A. P.

Electricals & Electronics

NANO ELECTRONICS

NANO ELECTRONICS

A.S. BHATIA

Foreword by

DR. S.M. ISHTIAQUE
Professor
Department of Textile Technology
Indian Institute of Technology, Delhi

NuTech Books

from

DEEP & DEEP PUBLICATIONS PVT. LTD.
F-159, Rajouri Garden, New Delhi - 110 027

NANO ELECTRONICS

ISBN 978-81-8450-330-2

Typeset by THE COMPOSERS
260 C.A. Apt., Paschim Vihar, New Delhi - 110 063

Printed in India at MAYUR ENTERPRISES
WZ Plot No. 3, Gujjar Market, Tihar Village, New Delhi - 110 018

NuTech Books are published by DEEP & DEEP PUBLICATIONS PVT. LTD.
F-159, Rajouri Garden, New Delhi - 110 027 • Phone : 25435369, 25440916
E-mail : ddpubs@gmail.com • ddpbooks@yahoo.co.in
Showroom :
2/13, Ansari Road, Daryaganj, New Delhi - 110 002 • Telefax : 23245122

Contents

डा. एस.एम. इश्तियाक
प्रोफेसर
Dr. S.M. Ishtiaque
Professor

वस्त्र प्रौद्योगिकी विभाग
DEPARTMENT OF TEXTILE TECHNOLOGY
भारतीय प्रौद्योगिकी संस्थान, दिल्ली
Indian Institute of Technology, Delhi
हौज़ खास, नई दिल्ली - 110 016
Hauz Khas, New Delhi- 110 016
Tel. : +91-11-26591401, 26591410 (O)
Tel. : +91-11-26515937, 26591940 (R)
Fax : +91-11-26581103, 26862037
e-mail:ishtiaque@textile.iitd.ernet.in
hodtextile@textile.iitd.ernet.in

Foreword

This book lays the foundations on which prospects in nano electronics may be pursued. This book is aimed at post-graduate students and young engineers, research scholars, scientists and teachers and provides a complete review of all relevant aspects in the field of nano electronics, from FETS, MOSFETS, LED'S, Quantum Dot Lasers to 3D-integrated circuits.

This book is well written using simple language and has illustrations which also help in understanding the topics covered.

I am sure that this book will prove quite useful to students who want to learn about Nano electronics.

Dr. S.M. ISHTIAQUE
Professor,
Department of Textile Technology
Indian Institute of Technology, Delhi
Hauz Khas, New Delhi-110016

Preface

This is believed to be the first book that takes a view of nanotechnology from a nanoelectronics perspective. Nanotechnology refers to the manipulation of materials at the atomic or molecular level. Nanotechnology is getting a lot of attention of late not only in academic settings and in laboratories around the world, but also in government and venture capitalists initiatives.

At the start of the century it was observed that we are just beginning to understand how to use nanotechnology to build devices and machines that imitate the elegance and economy of nature. The gathering nanotechnology revolution has made possible a huge leap in computing power, vastly stronger yet much lighter materials, advances in electronics as well as devices and processes with much lower energy and environmental costs.

Nanotechnology is a nanometer-level *bottom-up* assembly approach that allows developers to engineer particles at the molecular level, building them up to the "right size", with engineered functional properties. A nanometer is one billionth of a meter. Bottom-up process technology provides a control mechanism, over development of particles with respect to their size, shape, morphology, and surface conditions. Because of the challenges involved in working at this microscopic scale of a few nanometers, research and engineering efforts involving manipulation of components as "large" as 100 nm are typically included in the field of nanotechnology. Atoms are typically between one-tenth and one-half of a nanometer wide.

Research and development topics in nanotechnology range from molecular manipulation to nanomachines (microscopic devices that can themselves carry out takes at the atomic or sub-atomic level). While nanomachines represent futuristic initiatives nanoelectronics is already resulting or will do so in usable technologies.

In this book I have focussed on developments and technologies that have the potential to be used (or are already being used) in electronic environments. Such applications include Quantum Dot Lasers, LEDs, Magnetic Nanowires, 3-dimensional integrated Circuits and many more.

In 1965, Gordon Moore formulated his now famous Moore's law, which became the catalyst for advancements in the semiconductor industry. The semiconductor industry has brought us the sub-100-nm era with all the advancements we see today. With these advancements come difficulties in process control and subsequent challenges to circuit and physical design. As a result, the degrees of freedom in design methology are fast shrinking and will require a revolutionary change in the way we put together chips that are not only functional but also meet the design objectives and are high yielding.

However, the explosive growth of semiconductor models developed, has resulted in the isolation of process/device engineers from circuit design engineers, leading to some understanding of the impact of their design upon manufacturability, yield, and performance, due to the fundamental limitations of technology and device physics. As we enter the nano-CMSO era, knowing how to traverse these issues is critical to the success of products and companies. These communities of engineers must work together to fill each other's knowledge gaps, which are ever widening as we travel down the road of dimensional scaling. Only by doing this can goals be realised.

In Chapter 1, I have reviewed the electronic revolution. In Chapters 2 and 3 supportive topics such as semiconductor electronics and in Chapter 3 basic introduction to transistors, FETs, MOSFETs, etc. are covered. In Chapters 5 to 9, I look at Nano arrays and quantum dots, quantum dot lasers, Magnetic nanowires and Characteristics of Nanosilicon devices.

Relentless assaults on the frontiers of CMOS technology over several decades have produced a marvel of a technology. The world we live in has been changed by complex integrated circuits now containing a billion transistors with line widths of less than 100 nm, that are fabricated. This microelectronics revolution was made possible only through the dedication and ingenuity of many specialized experts with detailed knowledge of their crafts.

I have covered the atomic level manipulation of CMOS in Chapter 10. While the traditional approach to CMOS scaling are being pushed to their limits, new devices, materials, and designs are being developed to further extend the "silicon age". Chapter 11 is about organic-inorganic devices while in Chapter 12, I have tried to explain about 3-dimensional integrated circuits.

This book is intended as an introduction to the field of nanotechnology for nanoelectronics vendors, researchers, and students who want to start thinking about the potential opportunities afforded by these emerging scientific developments and approaches for the next-generation network to be deployed 5-10 years in the future. Advanced planning is a valuable and effective exercise.

I hope it will serve as motivation, by raising interest, to continue the line of investigation and research into the field. I have made every effort to make it relatively self-contained by discussing the introductory fundamental principles involved.

New Delhi

A.S. BHATIA

1

The Electronic Revolution

MINIATURIZATION OF ELECTRONIC DEVICES

We have entered another era, namely age of nanotechnology. Development in electronic devices means a race for a constant decrease in the order of dimension. The electronics revolution began with the realization of transistors i.e. hand-held radio. The transistor then meant bipolar junction transistor of BJT for short. While BJTs were the very springboard for launching the microelectronics age, these devices gave way to metal oxide semiconductor field effect transistor, MOSFETs for short. MOSFETs themselves laid the ground for complementary MOSFETs or CMOS for short. Due to their comparatively low power consumption, CMOS circuits have made possible microprocessors with ever-increasing speed and density. The generic term for miniaturized circuits which has become a household name in very culture is microchip. The microchip conquered the information age in a quarter of a century. It compacted the power of mainframe computers to desktop computers on everyone's desk. It is responsible for increased efficiency in the work place, a diagram of CMOS is shown in Fig. 1.1.

In silicon-based electronics it has been the reduction of dimensions which not only allows more devices circuits per unit area, but also increases the device/circuit speed. Functional silicon MOS devices with gate lengths as low as 15 nm have been produced with frequency performance in the terahertz regime. Development of pattern generation and pattern transfer tools, and related fabrication processes & a new generation of central processing unit (CPU) is developed every 3 years. Thus the memory density increases by about four times, and logic density increases by about 2.5 times (is very 6 years), the feature size decreases by 2 transistor current density, circuit speed, chip area, chip current and maximum I/O pins increases by 2. Trends and future estimates regarding the maximum dimensions, chip size, number of transistors per chip and many other important parameters e.g. the number of logic transistors per chip will be at 60 billion in the year 2016, and the chip size could be as large as 620 mm^2, whereas presently production CMOS lines are using 0.13 μm (or 130 nm) gate lengths.

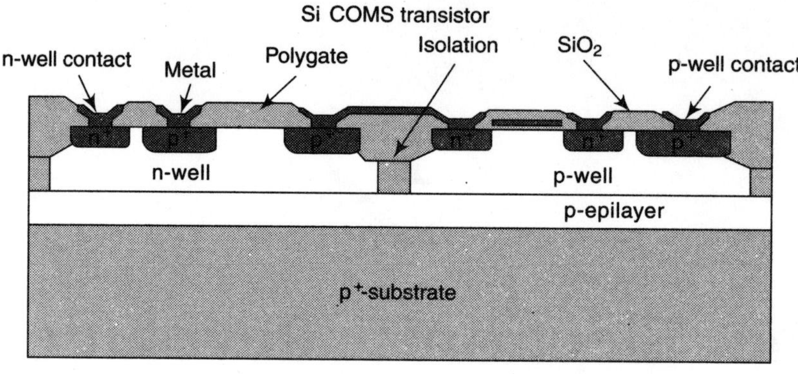

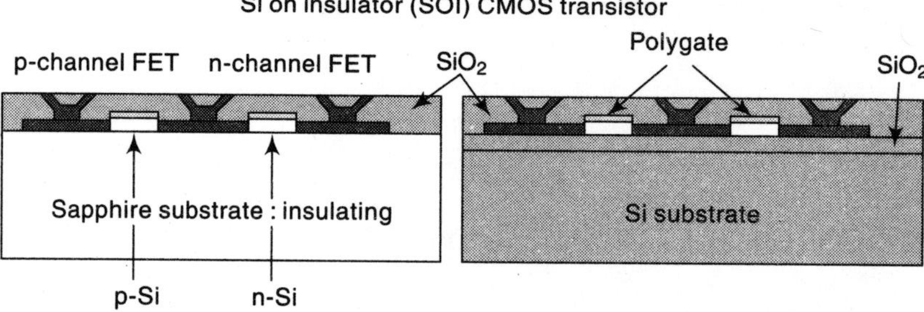

Fig. 1.1 *The building blocks of a CMOS circuits implemented in silicon (top), and silicon on insulator, such as SIMOX (bottom right) and silicon on sapphire (bottom left).*

The gate dimension in mass produced devices is projected to be 5 to 6 nm by the year 2016 the gate dimension in mass produced devices is projected to be less than 20 nm. This figure represents a factor of 40 reduction in sized inside of just three decades. The rate of progress is so fast that it was nearly impossible to foresee that a desktop computer, would pose a formidable challenge to mighty super-computers in the span of just a few years. CPUs with greater than 2 GHz clock speeds are already commercially available which utilize Cu interconnect technology for reducing the wire resistance and thus the RC time delay. With new lithographic tools in their arsenal, the chipmakers will be able to boost the speed of CPUs to 10 GHz.

In integrated circuits, the higher the current that the devices (switches) can produce at voltages as low as possible, the better the circuit performance. As the device dimensions are reduced for both high speed and high density, the number of carriers decreases.

One option is to simply increase the number of dopants which comes to the expense of increases leakage current, where unintended current paths are created. As feature sizes shrink, these dopant levels become an increasingly serous problem. Lithographic tools are developed for chip makers and also extreme ultraviolet (EUV) as the light source in photolithography; for the next generation lithography (NGL); X-ray lithography (PXL). Also, the lithographic PXL technology for 4-Gbit dynamic random access memory (DRAM) has been demonstrated, and a straightforward development path has been defined that leads all the way to the end of the silicon semiconductor era at about 25 nm channel lengths.

In the EUV approach, soft X-ray are emitted upon the excitation of a xenon gas streaming into the path of a high power laser. This emission is then focused on a reflective mask. To reduce this image, curved mirrors coated with 80 alternating layers of silicon molybdenum, polished with atomic-scale precision, were developed. Because air absorbs EUV radiation, the entire apparatus must be placed in a vacuum. The result is a 3m × 3m machine that stands some 4m high, which is quite tall compared to the lithographic tools in use today.

In the scale of laboratory research and exploration of exceedingly small dimensions in conjunction with devices and nanostructures, tools for imaging purposes have been demonstrated with some modification, to the creation of structures not achievable with the lithographic tools. These include scanning tunneling microscopy (STM) and atomic force microscopy (AFM) techniques. In addition to patterning a resist material by passing current through a nanometer-sized tip, as in the case of STM, mechanical definition of patterns, as in the case of AFM have been explored. Furthermore, multiple tips working in unison have been utilized in pattern generation in nanoscale.

The diffraction limit of photolithography has continuously been shrunk owing to the use of shorter and shorter wavelength UV sources which have been developed, and also the development of mask fabrication methods with phase correction played a pivotal role. However, it is clear to the Semi-conductor Industry Association that the UV sources (which are already down to 193 nm wavelength) will eventually be the limiting factor in scaling the technology to device features of smaller and smaller size, the efforts are underway to meet this challenge by developing alternative approaches which include variants of electron beam exposure and soft X-rays.

FETs are scaled down to sub-micrometer regime (see Fig. 1.2) concepts of drift and diffusion would not apply, as the device dimensions will become comparable to scattering lengths. Small

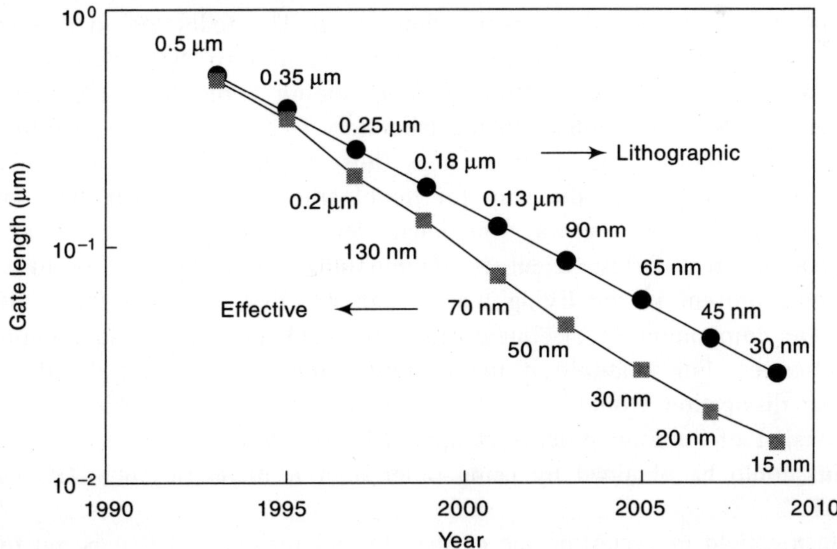

Fig. 1.2 *Evolution of the past and expected gate dimension, both lithographic and actual, in production Si FET technology.*

dimensions also mean that the carriers are subjected to high driving fields in the channel as well as high confining field at the gate. The aspect ratio consideration and the need to reduce the operation voltage require that the gate dielectric thickness be reduced which eventually runs into the direct tunneling regime, less than 4 nm. The combat this scaling problem, high dielectric constant oxides, dubbed the high K materials, are being explored. With high K dielectrics, the gate dielectric thickness requirement (while remaining within scaling rules) is relaxed. For example, by using $SrTiO_3$ for gate dielectric, thickness of about 100 Å is equivalent to about 10 Å of SiO_2 gate oxide.

The mind-boggling speed with which electronic devices have been scaled to first micrometers, then to submicrometers and recently to nanometer scale has fulled an unprecedented advancement in telecommunications and computers. While the devices bedded in silicon are scaled to nanometer size, dubbed the nanoelectronics, strides are being made on the fronts. Among them are nanowires based on carbon nanotubes FET-like devices based on carbon nanotubes that are operative at room temperature semiconductor nanowires and logic circuits based on carbon nanotubes and molecules. With ever-shrinking dimensions semiconductor silicon components cannot be shrunk much below 0.1μ in size for CMOS unless the issues of leakage current and gate oxide scaling area are addressed.

However, it is reasonable to expect that there will be a limit somewhere along the way to the conventional scaling of silicon circuitry. Once and if that limit is reached, entirely new technologies must be introduced to etch ever tinier transistors onto silicon wafers, and new materials such as molecules, carbon tubes or some other material may have to be introduced. It may be that components on a microchip will have to be made as small as the coil of a DNA molecule. Sooner or later, the elements of the microchip will become so small that their dimensions may be expressed in molecular dimensions, where the laws of quantum physics prevail. Luckily, other semiconductors such GaAs and related varieties have been aggressively used in the pursuit of very small geometry dimensions. The reduction in dimension has been extended from three to two, from two to one, and then to zero dimension to the point where quantum dots and allowed states within them are being exploited for potential computing machines. Already Si MOS technology in the laboratory is around 01 nm in critical dimensions. As mentioned earlier, other components such as those based on carbon nanotubes for devices and chemical storage may pave the way for novel applications which may someday get us around some log jams closing the way functional elements are produced. Along parallel paths, but in somewhat different fields, scientists are pursuing computation as performed by nature: using components present in the living world such as DNA. Even if these efforts are truly successful at the component level, integration and packaging of these components present enormous challenges. For example if the current trends were extrapolated to high-density microchips, heat dissipation would be sufficient to melt the package. Already, power densities in the latest version of Pentium processors approach 100 W cm^{-2}.

Similar gains could be obtained by using other high K dielectric such ZrO_2 used in Intel's THZ MOSFET.

The gate electric field may confine the carriers to nanometer scale that is, on the order of the de Brogile wavelength of current carriers. When the source to drain distance is lowered well below 100 nm, tunneling may take place creating an additional current path a very short channel n-MOSFET when treated with a simple and analytically solvable model; assumes ballistic dynamics of two-dimensional electrons in the channel. By including the source to gate tunneling effects the voltage gain drops sharply at L ~ 10 nm, while the conductance modulation remains sufficient form memory applications until L ~ 4 nm. In short, simple extrapolations fail when extended to nanometer scale, and quantum mechanical treatment of the problem becomes necessary. Traditionally, semiconductor materials, which are better suited for high-speed devices in terms of electrical properties, have led the way in terms of dimensions and speed. While these devices fulfill niche markets, such as higher end consumer products, military hardware and high-end communication gear, they also serve to advance the state of the knowledge and art in device physics and lithographical limits. In particular, a device which has been very successful is the modulation doped field effect transistor. With scaling down to very small dimensions and increasing density, leakage paths and other paths of interference become important.

One idea that has been on the table for many decades in the silicon-on-insulator (SOI) area is to produce a device with a thin and undoped silicon channel. SOI with a buried oxide layer is accomplished by either creating a buried oxide layer beneath the silicon surface by ion implantation or by grafting a silicon layer on top of an oxide layer. The other approach relies on deposition of silicon-on-sapphire (SOS) which lacks quality. Schematics for both of these approaches are shown in Fig. 1.1, which shows quite clearly the simplicity, afforded as opposed to conventional CMOS approaches. Availability of insulating layers paves the way for substantially reducing the capacitance under the source and drain tubs leaving only a parasitic bipolar device extending from the source to the drain. Simplicity renders the approaches very amenable to high density.

The SOI technology has remained as a laboratory curiosity. In addition to the standard device designs on insulators, novel approaches afforded by lateral epitaxial overgrowth have been explored. Example, the meaning of "depletion region" with variable gate-induced depth, created along the gate edge, is modified because the entire body of such transistors if fully depleted. Instead, the entire channel's electronic properties are modulated by the electric field across it, generated by two transistor gates.

As always, the dopant-level problem is connected to the channel thickness, because thickness affects how the electric field modulates channel behavior. This is achieved by a novel fabrication method, shown in Fig. 1.3, which relies on epitaxially growing silicon through a mask to create

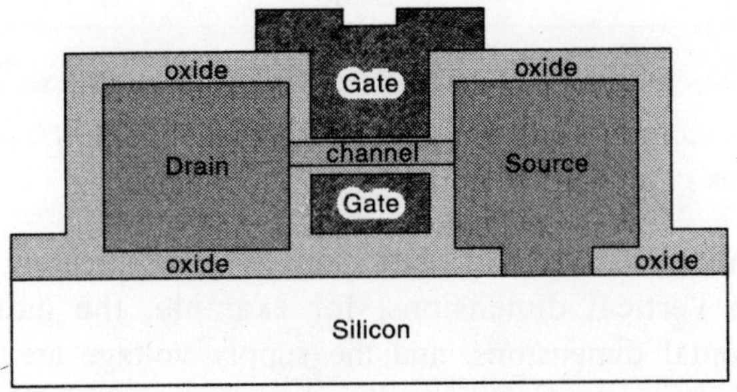

Fig. 1.3 *A nanoscale MOSFET, which reduces short channel effects.*

the source, channel, and drain regions. The device performance and most importantly, the method of fabrication—epitaxial growth—are scalable to the ultimate CMOS limit of 20–30 nm gate lengths. The processing steps, capitalizing on the continuing developments in silicon processing are as follows: first, sacrificial layers are built up and patterned to produce a thin, isolated layer of amorphous silicon. This is etched away to produce a "window" and the silicon substrate is exposed to create a "seed window" on which to grow the rest of the single crystalline sections such as a the source and drain regions and the channel layer. When the epitaxial layers have been completed, the sacrificial layers can be removed, resulting in a suspended silicon bridge around with the rest of the transistor can be fabricated.

It is really not very important that this device structure is the one that winds up being used. It will be years before we know which of the device designs that are out there will end up being used. Already device structures that are very amenable to production have been developed. One particular device that utilizes SOI technology and high *K* gate dielectric (ZrO_2), dubbed the depleted substrate transistor (DST), is shown in Fig. 1.4. It should be mentioned that the acronym DST has already been used for fully depleted SOI MOSFET. The source and drain resistance were reduced by raising the source and drain tubs, effectively increasing the thickness of the access area and thus the resistance. The device sports an effective gate length of 15 nm and the associated gate delays portend operation over 2 THz. Excessive power dissipation associated with high device density is alleviated somewhat by reduced power supply voltage which is facilitated by reduced device dimensions. However, increased gate leakage and off-state leakage currents, unless countered, eat away what is gained by supply reduction. The evolution of the supply voltage in nanometer gate devices is rather complex, but a figure of 0.75 V is given here for a ballpark value. Everything else being the same, any reduction in power supply voltage translates, through square of the reduction factor, to power density reduction.

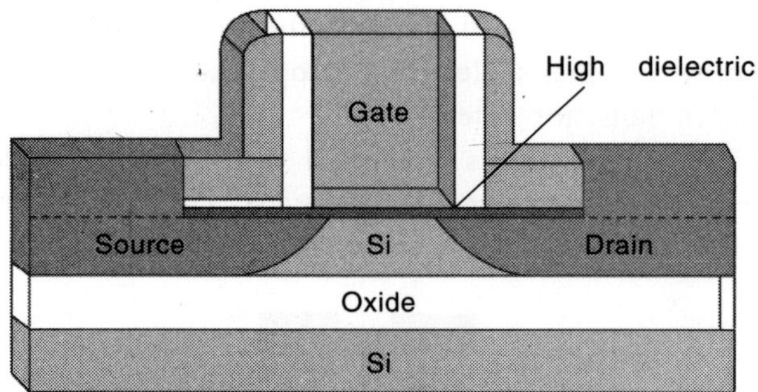

Fig. 1.4 *Schematic representation of Intel's depleted substrate transistor (DST) which recently has been demonstrated at the 15 nm gate length level.*

For keeping short channel effects in check, constant field scaling is commonly utilized. In this frame of mind, the vertical dimensions, for example, the gate insulator thickness and junction depth, the horizontal dimensions, and the supply voltage are reduced, and the substrate doping concentration is increased (decreasing the depletion width). In this approach, the device voltages and dimensions (both horizontal and vertical) are scaled by the same factor, κ (>1), to keep the electric field (E) unchanged.

All capacitances (including wiring load) scaling down by k since they are proportional to area and inversely proportional to thickness. The charge per device ($\sim C \times V$) scales down by κ^2 while the inversion-layer charge density (per unit gate area), Qi, remains unchanged after scaling. This means that the drift current scales down by κ, consistent with the behavior of both the linear and the saturation MOSFET currents. A key implicit assumption is that the threshold voltage also scales down by κ. However, the diffusion current does not scale down the same way as the drift current. This has significant implications on the MOSFET subthreshold currents not following the scaling rules which forces the operating voltage to not scale.

The circuit delay, which is proportional to RC to $(V/I)C$, scales down by κ or the circuit speed increases by the same factor. Moreover, the power dissipation per device is reduced by κ^2. Since the device density also increases by κ^2, the power density remains unchanged. Consequently, the power-delay product of the scales CMOS circuit shows a dramatic improvement by a factor of κ^3. Due to increased complexity of each generation of chips beyond the aforementioned scaling rules, power dissipation per chip has been steadily increasing which poses a formidable problem.

In parallel to developments in electronic devices, the digital recording an storage medium, which is the other component of computers, has made tremendous strides outspacing speed experienced in silicon electronics. Only a few decades ago 200-MB memory disk was almost about 50 cm in diameter, and it required tremendous mechanical stability at tremendous cost. Many desktop and laptop computers have magnetic discs capable of storing well over 40 GB in an tiny package.

This was made possible by the discovery of giant magnetoresistance (GMR) with its associated giant magnetoresistive sandwich materials, and rapid pace of technology transfer from the research phase to the production phase. These materials are beginning to shatter their traditional roles and venture into what used to be the domain of electronic devices based on capacitively charged storage devices which in turn are based on transistorized DRAM, and others as alluded to below.

Recent development sin GMR have led to the development of a new generation of metallic devices, such as spin value GMR heads and magnetic random access memory (MRAM) devices based on magnetic tunnel junctions, in which electronic spin plays a central role. The next area in which electron spin effects can make a tremendous impact is in semiconducting electronic devices, where spin-based devices, such as spin transistors, spin memory devices, and spin quantum computers, may revolutionize the industry.

Spin-based devices require spin junction from one material into another across an interface. Spin injection from one ferromagnetic metal into another metal can readily occur, however, spin junction into a semiconductor is altogether different. There are strong theoretical reasons that direct injection of spins from an ordinary ferromagnetic metal film deposited onto a semiconductor is not likely to be successful. The fundamental obstacle for spin injection in the diffusive mode is the large contrast in conductivity and carrier density between a metal and a semiconductor. The best prospect for spin injection into semiconductors, aside from indirect injection by optical means, is direct injection from a dilute magnetic semiconductor. Because of the lack of suitable magnetic semiconductors, direct injection has only been demonstrated at 20 K, with little prospects of realistic devices at room temperatures. Consequently, attainment of Curie temperature near room temperature is of paramount important for semiconducting spin-based devices.

Semiconductor Electronics

The reason that electronics have become the fundamental technique of the information age does not only lie in the discovery of the transistor effect and the miniaturization of these devices. It is also based on the special properties of the electron, as the lightest elementary particle accessible for technical applications, which are key factor in all chemical interactions. An electron contains a charge, and it is distinguishable from all ions and also the proton in its extremely large charge/mass ratio. The transfer rates are controllable by electric fields, and they are extremely sensitive to the medium, so that electrical conductivity can be varied by many orders of magnitude if changing the material.

The first technical devices for controlled electron transport were vacuum devices (tubes). Their replacement by semiconductor devices began in the 1950s. These devices become key components for controlled electron transport in the micron and sub-micron range, especially in the area of transistors. Semiconductor technology is also interesting to the nanometer range, and the technology of integrated circuits can now reach below the 100 nm for memory and processor applications.

Semiconductor behave as insulators at absolute zero temperature ($T = 0$) but at non-zero temperatures ($T > 0$) exhibit a relatively small electrical conductivity, the size of which increases rapidly with increasing temperature. Their electrical conductivity can be increased by adding small amounts of certain impurities (dopants) or by illumination with particular wavelengths of light whereas the properties like good conduction (metals whose electrical conductivity is many orders of magnitude larger) decreases relatively weakly with increasing temperature and, is not affected by small levels of impurities of illumination.

Figure 2.1 show the main features of the band structure of a semiconductor. At absolute zero temperature all states in the valence band are occupied by electrons and all states in the conduction band are empty. So, electrical conduction cannot occur. If the temperature is increased the electrons are excited form the valence band across the band gap E_g into the conduction and band below. Electrical conduction is possible via the small number of electrons in the conduction band and the large number of electrons which remain in the valence band, but whose motion is limited because there are only a small number of vacancies. Although

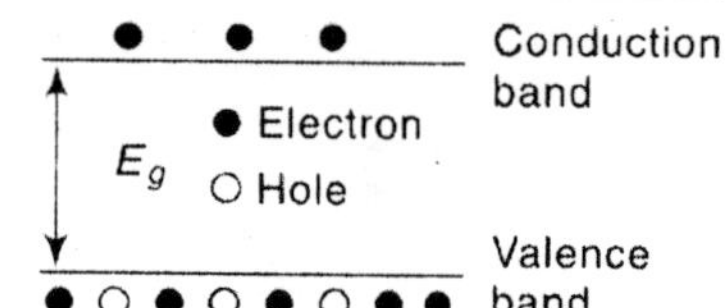

Fig. 2.1 *The electronic band structure of a semiconductor. Electrical conduction occurs in the conduction band by a small number of electrons and in the valence band by a small number of vacancies or holes.*

electrical conduction in the valence band is due to the movement of the large number of electrons, it is more convenient instead to consider the contribution to the electrical conductivity in terms of the much smaller number of vacancies. These vacancies, which are termed holes, move in the opposite direction to the electrons and hence they behave as carriers of opposite charge sign.

CHANGES IN SEMICONDUCTOR PHYSICS AND TECHNOLOGY

1. Artificial Atoms vs. Layers

Figure 2.2(i) compares the electronic levels of a single atom, a bulk semiconductor, and a quantum dot. It is well known that a single atom has discrete energy levels separated by forbidden energy gaps, as shown in Fig. 2.2(i). When the atom is excited, the electron goes to the higher energy level, and when it relaxes back to the ground state, a photon with strictly defined energy is emitted. The width of the emission or absorption line (ΔE) is defined by a fundamental relation involving the lifetime of the electron into the upper state. The uncertainty in the emitted energy is

$$\tau \, \Delta E \geq h,$$

where τ is the coherence time of the electron in the excited state.

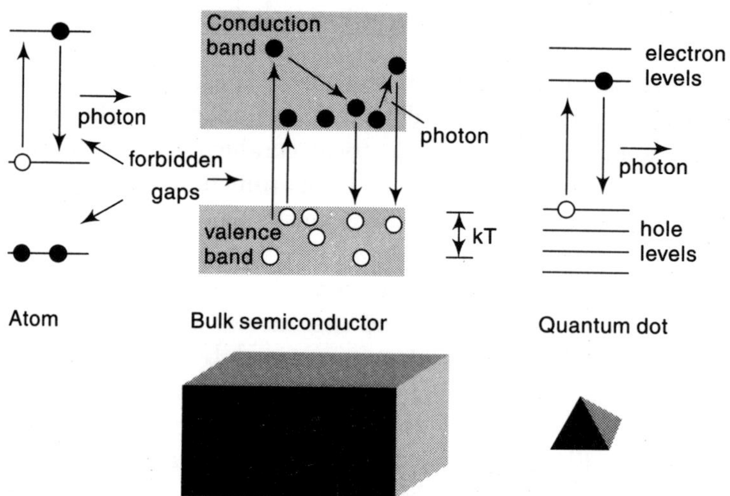

Fig. 2.2(i) *Representation of energy levels in a single atom, a bulk semiconductor and a quantum dot.*

Atoms in crystals are strongly bound to each other. Their high density ion crystals allows high absorption or gain coefficients, provide high conductivity and high density flow of charged carriers through the crystal; example:

A modern semiconductor laser with length 1 mm and cross-section 10^{-4} mm^2 can emit continuous light with a power of a few watts. Small separations between atoms make interactions between their electron levels unavoidable. This interaction results in the formation of wide bands of allowed states, in contrast to the discrete (δ-function-like) energy spectrum of single atoms. In semiconductors, the last filled band of allowed states is called the *valence band* and the next empty band is called the *conduction band.* Due to the broad spectrum of allowed states in these bands, a wire range of transition energies exists between electrons from the filled valence band to empty states in the conduction band. The absorption band becomes rather broad, of the order of a few electronvolts, in marked contrast to the sharp line absorption spectrum of single atoms. Excited electrons in the conduction band, as well as empty states in the valence band (so-called holes) can move in the crystal via tunneling between crystal lattice sites. Since the atomic potential profile in a crystal is periodic, electrons and holes can move freely through the crystal, as free carriers do in vacuum. However, the motion of charged carriers in crystals is described by a different mass to that of free electrons, defined by the crystal field, The carriers are thus called quasiparticles. In the widely used optoelectronic III-V materials, e.g., gallium arsenide (GaAs), idium arsenide (InAs), etc., electron effective masses lie in the range 0.0–0.1 of the electron mass in vacuum.

Wide bands of allowed states in the crystal provide sufficient opportunity for scattering of electrons and holes. Lattice vibrations easily stimulate transitions of charge carriers in the energy range defined by the lattice temperature and/or scatter the direction of motion of the carriers. The tails of the carrier distribution near the bottom of the conduction band and the top of the valence band increase remarkably with temperature. Thus, the concentration of carriers per energy interval near the band edge drops. For the same concentration of injected carriers, a broadening of their energy spectra results in a decrease in maximum gain, and degradation of laser performance, among other disadvantages.

The situation changes remarkably if the motion of the charged carrier in the crystal is limited to a very small volume, e.g., to a three-dimensional rectangular box. Localization of carriers can be provided by a surrounding (matrix) material. For laser applications, it is important for the matrix material to have a larger bandgap than the box material and also for the potential wells to be attractive for both electrons and holes. Since electrons exhibit both particle and wave properties, if the size of the box is small, the electron energy spectrum is quantized rather as it would be in the attractive coulomb potential of a nucleus.

Electrons in crystals usually have rather small effective mass, and an already large box size of about 10 nm can result in a large energy separation between electron sublevels (about 100 meV for a GaAs QD). The latter value significantly exceeds the thermal energy at room temperature (26 meV), so that population of excited states can be avoided. In this sense the optical spectrum of such a box will display no temperature dependence over a wide temperature range, and the realization of temperature-insensitive devices becomes possible.

Quantum dots combine the advantage of single atoms (discrete energy spectrum) and of solids (a rather large volume density). In view of their discrete spectrum, quantum dots may be called artificial atoms, although they may consist of a few 10^2 to a few 10^5 atoms.

The traditional approach in semiconductor technology involves planar epitaxial growth followed by batch lithographic processing. Traditional lithographic techniques do not allow fabrication of the required quantum dot structures due to size limitations and the high density of defects generated by lithography.

Advanced lithographic techniques are under development that employ radiation with wavelength shorter than visible light. These methods are electron beam lithography, X-ray lithography using synchronous light sources (a review may be found in, focused ion beam techniques and atom beam lithography. Although these techniques yield lateral resolution down to a few 10 nm and offer almost infinite design variations, the disadvantages are the high number of technological processes involved and the high cost. Moreover, these techniques still generate a high density of defects hindering possible applications of nanostructures in optoelectronics.

2. Self-Organised Nanoepitaxy vs. Lithography

Recently, two alternatives to the concept of optical lithography have been introduced. The first is based on scanning probe microscope (SPM) techniques that have been developed since the early 1980s. Nanostructures down to an atomic scale may be achieved either by manipulating single atoms or by using the SPM tip as a pen to 'write' nanoscale structure. Although the 'writing speed' of these procedures has been steadily increased, they are still very time-consuming because of their serial nature. To overcome this disadvantage, attempts have been implanted to operate more than one probe in parallel, resulting in linear arrays. Although some 2D arrays have been fabricated, the prospects for this 'writing' technique are still challenging.

The second, much more elegant and exciting alternatives to lithography uses the effects of self-organization. Spontaneous formation of spatial, temporal, or spatio-temporal patterns by self-organisation of individual constituents is a common phenomenon in nature. It covers a wide range of length scales from atomic to cosmic dimensions; examples are, lasers and heated fluids in physics, the Belousov-Zhabotinsky reaction in chemistry, dune patterns in earth science, and mor-phogenes in biology

The use of self-organisation phenomena at surfaces allows us to fabricate quantum dot structures within the context of planar technology; just the 'conventional' planar growth is used. The massive parallel process of spontaneous formation of nanostructures makes it possible to produce 10^{10} to 10^{12} quantum dots per cm^2 per second. Hence, the effects of self-organization can result in the formation of ordered nanostructures from an initially random distribution of atoms.

Figure 2.2(ii) shows a few examples of spontaneously forming nanostructures. When the phenomenon occurs in semiconductor materials, further overgrowth of the surface nanostructure results in the formation of wire-shaped or dot-shaped insertions in the matrix. In the case of narrow band gap insertions in a wide bandgap matrix, quantum wires or quantum dots are formed.

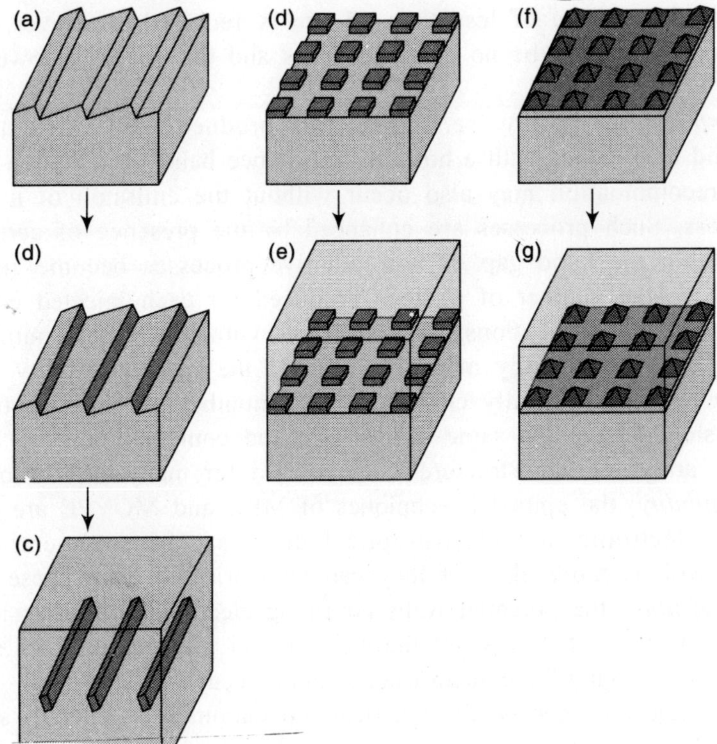

Fig. 2.2(ii) *Examples of spontaneously formed nanostructures. (a) periodically faceted surface. (b) Wires of a deposited material formed in the grooves of the periodicity faceted surface. (c) Capped structure (b) forming quantum wires within a matrix. (d) Array of two-dimensional islands in submonolayer heteroepitaxy. (e) Capped structure (d) forming two-dimensional quantum dots embedded in a matrix. (f) Array of three-dimensional coherently strained islands over a wetting layer on a substrate surface. (g) Capped structure (f) forming an array of three-dimensional quantum dots embedded in matrix.*

REQUIREMENTS FOR AN IDEAL SEMICONDUCTOR NANOSTRUCTURE

The following lists the main requirements for an ideal semiconductor nanostructure. In practice the relative important of these requirements will be dependent upon on the precise application being considered.

- *Size:* for many applications the majority of electrons and holes should lie in their lowest energy state, implying negligible thermal excitation to higher state. The degree of thermal excitation is determined by the ratio of the energy separation of the confined states and the thermal energy, k_BT. At room temperature $k_BT \approx 25$ meV and a rule of thumb is that the level separation should be at least three times this value (~75 meV). As the spacing between the states is determined by the size of the structure, increasing as the size decreases this requirement sets an upper limit on the size of a nanostructure. For electrons in a cubic

GaAs quantum dot, a size of less than 15 nm is required. However, below an certain quantum dot size there will be no confined states and this places a lower limit on the dot size.

- *Optical and structural quality:* semiconductors produce light when an electron in the conduction band recombines with a hole in the valence band—a radiative process. However, electron-hole recombination may also occur without the emission of a photon in a non-radiative process. Such processes are enhanced by the presence of certain defects which form states within the band gap. If non-radiative processes become significant then the optical efficiency—the number of photons produced for each injected electron and hole—decreases. For optical applications, nanostructures with low defect numbers are therefore required. Poor structural quality may also degrade the carrier mobility.
- *Uniformity:* device will typically contain a large number of nanostructures. Ideally each nanostructure should have the same shape, size and composition.
- *Density:* dense arrays of nanostructures are required for many applications.
- *Growth compatibility:* the epitaxial techniques of MBE and MOVPE are used for the mass production of electronic and electro-optical devices. The commercial exploitation of nanostructures will be more likely if they can be fabricated using these techniques.
- *Confinement potential:* the potential wells confining electrons and holes in a nano-structure must be relatively deep. If this is not the case then at high temperature significant thermal excitation of carriers out of the nanostructure will occur.
- *Electron and or hole confinement:* for electrical applications it is generally sufficient for either electrons or holes to be trapped or confined within the nanostructure. For electro-optical applications it is necessary for both types of carrier to be confined.
- *p-i-n structure:* the ability to place a nanostructure within the intrinsic region of a *p-i-n* structure allows the efficient injection or extraction of carriers.

Light Emission process in Nanostructures

Electrons and holes can be created in semiconductor either optically with incident photons of energy greater than the band gap, or by electrical injection in a pn junction. The electrons and holes are typically created with excess energies above their respective band edges. However, the time required to lose the excess energy is generally much shorter than the electron-hole recombination time, consequently the electron and holes relax to their respective band edges before recombining to emit a photon. Emission therefore occurs at an energy corresponding to the band gap of the structure, with a small distribution due to the thermal energies of the electrons and holes. The influence of rapid carrier relaxation is demonstrated in the emission spectrum of a structure containing quantum wells of five different widths. Only emission corresponding to the lowest-energy transition of each well is observed, even though the wider wells contain a number of confined states.

Higher-energy transitions in a nanostructure can be observed in emission if the density of electrons and holes is sufficiently large that the underlying electron and hole states are populated. This can occur under high-excitation conditions where the lower energy states become fully occupied and carriers are prevented from relaxing into these states by the Pauli exclusion principle. Figure 2.3 shows emission spectra of an ensemble of self-assembled quantum dots for different optical excitation powers. At low powers the average number of electrons and holes in each dot is very small, and consequently only the lowest-energy, ground state transition is observed. However with increasing power the ground state, which has a degeneracy of two, is fully occupied and emission from higher-energy, excited states is observed.

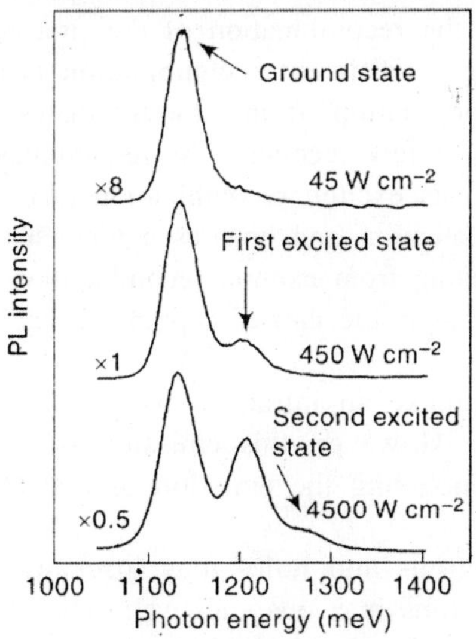

Fig. 2.3 *Emission spectra of an ensemble of InAs self-assembled quantum dots for three different laser power densities. At the highest power, emission from three different transitions is observed. The numbers by each spectra indicate the relative intensity sale factory.*

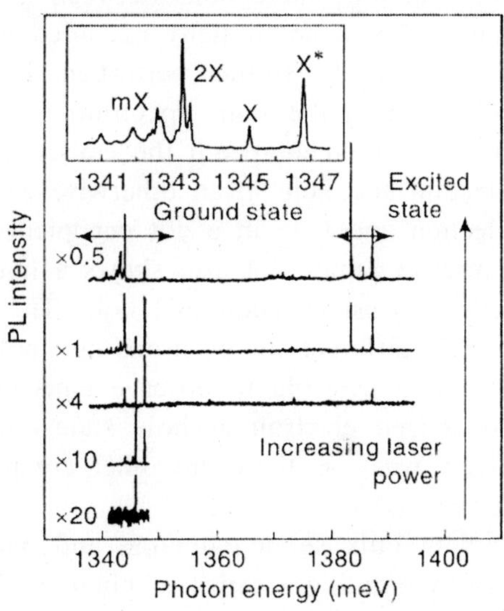

Fig. 2.4 *Emission spectra of a single InAs self-assembled quantum dot as a function of laser power. The inset shows emission from the ground state in greater detail. Emission lines are observed corresponding to an exciton recombining in an otherwise empty dot (X), the dot occupied by one additional exciton (the biexciton 2X), the dot occupied by > 1 additional excitons (the mX lines) and one additional hole (a charged exciton X*).*

Fluctuations in the size, shape and composition of quantum dots result in the significant inhomogeneous broadening of optical spectra recorded for large numbers of dots. Only by proving a small number of dots can the predicted very sharp emission be observed. Figure 2.4 shows emission spectra obtained from a single self-assembled InAs quantum dots as a function of the incident laser power. At low powders a single, sharp line is observed, arising from the recombination of an electron and hole from their respective ground states. At high powers these states are fully occupied and carriers are forced to occupy higher-energy excited states form which emission is then observed.

An added complication in the spectra of Figure 2.4 is that, at high excitation powers, multiple emission lines are observed for the ground state and excited state transitions. This is shown more clearly in the inset to Figure 2.4, which depicts the ground state emission for high laser power. These multiple emission lines result from interactions between the carriers confined within the dot. If the dot is occupied by a single electron and hole, one exciton, then emission will occur at a particular energy. If however, the dot is occupied by two electrons and two holes, two excitons, then the energy of the first recombining electron and hole will be perturbed by the Coulomb interactions between the two excitons, and the emission will be slightly shifted in energy. For a dot initially occupied by three excitons, the recombination of the first electron and holes will be further perturbed. Lines corresponding to different recombination processes are observed in the same spectrum because the carrier population of the dot fluctuates during the time required to record the emission spectra, typically a few seconds. The recombination of an electron and hole in an otherwise empty dot is known as exciton recombination (X), that of an electron and hole in a dot occupied by an additional electron and hole as a biexciton (2X). The inset to Figure 2.4 also shows a line labelled X*, arising from exciton recombination in the presence of just an additional hole. This process is possible if the carrier capture probability of the dot is different for electrons and holes.

It is also possible to observe emission when carriers make an intraband transition between the quantised electron or hole states of a nanostructure. However, this emission is generally relatively weak as there are other competing processes, including the emission of one or more phonons.

In both bulk semiconductors and nanostructures, electrons and holes may lose any excess energy by emitting a series of phonons. Figure 2.5(a) electrons in a quantum well. The electrons initially have zero energy as they are at the conduction band edge of the barrier, but on transferring into the well they are left with an excess energy. Associated with each confined well state is a continuum of states, resulting from the in-plane motion, and this allows the electron to lose energy by emitting a sequence of phonons, as shown in the figure. For III-V semiconductors the carriers interact most strongly with longitudinal optical (LO) phonons and it is these phonons which are emitted as the carriers lose energy. A typical LO phonon energy is ~30 meV and a typical time to emit a single LO phonon is ~150 fs. Once the carriers reach an energy less than one LO phonon energy from the band edge, further LO phonon emission becomes impossible and the final energy is lost by the emission of low-energy acoustic phonons, a slower process compared to LO phonon emission.

Figure 2.5(a) also applies to bulk semiconductors and quantum wires, both of which have continuum of states. However, the situation is very different for a quantum dot, where the energy level are discrete. Here emission by a series of LO phonons is only possible if the spacing between the energy levels equals the LO phonon energy, an unlikely coincidence. Hence carrier relaxation in a quantum, dot must occur by an alternative, slower process. Possibilities include the emission of multiple phonons, for example an LO plus an acoustic phonon, or an Auger process where the energy released by one carrier as it relaxes is transferred to a second carrier, which is excited to a higher energy state, for example the continuum states associated with the barrier. These processes are shown schematically in Figure 2.5(b) and (c). The slow carrier relaxation predicted to result from the discrete energy levels of a quantum dot is known as the phonon bottleneck. In severe cases this slow relaxation may affect the performance of quantum dot devices, particularly for high-speed applications.

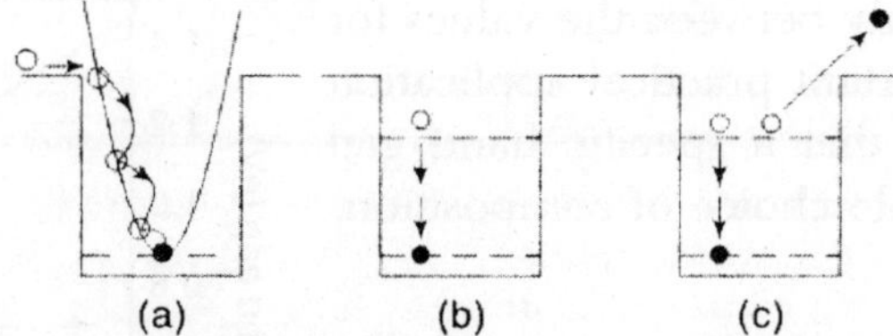

Fig. 2.5 *(a) Electron relaxation in a quantum well. The electron is injected into the well with excess energy but the continuum of well states allows this energy to eb lost by the emission of a sequence of LO phonons. In a quantum dot, the separation between the discrete, confined states does not generally match the LO phonon energy and the electron relaxes by (b) simultaneously emitting two different phonons or (c) transferring energy to a second electron that is excited into the continuum states of the barrier.*

Types of Semiconductor

The majority of purely electronic devices are based on the elemental semiconductor silicon (Si). However Si has an indirect band gap, with the lowest energy state in the conduction band occurring at a different wavevector to the lowest energy state in the valence band. When an electron and hole in Si recombine, this wavevector difference must be conserved, in addition to energy conservation. A photon is unable to conserve both energy and wavevector, so a second particle, usually a phonon, must be created in addition to the photon. This two particle, photon plus phonon, recombination occurs relatively slowly, hence it allows other processes to occur in which a photon is nor created.

For example, the electron may return to valence band by relaxing via phonon emission through a series of impurity states formed within the band gap, or it may transfer is energy to a second electron which is excited to a higher state in the conduction band. As a result of these non-radiative processes the majority of electrons and holes recombine without the emission of a photon; consequently, the light production efficiency of Si is very poor, making it unsuitable for many electro-optical applications.

Light production efficiency is much greater in direct band gap semiconductors where the recombining electron and hole have the same wavevector, and only a photon is required to satisfy energy conservation. For electro-optical applications, binary semiconductors consisting of elements from columns three and five of the periodic table are typically used, the majority of which have direct band gaps. Examples of III-V semi-conductors include gallium arsenide (GaAs), indium phosphide (InP) and gallium nitride (GaN). It is also possible to form semiconductors by combining elements from columns two and six, although these II-VI semiconductors, which include cadmium telluride (CdTe) and zinc selenide (ZnSe), are technologically less important. Furthermore, it is possible to combine two semiconductors to form an allow semiconductor. For example, InAs and GaAs can be combined to form the ternary semiconductor gallium indium arsenide ($Ga_xI_{1-x}As$), where the variable x ($0 \leq x \leq 1$) indicates the relative proportions of InAs and GaAs. The properties of an alloy semiconductor are approximately equal to the appropriate weighted average of the constituent semiconductors. Figure 2.6 show the variation of the band gap and lattice constant of $Ga_xI_{1-x}As$ as a function of x.

Both quantities vary smoothly between the values for InAs and GaAs. One important practical application of alloy semiconductors is that a specific band gap can be obtained by a suitable choice of composition.

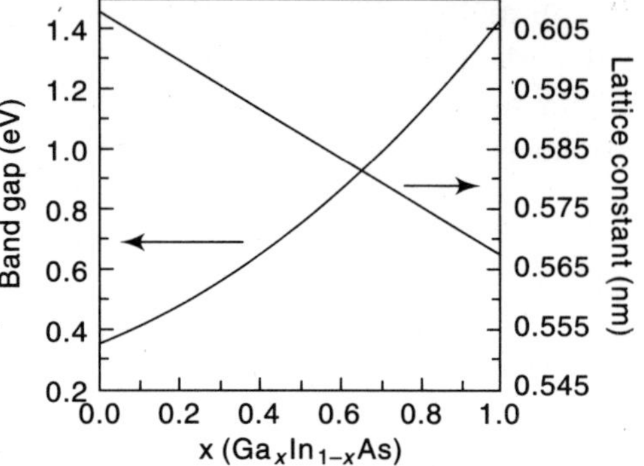

Fig. 2.6 *The composition variation of the lattice constant and band gap of the ternary alloy semiconductor $Ga_xIn_{1-x}As$*

Size of Semiconductor

At the present time, sizes are going down from the submicrometer region to the nanometer region, i.e., decreasing below 100 nm. Semiconductor nanostructures are structures having characteristics features with a typical size of 1–10 nm in the lateral plane.

Nanostructures have already appeared in the form of quantum wires or quantum dots (Fig. 2.7). The use of such structures to become mainstream throughout semiconductor technology is there.

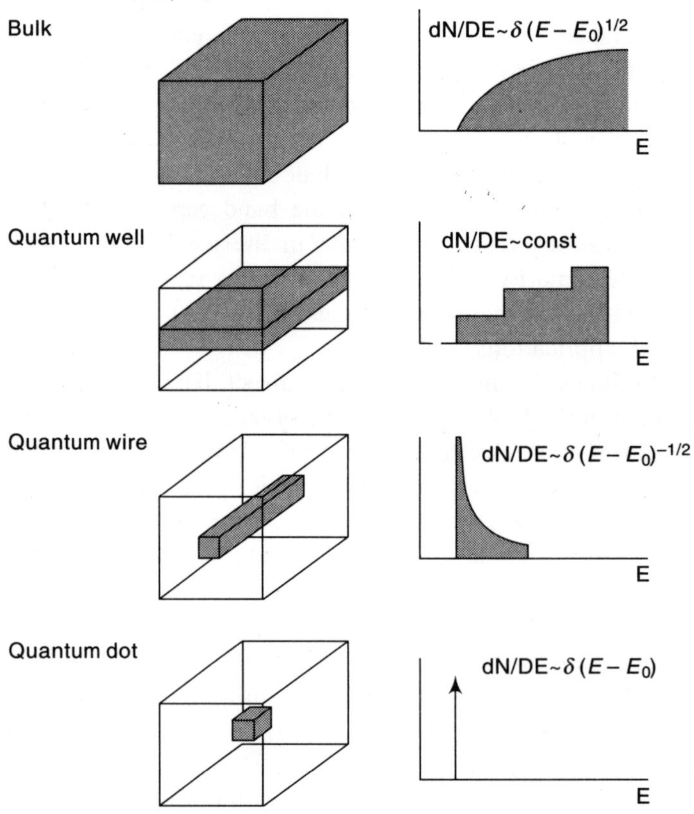

Fig. 2.7 *Semiconductor structures and respective electronic density of states near the edge of electronic band.*

The major advantages from QDs as compared with quantum wells.

Due to the discrete nature of their electronic spectrum, QD-based devices should exhibit a higher temperature stability and allow temperature-insensitive device operation. Quantum dot structures should obey certain requirements.

- The QD should not be too small, other wise it will not have localized states.
- The QD should not be too large, otherwise the spacing between energy levels becomes too small and may hinder the temperature stability of the structure. An estimate made by Ledenstov [1.4] for InAs QDs in a GaAs matrix suggests that the QD size should lie in the interval

$$4 \text{ nm} < L < 20 \text{ nm} \tag{1.4}$$

- The density of QDs should be rather high to ensure a high modal gain for lasers.
- QDs should be uniform in shape and size.
- A QD heterostructure should contain a low density of defects which produce centers of irradiative recombination.

The pn Junction

The majority of semiconductor devices are based on the pn junction, which is formed at the interface between two regions, one doped n-type the other *p*-type. In equilibrium a potential step is formed at the interface which prevents the net movement of electrons from the n-type region into the *p*-type region, and vice versa for holes. In addition, free carriers are absent from regions either side of the junction, forming a depletion region. A schematic diagram of a pn junction under equilibrium conditions is shown in Figure 2.8(a). If an external voltage of the correct sign if applied (Figure 2.8(b)) the potential step is reduced, allowing electrons and holes to move across the junction, a process known as injection. In a pn junction designed for optical applications, an undoped or intrinsic (i) region may be placed between the *n*- and *p*-type regions to form a *p-i-n* structure. Electrons and holes meet in the intrinsic region, where they recombine to produce photons. A nanostructure may be incorporated within the intrinsic region, providing a convenient and efficient mechanism for injection of electrons and holes into the nanostructure.

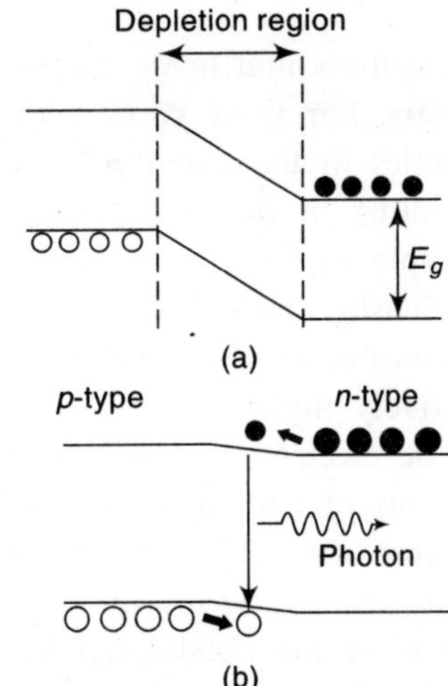

Fig. 2.8 *Schematic band diagrams of a pn junction (a) under equilibrium conditions and (b) with an external voltage applied to reduce the potential step, resulting in carrier injection across the junction.*

The bias condition for current injection is referred to as forward bias. Changing the polarity of the applied voltage produces reverse bias. In this case the potential step is increased and there is negligible current flow. However, electrons and holes created in the intrinsic region by photon absorption may be swept out into the *n*-and *p*-type regions, respectively, resulting in an electrical current that can be measured by an external circuit. This process allows a semiconductor to act as a photon detector.

Phonons

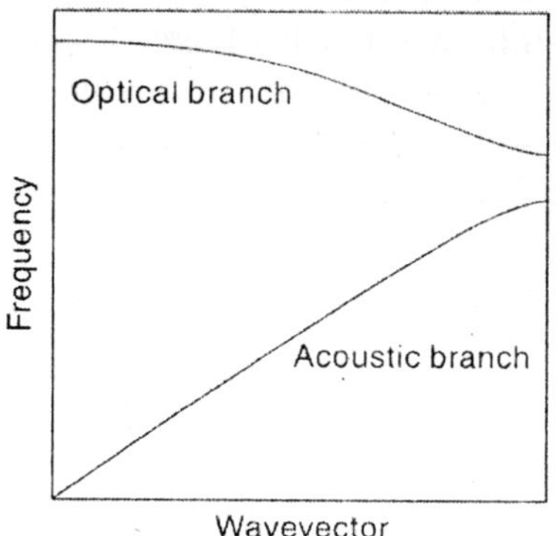

Fig. 2.9 *Schematic diagram of the phonon dispersion relationship for a non-dimensional linear chain consisting of atoms of alternating mass.*

Carriers in a solid may lose or gain energy by emitting or absorbing a phonon. Figure 2.9 shows the phonon dispersion—the frequency-wave vector relationship–calculated for a one-dimensional chain consisting of atoms of two alternating masses. This model provides a good approximation to real semiconductor. Of the two calculated branches, the lower or acoustic branch corresponds to the propagation of sound, and has a frequency which tends to zero for small wave vectors. The upper or optical branch corresponds to phonons which can interact with electromagnetic radiation, and has a frequency which remains non-zero for small wave vectors. For three-dimensional solids each branch consists of three sub-branches, corresponding to the three possible directions of the lattice vibrations with respect to the propagation direction, two transverse and one longitudinal. For GaAs and related semiconductors the strongest carrier-photon interaction occurs for longitudinal optical (LO) phonons, and it is these phonons that are preferentially emitted as carriers lose energy.

DOPING PROPERTIES OF SEMICONDUCTORS

Electrons and holes created by thermal excitation across the band gap are known as intrinsic carriers. For these carriers the density of electrons in the conduction band n equals the density of holes in the valence band p. Although n and p increase rapidly with increasing temperature (resulting in increased electrical conductivity), their absolute values are relatively small. For example, in Si (E_g = 1.12 eV) $n = p \sim 10^{15}$ cm^{-3} at 300 K, many orders of magnitude less than the density of electrons in the conduction band of a typical conductor ($\sim 10^{22}$ cm^{-3}). Consequently, a semiconductor's intrinsic electrical resistance, which is inversely related to its conductivity, is relatively high.

The electron or hole densities in a semiconductor can be increased by the addition of small amounts of certain impurities, a process known as doping. If atoms with one additional valence electron are added to the host semiconductor, such as phosphorus to silicon, the impurity atoms form a series of new states within the forbidden band gap, located slightly below the bottom of the conduction band. At $T = 0$ these states are occupied by the additional electrons but for $T \neq 0$ these electrons may be thermally excited into the conduction band, increasing the free electron density. Because the impurity states are relatively close to the conduction band, this excitation requires relatively little thermal energy, and at moderately high temperatures all the impurity atoms will lose their electrons to the conduction band. Therefore n is increased by an amount approximately equal to the density of impurity atoms, which are known as donors; this process is called *n-type doping*. Similarly, the introduction of impurity atoms with one fewer valence electron that the host semiconductor, such as boron to silicon, forms a series of levels slightly above the top of the valence band, which are unoccupied at $T = 0$. For $T \neq 0$ electrons may be excited into these states, leaving free holes in the valence band. This is called *p-type doping* and increases the free hole density by an amount approximately equal to the density of the impurity atoms, which are known as acceptors. Electrons or holes produced by doping are known as *extrinsic carriers* and for a doped semiconductor $n \neq p$. The concept of doping is summarised in Figure 2.10.

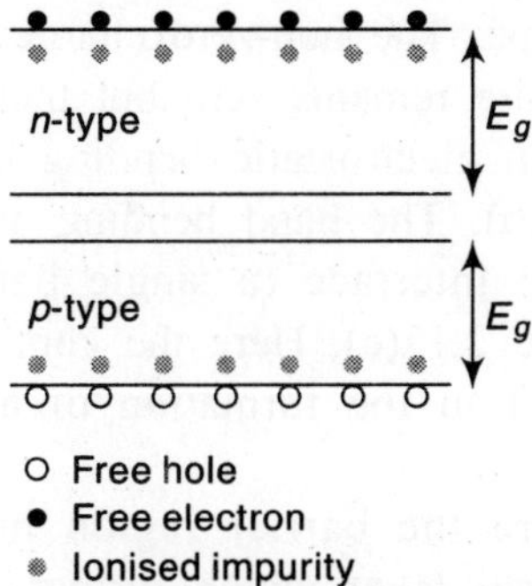

Fig. 2.10 *n- and p-type doping. Impurity states are formed just below the bottom of the conduction band and just above the top of the valence band for n- and p-type doping, respectively. The close proximity of these states to the respective band edges means the thermal excitation of extrinsic carriers is much more probable than intrinsic carriers excitation across the band gap.*

Modulation Doping (n-type Doping of a Quantum Well)

The low-temperature carrier mobility of a bulk semiconductor is limited by scattering from impurities. This mechanism is particularly efficient in doped semiconductors where the scattering results from the charged dopant atoms. Consequently, the low-temperature carrier mobility in a bulk doped semiconductor is very low. However, in a semiconductor nanostructure it is possible to spatially separate the dopant atoms and the resulting free carriers. This reduces impurity scattering, resulting in an extremely high carrier mobility at low temperature. A spatial separation of the dopant atoms and free carriers is achieved through remote or modulation doping as shown schematically for *n*-type doping of a quantum well structure in Figure 2.11(a). Here, the donor atoms are placed only in the wider band gap barrier material, the quantum well remains

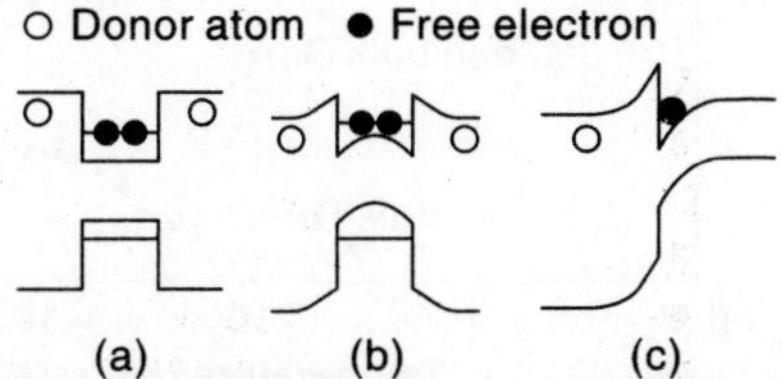

Fig. 2.11 *(a) n-type modulation doping of a quantum well showing the transfer of electrons from the barriers to the well; (b) the band edge profile of a modulation-doped quantum well, including the effects of band bending which results from the non-zero space charges in the well and barriers; (c) the production of a 2DEG in a modulation-doped single heterojunction.*

undoped. This is achieved during MBE growth by only opening the shutter in front of the cell containing the dopant atoms during growth of the barriers. In MOVPE the gas carrying the dopant atoms is similarly switched. Following thermal excitation of the electrons from the donors into the conduction band of the wider band gap semiconductor, these free electrons transfer into the lower-energy quantum well states, resulting in a spatial separation of the free electrons and the charged donor atoms. The confined electrons in the quantum well are said to form a two-dimensional electron gas (2DEG); a two-dimensional hole gas can similarly be formed by doping the barriers *p*-type. The non-zero charge present in both the barriers and the well (the total charge in the structure remains zero but there are equal and opposite charges in the well and barriers) results in an electrostatic bending of the conduction and valence band edges, as indicated in Figure 2.11(b). The band bending allows the formation of a modulation doping induced 2DEG at a single interface (a single heterojunction) between two different semiconductors, as shown in Figure 2.11(c). Here the combined effects of the conduction band offset and the band bending result in the formation of a triangle-shaped potential well that provides confinement of the electron.

In a modulation-doped structure the barrier region immediately adjacent to the well is generally undoped, forming a spacer layer which further separates the charged dopant atoms and the free carriers. By optimising the width of this spacer layer and the structural uniformity of the interface, and by minimising unintentional background impurities, it is possible to achieve an extremely high carrier mobility at low temperatures. Figure 2.12 compares the temperature variation of the electron mobility of standard bulk GaAs, a very clean bulk specimen of GaAs and a series of GaAs-AlGaAs single heterojunctions. At high temperatures, where mobility is limited by phonon scattering, the mobility of the different structures are very similar. At low temperatures the mobility in bulk GaAs is increased in the cleaner material,

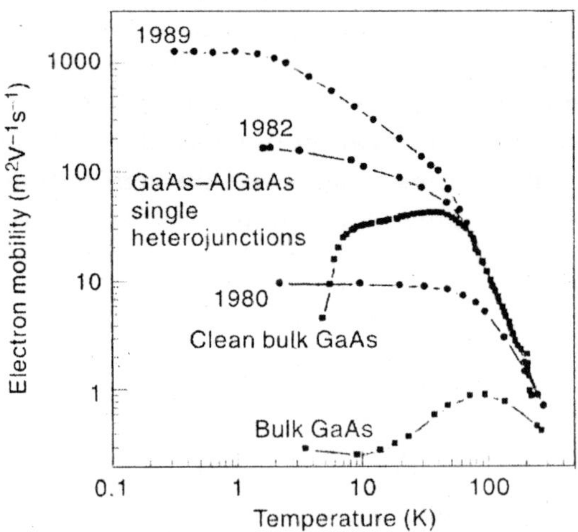

Fig. 2.12 *The temperature dependence of the electron mobility of two bulk GaAs samples having different purity, plus a series of n-type modulation-doped GaAs-AlGaAs single hetero-structures. The labels for the heterostructures give the year of growth and demonstrate the improvement of the low-temperature mobility with time.*

where a lower impurity density reduces the charged impurity scattering. However, the absence of doping results in a low carrier density and, as a consequence, a low electrical conductivity. In contrast, modulation doping result in high free carrier densities and a low-temperature mobility more than two orders of magnitude than that of the clean bulk GaAs sample. The data for the different heterojunctions presented in Figure 2.12 demonstrates how the low-temperature mobility has increased over time, reflecting optimisation of the structure, the use of purer source materials and improved cleanliness of the MBE growth reactor. A low-temperature mobility of 1200 $m^2V^{-1}s^{-1}$ is achieved for a GaAs-AlGaAs single heterojunction. e.g. 3000 $m^2V^{-1}s^{-1}$ for a modulation-doped 30 nm wide GaAs-AlGaAs quantum well is achieved.

The ability to produce 2DEGs exhibiting an extremely high mobility has allowed the observation of a range of interesting and novel physical processes. The modulation doping can be used to provide the channel of field effect transistors (FETs), where it is particularly useful for high-frequency applications. FETs that incorporate modulation doping are known as high electron mobility transistors (HEMTs) or modulation-doped field effect transistors (MODFETs). Although modulation doping provides only a minor enhancement of the room temperature carrier mobility, it produces free carriers that are confined within a two-dimensional sheet, in contrast to layer of non-zero thickness produced by conventional doping. This positioning of the carriers in the channel results in FETs which exhibit improved linear characteristics and, lower noise.

Effective Mass

Electrons and holes in a semiconductor are not free particles, possessing, in addition to kinetic energy, potential energy due to their electrostatic interaction with the charged ions. Particles with potential energy are considerably more difficult to describe mathematically then free particles, but in a solid this problem can be simplified by using the concept of effective mass. In this model the electrons and holes are treated as free particles by assigning them a modified mass, the effective mass, which combines their potential and kinetic energies into a single kinetic-like energy. The effective mass must be used in all equations describing the dynamical properties of carriers in a solid. The symbol for effective mass is m^* with, in general, a subscript e or h to indicate the effective mass of the electron or hole, respectively. Effective masses are expressed as a multiple of the free electron mass, and holes and electrons typically have different effective masses. As an example, the semiconductor GaAs has an electron effective mass $m_h^* = 0.067m_e$ and a hole effective mass $m_e^* = 0.35m_e0$, where m_e0 is the free electron mass.

CARRIER TRANSPORT, MOBILITY AND ELECTRICAL CONDUCTIVITY IN SEMICONDUCTORS

An externally applied voltage produces an electric field within a semiconductor and the results in an electrostatic force that acts on the charge carriers. This force produces an acceleration and hence motion of the carriers along the field direction; it is this motion which constitutes an electrical current. Carriers are accelerated by the electric field until they hit an obstacle, at which point their velocity is randomised. Following this collision, the acceleration recommences. This process is shown in Figure 2.13. If average time between any two collisions is τ, and mean velocity, (the carrier drift velocity) is v_d; the electrical conductivity is defined to be directly proportional to v_d; this is a measure of how easily the carriers are able to move through the semiconductor. At low fields v_d is proportional to the size of the electric field, hence a measurement-independent quantity can be obtained by dividing v_d by the field; the resultant quantity is the carrier mobility μ. It can be shown that the mobility is related to be scattering time by $\mu = e\tau/m^*$, where e is the electronic charge.

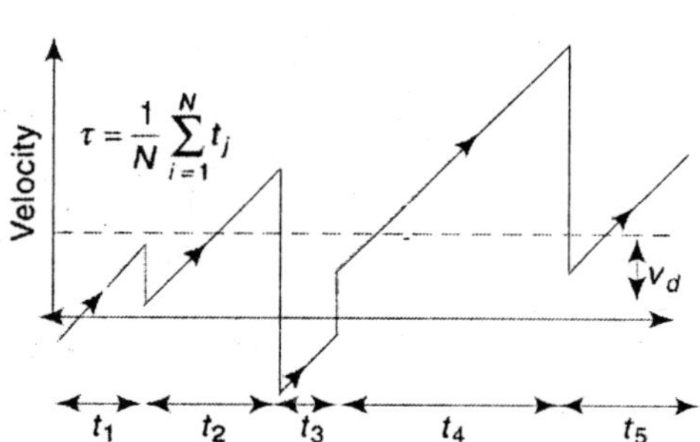

Fig. 2.13 *The dynamics of a charge carrier subjected a constant electric field. A mean scattering time, τ, can be defined by averaging over a large number of events, leading to an average drift velocity v_d*

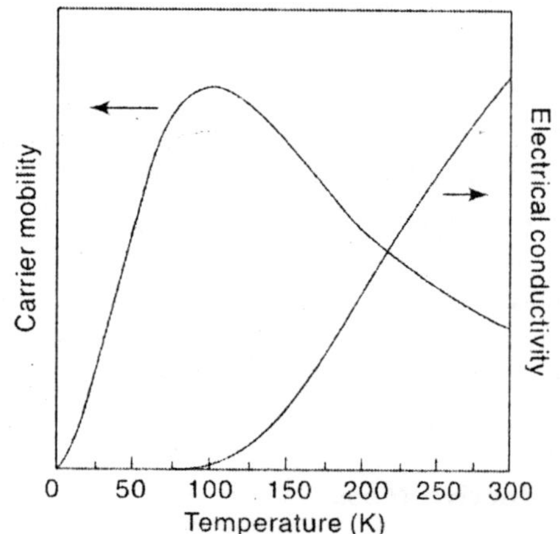

Fig. 2.14 *The temperature variation of the electrical mobility and electrical conductivity for a semiconductor.*

The mean time between collisions is dependent on the nature of the collisions. A carrier travelling through a periodic crystal lattice only experiences a collision if there is a local departure from the crystal periodicity. This can be result from the presence of an impurity atom, such as a dopant atom, or by thermal vibrations, of the lattice, the quanta of which are termed phonons. Scattering by impurity atoms is important at low temperatures but decreases with increasing temperature. In contrast, scattering by phonons increases with temperature, reflecting the increasing amplitude of the lattice vibrations. The combined effect of these two processes to give a mobility which, at low temperatures, increases with increasing temperature, followed by a decrease at high temperatures. This behaviour is shown schematically in Figure 2.14.

Ballistic Carrier Transport

The carrier transport is controlled by a series of random scattering events. However the high carrier mobilities which can be obtained by the use of modulation doping correspond to very long distances between successive scattering events, distances that can significantly exceed the dimensions of a nanostructures. Under these conditions a carrier can pass through the structure without experiencing a scattering events, a process known as ballistic transport. Ballistic transport conserves the phase of the charge carriers and leads to a number of novel phenomena, two of which will now be described.

When carriers travel ballistically along a quantum wire there is no dependence of the resultant current on the energy of the carriers. This behaviour results from a cancellation between the energy dependence of their velocity ($v = \sqrt{2E/m^*}$) and the density of states, which in one dimension varies as $1/\sqrt{E}$. For each occupied subband a conductance equal to 2r2/h is obtained, a behaviour known as quantised conductance. If the number of occupied subbands is varied then the conductance of the wire will exhibit a step-like, behaviour, with each step corresponding to a conductance change of $2e^2/h$. Quantum conductance is readily observable in electrostatically induced quantum wires. The gate voltage determines the width of the wire, which in turn controls the energy spacing between the subbands. For a given carrier density, reducing the subband spacing results in the population of a greater number of subbands, and hence increased conductance. Figure 2.15 shows quantum conductance in a 400 nm long, electrostatically induced quantum wire. The structure of the device is shown in the inset. Such measurement are generally performed at very low temperatures to obtain the very high mobilities required for ballistic transport conditions. In contrast to the highly accurate values observed for ρ_{xy} in the quantum Hall effect, which are independent of the structure and quality of the device, the quantised conductance values of a quantum wire are very sensitive to any potential fluctuations, which may result in scattering events. This sensitivity prevents the use of quantum conductance as a resistance standard.

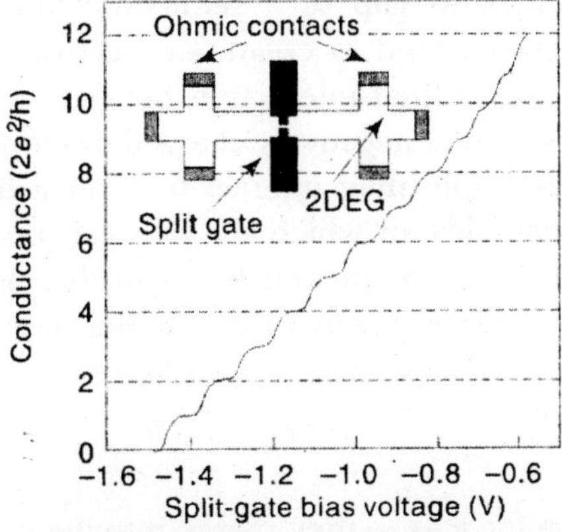

Fig. 2.15 *Quantised conductance steps in a 400 nm long electrostatically defined quantum wire measured at a temperature of 17 mK. The wire is produced by a split gate formed on the surface of a modulation-doped GaAs-AlGaAs heterostructure. The inset shows the form of the electrical contacts.*

The inset to Figure 2.16 shows a structure where a quantum, wire splits into two wires, which subsequently rejoin after having enclosed an area A. under ballistic transport conditions the wavefunction of an electron incident on the loop will split into two components which, upon recombining at the far side of the loop, will interfere. This process requires that the phase of the electron wavefunction is conserved as it transits the structure. If a magnetic field is now applied normal to the plane of the loop, the wavefunctions acquire or lose an additional phase, depending on the sense in which they traverse the loop. The phase difference between the two paths increases by 2π when the magnetic flux through the loop, given by the area multiplied by the field ($=BA$) changes by h/e. As the magnetic field increases, the system oscillates between conditions of constructive interference (corresponding to a high conductance) and destructive interference (corresponding to low conductance). The change in field, ΔB, between two successive maxima (or maxima) is given by the condition $\Delta BA = h/e$, resulting in the conductance of the system oscillating periodically with the field. An example of this behaviour, known as the Aharonov-Bohm effect, is shown in Figure 2.16 for a loop of diameter 1.8 μm, formed from the 2DEG of a GaAs-AlGaAs single heterostructure by electron beam lithography.

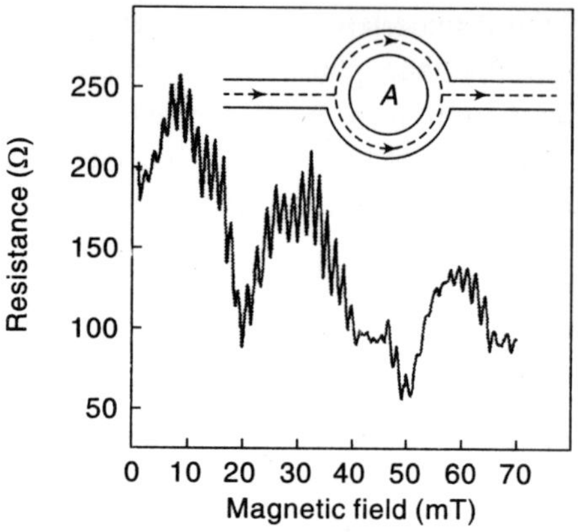

Fig. 2.16 *The aharonov-Bohm effect in a 1.8 μm diameter ring, measured at a temperature of 280 mM. The inset shows the geometry of the structure.*

Band Gap

The band gap of a semiconductor represents the energy required to create an electron and hole when there is no final interaction between the two carriers. However, the negatively charged electron and positively charged hole may interact to form a hydrogen-atom-like complex in which the two carriers orbit each other, a system known as an exciton. The electrostatic interaction between the electron and hole reduces their energy compared to the non-interacting case, resulting in a series of energy levels just below the conduction band edge. These excitonic states have discrete energies.

$$E_n = E_g - E_b/n^2 \quad (n = 1, 2, 3, \ldots, \infty)$$

where for $n \to \infty$ they merge into the continuum states of the conduction band. The binding energy of the exciton E_b is the energy difference between the lowest exciton state ($n = 1$) and the conduction band edge ($n = \infty$). In addition to the states formed below the conduction band edge, there is a modification of the states above the band edge, referred to as the Sommerfeld enhancement. Absorption into excitonic states is possible, and the inset of Figure 2.17 shows how the absorption of a bulk semiconductor is modified by the inclusion of excitonic effects. Although there are an infinite number of exciton states, their adsorption strength and separation both decrease rapidly with increasing n, and hence experimentally only absorption into the $n = 1$ state is generally observed. Figure 2.17 shows absorption spectra for the semiconductor gallium arsenide (GaAs) at low temperature and room temperature. At low temperature, excitonic effects are clearly visible in the spectrum as an enhanced

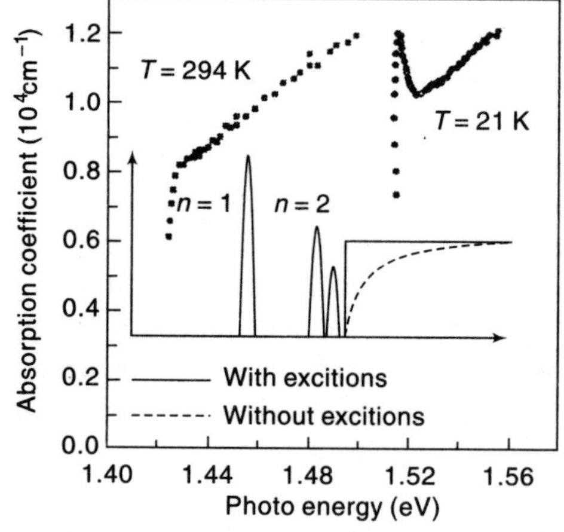

Fig. 2.17 *Low-temperature and room temperature absorption spectra of bulk GaAs. The energy shift between the two spectra results from the temperature variation of the band gap. The inset shows the density of states with and without the inclusion of exciton of excitonic effects.*

absorption close to the band gap. However, because the exciton binding energy in GaAs is only 4.2 meV, at room temperature there is sufficient thermal energy (k_BT = 25 meV) to ionise the majority of excitons. Hence excitonic effects are absent, or at most extremely weak, in GaAs and most other bulk semiconductors at room temperature. It is possible to significantly increase the exciton binding energy in a nanostructure, allowing the observation of excitonic effects at higher temperatures.

The Quantum Hall Effect

The Hall effect is a standard characterization technique that can be used to determine both the density and type of majority carrier in a semiconductor. The inset to Figure 2.18 shows a schematic diagram of a Hall measurement, which can be applied to either a bulk semiconductor or a suitable nanostructure. An external voltage source causes a current I_x to flow along the bar, resulting in a current density J_x. On application of a magnetic field B_z, applied normal to the plane of the sample, a lateral electric field E_y is produced, which appears as a voltage V_y measured across the sample. The quantity E_y/B_zJ_x is known as the Hall coefficient, R_H, and for a bulk sample in which transport is dominated by one tube of carrier (electrons or holes) R_H = $1/ne$, which n is the free carrier density. The magnitude and sign of R_H allow determination of the free carrier density and the type of majority carrier, respectively.

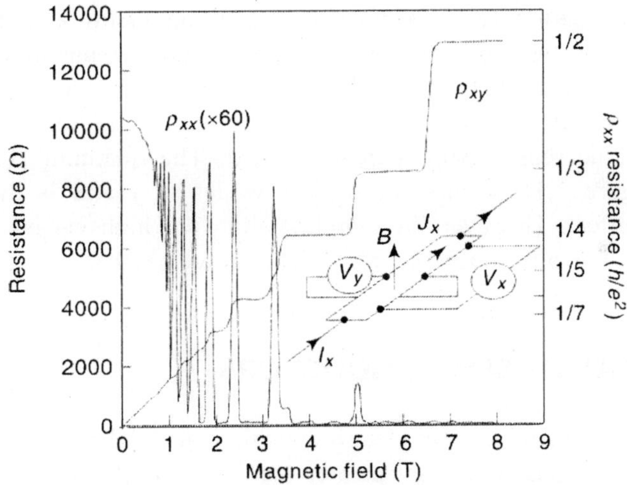

Fig. 2.18 *The integer quantum Hall effect as measured for a 2DEG in a GaAs-AlGaAs single heterojunction at a temperature of 50 mK. The inset shows the experimental geometry used for Hall effect measurements.*

Experimentally the electric field along the sample, E_x, can also be determined by measuring V_x as shown in Figure 2.15. This allows two resistivities to be defined and measured:

$$\rho_{xx} = \frac{E_x}{J_x}, \qquad \rho_{xy} = \frac{E_y}{J_x}. \tag{1}$$

For a bulk semiconductor $R_H = E_y/B_zJ_x$, hence $\rho_{xy} = R_HB_z \cdot \rho_{xy}$ therefore increases linearly with increasing magnetic field, whilst ρ_{xx} remains constant. However, for a two-dimensional system, a very different behaviour is observed, as shown in Figure 2.18. In this case, although ρ_{xy} increases with increasing field, it does so in a step-like manner. In addition, ρ_{xx} oscillates between zero and non-zero values, with zeros occurring at fields where ρ_{xy} forms of plateau. The surprising behaviour is known as the quantum Hall effect.

The quantum Hall effect arises as a result of the form of the density of states in a two-dimensional system in a magnetic field. This corresponds to that of a fully quantised system, with quantization in one direction resulting from the physical structure of the sample and in the remaining two directions by the magnetic field. As full discussion of the physics underlying the quantum Hall effect, including the importance of disorder, is beyond the scope of this book. Suggestions for further reading are given at the end of this chapter.

An importance practical application of the quantum Hall effect arises from the plateau values of ρ_{xy}. It can be shown that these are given by

$$\rho_{xy} = \frac{1}{j}\frac{h}{e^2} = \frac{25813}{j}\,\Omega, \tag{2}$$

where j is an integer whose value decreases with increasing magnetic field. ρ_{xy}, which is independent of the sample, can be measured to be high accuracy and is now used as the basis of the resistance standard and also to calculate the fine structure constant $\alpha = \mu_0 ce^2/2h$, where the permeability of free space, μ_0, and the speed of light, c, are defined quantities.

The parameter j in Equation (2) is known as the filling factor and denotes the number of different states occupied by the free carriers. The degeneracy (the maximum number of electrons or holes a given state can contain) of the states formed in a magnetic field increases with increasing field; consequently, for a constant total electron number, the number of states occupied, and hence j, decreases with increasing field. The quantum Hall effect discussed so far, and shown in Figure 2.18 occurs for integer values of j and is therefore known as the integer quantum Hall effect. However, in samples with very high carrier mobilities, plateaue in ρ_{xy} and minima in ρ_{xx} are also observed for fractional values of j, giving rise to the fraction quantum Hall effect.

OPTICAL PROPERTIES OF SEMICONDUCTORS

The application of a semiconductor in electro-optical devices relies on their ability to efficiently emit or detect light. If photons of energy greater than or equal to the band gap are incident on a semiconductor, they may excite an electron from the valence band to the conduction band. In this process the photon is destroyed (absorbs) and an electron and hole are created. In the reverse process an electron in the conduction band may return to the valence band and recombine with a hole; the energy lost by the electron creating a photon. As the energies of the electron and hole will generally be very close to the bottom of the conduction band and the top of the valence band respectively, the emitted photon will have an energy approximately equal to the band gap of the semiconductor.

Optical and Electrical Characterization

(A) The most commonly applied optical characterisation technique is *photoluimunescence (PL)*. PL involves creating electrons and holes by illuminating the structure with photons of sufficient energy, generally using light from a laser. Typically photon energies are used such that absorption occurs in the barriers, as this produces a relatively high density of electrons and holes and therefore results in a strong PL signal. After creation the carriers diffuse spatially and are captured and localised in different parts of the structure. This is followed by relaxation as the carriers lose any excess kinetic energy, generally reaching the lowest possible energy states before recombining to emit a photon. The energies of the emitted photons are determined using a spectrometer and suitable detector.

The carrier transport efficiency between different parts of a nanostructure affects the regions which contribute to the PL. In quantum wells and self-assembled quantum dots there is a very rapid transfer of carriers from the barriers into the lower energy states provided by the wells or dots. As a result, emission from the barriers, and from the wetting layer in quantum dots structures, in generally not observed. However, in nanostructures where the different regions are spatially well separated, efficient carrier transport may not be possible, and a number of regions may contribute to the PL. For example, in V-groove quantum wire structures the quantum wires, the quantum wells formed on the sides of the grooves and between the grooves, and the bulk GaAs may all give distinct emission, as shown in Figure 2.19. This behaviour complicates the interpretation of the emission spectrum and is a serious disadvantage for optical devices where emission at a single energy is generally required.

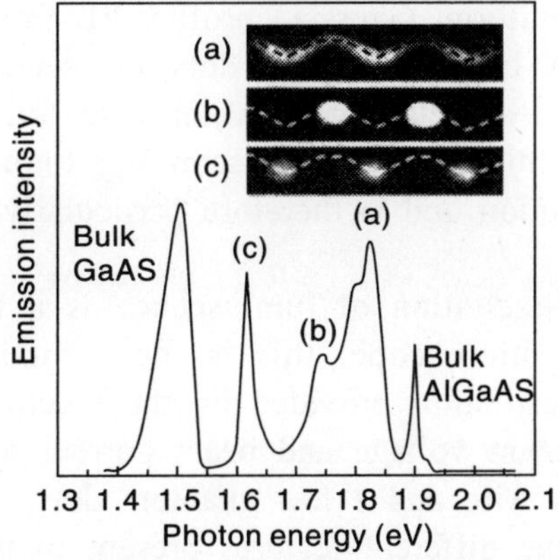

Fig. 2.19 *Cathodoluminescence (CL) spectrum of a V-groove quantum wire structure showing emission from the different regions of the structure. Scanning CL images are shown recorded for the detection of emission lines (a), (b) and (C). The forms of these images are consistent with line (a) arising from carrier recombination in the quantum wells formed on the side walls of the grooves, (b) quantum wells formed on the planar regions between the grooves, and (c) the quantum wires at the bottom of the grooves. The dashed lines in the images indicate the approximate position of the pre-growth surface.*

In conventional PL measurements the exciting laser beam is focused to a spot of diameter ~ 100 μm. Because of the high density of dots or wires in a typical structure, this results in the simultaneous measurement of many nanostructures. For example, V-groove quantum wires are typically spaced by ~ 1 μm and self-assembled quantum dots may have a density ~ 10^{11} cm^{-2}.

For these densities a spot of diameter 100 μm excites 100 wires or 100 million dots. Consequently, the PL spectra are inhomogeneously broadened due to unavoidable fluctuations in the wire or dot shape, size and composition. Although the magnitude of this broadening provides information on the homogeneity of the nanostructure, it prevents processes that occur on smaller energy scales, such as the perturbation of the emission energy of a quantum dot as additional excitons are added.

In order to study physical processes occurring on an energy scale smaller than the inhomogeneous broadening, it is necessary to probe individual dots or wires. This can be achieved through reducing the size of the focused laser beam, either by using-aperture microscope objective, for which a diffraction-limited spot size of ~1μm is possible, or by using a scanning near-field optical microscope (SNOM) that allows the diffraction limit to be circumvented. The typically large quantum dot densities, necessitates the use of additional steps to prove single quantum dots. This generally involves the reduction of the dot density, achieved by modifying the growth conditions, followed by the physical isolation of a single dot, either by etching submicron mesas of forming small holes in an otherwise opaque metal surface mask. Example: single quantum dot spectroscopy.

Carriers may also be created electrically by placing a nanostructure in a *p-i-n* device. In this case the process of light emission is known as electroluminescence (EL). EL has the advantage that the rate of carrier injection is uniform across the area studied, in contrast to PL, where the incident laser beam has a non-uniform, Gaussian profile. EL from a single quantum dot may be observed by exciting a relatively large number of dots and selecting the emission from a single dot by using a small aperture in a metal mask, which also forms one of the contacts. EL can also be excited using current injection from a scanning tunneling microscope (STM). This method offers high spatial resolution and is therefore particularly suitable for the study of single quantum dots.

A third mechanism for the excitation of luminscence is a beam of high-energy electrons, such as found in an electron microscope: this is the cathodoluminscence (CL) technique. Although the nominal spatial resolution provided by the electron beam is degraded by carrier diffusion, a careful choice of beam voltage and beam current makes it possible to observe the emission from a single quantum wire and a few quantum dots. One powerful application of CL is in identifying the origin of the different features present in the emission spectra of complex structures; for example V-groove quantum wires. Initially the emission spectrum for excitation of a large area is recorded. The system is then set to detect photons corresponding to one of the emission features, and the electron beam is raster scanned over the sample. Regions of the structure responsible for the selected emission appear bright in the resultant image, allowing their position and shape, and hence their origin, to be determined. This procedure is then repeated for the different emission features. Scanning CL images from the cleaved edge of a V-groove quantum wire structure are shown in Figure 2.19. Detecting photons corresponding to line (c) results in emission located at the bottom of the grooves, and therefore originating from the quantum, wires. Lines (a) and (b) result in emission from the side walls of the grooves and the flat surfaces between the grooves, consistent with emission from the side and top quantum wells, respectively. These wells have different thickness, hence they emit at different energies, a result of the dependence of the GaAs growth rate on surface orientation.

Carrier relaxation is generally much faster than radiative recombination hence only the lowest-energy, ground state transition is observed in emission. Emission from higher energy states can be observed by increasing the carrier excitation or injection rate so that the population of carriers in the ground state is sufficient to block relaxation from the excited states. Although high carrier injection allows the excited states to be observed, the system is highly occupied and the interaction between the carriers may significantly perturb the energies of the transitions. In addition, emission techniques do not allow the true relative strengths of the transitions to be determined, as the emission intensity is dependent not only on the intrinsic transition strength but also on the carrier population in the initial and final states.

A determination of the unperturbed energies and the strengths of optical transitions therefore require the use of an absorption technique. However, it is very difficult to measure the direct absorption of a single nanostructure because only a very small fraction of the light is absorbed in comparison to the majority of the light that simply passes through the sample. Measuring this small change in transmitted light against the large background is technically very difficult. For quantum wells it is possible to use multiple-well structure to increase the absorption to a measurable level, and for colloidal and nanocrystal dots it is possible to produce films or solutions containing the dots of sufficient thickness to allow direct absorption measurements. However, absorption measurements of epitaxially grown wires and dots are more difficult because, as they occupy only as relatively small fraction of the cross-sectional area, their intrinsic absorption is very low. The absorption spectrum of a single layer of self-assembled quantum dots can be measured, but this requires the use of very sensitive and expensive commercial systems. Even with such systems the very small absorption signal requires the use of extremely long integration times, typically a few hours. By focusing a laser beam to a spot size <1 μm, using a large-aperture microscope objective, it is possible measure the absorption spectrum of a single-assembled quantum dot, whose physical cross sectional is now a reasonable fraction of the laser beam area. However, the absorption is still too low to allow the direct measurement of the absorption. Consequently, a modulation technique consisting of an oscillating electric field that varies the transition energies via the quantum-confined Stark effect is used. The resulting spectra correspond to the first derivative of the absorption.

Because of these experimental difficulties, absorption studies of quantum wires and dots generally makes use of an absorption-related technique photocurrent (PC) spectroscope and photoluminescence excitation (PLE) are the main examples. In PC, incident photons create electrons and holes in a nanostructure, which is placed in the intrinsic region of a p-i-n device. Under suitable conditions the carriers are able to escape from the nanostructure before be recombining , giving rise to a current that can be measured by an external circuit. As the energy of the incident photons is varied, a change in absorption will alter the number of carrier created, and hence the magnitude of the photocurrent. In PLE the intensity of the photoluminescence is monitored as the energy of the exciting photons is varied. A change in the absorption alters the number of carriers created, and hence the intensity of the photoluminscence. Both PC and PLE measure a small signal against a zero background; although the majority of incident light still passes through the structure, this does not contribute to the measured signal. PC and PLE and therefore referred to as background-less technique and are particularly suited to the study of nanostructures that absorb only very weakly. However, both techniques involves an additional step beyond photon absorption. In PC the photo-carriers must escape from the nanostructure and in PLE they must relax to the transition being monitored. If the probability of these processes is not constant, but depends on the initial energy of the carriers, then the form of the resultant spectra will not reflect the true absorption.

(B) A further important class of optical characterisation techniques is *time-resolved spectroscopy.* Here a structure is excited by a very sort pulse of light, and the subsequent temporal changes in its properties are determined as carriers recombine or relax in energy. The pulse length and time resolution of the measurement system are typically in the range 1 ns to 0.1 ps; the precise value is dependent on the physical processes being studied. Time-resolved spectroscopy has been used to study a range of carrier processes in semiconductor nanostructures, including the rate at which carriers are captured from the barriers into the nanostructure, the rate at which carriers relax between confined levels, dephasing times and recombination times. Figure 2.20 shows results form a study of carrier relaxation mechanisms in InGaAs self-assembled quantum dots. Electrons and holes are initially created in the GaAs barriers by 1.5 ps pulses of light. The PL from the ground state of the quantum dots is detected and the rise time of this emission provides an indication of the time taken for carriers to be captured into the dots, followed by relaxation to their ground state. In Figure 2.20 the photoluminesence rise time is plotted as a function of the laser power, which is used to vary the density of carriers created in the structures. For low powers, equivalent to less than one electron and hole per dot, the rise time is relatively long, and reflects the slow relaxation of carriers between the discrete states of the dots by the emission of multiple phonons. This slow rise time is retained until the laser power becomes sufficient to excite an average of one electron and hole per dot, indicated by the vertical dashed line in the figure. Above this power the rise time starts to decrease, reflecting carrier capture and relaxation by a much faster process in which the energy lost by one carrier is transferred to a second carrier. This mechanism, known as an Auger process, requires an electron or hole in addition to the relaxing carriers, hence it only becomes possible above an average electron-hole number of one per dot. Further increase in the number of carriers per dot increases the efficiency of this process, resulting in the continuous decrease of the rise time observed at high laser powers. The insets to Figure 2.20 show the carrier relaxation process relevant to the low and high carrier density regimes.

Electrical measurements in their basic form involve the determination of currents and voltages from which samples resistance and electrical conductivity can be obtained. With the addition of a magnetic field, the integer and fraction quantum Hall effects can be studied, and measurements at very low fields allow a determination of carrier mobility. Two-dimensional carrier densities are determined from the period of the ρ_{xx} oscillations in the integer quantum Hall regime; this is also known as the Shubnikov-de Hass effect. Information on the dominant carrier scattering mechanism may be obtained by studying the temperature variation of the carrier mobility.

The capacitance of a quantum dot structure is affected by the charge state of the dots and this provides a method for probing the number of electrons or holes which have been loaded into a dot. A refinement of this technique is deep-level transient spectroscope (DLTS), which measures the temporal evolution of the capacitance following the application of a voltage pulse.

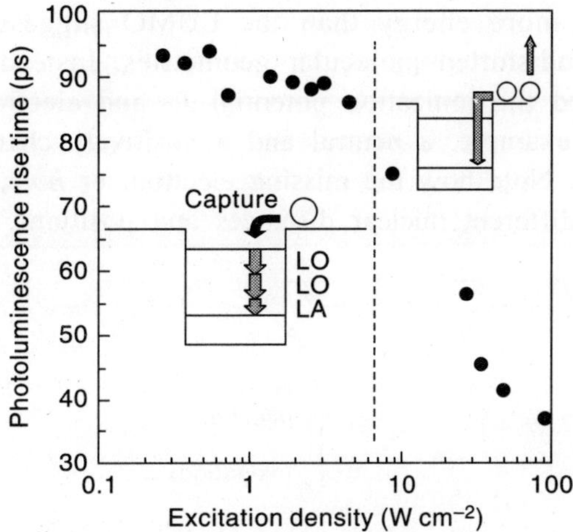

Fig. 2.20 *The rise time of the PL from InGaAs self-assembled QDs measured as a function of laser power density. The rise time indicates the time required for the carriers to be captured by the dots followed by energy relaxation to the ground state. Two possible realization processes are shown schematically. The vertical dashed line indicates the laser power corresponding to the creation of an average of one electron and hole per quantum dot.*

If the size and sign of the pulse are chosen such that excess carriers are loaded onto the dot, then a transient change in the capacitance will occur. Following removal of the pulse, the capacitance will revert back to its original value as the excess carriers escape from the dot. By measuring the rat eat which the capacitance recovers it is therefore possible to determine the carrier escape rate, and measurements as a function of temperature allow the height of the potential barrier confining the carriers to be determined.

Theory of Semiconductors

In any semiconductor application, the material will not be in its ground state. To transport charge and/or emit light, the semiconductor needs to sustain excitations, and in the case of charge transport, these *excitations* also need to be mobile.

When an electron is taken away from the HOMO or added to the LUMO of the molecule, the resulting molecule is termed a radical ion; namely a radical cation for positive charge, and radical anion for negative charge. The word 'radical' refers to the net spin the molecule will have due to the unpaired remaining (or added) electron. After removal or addition of the electron, molecular orbitals and the positions of nuclei will respond by a relaxation to a new position of minimum energy. Radical ions are often called *polarons*, analogous top the term used for inorganic semiconductors. Again, however the inorganic polaron is delocalized with an associated wavevector **k** describing its coherent movement; the radical ion is not delocalized.

Due to the localized character of the polaron, its charge strongly couples to molecular geometry. Bond distances and angles will be distorted compared to the neutral molecule. This distortion will always reduce the energetic cost of forming a polaron. Therefore removing an electron costs somewhat less energy than the HOMO suggests, and an electron joining the molecule gains somewhat more energy than the LUMO suggests, because HOMO/LUMO levels are calculated for undistorted molecular geometries. Instead, the energies required for polaron formation are called the ionization potential I_p, and electron affinity E_a, respectively. As a practically important example, a neutral and a positively charged polythiophene segment are sketched in Figure 2.21. Note how the missing electron, or *hole*, leads to a redistribution of the π-bonds and hence to different nuclear distances and positions.

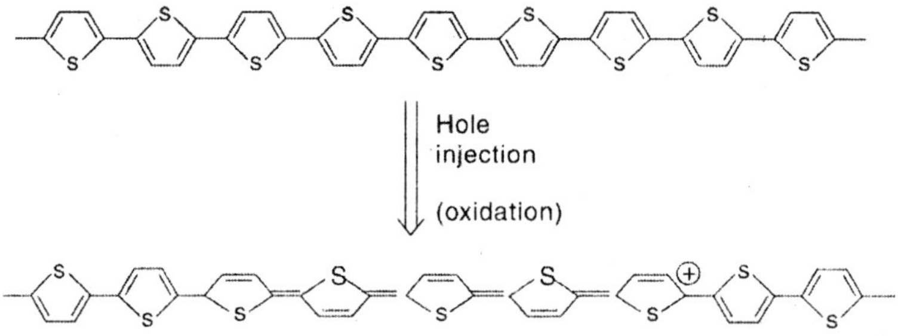

Hole injection (oxidation)

Fig. 2.21 *A polythiophene segment and the derived radical cation*

What an electron is removed from the HOMO but is placed into the LUMO instead of being removed entirely from the molecule, we arrive at an electrically neutral excitation, the so-called *exciton*. A typical way of lifting an electron from the HOMO into the LUMO if via the absorption of a photon. The π electrons redistribute into the excited π^* orbitals, which are also known as antibonding orbitals, as they destablize the molecule. Nevertheless, the strong σ bonds are crucial in keeping the molecule intact. Again, the excitation leads to a related relaxation of the surrounding crystal lattice or molecule. Figure 2.22 shows the geometric relaxation and redistribution of electron density in an excited phenylene-vinylene segment.

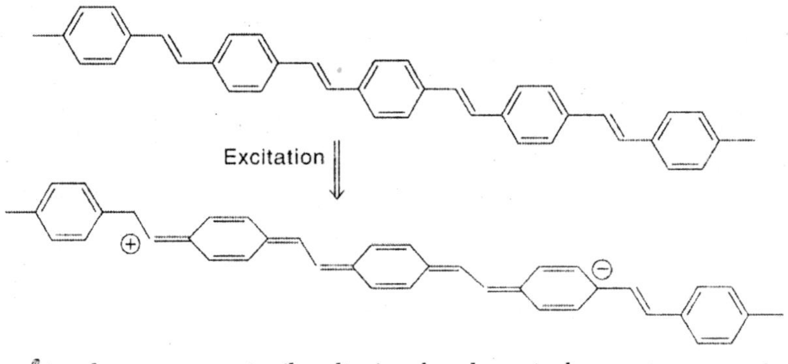

Excitation

Fig. 2.22 *The transition from an aromatic, 'bonding' π phenylene-vinylene system to a quinoidal, 'antibonding' π system on optical or electrical excitation.*

The size of the exciton is about three repeat units, or 10 nm, and the exciton has clearly intramolecular, one-dimensional character. This makes organic excitons *Frenkel* excitons. Typical exciton binding energies E_b are in the range 0.2–0.5 eV. Note how considerable ambiguity arises in the term 'band gap', which can mean either the energy difference between LUMO and HOMO, or $I_p - E_a$ or $(I_p - E_a) - E_b$.

Figure 2.23 is an alternative, more schematic representation of the electronic ground state, radical ions (here called polarons), and excitons in organic semiconductors. In shows two different types of exciton: *singlet* and *triplet* excitons. There are three ways in which hole and electron spin can combine so that the resulting overall spin part of the wavefunction is symmetric under particle exchange, and has total spin $S = 1$, namely $|\uparrow\uparrow\rangle$, $|\downarrow\downarrow\rangle)$, and $(1/\sqrt{2})(|\uparrow\downarrow\rangle + |\downarrow\uparrow\rangle)$. Excitons with that property are called triplet excitons. The combination of spins $(1/\sqrt{2})(|\uparrow\downarrow\rangle - |\downarrow\uparrow\rangle)$ results in a spin party of the wavefunction that is antisymmetric under particle exchange, and total spin $S = 0$. This combination is called a singlet exciton.

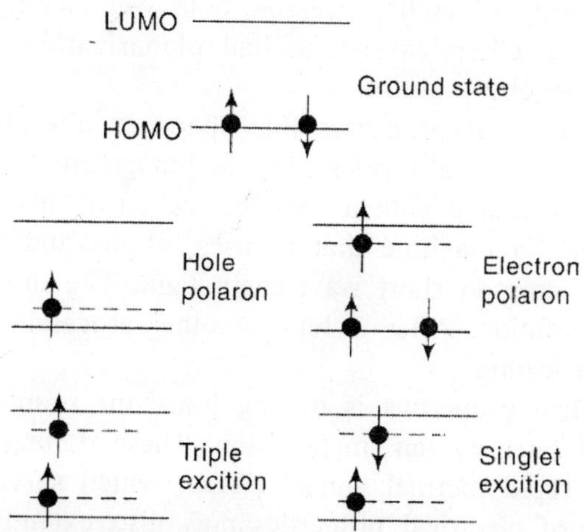

Fig. 2.23 *Energy level diagrams for excitations in organic semiconductors*

SEMICONDUCTOR DEVICE FABRICATION

Semiconductor Device Fabrication is the process used to create chips, the integrated circuits that are present in everyday electrical and electronic devices. It is a multiple-step sequence of photographic and chemical processing steps during which electronic circuits are gradually created on a wafer made of pure semiconducting material. Silicon is the most commonly used semiconductor material today, along with various compound semiconductors.

A typical wafer is made out of extremely pure silicon that is growth into mono-crystalline cylindrical ingots upto 300 nm in diameter. These ingots are then sliced into wafers about 0.75 mm thick and polished to obtain a very regular and flat source.

Once the wafers are prepared, the desired semiconductor integrated circuit are obtained by the following four steps.

- Front end processing
- Back end processing
- Test
- Packaging

PROCESSING

In semiconductor device fabrication, the various processing steps fall into four general categories: deposition, removal, pattering, and modification of electrical properties.

(a) Deposition is any process that grows, coats, or otherwise transfers a material onto the wafer. Available technologies consist of physical vapor deposition (PVD), chemical vapor deposition (CVD), electrochemical deposition (ECD), molecular beam epitaxy (MBE) and more recently, atomic layer deposition (ALD) among others.

(b) Removal processes are any that remove material from the wafer either in bulk or selective form and consist primarily of etch processes, both wet etching and dry etching such as reactive ion etch (RIE). Chemical-mechanical planarization (CMP) is also a removal process used between levels.

(c) Pattering covers the series of processes that shape or alter the existence shape of the deposited materials and is generally referred to as lithography. For example, in conventional lithography, the wafer is coated with a chemical called a "photoresist". The photoresist is exposed by a "stepper", a machine that focuses, aligns and moves the mask, exposing select portions of the wafer to short wavelength light. The unexposed regions are washed away by a developer solution. After etching or other processing, the remaining photoresist is removed by plasma ashing.

(d) Modification of electrical properties is doping transistor sources and drains originally by diffusion furnaces and later by ion implantation. These doping processes are followed by furnace anneal or by rapid thermal anneal (RTA) which serve to activate the implanted dopants. Modification of electrical properties now also extends to reduction of dielectric constant ion low-k insulating materials via exposure to ultraviolet light in UV processing (UVP).

It is the formation of transistors directly on the silicon. The raw wafer is engineered by the growth of an ultrapure, virtually defect-free silicon layer through epitaxy. In the most advanced logic devices, *prior* to the silicon epitaxy step, tricks are performed to improve the performance of the transistors to be built. One method involves introducing a "straining step" where in a silicon variant such as "silicon germanium" (SiGe) is deposited. Once the epitaxial silicon is deposited, the crystal lattice becomes stretched somewhat, resulting in improved electronic mobility. Another method, called "silicon on insulator" technology involves the insertion of an insulating layer between the raw silicon wafer and the thin layer of subsequent silicon epitaxy. This method results in the creation go transistors with reduced parasitic effects.

Silicon Dioxide

Front end surface engineering is followed by: growth of the gate dielectric, traditionally silicon dioxide (SiO_2), patterning of the gate, pattrerning of the source and drain regions, and subsequent implantation of diffusion of dopants to obtain the desired complementary electrical properties. In memory devices, storage cells, conventionally capacitors, are also fabricated at this time, either into the silicon surface or stacked above the transistor.

Once the various semiconductor devices have been created they must interconnected to form the desired electrical circuits. This "Back End of Line" (BEOL–the latter portion of the front end of water fabrication involves creating metal interconnecting wires that are isolated by insulating dielectrics. The insulating material is traditionally a form of SiO_2 or a silicate glass, but recently new low dielectric constant material are being used. These dielectrics take the form of SiOC and have dielectric constants around 2.7 (compared to 2.9 for SiO_2). The metal wires consist of aluminum. In this approach to wiring often called "subtractive aluminum", blanket atoms of *aluminum are deposited first*, patterned, and then etched, leaving isolated wires. Dielectric material is then deposited over the exposed wires. The various metal layers are interconnected by etching holes, called "vias", in the insulating material and tungsten is deposited in them with a CVD technique. This approach is used in the fabrication of dynamic random access memory (DRAM) chips; as the number of interconnect levels is small, not more than four.

Recently, as the number of interconnect levels for logic has increased due to the large number transistors that are interconnected in a modern microprocessor, the timing delay in the wiring has become significant; prompting a change in wiring material from aluminum to copper and from silicon dioxides to newer low-K material.

This performance enhancement eliminates processing steps. Here, *the dielectric material is deposited first* as a blanket film and is patterned and etched leaving holes or trenches. In a "single" processing, copper is then deposited in the holes or trenches surrounded by a thin film resulting in filled vias or wire "lines" respectively. In "dual" technology, both the trench and via are fabricated before the deposition of copper resulting information of both the via and line simultaneously, further reducing the number of processing steps. The thin barrier film, called Copper Barrier Seed (CBS), is necessary to prevent copper diffusion into the dielectric. The ideal barrier film is effective, but is barely there. As the presence of excessive barrier film competes with the available copper wire cross section, formation of the thinnest yet continuous barrier represents one of the greatest ongoing challenges in copper processing today.

As the number of interconnect levels increases, planarization of the previous layers is required to ensure a flat surface prior to subsequent lithography. Without it, the levels would become increasingly crooked and extent outside the depth of focus of available lithography, interfering with the ability to pattern. CMP (Chemical Mechanical Polishing) is the primary processing method to achieve such planarization although dry "etch back" is still sometimes employed if the number of interconnect levels is no more than three.

Once the Front End Process has been completed, the semiconductor devices are subjected to a variety of electrical tests to determine if they function properly. The proportion of devices on the wafer found to perform properly is referred to as the yield.

The fab tests the chips on the wafer with an electronic tester that presses tiny probes against the chip. The machine marks each bad chip with a drop of dye. Chips are often designed with "testability features" to speed testing.

A good chip design made by a good process will have more than 90% yield.

Semiconductor Processing by TEM

During the processing of a silicon wafer into devices. Specific microstructures must be formed repetably with a high degree of precision. Metal, insulator, and semiconductor layers of well-controlled thickness and composition must be deposited and patterned to create a composite structure with appropriate properties such as electrical resistivity or diffusion resistance. As the critical dimensions for devices become smaller, it is becoming necessary to specify precisely the crystal structure, the interface morphology, and the shapes and sizes of individual features, and control at the atomic scale is essential.

Reaction mechanisms and the relationship between structure and properties at all levels of integrated circuit fabrication are investigated the example; consider the relationship between the quality of an ultrathin gate oxide layer and the initial silicon surface on which it is grown. It is found that interface steps do not move during oxidation it has therefore become clear that the initial silicon surface must be extremely flat if a uniform oxide with a smooth interface is to be created. In the field of Al- and Cu-based chip metallization, study of the dynamics of voids has led to designs of the lines and vias which mitigate the effect of electromigration. For novel devices such as single-electron transistor or solid-state lasers, an understanding of the growth mechanism of nanosize islands, or "quantum dots," allows these islands to be formed controllably and devices to be fabricated.

Many analytical techniques are available for studying growth and processing during the fabrication of electronic devices. Transmission electron microscopy (TEM), in combination with advanced specimen preparation methods is an indispensable technique for examining the layered structures which make up semiconductor devices. TEM has been used successfully, to identify phase in polycrystalline films, to measure film texture, and in the atomic scale imaging of relevant interfaces, most TEM investigations have relied on postprocessing examination of device structure. A full understanding of processing must also include the transient effects which occur during heating, exposure to gases, or deposition, and these effects may be time-consuming or impossible to capture by conventional Tem analysis. Therefore a transmission electron microscope in which many of the processing steps important in device fabrication can be reproduced is used, to observe growth and phase transitions and understand their mechanism and processes.

Consider the following two experiments. In the first the growth of nanosize semiconductor islands, or "quantum dots," by self-assembly during strained-layer epitaxy is described and shows how grow islands in the TEM allows us to observe the phenomenon of self-assembly in real time. By observing the evolution of islands sizes shapes, one can determine how to optimize processing conditions to achieve the narrow size distribution required for quantum dot devices. Furthermore, observations of islands nucleation suggested ways to pattern the islands in particular regions of the specimen. Together these results indicated that self-assembly is a viable technique for forming quantum dots for novel devices. In the second example, a phase transition in the Ti–Si system is described. The transition from the C49 phase of TiSi2, which has a high resistivity, to the C54 phase, having a much lowest resistivity, much occur consistently during processing of low-resistance contacts. Real-time observations of silicide phase formation in Ti films deposited *in situ* and *ex situ* allows us to measure the growth kinetics of the describes C54 phase and to show that the propagation of the phase boundary is rate-limited by oxygen and other impurities at grain boundaries in the precursor C49 phase. These results suggest changes in processing conditions which encourage the phase transition and thereby improve contact properties.

Resonant Tunneling

Quantum mechanical tunnelling, which a particle passes through a classically for bidden region, is the mechanism by which α-particles escape from the nucleus during α-decay and electrons escape from a solid in thermionic emission. In both cases the tunneling probability is a very sensitive function of the energy of the particle and the thickness and height of the potential barrier. Carrier tunnelling can also be observed in nanostructures where a tunneling barrier is formed by sandwiching a thin layer of a wide band gap semiconductor between layers of a smaller band gap semiconductor. Incorporated into a suitable device, this allows the behaviour of electrons incident on the barrier to be studied. The energy of the electrons is determined by the voltage applied to the device, and the probability of tunneling through the barrier is reflected by the magnitude of the current that flows. Of greater interest, however, is the case of two barriers separated by a thin quantum well, double-barrier resonant tunneling structure (DBRTS). A schematic diagram of a DBRTS is shown in Figure 2.24. Quantised energy levels are formed in the quantum well, A DBRTS is generally grown between two doped layers (n-type in Figure 2.24) which provide reservoirs of carriers.

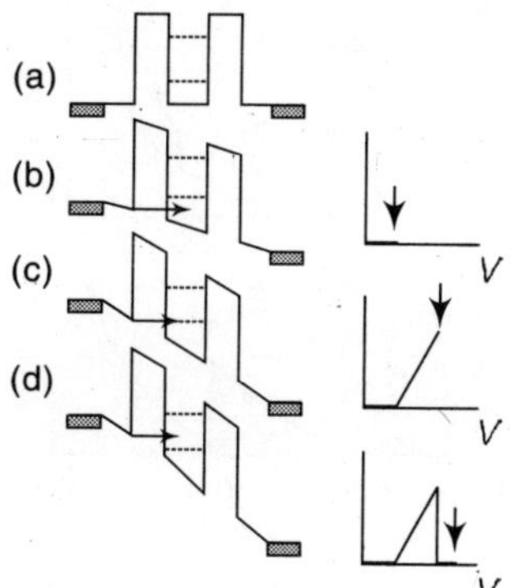

Fig. 2.24 *Schematic conduction band diagram of double-barrier resonant tunneling structure. The band structure is shown for different applied voltages and with the corresponding current, (a) no applied voltage, (b) voltage below the first resonance, (c) in resonance with the lowest energy state in the quantum well, and (d) above the first resonance.*

Figure 2.24 shows a DBRTS for various applied voltages. For the sign of voltage shown, electrons travel from left to right. Electrons are first incident on the leftmost barrier, through which they attempt to tunnel into the well, followed by tunnelling out of the well via the second barrier. At low voltage, condition (b) the electron energy following tunnelling into the well is below that of the lowest confined state. Hence there are no available states in the well and it acts as a further barrier. For this condition (b) the two barriers plus the well act as one effective thick barrier, as a consequence the tunneling probability, and hence the current, is very low. As the voltage is increased to condition (c), the energy of the electrons tunnelling through the first barrier comes into resonance with the lowest state in the well. The effective barrier width is now reduced and it becomes much easier for the electrons to pass through the structure. As a result, the current increases significantly. For a further increase in voltage, condition (d), the resonance condition is lost and the current decreases. Although for condition (d) the energy of the tunnelling electrons coincides with higher energy states in the quantum well subband, these states correspond to non-zero in-plane motion. The tunnelling electrons are moving parallel to the growth direction only, so tunneling into these subband states is not possible. However, for high applied voltages, additional resonance may be observed, corresponding to higher confined states. Figure 2.24 also shows the expected current-voltage characteristics of a DBRTS, indicating the relation between specific points on the characteristics and the different bias conditions of the structure. Figure 2.25 shows experimental results obtained for a DBRTS consisting of a 20 nm GaAs quantum well confined between 8.5 nm AlGaAs barriers. Resonances with five confined quantum well states are observed. Beyond each resonance a DBRTS exhibits a negative differential resistance, a region where the current decreases as the applied voltage is increased. This characteristic has a number of applications, including the generation and mixing of microwave signals.

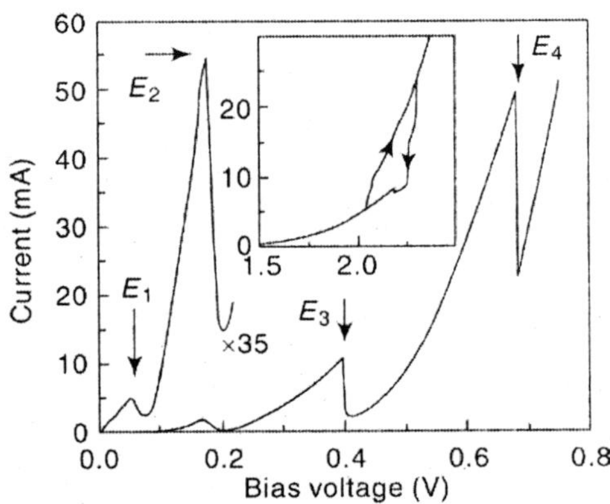

Fig. 2.25 *Current-voltage relationship for a double-barrier resonant tunneling structure consisting of a 20 nm GaAs quantum well with 8.5 nm AL0.4Ga0.6 As barriers. Resonances with four confined well states are observed. The inset shows the hysteresis observed for an asymmetrical devices with 8.5 and 13 nm Al0.33Ga0.67As barriers and a 7.5 nm In0.11 Ga0.89As quantum well.*

DBRTS devices may also exhibit hysteresis in their current-voltage characteristics, particularly when the two barriers have unequal thickness. A thinner first barrier allows carriers to tunnel easily into the well whilst a thicker second barrier impedes escape, resulting in charge build-up in the well. This charge build-up modifies the voltage dropped across the initial part of the structure and maintains the resonance condition to higher voltages than would occur in an empty well. This broadened resonance is only observed as the voltage is increased, allowing charge to accumulate in the well. When the voltage is finally taken above the resonance condition, the well empties. If the voltage is now decreased, a narrower resonance is observed as there is now no charge accumulation. For such a structure the current follows a path that is dependent upon the direction in which the voltage is swept; the current-voltage characteristics exhibit hysteresis. The inset to Figure 2.25 shows the characteristics of an asymmetrical DBRTS with 8.5 and 13 nm thick $Al_{0.33}Ga_{0.67As}$ barriers and a 7.5 nm $In_{0.11}Ga_{0.89}As$ quantum well.

Charging Effects

A charge carrier in a quantum dot is highly spatially localised. If a quantum dot already contains on or more carriers, then significant energy is required to add an additional carrier, as a result of the work that must be done against the repulsive electrostatic force between like charges. This charging energy modifies the energies of the dot states relative to their energies in the uncharges system.

The inset to Figure 2.26 shows the conduction band profile of a structure consisting of a quantum dot placed close to a reservoir of free electrons. Applying a voltage to a metal contact on the surface of the structure allows the energy of the dot to be varied with respect to the reservoir. If a given energy level in the dot is below the energy of the reservoir, then electrons will tunnel from reservoir into the dot level. Alternatively, if the energy level is above the reservoir, then the level will be unoccupied. Hence by varying the gate voltage, the dot states can be sequentially filled with electrons. This filling can be monitored by measuring the capacitance of the device, which will exhibit a characteristics feature each time an addition electron is added top the dot.

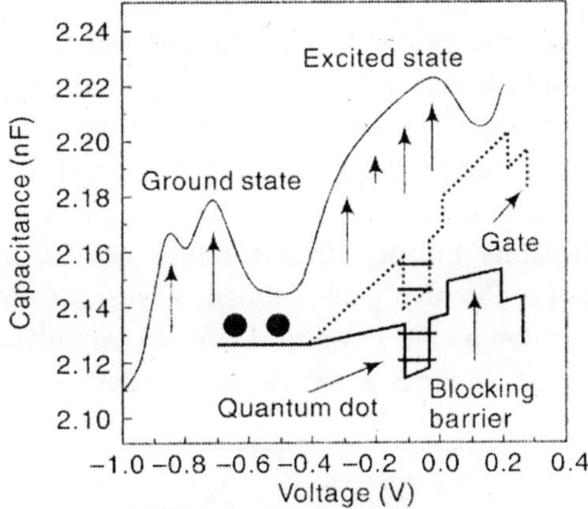

Fig. 2.26 *Capacitance-voltage profile of approximately 2 million self-assembled quantum dots showing features as successive electrons are added to the dots. The vertical arrows indicate the positions of the features corresponding to the loading of electrons into the ground state (two features) and the excited state (four unresolved features). The inset shows the conduction band profile for two bias voltages.*

Figure 2.26 shows the capacitance of a device containing approximately 2 million self-assembled quantum dots. These dots have two confined electron levels: the lowest level (ground state) is able to hold two electrons with the excited level able to hold four electrons. In the absence of charging effects, only two features would be observed in the capacitance trace, one at the voltage corresponding to the filling of the ground state, the other when the voltage reaches the value required for electrons to transfer into the excited state. However, once one electron has been loaded into the ground state, charging effects result in an additional energy, hence a higher voltage, being required to add the second electron. This leads to two distinct capacitance features corresponding to the filling of the ground state. Similarly, four distinct features are expected as electrons are loaded into the excited state, although in the present case inhomogeneous broadening prevents them being individually resolved. This charging behaviours is known as Coulomb blockade and is observed experimentally when the charging energy exceeds the thermal $k_B T$.

Coulomb blockade effects may also be observed in transport processes where carriers tunnel through a quantum dot. Suitable dots maybe formed electrostatically using split gates to define the dot and to provide tunnelling barriers between the dot and two 2DEGs which act as carrier reservoirs. The device is analogous to resonant tunnelling structures but with the quantum well replaced by a quantum dot. An additional gate electrode allows the energy of the dot to be varied with respect to the carrier reservoirs. The relatively large dot size (>100 nm) results in Coulomb charging energies – given by $e^2/2C$ where C is the dot capacitance – much larger than the confinement energies. The Coulomb charging energies therefore dominate the energetics of the system The inset to Figure 2.27 shows a schematic diagram of the structure with a small bias voltage applied between the left and right reservoirs. The dot initially contains N electrons, resulting in a dot energy indicated by the lower horizontal line. An additional electron can tunnel into the dot from the left-hand reservoir but this increases the dot energy by an amount equal to the charging energy. This process is therefore only energetically possible if the energy of the dot with $N + 1$ electrons lies below the maximum energy of the electrons in the left-hand reservoir. Tunneling of this additional into the right-hand reservoir may subsequently occur, but only if the $N + 1$ dot energy lies above the maximum energy of this reservoir, as the electron can only tunnel into an empty state. If these two conditions are satisfied, requiring that the dot energy for N + 1 electrons lies between the energy maxima of the two reservoirs, a sequential flow of single electrons through the structure occurs. For this condition the system exhibits a zon-zero conductance. As the gate voltage varies the dot energy, rigidly shifting the different confined states up or down, the condition for sequential tunnelling with be successively satisfied for different values of N, and a series of conductance peaks will be observes. An

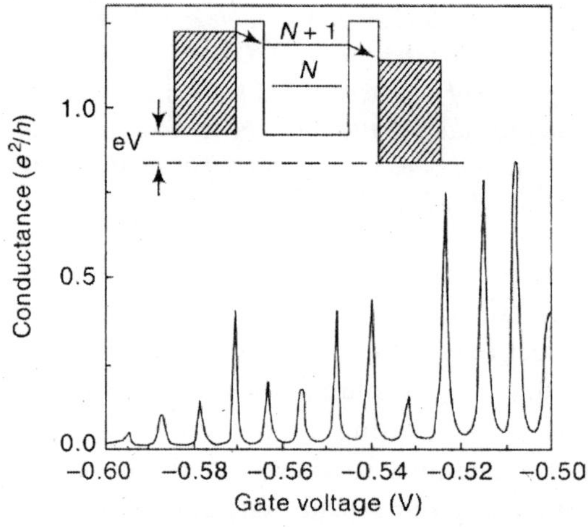

Fig. 2.27 *Coulomb blockade oscillation for an electrostatically defined quantum dot measured at a temperature of 10 mK. The inset shows the transport of a single electron through the structure.*

example is shown in Figure 2.27 for a dot of radius 300 nm. This large dot size results in a large capacitance and correspondingly small charging energy (0.6 meV for the present example). Hence measurements must be performed at very low temperatures in order to satisfy the condition $e^2/2C >> k_B T$.

Interband Absorption in Semiconductor Nanostructure between VB & CB

A semiconductor can absorb a photon in a process where an electron is promoted between the valence and conduction bands. The strength of this absorption is proportional to the density of states in both bands–the joint density of states. The joint density of states has a form similar to the individual density of states and is therefore a strong function of the dimensionality of the system. In addition, the absorption will be modified by the quantised energy levels of a nanostructure, resulting in a number of different energy transitions occurring between the confined hole and electron states.

A further modification arises from the influence of excitonic effects. In a nanostructure the electron and hole are prevented from moving apart in one or more directions and, as a result, their average separation is decreased and the exciton binding energy is increased. For an ideal two-dimensional system, the exciton binding energy is increased by a factor of four compared to the bulk value. However, in a real quantum, well the electron and hole wavefunctions penetrate into the barriers, and the ideal two dimensional case is never achieved, although up to an approximately twofold enhancement is possible. A larger binding energy decreases the probability of exciton ionisation at high temperatures and, as a consequence stronger excitonic effects are observed in a nanostructure at room temperature than in a comparable bulk semiconductor.

Figure 2.28 shows the absorption spectrum of a 40-period GaAs multiple quantum well structure with wells of width 7.6 nm and AlAs barriers. The gross form of the spectrum consists

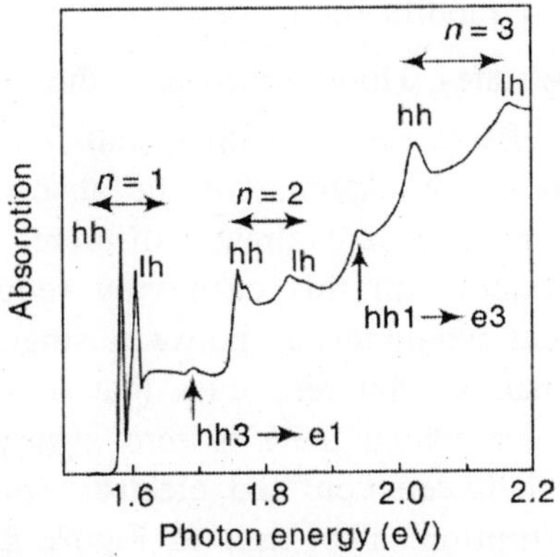

Fig. 2.28 *Low-temperature absorption spectrum of a 40-period GaAs-AlAs multiple quantum well structure with 7.6 nm wide wells. The most intense features result from transitions between the nth (n = 1, 2, 3) confined light hole (1h) and heavy (hh) hole states and identical index elecytron states. IN addition, two weaker transitions are observed between the first and third heavy hole and electron states (hh3 ↑ el and hhl → e3).*

of a series of, steps, representing the density of states of a two-dimensional system with an excitonic enhancement at the onset of each step. A number of transitions are observed between the *n*th confined hole state and *n*th confined electron state. These transitions are subject to selection rules that result in transitions between identical electron and hole index states ($m = n$) being the most intense. Three orders of such transitions are observed in the spectrum of Figure 2.28. Weaker transitions occur when the index changes by an even number ($|n - m| = 2, 4$, etc.). Transitions where the index changes by an odd number ($|n - m| = 1, 3$, etc.) are forbidden by parity considerations. The spectrum is further complicated by the presence of two different valence bands, known as the heavy and light hole bands, which result in two distinct series of confined valence band states. Transitions are possible from both series to the conduction band.

Figure 2.29 compares the room temperature absorption of bulk GaAs, and a GaAs multiple quantum well structure with 10 nm wide wells. The enhancement of the excitonic strength in the quantum well, due to the increased exciton binding energy, is clearly visible. The presence of an exciton provides a sharp onset of the band edge absorption, and this has a number of practical applications, for example in optical modulation. Nanostructures allow excitonic effects to be exploited more readily as these effects persist to considerably higher temperatures than in bulk semiconductor.

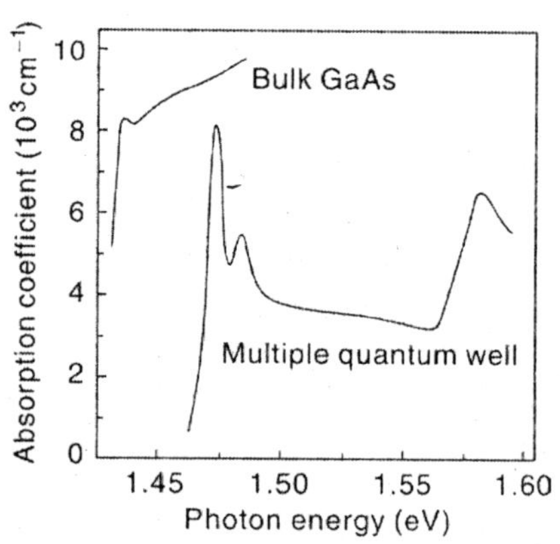

Fig. 2.29 *Comparison of the room temperature absorption spectra of bulk GaAs and a 77-period GaAs-AL0.28Ga0.72As multiple quantum well with 10 nm wide wells.*

The absorption spectrum of a quantum wire should be similar to that of a quantum well, but with a further enhancement of the exciton binding energy provided by the additional quantum confinement. There will also be a modification due to the $1/\sqrt{E}$ form of the density states. However, to date the quality of available quantum wires is not sufficient to observe these effects clearly, with spatial fluctuations of the wire cross section leasing to significant inhomogeneous broadening of the absorption spectrum.

The energy levels of a quantum dots are already discrete, hence excitonic effects are less obvious in zero-dimensional systems. In this case they reduce the energies of the optical transitions compared to those that would occur between single-particle, non-interaction states. The absorption spectrum of a quantum dot resembles that of an atom, consisting of a number of discrete absorption lines between which there is zero absorption.

In nanostructures, absorption between confined electron (or hole) levels can also occur, a process known as interband absorption. The inset to Figure 2.30 shows interband absorption for the conduction band of a quantum well. Electrons are required in the initial state and these are generally provided by doping. Photon absorption involves the excitation of electrons between the $n = 1$ and higher confined electron states. In some structures excitation between confined well states and the unconfined states of the barrier is also possible. The energy separation between the confined states is typically ~10–20 meV, corresponding to wavelengths

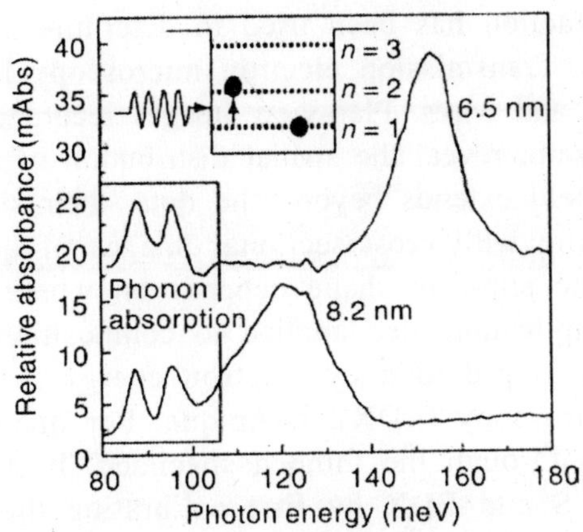

Fig. 2.30 *Intraband absorption spectra of two 50-period GaAs multiple quantum well structures with wells of width 6.5 and 8.2 nm. Transitions occur between the n = 1 and n = 2 confined electron levels. The inset shows a schematic diagram of absorption process.*

~6–120 μm and resulting in absorption in the infra-red region of the electromagnetic spectrum. The energies of the transitions can be varied over this wide range by altering the well width, allowing the absorption to be tuned to a specific wavelength. Interband absorption spectra are shown in Figure 2.30 for two quantum wells of widths 6.5 and 8.2 nm. With decreasing well width, the separation between the $n = 1$ and $n = 2$ states increases and the absorption shifts to higher energy. There are no excitonic effects associated with Interband absorption, because only one type of carrier is involved; excitons require both an electron and a hole, which are only present in interband processes.

Intraband absorption is also served between the confined states of quantum wires and dots. For quantum wells, one important selection rule for intraband absorption requires that the incident radiation has an electric field component normal to the plane containing the wells. As a consequence, for light incident along the growth direction, which is the most convenient experimental geometry, intraband absorption does not occur. Intraband absorption in a quantum well therefore requires the use of less convenient geometries, including light incident on the edge of the structure or normally incident light that is bent into the structure by a diffraction grating deposited on the surface. In contrast, for quantum dots this selection rule is modified and intraband absorption is possible for normally incident light. This difference is particularly advantageous for the practical application of intraband absorption infra-red detectors.

STRUCTURAL CHARACTERISATION

A number of techniques are available for the direct study of the physical structure of quantum wells and superlattices. These include: X-ray diffraction, which can determine periodicity, layer uniformity and in some cases composition; transmission electron microscopy, which can determine layer thicknesses and compositions which enables the nature of interfaces to be directly imaged. X-ray studies of quantum wires and dots are more limited, although shallow incidence X-ray diffraction has been used to determine the distribution of indium in self-assembled quantum dots. Transmission electron microscopy has been applied extensively to the study of quantum dots and wires. Plan-view images recorded along the growth axis, and for a geometry sensitive to strain, reveal the spatial distribution of self-assembled quantum dots, although because the strain field extends beyond the dots, their size and shape is not directly imaged. A complication arising with cross-sectional images of quantum dots is that for dot shapes other than cuboidal the apparent shape depends on where the dot is sectioned during sample preparation. This complication also applied to compositional studied.

Compositional information is provided by electron energy loss spectroscopy (EELS) and energy-dispersive X-ray spectroscopy (EDX), technique. For quantum wells and wire, which have a uniform cross section through the thinned specimen, it is possible to obtain absolute compositions from both EELS and EDX by first calibrating the system with samples of a known composition. However, for quantum dots the uncertainty in how the dot has been sectioned makes it difficult to achieve absolute compositional determine. Figure 2.31 shows an EELS image of the gallium distribution in an INAs self-assembled quantum dot grown within a GaAs matrix. In addition, the two traces show EDX line scans of the indium distribution along directions parallel to the growth axis and passing through either just the wetting layer or both the dot and wetting layer. The profile of the latter trace is broader, reflecting the additional indium-containing region of the quantum dot.

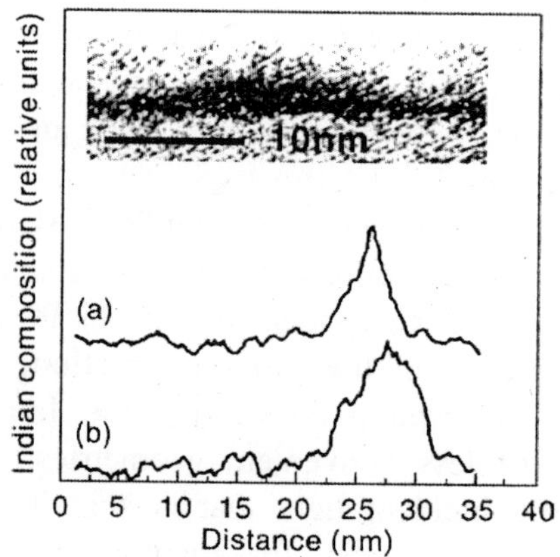

Fig. 2.31 *EDX profiles of a self-assembled InAs quantum dot structure recorded for line scans (a) passing through the wetting layer and (b) passing through the wetting layer and a quantum dot. Also shown is an FELS image of the Ga composition in an InAs self-assembled quantum dot structure with GaAs barriers.*

Atomic force microscopy (AFM) is routinely used to study nanostructures as no complicated sample preparation is required. AFM allows the shape, size and distribution of self-assembled quantum dots to be determined but these dots must be situated on the sample surface, requiring the growth to be terminated immediately after the dots have been formed. In contrast,. Dots for optical and other applications musty be covered by a relatively thick 'protective' layer to prevent the strong non-radiative carrier recombination which otherwise occurs at a free surface.

As there is some evidence that material diffuses into and out of the dots during the growth of the capping layer, altering their shape, size and composition, it is possible that dot parameters determined by AFM may be different to those of covered dots.

AFM may also be used to determine alloy compositions in aluminium-containing nanostructures. This requires the nanostructure to be cleaved and exposed to air, causing the aluminium-containing materials to oxidise and bow outwards slightly from the surface. The degree of bowing is proportional to the local aluminium content and can be measured by an AFM. By calibrating the system with samples of known composition it is possible to produce a two-dimensional image of the aluminium composition.

Lattice-Mismatched vs. Lattice-Matched Growth

The classical approach to heteropitaxial growth focused mainly on lattice-matched or nearly lattice-mentioned growth. Lattice-mismatched systems grow coherently (dislocation-free) only below a certain critical thickness, beyond which the structure becomes dislocated. This deteriorates device characteristics for both microelectronic applications. One disadvantage was a severely limited range of materials. Only a few material combinations, e.g. $GaAs/Ga_{1-x}A_{1-x}As$, $InP/Ga_{0.52}Al_{0.48}As$, GaAs/ZnSe, were possible.

Additionally, in the classical approach, the goal was to stay with a planar morphology on the surface. Islands were highly undesirable, as they were though to be necessarily dislocated.

In the self-organised growth of quantum dots, lattice-mismatched growth is the main route, and lattice-matched growth is in many cases inappropriate. Islands are widely used as quantum dots, since they are dislocation-free below a certain critical volume. A wide variety of material combinations may thus be used to fabricate quantum dots.

Figure 2.32 illustrates the formation of QDs in a wide variety of highly-mismatched systems. Plan-view transmission electron microscopy (TEM) images of InAs/GaAs QDs and GaSb/GaAs QDs show the strain contrast representing strained insertions. Cross-sectional high resolution transmission electron microscopy (HRTEM) images of InGaN/GaN and CdSe/ZnSe structures have been processed by the digital analysis of lattice imaged (DALI) technique which reveals the local map of the lattice parameter in the vertical direction. This map allows us to estimate the local content of In (Fig. 2.32(c)) or Cd (Fig. 2.32(d)).

Design of a Transmission Electron Microscope for the study of growth processes

Experiments are performed in a 300-kV ultrahigh-vacuum transition electron microscope (UHV TEM) having a base pressure of 2×10^{-10} mmH. The microscope is modified to enable several types of growth process to be carried out during observation of the specimen (Figure 2.33). Gases are introduced to the specimen area through capillary tubes, allowing oxidation or growth by chemical vapor deposition (CVD). A pressure of upto to 10^{-5} mm Hg can be sustained with the beam on. An electron-beam evaporator mounted directly above the polepiece allows us to carry out evaporation *in situ*. For further analysis and preparation, and Auger spectroscopy system, an ion gun, and two additional 5-kW evaporators are located in adjacent chambers to which the specimen can be transferred without breaking vacuum. Specimen heating is carried out by direct current using a double-tilt heating holder designed especially for the system.

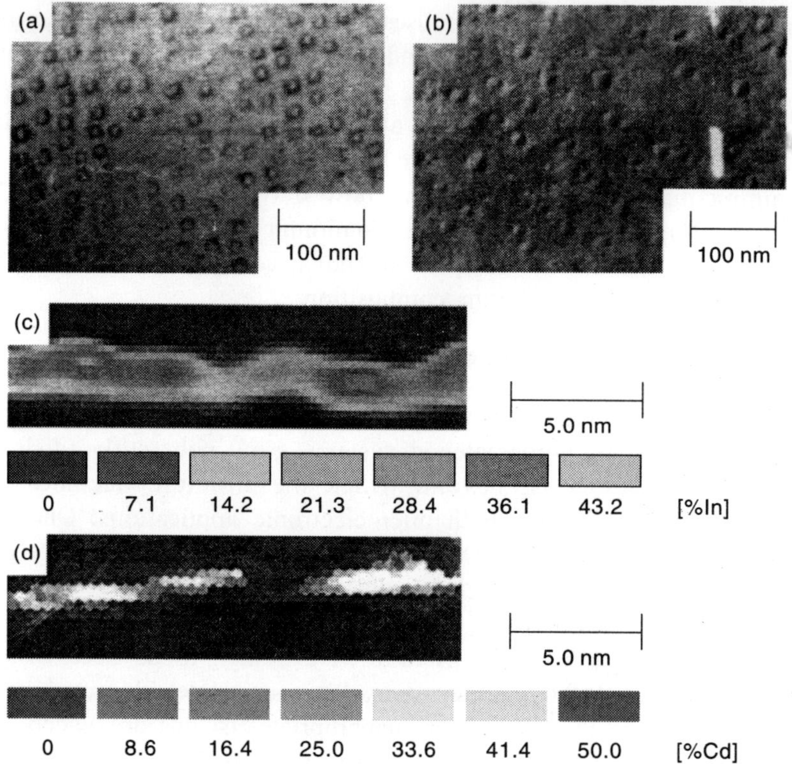

Fig. 2.32 *Spontaneously formed arrays of quantum dots (QDs) in various material systems. (a) plan view transmission electron microscopy (TEM) image of InAs/GaAs QDs. (b) Plan view TEM image of GaSb/GaAs QDs. (c) Cross-sectional high resolution transmission electron microscopy (HRTEM) image of In GaN/GaN QDs processed by the digital analysis of lattice images (DALI) evaluation technique. (d) Cross-sectional HRTEM image of CdSe/ZnSe QDs processed by the DALI technique.*

Silicon substrate specimens were used in the present study. The specimens consisted of 2 × 4-mm rectangles cleaved from 100-μm-thick Si(001) or Si(111) wafer which had been polished on both surfaces. They were thinned to electron transparency with an $HF/HNO_3/CH_3COOH$ mixture. Final thinning was carrier out by etching in the microscope polepiece using a low pressure of oxygen (10^{-6} mm Hg) at about 1173 K until a desired thickness was achieved. The specimens were then flash-heated to above 1523 K form a clean surface on which growth could take place. Dynamic processes were recorded in real time using an image intensifier linked to a video system, or on photographic plates which were subsequently digitized for computer analysis.

The Growth and Patterning of Quantum Dots of Ge on Si

Considerable interest has recently centered around the properties of quantum dots and their potential use in novel microelectronics devices. The term quantum dot refers to a volume of material which is small enough, of the order of a few nanometers or less in diameter, that quantum confinement effects lead to unusual electronic properties. Several exciting applications

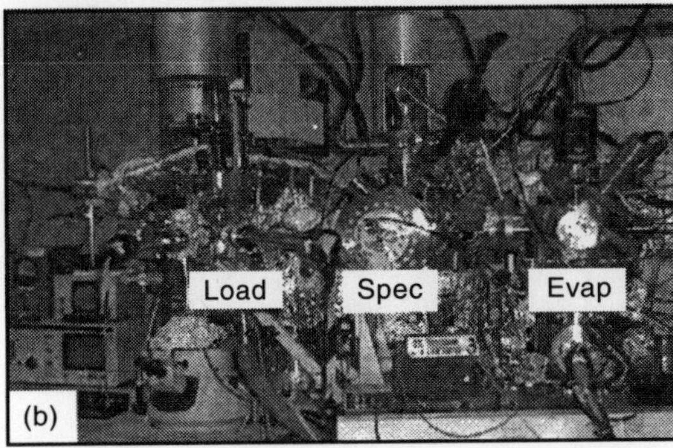

Fig. 2.33 *The UHN transmission electron microscope: (a) View showing the microscope column and the attached loading chamber. Specimens can be transferred from the loading chamber either into the microscope or into the adjacent specimen preparation chambers. (b) View of the loading and preparation chambers. Attached to the central "spectroscopy" chamber is an electron gun for SEM and Auger analysis and an optical pyrometer for calibration of specimen temperature. The specimen can be heated by direct current while in the preparation chambers or in the microscope, and a current-temperature curve can be obtained for each specimen. In the right-hand "evaporation" chamber are two 5-kW electron-beam evaporators and an ion gun used for thinning specimens.*

for such dots have been envisaged, including single-electron transistors and lasers. For these devices to function efficiently, a well-organised group of identically sized dots is necessary. However because of their small size, the fabrication of such dots is at present beyond the limits of conventional lithography, and an alternative technique for growth is necessary.

One very promising route to the fabrication of nanosize volume of material is to make use of the phenomenon of island self-assembly during strained-layer epitaxial growth. To illustrate how this occurs, consider the case of Ge grown epitaxially on Si (Figure 2.34). The Ge lattice spacing is about 4% larger than that of Si, so significant strain energy is stored in the Ge lattice as it grows. As Ge is deposited beyond about three monolayers (ML), the growing surface becomes nonplanar and islands form relieving strain by allowing the Ge lattice to expand at the island peaks. Depending on the growth conditions, the islands may be up to 100 nm in diameter and can have a very uniform size distribution.

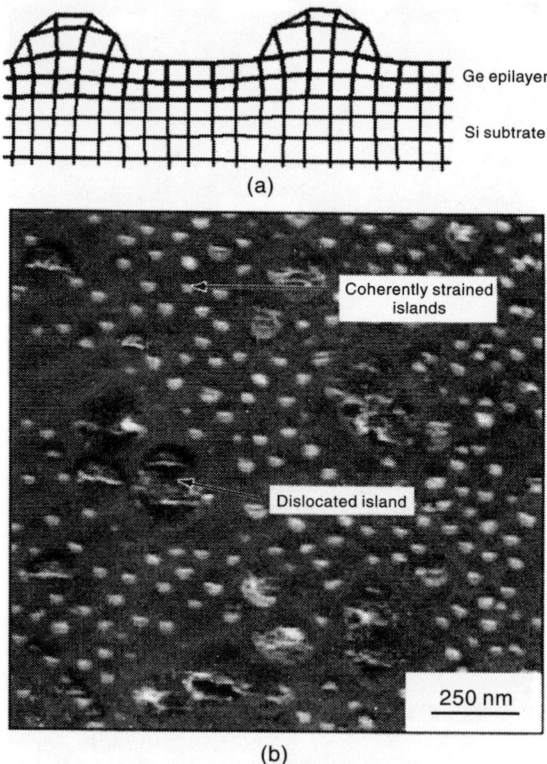

Fig. 2.34 *Schematic diagram showing epitaxial growth of Ge on Si(001). Islands from because strain can be relieved by allowing the Ge lattice to expand at the tops of the islands. (b) Ge islands formed after deposition of approximately 10 ML of Ge on Si(001). In this example the islands were not of uniform size. The small islands were coherently strained, as shown in (as), while in the largest islands a second more of strain relief was visible: the introduction of misfit dislocations at the Ge/Si interface.*

The spontaneous formation of nanosized islands during strained-layer growth, if it can be sufficiently well controlled, provides a means to fabricate quantum dots which bypasses the limitations of lithography. However, the self-assembly processes has its own limitations: In particular, the placement of individual islands is not defined at the start of growth, and island sizes may show large variation.

Postgrowth examination of island populations has revealed several intriguing aspects of the self-assembly process. One particularly important result is that in many cases islands can be grown with a very narrow size distribution. Although this is very convenient for device applications, the reason for the narrowness of the distribution is not well understood. The distribution is generally considered to narrow to be due to Ostwald ripening, and several other growth mechanisms have been proposed. Some models are based on thermodynamic arguments, postulating the existence of a minimum energy size to which islands will tend to grow. Other models suggest that kinetic effects slow down the growth of larger islands, allowing the smaller ones to catch up. Possible reasons for this may be the slowing of diffusion to an island due to the strain field around it or the reduced probability of nucleating a step on the surface of a larger island, again due to strain.

A interesting result, is observed in the Ge/Si(001) system. It is the observation of a bimodal island size distribution, with the two peaks corresponding to two different island shapes. Smaller islands (less than about 600 nm in diameter) have been found to be square-based pyramids, while larger ones adopted steeper "dome" shapes with additional facets. The origin, evolution, and distribution of the pyramids and domes are important for the development of quantum dot devices.

We therefore examined the process of island self-assembly in Ge on Si in real time using the *in situ* TEM system. Ge was grown by CVD on Si(001) at a specimen temperature of 650°C using digermane gas (Ge_2H_6) at a pressure between 10^{-8} and 10^{-6} Torr, resulting in a growth rate of 0.1-10 monolayers of Ge per minute. A weak-beam imaging conditions was chosen which showed the strain field around the islands. The allowed island positions and sized to be resolved, although the images did not show the shapes of individual islands.

Representative results form the experiments are shown in Figure 2.35 and 2.36. The *in situ* observations revealed a rich evolutionary process. Figure 2.35 indicate how the bimodal size distribution develops in a large population of islands. Shortly after islands become visible, a broad bimodal distribution develops, with two peaks corresponding (presumably) to the pyramid- and dome-shaped islands. At laser times, the smaller (pyramid) peak weakens and almost disappears. While a single postgrowth observation made, for example, after 50 seconds [as in Figure 2.35(b)] might suggest two stable population, the complete series shows that the pyramids are transient, with no particular size preferred, and vanish at later times. Note that the domes show their narrowest size distribution just as the pyramids are disappearing [Figure 2.35(c)].

To explain these results, we have developed a simple model based on the existence of the two known island shapes. During growth, islands can exchange Ge atoms, since at the relatively high growth temperature of 650°C the equilibrium concentration of Ge adatoms on the surface is large (~10%), and Ge atoms can easily detach from an island, diffuse along the surface between the islands, the reattach to another island. Calculation show that this process occurs so readily that the flux of detaching and attaching Ge adatoms is more important in controlling the development of an island that the flux of Ge arriving from the gas phase. If one island, because of its shape or size, has lower energy per atom then another island, Ge atoms will preferentially detach from higher-energy islands and attach to the lower-energy island. Islands can thus shrink or grow depending on their energy per atom in comparison to the other islands.

If all of the islands were the same shape, we would expect that the larger islands would always have a lower energy per atom that the smaller ones by virtue of their lower surface energy/volume ratio. The larger islands would grow at the expense of the smaller ones, leading to the well-known classical phenomenon of Ostwald ripening. However, if two different island shapes are allowable, the process becomes much more complex and interesting. For the Ge pyramid and dome shapes described above, simple calculations of the energy per atom show that the pyramid shape is preferred for smaller islands below a critical volume V_c, while the dome shape is favored at volumes greater than V_c. This islands formed during the early stages

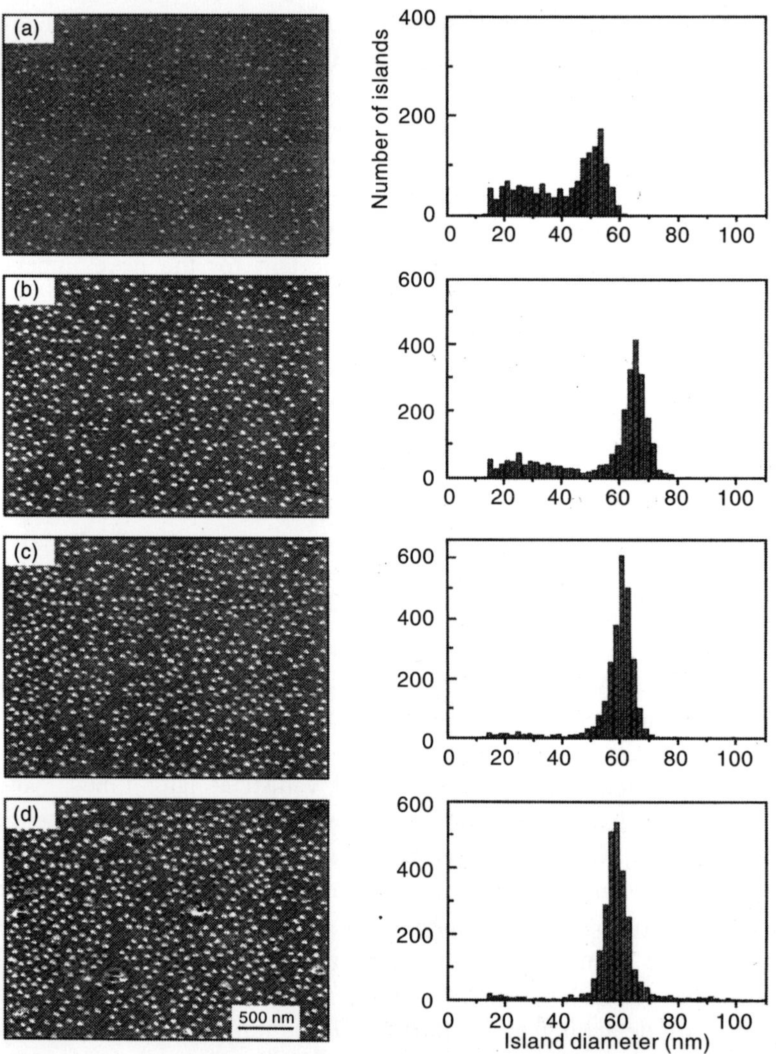

Fig. 2.35 *Weak-beam (220) dark-field images and corresponding histrograms obtained during Ge deposition from 2 × 10–7 Torr Ge2H6 at a substrate temperature of 873 K. The images show the same area and were taken (a) 21 s, (b) 51 s, (c) 98 s, and (d) 180 s after "nucleation" (defined as the time at which the island strain contrast if first seen). Nucleation occurred after a dose of ~50 L digermane. Note that the two island shapes (pyramids and domes) cannot be distinguished in these images. The histograms were obtained by digitizing the images and using particle-counting algorithm on the ~2600 islands visible. The distribution appears to cut off at ~15 nm diameter because the smallest islands were not detected by the algorithm owing to weak contrast. In (d) some large islands with multiple dislocations formed, creating a long tail (not shown) in the size distribution.*

of Ge deposition are all small and therefore pyramid-shaped and undergo Ostwald ripening, with the larger islands growing at the expense of smaller ones. As the successful pyramids grow, the first few to reach the volume V_c make the transition to the dome shape. Their energy per atom rapidly decrease, making them very attractive sites for adatoms and increasing their growth ate still further at the expense of the smaller islands. The result of this process is shown in Figure 2.36(a), while in Figure 2.36(b) this model is compared with experimental date. The model reproduces the correct trends here as well as for larger ensembles of islands such as in Figure 3, where it predicts that population of unstable small islands will disappear to leave a relatively narrow distribution larger islands.

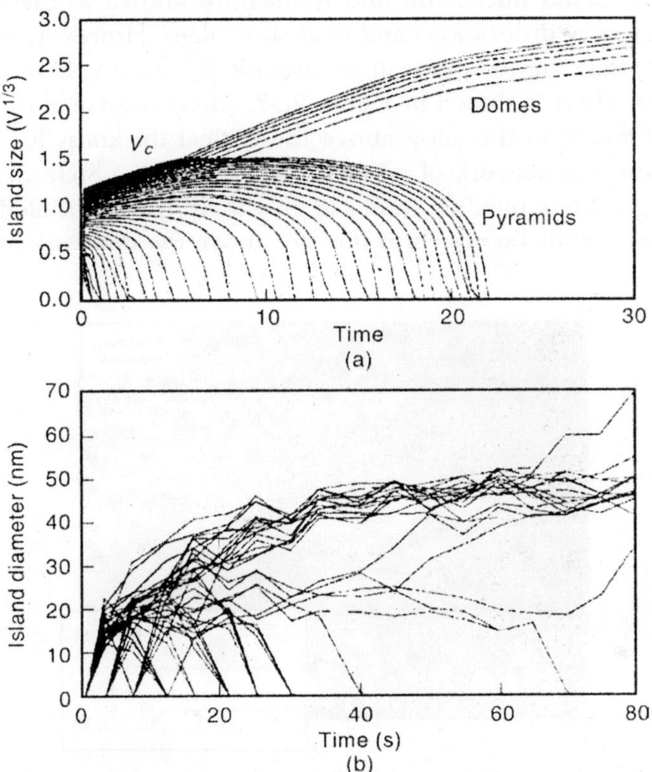

Fig. 2.36 *Simulation showing the fate of a group of islands with different initial sizes. We have assumed that islands make the transition from pyramids to domes at a critical volume V_c. The rapid growth rate of the first islands to exceed V_c leads to the subsequent bimodal size distribution. At later times, note the narrowness of the size distribution for the domes compared with the initial distribution. The scale of the plot is chosen to match the data shown below. (b) Experimental data showing the growth of all islands in a 0.5-µm × 0.5-µm area. Data were measured from a video obtained during growth at 5 × 10^{-7} mm Hg Ge_2H_6 and 923 K.*

In this simple model, islands have no preferred size, but instead grow or shrink at a rate determined by their environment. Islands continue to grow as long as smaller islands are available to supply adatoms, until the final configuration, one very large island, is reached. In contrast to the models described previously, there is no factor which tends to equalize island sizes; the narrow size distribution is the result of selection of the largest islands at the volume V_c. It is clear that for best uniformity, growth should be stopped just after the last pyramids have disappeared; further growth or annealing does not improve the inland distribution.

Narrowing the island size distribution, even below the width shown in Figure 2.35(c), is be advantageous to the performance of real device. One prediction from the model is that the distribution at later times can be improved by controlling the initial nucleation of the islands. If all of the islands were to nucleate at the same time and be equally spaced, they should develop similarly and have a very narrow distribution.

Any irregularity on the initial Si surface which changes the local sticking probability of Ge adatoms can influence island nucleation and islands are known to nucleate preferentially on mesa edges, near clusters of dislocations and near step edges. However, we have found that the strain fields from individual buried dislocations provide the most convenient way of controlling island nucleation. This effect is shown in Figure 2.37, where a SiGe alloy was used to generate buried dislocations. Growth of the alloy above the critical thickness for dislocation formation resulted in the creation of a network of misfit dislocations at the SiGe/Si interface (see inset in Figure 2.37). Further growth of pure Ge results in growth in rows associated with the dislocations. Figure 2.37 shows that islands do not form directly above the dislocations but are offset to one side or the other.

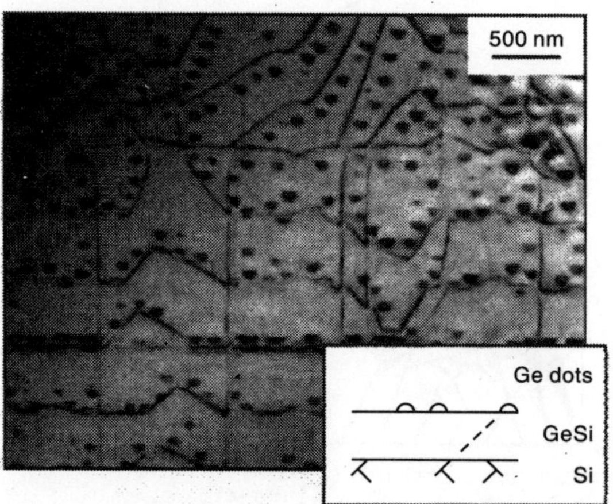

Fig. 2.37 *Ge islands formed on a 300-nm Si0.8Ge0.2 alloy layer on a Si substrate. The alloy layer has partially relaxed, forming misfit dislocations at the SiGe/Si interface, and the Ge islands have grown in rows associated with these dislocations.*

The nucleation of these islands is influenced by the strain field of the dislocation which extends up to the specimen surface. A calculation of the surface strain field due to a single dislocation shows that the region of maximum tensile strain, where we might atoms are expected to stick preferentially, is offset to one side of the dislocation; the islands formed in a row at the expected location (Figure 2.38).

These experiments have therefore shown that it is possible to pattern single rows of islands along individual dislocations. A two-dimensional array of dislocations allows islands to adopt a regular pattern, resulting in a structure which is conductive to information storage. Techniques for patterning arrays of dislocations are developed and the fabrication of devices requiring regular arrays of quantum dots are therefore be feasible.

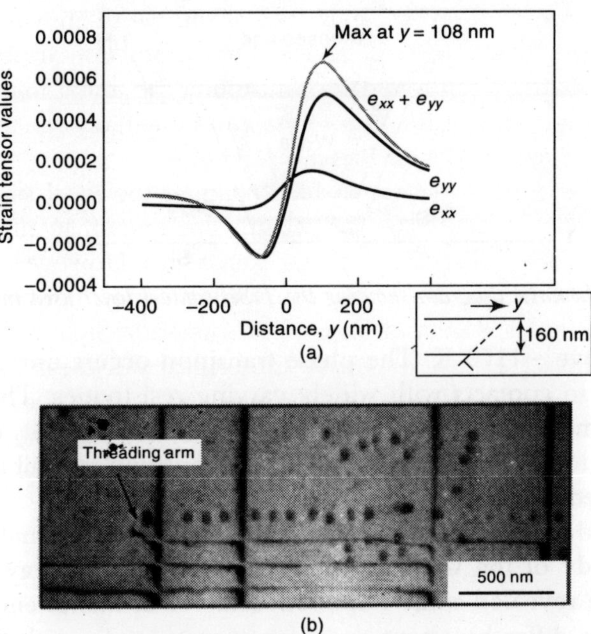

Fig. 2.38 *(a) Calculation of the strain field at the surface associated with a single 60° misfit dislocation which is parallel to and 160 nm below the surface of a Si crystal. The maximum tensile strain ($e_{xx} + e_{yy}$) occurs approximately at the point where the slip plane intersects the surface, (b) Ge islands grown on a 160-nm $Si_{0.85}Ge_{0.15}$ alloy layer on a Si substrate. A single row of islands formed at the location expected from the calculation. The end of the threading arm shows where the slip plane for this particular dislocation intersected the surface.*

Phase Transitions in $TiSi_2$

$TiSi_2$ is an important component in high-density integrated circuits such as those found in random-access memory. Its use in low-resistance contacts is shown schematically in Figure 2.39. The purpose of the structure shown is to form an electrical contact between the Al-based line and the Si substrate, and this is achieved by patterning a vertical-walled via into the dielectric layer and filling it with tungsten. A thin TiN diffusion barrier is added to prevent the tungsten from diffusing into the silicon. However, electrical contact between the TiN and Si is poor, so a low-resistivity $TiSi_2$ layer is also added. The layer is formed by depositing a blanket film of ~30 nm Ti onto areas patterned with polysilicon and SiO_2 and then annealing to form TiSi2 in the polysilicon areas. The annealing, carried out at 873-973 K, actually forms a high-resistivity metastable phase of $TiSi_2$ known as C49; the final step is conversion of the C49 phase to the stable C54 phase, which has a lower resistivity, a by a rapid anneal at 1123–1173 K.

The contact scheme described above is complex but works well for circuit designs with contact areas greater than about 1 μm^2. However, the miniaturization of components in integrated circuit design has reduced the area available for such contacts and led to an unexpected problem: In very small areas, the conversion of C49 $TiSi_2$ to C54 $TiSi_2$ is retarded, occurring

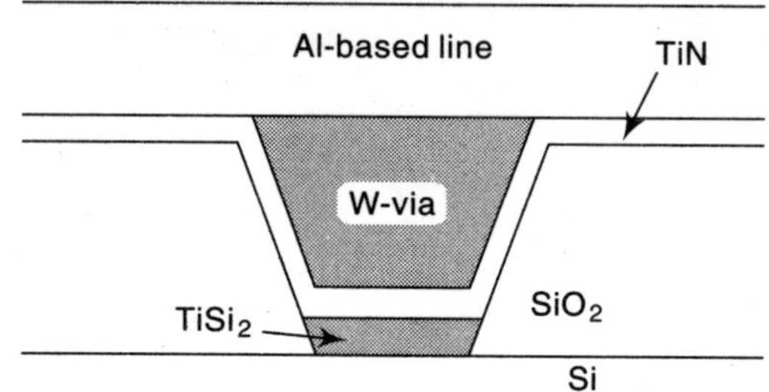

Fig. 2.39 *Schematic diagram showing the $TiSi_2$ contact layer used in a typical via.*

only a temperatures above ~1173 K. The phase transition occurs unreliably during the 1123-1173 K anneal, leading to contacts with widely varying resistivities. This problem has slowed progress in the development of future generations of circuits with $TiSi_2$ contact layers, and it is therefor very important to understand why the phase transition is inhibited in small areas, and how it can be encouraged to occur at lower temperatures.

At might be expected for such an important problem, several analytical techniques have been applied to the study of the C49-C54 phase transition and X-ray diffraction and *ex situ* microscopy studies have revealed some important features of this phenomenon. Grain sizes in the C49 and C54 phases differ by about an order of magnitude, with C49 grains around 100 nm in diameter transforming to C54 grains more than 1 μm across. The very large size of C54 grains suggests that the nucleation of the C54 phase is difficult, and the nucleation sites for C54 have not yet been unambiguously identified. Further studies have shown that alloying Ti with other metals, such as Mo or NMb, before permitting it to react with Si can reduce the C54 grain size and improve the uniformity of the reaction in small regions. Synchrotron X-ray experiments following the progress of the phase transition in blanket and patterned films have revealed subtle changes in reaction temperatures and film texture which reflect the underlying mechanism of the effect of these alloying elements.

However, one important factor which has not been addressed in these studies is the effect of impurities other than alloying metals in the films. Because Ti is such an efficient gettering agent, sputtered films of Ti contain substantial amounts of oxygen, and even relatively clean films oxidize further between growth and annealing. This effect is particularly important for the thinnest films of most interest in integrated circuit manufacturing. We have therefor used the *in situ* cleaning and evaporation capabilities of the UHV-TEM to examine the effect of oxygen impurities on the $TiSi_2$ phase transition. We have compared the transition in clean, UHV-deposited Ti with the transition in sputtered films which contained appreciable amounts of oxygen. The differences in reaction kinetics were striking and showed that the presence of oxygen at grain boundaries in the C49 phase causes a significant reduction in the rate of the C49-C54 phase transition.

To observe the transition *in situ*, a clean (001) Si surface was formed by repeated flashing of a thinned Si specimen to 1573 K, and a 10-mm-thick Ti film was deposited at room temperature at a pressure of less than 5×10^{-9} mm Hg. The specimen was then heated progressively in the microscope to 1123 K. The phases formed were identified from images and diffraction patterns, and the reaction kinetics were recorded at video rate. A similar heating sequence was applied to a 10-nm-thick Ti film on Si(001) which had been deposited by sputtering *ex situ*: the specimen was then thinned to electron transparency by etching from the back surface.

Figure 2.40 shows the phases formed during the heating of the UHV-deposited T film. As-deposited, the Ti film had a grain size of around 10 nm [Figure 2.40(a)], and annealing at low temperatures resulted in the formation of a small grained TiSi phase (not shown). The C49 TiSi, phase appeared after heating to 973 K, while the C53 phase appeared at 1123 K.

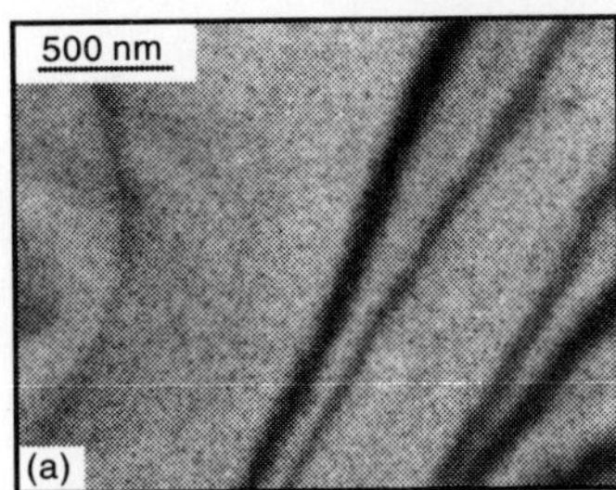

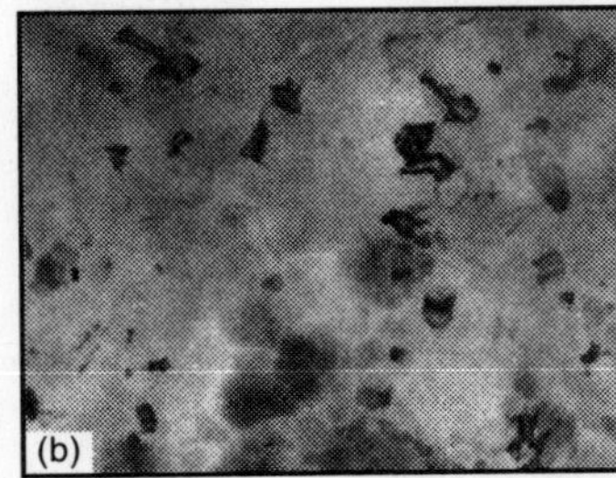

Fig. 2.40 *Some of the phases formed during heating of a 10-nm-thick Ti film deposited under UHV on Si(001): (a) The Ti film as-deposited was microcrystalline. Heating to 573 K caused development of a fine-grained TiSi phase as well as other phases of intermediate stoichiometry (not shown). (b) After heating to 973 K the C49 phase was identified by its faulted structure and 100-nm grain size. (c) Above 1123 K, micron-sized grains of C54 were visible. The dark lines were stacking faults, and the appearance of pinholes in the film showed that it was starting to agglomerate.*

The phases formed during heating of the *ex-situ*-sputtered film appear indistinguishable from those for the UHV-deposited film, with similar grain sizes and diffraction patterns. However, the kinetics of the C49-C54 phase transformation were quite different. In Figure 2.41 it can be seen that the transition occurred smoothly in the UHV-deposited film, with a large C54 grain steadily consuming the smaller C49 grains. The interface velocity was independent of the orientation of the C49 grains. In contrast, the sputtered film underwent a very jerky transformation, pinning at grain boundaries in the C49 structure, then moving rapidly across C49 grains. We attribute the differences in reaction kinetics to oxygen segregation at the grain boundaries. Since C49 and C54 have the same stoichiometry, the reaction in the UHV-deposited film could occur relatively easily without any need for long-range diffusion. However, the reduction in the number of grain boundaries which also occurred during the transformation requires diffusion of any grain-boundary segregants to other sites on the surface or interface. It is this process which appeared to reduce the reaction rate in the *ex-situ*-sputtered film.

This result is significant in our attempts to enhance the C49-C54 reaction rate during integrated circuit processing. For a rapid transformation, low oxygen concentrations are preferable. However, it is also advantageous to increase the nucleation density of the C54 phase so that individual reaction fronts have less distance to cover before the reaction is completed. Since the C54 nucleation density can be increased by alloying, we are now conducting further experiments to determine how oxygen impurities affect the transformation kinetics in alloyed films.

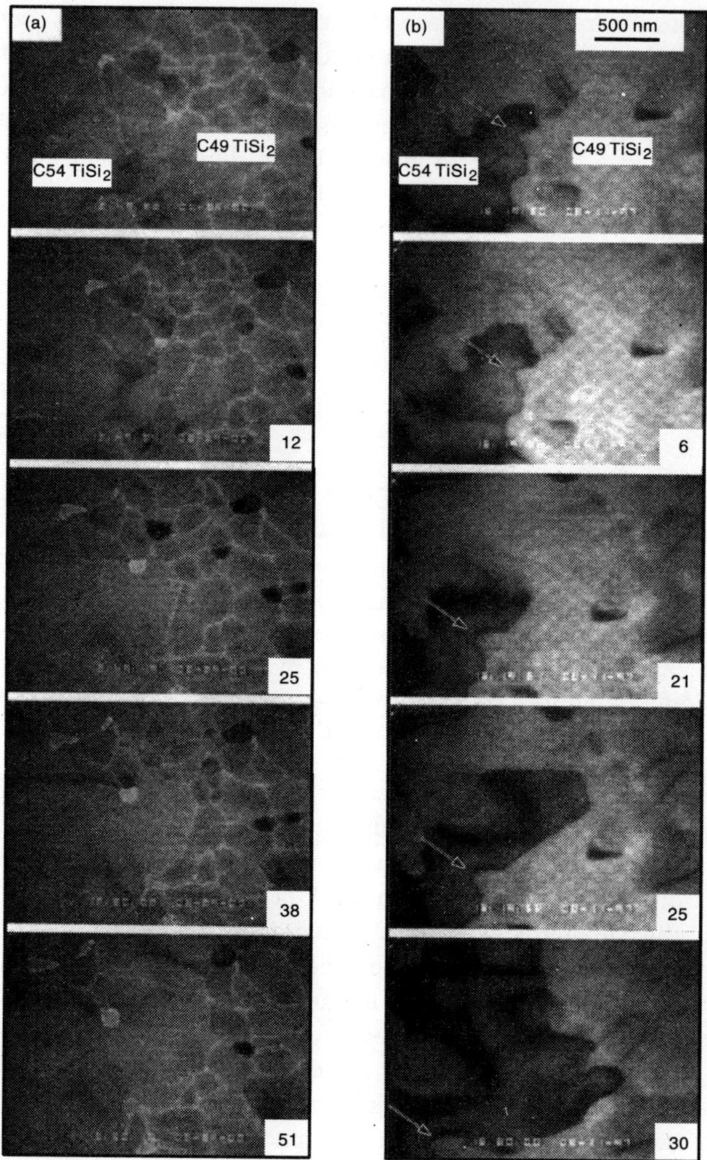

Fig. 2.41 *Video frames showing the transformation from the C49 to C54 phase. The numbers indicate the time elapsed in seconds since the first frame. (a) The phase transition in a 10-nm-thick UHV-deposited Ti film, recorded during heating at 1098 K. The reaction front moves from left to right smoothly with no pinning. The bright feature is a pinhole. (b) Phase transition in a 10-nm film sputtered ex situ and containing about 33% oxygen, as measured by Auger electron spectroscopy. A single C54 gain is in a strong diffracting condition and appears dark. The C49 phase has a grain structure similar to that shown in (a), through the grain boundaries are less visible due to a greater substrate thickness. The arrow marks a reference point on the specimen. The reaction occurs in a jerky fashion with pinning events at the grain boundaries three frames) followed by rapid jumps across several grains (fourth and fifth frames).*

Studying the Liquid/Solid Interface by *in situ* Microscopy

The preceding examples have shown how processing-related reaction mechanism can be observed and growth phenomena understood through *in situ* experiments. Many phenomena of interest during processing take place at gas/solid or solid/solid interfaces and are therefore amenable to study *in situ* in the TEM. However, there is an important class of growth processes which occur at a liquid/state interface, and these have not received corresponding microscopic study. One example of topical interest is the deposition of copper by electroplating. The success of copper as a replacement for aluminum in metal lines has encouraged the study of several different copper-deposition techniques. Electroplating is the method of choice for integrated circuit fabrication, but important issues remain in applying this technique. Examples include the problem of reliably depositing copper into small vias, and the control of the microstructure of the lines (grain size and texture) to ensure good coverage and reduce electromigration.

Several phenomena are unique to the electroplating process used in integrated circuit fabrication. These include the effect of the underlying "seed layer" on grain nucleation and orientation, and a "transient grain growth" process in which grain size increase, even at room temperature, in the minutes or hours after deposition. Observations made in real time during and immediately after deposition would be helpful in understanding these phenomena. We are therefore developing a liquid cell in which electroplating, or other liquid processes, can be reproduced while under TEM observation.

The cell, shown schematically in Figure 2.42, is designed to fit into a single-tilt holder which is of standard design except for the addition of electrical connections. It is made up of two silicon-nitride-covered Si wafers which have been etched to form small windows. The wafers are glued together face to face so that these "imaging window" are aligned, while a spacer layer keeps the windows a fixed distance apart. Two larger windows are also etched into the upper wafer, and the nitride is removed from these windows. The two etched volumes act as reservoirs for the liquid, so that a drop of liquid can be placed in one reservoir and can flow by capillary action across the imaging window. After liquid is introduced, the reservoirs are sealed by gluing sapphire squares on top of the cell.

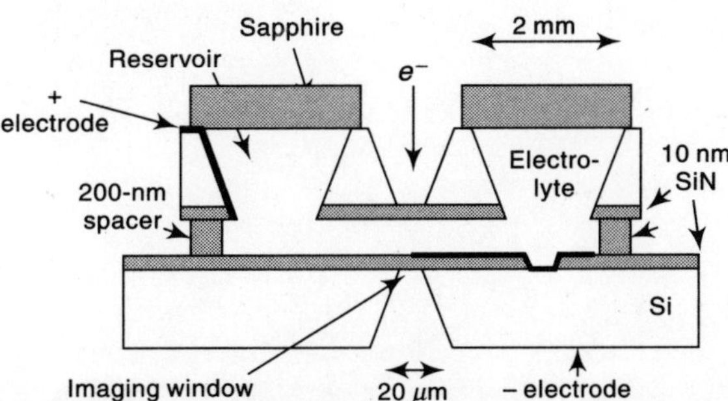

Fig. 2.42 *Schematic diagram of experimental liquid cell. The spacer layer is formed from SiO_2, and the positive electrode is formed from deposited copper. The contacts to the negative elctrode is achieved through a hole etched in the SiN.*

To study electroplating, electrical contacts are patterned on the wafers before the cell is assembled (Figure 2.42), and a copper sulphate/sulphuric acid solution is used as the electrolyte. It is intended that plating will take place onto an anode which is visible through the imaging window. Calculations suggest that application of a potential of around 2 V will allow a current density of 10^{-6} A/mm^2 to flow, depositing 200 nm of Cu in several minutes. Reserving the polarity allows the copper to be dissolved off the anode so that the experiment can be repeated. It is important to note that the volumes of copper which are to be plated are sufficiently small that the finite amount of electrolyte present is not significant.

Although the full system is still undergoing testing, we have already shown that cells of this design can hold liquid in the vacuum environment of the microscope. However, due to curvature of the SiN, the total specimen thickness (liquid, metal, and SiN) is much greater than desirable for imaging. To improve the image quality, we have recently added an energy filter to the microscope.[4] This improves the image contrast significantly by removing electrons which have undergone inelastic scattering, allowing only elastically scattered electrons to contribute to the image. The development of the liquid cell should enable us to study many phenomena other than electrodeposition, including switching in liquid crystals and crystal growth from a solution.

FETs, MOSFETs, MODFETs and SETs Details of Semiconductor Electronic Devices

(I) IN NANOSCALE ELECTRONIC DEVICES

(a) Electrical Contacts and Nanowires

Semiconductor devices rely on the formation of spatial charge domains. Miniaturization of the devices requires a minimization of the dimensions of these domains. So combinations of high and extremely low conductive materials gains importance. Beside classical semiconductors, the ongoing miniaturization has lead to highly doted semiconductor materials as well as nanostructured metal-isolator systems. The latter are of particular interest for the controlled transport of individual electrons by tunneling barriers and via conductive islands. In comparison with conventional semiconductor devices, such single electron devices would have the advantage of much less heat production at the same switch rates because of their extremely low power dissipation.

Circuits path with nanometer dimensions are elementary components of nanoelectronics in general, and especially for switches and transducers that control the transport of single electrons.

Contacts with widths in the medium nanometer range can be fabricated by electron beam lithography (EBL) wiring using a molecular monolayer as the positive resist. If a conductive substrate is utilized, the EBL-structured lines in the monolayer can be enhanced galvanically. Copper contacts of 50 nm widths fabricated by a combination of EBL, dry etching and chemical-mechanical polishing. Therefore, trench structures were EBL-written into a thin PMMA layer prior to transfer by plasma etching into an SiO_2 layer. After removal of the resist layer, these trenches were completely filled with a TaSiN or TaN adhesion layer prior to thicker copper layer. Chemical-mechanical polishing removed the copper on the planar SiO_2 surface, resulting in metal contact structure in the pattern of the trench structure.

A simple technique for the fabrication of narrow metal contacts relies on the nano-epitactical deposition of metal strips along monoatomic steps of single crystalline dielectric materials, which are cut at a defined angle to the crystal planes (Fig. 3.1). These strips can be miniaturized down to a width of 2–3 nm. Unfortunately, the geometries of such contact lines are predetermined by the arrangement of the crystal planes, so that they cannot be chosen arbitrarily.

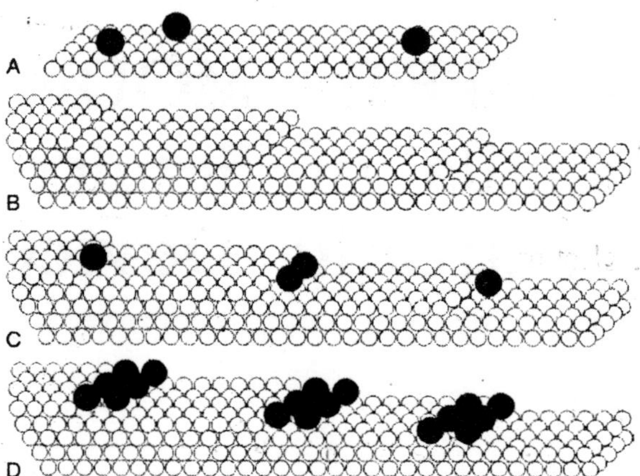

Fig. 3.1 *Preparation of metal lines with nanometer dimensions by decoration of elementary steps on single-crystalline surfaces*

The electrical conductivity of molecules with mobile electrons is a required condition for the preparation of nanowires using single molecules. The macromolecules typically prepared by polymerization or polycondensation often exhibit a linear geometry with a greater number of rotating bonds, so that they can form a variety of unpredictable conformations.

For nanoconstruction and the fabrication of nanowires with predetermined geometries, rigid molecules are required. Therefore, compact rod-like molecules have been developed with conjugated ring-shaped structures, which combine increased rigidity with high intermolecular electron mobility.

In DNA conductivity, the experimental evidence points to electron transport over short (lower nanometer scale) distances. A single poly(G)-poly(C) double strand of 10.4 nm length positioned in a nanogap of 8 nm yields a non-linear current-voltage plot. At room temperature it points to the transport of individual electrons, for a molecular wire. The current is increased up to 100 nA, indicating a high conductivity. The transport rate of the charge is about 100 electrons per μm and picesecond. A theoretical and experimental study indicates that conductivity is achieved after a certain exposure to energetic electrons.

Instead of the direct use of individual macromolecules, in nanolithographic processes molecules are applied as masks in order to form single molecule-based nanowires. One example is the fabrication of Au nanowires by dry etching of a thin Au layer masked by an absorbed microtubule molecule. Such a microtube molecule acts as the template for a nanowire for an electroless Ni or Pd metallization. This principle is also applicable to DNA.

In the combination of chip surfaces or lithographically fabricated microelectrodes with molecules and metal nanoparticles; an assembly of surface functionalized metal nanoparticles yield electrical contacts with nanometer dimensions. With the application of an electric field, rod-like metal macro- and nano-particles are oriented and positioned on substrate surfaces. Hence gold particles of 70–350 nm diameter and 8 μm length are positioned between gold electrodes out of a dielectric solvent. Layers of chemically functionalized gold nanoparticles are generated in a stepwise process. The deposition of ensembles of metal nanoparticle on immobilized molecules is applied for the detection of molecular interactions. DNA molecules individually positioned between microelectrodes are an interesting objective for such an approach.

Nanowires can be based on molecules by using an extended chain-like molecule as a template for the binding of metal nanoparticles. One example utilized super molecular aggregates of DNA streptavidin conjugates that are decorated with biotinylated gold nanoparticles, resulting in chains of gold nanoparticles. Through the integration of branched DNA, more complex supermolecules are accessible. The applied supermolecular streptavidin-DNA aggregates are also interesting for the highly sensitive detection of bimolecules by immuno-PCR.

Free-standing nanowires with a width of about 20 nm are fabricated by the electron beam induced deposition of carbon needles at the edge of a prestructures gold layer, resulting in free-standing nanodiodiodes with a gap with of about 5 nm.

Electrical contacts consisting of chains of conductive nanoparticles are of both technological and functional interest. Particles below 100 nm diameter are comparable to molecules with respect to their binding behavior, which means that their non-specific binding to surfaces can be suppressed and specific bonds can be generated through complementary coupling groups on the surface of the substrate or other nanoparticles.. Hence contacts can be formed by chemical self-organization, e.g., binding via thiol groups (Fig. 3.2). The conductive particles present barriers to the electron transport at their contacts points. Dielectric molecular surface layers on gold nanoparticles transport at their contact points. Dielectric molecular surface layers on gold nanoparticles can further increase the barrier effect, and the barrier height can be controlled by the choice of thickness and composition of the molecular monolayer. Therefore, such particle chains are not only interesting as simple wiring elements, but also for controlled electron transport. Chains of assembled nanoparticles of 100 nm diameter span gaps of 30 nm and 150 nm width, respectively. On the other hand, fairly large electrode gaps of several millimeters can also be connected through the dielectrophoretic assembly of metal nanoparticles.

(b) Tunneling Barriers

Well-defined tunneling barriers are necessary for all devices with electron tunneling processes. The requires superconductive devices they rely on the mutual tunneling of electron pairs over barriers single electron tunneling (SET) devices. Such barriers should be thinner than 2 nm, because the probability of electron transfer decreases steeply with the barrier thickness due to the exponential decay of the tunneling conductivity. So thicker barriers provide no applicable switches or transducers. The adjustment of the barrier thickness is usually realized by the

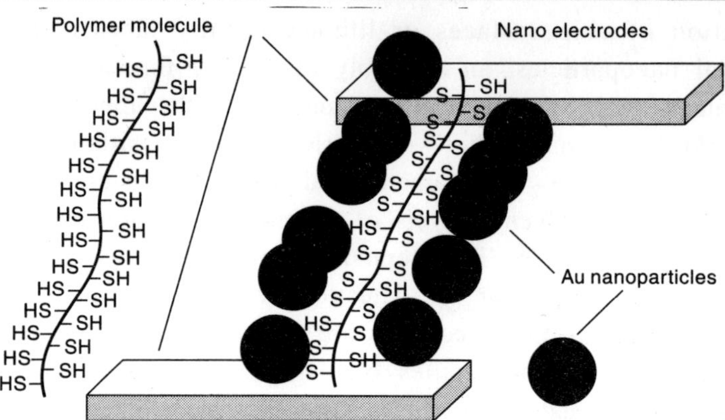

Fig. 3.2 *Single-molecule based preparation of conducting paths of a chain of nanoparticles by immobilization of a chain-like molecule between lithographically structured electrodes and coupling of particles by thiol groups*

adjustment of a thin layer. The direct deposition of dielectric layers with a highly defined thickness in the range of 1–2 nm inside a stack of layers is challenging. Therefore, instead of directly deposited layers, of a less noble metal are often oxidized in a spontaneous reaction with the oxygen or water contained in the air into a dielectric layer that can be used as a tunneling barrier. Such as chemically formation of the tunneling barrier layer can be regulated by the oxygen or particle pressure of water, the temperature and the duration of the incubation in the reactive atmosphere.

In addition to the thickness, the lateral extension of tunneling barriers is also important for SET devices. It should be low to achieve high integration densities. The individual switching process depends on the capacity of the contact, which should be low. The capacity depends on the barrier thickness, but this thickness has upper limits for which tunneling is still possible. Another parameter influencing the capacity is the area. Thus to realize low capacitance–and the applies especially to room temperature devices–nanostructure barriers are required.

Small lateral dimensions of tunneling barriers are achieved with metal nanoparticles or clusters that are separated by ultrathin dielectric layer from metal or semiconducting contacts. The isolating layer can cover a larger area, but it acts as a tunneling barrier only in the contact region with the nanoparticle. Barriers with an efficient diameter only 4 nm have been prepared, for example for the couple of 4 nm. Au clusters via a monolayer o-dithiolxylol layer onto a GaAs layer. SEM imaging presents the contacts as small flat structures.

Another approach to the fabrication of nanostructured tunneling barriers utilized local oxidation of a metal contact by STM conductive AFM tips (Fig. 3.3). Therefore, the metal layer is anodically polarized. After the approach of the scanning tip, the metal is locally oxidized and generates a localized barrier to the electron transportation. This approach requires easily oxidizable metals, such as aluminum. To achieve electron confinements for controlled single electron transport, a metal layer is pre-structures as a narrow contact prior to writing an island through local oxidation.

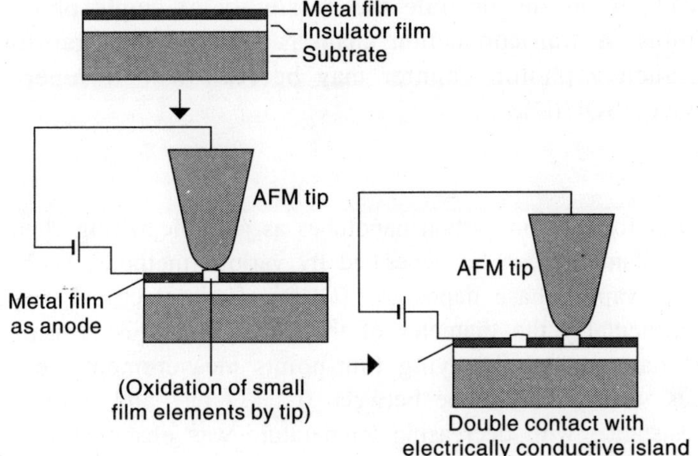

Fig. 3.3 *Preparation of a pair of tunneling contacts by local anodic oxidation of a metal layer using a conductive AFM.*

Nanostructures semiconductors can also be utilized as barriers for the electrostatic single electron transport. So a freestanding nanolever of highly doted silicon can be fabricated with a length of 800 nm and a cross-section of 24 nm × 80 nm. A gated electrode in close proximity allows for the possibility of a controlled single electron transport up to temperature of about 100 K.

Molecular tunneling barriers can be prepared by the generation of double gaps and the subsequent filling of the gaps with molecular mono-or double layers (Fig. 3.4). The preparation of vertical molecular tunneling barriers was carried out with the formation of an SAM monolayer of p-dithiohydrochinon on a thin gold wire. After mono layer formation, the wire was mechanically broken in one location. A reconstruction of the monolayer on both sides of the gap occurred, prior to reconnection of the two parts of the wire by evaporation of the solvent. In this manner an SAM double layer is formed. This breaking technique probably facilities the fabrication of contacts with charge transfer via individual atoms.

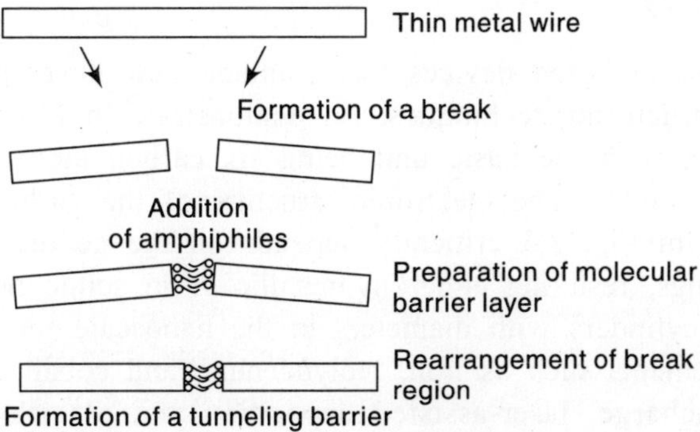

Fig. 3.4 *Preparation of molecular tunnelling barriers by the formation of SAMs in break junction of a wire.*

Theoretical considerations demonstrate the possibility of single photon detection with the help of nanostructures. A transconduction chain is required that transforms photons into an electronic signal. Such a photon counter may be read with a super-conducting quantum interferometric device (SQUID).

(c) Nanotubes

Specific interest has focused on carbon nanotubes as possible wiring elements in the nanometer range. These carbon filaments can be generated by various methods, such as laser ablation, arc discharge or chemical vapor phase deposition (CVD). Their electrical properties depend greatly on the geometry, particularly the diameter of the tubes, which is typically between about one and a few tens of nanometers. Applying four-points measurements, electrical resistances of individual nanotubes were found to be between 0.2 $k\Omega\ m^{-1}$ and more than 500 $M\Omega\ m^{-1}$. A great increase in resistance with decreasing temperature was observed for selected tubes, others exhibited a linear decrease of the resistance between 4 and 300 K only 10%. In general, semiconductor-like temperature dependence of the electrical behavior is observed. Logic circuits and a room-temperature single electron transistor were fabricated based on carbon nanotubes.

Carbon nanotubes exhibit several fairly different conformations: tubes can have a single wall, but they can also be double- or multi-walled, or narrower and sometimes helical tubes integrated into tubes of greater diameter. In addition to pure carbc.ı tubes, synthetic monofilament with conjugated bond backbones and with aromatic- and aromatic-heterocyclic units, such as polyphenylenes, polythiophenes or bisalkylthiophen-bridged ureas, are also candidates for organic conducting wire with nanometer dimensions.

Two Fe/Ti contacts separated by 1 μm gap can be bridged by individual carbon nanotubes prepared by gas phase deposition. The thin Fe layer thereby acted as a catalyst for the nanotube generation. Most of the nanowires of about 1 μm length and a diameter of 40–50 nm yielded electrical resistances of between 50 and 80 Ω. Directed growth of carbon nanotubes results in freestanding nanowires between microstructure silicon columns.

(II) CARBON NANOTUBE DEVICES

Structure and Technology

Carbon nanotube (CNT) based devices that combine new developments in material science with innovative nanostructuring techniques. As demonstrate in Fig. 3.5 carbon nanotubes are made out of a network with the basic unit being six carbon atoms in ring configuration and arranged in form of cylinder. The electronic structure of the carbonnanotubes as represented by the band diagrams in Fig. 3.5 critically depends on the geometry of the interconnnection between the carbon rings, resulting either in metallic or in semiconducting behavior.

The growth of the cylinders with diameters in the nanoscale range is generally induced by the use of catalytic elements such as iron, molybdenum, and cobalt. The most common growth techniques are arc-discharge, laser-assisted deposition and plasma-enhanced chemical vapor deposition (PECVD), using a methane plasma at relatively high temperatures. Depending on the growth parameters, the deposition processes result either in formation of multi wall nanotubes (MWNTs) or single wall nanotubes (SWNTs).

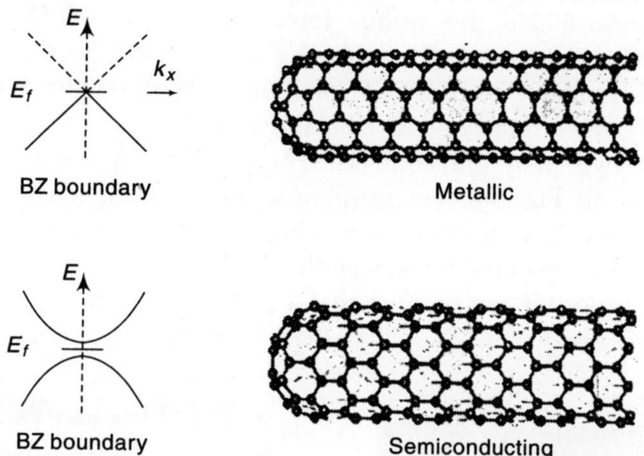

Fig. 3.5 *Structure and electronic band diagram of metallic and semiconducting carbon nanotubes. Note the different orientation of the rings.*

The particular interest in CNTs is due to their very low specific resistivities for metallic carbon nanotubes and high hole mobilities for semiconducting nanotubes. The low density of surface states explain these interesting electronic properties. The material forms a two-dimensional network of carbon atoms without the presence of dangling bonds. When assembling in cylindrical form the recombination at the edges of the semiconductor is avoided.

First applications of metallic CNTs are wiring of microelectronic circuits and the use as field emitters for high resolution flat panel displays. As an example of the latter application, the manufacturing of a gated 3 × 3 field emitter cathode array (FEA) is depicted in Fig. 3.6. A silicon surface is covered with a 1-nm thick iron layer as catalyst on top of a 10-nm thick aluminum layer and subsequently with a SiO_2 layer. After deposition of the molybdenum gate electrode and the opening of the single cathode windows by reactive ion etching, metallic multiwall carbon nanotubes are grown on top of the Al/Fe metallization as cathode electrodes using a CVD process with an acetylene plasma at 1173 K. To give an idea of the dimensions of the device: the lengths of the white marks are 50 μm in Fig. 3.6(g) and 2 μm in Fig. 3.6(h).

Carbon Nanotube Transistors

The very high values of the charge carrier mobilities in semiconducting carbon nanotubes together with the small device dimensions make CNT based devices very interesting for microelectronic applications. Field effect transistors have mostly been implemented because carbon nanotubes exhibit very high hole mobilities in particular. It should, however, also be mentioned that first experiments to realize a bipolar *p-n-p* transistor were successful.

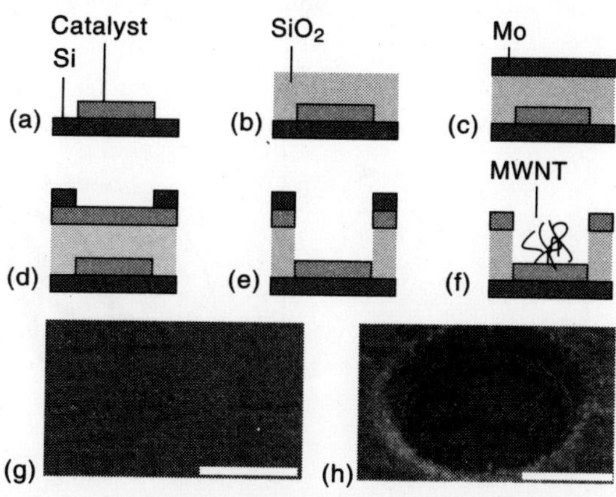

Fig. 3.6 *Fabrication process and SEM images of a carbon nanotube based gated field emitter array*

In Fig. 3.7, the structure and the atomic force microscope (AFM) image of a so-called *TUBEFET* are shown. On a silicon wafer covered with a thermal SiO_2 layer, which serves as a backside gate and gate insulator respectively, platinum (Pt) electrodes are deposited that form the source and drain contact. Subsequently the single wall carbon nanotubes are deposited or grown connecting the two platinum contacts. The room temperature characteristics of this TUBEFET are illustrate in Fig. 3.8. For positive voltages applied between drain and source contact, a clear threshold voltage for conduction has been observed whose value increases with increasing gate voltage. For negative voltage applied between drain and source, ohmic behavior has been found. Furthermore it can be seen that a 10 V change of the gate voltage results in a variation of the channel conductance of more than six orders of magnitude.

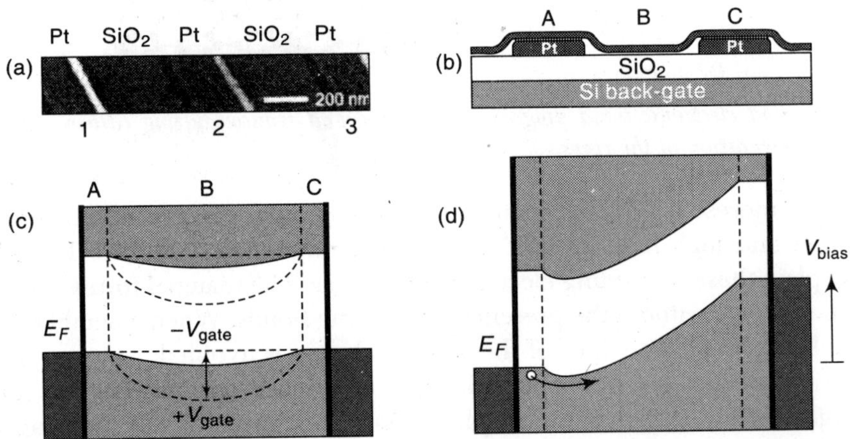

Fig. 3.7 *(a) AFM image, (b) schematic structure, and band diagrams without (c) and with (d) applied source-drain voltage of a TUBEFET.*

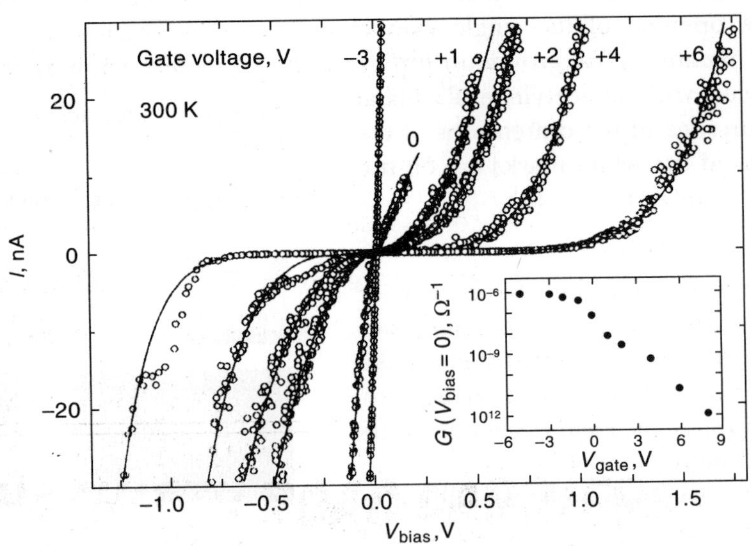

Fig. 3.8 *Electrical device characteristics of a TUBEFET measured at 300 K. Inset: channel conductance vs. gate voltage.*

One of the main problems regarding the fabrication integrated circuits using CNT transistor is the limited reproducibility of the CNT growth process. An alternative approach to lateral integration is the manufacturing of arrays of CNTs based on vertical structures. Very homogeneous and reproducible growth of vertical CNT arrays by porolysis of acetlyene on cobalt coated alumina substrates has been achieved. The structure and distribution of the diameters are shown in Fig. 3.9. The hexagonal cells, being open on top, have an average radius of 47 nm. They are positioned very symmetrically with an average distance of 98 nm. The manufacturing of vertical CNT transistors is also achieved. However, they operate only at cryogenic temperature of 4 K.

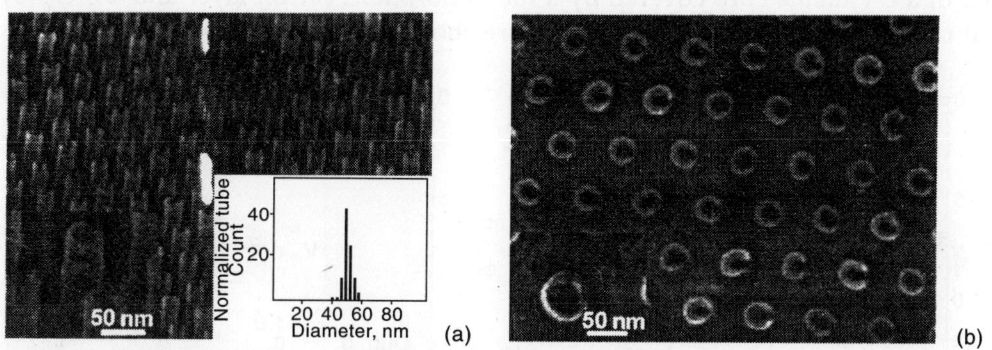

Fig. 3.9 *SEM image and histogram of the diameters of the nanotubes of a vertical carbon nanotube array.*

The first successful integration of CNT field effect transistors is done using lateral structures similar to the above-shown TUBEFT but with other materials. A structure using gold drain and source contacts is shown in Fig. 3.10. In this case, an aluminium gate contact covered by a

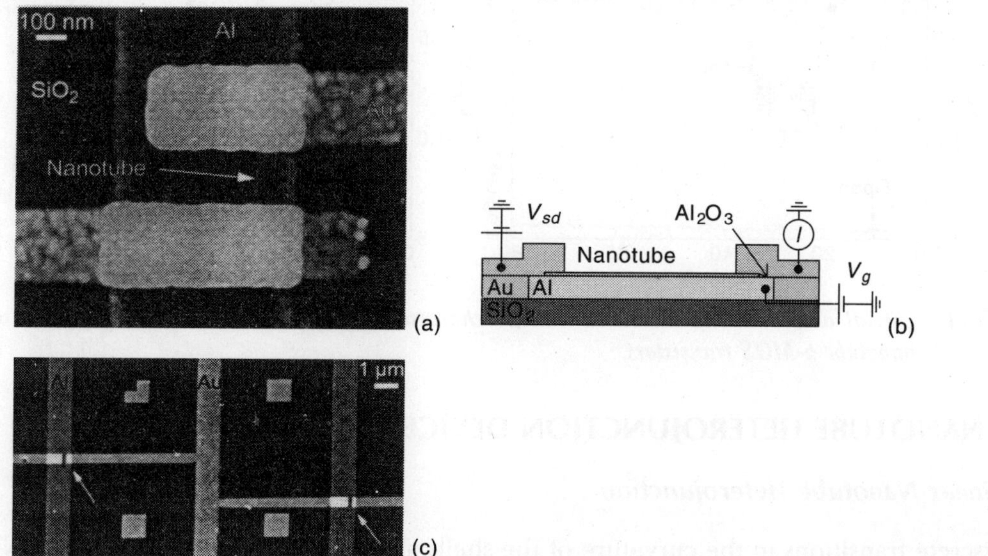

Fig. 3.10 *(a) AFM image of a single SET, (b) schematic structure, and (c) AFM image of an integrated circuit structure using single wall carbon nanotube transistors.*

100 nm thick Al_2O_3 layer is a gate isolater. Both layers are deposited on top of a SiO_2 layer as substrate. The resulting enhancement-type *p*-channel MOSFETs with a voltage gain exceeding 10 are wired together by gold metallization. In Fig. 3.11, the transfer characteristics of various digital functional circuits implemented by this technique are shown. Up to three interconnected transistors are used to implement an inverter, a NOR gate, a static RAM cell, and a ring-counter–with a long switching time.

Other applications of carbon nanotubes for electronic devices, can be made as heterojunctions between CNTs and silicon quantum wires are achieved. Here, the silicon quantum wires are grown by CVD deposition in a silane atmosphere selectively on top of the CNTs. They consisted of a crystalline core covered by a thin amorphous silicon layer and a SiO_2 layer. The electrical characterization of this heterostructure show a behavior similar to a Schottky diode.

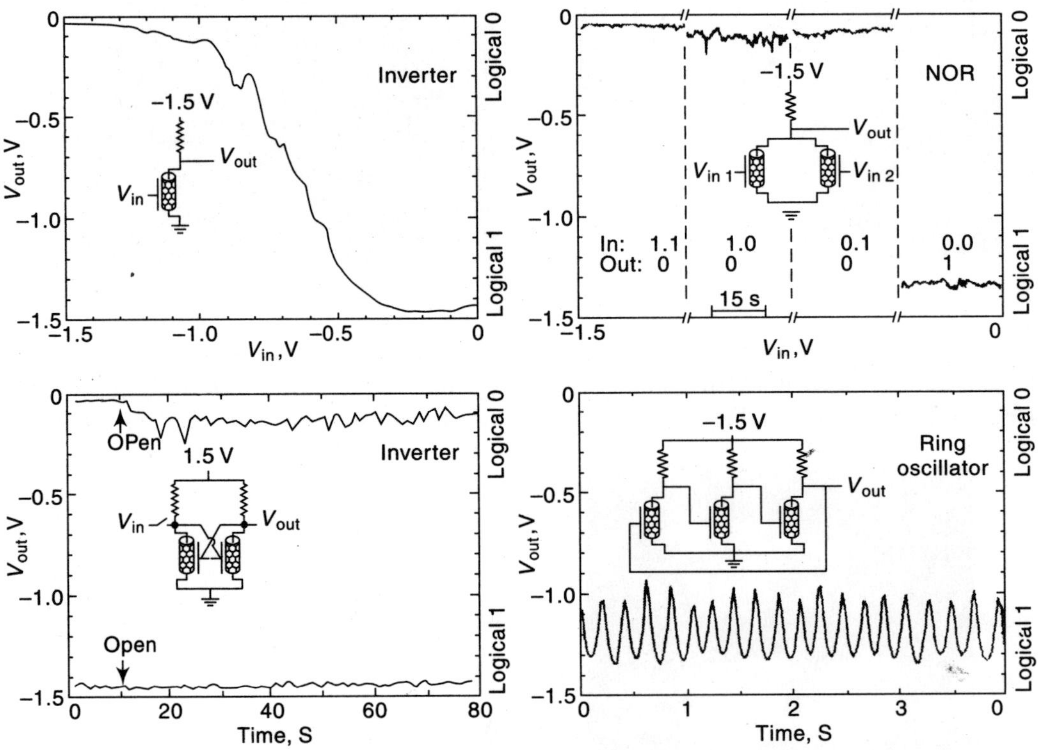

Fig. 3.11 *Circuit diagrams and measured transfer characteristics of integrated circuit produced with carbon nanotube p-MOS transistors.*

(III) NANOTUBE HETEROJUNCTION DEVICES

(a) Linear Nanotube Heterojunction

Discrete transitions in the curvature of the shells are observed while making MWNTs, and is due to the insertion of a pentagon or heptagon in the graphene lattice. Thus, nanotubes of different diameter and/or chirality maybe joined seamlessly by the introduction of pairs of pentagons and heptagons into the graphene lattice. All carbon metal-semiconductor diodes are made by joining a metallic and a semiconducting SWNT by a single pentagon-heptagon pair. So, an on-tube quantum dot is produced by a pair of pentagon-heptagon defects; which connects a metallic (5,5) SWNT to two semiconducting (6.4) SWNT leads.

Colloins *et al.* probed a tangled mat of laser ablation produced SWNT bundles with an STM tip. The tip when brought into electrical contact with the mat, is found to retract for several micro meters before loosing the electrical contact. They proposed a model of a sliding contact between tip and nanotube bundle as the tip is retracts. During retraction of the tip, current-voltage (*I–V*) characteristics of the tip-mat junction are acquired at various retraction distances. It is found that the I-V *characteristics* sometimes abruptly change from symmetric to strongly rectifying due to the tip sliding past a defect in a nanotube which acted as an on-tube device. The experiments of Collins et al. the STM was not used in imaging mode.

In fabricating tube-on-top electrical devices occasionally observed individual SWNTs with sharp kinks along their length. Out of 500 devices, four were found with single kinks, and one with two kinks. These kinks, interpreted as individual pentagon-heptagon pair defects causing an abrupt change of chirality of the SWNT at the defect. Electrical measurements on two kinked SWNTs are noted. The first device spanned only three electrodes, so independent two-terminal measurements could only be performed on one side of the kink. The portion of the nanotube on this side of the kink junction had a two-probe conductance of 110 kΩ, with no gate voltage dependence, indicating a metallic SWNT. Two-probe measurements across the kink, however, give an immeasurably low conductance (< 4 pS) at zero bias, but a strong non-linear onset of conduction when around +1 to 2 V was applied to the metallic SWNT. The onset of conduction shifted significantly with the application of a gate voltage, becoming more conducting at positive bias for negative gate voltages. Because of the gate voltage dependent conductivity of the kink segment, the side of the kink which contact only a single electrode is assumed to be a semiconducting SWNT. The rectifying behaviour attributes to the formation of a metal-semiconductor Schottky barrier at the kink, although the very high threshold voltage 1–2 V) and the incorrect sign of the rectification indicates that the device is somewhat complicated.

(b) Nanotube T and Y Junctions

The band structure of 1D, 2D, and 3D superstrctures formed by the joining of SWNTs with pentagons and heptagons, envisioned an all-carbon circuit network. Also, two T junctions connecting metallic and semiconducting SWNTs are found stable under molecular dynamics relaxation, and that Y junctions between SWNTs, rectify currents. Such ideal junctions between SWNT have been difficult to realize experimentally. However, branching structures of large; tubular carbon nanofibers, synthesized by templated pyrolysis of acetylene in nanochannel alumina, pyrolysis of nickelocene and thipphene, pyrolysis of methane over cobalt supported on MgO, pyrolysis of methane over iron particles on roughened silicon are achieved. These structures are all tens of nanometers in diameter and thus not electronically one dimensional at room temperature. Moreover, those structures for which TEM images exist show poor graphitization and thus do not fit the strict definition of nanotube. However, electrical measurements of individual examples as well as ensembles of Y junction fibers synthesized inside nanochannel alumina templates show interesting rectifying Y junctions.

(c) Crossed-Nanotube Junctions

Devices consisting of single junctions between individually electrically contacted nanotubes are fabricated using a metal-on-top method. Each crossed-SWNT device consists of two crossed individual SWNTs or small bundles (diameter < 3 nm) of SWNTs with four electrical contacts, one on each end of each SWNT or bundle. The inset of Fig. 3.12 shows an AFM image of such a crossed nanotube device; two crossed SWNTs interconnect Cr/Au contacts.

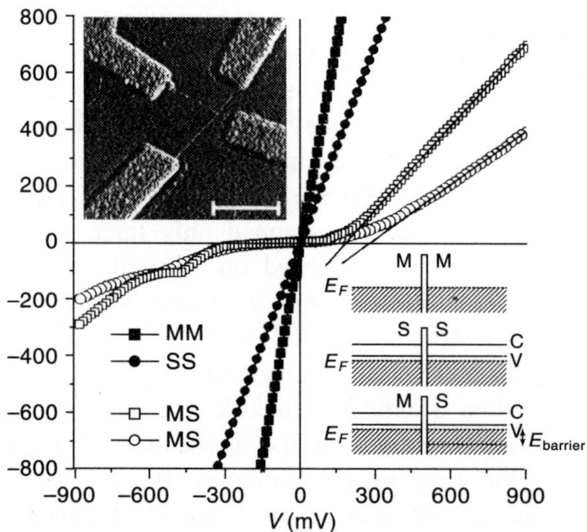

Fig. 3.12 *Current-voltage characteristics of crossed carbon nanotube junctions. The I-V curves of a metal-metal junction (filled squares), a semiconductor junction (filled circles), and two metal-semiconductor junctions (open squares and circles) are shown. The lower right inset shows the expected band profiles on either side of the junction for each case. The non-linear behavior of the metal-semiconductor junctions is due to the formation of a Schottky barrier at the junction, as shown in the inset. The solid lines in the main panel extrapolate the forward bias region to zero current, giving an estimate of the barrier height of 190–290 meV. The upper left inset show an AFM image of a typical crossed nanotube device. The narrow crossing lines are SWNTs, the larger blocks are Cr/Au electrodes. The scale bar is 1 μm.*

In each SWNT the junction is measured independently to determine its properties. At room temperature two-terminal conductance, as a function of gate voltage $G(V_g)$ characteristics is measured across the individual SWNTs; to find each, as metallic or semiconducting. Each crossed-SWNT device is composed of two metallic *SWNTs (MM)*, one metallic and one semiconducting *SWNT (MS)*, or two semiconducting *SWNTs (SS)*.

Figure 3.12 shows the four-terminal current-voltage (*I–V*) characteristic of an MM junction at 200 K (filled squares). The slope of the *I-V* curve corresponds to a resistance of 200 kΩ, or a conductance of $0.13e^2/h$. Similar measurements of three other MM junctions gave conductances of $0.086e^2/h$, $0.12e^2/h$, and $0.26e^2/h$. The measurements of SS junctions are often complicated by the presence of high resistance barriers in the laser-ablation synthesized semiconducting SWNT. Nevertheless, two-terminal conductance of SS junctions as high as $0.11e^2/h$ and $0.06e^2/h$ (the higher conductance curve is represented by the filled circles in Fig. 3.12) are observed.

The measured conductances of MM junction correspond to a transmission probability for the junction $T_j = G/(4e^2/h) \approx 0.02–0.06$. Thus, an electron arriving at the junction in one SWNT has a few percent chance of tunneling into the other SWNT. MM junctions make surprisingly good tunnel contacts, despite the extremely small junction area on the order of 1 mm^2. In order to understand these results, first principle density functional calculations of the conductance of MM junctions are performed. For two (5,5) SWNTs separated by the van der Waals distance of 0.34 nm, a transmission $T_j \approx 2 \times 10^{-4}$ is found. However, when the contact force between the nanotubes due to interact with the SiO_2 substrate is included, the nanotubes deformed significantly at the junction. The high conductance of MM junctions indicates that metallic SWNT junctions is useful for branching interconnects.

The MS case is qualitatively different from the MM and SS cases. Charge transfer at the junction between a doped semiconducting SWNT and a metallic SWNT is expected to form a Schottky barrier with the Fermi level E_F of the metallic SWNT aligned with the center of the band gap of the semiconducting SWNT at the junction (see lower inset, Fig. 3.12). The total barrier transmission probability T_{MS} is then given by $T_{MS} \approx T_j\, T_d$, where T_d is the transmission probability for tunneling through the depletion region to the location of the metal SWNT and T_j is the probability of tunneling between the SWNTs. $T_{MS} \approx 2 \times 10^{-4}$ for both MS devices in Fig. 3.12. If $T_j \approx 0.04$ for the MS junctions (comparable to the MM value), then $T_d \approx 5 \times 10^{-3}$.

A Schottky barrier also shows asymmetric transport; as is the case for the MS junction barrier (Fig. 3.12). The V-intercept of the linear positive-bias region gives a gross measure of the barrier height: $E_{barrier}$ = 190 and 290 meV for the two devices. This agrees reasonably well with the expected barrier height $E_{barrier} = E_g/2 \sim 250–350$ meV for 1–1.5-nm semiconducting SWNTs ($E_g \sim 500–700$ meV).

Poor rectification is obtained as the MS Schottky barrier is leaky. Better rectification is achieved with a three-terminal device which takes advantage of the double-width depletion barrier in the semiconducting nanotube (Fig. 3.13). A voltage is applied to one end of the semiconducting SWNT in an MS device, while the other end is grounded through a current-measuring amplifier. If the metal SWNT as well as one end of the semiconducting SWNT are grounded, the barrier to holes remain intact when a negative voltage is applied, because the metallic SWNT remains at roughly the same potential as the grounded end of the semiconducting SWNT [Fig. 3.13(c)]. However, when a positive bias is applied such that the potential difference between the metallic and semiconducting SWNT is greater than the barrier height, holes may pass the barrier, and current will flow through the semiconducting SWNT [Fig. 3.13(d)]. Such a response is observed in the measured current leaving the semiconducting SWNT [I_S in Fig. 3.13(a)].

The current from the semiconducting SWNT is approximately 100 times greater for a bias of +700 mV than for –700 mV. It is noted that the ineffective screening inherent to 1D systems

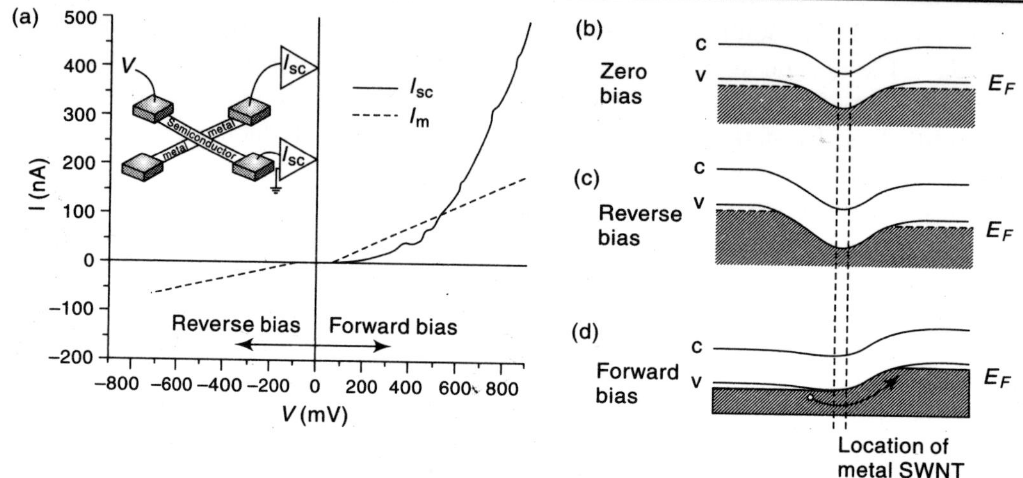

Fig. 3.13 *Three-terminal characteristics of a metal-semiconductor crossed-nanotube junction. In (a), the bias is applied to one end of the semiconducting nanotube, the current flowing to ground is measured at the other end (see inset). The metallic nanotubes is also grounded through a current amplifier. The solid curve in a (a) shows the current in the semiconducting nanotube as a function of bias voltage. The dotted curve shows the current flowing from the metallic nanotube. The schematic operation of the device is shown in (b)-(d). At zero bias (b), a depletion barrier exist in the semiconducting nanotube on either side of the metallic nanotube. On reverse bias (c), the barrier on the biased side increases, and no current flows. On forward bias (d), the barrier on the biased side is decreased, and holes may pass the barrier.*

poses problems for nanotube Schottky devices nanometer-scale depletion regions are likely to be leaky barriers to tunneling. This three-terminal device points out at least one solution to this problem; a good rectifier is constructed from narrow Schottky barriers. The active length of this device is on the order of 15 nm, demonstrating that useful devices consisting of only a few thousands of atoms can be constructed from SWNTs.

(d) Mechanical Devices

Carbon nanotubes are exceedingly light and stiff. This suggests that nanomechanical devices based on carbon nanotubes can operate at very high speed. Indeed, simple oscillators made form sub-micro meter nanotube beams have characteristic mechanical frequencies in the Gigahertz range.

In addition, nanotubes, are exceptionally tough; the elastic limit (elongation) for SWNT as is ~6%. Nanotubes may be bent through large angles, and even buckled, and return to their original shape elastically.

This suggests that nanotube mechanical devices can operate reversible for large numbers of cycles with no fatigue. A mechanical bistable nanotube device which acts as a memory or programmable logic elements proposed. The device consists of two nanotubes; one lies on a substrate, while the other nanotube crosses it, suspended above it at a distance of a few nanometers (see Fig. 3.14). The total energy of the device is analyzed as a function of the

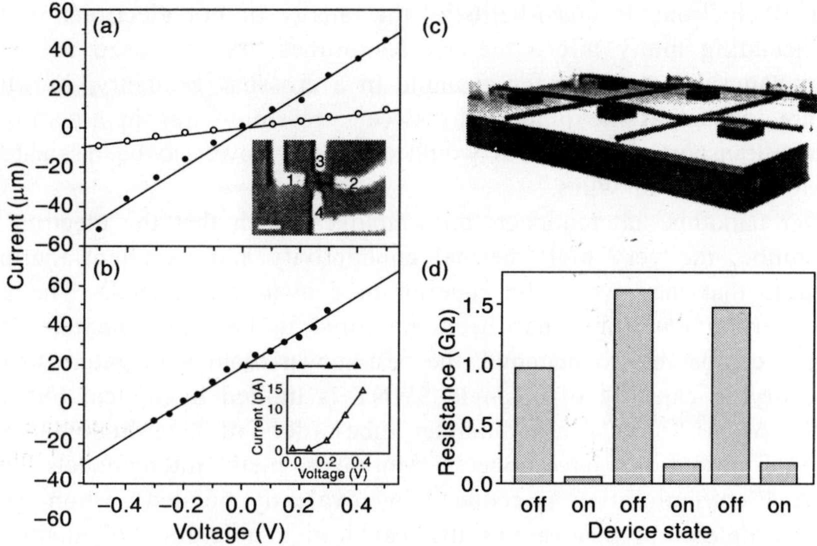

Fig. 3.14 *Carbon nanotube mechanical memory. The inset of (a) shows an optical micrograph (scale bar 4 μm) of a device consisting of two crossing bundles of SWNTs (dark lines) deposited on 150 nm high Cr/Au electrodes (while). The I-V characteristics measured from contacts 1 to 2 (solid circles) and 3 and 4 (open circles) are shown in (a). The I-V curves of the junction in the OFF state (open triangles) and ON state (solid circles) are shown in (b). The junction was switches to the ON state by the application of ±2.5 V. A possible nanotube mechanical memory array is shown in (c). (d) shows the ON/OFF resistance for a device similar to that shown in (a). This device could be switched ON with application of ±5 V, and OFF at 40 V.*

nanotube separation for a range of initial separations and nanotube lengths, and bistability of the device is found for a broad range of parameters. The two stable states correspond to the nanotubes in close contact (held by van der Walls forces), and relaxes. In order to switch between the two states, opposite polarity voltages are applied between the two nanotubes to cause them attract each other while like polarity voltage (relative to a third electrode) causes the nanotubes to separate.

By using bundles of SWNT deposited onto pre-fabricated gold electrodes with a height of 150 nm and a separation of ~4 μm is demonstrated (see Fig. 3.14). Reversible switching of one device from a separated state (>10 GΩ) to a contacted state (112 kΩ) at a voltage of 2.5 V i smade. Another device switched reversibly from separated (1.36 WΩ) to contacted (140 MΩ) with the application of ±5V to contact and 40 V to separate the SWNT bundles. Thus the, bistable devices can be made as small as 5–20 nm, with switching speeds of 100–200 GHz.

(e) Nanotube Interconnects

Nanotubes are excellent candidates for interconnects given their low electrical and thermal resistance, and stability at high temperature and current. The resistivity of metallic single-walled carbon nanotubes is on order 10^{-6} Ω cm, comparable or superior to copper. Furthermore, the conduction of electrons is quasi-ballistic; the energy of hot electrons is dissipated on a length scale exceeding many micro meters. Nanotubes may be used as exceedingly fine interconnects to nanoscale devices, for example in a crossbar geometry, in which molecular-sized devices are located at the junctions between crossing wires in an array. The ballistic nature of electron transport in nanotubes would allow the power to be dissipated far from the devices in macroscopic metal wires.

Even if longer nanotube interconnects are employed, such that the electron energy relaxes within the nanotube, the very high thermal conductivity and excellent thermal stability of nanotubes suggests that they would be superior to conventional metals. The phonon thermal conductivity of carbon nanotubes has been measured to be approximately 2000 W/m/K at room temperature, comparable to diamond, the best-known room temperature thermal conductor.

The current-carrying capacity of a single SWNT is limited by optical phonon emission to about 2.5×10^9 A/cm^2 for a 1 nm diameter tube-orders of magnitude larger than current densities found in present-day interconnects. Moreover, metal interconnects typically fail via electromigration. Electromigration is reduced in covalently bonded carbon nanotubes—there are no low-energy defects or dislocations that can lead to migration of atoms. Nanotubes are observed to carry current densities of 10^9 A/cm^2 for periods of weeks without failure.

(IV) RESONANT TUNNELING DIODE

Operation Principle and Technology

Resonant tunneling diodes (RTD) are the simplest devices based on quantum effects in semiconductor nanostructures. The basic RTD device incorporates a double quantum well (DBQW) structure (Fig. 3.15). The contact layers in the areas I, II, VI and VII consist of a heavily doped semiconductor with a relatively small band gap, for instance, GaAs. The layers III and IV are the tunneling barriers implemented with semiconductor of a relatively large gap and in particular with a large conduction band offset relative to the neighbouring regions like AlGaAs. The quantum well layer confined by the two tunneling barriers again consists of a material with a relatively small band gap.

For the operating principle see (Fig. 3.15), a local distribution of the electron energy shows that when a bias voltage is applied to a DBQW structure; the energy distribution of the electrons in the heavily doped region I, is in thermal equilibrium. At the boundary surfaces, there are multiple reflections of the electrons due to their wave nature, leading to destructive and constructive interferences as a function of the electron energy. Thus, the tunneling of those electrons is favored which hit the left barrier with an energy E_1 corresponding to the energy E_0 in the quantum well. The tunneling probability decreases drastically with both higher and lower electron energies. However, since the maximum of the electron energy distribution in the regions I and II can be tuned by changing the applied bias voltage, a local maximum (peak) is found in the current-voltage characteristics of the resonant tunneling diode, followed by a local minimum (valley). In Fig. 3.16, this is shown for an InGaAs/AlAs based RTD at 300 and 77 K.

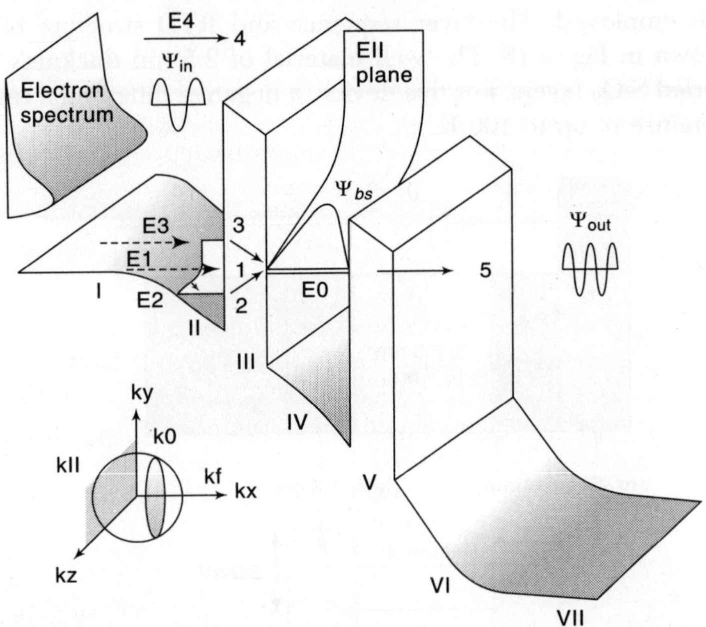

Fig. 3.15 *Structure and local electron energy distribution of a resonant tunneling diode with applied bias voltage.*

Even at room temperature, a region with negative differential resistance (NDR) identified with a peak-to-valley ratio (better than 20). Such characteristics suggest both distable and astable applications. The most common applications of RTDs are microwave oscillators operating at extremely high frequencies and very fast digital electronic circuits.

A negative differential resistance is also found in InAs/AlSb/GaSb resonant tunnel structures with AlSb barriers and GaSb quantum wells. Here, the special band structure favors electron tunneling between the energy levels within the valence band of the AlSb barrier and the energy levels within the conduction band of the GaSb quantum well layer. The resulting device is referred as *resonant interband tunnel diode* (RITD).

As in the area of optoelectronics, III-V materials are replaced with silicon for the production of resonant tunneling devices for integrating them with VLSI silicon ICs. As an example, the schematic structure and band diagram of an RTD with silicon quantum wells and CaF_2 barriers are depicted in Fig. 3.17. Alternatively, Si/Ge or silicon-on-insulator

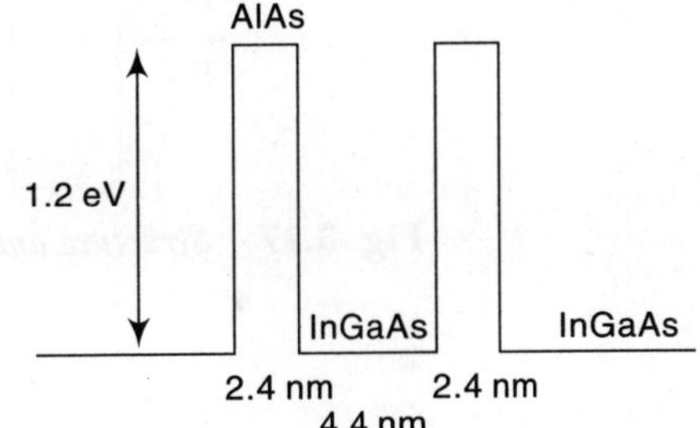

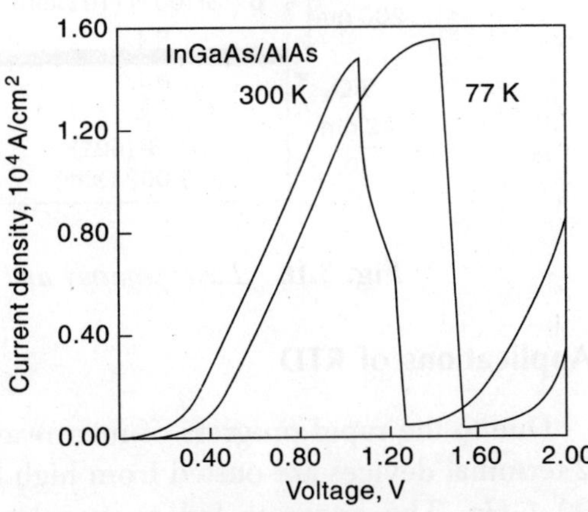

Fig. 3.16 *Conduction and diagram and current-voltage characteristics at 77 and 300 K of an ALAs/InAs RTD.*

(SOI) technology is employed. The layer sequence and RTD structure of an Si technology-based RTD are shown in Fig. 3.18. The well material of 2.5 nm thickness is confined by two ultrathin (2 nm) buried SiO_2 layers. For this device, a negative differential resistance is observed at operating temperature of up to 100 K.

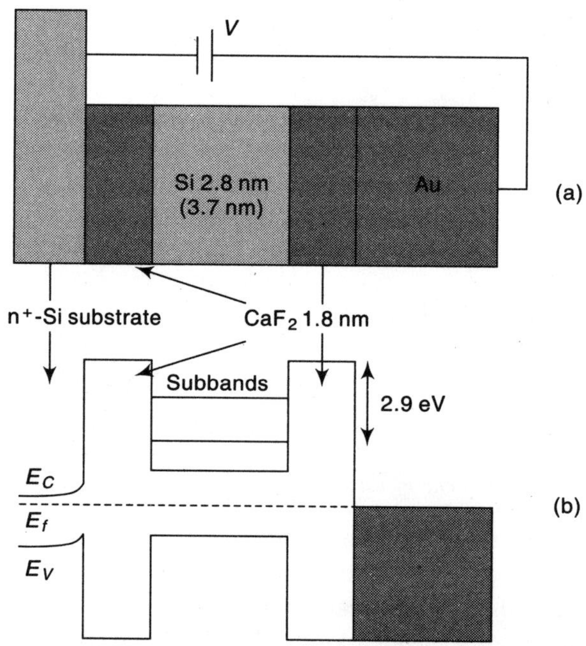

Fig. 3.17 *Structure and band diagram of a Si/CaF2 double barrier RTD*

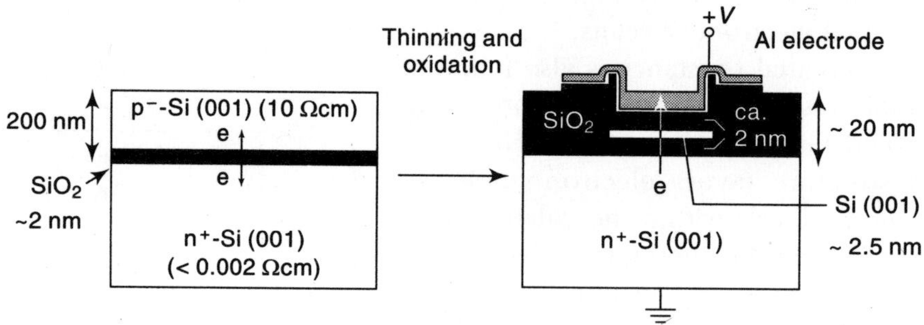

Fig. 3.18 *Layer sequence and device structure of a Si/SO_2 double barrier RTD*

Applications of RTD

Due to the rapid progress of microwave transistors regarding their high frequency behavior, 2-terminal devices are ousted from high frequency oscillator applications for frequencies below 30 GHz. The progress led to transistor cut-off frequencies of 350 GHz for InP/InGaAs heterobipolar transistors (HBT), 42 GHz for SiGe FETs and 85 GHz for GaN/AlFaNB high electron mobility transistors (HEMTs). For higher frequencies (30 GHz-l THz), IMPATT, Gunn diodes, and resonant tunneling diodes are still being considered for applications as microwave oscillators.

For transit time diodes (IMPATTs), a maximum oscillator frequency of 400 GHz is obtained with an output power of 0.2 mW, and at 44 GHz an output power of 1 W has been measured. Similar values are obtained by means of transferred electron devices (TEDs), also known as Gunn diode. In the latter case, an output power of 34 mW at 193 GHz and of about 96 mW at 94 GHz is achieved. Despite the somewhat lower performances at frequencies below 100 GHz, Gunn diodes are an important alternative to IMPATT diodes due to their low noise operation.

In comparison to the two device treated so far, lower, output powers are achieved with resonant tunneling diodes. At 30 GHz and 200 GHz, for example, output powers of about 200 mW and 50 μW, respectively, have been reported. However, with InAs/AlSb RTDs, a record frequency of 712 GHz has been obtained, achieved by an InAs/AlSb RTD with an output power 0.3 μW. The limitation of the maximum output power RTD based oscillators is mainly caused by the relatively high series inductance. A further advantage of RTDs compared to IMPATT and Gunn diodes is the fact that they can be easily integrated with other electronic devices, like modulation doped field effect transistors (MODFETs) and heterobipolar transistors. Thus, they are attractive for microwave integrated circuities (MMICs) even at moderate frequencies in the GHz range. Another reason of making resonant tunneling diodes appealing for digital circuit applications is the possibility of implementing very compact logical circuits since the number of active devices can be reduced as compared to conventional digital electronic circuits. In Table 3.1, the number of active devices required for the implementation of several digital functions using RTDs in comparison with TTL, CMOS, and ECL technology is listed.

Resonant tunneling diodes intrinsically have very short switching times of about 1.5 ps. As stated earlier, they can be easily integrated with ultrafast transistors like MODFETs and HBTs. Moreover, they can operate room temperature, which is a clear advantage as compared to a superconducting integrated circuit—another possible competitor for the implementation of ultrafast digital electronic circuits. A further advantage is the possibility of quite easily implementing multi-value-logic systems using multi-peak resonance RTDs. In Fig. 3.19, the circuit of a digital counter implemented with just three HBTs and one RTD and the corresponding output characteristics of the transistor $Q1$ with the multi-peak RTD as a load are depicted.

Table 3.1 *Number of devices required for the implementation of digital functions using RTD, TTL, CMOS and ECL technologies.*

Logical function	*TTL*	*CMOS*	*ECL*	*RTD*
bistable XOR	33	16	11	4
9-state memory	24	24	24	5
NOR2 + flip-flop	14	12	33	4
NAND2 + flip-floop	14	12	33	4

Other examples of digital circuits are a 50 GHz frequency divider fabricated with one resonant tunneling diode and one HEMT as well as NAND and NOR gates consisting of a

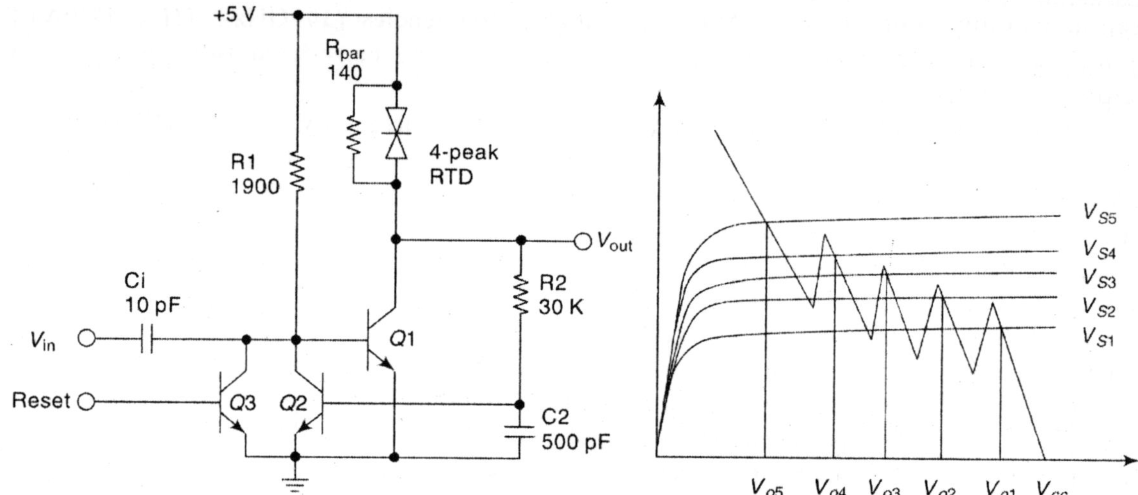

Fig. 3.19 *Counting circuit implemented with HBTs and a single RTD*

single resonant tunneling bipolar transistor with an integrated RTD structure (RTBT). Besides digital circuit applications, analog applications have also been proposed, e.g., analog/digital converters using an RTD as a multi value comparator. In optoelectronic applications, the intrinsic bistability of RTDs is used in particular The combination of an RTD growth on top of a multi quantum-well electro-optic modulator exhibited bistable operation at room temperature with switching in the mW range. The monolithic integration of an InAlAs/InGaAs RTD with an InGaAs/InGaAsP traveling wave photodiode on the same InP substrate enabled the production of an opto-electronic flip-flop operating at a clock rate of 80 Gb/s.

Further nanoelectronic digital circuits can be manufactured using chemical self-organized growth of quantum dot arrays. It would be much easier to use two-terminal devices rather than transistor-like three-terminal devices. The RTD is an attractive candidate for such nanodevice circuits because it combines the ability to implement complex logic functions with a relatively simple structure.

HETEROJUNCTION BIPOLAR TRANSISTOR

The **heterojunction bipolar Transistors (HBT)** can handle signals of very high frequencies (up to several hundred GHz) eg: in modern ultrafast circuits, mostly radio-frequency (RF) systems.

The use of different semiconductor materials for the emitter and base regions is made to, create a heterojunction. The effect is to limit the injection of holes into the base region, since the potential barrier in the valence band is so larger. This allows high doping to be used in the base, creating higher electron mobility while maintaining gain. The efficiency of the device is measured by the Kroemer factor.

Material systems such as AlGaAs/GaAs, GaInP/GaAs, InP/InGaAs. AllnAs/.InGaAs, and SiGe/Si are used for a wider band-gap emitter layer to make heterojunction bipolar transistor or HBT.

The GaAs- and SiGe-based one are used for power amplifiers in digital mobile telephones. The nominal devices show current gain cut off frequencies in excess of 100 GHz and maximum oscillation frequencies of about 200 GHz in compound semiconductor as well as SiGe-based ones. The InP-based varieties, when all the excess semiconductor is removed and contact pads are placed on dielectric layers with small dielectric constants, exhibit maximum oscillation frequencies in excess of 500 GHz.

HBT content is fueled by improved crystal growth methods, such as molecular beam epitaxy (MBE), metal-organic chemical vapor deposition (MOCVD), and ultra-thin-vacuum chemical vapor deposition (UHV-CVD). These technologies provided atomic level precision in layer thickness and doping concentrations with unprecedented control ensuring improvements in material quality thus new device structures, and non-equilibrium transport mechanisms such as ballistic transport are explored. The current gain h_{FE} of an HBT is sensitive to the material quality, and as the quality improves, the HBT current gain increases. GaAs/AlGaAs layers grown by MBVE and MOCVD are commercially obtained for HBT fabrication. In fact, the power amplifiers (PAs) of cellular digital telephone have been very successful.

A pseudomorphic heterojunction bipolar transistor is built from indium phosphide and indium gallium arsenide and designed with compositionally graded collector, base and emitter, has achieved a cut off speed of 710 gigahertz.

Besides being record beakers in terms of speed, HBTs made of InP/GaAs are ideal for monolithic optoelectronic integrated circuits. They bandgap of InGaAs fits for detection of 1.55μm-wavelength signal used in optical communication systems. Among other HBT applications are mixed signal circuits such as analog-to-digital and digital-to-analog converters.

In terms of process leading the speed of performance, HBT based on InP substrate where ever step is taken to reduce the parasitics and the transit times involved, cut-off frequencies approaching 1 THz are possible in Fig. 3.20.

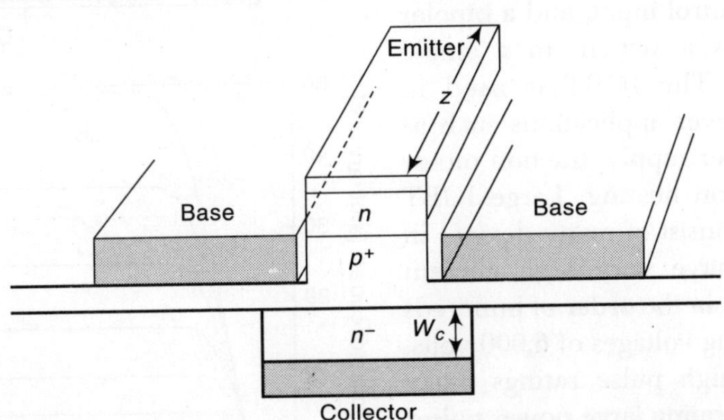

Fig. 3.20 *HBT with all the excess semiconductor removed and collector and emitter junction aligned at the sub-micrometer scale*

Substrate transfer processes are used to define narrow emitter-base and collector-base junctions, aligned are used to define to one another on opposing sides of the base epitaxial layer. Because the collector-base junction lies only directly below the emitter-base junction, excess collector capacitance is eliminated, thereby increasing device bandwidth. By scaling both the emitter and collector stripe widths to approximately 0.1 μm, base-spreading resistance is greatly decreased. In addition, transferring the device structure onto a metal substrate with reduced resistance aids in enhancing the radio frequency performance. Recent devices have consequently led to maximum oscillation frequencies of approximately 750 GHz and it is estimated that current gain cut-off frequencies above 350 GHz and power current gain cut-off frequencies above 1000 GHz be obtained.

In the transferred substrate method of fabrication, the wafer is covered with a 5 μm layer of BCB (wiring dielectric), and Au/Ni/Cu substrate is then platen. The wafer is inverted, and the InP growth substrate removed by selective etching in HCl. The Schottky collectors are then patterned, completing the process.

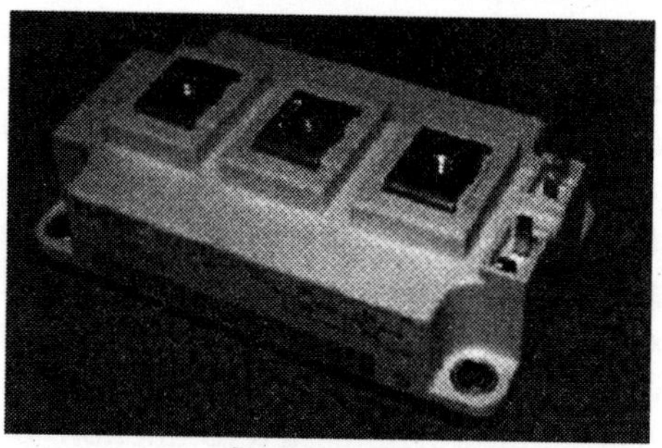

A power IGBT

INSULATED GATE BIPOLAR TRANSISTOR

The **Insulated Gate Bipolar Transistor** is a three-terminal Power semiconductor device that combines the simple gate drive characteristics of the MOSFETs with the high current and low saturation voltage capability of bipolar transistors by combining an isolated gate FET for the control input, and a bipolar power transistor as a switch, in a single devices Fig. 3.21. The IGBT is used in medium to high power applications such as switched-mode power supply, traction motor control and induction heating. Large IGBT modules typically consist of many devices in parallel and can have very high current handling capabilities in the order of hundreds of amps with blocking voltages of 6,000 volts.

The extremely high pulse ratings make them useful for generating large power pulses in areas like particle and plasma physics, which they are starting to supersede older devices like thyratrons and triggered spark gaps.

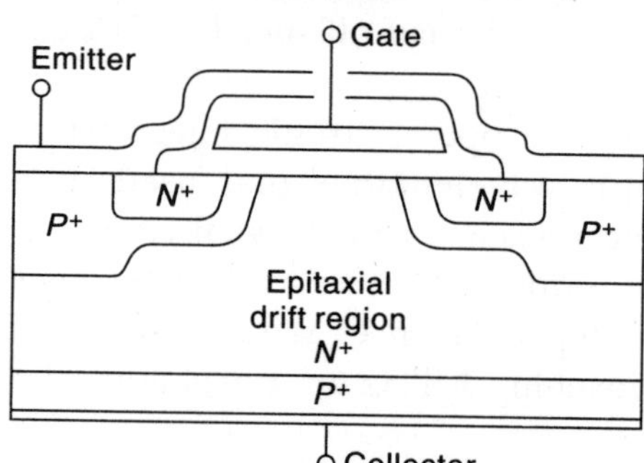

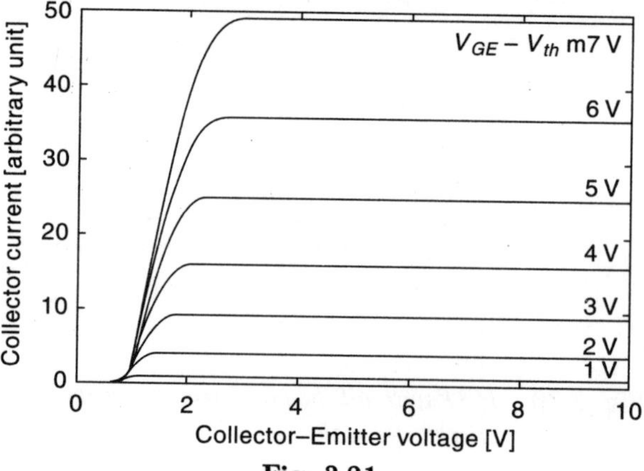

Fig. 3.21

Their high pulse rating, make them attractive to the high-voltage for generating large amounts of high-frequency power to drive experiments like Tesla coils.

Availability of affordable, reliable IGBTs is a key enabler for electric vehicles and hybrid cars and as an inverter to control two AC motor/generators connected to the DC battery pack. In addition, IGBTs are also used as audio amplifiers.

HEMT

HEMT stands for **High Electron Mobility Transistor,** and is also called **heterostructure FET (HFET).** A HEMT is a field effect transistor with a junction between two materials with different band gaps (i.e. a heterojunction) as the channel instead of an n-doped region. A commonly used combination in GaAs with AlGaAs. The effect of this junction is to create a very thin layer of conducting electrons with rather high concentration, giving the channel very low resistivity (or to output another way, "high electron mobility"). This layer is sometimes called a two-dimensional electron gas. As with all the other types of FETs, a voltage applied to the gate alters to conductivity of this layer.

Accordingly to semiconductor theory (such as n-CMOS), the semiconductor layer needs to be doped with n-type impurities (at Source and Drain) to generate electrons in the layer. However, this causes electron to slow down because they end up colliding with the impurities, residing in the same region, which were used to generate them in the first place. HEMT, however, is a smart device to resolve this seemingly inherent unsolvable contradiction.

HEMT accomplishes this by use of high mobility electrons, generated using the hetero-junction of a highly-doped n-type AlGaAs thin layer and a non-doped GaAs layer (no impurities). The electrons generated in n-type AlGaAs thin layer drop completely into the next GaAs layer to form a depletion AlGaAs layer, because the hetero-junction created by different band-gap material forms a steep canyon in the GaAs side where the electrons can move quickly without colliding with any impurities (because GaAs is undoped).

Ordinarily, the two different materials used for a heterojunction musthave the same lattice constant (spacing between the atoms). An analogy imaging pushing together two plastic combs with a slightly different spacing – at regular intervals, you'll see two teeth clump together. In semiconductors, these discontinuities are a kind of "trap", and greatly reduce device performance.

A HEMT where this rule is violated is called a **PHEMT** or *pseudomorphic* HEMT. This feat is achieved by using an extremely thin layer of one of the materials – so thin that it simply stretches to fit the other material. This techniques allows the construction of transistors with bigger bandgap differences than otherwise possible. This gives them better performance.

Another way to use materials of different lattice constants is to place a buffer layer between them. This is done in the **mHEMT** or *metamorphic* HEMT, an advancement of the PHEMT developed in recent years. In the bugger layer made of AlInAs, the indium concentration is graded, so that it can match the lattice constant of both the GaAs substrate and the GaInAs channel. This bridge the advantage that practically any Indium concentration in the channel can be realized, so the devices can be optimized for different applications (low indium concentration provides low noise, high indium, concentration gives high gain).

Applications are similar to MESFETs – microwave and millimeter wave communications, radar, and radio astronomy. (Heterojunction bipolar transistors have been demonstrated at frequencies over 600 GHz). Numerous companies worldwide develop and manufacture HEMT-based devices. These can be discrete transistors but more usually in the form of an integrated circuit called a MMIC standing for 'monolithic microwave integrate circuit'. HEMT devices are found in many types of equipment ranging from cellphones and DBS receivers to electronic warfare systems such as radar and for radio astronomy.

JFET

The junction gate field-effect transistor (JFET) is the simplest type of field effect transistor Fig. 3.22. Like other transistors, it can be used as an electronically controlled switch. They are also used as voltage-controlled resistance. An electric current flows from one connection, called the source, to a second connection, called the drain. A third connection, the gate, determines how this current flows. By applying an increasing negative (for an n-channel JFET) bias voltage to the gate, the current flow from source to drain can be impeded by pinching off the channel, in effect switching off the transistor.

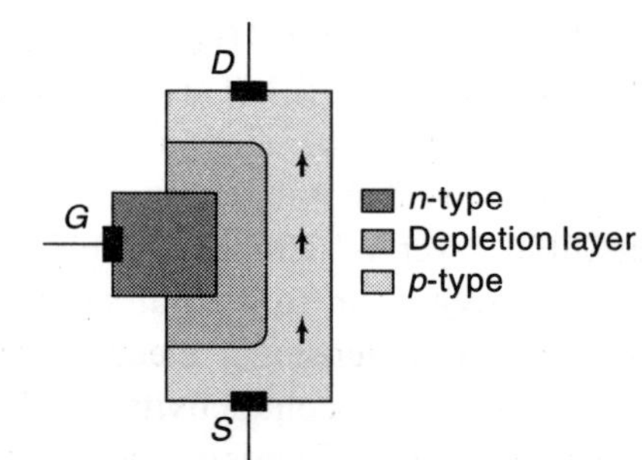

Fig. 3.22 *Electric current flow from source to drain in a p-channel JFEt is restricted when a voltage is applied to the gate.*

The JFET consists of a long channel of semiconductor material. The material is doped so that it contains an abundance of positive charge carrier (*p*-type), or of negative charge carriers (*n*-type). There is a contact at each end; these are the source and drain. The third control terminal, the gate, surrounds the channels and is doped opposite to the doping-type of the channel. Then, a pn junction is formed at the interface of the two types of the material and one has toe make sure that the terminal made with the semiconductor are usually made Ohmic.

The operation of a JFET can easily be understood by considering a garden hose. The flow of water through a garden hose can be controlled by squeezing it and reducing its cross section; the flow of electric charge through a JFET is controlled by construction the cross section of the current-carrying channel.

Sometimes the JFET gate is drawn in the middle of the channel instead of at the drain/source electrode as in these examples. This symmetric variation is hinting at that the channel is indeed symmetric in sense that *drain* and *source* are interchangeable physical terminals Fig. 3.23 (a) and (b). So this symbol variation should be used for only for JFEts where drain and source indeed are interchangeable, which is not true for all JFETs.

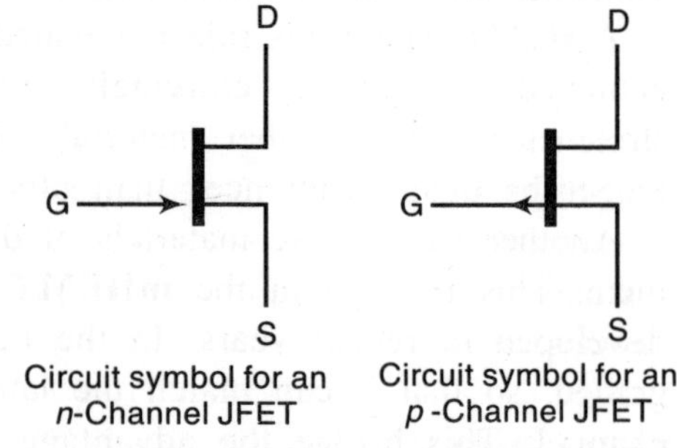

Fig. 3.23 *(a) and (b) Circuit symbol for a p-channel JFET*

Traditionally, the US style of the symbol was drawn with the whole component in side a circle, although this has been simplified in favor of the European style to draw it without a circle. In every case the arrow head indicates the polarity of the P-N-junction of the gate in relationship of the channel. (As with a diode, the arrow points from P to N, indicating the direction of current flow when forward-biased.) In order to pinch off the channel, one must produce a certain voltage in reverse direction (V_{GS}) of that junction. The precise value of this *pinch off voltage varies* with individual JFETs, even with JFETs of the same type, typical values ranging between 0.5 to 10 V.

The appropriate voltage bias can be remembered easily, since the n-channel device requires a negative gate-source voltage (V_{GS}) to switch off the JFET, while the p-channel device requires a positive gate-source voltage (V_{GS}) to switch off the JFET.

MESFET

MESFET stands for **Metal-semiconductor Field Effect Transistor.** It is quite similar to a JFET. The difference is that instead of a using p-n junction for a gate, a Schottky (metal-semiconductor) junction is used. MESFETs are usually constructed in compound semiconductor technologies lacking high quality surface passivation such as GaAs, InP, or SiC, and are faster but more expensive than silicon-based JFET or MOSFETs. Production MESFETs are operated up to approximately 30 GHz, and are used for microwave frequency communications and radar. From a digital circuit design perspective, it is increasingly difficult to use MESFETs as the basis for digital integrated circuits as the scale of integration goes up, compared to CMOS silicon based fabrication.

The MESFET differs from the common insulated gate FET in that there is no insulator under the gate over the active switching region. This implies that the MESFET gate should, in transistor mode, be biased such that one does not have a forward conducting metal semiconductor diode instead of a reversed biased depletion zone, controlling the underlying channel. This restriction inhibits certain circuit possibilities. MESFET analog and digital devices work well if kept within the confines of design limits. The most critical aspect of the design is the gate metal extent over the switching region. Generally, the narrower the gate modulated carrier channel the better the frequency handling abilities. Spacing of the source and drain with respect to the gate, and the lateral extent of the gate are important. MESFET, current handling ability improves as the gate is elongated laterally, keeping the active region constant, however is limited by phase shift along the gate due to the transmission line effect. As a result most production MESFETs use a built up top layer of low resistance metal on the gate.

Numerous MESFET fabrication possibilities are explored for a wide variety of semiconductor systems. Some of the main application areas are: military communications, military radar devices, commercial optoelectronics, satellite communications

THIN FILM TRANSISTOR

A **thin film transistor (TFT)** is a special kind of field effect transistor made by depositing thin films for the metallic contacts, semiconductor active layer, and dielectric layer. The channel region of a TFT is a thin film that is deposited onto a substrate (often glass, since the primary application of TFTs is in liquid crystal displays).

In *Displays technology* (as in laptops), a filtered background light is allowed to pass through to the screen or is blocked. This is accomplished by thin film transistors (TFTs) in conjunction with the approach of either applying or not applying a field on liquid crystal displays. These structures are prepared on transparent substrates of large sizes such as glass. Again, because they have to be on quartz, these transistors are of hydrogenated amorphous Si (a-Si:H) and bandgap is assumed to be less than that of crystalline silicon (1.112 eV). However, as indium tin oxide (ITO) are used as electrodes, so, terminal and interconnections are often transparent. As a-Si is replaced with poly-Si (higher electron mobilities) it allows smaller and faster devices operating at lower voltage for a larger display. With organic semiconducting materials, these plastics are exploited as they are easy to deposit on transparent substrates. The generic term for FET where the channel is separated from the gate by an insulating material, other than SiO_2, is insulated gate FET (or IGFET for short). IGFETs based on conjugated polymers, oligomers, or other molecules have been fabricated. Typical device configuration used is depicted in Fig. 3.24. The fabrication process is as follows: pentacene films are deposited by means of vapor deposition. The barium zirconium titanate (BZT) gate dielectric films are deposited on oxidized silicon wafers or thin polycarbonate sheet, provided with Pt gate lines, with ratio frequency (RF) magnetron sputtering of a sintered powder target of BZT, in an Ar/O_2 gas mixture. Gold source and drain electrode are vapor deposited on top of the pentacene layer through a shadow mask. To create devices with channel length (L) on the order of 10 μm, special silicon membrane masks are used as conventional silicon device processing methods.

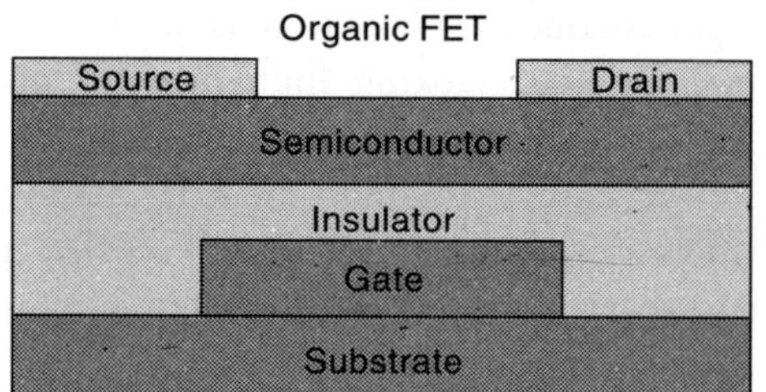

Fig. 3.24 *Schematic representation of layer structure of plastic FET.*

Owing to the relatively low mobility of the organic semiconductor layers, organic IGFETs will not compete for CPUs which require high switching speeds. The performance of organic IGFETs suggest that they have good applications in large areas; mechanical flexibility and low-temperature processing. These applications include TFT switching devices for active matrix liquid crystal displays (AMLCD), where hydrogenated amorphous silicon (a-Si:H) TFTs are used, and active matrix organic light emitting diode displays (AMOLED), and low- end date storage (such as inexpensive smart cards and identification tags that are dispensable).

So far, two classes of organic semiconductors, thiophene oligomers and pentacene have exhibited the best IGFET performance with good stability, α, ω-Alkyl-substitued hexathiophenme oliogomers and their derivatives are conjugated oliogomners that exhibit a pronounced capability for self-assembly in close-packed configurations. The highest value of field-effect mobility (μ) reported for α, ω-dihexyl-hexathienylene (DH6T) thin films is 0.13 $cm^2\ V^{-1}\ s^{-1}$. Pentacene TFTs have produced the largest field-effect mobility values reported for organic IGFET, approaching or in some cases greater than 0.6 $cm^2\ V^{-1}s^{-1}$. This mobility depends on the deposition conditions of the pentacene layer and the device configuration. The mobility measured in solution-processed pentacene IGFETs is some 50 times smaller than this measured in vapor-deposited, and purified pentacene which is as high as 0.62 cm^2 V^{-1} s^{-1}. These devices require large voltage to operate.

But, Pentacene-based TFT on very transparent plastic substrates (polycarbonate) and on quartz or SiO_2/Si substrates fabricated with BZT gate insulator (dielectric constant 17.3) operate at gate voltages of 4 V.

The best known application of thin-film transistor is in TFTLCDs, a variant of LCD technology. Transistors are embedded within the panel itself, reducing crosstalk between pixels and improving image stability. Presently, all color LCD screens use this technology. TFT panels are heavily used in digital radiography applications including Mammography and General Radiography.

The new AMOLED (Active-Matrix OLED) screens also contains a TFT layer.

FETs, MOSFETs, MOSFETs, AND SETs

I. FETs: The **field-effect transistor** is a type of transistor that relies on an electric field to control the shape and hence the conductivity of a 'channel' in a semiconductor material. The *concept* of the field effect transistor predates the bipolar junction transistor (BJT), though it was not physically implemented until *after* BJTs, due to the limitations of semiconductor materials and relative ease of manufacturing BJTs compared to FETs at the time.

All FETs (see types of FEt below) except J-FETs have four terminals, which are known as the *gate, drain, source* and *body/base/bulk/substrate*. Compare these to the terms used for BJTs: *base, collector and emitter*. BJTs and J-FETs have no body terminal.

The names of the terminals refer to their function. The gate terminal may be thought of as controlling the opening and closing of a physical gate. This gate permits electrons to flow through or blocks their passage by creating or eliminating a channel between the source and drain. Electrons flow form the source terminal towards the drain terminal if influenced by an applied voltage. The body simply refers to the bulk of the semiconductor in which the gate, source and drain lie. Usually the body terminal is connected to the highest or lowest voltage within the circuit, depending on type. The body terminal and the source terminal are sometimes connected together since the source is also sometimes connected to the highest or lowest voltage within the circuit, however there are several uses of FETs which do not have such a configuration, such as transmission gates and cascade circuits.

The FET can be constructed from a number of semiconductors, silicon being by far the most common. Most FETs are made with conventional bulk semiconductor processing techniques, using the single crystal semiconductor wafer as the active region, or channel.

The Principal of FET Operation

Figure 3.25 is a sketch of a thin film FET with electric connections. A voltage between source (S) and drain (D), called the *drain or source-drain voltage* V_D, attempts to drive a current through the semiconducting transistor channel. However, a drain current I_D will only flow if there are mobile charge carriers in the channel.

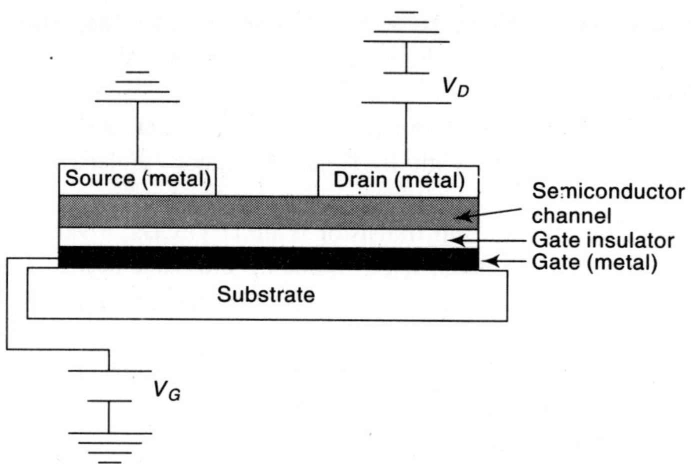

Fig. 3.25 *A Fet with bottom-gate architecture. Transistors may also be built the other way round, with top-gate architecture.*

A gate voltage V_G will extract carriers form the source into the semiconducting channel. However, since the gate *dielectric* is an insulator, these carriers cannot reach the gate metal. Instead, they will form an *accumulation* layer at the channel-insulator interface. The channel semiconductor thus gets doped by applying V_G. Unlike chemical doping, this quickly reversible, simply by switching off V_G. V_G switches channel conductivity, in some (desirable) cases by several orders of magnitude. Accordingly, I_D at a given V_F will change with V_G; *the FET is an electronic switch.* Note that an OFET is switched on by applying V_D and V_G of equal polarity *opposite* to the sign of the mobile carriers. Thus, OFETs can easily be used to determine the type of carriers a particular material sustains. If positive V_G, V_D switch the transistor on, then the carriers are negative (electrons); if negative V_G, V_D are required, then the carriers are holes. To minimize injection barriers from source to semiconductor, high work function metals such as gold or PEDOT/PSS are used for hole channel semiconductor, and low work function metals such as Ca for electron channel semiconductors.

To characterize the gate and source – drain voltage-dependent drain current in a FET beyond the qualitative on/off description, usually two types of measurements are carried out: the *output* and *transfer* characteristics. The output characteristic of a FET is the family of curves I_D against V_D at different V_G, but V_G fixed during every single V_D scan. The transfer characteristic is the family of I_D against C_G curves at different V_D, but V_D fixed during every single V_G scan.

The following formulae apply

$$I_D = \frac{Z}{L} C_i \mu (V_G - V_T)\, V_D \text{ for small } V_D \; (|V_T| < |V_D| << V_G) \tag{i}$$

$$I_D = I_{sat} = C_i \mu (V_G - V_T)^2 \text{ for large } V_D \; (V_D \geq V_G) \tag{ii}$$

where Z/L is the channel width/length, C_i is the capacitance per unit area of the insulator, V_T is a threshold voltage, and μ is the charge carrier mobility. For a dielectric, $C_i = \varepsilon_r \varepsilon_0 / d$, where ε_r is the dielectric constant and d is the film thickness. ε_r describes how much a dielectric shields the applied field. The large V_D regime is known as the *saturation regime*, I_D does not rise with higher

V_D any more but levels off at a (V_G-dependent) saturation current I_{sat}. Saturation will be reached at $V_D \approx V_G$, because at $V_D = V_G$ the field between gate and drain is zero, and the accumulation layer pinches off. The theory of the threshold V_T is rather intricate, and has a specific differences between organic and inorganic FETs.

From experimental transistor characteristics, the charge carrier mobility can be determined with the help of Equation (i) and (ii). A robust method is to plot against V_G (for $|V_D| > |V_G|$), which should produce a straight line with a slope directly related to μ.

For a low-noise measurement, we wish to have a relatively high drain current, which at given mobility, dielectric and operating voltages is determined by the geometry factor, written as Z/ L. The maximum geometry factor leads to higher currents that can eb measured more comfortably. Often, Source and drain and therefore evaporated through an interdigiting comb shadow mask, with many channels in parallel. Practical transistors, however, have to be small and should work with only one channel. Also, L controls the time τ that a carrier requires to cross the channel:

$$\frac{1}{\tau} = f = \frac{\mu V_D}{L^2} \tag{iii}$$

L enters the equation L^2; one factor of L come s from the linear dependence of the transit time of carriers through the channel on L, the other factor of L arises from the dependence of the electric field on $1/L$ τ is one of the limiting factors of switching speed. A genuine engineering challenge is therefore to achieve, at low cost, the smallest possible L without inadvertently creating electrical short circuits.

NANOTUBE NANOELECTRONIC DEVICES

Nanotube field-effect transistor

Transistors are the basic building blocks of integrated circuits. To use nanotubes in future circuits it is essential to be able to make transistors from them. These nanotube transistors have been successfully fabricated and tested using individual multi-wall or single-wall nanotubes as the channel of a field-effect transistor (FET). (See fig. 3.26).

The electrical characteristics of our nanotube-FETs are measure and it is found the amount of current (I_{SD}) flowing through the nanotube channel by a factor of 100,00 can be changed by changing the voltage applied top a gate (V_G), (Fig. 3.27a to d). G is the low bias conductance of the tube.

On cooling the FET down from room temperature to 4 degree Kelvin the device behavior changes dramatically. While the device acts like a field-effect transistor at room temperature, at 4 K it behaves like a single-electron transistor (SET).

The three plots below show this change in device behavior for different temperatures.

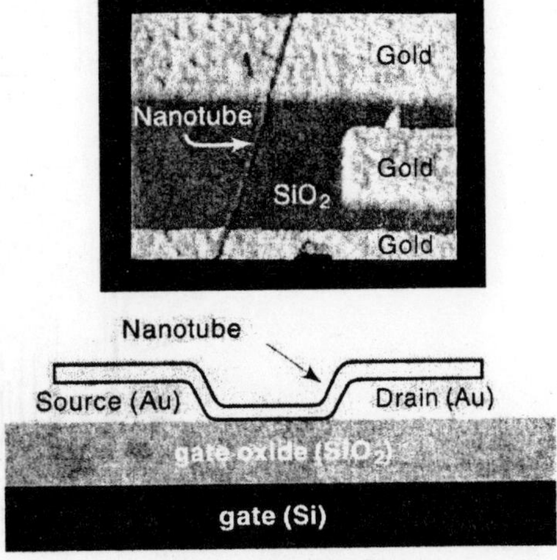

Fig. 3.26 *(FETs)*

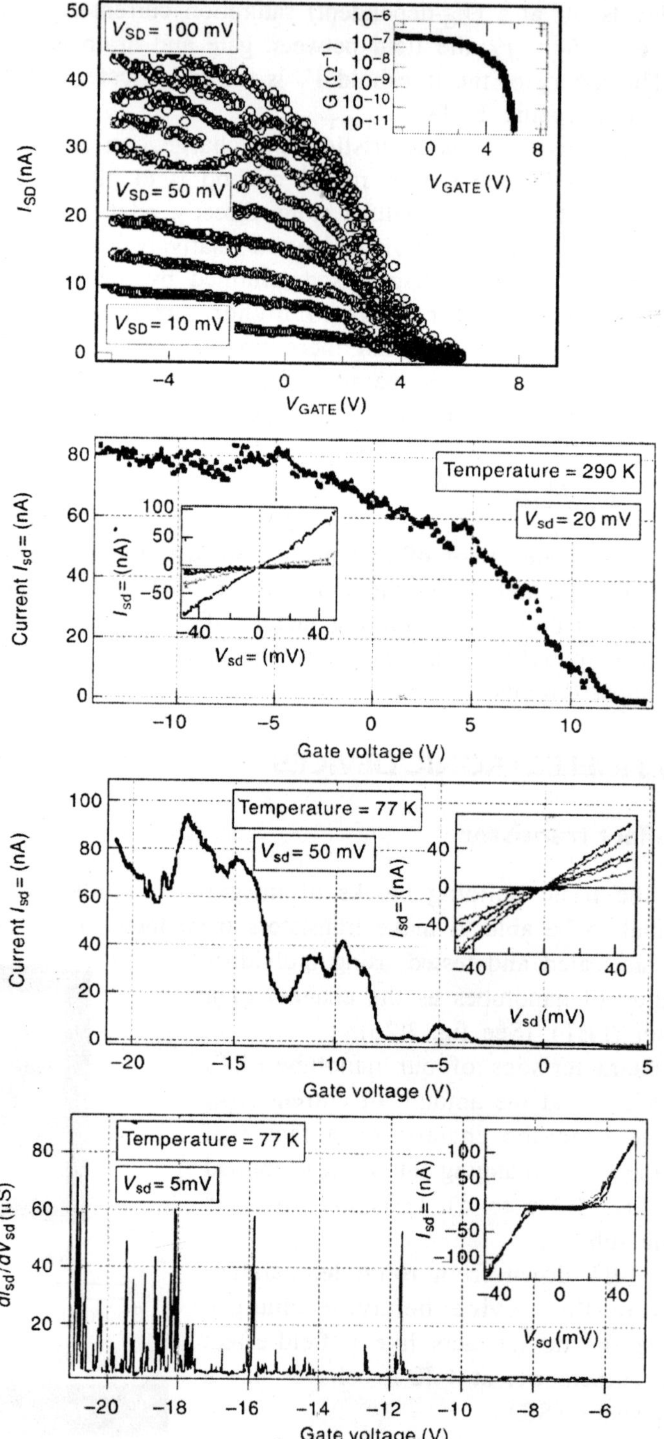

Fig. 3.27 *Electrical characteristics of FETs*

The semiconducting SWNT channel acts like an FET; positive gate voltage shuts off conduction, while the conduction is enhanced by negative gate voltage first electrical transport measurements on an individual semiconducting SWNT was by using the laser ablation tube-on-top method. That the semiconducting SWNT acts as an intrinsic *ballistic* channel between the source and drain electrodes. Because of difference in work function ϕ between the metal electrodes (platinum, with a work function ϕ = 5.7 eV) and the nanotube (with ϕ = 4.5 eV), the Fermi level in the nanostructure in pinned in the valence band at the contacts. Far from the electrodes, the nanotube is intrinsic. The bands bend towards the intrinsic position over the electrostatic screening length, which may be very long in one dimension. This creates a barrier for hole transport through the nanotube, which may be lowered by applying a negative voltage to the gate electrode. The analogous semiconductor device is termed a barrier impact transit time (BARITT) diode. Since the barrier height depends strongly on the ratio of the device length to the electronic screening length, the BARITT model predicts a large sensitivity of the device characteristics to device length, which was not observed in subsequent research.

Also fabrication of semiconducting SWNT devices from laser ablation derived material using the tube-op-top technique. Here in the transistor behavior conduction through the semiconducting SWNT in *diffusive*. The semiconducting SWNT acts as the channel of a MOSFET, apparently intrinsically doped slightly *p*-type, and the device operates in a *p*-channel depletion mode (positive gate voltage shuts off the conduction, and negative gate voltage enhances the conduction). The conductance saturates at large negative V_g due to the series resistance of the contacts. The linear region in the conductance as a function of gate voltage was used to extract a hole mobility with a rather low value of 20 cm^2/V · s.

The electronic transport through laser ablation derived semiconductors SWNTs is dominated by a number of large conductance barriers which explains the low measured mobility of the devices.

In CVD-grown semiconducting SWNTs, the intrinsic resistance is much lower, the mean-free path is several hundred nanometers, and the hole mobility may be as high as 20000 cm^2/ V · s.

Transconductance

The conductance of an FET is given by G = $C_g(V_E - V_g)\mu/L$. The transconductance dI_{SD}/dV_g is then maximized by increasing the gate capacitance and the carrier mobility. The transconductance of the first SWNT FETs were very low, as they used gate oxide thicknesses of hundreds of nanometers. Gate coupling is increased by depositing SWNTs on top of the oxidized Al wires which act as gate electrodes. Thus, Fets are constructed with dielectric thicknesses of a few nanometers. Although low-mobility laser ablation SWNTs are used, transconductance of up to 0.3 μS are is obtained, and transistor-resistor logic devices showing grain are constructed.

An electrolytic solution is used to gate CVD-grown semiconducting nanotube devices (see Fig. 3.28). Here, the effective dielectric thicknesses is made very small, (a few nanometers). A maximum transconductance of 20 μS is observed. Normalizing this to the device width of ~2 nm, gives a transconductance per unit width of ~10 mS/μm; better than current-generation MOSFETs, but reasonable in light of the high hole mobility of semi-conducting SWNTs.

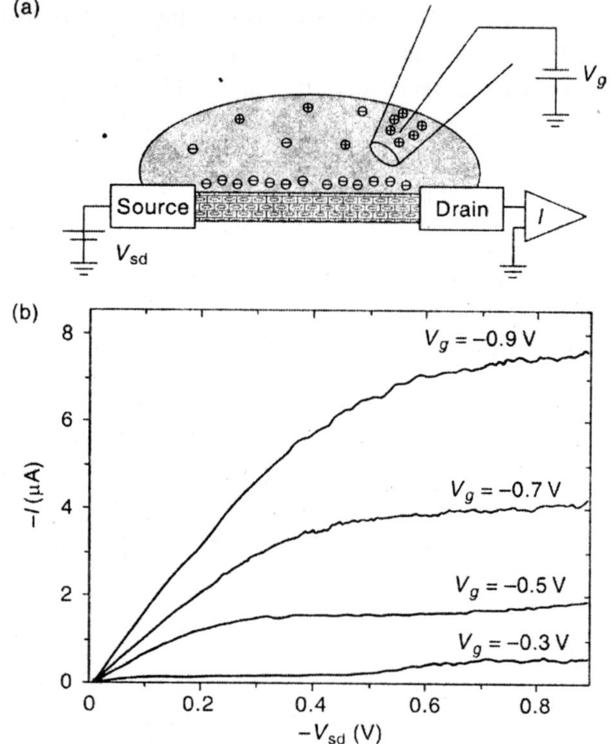

Fig. 3.28(a) *Gating an SWNT FET using an ionic solution. A schematic of the device is shown in (a). A pipette containing a NaCl solution (1–100 mM) is used to place a top of solution over the device. An electrode inside the pipette acts as a gate terminal. The device characteristics are shown in (b). The maximum transconductance of the device is 20 μS.*

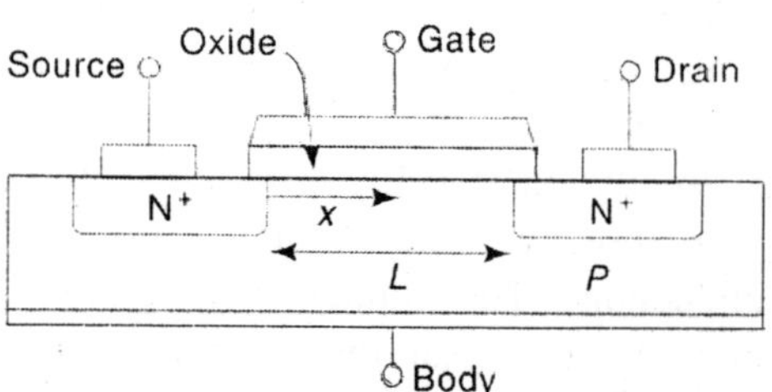

Fig. 3.28(b) *Cross section of an n-type MOSFET*

Types of FET

The channel of a FET is doped to produce either an N-type semiconductor or a P-type semiconductor. The drain and source may be doped of opposite type to the channel, in the case of enhancement mode FETs, or doped of similar type to the channel as in depletion mode FETs. Field-effect transistors are also distinguished by the method of insulation between channel and gate. *Types of FETs are:*

- The *MOSFET* (Metal-Oxide-Semiconductor Field-Effect Transistor) utilizes an insulator (typically SiO_2) between the gate and the body.
- The *JFET* (Junction Field-Effect Transistor) uses a reverse biased p-n junction to separate the gate from the body.

- The *MNESFET* (Metal-Semiconductor Field-Effect Transistor) substitutes the p-n junction of the *JFET* with a Schottky barrier; used in GaAs an other III-V semiconductor materials.
- Using bandgap engineering in a ternary semiconductor like AlGaAs gives a *HEMT* (High Electron Mobility Transistor), also called an *HEMT* (heterostructure FET). The fully depleted wide-band-gap materials forms the isolation between gate and body.
- The *MODFET* (Modulation-Doped Field Effect Transistor) uses a quantum well structure formed by graded doping of the active region.
- The *IGBT* (insulated-gate bipolar transistor) is a device for power control. It has a structure akin to a MOSFET coupled with a bipolar-like main conduction channel. These are commonly used for the 200-3000 V drain-to-source voltage range of operation. Power MOSFETs are still the device of choice for drain-to-source voltages of 1 to 200 V.
- The *FREDFET* is a specialized FET designed to provide a very fast recovery (turn-off) of the body diode.
- The *DNAFAD* is a specialized FET that acts as a biosensor, by using a gate made of single-strand DNA molecules to detect matching DNA strands.

The FET controls of the flow of electrons (or electron holes) from the source to drain by affecting the size and shape of a "conductive channel" created and influenced by voltage (or lack of voltage) applied across the gate and source terminals. (For ease of discussion, this assumes body and source and connected). This conductive channel is the "stream" through which electrons flow from source to drain.

In n-channel *"depletion-mode"* device, a negative gate-to-source voltage causes a *depletion region* to expand in width and encroach on the channel from the sides, narrowing the channel. If the depletion region expands to completely close the channel, the resistivity of the channel from source to drain becomes large, and the FET is effectively turned off like a switch. Like wise a positive gate-to-source voltage increases the channel size and allows electrons to flow easily.

In n-channel *"enhancement-mode"* device, a positive gate-to-source voltage is necessary to create a conductive channel, since one does not exist naturally within the transistor. The positive voltage attracts free-floating electrons within the body towards the gate, forming a conductive channel. But first, enough electrons must be attracted near the gate to counter the dopant ions added to the body of the FET; this forms a region free of mobile carriers called a depletion regions, and the phenomenon is referred to as the *threshold voltage* of the FET. Further gate-to-source voltage increase will attract even more electrons towards the gate which are able to crate a conductive channel from source to drain; this process is called *inversion.*

For either enhancement- or depletion-mode devices, at drain-to-source voltages much less than gate-to-source voltages, changing the gate voltage alters the channel resistance. In this mode the FET operates like a variable resistor and the FET is said to be operating in a *linear mode.* This mode is not employed when amplification is needed.

If drain-to-source voltage is increased, this creates a significant asymmetrical change in the shape of the channel due to a gradient of voltage potential from source to drain. The shape of the inversion region becomes "pinched-off" near the drain end of the channel. If drain-to-source voltage is increased further, the pinch-off point of the channel begins to move away from the drain towards the source. The FET s said to be in *saturation mode.*

Even though the conductive channel formed by gate-to-source voltage no longer connects source to drain during saturation mode, carriers are not blocked from flowing. Considering again an n-channel device, a depletion region exists in the p-type body, surrounding the conductive channel and drain and source regions. The electrons which comprise the channel are free to move out of the channel through the depletion region if attracted to the drain by drain-to-source voltage. The depletion region is free of carriers and has a resistance similar to silicon. Any increase of the drain-to-source voltage will increase the distance from drain to the pinch-off point, increasing resistance due to the depletion region proportionally to the applied drain-to-source voltage. This proportional change causes the drain-to-source current to remain relatively fixed independent of changes to the drain-to-source voltage and quite unlike the linear mode operation. Thus in saturation mode, the FET behaves as a constant-current source rather than as a resistor and can be used most effectively as a voltage amplifier. In this case, the gate-to-source voltage determine the level of constant current through the channel.

The most commonly used FET is the MOSFET. The CMOS (complementary-symmetry metal oxide semiconductor) process technology is the basis for modern digital integrated circuits. This process technology uses an arrangement where the (usually ("enhancement-mode") *p*-channel MOSFET and n-channel MOSFET are connected in series such that when one is on, the other is off.

The fragile insulating layer of the MOSFET between the gate and channel makes it vulnerable to electrostatic damage during handling. The is not usually a problem after the device has been installed.

In FETs electrons can flow in either direction through the channel when operated in the linear mode, and the naming convention of drain terminal and source terminal is somewhat arbitrary, as the devices are typically (but not always) built symmetrically from source to drain. This makes FETs suitable for switching analog signals between paths (multiplexing). With this concept, one can construct a solid-state mixing broad, for example.

Organic Field Effect Transistors

The aim of organic FET (OFET) research is not to outperform current inorganic semiconductor processors in terms of integration density of processing speed. Instead, the idea is to make low-performance integrated circuits based on organics at extremely low cost, cheap enough to be discarded after single use (disposable electronics). One target application is an electronic RF price tag on a food wrapper that a supermarket checkout can read remotely. The target of extremely low cost imposes strict limits on the processing technologies that can be used practically. For example, any vapors deposition step in the device manufacture has to be avoided; vacuum is too expensive.

Good OFETs should display high drain current at low drain and gate voltages, without relaying on the optimized geometry factor; they should also have high on off ratios, which means drain current at $V_G = 0$ should be extremely low.

Requirements on OFET Semiconductors

Equation (i), (ii) and (iii) (in FET above) underscore the importance of high carrier mobilities for good drain current at moderate gate and drain voltages High much effort in synthetic chemistry and physical chemistry is geared towards high mobility materials.

In an OFET, carriers are confined to a very thin interface layer between gate insulator and semiconductor, typically only ~5 nm thick. The resulting interface mobility differs from the bulk mobility, as measured, by the time-of-light (ToF) method. The approach is thus to build an OFET, measure saturated transfer characteristics, and determine μ from the versus V_G plot.

Generally, it is fair to say that *p*-type (hole-transporting) OFET materials are now developed to a much higher state of the art then *n*-type (electron-transporting) materials. C_{60} will not plug that gap, owing to its oxygen sensitivity. A single OFET can operate with a p-type channel, However, the CMOS (complementary metal oxide semiconductor) inorganic semiconductor circuits require *n*-type and *p*-type transistors on the same chip.

With high carrier mobility, the operation of an OFET as a switch, moderates the values of V_G/V_D that lead to a decent I_D, and the switch can be switched on easily. However, this is not the only requirement on an OFET. A switch is useless if it cannot be switched off as well as on. An off-current may result if the OFET semiconductor is permanently (chemically) doped, thus giving channel conductivity even at $V_G = 0$. The on/off ratio can be read directly from the transfer characteristic with V_D applied at the level at which the respective device is meant to operate, and V_G applied or switched off again at whatever level is available. The required of/off ratios depend on the application; e.g., an on/off ratio of 10^6 is required for active matrix addressed liquid crystal displays. For high on/off ratios, OFET semiconductors have to show high mobility μ at very low (in intrinsic) conductivity σ. However, since $\sigma = nq\mu$, where n is the charge carrier density, a high mobility implies a high conductivity unless *n* is extremely low. Here vapour-deposited films have the edge over solution processing, because vapor deposition is a purification step. Ultrapore preparation is a great challenge to synthetic chemistry. As an example, the conventional preparation of 6T couples two 3T by oxidative dimerization with ferric-chloride ($FeCl_3$). However, traces of ferric chloride make it into the final product, and chloride is known to be an oxidative dopant; for example, Cl^- can turn PA into a synthetic metal. This intrinsic doping leads to unwanted conductivity, and OFET grade 6T is now made by coupling lithiated 3T with a ferric acetylacetonate instead. Thiophenes also suffer from oxygen exposure, but in a different way from C_{60}; thiophenes form a weakly bound charge transfer complex with molecular oxygen that acts as dopant, leading to unwanted conductivity.

Integrated Circuits based on OFETs

Logic circuits always require a large number of transistors wired up to each other. The desire for high packing density is one driving force to achieve OFET architectures as small as possible. The other is a direct consequence of Equation (i), (ii) and (iii), which show that a short channel *L* is required for fast OFETs with high drain current at moderate V_G and V_D.

The basic building block of a logic circuit based on FETs is the inverter. It can be built from two FETs, shown in Figure 3.29. $-V_{DD}$ is a supply voltage. When V_{in} is high, V_{out} will be low, and vice versa; the inverter performs the logical operation NOT. Note that the drain of the second FET is connected to its gate. This introduces a key feature of circuits as compared to individual transistors, which poses a considerable challenge to device engineering: to make useful circuits, there needs to be interconnects from the source/drain level to the gate level, going through the semiconducting channel/and through the gate insulator. These interconnects are called vias. The preparation of via interconnects represents another great engineering challenge for the manufacture of practical organic circuits. Patterning on the source/drain and the gate level are also necessary. Currently, two approaches to plastic circuit engineering are competing: photolithography and inject printing techniques.

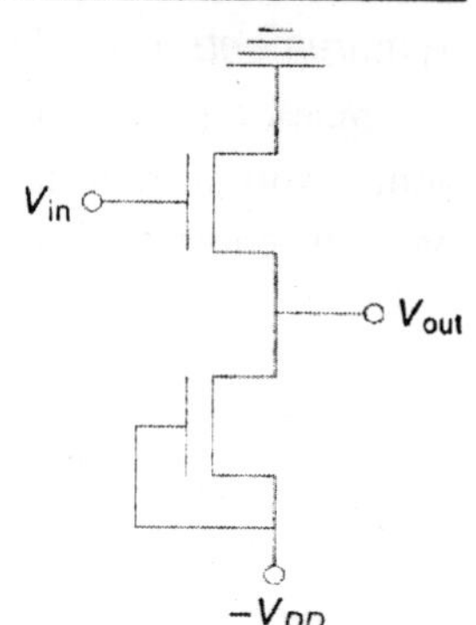

Fig. 3.29 *The principle of an inverter*

Use is made of Camphor sulphonic acid doped PANi as the source/drain and gate metal. To pattern the source/drain level, a photoinitiator molecule is added; on exposure to UV light, it cross-links the PANi and leads to an increase in PANi resistivity of 10 orders of magnitude thus, practically turning PANi into an insulator. Hence the negative of a mask can be patterned into the PANi. The insolubility of PANi in organic solvents is crucial for deposition of the semiconductor poly(thienylene vinylene) (PTV) and the gate insulator PVP. Vias are punched mechanically with pins, providing sufficient mixing between the gate and source/drain PANi to give electrical contact. A 15-bit all-plastic programmable code generator on demonstration still worked when the device is sharply bent.

The photolithography approach to organic electronics has its limitations. By using plastic substrates, dimensional stability is not given over large areas; there is some warp between the gate and source/drain level, which leads to a loss of registration. Al alternative approach is to use inkjet printing to put an inverter in the top-gate. The source/drain and gate are made from PEDOT/PSS, with F8T2 as channel semiconductor and PVP as gate insulator. Since the inkjet printer can be mounted with an optical system, local registration between source/drain and gate is possible and, in principle, circuits of unlimited size can be manufactured. Importantly, an effective method formatting via holes is also devised. When a droplet of plain solvent is printed onto an existing semiconductor, it dissolves the semiconductor. When the solvent evaporates, semiconductor is deposited mainly at the edges of droplet ('coffee stain effect'). Effectively, the solvent droplet etches a hole into the semiconducting film which can be filled with a synthetic metal droplet to make a via interconnect.

II. MOSFET: *The Metal-oxide-semiconductor field-effect transistor (MOSFET)* is the most common field-effect transistor in both digital and analog circuits. The MOSFET is composed of a channel of n-type or p-type semiconductor material and is accordingly called nMOSFET, pMOSFET.

The 'metal' in the name is often incorrect as current process technologies usually use polysilicon gates. Aluminium was used as the gate material until the 1980s when polysilicon became dominant owing to its capability to form self-aligned gates. Lately, at the 65nm mode and smaller, metal gates are again in use. **IGFET** is a related, more general term meaning *insulated-gate field-effect transistor*, and is almost synonymous with MOSFET, though it can refer to FETs with a gate insulator that is not oxide. With the new generation of high-k technology, metal gates in conjunction with the high-k dielectric material replacing the silicon dioxide are making a come back replacing the polysilicon

Usually the semiconductor is of silicon, but now a mixture of silicon and germanium (SiGe) in MOSFET channels is used. Unfortunately, many semiconductors with better electrical properties than silicon, such as gallium arsenide, do not form good gate oxides and thus are not suitable for MOSFETs.

The *gate* terminal in the current generation (65 nanometer node) of MOSFETs is a layer of polysilicon (polycrystalline silicon; placed over the channel, but separated from the channel by a thin insulating layer of silicon dioxide or silicon oxynitride. A high-k + metal gate combination in the 45 nanometer node is also being used. When a voltage is applied between the gate and source terminals, the electric field generated penetrates through the oxides and creates a so-called "inversion channel" in the channel underneath. The inversion channel is of the same type (P-type or N-type) as the source and drain, so it provides a conduit through which current can pass. Varying the voltage between the gate and body modulates the conductivity of this layer and makes it possible to control the current flow between drain and source.

The growth of digital technologies like the microprocessor has provided the motivation to advance MOSFET technology faster than any other type of silicon-based transistor. The principal reason for the success of the MOSFET is the development of digital CMOS logic, which uses p- and n-channel MOSFETs as building blocks. The great advantage of CMOS logic is that they allow no current to flow (ideally), and thus no power to be consumed, except when the inputs to logic gates are being switched. CMNOS accomplishes this by complementing every nMOSFET with a pMOSFET and connecting both gates and both drains together. A high voltage on the gates will cause the nMOSFET to conduct and the pMOSFET not to conduct and a low voltage on the gates cause the reverse. During the switching time to voltage goes from one state to another and both will conduct briefly. The arrangement greatly reduces power consumption and heat generation. Overheating is a major concern in integrated circuits, since ever more transistors are packed into ever smaller chips.

Another advantage of MOSFET for digital switching is that the oxide layer between the gate and the channel prevents DC current from flowing through the gate, further reducing power consumption and giving a very large input impedance. The insulating oxide between the gate and channel effectively isolates a MOSFET in one logic state from earlier and consequent stages, which allows to drive a considerable number of MOSFEt inputs from a single MOSFET output. Bipolar transistor-based logic do not have such a high capacity. This isolation makes it easier for the designers to ignore loading effects between logic stages independently i.e. the operating frequency: as frequencies increases, the input impedance of the MOSFETs decreases.

Analog

The MOSFET's strengths as transistor in most digital circuits do not translate in all analog circuits. The bipolar junction transistor (BJT) is the analog designer's transistor, due largely to its high transconductance and unique properties. MOSFETs are widely relied upon in many types of analog circuits. The characteristics and performance of many analog circuits are designed by changing the sizes (length and width) of the MOSFETs used. Only in specialized bipolar circuits, the size of the bipolar device used, significantly affect the performance. MOSFET's ideal characteristics regarding gate current (zero) and drain-source offset voltage (zero) also make them ideal switch elements and switched capacitor analog circuits. In their linear region, MOSFETs are used as precision resistors, which can have a much higher controlled resistance than BJTs. In high power circuits, MOSFETs do not suffer from thermal runaway as BJTs do. Also, they can be formed into capacitors and specialized circuits, allow-op-amps made form them to appear as inductors, thereby allowing all of the normal analog devices, to be built entirely of MOSFEts. This allows for complete analog circuits to be made on a silicon chip in a much smaller space. Some ICs combine analog and digital MOSFET circuitry on a single chip, making the broad space even smaller. Thus, the analog circuits from the digital circuits on a chip level are isolated and lead to the use of isolation rings and Silicon-On-Insulator (SOI). The main advantage of *BJTs vs MOSFEts* in the analog design process is the ability of BJTs a handle a larger current in a smaller space. Fabrication processes exist that incorporate BJTs and MOSFETs into a single device, called Bi-FGETs (Bipolar-Fets) if they contain just one BJT-FET and BiCMOS (bipolar-CMOS) if they contain complementary BJT-FETs. This device provides for the advantages of both the insulated gate and the higher current density.

The BJT also has some advantage over the MOSFET in certain digital circuits i.e. for at least two digital jobs. The first is in the high speed switching because they don't have the "larger" capacitance from the gate, which when multiplied by the resistance of the channel gives the intrinsic time constant of the process. The intrinsic time constant places a limit on the speed a MOSFET can operate at because higher frequency signals are filtered out. Widening the channel reduces the resistance of the channel, but increases the capacitance by the exact same amount. Reducing the width of the channel increases the resistance, but reduces the capacitance by the same amount. There is no way to minimize the intrinsic time constant for a certain process. Different processes using different channel lengths, channel heights, gate thicknesses and materials have different intrinsic time constants.

Secondly, when driving many other gates, the resistance of the MOSFET is in series with the gate capacitance of the other FETs, creating a secondary time constant. Delay circuits use this fact to create a set signal delay by using a small CMOS device to send a signal to many other, many other larger CMOS devices. The secondary time constant is minimized by increasing the driving FETs channel width to decrease its resistance and decreasing the channel width of the FETs being driven, decreasing their capacitance. This does have a drawback because it increases the capacitance of the driving FET and increases the resistance of the FETs being driven, but usually those drawbacks are a minimal problem when compared to the timing problem. BJTs are better to drive the other gates because they can output more current than MOSFETs, allowing for the FETs being driven to charge faster. Many chips employ MOSFET inputs and BiCMOS outputs.

Circuit Symbols

A variety of symbols (Fig. 3.30) are used for the MOSFET. The basic design is generally a line for the channel with the source and drain leaving it at right angles and then bending back

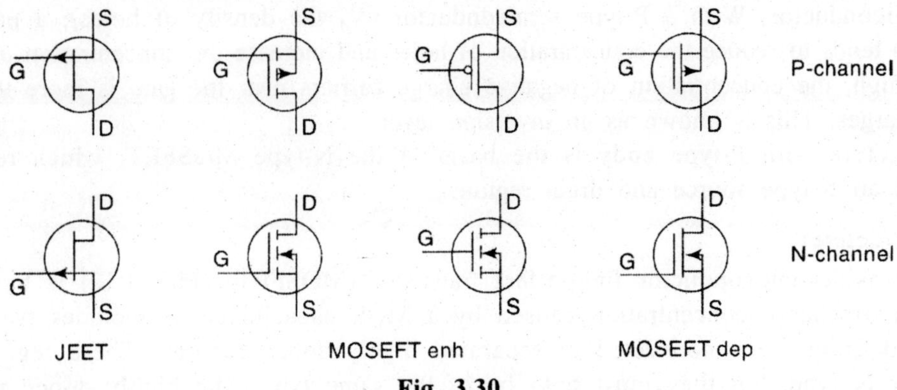

Fig. 3.30

into the same direction as the channel. Sometimes a broken line is used for enhancement mode and a solid one for depletion mode, but the awkwardness of drawing broken line means this distinction is often ignored. Another line is drawn parallel to the channel for the gate.

The bulk connection, if shown, is shown connected to the back of the channel with an arrow indicating PMOS or NMOS. Arrows always point from P to N, so an NMOS (n-channel in P-well or P-substrate) has the arrow pointing in. If the bulk is connected to the source (as is generally the case with discrete devices) it is angled to meet up with the source leaving the transistor. If the bulk is not shown (as is often the case in IC design as they are generally common bulk) an inversion symbol is sometimes used to indicate POMOS, alternatively an arrow on the drain may be used in the same way as for bipolar transistors (our for NMOS in for PMOS).

For the symbols in which the bulk, or body terminal is shown it is here shown internally connected to the source. In general, the MOSFET is a four-terminal device, and in integrated circuits many of the MOSFETs share a body connection, not necessarily connected to the source terminals of all the transistors.

MOSFET Operation

Metal-oxide-Semiconductor Structure

A traditional metal-oxide-semiconductor (MOS) structure Fig. (3.31) (a) and (b) is obtained by depositing a layer of silicon dioxide (SiO_2) and a layer of metal (polycrystalline silicon is commonly used instead of metal) on top of a semiconductor die. As the silicon dioxide is a dielectric material its structure is equivalent to a plane capacitor, with one of the electrodes replaced by a semiconductor. When a voltage is applied across a MOS structure, it modifies the distribution of charges

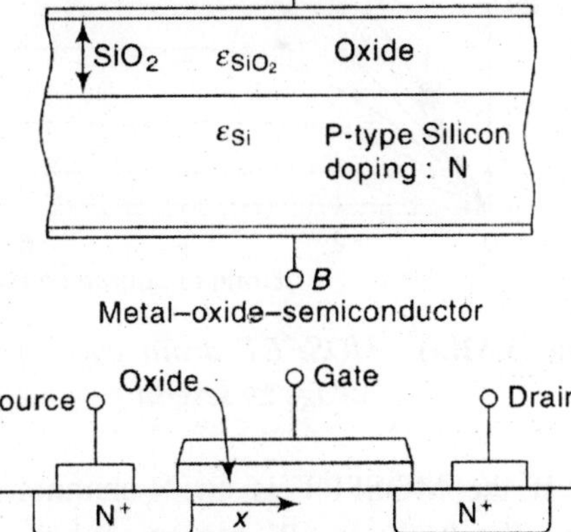

Fig. 3.31 *(a) Metal–oxide–semiconductor structure (b) Cross section of an NMOS*

in the semiconductor. With a P-type semiconductor (N_A the density of holes); a positive V_{GB} (see figure) tends to reduce the concentration of holes and increase the concentration of electrons. If V_{GB} is high, the concentration of negative charge carriers near the gate is more than that of positive charges. This is known as an inversion layer.

The structure with P-type body is the basis of the N-type MOSFET, which requires the addition of an N-type source and drain regions.

MOSFET Structures

A metal-oxide-semiconductor field-effect transistor (MOSFET) (Fig. 3.32) is based on the modulation of charge concentration caused by a MOS capacitance. It includes two terminals (source and drain) each connected to separated highly doped regions. These regions can be either P or N type, but they must both be of the same type. The highly doped regions are typically denoted a '+' following type of doping (see the image at the right). These two regions are separated by a doped region of opposite type, known as the body. This region is not highly doped, denoted by the lack of a '+' sign. The active region constitutes a MOS capacitance with a third electrode, the gate, which is located above the body and insulated from all of the other regions by an oxide.

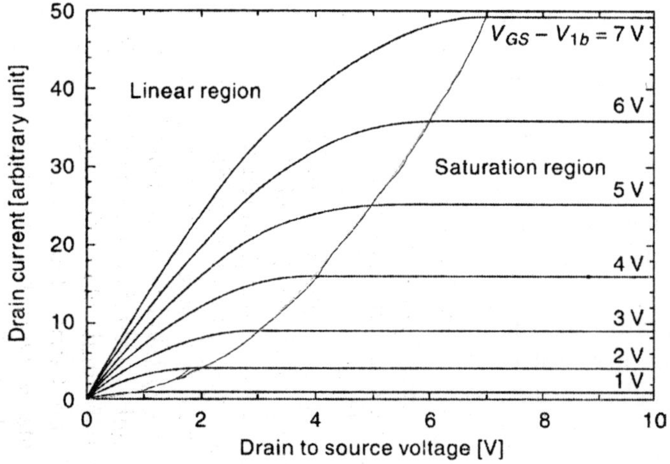

Fig. 3.32(a) *MOSFET drain current vs. drain-to-source voltage for several*

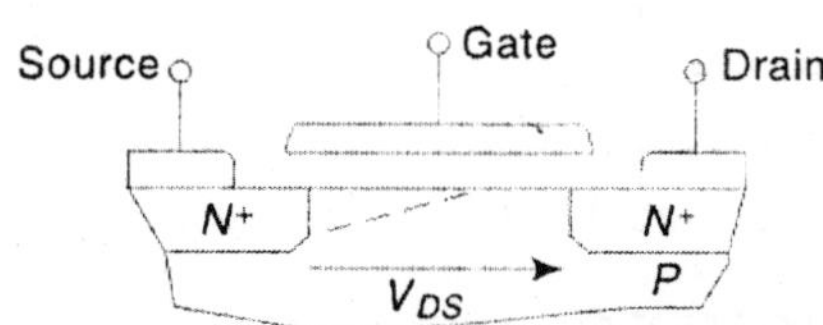

Fig. 3.22(b) *Cross section of a MOSFET*

If the MOSFET is an N-channel or nMOSFET, then the source and drain are 'N+' regions and the body is 'P' region. When a positive gate-source voltage is applied, it creates an *N-channel* at the surface of the P region, just order the oxide, by depleting this region of holes. This channel extends between the source and the drain, but current is conducted through it only when that potential is high enough to attract electrons form the source into the channel. When zero or negative voltage is applied between gate and source, the channel disappears and no current can flow between the source and the drain.

If the MOSFET is an P-channel or pMOSFET, then the source and drain are 'P+' regions and the body is 'N' region. When a negative gate-source voltage (positive source-gate) is applied, it creates a *P-channel* at the surface of the N region, just order the oxide, by depleting this region of electrons. This channel extends between the source and the drain, but current is conducted only when the gate potential is low enough to attract holes form the source into the channel. When a near-zero or positive voltage is applied between gate and source, the channel disappears and no current can flow between the source and the drain.

The source is so named because it is the source of the charge carriers (electrons for n-channel, holes for P-channel) that flow through the channel; similarly, the drain is where the charge carriers leave the channel.

Modes of Operation

The operation of MOSFET is into three different modes, depending on the voltages at the terminals. For an *enhancement-mode, n-channel MOSFET* the modes are:

1. Cut-off or sub-threshold mode

   When $V_{GS} < V_{th}$ where V_{th} threshold voltage of the device.

   According to the basic threshold model, the transistor is turned off, and there is no conduction between drain and source. In reality the Boltzamann distribution of electron energies allows some of the more energetic electrons at the source to enter the channel and flow to the drain, resulting in a subthreshold current what is an exponential function gate-source voltage. While the current between drain and source should ideally be zero when the transistor is being used as a turned-off switch, there is a weak-inversion current, sometimes called subthreshold leakage.

2. Triode or linear region

   When $V_{GS} > V_{th}$ and $V_{DS} < (V_{GS} - V_{th})$

   The transistor is turned on, and a channel has been created which allows current to flow between the drain and source. The MOSFET operates like a resistor, controlled by the gate voltage relative to both the source and drain voltages. The current from drain to source modeled as:

   $$I_D = \mu_n C_{ox} \frac{W}{L}\left((V_{GS} - V_{th}) - \frac{V_{DS}^2}{2}\right)$$

   where μ_m is the charge-carrier effective mobility, W is the gate width, L is the gate length and C_{ox} is the gate oxide capacitance per unit area. The transition from the exponential subthreshold region to the triode region is not sharp.

3. Saturation

   When $V_{GS} > V_{th}$ and $V_{DS} > V_{GS} - V_{th}$

   The switch is turned on, and a channel is created, which allows current to flow between the drain and source. Since the drain voltage is higher then the gate voltage, a portion of the channel is turned off. The onset of this region is known as pinch-off. The drain current is now relatively independent of the drain voltage and the current is controlled by only the gate-source voltage, as:

$$I_D = \frac{\mu_n C_{ox}}{2} \frac{W}{L} (V_{GS} - V_{th})^2$$

this equation can be multiplied $(1 + \lambda V_{DS})$ to take into account the channel length modulation.

When the channel length becomes very short, carrier transport is quasi-ballistic transport. When short-channel effects dominate, the I-V characteristics are no longer approximated by the above equations. Rather, the saturation drain current is linear than quadratic in V_{GS}.

The changes in the threshold voltage *(body effect)* is described by the change in the source-bulk voltage, is given by the following equation:

$$V_{TN} = V_{TO} + \gamma(\sqrt{V_{SB} + 2\phi} - \sqrt{2f}),$$

where V_{TO} is the zero substrate bias, γ is the body effect parameter, and 2φ is the surface potential parameter.

The body can be operated as second gate, and is sometimes referred to as the "back gate"; the body effect is sometimes called the "back-gate effect."

MOSFET Scaling

Over the past decades, the MOSFET has continually been scaled down in size; typical MOSFET channel lengths are reduced from several micrometers, to less than a tenth of a micrometer.

Small MOSFETs are desirable for several reasons. The main reason to scale the transistors is to pack more and more devices in a given chip area. This results in either small chips or chips with more computing power in the same area. Since fabrication costs for a semiconductor wafer are relatively fixed, the cost per integrade circuits is mainly related to the number of chips that can be produced per wafer. Hence, smaller ICs allow more chips per wafer, reducing the price per chip. In fact, over the past 30 years the number of transistors per chip has been doubled every 2-3 years once a new technology node is introduced. For example the number of MOSFETs in a microprocessor fabricated in a 45 nm technology is twice as large as in a 65 nm chip. This doubling of the transistor count, first observed by Gordon Moore in 1965 is commonly referred to as Moore's law.

Smaller transistors switch faster. MOSFETs scaling of the transistor dimensions does not necessarily translate to higher speed. Commensurate scaling of the MOSFET requires that all device dimensions are scaled with the same pace. The main device dimensions are the transistor length, width, and the oxide thickness, each (used to) scale with a factor of 0.7 per node. This way, the transistor channel resistance does not change with scaling, while gate capacitance is cut by a factor of 0.7. Hence, the RC delay of the transistor scales with a factor of 0.7.

Because of small MOSFET geometries, the voltage applied to the gate must be reduced to maintain reliability. To maintain performance, the threshold voltage of the MOSFET is also to be reduced. As threshold voltage is reduced, the transistor cannot be completely turned off; that is, the transistor operates in weak-inversion mode, with a subthreshold leakage, or subthreshold conduction, between source and drain. Subthreshold conduction, which has ignored in the past, now can consume upwards of half of the total power consumption modern high-performance VLSI chips.

Some micropower analog circuits are designed to take advantage of subthreshold conduction; by working in the weak-inversion region, the MOSFETs in these circuits deliver the highest possible transconductance-to-current ration.

Traditionally switching time was roughly proportional to the gate capacitance of gates. However, with transistors becoming smaller and more transistors being placed on the chip, interconnect capacitance (the capacitance of wires connecting different parts of the chip) is becoming a large percentage of capacitance. Signals have to travel through the interconnect, which leads to increased delay and lower performance.

The ever-increasing density of MOSFETs on an integrated circuit is creating problems of substantial localized heat generation that can impair circuit operation. Circuits operate slower at high temperature, and have reduced reliability and shorter lifetimes. Heat sinks and other cooling methods are now required for many integrated circuits including.

Power MOSFETs are at risk of thermal runaway. As their on-state resistance rises with temperature, if the load is approximately a constant-current load then the power loss rises correspondingly, generating further heat. When the heatsink is not able to keep the temperature low enough, the junction temperature may rise quickly and uncontrollably, resulting in destruction of the device.

The gate oxide, which serves as insulator between the gate and channel, should be made as thin as possible to increase the channel conductivity and performance when the transistor is on and to reduce subthreshold leakage when the transistor if off. However, with current gate oxides with a thickness of around 1.2 nm (which in silicon is ~5 atoms thick) the quantum mechanical phenomenon of electron tunneling occurs between the gate and channel, leading to increased power consumption.

Insulators (referred to as high-k dielectrics) that have a larger dielectric constant than silicon dioxide, such as group IVb metal silicates e.g. hafnium and zirconiuuim silicates and oxides are going to be used to reduce the gate leakage from the 45 nanometer technology node onwards. Increasing the dielectric constant of the gate oxide material allows a thicker layer while maintaining a high capacitance. The higher thickness reduces the tunneling current between the gate and the channel. An important consideration is the barrier height of the new gate oxide; the difference in conduction band energy between the semiconductor and the oxide (and the corresponding difference in valence band energy) will also affect the leakage current level. For the traditional gate oxide, silicon dioxide, the former barrier is approximately 8 eV. For many alternative dielectrics the value is significantly lower, somewhat negating the advantage of higher dielectric constant. With MOSFETs becoming smaller, the number of atoms in the silicon that produce many of the transistor's properties is becoming fewer.

MOSFET Construction

Gate Material

The primary criterion for the gate material is that it is a good conductor. Highly-doped polycrystalline silicon is an acceptable, but certainly not ideal conductor, and it also suffers from some more technical deficiencies in its role as the standard gate material. Nevertheless, there are several reasons favoring use of polysilicon as a gate material:

1. The threshold voltage (and consequently the drain to source on-current) is modified by the work function difference between the gate material and channel material. Because polysilicon is a semiconductor, its work function can be modulated by adjusting the type and level of doping. Furthermore, because polysilicon has the same bandgap as the underlying silicon channel, it is quite straighforward to tune the work function, so as to achieve low threshold voltages for both NMOS and PMOS devices. By contrast the work functions of metals are not easily modulated, so tuning the work function to obtain low threshold voltages becomes a significant challenge. Additionally, obtaining low threshold devices on both PMOS and NMOS devices would likely require the use of different metals for each device type, adding additional complexity to the fabrication process.
2. The Silicon-SiO_2 interface is known to have relatively few defects. By contrast many metal-insulator interfaces contain defects which lead to fermi-level pinning, charging, that ultimately degrade device performance.
3. In the MOSFET IC fabrication process, it is preferable to deposit the gate material prior to certain high-temperature steps in order to make better performing transistors. Such high temperature steps would melt some metals, limiting the types of metals that could be used in a metal-gate based process.

While polysilicon gates have been the standard for the last twenty years, they do have some disadvantages. They are replacemed by metal gates because:

1. Polysilicon is not a great conductor (approximately 1000 times more resistive than metals). It reduces the signal propagation speed through the material. The resistivity is lowered by increasing the level of doping, but even the highly doped polysilicon is not as conductive as most metals. To improve conductivity further, sometimes a high temperature metal such as tungsten, titanium, cobalt, and nickel, is alloyed with the top layers of the polysilicon. Such blended material is called silicide. The silicide-polysilicon combination has better electrical properties than polysilicon along and still does not melt in subsequent processing. Also the threshold voltages is not significantly higher than polysilicon alone, because the silicide material is not near the channel. The process in which silicied is formed on both the gate electrode and the source and rain regions is sometimes called silicide, self-aligned silicide.
2. When the transistors are extremely scaled down, it is necessary to make the gat dielectric layer very thin, around 1 nm in state-of-the-art technologies. A phenomenon observed there is the so-called poly deletion, where a depletion layer is formed in the gate polysilicon layer next to the gate dielectric when the transistor is in the inversion. To avoid this problem a metal gate is desired. A variety of metal gates such as tantalum, tungsten, tantalum nitride, and titanium nitride, usually in conjunction with high dielectrics. An alternative is to use fully-silicoded polysilicon gates, and the process is referred to as FUSI.

Types of MOSFET

Dual Gate MOSFET

The dual gate MOSFET has a tetrode configuration, where both gates control the current in the device. It is commonly used for small devices in radio frequency applications where the second gate is normally used for gain control or mixing and frequency conversion.

Depletion-mode MOSFETs

There are *depletion-mode* MOSFET devices, which are less commonly used than the standard *enhancement-mode* devices already described. These are MOSFET devices that are doped so that a channel exists even with zero voltage from gate to source. In order to control the channel, a negative voltage is applied to the gate (for an n-channel device), depleting the channel, which reduces the current flow through the device. In essence, the depletion-mode device is equivalent to a normally closed (on) switch, while the enhancement-mode device is equivalent to a normally open (off) switch.

Due to their low noise figure in their RF region, and better gain, these devices are often preferred to bipolars in RF front-ends such as in TV sets.

NMOS Logic

n-channel MOSFETs are smaller than p-channel MOSFETs and either of them is produced on a silicon substrate. These are the driving principles in the design of NMOS logic which uses n-channel MOSFETs exclusively. However, unlike CMOS logic, NMOS logic consumes power even when no switching is taking place. With advances ion technology. CMOS logic has displaced NMOS logic to become the preferred process for digital chips.

Power MOSFET

Power MOSFET fig. 3.33 have a different structure than the one presented above. As with all power devices, the structure is vertical and not planar. Using a vertical structure, it is possible for the transistor to sustain both high blocking voltage and high current. The voltage rating of the transistor is a function of doping and thickness of the N epitaxial layer while the current rating is a function of the channel width (the wider the channel, the higher the current). In a planar structure, the current and breakdown voltage ratings are both function of the channel dimensions (respectively width and length of the channel), resulting in efficient use of the "silicon estate". With the vertical structure the component area is roughly proportional to the current it can sustain, the component thickness (actually N-epitaxial layer thickness) is proportional to the breakdown voltage.

It is worth noting the power MOSFETs with lateral structure are mainly used in high-end audio amplifiers. Their advantage is a better behaviour in the saturated region (corresponding to the linear region of a bipolar transistor) than the vertical MOSFETs. Vertical MOSFETs are designed for switching applications.

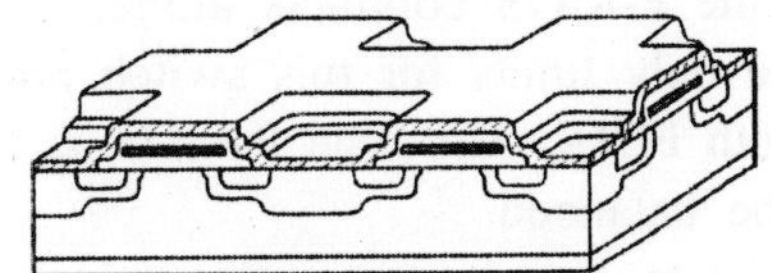

Fig. 3.33 *Cross section of a Power MOSFET, with square cells. A typical transistor is constituted of several thousand cells.*

DMOS

DMOS stands for double-Diffused Metal Oxide Semiconductor. Most of the power MOSFETs are made using this technology

MOSFET Analog Switch

MOSFET analog switching use the MOSFET channel as a low-on resistance switches to pass analog signals when on, and as a high impedance when off. Signals flow in the both directions across a MOSFET switch. In this application the drain and source of a MOSFET exchange places depending on the voltages of each electrode compared to that of the gate. For a simple MOSFET without an integrated diode, the source is the more negative side of an N-DMOS or the more positive side for a P-MOS. All of these switches are limited on what signals they can pass or stop by their gate-source, gate-drain and source-drain voltages, and source-to-drain currents; exceeding the voltage limits will potentially damage the switch.

Single-type MOSFET Switch

This analog switch uses a four-terminal simple MOSFET of either P or N type. In the case of N-type switch, the body is connected to the most negative supply (usually GND) and the gate is used as the switch control. Whenever the gate voltage exceeds the source voltage by at least a threshold voltage, the MOSFET conducts. The higher the voltage, the more the MOSFET can conduct. An N-MOS will pass through all voltages less than (Vgate-Vtn). When the switch is conducting, it typically operates in the saturation region, since the source and drain voltages will typically be nearly equal.

In the case of a P-MOS, the body is connected to the most positive voltage, and the gate is brought to a lower potential to turn the switch on. The P-MOS switch passes al voltages higher than (Caget+Vtp). Threshold voltage (Vtp) is typically negative in the case of P-MOS.

A P-MOS switch will have about three times the resistance of an N-MOS device of equal dimensions because electrons have three times the mobility of holes in silicon.

Dual-type (CMOD) MOSFET Switch

The "complementary" or CMOS type of switch uses one P-MOS and one N-MOS FET to counteract the limitations of the single-type switch. The FETs have their drains and sources connected in parallel, the body of the P-MOS is connected to the high potential (V_{DD}) and the body of the N-MOS is connected to the low potential (Gnd). To turn the switch on the gate of the P-MOS is driven to the low potential and the gate of N-MOS is driven to the high potential For voltages between (VDD – Vtn) and (Gnd + Vtp) both FETs conduct the signal, for voltages less than (Gnd + Vtp) the N-MOS conducts alone and for voltages greater then (V_{DD} – Vtn) the P-MOS conducts alone.

The only limits for this switch are the gate-source, gate-drain and source-drain voltages limits for both FETs. Also, the P-MOS is typically three times the width of the N-MOS so the switch will be balanced.

Tri-state circuitry sometimes incorporate a CMOS MOSFET switch on its input to provide for a low ohmic, full range output when on and a high ohmic, amid level signal when off.

Double Gate MOSFET

Though several silicon transistors with gate length near or below 20 nm, and 10-nm-scale MOSFETs devices have been developed yet the picture of all possible options is not clear. Consider, a double-gate MOSFET with undoped, ultrathin SOI channel connecting highly doped (degenerate) source and drain Fig. 3.34 (a) and (b). The reasons for this particular choice are as follows:

1. Transistors with planar geometries, have been implemented although their fabrication is complex than that of the usual bulk MOSFETs.
2. This theoretical model can be explored in considerable detail.
3. Most importantly although the double-gate MOSFET device is still not the "ultimate" FET, it is a good first approximation to the detail.

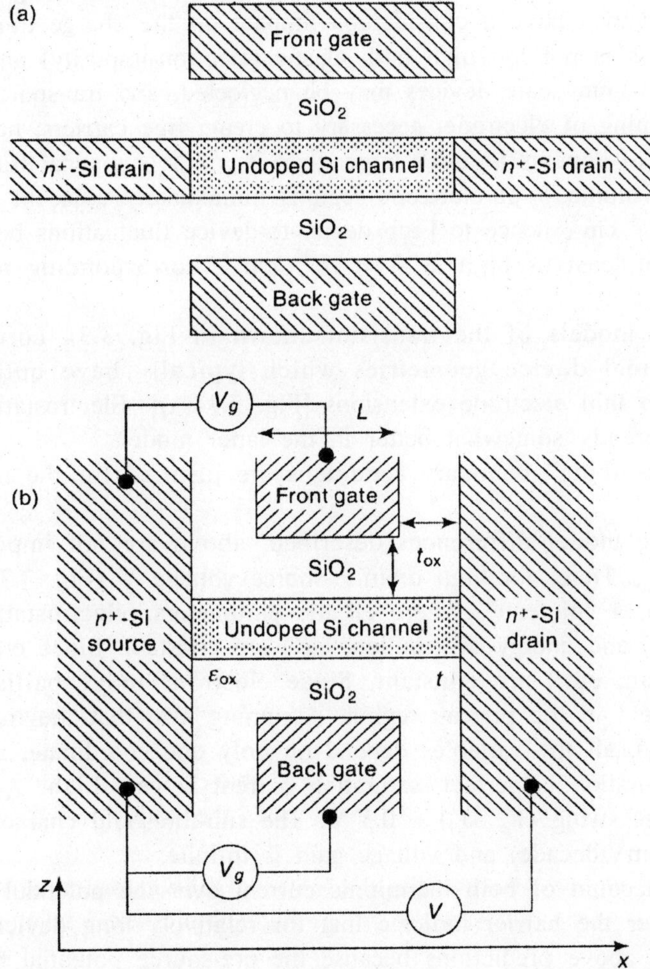

Fig. 3.34 *(a) and (b) Two simple models of double-gate MOSFETs with ultrathin intrinsic channel*

Indeed, double-gate or very close "surrounding-gate" or wrap-around-gate" transistors with ultrathin channel allow to avoid most short-channel effects for bulk MOSFET, because of the electrostatic potential. In addition, lateral quantum confinement of electrons in a channel with a few nanometers thickness 't' increases the effective bandgap by

$$\Delta E_g = E_z + E_z' ; \quad E_z = \pi^2 h^2 / 2 m_h t^2,$$

(where $m_h \approx m_0$ is the heavy electron mass, and E_z' is a similar valence band edge shift) and ence decreases current leakage via midgap traps.

For $t = 2$ nm, the increase is quite considerable (~0.1 eV).

The assumptions concerning doping come from the natural requirement of the device reproducibility: in a 2-nm thick Si channel, each dopant would create a randomly located potential bump with height about 0.1 eV, that is, quite comparable with typical scale of potential profile in the channel (see Fig. 3.35). Thus, any channel doping above $\sim 10^{17}$ cm^{-3} is unacceptable. This fact may have a considerable impact on the charge transfer in the channel: if the surface roughness is not too high both elastic (electron-impurity) and inelastic (electron-phonon) scattering in 10-nm scale devices may be neglected, and transport considered *ballistic.*

On the contrary, doping of electrode, necessary to create free carriers, has to be so high that the average number N of dopants is much larger than its r.m.s. statistical fluctuation $\sqrt{N}$. For 10-nm-scale devices the volume V of electrode regions immediately adjacent to the channel is of the order of 0.3×10^{-18} cm^3; hence to keep device-to-device fluctuations below 10% the doping rate N/V should be at least as high as 3×10^{20} cm^{-3}, corresponding to deeply degenerate silicon.

The two particular models of the transistor shown in Fig. 3.34 correspond to different idealizations of the real device geometries which typically have bulk source and drain [Fig. 3.34(b)], but also thin electrode extensions [Fig. 3.34(a)]. Electrostatic properties of very short devices are apparently somewhat better in the latter model.

The "completely absorbing" boundary conditions are justified for the model shown in Fig. 3.34(b).

It is seems that the model differences described, above are not important in the "long" device limit $L >> t$, t_{ox}. Here, for high drain-to-source voltage V (Fig. 3.34(b), channel has to feature a long section, a "plateau", where all the parameters (electrostatic potential, current, electron density, speed, and energy distribution, etc.) are constant. Since electron density, speed and energy distribution, etc.) are constant. Since electrons move ballistically, there is no longitudinal electric field in the plateau region. Assuming also that the transistor width is not too small (W >> t, t_{ox}), all the variables depend on only one coordinate, z across the channel. That the transistor saturation is perfect, saturation current is very high: J_s above 2000 μA/μm for very modest voltage swing $(V_g - V_t) \sim 0.3$ V. The sub-threshold characteristics have perfect slope (at 300 K, ~60 mV/decade) and voltage gain is infinite.

From Fig. 3.34(b) account of both thermionic current over the potential barrier, and source-to-drain tunneling under the barrier indicate that for relatively long devices (L > 10 nm) the current is close to the above predictions because the pre-source potential bump height U_{max} is close to the position of the plateau.

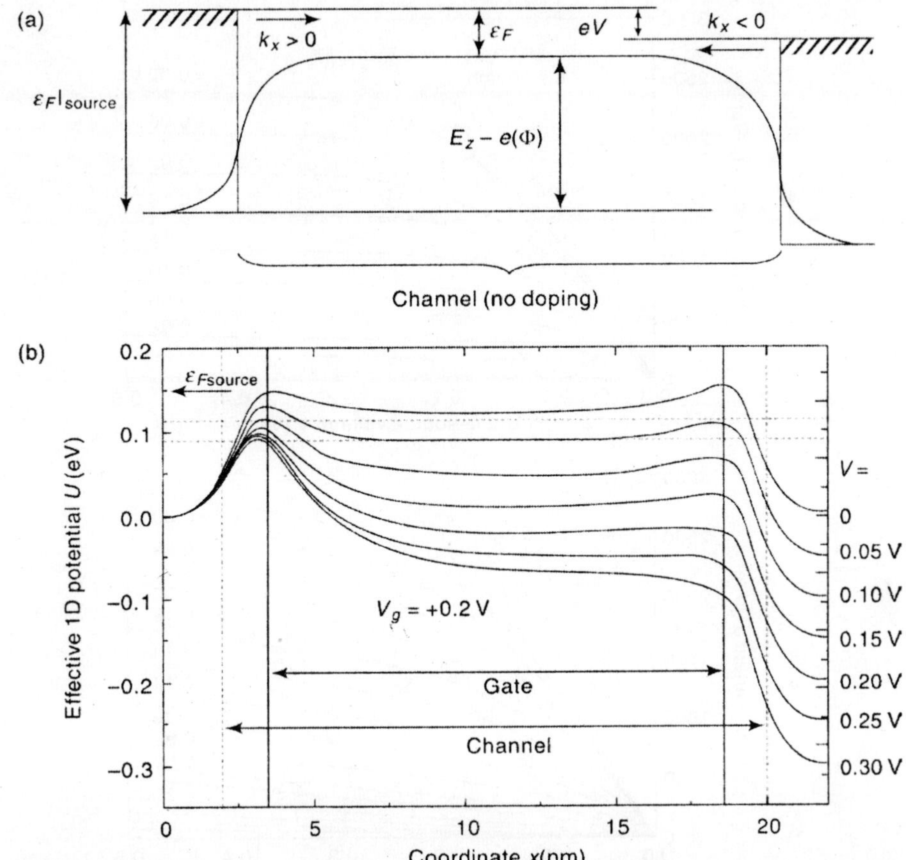

Fig. 3.35 *The effective 1D potential profile $U(x) = E_z - e\langle\Phi\rangle$ (i.e., the conduction band edge diagram) of a relatively log ($L >> t, t_{ox}$) ballistic transistor according to: (a) the 1D theory and (b) the numerical solution of the problem for the model shown in Fig. 3.32, for the case L = 15 nm, t = 2 nm ($E_z \approx$ 0.1 eV), t_{ox} = 1.5 nm, $N_D = 3 \times 10^{20}$ cm^{-3} ($\varepsilon_{F|source} \approx 0.15$ eV), T = 300 K, and several values of drain-to-source voltage V In (b), the dashed and dotted lines indicate the potential plateau values for low and high V, respectively, according to the 1D theory.*

These numerical results show that transistor performance degrades as the gate length L is reduced below approximately 5 nm (channel length, below ~8 nm). Fig. 3.36 shows that the drain current saturation becomes less pronounced and hence the voltage gain $G_V \equiv \partial V/\partial V_g|_{J=\text{const}}$ (Fig. 3.37) becomes finite. The sub-threshold curve slope (Fig. 3.38) becomes considerably lower than the perfect, thermally determined value (for 300 K, ~60 mV/decade) and shows increasing DIBL ($\sim 1/G_V$) as L decreases. Moreover, the sub-threshold curves band upward for strongly closed devices (large negative values of V_g), due to the contribution from source-to-drain tunneling along the channel.

For larger values of (t and t_{ox}); two effects contribute comparably to the device degradation at $L \to 0$: a loss of electrostatic control of the bottleneck potential U_{max} by the gate voltage V_g, and source-to-drain tunneling along the channel.

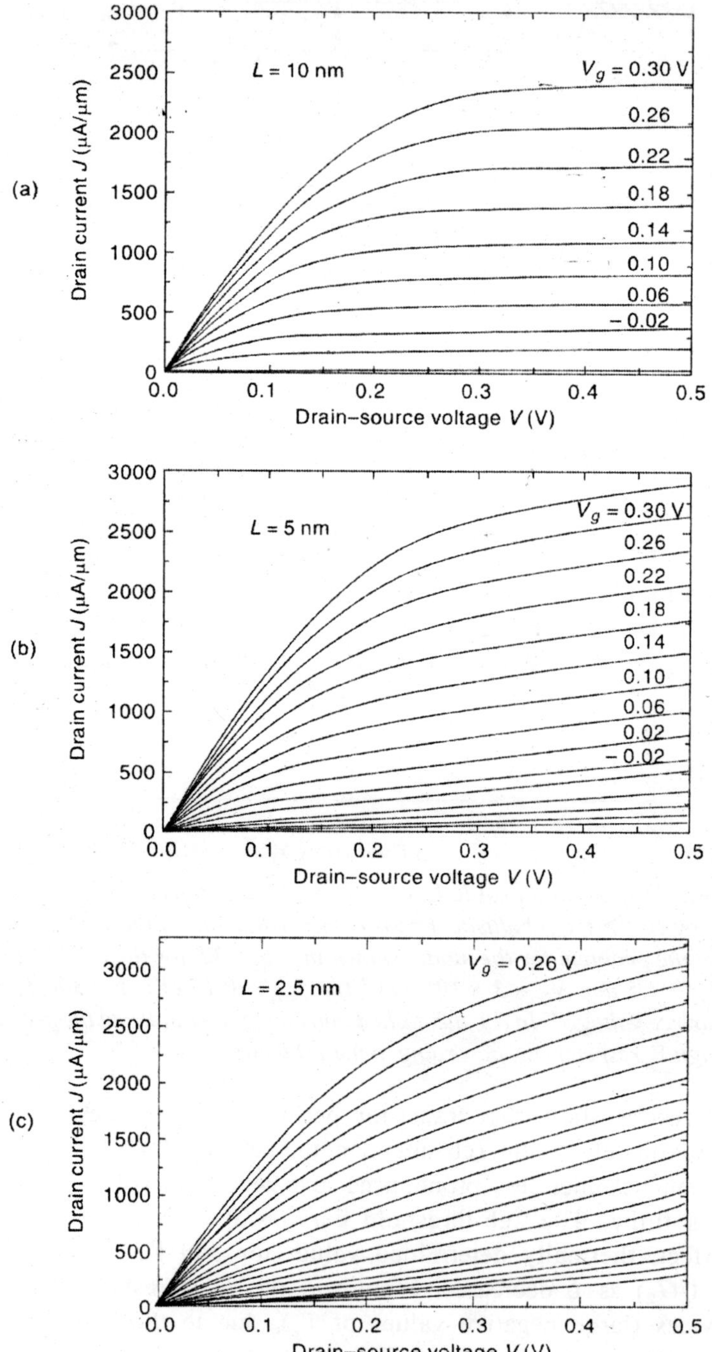

Fig. 3.36 *I-V characteristics of double-gate, ballistic MOSFETs for three values of gate length L. numerically calculated using the model shown in Fig. 3.31(b). Besides L, all the parameters are the same as for Fig. 3.32(b). The gate-source voltage V_g is changed with a step of 0.04 V for all frames.*

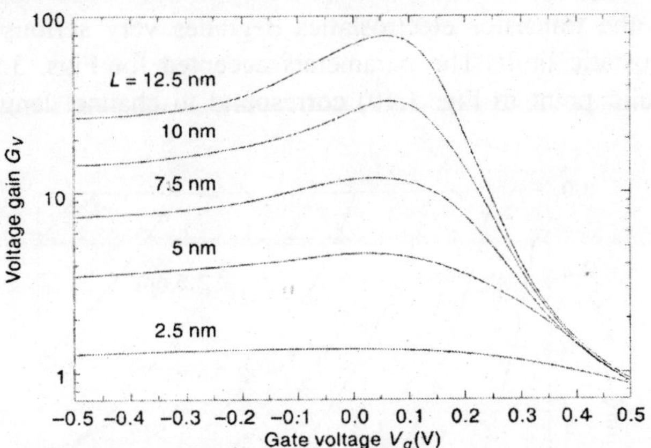

Fig. 3.37 *Voltage gain $G_V \equiv \partial V/\partial V_{g/J=const}$ of nanoscale MOSFETs as a function of V_g for $V = 0.3$ V and several values of gate length L. The other parameters are similar to those used in Fig. 3.32*

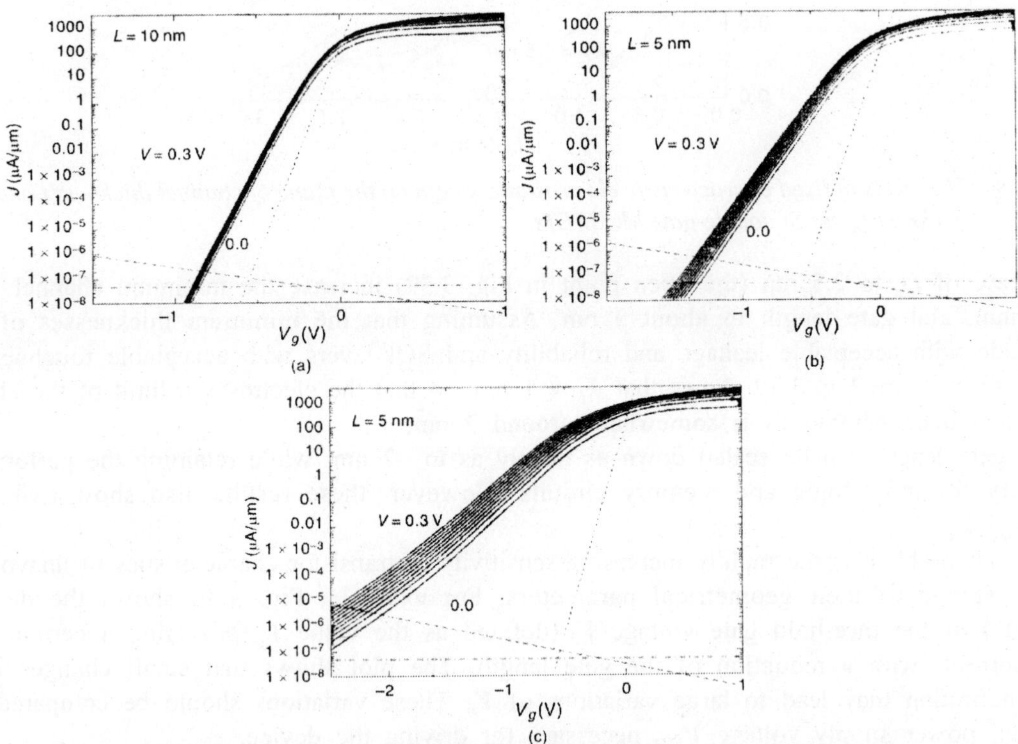

Fig. 3.38 *Sub-threshold curves of nanoscale ballistic MOSFETs [Fig. 3.31(b)] for three values of gate length L, and for 10 values of drain-source voltage V (with a step of 0.03 V). Transistor parameters are the same as for Figs. 3.32(b), 3.33 and 3.34, besides a thicker oxide thickness is taken (t_{ox} = 2.5 nm), because sub-threshold curves are mostly important for memory applications where I-V curve saturation maybe sacrificed. Dotted lines show the ideal slope of the curves (60 mV/decade). Dashed lines show estimated gate oxide leakage current, while the dash-dotted lines indicate the upper boundary of the region where hole current cannot be neglected.*

One can expect that the transistor electrostatics degrades very seriously if the channel length is reduced below electrostatic limit. The parameters accepted for Figs. 3.36 and 3.37 (t = 2 nm, t_{ox} = 1.5 nm, see the solid point in Fig. 3.40) correspond to channel length about 5.5 nm, that is $L \approx 2.5$ nm.

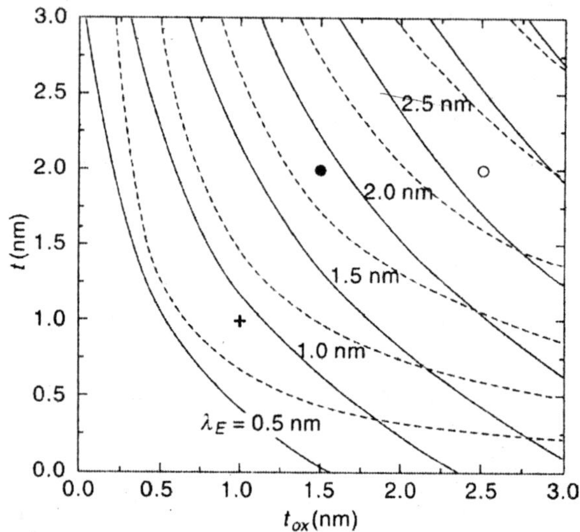

Fig. 3.39 *The levels of fixed characteristic electrostatic length on the plane of channel thickness t and oxide thickness t_{ox} for Si double-gate MOSFETs.*

Increase of t_{ox} to 2.5 nm (the open point in Fig. 3.39) increase the minimum channel length to ~8 mm, and gate length to about 3 nm. Assuming that the minimum thicknesses of SiO_2 gate oxide with acceptable leakage and reliability and SOI layers with acceptable roughness are both close to 1 nm Fig. 3.39 shows that $\lambda_E \approx 1$ nm, so that the electrostatic limit of the channel length of silicon MOSFETs is somewhere around 3 nm.

FET gate length can be scaled down as deeply as to ~2 nm, while retaining the performance necessary for most logic and memory circuits. However, these results also show two major challenges.

The first problem is the rapidly increasing sensitivity of transistor characteristics to unavoidable random spread of their geometrical parameters. For example, Fig. 3.40 shows the decrease ("roll-of") of the threshold gate voltage V_t (defined as the value V_g providing a certain small drain current) with a reduction of the gate length. The plot shows that small changes in the length definition may lead to large variations of V_t. These variations should be compared with minimum power supply voltage V_{DD} necessary for driving the device.

Consider a relatively long device with physical gate length L = 9 nm, then the critical dimension control accuracy will reach ~0.7 nm. Figure 3.40 shows that this control enables variation of V_t to be kept within approximately 50 mV. In order to keep ON current density at the standard 600 μA/μm the voltage swing V_{DD} should be above 20 mV; hence, the 50 mV variation seems acceptable.

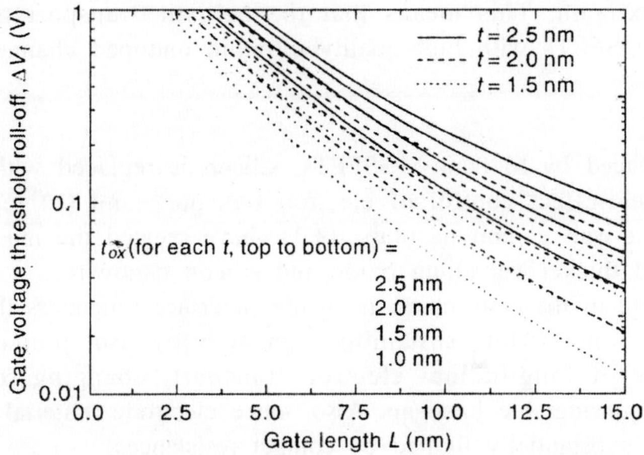

Fig. 3.40 *Threshold voltage roll-off (relative to that for L → ∞) as a function of gate length L, for all combinations of three values for oxide thickness t_{ox} and the four values for channel thickness t. The drain-to-source voltage V is 0.3 V; other parameters are the same as for Fig. 3(b).*

Now consider a shorter device with L = 5 nm. According to Fig. 3.37, in order to keep fluctuations of V_t below 50 mV, the critical dimension should be controlled better than ~0.15 nm.

The second factor which may seriously affect the practical use of sub-20-MOSFETs is power scaling. The standard MOSFET scaling leads to specific power (per unit channel width) *decreasing* with L, since at drift-diffusion transport the ON current density depends on the longitudinal electric field $E \approx V_{DD}/L$. Hence, for fixed current density J_{ON}, reduction of L allows the voltage swing V_{DD}, and hence the specific power, to be reduced. As soon as the transistor is scaled down to the ballistic transport mode, this scaling, however, no longer holds, and at relatively large L ($\geq$ 10 nm) does not depend on the device length at all.

Next, the MOSFET length is scaled down below ~10 nm, its deterioration due to electrostatics and source-to-gate tunneling leads to a gradual increase of the specific power per unit width.

Ballistic Transport

This transport regime is observed at low temperature so the issue is whether the scattering can be low enough in ultrathin silicon layers at room temperature. The mobility in undoped SOI layers (as thin as 5 nm) is very high (>400 cm^2/V/s), that is, essentially the same as in bulk MOSFETs in the equivalent perpendicular field. If similar surface roughness is sustained down to t ~ 2 nm, the mobility is about 200 cm^2/V/s (because of electron scattering at both interfaces). This mobility (μ) corresponds to an elastic scattering time $\tau = m_1\mu/e$ close to 25 fs.

For the most important transport region (where the effective potential $U(x)$ is close to its maximum), the electron kinetic energy is of the order of thermal energy k_BT ~ 30 meV, that is, their average speed $c \approx (2\ k_BT/m_t)^{1/2}$ is close to 2 × 10^7 cm/s; for this speed the elastic mean free path $l = \tau v$ should be about 5 nm. When this value of l may be lower than the total channel length, it is still larger than that of the transport bottleneck, where electrons overcome the potential barrier maximum. This means that the ballistic transport may be a reasonable approximation for MOSFETs with high-quality ultrathin undoped channel.

Alternative Ways

For, the problem faced by Si nano-MOSFETs, silicon is replaced with another material by (a) still using the field-effect control of electron transport; and (b) replacing it with other physical effects. Single-wall carbon nanotube FETs, have created the most publicity multi-wall carbon nanotubes and thicker vanadium oxide and silicon nanowires.

A possible advantage of these structures is smaller interface roughness that allows to use very thin channels (t ~ 1.5 nm) while sustaining high mobility. But production methods affect nanotubes parametrs of longitudinal electron transport, changing them from metals to semiconductors with considerable bandgap. Also, since electrode material is different from the tube, current may be substantially limited by contact resistances.

To change the field effect, Mott metal-insulator transition is controlled by the electric field of the gate, Mott-transition FETs based on perovskite capture films give very low current and voltage gain much below unity. The development of organic FETs mobilities exceeding 1 cm^2/V/s are achieved for both n- and p-type transistors, and hole mobility at low temperatures (105 cm^2/V/s). Thus there is development of organic MOSFETs large-area circuits (e.g., flat panel displays) & for high-performance electronics.

Ultrashort transistors using a single organic monolayer as a channel are achieved. Fig. 3.41. A drawback of such a structure is that the electric field applied to the doped Si substrate (serving as a gate) cannot penetrate the organic monolayer much more than by the organic layer thickness (in this particular case, 1-2 nm), so that effective transport control is possible only if layer width W is reduced to a similar scale.

However, devices with W as large as 100 nm has shown voltage gain $G_V > 1$, and the ON/OFF current ratio above 10^5.

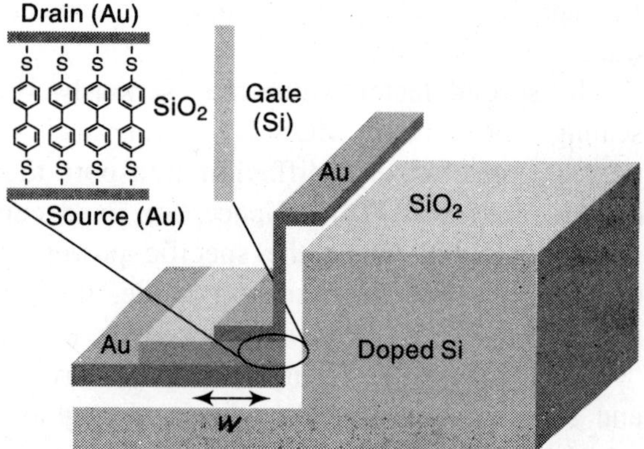

Fig. 3.41 *Schematic structure of the ultrashort organic transistor*

(III) MODULATION-DOPED FIELD EFFECT TRANSISTOR (MODFET)

The modulation-doped field effect transistor (MODFET) is the compound semiconductor analog of MOSFET which has been very successful and represents almost all the volume of semiconductor electronics.

The modulated doping field effect transistor or modulation-doped field effect transistor (MODFET) is a type of field-effect transistor. It is manufactured by epitaxial growth of strained SiGe layer, with the germanium content linearly increasing to about 40-50%. The high content of germanium allows formation of a quantum well structure with high conduction band offset and high density and mobility of charge carriers, resulting in parts with ultra-high speed and low noise. In GaAs/AlGaAs, AlGaN/InGaN, and other structures are also used and are known as HEMTs (High Electron Mobility Transistors). They basically consist of a wide and a narrow band gap material aligned adjacent to each other. This is known as Heterojunction structure as two different materials are used to make this device. The presence of two dissimilar material side by side results in bending of the conduction and Valence band. Electrons from a wide band n-type material, start diffusing to the conduction band of the narrow band material due to the availability of states in there. As the transition is to a conduction band, it should be no surprise that there area lot of states available in there. The diffusion process continues until the charges are balanced and a junction is formed. There is a band discontinuity in both the conduction and the valence band which acts as a barrier for any carrier to move across the junction. Band discontinuity across the bands is modified separately. This leads to the type of carriers for the required device. As HEMTs requires electrons as the main carriers, a graded doping is applied in one of the material making the conduction band discontinuity smaller, and keeping the valence band discontinuity the same. The diffusion of carriers leads to accumulation of electron along the boundary of the two regions inside the narrow band gap material. This accumulation of electron leads to a very high current in these devices. The accumulated electrons are also known as 2 DEG or two dimension electron gas.

Thus, MODFETs utilize a pseudo-two-dimensional carrier gas, the concentration of which is modulated by a gate potential. In a MODFET, a larger bandgap material with a high doping concentration (assumed to be n-type in this discussion) is grown on a lower bandgap intrinsic material. The growth of a thin intrinsic layer of the high bandgap material, setback layer or spacer precedes growth of the rest of the high bandgap material. Electrons diffuse from the doped larger bandgap material to the smaller bandgap material where they are confined and form a conducting sheet, two-dimensional electron gad (2 DEG). When the gate voltage is adequately high so that the source drain channel is no longer depleted, the 2DEG is free to conduct in the intrinsic material. Because the undoped low bandgap material has no donor atoms cluttered about, *in situ* impurity scattering no longer inhibits the carrier mobility and saturation velocity. A schematic diagram of a modulation-doped structure is shown in Fig. 3.42. The spacer mentioned above serves to increase the channel mobility further by shielding the 2DEG from ionized impurities, though at the cost of decreased sheet carrier density. Early versions of MODFETs relied on GaAs as the conducting medium and AlGaAs as the do not layer of larger bandgap in relation to GaAs. Shortly thereafter, the material system expanded to include the InGaAs/InAlAs (InP) structures on InP substrates and SiGe strained layers on Si substrates. Although, the InP based heterostructures operate at higher frequencies, the structure that has the widest range of applications is the pseudomorphic AlGas/InGaAs/GaAs PMODFET.

Advantages of HEMTs are as follow: Firstly, they have high gain. They have high switching speeds, which is achieved because the main charge carriers in these devices are the majority carrier, and minority carriers are not significantly involved. They have extremely low noise value, as the current variation in these devices is low as compared to FETs. Earlier, they were made using GaAs as the base material, however, InP and GaN based HEMTs becoming more and more popular nowadays, being superior to their GaAs based devised, both in power and

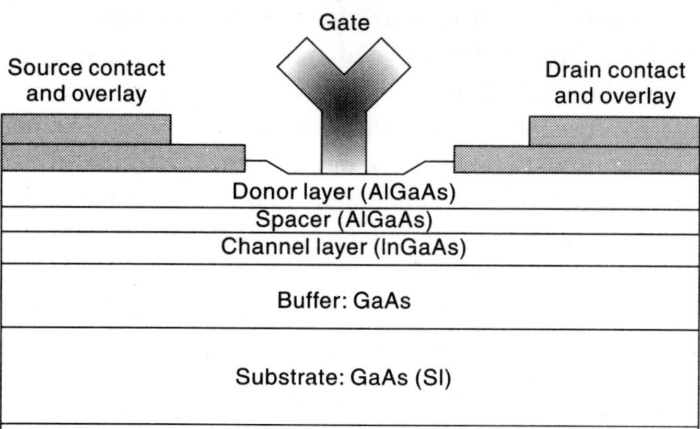

Fig. 3.42 *Cross-sectional view of a MODFET implemented in the AlGaAs/InGaAs/GaAs system.*

noise ratios. HEMTs are widely being used satellite receivers, low power amplifiers and have a wide range of military applications.

I. GaAs-based Pseudomorphi MODFETs

The sheet carrier concentration is improved by introducing MODFET; through increased band discontinuity, ΔE_C, by increasing the InAs mole fraction in the channel. The conventional MODFET uses AlGaAs and GaAs for the high and low bandgap materials, respectively, on a GaAs buffer. Carrier density and its confinement in the GaAs layer improved with increasing ΔE_C, induced by a larger Al mole fraction in$Al_xGa_{1-x}As$.

The PMODFET, or pseudomorphis MODFET provides a channel in InGaN with superior transport properties over GaAs. The structure is achieved by inserting a thin InGaAs layer between the GaAs buffer and the AlGaAs setback layer in otherwise conventional MODFET. The higher the in mole fraction in $In_xGa_{1-x}As$, the higher the electron mobility within the 2DEG. Additionally, greater carriers confinement are is achieved with pseudomorphic systems, because large conduction band discontinuities are obtained between InGaAs and low AlAs mole fraction AlGas. The thickness of the lattice mismatched InGaAs is below the critical thickness beyond which misfit dislocations occur. The lattice constant, lattice mismatch, and the critical thickness are functions of the In mole fraction of the strained InGaAs layer. The In mole fraction content is about 30% on GaAs. There is also the lattice matched and pseudomorphic MOSFETs based on InP substrates which take advantage of the superior mobility and confinement properties of the AllnAs/InGaAs. Devices based on InP have exhibited the lowest noise figures.

PMODFETs is galvanized with the development of 0.25μm gate length devices. It is found that as the In mole fraction is increased, the noise and gain performance for a given gate length and frequency improves. As the current gain cut-off frequency increases by more than a factor of 2, from about 110 to 260 GHz, as the InAs mole fraction is increased from zero to 65%. A compilation of minimum noise figure for GaAs-and InP-based MODFET technologies are shown in Fig. 3.43 where the noise figures obtained at about 100 GHz is less than what the FET technology provides at 4 GHz.

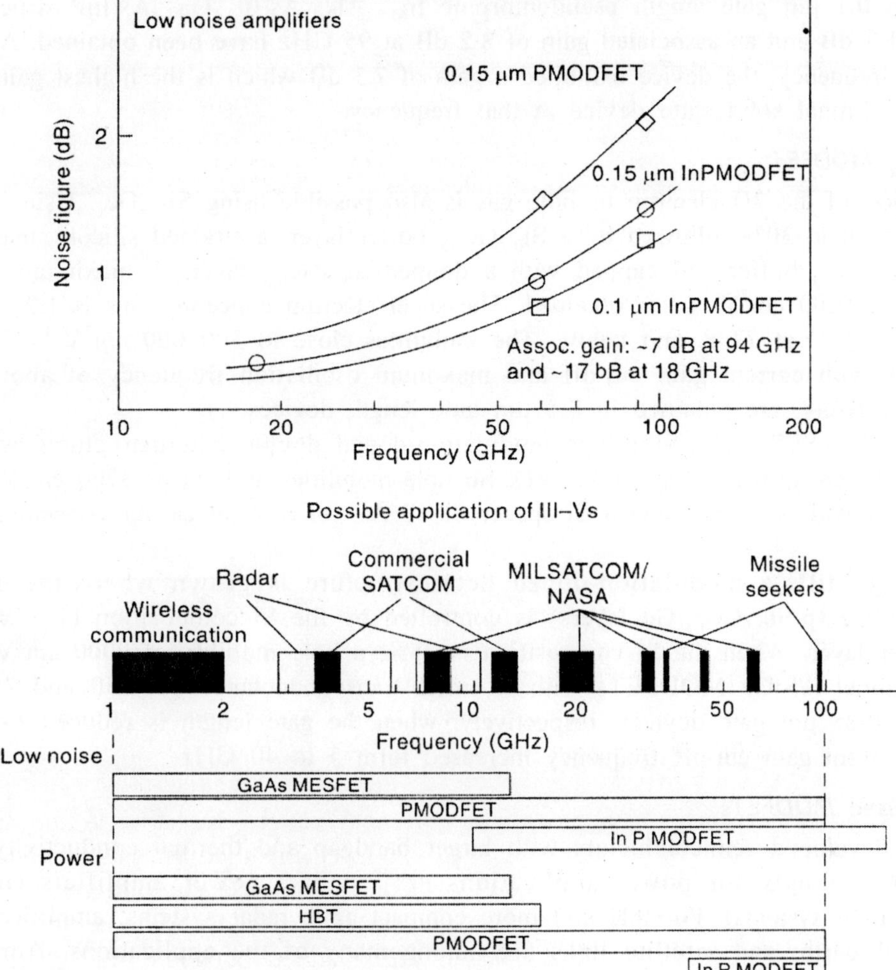

Fig. 3.43 *(a) Minimum noise figure as a function of frequency obtained from PMODFETs, based on GaAs substrates and MODFETs based on InP substrates., (b) Application areas of MODFETs, GaAs metal semiconductor FETs (MESFETs), and HBTs.*

The unity current gain cut-off frequency, where the current gain goes to unity, is an important parameter in logic gates. The maximum oscillation frequency (f_{max}), the frequency at which the maximum available gain of the device goes to unity, is also an important factor in determining performance. This figure too shows a marked improvement as the In mole fraction is increased. For example, the f_{max} for GaAs channel devices is increased by about 30% to about 350 GHz. With 53% mole fraction a remarkable 405 GHz was obtained with 0.15 μm gate devices. The minimum noise levels from MOSFETs based on GaAs are shown in Fig. 3.43. Performance above 100 GHz has been measured.

Using a 0.1 μm gate length pseudomorphic $In_{0.52}Al_{0.48}As/In_{0.6}Ga_{0.4}As/InP$ structure, a noise figure of 1.3 dB and an associated gain of 8.2 dB at 95 GHz have been obtained. At 141.5 GHz, a D band frequency, the device exhibited a gain of 7.3 dB which is the highest gain reported for any three-terminal solid state device at that frequency.

2. $Si_{1-x}Ge_x$ MODFET

Formation of the 2D electron or hole gas is also possible using $Si_{1-x}Ge_x$. Using a SiGe layer graded from 0 to 30% followed by a $Si_{0.7}Ge_{0.3}$ buffer layer, a strained silicon channel is grown on the $Si_{0.7}Ge_{0.3}$ buffer and capped with a dopped $Si_{0.7}Ge_{0.3}$ layer. A maximum Hall mobility at 4 K of 125000 $cm^2V^{-1}s^{-1}$ is obtained. The sheet electron concentrations is 1.2×10^{12} at 300 K, to 7.8×10^{11} at 77 K and below. The mobilities close to 300 000 $cm^2V^{-1}s^{-1}$ are obtained. MODFETs with current gain cut-off and maximum oscillation frequency of about 32 and 40 GHz, respectively, are obtained in 0.5 μm gate length devices.

High-quality $Si/Si_{1-x}Ge_x/Si$ p-type modulation-doped double heterostructures with $x = 0.12$ and $x = 0.15$ are grown using UHV/CVD. So hole mobilities as high as 3700 $cm^2V^{-1}s^{-1}$ at 14 K are obtained with $x = 0.12$ and a Si spacer of 60 Å—for a sheet carrier concentration of $\sim 8 \times 10^{11}$ cm^{-2}.

Utilizing MBE a modulation-doped heterostructure is grown where the strain at the heterointerface ($p\text{-}Si_{0.5}Ge_{0.5}/Ge$ layers) is controlled by the Si composition $(1 - x)$ of the $Si_{1-x}Ge_x$ buffer layer. When the Si composition is 25%, a hole mobility of 9000 $cm^2V^{-1}s^{-1}$.

For channel S/SiGe MODFETs with extrinsic transconductances of 150 and 250 mS mm^{-1} for 1 and 0.25 μm gate devices, respectively, when the gate length is reduced from 1 to 0.25 μm, the current gain cut-off frequency increased form 5 to 40 GHz.

3. GaN-based MODFETs

Newly developed semiconductors with larger bandgap and thermal conductivity are coming along very strongly for power applications eg: compact power amplifiers (in radar, and communication systems). Portable and more compact agile radar systems, amplifier for satellite-earth, earth-earth, and satellite links are among many of the applications. Among the new power devices are FETs based on GaN and SiC. The latter as a good thermal conductivity of 4.5 W K^{-1} cm^{-1}. The GaN-based devices are MODFET type. Power applications are divided into RF power at high frequencies and switching power at relatively lower frequencies. The former is served increasingly well by GaN and related heterostructures. The latter by developing diode like switches for the electric power industry. This semiconductor is also being explored for smart power for large electric motors, etc.

Gate dimensions of approximately 2 μm used to optimize GaN MODFET structure for contact and Schottky barriers. Short gate lengths reduce the time required for carriers to traverse the length of the channel Thus higher operating frequencies and higher gains or both are obtained. Unprecedented power levels are achieved with near-half-macrometer gate lengths. Sapphire has a poor thermal conductivity whereas SiC has a good thermal conductivity. To operate the device at high voltages imperative for large power, the heat dissipation in the case of sapphire substrate is managed by flip chip mounting the device on AlN ceramic grade boards with good thermal conductivity. In this approach, heat diffusion is through GaN with its three times better heat conductivity as compared to that for sapphire substrates that 0.45 μm gate, high-power MODFEts, on SiC substrates exhibited a power density of 6.8 W mm^{-1} in a 125 μm wide device and a total power of 4 W (with a power density of 2 W mm^{-1}) at 10 GHz superior performance GaN-based MODFETs on SiC and sapphire substrates example, GaN/AlGaN MODFETs prepared by MBE on SiC substrates, exhibited a total power level of 6.3 W at GHz from a 1 mm wide device. (the power density extrapolated from a 0.1 mm device is 6.5 W). When four of these devices are power combined in a single stage amplifier, an output power of 22.9 W with a power added efficiency of 37% is obtained at 9 GHz. Equally impressive is the noise figure of 0.85 dB at 10 GHz with an associated gain of 11 dB. The drain breakdown voltages in these quarter micrometer gate devices are about 60 V.

(IV) SINGLE-ELECTRON DEVICES

Single-electron device is also called the "single-electron box". The device consists of just one small conductor (island") separated from an eternal electrode by a tunnel barrier with high resistance, $R >> R_Q$.

An external electric field is applied to the island using a capacitively coupled gate electrode. The field changes the local Fermi level of the island and thus determines the conditions of electron tunneling. Elementary electrostatics show that energy W of the system may be presented as

$$W = (Q_0 - n_e)^2/2C_\varepsilon + \text{const}, \quad Q_0 = C_g V_g \tag{i}$$

where $Q = -ne$ is the island charge (n is the number of uncompensated electrons), V_g is the island-gate capacitance, while C_ε is the total capacitance of the island (including C_g) parameter Q0 is usually called the "external charge". From its definition, it is evident that in contrast with the discrete total charge Q of the island, the variable Q_0 is continuous, and may be a fraction of the elementary charge e.

An elementary calculation using Eq. (i) shows that Q is a step-like function of Q_0, i.e. of the gate voltage [Fig. 3.44(b)], jumping by e when

$$Q_0 = \left(n + \frac{1}{2}\right), \qquad n = 0, \pm 1, \pm 2,$$

At sufficiently low temperature

$$k_B T \approx E_C, \qquad E_C \equiv e^2/C_\varepsilon, \tag{iii}$$

the stationary number n of electrons in the island corresponds to the minimum of W.

If the temperature is increased to $k_B T \sim E_C$, the system has non-vasnishing probability p_n to be in other states as well, with $p_n = \exp\{-W(n)/k_B T\}/\Sigma_n \exp(-W(n)/k_B T\}$. A straight forward calculation of the average charge $\langle Q \rangle = -\Sigma_n enp_n$ yields the pattern shows in Fig. 3.44(b): the step-like dependence of charge on gate voltage is gradually smeared out by thermal fluctuations. This is typical for all single-electron devices, so that the operation temperature of most of them should satisfy Eq. (ii). Increasing gate voltage V_g attracts more and more electrons to the island. The discreteness of electron charge low-transparency barriers makes this increase step-like.

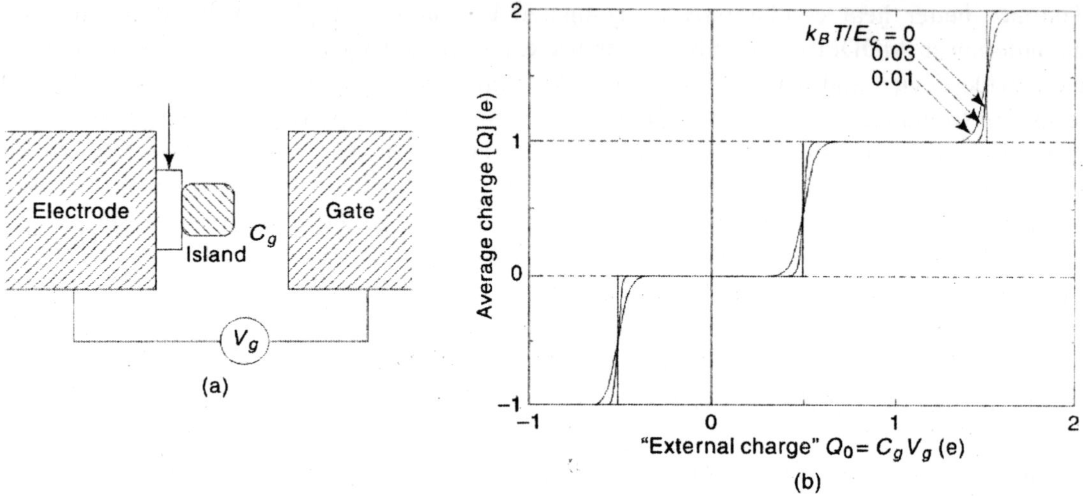

Fig. 3.44 *Single-electron box; (a) schematics, and (b) the "Coulomb staircase", that is, the step-like dependence of the average charge Q of the island on the gate voltage V_g of several values of temperatures.*

Quantum Dot Nanostructure for Single-Electron Devices

The idea of using quantum dots to control electric charges in the system at the ultimate level of one electron is related to the Coulomb blockade phenomenon. The classical electrostatic energy of a quantum dot with capacity C which is a capacitively coupled to a gate at a bias voltage V_g is given by

$$E = \frac{Q^2}{2C} - Q\,\alpha V_g,$$

where α is a dimensionless factor relating the gate voltage to the potential of the island and Q is the charge. Mathematically, the minimum energy is reached for a charge $Q_{\min} = \alpha C V_g$. However, the charge has to be an integer multiple of the elementary charge e, i.e., $Q = Ne$. If V_g has a value such that $Q_{\min}/e = N_{\min}$ is an integer, the charge cannot fluctuate, as long as the temperature is low enough, i.e.,

$$k_B T << \frac{e^2}{2C}.$$

Tunneling in or out of the dot is suppressed by the Coulomb barrier $e^2/(2C)$, and conductance is very low. Similarly, the differential capacitance is small. Peaks in the tunneling current occur when the gate voltage is such that the energies for N and $N + 1$ electrons are degenerate, i.e., $N_{\min} = N + 1/2$. The expected level spacing is

$$e\alpha\Delta V_g = \frac{e^2}{C} + \Delta\varepsilon_N,$$

where $\Delta\varepsilon_N$ denotes the change in lateral (kinetic) quantization energy for the added electron and e^2/C corresponds to the charging energy.

The Coulomb charging energy is calculated for self-organized InAs/GaAs QDs. It is of the same order of magnitude as the energy spacing between in individual energy levels. For lens-shaped InAs/GaAs QDs with a radius of 25 nm, charging energy is $\approx$ 30 meV. For pyramid-shaped INAs/GaAs QDs, charging energy varyies from 25 meV for a pyramid base length of 10 nm to 18 meV for a pyramid base length of 20 nm. However, the localization energy of electrons in InAs/GaAs QDs is only about 200–300 meV at maximum. This corresponds to about $10k_BT$. A localization energy of about $100k_BT$ is required for data storage applications.

Hence, single-electron devices operating at room temperature can hardly be constructed on the basis of individual QDs because of the string thermal evaporation of carriers. Systems based on ensembles of a moderate number say, 10 to 100 coupled QDs are more realistic, so that the probability of simultaneous escape of electrons is negligible.

The impact of electronic coupling of dots on the Coulomb blockade of an array of QDs is analyzed at three zero-temperature phases depending on the strength of inter-dot tunneling which are

(a) Coulomb blockade of individual dots (regime of small coupling),
(b) collective Coulomb blockade (intermediate coupling, where at least separate minibands exist),
(c) breakthrough of Coulomb blockade (strong coupling regime).

The collective Coulomb blockade is used to construct logic elements (operating at room temperature). Let a set of $\mathcal{N}$ electronically coupled dots be capacitively coupled to a gate, and the charging energy of every dot be of the order of thermal energy, $e^2/(2C) \sim k_BT$. Let the voltage be such that, at $T = 0$, dots are not charged. Then at room temperature the number of dots changed due top thermal fluctuations will be about $\sqrt{\mathcal{N}}$. In order to have all or nearly all dots charged by one electron, voltage which overcomes the Coulomb barrier must be applied.

Now let the voltage be such that, at $T = 0$, every dot contains one electron. Then the charge of ensemble of QDs fluctuates mainly within the interval from $(\mathcal{N} - \sqrt{\mathcal{N}})e$ to $(\mathcal{N} + \sqrt{\mathcal{N}})e$. With a sufficient number of dots, say a few tens of dots, in the ensemble, it is possible to ensure distinct switches between logical 0 and logical 1 at room temperature.

(V) SINGLE ELECTRON TRANSISTOR (SET)

Modifying the circuit of a single electron device by splitting its electrode into two parts a very important device, the SET Fig. 3.45 is obtained. The single electron transistor (SET) is an example of an electronic device where a final limit of electronics has already been reached: the switching of a current carried by just one electron. This device is clearly reminiscent of an FET, but with a small conducting island connecting source and drain.

Operating Principle

The SET consists of two tunnel contacts with associated capacitances (C_s and C_d) and an intermediate island to which an operation voltage (V_g) is capacitively coupled via C_g Fig. 3.46. By varying V_g, the electrical potential of the island is changed. An insulator is embedded

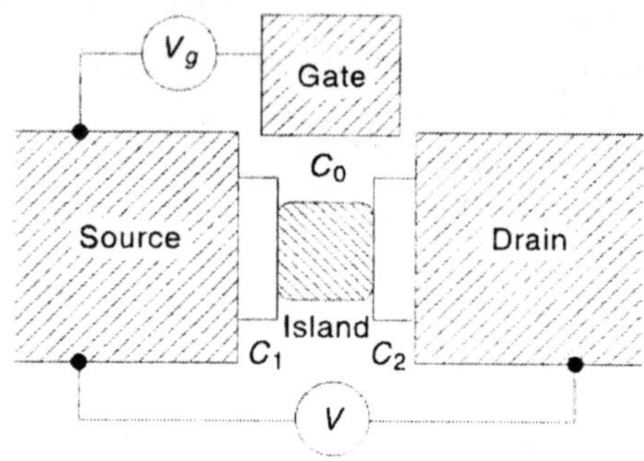

Fig. 3.45

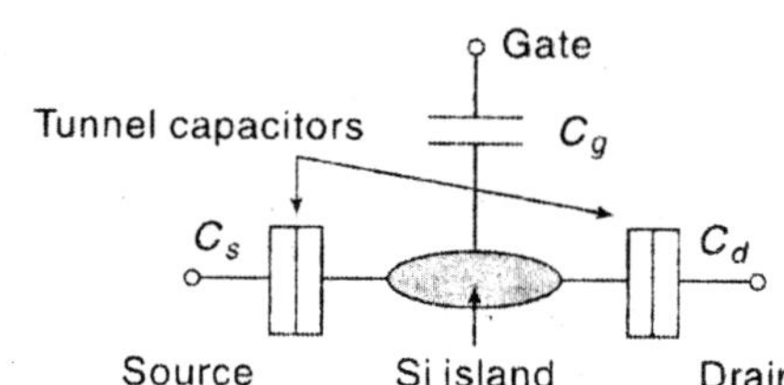

Fig. 3.46 *Schematic representation of the double barrier structure of a single electron transistor*

between two electrical conductors or semiconductors partly loses it insulating characteristics for a layer thickness which is in the lower nanometer range about (5 nm) due to charge carrier tunneling. Thus, the transmission of charge carriers increases exponentially with decreasing thickness of the insulator layer. This quantum mechanical effect limits further miniaturization of the MOS transistor due to increasing gate oxide leakage, thus, leading to stability problems due to the charge carrier capture within the gate oxide.

Also, this tunneling current enables the operation of the single electron transistor. Besides the capacitances C_s and C_d, the electrical conductivities $1/R_s$ and $1/R_d$ are associated with the two tunnel contacts due to the tunneling current. So a current flows between drain and source if an external voltage V_d is applied between these contacts. Assume that there are already n electrons on the island. One further electron can tunnel though the left barrier, if it has a charge energy of

$$(n + 1)\frac{e^2}{2(C_s + C_d + C_g)}$$

At the same time, this electron gains energy (ΔE) by its transfer from the left electrode to the island due to energy difference between these two points:

$$\Delta E = e\frac{V_d C_s + V_g C_g}{C_s + C_d + C_g}$$

Therefore, the electron will arrive on the island if the total energy difference is negative. In the opposite case, a coulomb blockade is given which can be overcome by a further increase of the applied voltage V_d. A possible implementation of a SET is shown in Fig. 3.47. The basic element is an ultrathin silicon-on-insulator (SOI) film. Thickness modulation due to anisotropic etching of the center part of the SOI film results in the formation of a series of quantum dots. The gate contact is formed y polysilicon deposition on the upper oxide layer. The resulting current-voltage characteristics of such a device are presented in Fig. 3.48. As can be seen characteristics levels are formed with constant gate voltage and varying V_d (Fig. 3.48(a)), while the SET current oscillates at a constant voltage V_d and a varying gate voltage (Fig. 3.48(b). The oscillation is due to the fact that, tunneling from the right contact, to island during increasing

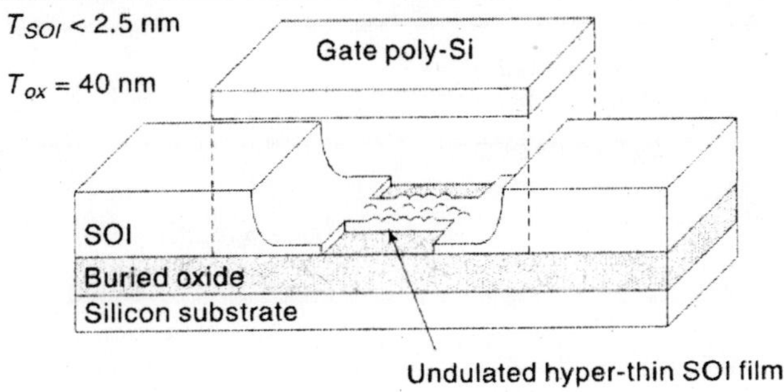

Fig. 3.47 *SOI technology based single electron transistor*

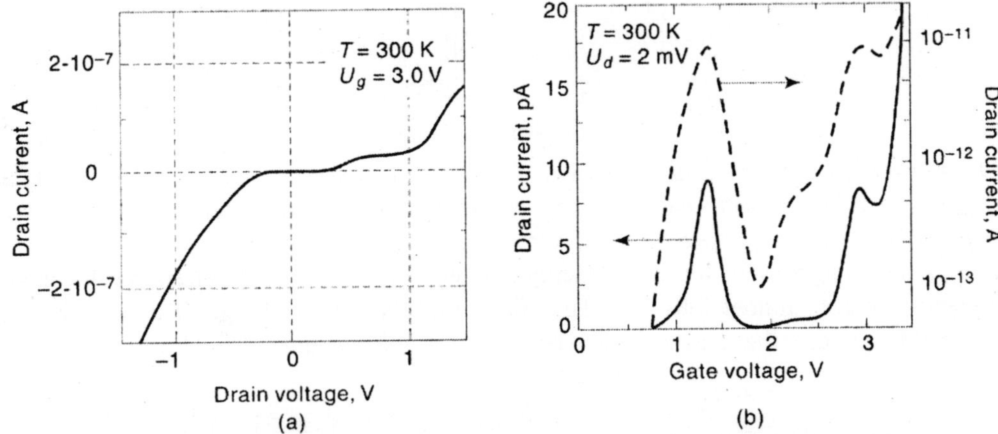

Fig. 3.48 *Current of a SET measured at room temperature (a) as a function of the applied drain-source voltage and (b) as a function of the applied gate-source voltage*

gate voltage is possible and that the transistor is again switched into the blocking state. This process is repeated by further increasing the gate voltage and leads to the observed oscillations of the electrical current.

The plots presented in Fig. 3.48 have been taken at room temperature. As a condition for the operation of the SET, the charge energy

$$E_C = \frac{e^2}{2(C_s + C_d + C_g)}$$

is large than the thermal energy kT. This means that the maximum operating temperature decreases linearly with increasing device capacitance. While a capacitance of about 1 aF is tolerable at room temperature, this value rises to about 60 aF at 4.2 K. i.e., the smaller the device dimensions or electrical capacitances, the higher the maximum allowed operating temperature.

Figure 3.49 shows a typical set of dc *I-V* curves of such transistor, for several values of the "external charge" Q_0, defined in the same way as in the single-electron box. At small drain-to-source voltage V, there is virtually no current, besides the special values of Q_0 even if $V > 0$,

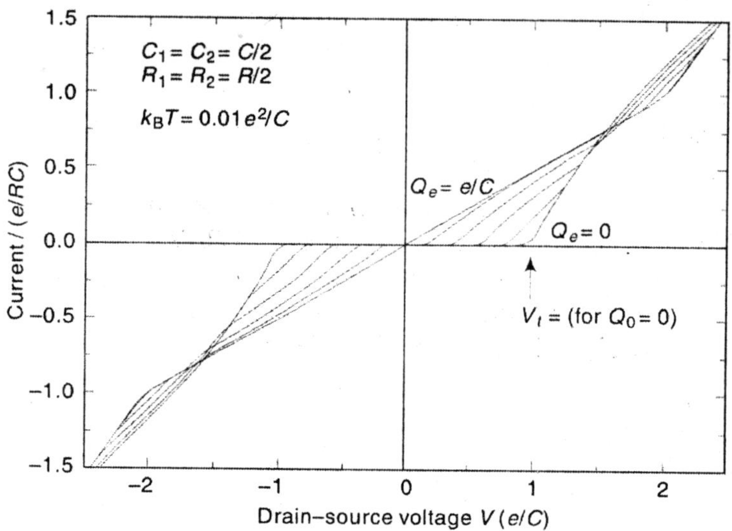

Fig. 3.49 *A typical set of source-drain I-V curves of a symmetric transistor for several values of Q_0, that is, of the gate voltage V_g calculated using the "orthodox" theory of single-electron tunneling.*

and thus it is advantageous in terms of energy for an electron to go from source to drain, on its way the electron has to tunnel into the island first, and change its charge Q it by $\Delta Q = -e$. Such charging would increase the electrostatic energy W of the system

$$W = (Q_0 - ne)^2/2C_\Sigma - eV(n_1C_2 + n_2C_1)/C_\Sigma + \text{const},$$

$$C_\Sigma \equiv C_0 + C_1 + C_2,$$

(where n_1 and n_2 are the number of electrons passed through the tunnel barriers 1 and 2, respectively, so that $n = n_1 - n_2$), and hence at low enough temperatures ($k_BT \approx E_C$) the tunneling rate is exponentially low.

At a certain threshold voltage V_t, currents starts to grow with V. The most important property of the SET is that V_t is a periodic function of V_g, vanishing in special values of gate voltage, the reason for these oscillations is that one electron maybe transferred to the island form either drain or source without changing the electrostatic energy of the system even at $V = 0$. Hence, an electron tunnels form the source of the island and then to the drain even at negligible V, so that $V_t = 0$. At low temperatures the dependence of V_t on V_g is piecewise-linear, with its lower and upper branches forming the so-called "diamond diagram" (Fig. 3.50).

Since the *I-V* curves of the transistor are continuous [Fig. 3.49], if a small current is fixed by an external circuit (say, by a large resistor in series with the SET), V is close to V_t and also follows the diamond diagram, for example, Fig. 3.50(b). Thus, the voltage gain and transconductance of SETs may changing sign depending on the gate voltage—an important difference in comparison with usual FETs. This property is very convenient for the design of SET analogs of CMOIS circuits Fig. 3.51 shows, SETs of only one type.

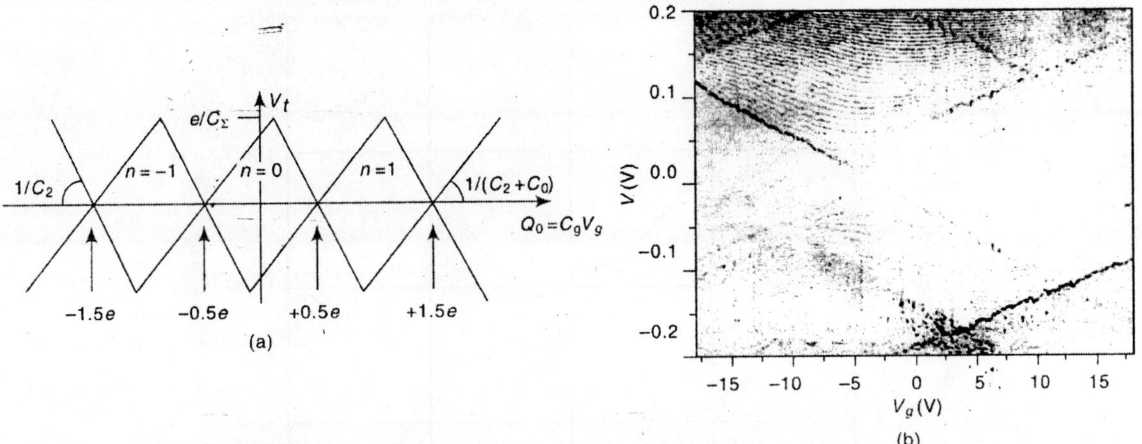

Fig. 3.50 *SET: (a) theoretical dependence of the Coulomb blockade threshold voltage V_t as a function of Q_0 at $T \to 0$, and (b) experimental contour plots of current on the $[V, V_g]$ plane for an aluminum SET with $e/C_\Sigma \approx 0.1V$ at $T = 4.2$ K.*

On the other hand, the same diamond diagram shows that the voltage gain is limited by a capacitance ratio: $(G_V)_{max} = C_0/C_2$.

It may be larger than unity but can hardly higher than ~3 in room temperature transistors.

Figure 3.52 shows the energy E_a necessary to put an additional electron on a transistor island of a certain radius, calculated within the framework of a simple model. This energy is crudely a sum of the electrostatic contribution $E_C = e^2/C$, where C is the island capacitance, and quantum confinement energy E_k. While E_C dominates for relatively large islands, for 1-nm-scale islands with size comparable with the electron de Brogile wavelength, E_k becomes substantial. For operation of most digital single-electron devices the single electron addition energy should be at least 100 times larger than k_BT. This means that for room temperature operation, E_a should be as large as ~3 eV. Accordingly to Fig. 3.52, this value corresponds to island size about 0.7 nm, although insulating materials with lower dielectric constant may relax their requirement by a factor of 2 or so.

Thus, fabrication of room temperature SETs which has only been partly achieve, despite the fact that single-electron tunneling is achieved at

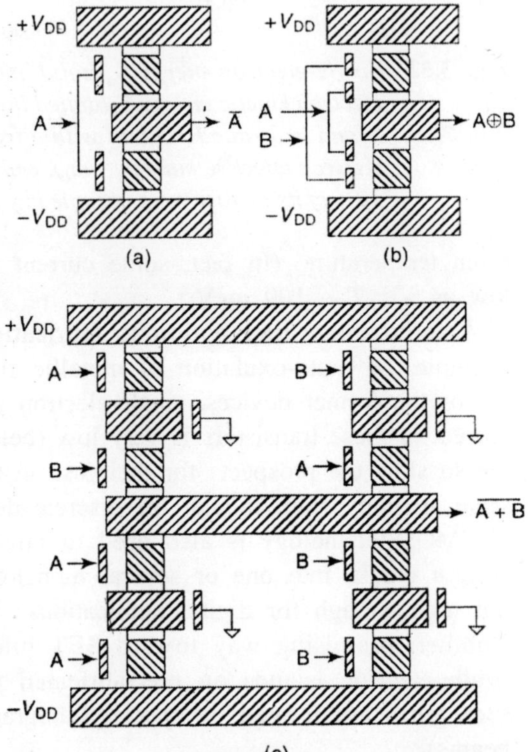

Fig. 3.51 *Basic gates of the complementary SET logic family using capacitively coupled SETs (a) inverter, (b) COR and (c) NOR/NAND.*

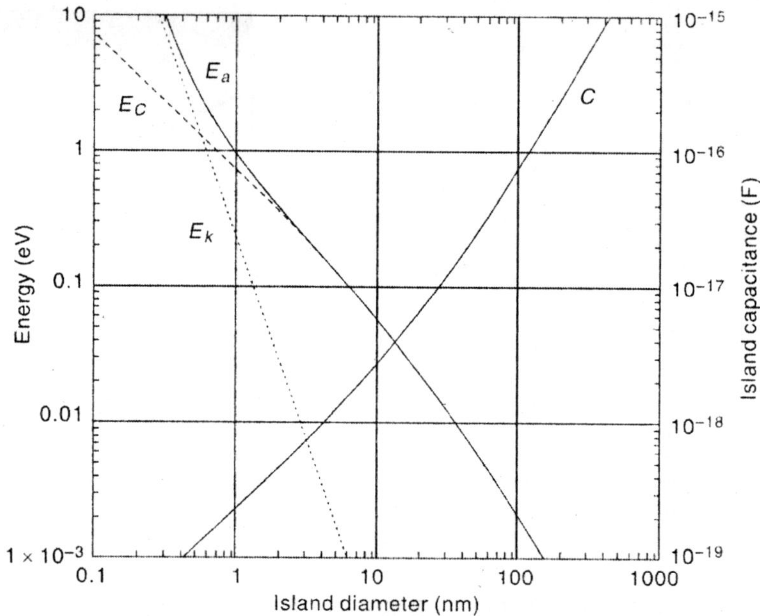

Fig. 3.52 *Single-electron energy E_a (solid line), and its components: charging energy E_C(dashed line) and electron kinetic energy E_k(dotted line), for a simple model of conducting island., In this model the island is a round 3D ball with a free, degenerate electron gas (electron density $n = 10^{22}$ cm ^{3}, electron effective mass $m = m_0$), embedded into a dielectric matrix (dielectric constant $\varepsilon = 4$), with 10% of its surface area occupied by tunnel junctions with a barrier thickness $d = 2$ nm.*

room temperature. (In fact, some current modulation by gate voltage becomes visible at E_a as low as ~$3k_BT$ ~ 100 meV.)

A more interesting option is fabrication of discrete transistors with scanning probes, for example, by nano-oxidation of metallic films or manipulation with carbon nanotubes.

For the former devices, single-electron addition energies as high as 1 V is obtained, but, the current in these transistors is very low (below 10^{-11} A). Moreover, the scanning probe techniques are so slow the prospects for their use in the fabrication of circuits of any noticeable integration scale are very slim, though for discrete devices this approach is promising.

CMOS technology is also used to fabricate SETs, mostly by etching a thin silicon channel until it breaks into one or several tunnel-coupled islands. The highest values of E_a reached are not high enough for digital applications. The necessary values E_a ~ 3 eV is still not reached. Another, promising way toward SET integrated circuits is chemically directed self-assembly single-electron islands on pre-fabricated templates of metallic nanowires. (Periodic arrays of such wires are formed by special lithographic methods, example, interference patterns of EUV beams).

The resistance R of a tunnel barrier shows that the output resistance of these transistors is high, (above ~100 kΩ). This is a drawback of SETs. However, what is really important for applications (e.g., interconnect recharging speed) is the ON current density per unit width. For a room temperature SET with $V \sim V_t \sim E_a/e \sim 1$ V, the total ON current would be below

10 μA. However, since width of such transistor may be as small as a few nanometers, the available current density may be above 1000 μA/μm, that is, even higher than that of standard silicon MOSFETs.

Figure 3.53 illustrates sensitivity to electric charge of the SET. Let a single charged impurity be trapped in the insulting environment, say on the substrate surface, at a distance r from the island of size a. The impurity polarizes the island, creates on its surface a polarization charge of the order of *e*. This charge is subtracted from the external charge Q_0 which determines threshold V_t—see Fig. 3.50. For $r \sim a$, this is large, of the order of $(V_t)_{max}$, even for a simple impurity, 10^9 cm^{-2}, for the maximum concentration of charged impurities, and assuming 1-nm island size, at least 10^{-3} part of the chip area is ineffective, so that the same fraction of single-electron devices have an unacceptably large background charge fluctuation, $\delta Q_0 \geq 0.1e$.

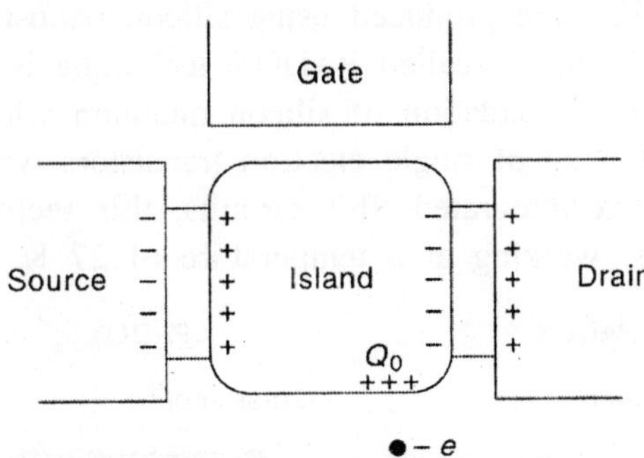

Fig. 3.53 *Random background charge effect: a single charged impurity induces on a single-electron island a polarization charge of magnitude $Q_0 < e$, which depends on the impurity position. This bound charge drops off the valence of tunnel junction charges and thus shifts the Coulomb blockade threshold by $\delta V_t \sim (Q_0/e)(V_t)_{max}$*

For the fabrication of single electron transistors some structures, implemented are specified in Table 3.2. The island and barrier materials are named for the respective manufacturing methods. Also, energy E_a is listed in this table. The maximum allowed operating temperature (*T*) can be estimated from the relation;

Table 3.2 *Data on selected single electron transistor technologies*

Materials (Island; barrier)	*Fabrication methods*	E_a, *meV*
Al; AlO_x	Evaporation by means of an e-beam produced mask	23
CdSe; Organic	Nanocrystal bond to structured gold electrodes	60
Al; AlO_x	Evaporation on a structured Si_3N_4 membrane	92
Ti; Si	Metal evaporation on a structured Si substrate	120
Carbonate molecule	E-beam structured thin layer gate + STM electrode	130
Si; SiO_2	E-beam structuring + oxidation on SIMOX layer	150
Nb; NbO_x	Anodic oxidation by means of STM	1000

$$kT < \frac{E_a}{10}$$

with k being the Boltzmann constant.

It is observed from Table 3.2 that mostly metal/metal oxide and semiconductor/insulator systems are used for the fabrication of SETs. The growing use of organic layers and molecules is also remarkable. Regarding SET manufacturing, evaporation techniques using electron-beam structured masks and structuring with the scanning tunneling microscope (STM) are used. Single electron transistors are fabricated using superconducting Nb/Al structures with AlO_x barriers self-organized growth of GaN quantum dots or Co nanoparticles. The most advanced SETs, are devices manufactured by microstructuration techniques and based on silicon or III-V semiconductor technology. Coulomb blockade oscillations are observed in multi-gate SET structures using GaAs/InGaAs/AlGaAs or AlGaAs/GaAs/AlGaAs as semiconductor. Integrated structures of more than one SET are produced using silicon nanostructures. As an example, the production of a SET employing the so-called PADOX technique is illustrated in Fig. 3.54. This technique is based on the thermal oxidation of silicon quantum wires with a trench structure. It has enabled the fabrication of pairs of single electron transistors with good reproducibility. As a first step toward more complex integrated SET circuits, this technology has also enabled the production of simple inverters, working at a temperature of 27 K.

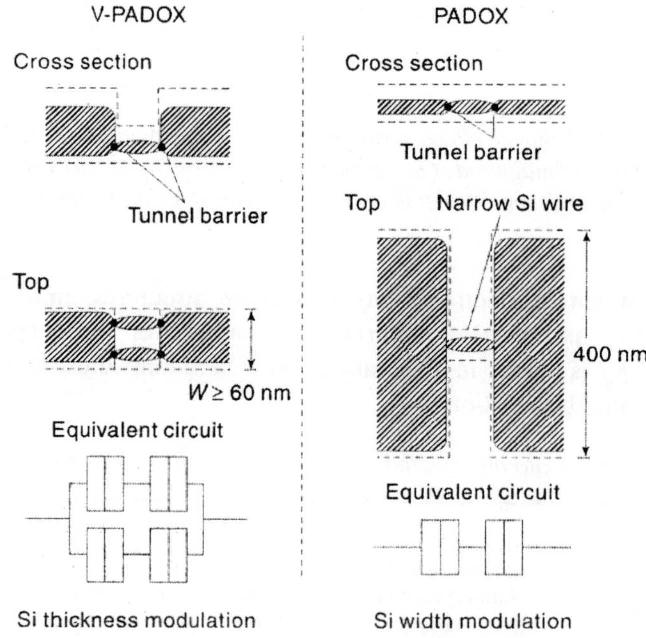

Fig. 3.54 *Principle of SET fabrication based on PADOX and V-PADOX technology*

Semiconductor single electron FETs based on the Coulomb blockade have small size, low power consumption, and unique functionality. A silicon-based FETs shown in Fig. 3.54 relies on a SIMOX channel that is etched followed by an oxidization and etching process to reduce the dimension even further. The electrons tunnel from the source to the drain provides the

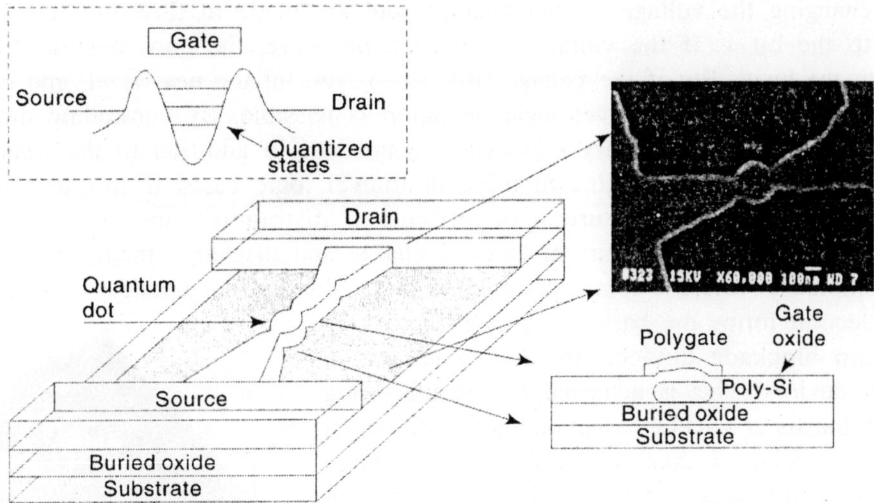

Fig. 3.55 *Single electron transistor (SEL) relying on a SIMON wafer*

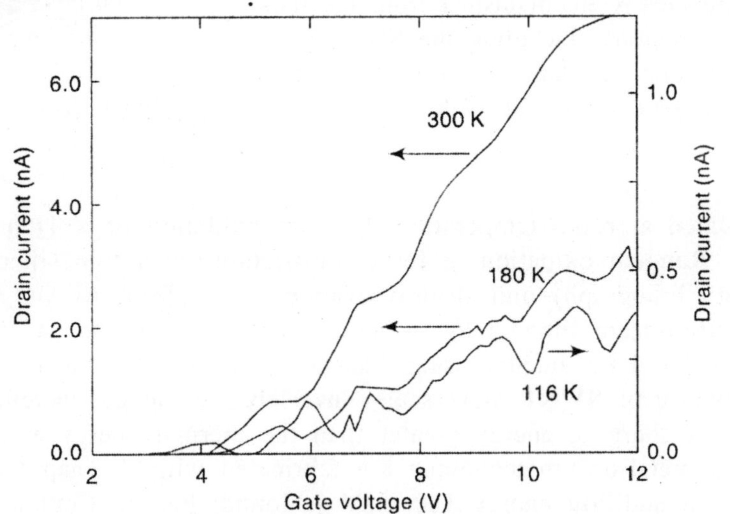

Fig. 3.56 *Drain current in a single electron transistor with features caused by quantum blockade.*

electron energy level in the access arm which resonates with one of the levels in the dot itself. The states under the dot are changed with gate bias and the aforementioned resonance is achieved, the electrons tunnel only one at a time due to Coulomb blockade into the dot under the gate. These Coulomb blockade oscillations give rise to discrete current spikes in the source or drain current versus the gate voltage. The smaller the dot dimension, the larger the separation of the quantum states and higher the temperatures at which they are observed. Thus smearing due to thermal energy is not as detrimental as it is in large dots. This allows the observation of features associated with single electron transport in the drain current at room temperature as shown in Fig. 3.56. Hence, the size of the quantum dots has to be smaller than 15 nm in order for the devices to show distinct resonances at room temperature. Progress is made in this front in that single-electron addition energies up to 1eV is achieved.

Thus, by changing the voltage on the quantum dot are made to flow through the dot. This corresponds to the bit 1. If the voltage is raised a bit more, the current stops flowing. This corresponds to the bit 0. But if the voltage rises again, you hit the next level, and current flows once more. In this way, a multilevel logic operation is possible. By controlling the voltage on the quantum dot, a series of binary messages are created. In addition to the issue of current touched upon a little, noise margins in these multilevel logic cases is low and with thermal smearing due to device temperature it is difficult to distinguish ones from zeros. So, the capacitances and need for large current level to charge and discharge them, are eliminated, by adopting a completely different circuit design, and hence the world's smallest transistor consisting of a single electron forms the basis for computers.

The Column blockade devices are also extended to hole transport devices. This is achieved by manufacturing a single hole transistor (SHT) based on the modification of a hydrogen terminated diamond surface by means of an atomic force microscope (AFM). An AFM image of such a transistor is depicted in Fig. 3.57. The bright local oxidized regions are clearly distinguished from the dark, hydrogen terminated regions (including the SHT island). This SHT shows Coulomb oscillations of the drain current with variations of the gate voltage at 77 K.

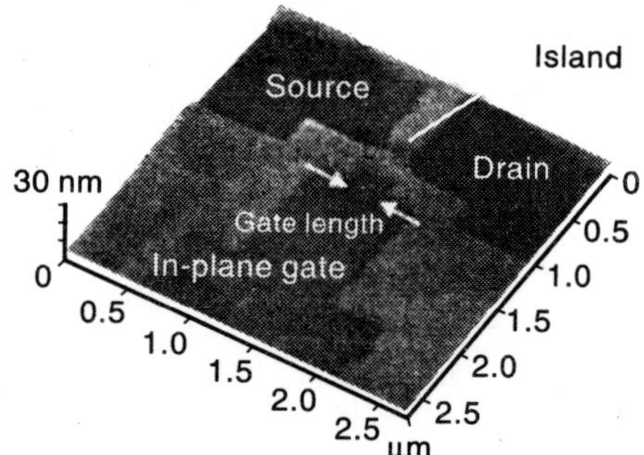

Fig. 3.57 *Atomic force microscope image of a diamond single hole transistor*

Fabrication

SETs are fabricated at room temperature, by nano-oxidation of a Ti metal film using a scanned-probe tip. Random oxidation to form constrictions in a thin silicon wire and even using electron-beam lithography and shadow evaporation to form $Al/AlO_x/Al$ structures; by modifying the electron-beam lithography to achieve a fine resolution. A charging energy of over 150 meV is achieved for an SET using a single C_{60} molecule as an island.

Since the capacitance of SWNT SETs scales inversely with length therefore lengths shorter than ~50 nm have a charging energy greater than the thermal energy at room temperature. SWNT SETs via a metal-on-top technique are fabricated with the gap between electrodes. Although an electron addition energy 100 meV is found, but no Coulomb oscillations are observed at room temperature, due to thermal activation of electrons over the tunnel barrier being formed at the nanotube-electrode interface.

That tunnel junctions are produced in metallic SWNTs through AFM manipulation while producing a kink pairs of tunnel junctions are produced in SWNTs to form SWTs with island sizes less than 100 nm. These small islands act as SETs. One island with length ~50nm had an addition energy of ~70 meV, and showed Coulomb oscillations up to a temperature of about 165 K. By using the double-kink technique a device with a length of ~25 nm, is produced which show Coulomb oscillations at room temperature (see Fig. 3.58). This device has an addition energy of ~120 meV.

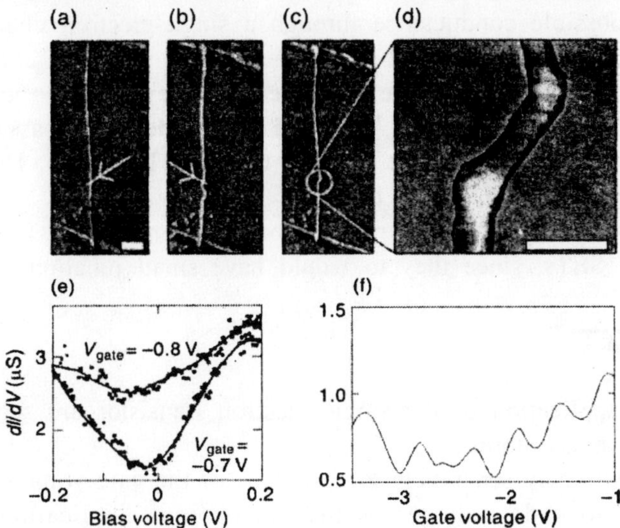

Fig. 3.58 *Room temperature nanotube SET. An AFM tip was used to modify a nanotube device by dragging over the device along the lines shown in (a) and (b). This formed two kinks in the nanotube shown in (c) and (d). The scale bar in (a)–(c) is 200 nm, and the scale bar in (d) is 20 nm. The differential conductance as a function of bias voltage (e) at 3980 K shows significant variation with gate voltage. The differential conductance as a function of gate voltage (f) at 260 K shows periodic oscillations due to Coulomb blockade.*

The AFM manipulation technique used to fabricate double-kink SWNT SETs is uncontrollable, and difficult to envision as a method of mass production. However, these devices show in general some of the advantages of making SETs from SWNTs, which in principle may be generalized to other fabrication techniques.

The reason that scaling of SWNT SETs to small lengths produced high charging energy devices is that the capacitance is dominated by the length of the nanotube, not the area of the contacts. In SETs fabricated by, for example, the $Al/AlO_x/Al$ process the total capacitance is typically dominated by the capacitance of the tunnel junctions. In order to produce smaller capacitance devices, one must make smaller area tunnel junctions.

Fabricating small tunnel junctions by brute-force lithography is extremely difficult, as the resistance of the tunnel junction increases as the area decreases. With SWNT SETs, a small-area tunnel junction is self-assembled; the area of the SWNT controls the area of the junction. Lithography is thus used for long dimension of the SET, of the order of 25 nm for room temperature operation.

This advantage is seen in the ratio of gate capacitance to total capacitance C_g/C_ε for SWNT SETs. Both double-kink devices discussed above have ratios of C_g/C_ε of around 0.3, while the room temperature SET has a ratio C_g/C_ε of about 3×10^{-3}, two orders of magnitude smaller. This ratio determines the maximum transconducatance of an SET. The transconductance is the change in current through the device divided by the charge in gate voltage. The maximum current modulation will be produced if the device operates at a source-drain bias $V_{sd} \approx E_c =$

$e/2C_\varepsilon$. The maximum possible conductance through a single-electron charging device is $G_{max} \approx e^2/h$. Thus, the maximum possible current is $I_{max} = G_{max} V_{sd} = e^3/2hC_\varepsilon$. The current varies from minimum to maximum over a change in gate voltage $\Delta V_g/2 = e/2C_g$. Therefore, the transconductance is $2I_{max}/\Delta V_g \approx (e^2/h)(C_g/C_\varepsilon)$. The simple derivation assumes that the device is classical, that is $\Delta E << k_B T$, which is not true for the SWNT devices. However, incrreasing C_g/C_ε increases the transconductance of the device.

Hence, not only would SWNT islands make good SETs, but metal islands with SWNT leads would also make good SETs, since they to would have small junction capacitances.

APPLICATION OF SET

The main fields of application of the single electron transistor are sensor technology, digital electronic circuits, and mass storage.

As the SET reacts extremely sensitively to variations of the gate voltage V_g if the voltage V_d is adjusted as Coulomb blockade voltage, so that an obvious application is a highly sensitive electrometer.

The Coulomb blockade is only effective if the thermal energy is lower than the charge energy of the island. Therefore, the differential electrical conductivity within this area is also a measure for the ambient temperature and enables the use of the SET as a temperature probe, particularly in the range of very low temperatures.

Moreover, the SET is a suitable measurement setup for single electron spectroscopy. For this purpose, the island of the SET structure, for instance, can be taken as individual quantum point.

Apart from the applications as sensors, further applications as direct current normal is interesting. Since exactly one electron is transported in each period when applying an alternating voltage of suitable amplitude to the gate of the SET, the current between source and drain which flows through a single electron transistor is directly proportional to the frequency of the applied alternating voltage V_g. Since frequencies can be measured with high accuracy, a precise direct current measurement standard can thus be implemented.

The applications of the SET as a switch and memory in digital electronics, operating at room temperature, is of great interest. But, the operation of a single electron transistors at room temperatures requires extremely small island capacities of about 1 aF and structure widths around 1 nm. This is achieved in single structures. The operation of a SET at room temperature and at higher frequencies is difficult. Very low device capacitances are obtained with small dimensions, but the electrical resistances associated with the tunneling barriers increase and reach values in the MΩ range. As a result, the critical frequencies of the SET are limited by relatively large RC constants. Also, a variety of digital logic functions, including AND and NOR gates, are obtained with the SOI technology based single electron transistor operating at room temperature (See 3.46.)

For, the potential use of the SET as electronic mass memory, the number electrons on a neighbouring conducting island containing the stored information is achieved by means of a SET. Memory densities around 10^{12} bit/cm^2 is reached, which is higher than the values achieved by MOS memories.

Two different concepts are used for the SET for data storage;

(1) SET-FET hybrid approach—up to 100 SET based memory cells are read out by a field transistor (FET) based amplifier. This type of memory requires—a refreshing of the memory content after each reading. This approach gives memory densities up to Gbit/cm^2 at room temperature. The recording procedure, uses tunneling, with a typical delay time of approximately 10 ns, which is relatively slow. An illustration of the SET-FET hybrid concept is shown in Fig. 3.59. It is interesting to note that this type of memory is not sensitive to background charges and requires structure lengths of about 3 nm.

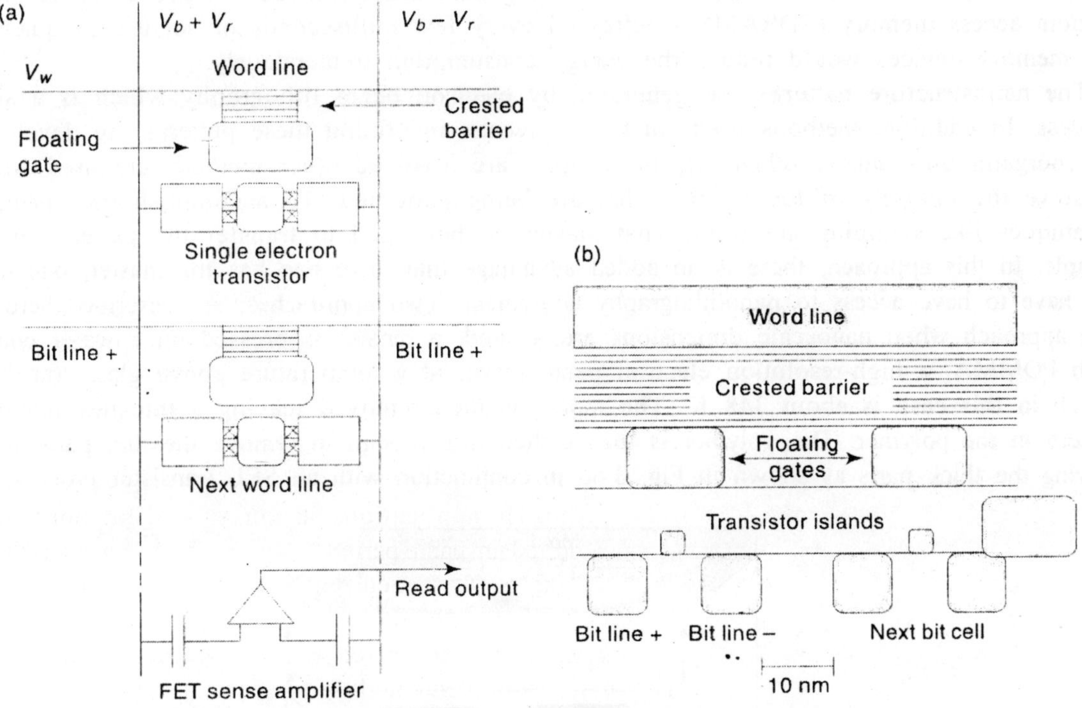

Fig. 3.59 *(a) Principle and (b) implementation of a SET-FET hybrid circuit*

(2) The magnetics recording technique is replaced by electrostatic storage (ESTOR). The information is kept in loaded grains in a memory layer separated by means of tunnel barriers and a metallic layer form an insulating substrate. The process of writing and reading takes place by using a head floating about 30 m over surface with the island of the SET at the top. Memory densities of approximately 1 Tbit/inch^2 are achieved. Which is about 30 times higher than the theoretical limit calculated for magnetic memories.

The SET is also a good choice for multi-value logical circuits due to the periodical oscillations in the current-voltage characteristics, example is, the combination of a SET, fabricated with PADOX process, with a conventional MOSFET that results in a completed current-voltage characteristics, with multiple hyteresis. Also, silicon based SETs are very attractive for digital applications as they are easily interfaced to conventional electronics. An example of an analog applications of the single electron transistor is the fabrication of a radio frequency mixer. It uses the nonlinear gate voltage-drain current characteristics of the SET, for the fabrication of a homodyne receiver operating at frequencies between 10 and 300 MHz.

Single electron transistors are used to scan and map out lateral impurity distribution in semiconductors.

The same concept is used to store charge for storage (memory) applications. When there is charge in the potential well under the gate, the potential measured at the gate is different from the case when there is not. The amount of charge or energy needed to charge the memory is meager compared to present approaches utilizing standard MOSFETs. Since dynamic static random access memory (SDRAM) is refreshed every few milliseconds or so, use of quantum dot memory devices would reduce the energy consumption tremendously.

The nanostructure patterns are generated by electron beam lithography which is a slow process. In addition, methods used, such as solvents, indefining these patterns are developed for inorganic substances. When organic samples are involved other methods are used which preserve the integrity of the samples that are being patterned. To accomplish this, printing techniques like stamping are used. That master is then used to transfer the pattern on the sample. In this approach, there is an added advantage that once one has the master, one does not have to have access to nanolithography equipment. Two approaches, are described here. In one approach when nanoscale dimensions are desired, a master is pressed on a wafer coated with POMMA, a high-resolution electron beam resist, at a temperature above glass transition which in this case is about 348 K. The model is then removed leaving a thickness contrast pattern in the polymer. The polymer is then etched just enough to remove the thin parts while leaving the thick parts as shown in Fig. 3.60 in conjunction with an SEL transistor process.

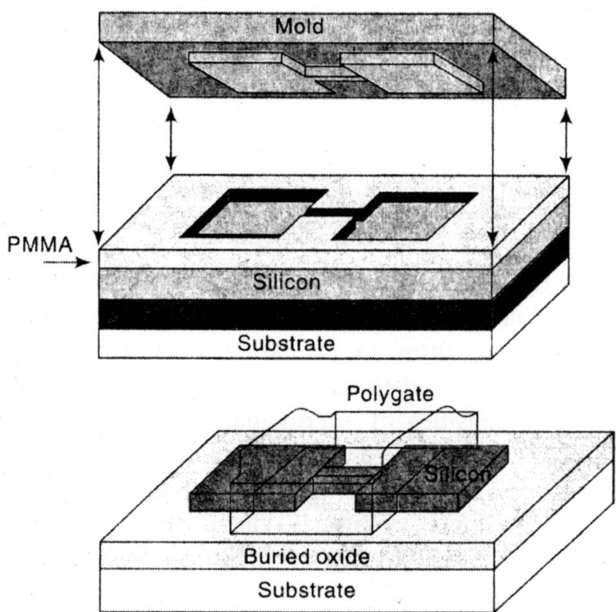

Fig. 3.60 *Imprint technology used for producing nanoscale single electron transistor.*

Similar in concept but designed for biological samples are what is called micro contact printing (MCP) which is compatiable with biological samples. Using microprinting, a diffraction grading of antibody has been pressed on a silicon wafer insignificant diffraction occurs with the antibody alone, but when bacterial cells are immunocaptured, significant diffraction occurs, the details of which are used to detect presence of various toxins. In this approach an elastomer (epoxy) stamp is generated from a Si master, and the stamp is used repeatedly until it is damaged.

Light Emitting Diodes (LEDs)

In optoelectronics, tremendous developments have been made in light emitting diodes (LEDs) on the display and lighting side and lasers on the telecommunications and compact disk side. On the LED side, blue and true green LEDs were not available until the advent of GaN and related heretojunctions. At present, almost the entire range of colors are obtained with use of nitride-based LEDs covering the blue and green colors and GaAlAs or GaInAlp LEDs covering the red end of the spectrum. Organic LEDs (polymers) and organic crystals are used as LED materials, to produce all three primary colors, red, green, and blue. The difficulty has until recently been in obtaining blue with sufficient brightness and longevity. White light is generated by pumping an inorganic crystal, or an organic medium with appropriate dyes, with nitride based blue LEDs.

These emitters are used for indicator lights, background lights, and emergency flashlights. Being very efficient, battery usage is very economical. White light generation By LEDs has become attractive.

Theory

An LED is a semiconductor device which emits visible, infrared (IR), or ultraviolet (UV) radiation due to flow of electric current through it. Essentially, it is a p-n junction device with p- and n-regions made from the same or different semiconductor. The color of the emitted light is determined by the energy of the photons, and in general, this energy is usually approximately equal to the energy bandgap E_g of the semiconductor material in the active region of the LED. III-V semiconductors such as GaAs, GaP, AlGaAs, InGaP, GaAsP, GaAsInP, AlInGaP, etc. are the common constituents of an LED. The materials used for a particular LED, depends on the choice of color, performance, and cost. Typically, a semiconductor chip embodying LED is 250 $\times$ 250 μm^2 which is mounted on one of the electrical leads. The top of the chip is electrically connected to the other leads through a bound wire. The epoxy dome serves as a lens to focus the light and as a structural member to hold the device together. Operating currents at a forward voltage of about 2V are usually in the range of 1–50 mA. Typical LED materials against a wavelength/colour bar are shown in Fig. 4.1.

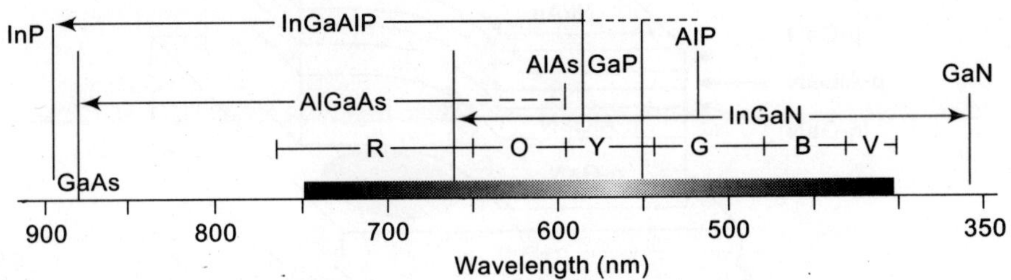

Fig. 4.1 *Typical LED materials against a wavelength/colour bar.*

The external quantum efficiency, is defined is the number of photons per each electron passing through the device. It is roughly equal to the power efficiency of LEDs, and ranges from less than 0.1% to more than 10%.

LEDs are obtained by multiplying the power efficiency by the eye sensitivity curve. The performance of LEDs typically in the rage of 1–10 lmW^{-1}, although performance as high as 20 lm W^{-1} can also be achieved. This is comparable with 10–15 lmW^{-1} performance of an incandescent bulb. LEDs are suitable for room illumination because at lower level, they can operate even at less than 0.1 W. The wider applicability of LEDs is for lighting and indicator lighting in vehicles.

Structure

Structure of an LED is shown in Fig. 4.2. Both GaAsP and GaP doped with Zn or OLEDs exhibited an efficiency of about 0.1 lmW^{-1}. That nitrogen provides an efficient recombination center in both GaP and GaAsP led to the introduction of red, orange, yellow, and green in GaAsP:N and GaP:N LEDs with an improved performance of about 1 lmW^{-1}. It is found that both homostructure AlGaAs and heterostcutre AlGaAs LEDs have advantages over GaAsP and GaP homojunction LEDs. The evolution of visible LEDs is shown schematically in Fig. 4.3. For comparison, Edison's incandescent light bulb and fluorescent light sources are also shown. The luminous efficiency or the wall plug efficiency represents the optical power generated for each watt of electrical power spent. While the liquid phase epitaxy (LPE) is instrumental in developing high-performance LEDs; growing high-quality multilayered device structures and producing large quantities of LED material is not advised. The performance of these LEDs is improved, from 2 to 10 lmW^{-1} depending upon the structure. Thus, LEDs broke the efficiency barrier of filtered incandescent bulbs, enabling them to replace light bulbs in many outdoor display applications. The development of metal-organic vapor phase epitaxy (MOVPE) lead to high-performance AlInGaP LEDs. This material system produced bright LEDs in the red-yellow range. SiC-based LEDs emerged as blue color emitters despite their low performance simply because there was no other alternative. Green and blue LEDs using ZnSe, have shown promise in laboratory and even in production but GaN-based LEDs successfully cover the violet to green, even to amber, part of the spectrum made the SiC and ZnSe based LEDs obsolete.

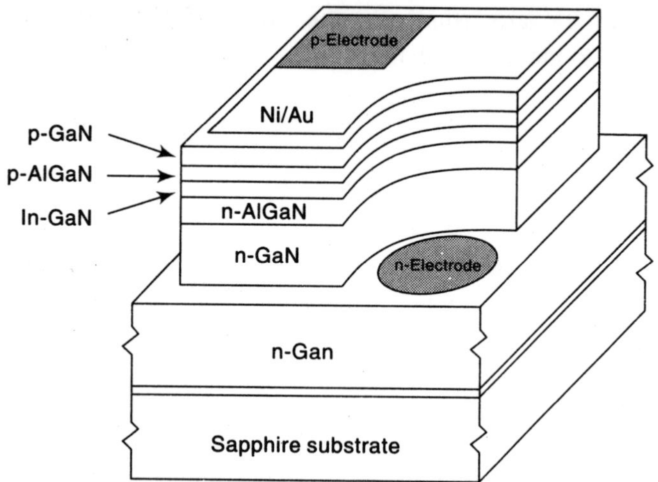

Fig. 4.2 *A schematic structure of an InGaN LED*

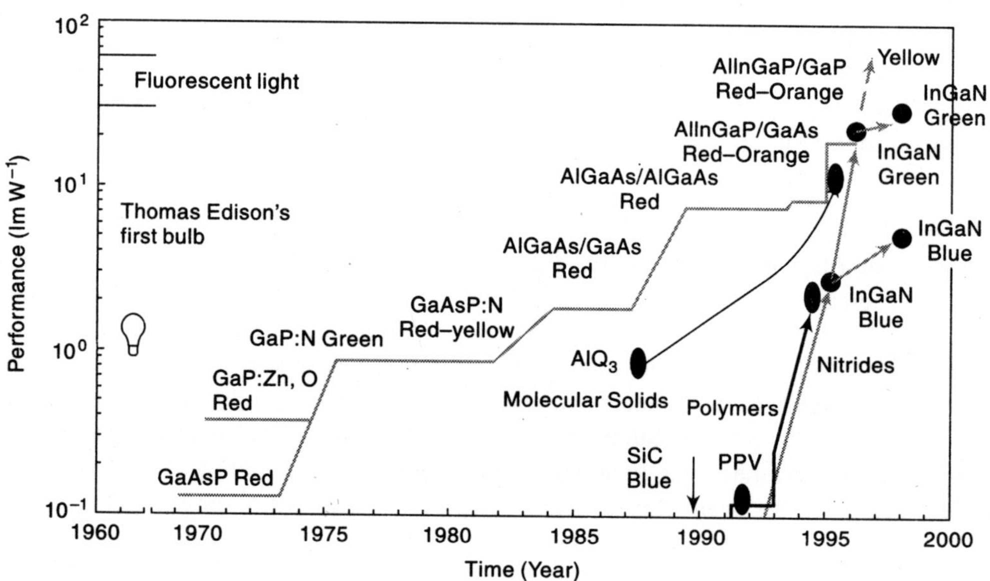

Fig. 4.3 *The evolution of LED OLED.*

LIGHT EMITTING DIODES

Si-Based

The requirement for a light emitting diode (LED) are small size, low power consumption, and reliable ways to connect in limited space optical and electronic devices.

Si is the material of microelectronic industry; has indirect energy bandgap, which leads to very long radiative lifetime. Electroluminescence (EL) from bulk crystalline Si are obtained by biasing a *p-n* junction, either in forward or reverse mode. The power efficiency (the ratio of the emitted optical power to the driving electrical power) is typically about 10^{-4} in forward bias, and about 10^{-8} in reverse bias while in avalanche breakdown regime.

LEDs are "bulk-Si" if line space and peak position of the EL spectra are consistent with the bandgap (1.1 eV) and the density of states of bulk Si.

In Figure 4.4, the emission peak has the energy hv = 1.1 eV (or, equivalently, the wavelength = 1127 nm) at 77 K. At RT, several changes are observed. The first is a lowering of the peak intensity and of the internal quantum efficiency (IQE, ratio of the number photons internally generated versus the number of electron-hole pairs excited). The drop is due to raise of nonradiative recombination rate and exciton ionization. The second change is a shift to lower energy because of thermal bandgap shrinkage. The third effect is the broadening of the luminescence band, as a consequence of the rise in the free carrier kinetic energy. At RT, various spectral features like interband transition, bremstrahlung, intraband transistors, and indirect interband recombination under high-field conditions (hot carriers) are observed. The peaks at hv_3 = 1.038 eV and hv_2 = 1.072 eV observable at 77 K are extrinsic peaks, to the presence of B and As impurities, although the latter might also be due to B. The position of extrinsic peaks range between 0.96 and 1.1 eV depending on doping impurities.

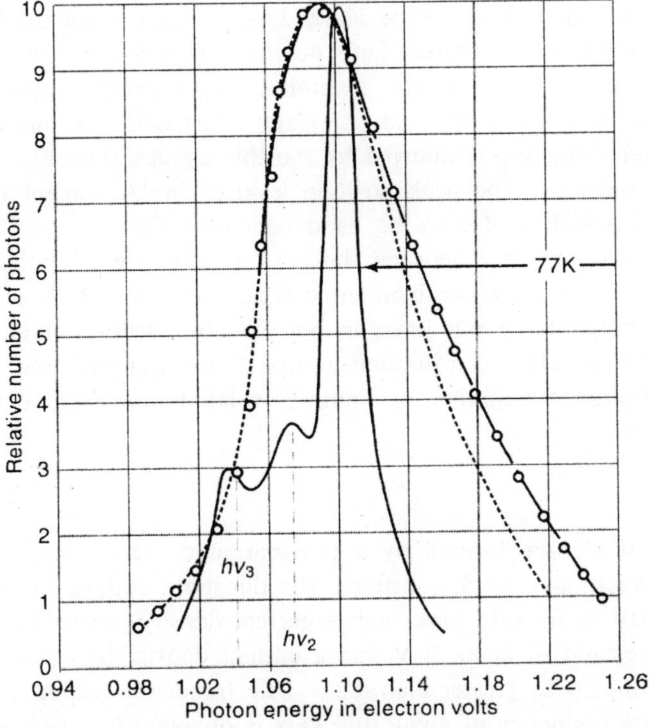

Fig. 4.4 *Electroluminescence spectrum due to recombination of electrons and holes in Si, the dashed line and the line with circle markers show emission at room temperature.*

Efficiency

To enhance LED efficiency in bulk-Si LEDs first increase the non-radiative lifetime, τ_{nr}. In the ideal limit, condition $\tau_{nr} >> \tau_r$, where τ_r is the radiative lifetime, implies $\eta_{\text{int}} \approx 1$. The non-radiative rates are reduced by using (1) high-quality Si substrates, (float-zone being preferred over CZ) (2) passivation of surfaces by high-quality thermal oxide, (to reduce surface recombination) (3) small metal areas, and (4) high doping regions (to reduce the SRH recombination in the junction region). Recombination lifetimes of the order of millisecond have been demonstrated in forward biased *p-n* junctions.

A second strategy to enhance LEDs efficiency is to reduce parasitic absorption of photons once they have been generated. For example, the reabsorption can be minimized by keeping the doping level to moderate value, such as about 10^{16} cm^{-3}.

In third strategy, light emission is enhanced by suitably texturing the Si surface. The effect of a texture on the Si/air interface is to increase the system absorptivity, which is equal to its spectral emissivity. By acting as light trapping elements, textures surfaces arise emissivity by over one order of magnitude, compared to planar surfaces. Textured surfaces are used for Si photovoltaic cells.

Another approach is by using dislocation loops. Field-induced carrier confinement in quasi-2D structures are observed in MOS LED structures. The LEDs are essentially MOS capacitors with thin gate oxide (2.7 nm). Both *n*-type and *p*-type substrates are used. By biasing the LED in inversion mode (in p-type structures, gate positive with respect to the substrate and vice versa in *n*-type) two relevant effects are generated: (1) quasi-2D minority carrier layers are created at the interface SiO_2/substrate, and (2) majority carriers are injected by tunneling from the gate. Nondestructive tunneling is allowed by the thin oxides. The EL line shape is similar to that observed Si *p-n* junctions. The peak position is also similar, except for a small low-energy shift (about 100 meV), which is due to the band bending effect.

Given the EL spectral similarity between the devices they are essentially based on the same mechanism, despite the quasi-2D confinement claimed for the MOS. However, EL at the Si band edge is excited not only in p-n junction but also by carrier tunneling through thin oxide MOS structures. Additionally, diminished temperature dependence is observed in such MOSLEDs, due to limited availability of impurity related nonradiative recombination centers in the quasi-2D region.

Porous SI LED

The LEDs based on PS are formed by a transparent or semitransparent top contact, a PS layer about 1–10 μm thick, and the Si substrate. The threshold voltage for the electroluminescence (EL) is greater than 10V in forward bias, with a current density greater than 10 mA·cm^{-2}, which when reduced to a threshold of about 2 V and a current density less than 1 mA·cm^{-2} improved the efficiency from 0.0001% to greater than 0.2%. The frequency response of a PSLED is shown in Figure 4.5. A cut-off frequency of about 300 kHz is obtained by using a thin layer of PS or a *p-n* porous junction.

The emission spectrum of PS LEDs is broad, due to the size dispersion of the nanocrystals in the porous layer (Figure 4.6). The emission wavelength is tuned in a very wide range by changing the porosity of the PS or by introducing impurities.

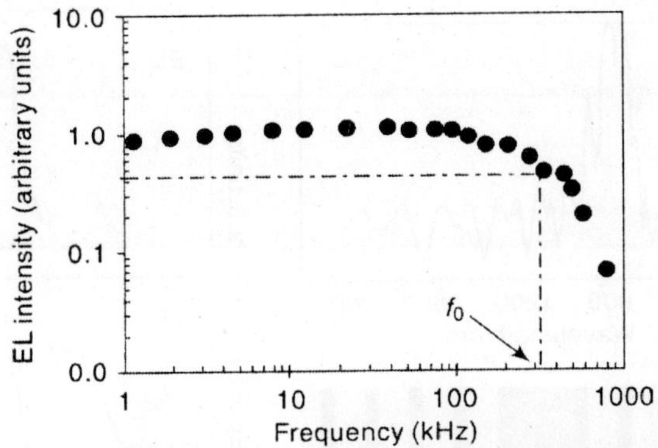

Fig. 4.5 *Small signal amplitude of the electroluminescence as a function of the modulation frequency in a P-N junction with PS. The cut-off frequency to –3 dB is equal to about 300 kHz.*

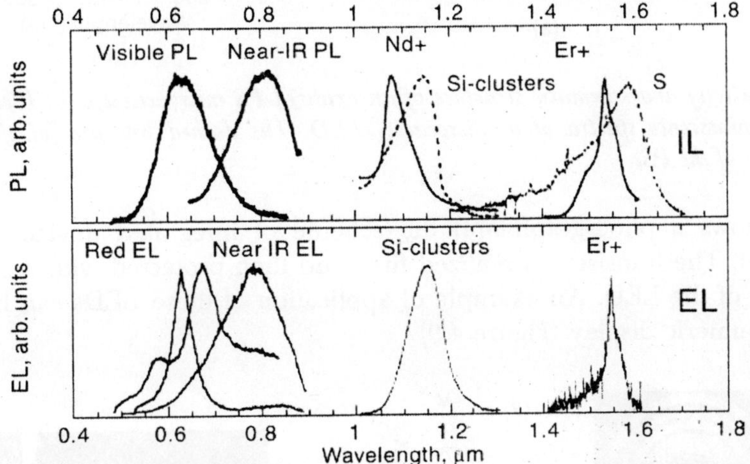

Fig. 4.6 *Photoluiminescence and electroluminescence spectrums of PS, prepared with less than 5 nm sized Si nanocrystals (luminescence in the visible and the infrared), with bigger crystal (>> 10 nm) (Si-clusters),. Or doped with Neodymium (Nd), Erbium, (Er), or Sulphur (S).*

For some applications, there is the need of a higher purity in the emission colours. A narrowing of the luminescence spectrum is obtained by using an integrated microactivity. This is formed by growing two Bragg mirrors around the active layer. Bragg mirrors are fabricated by alternating PS layers of high and low porosity. The width of the electroluminescence spectrum is reduced to 10 meV or less. In Figure 4.7, the reflectivity spectrum of a microactivity and the electroluminescence spectrum of a device are shown.

For a PS LED a microelectronic circuit, the anodization process must be small and it should not interface with the other CMOSD process steps. Various techniques are developed to allow masking of Si and allow opening a window to selectively form PS on small are of 1 μm^2 or less. These include local amorphization through, and the protection, low energy ion bombardment to drive the pores formation, and the protection of the selected areas through a nitride Si and photoresist combination. Using a combination of these techniques, PS based LED integrated with the driving transistors are produced.

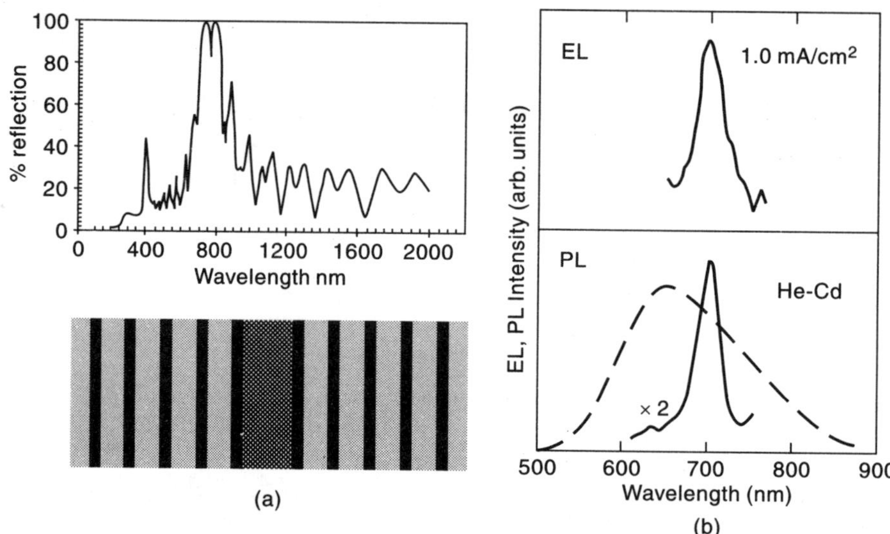

Fig. 4.7 *(a) reflectivity and schematic structure of an oxidized PS microactivity. (b) Electroluminescence and photoluminescence spectra of a microcavity LED. The dashed line are for the photoluminescence spectrum of the PS.*

Figure 4.8 shows a photograph and the scheme of integrated device, together with its equivalent circuit. The transistor is realized first, and then protected with a Si_3N_4 layer before the manufacture of the LED. An example of application of these LEDs can be a simple seven segments alphanumeric display (Figure 4.9)

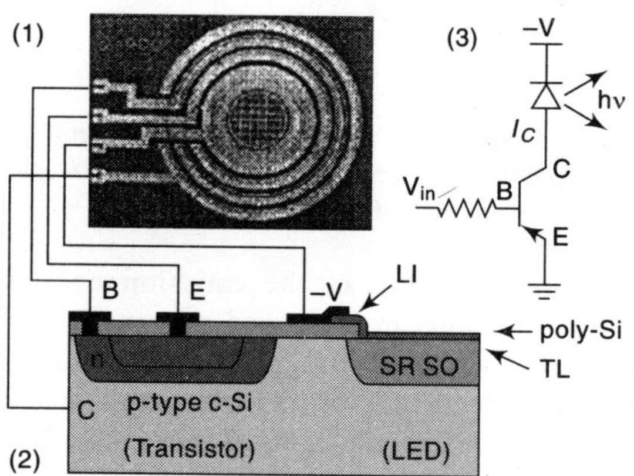

Fig. 4.8 *Integrated structure of a bipolar transistor and a LED in PS. (1) top view of a LED with an active area of 400 μm^2; (2) Cross section view (SRSO it is the oxidized PS light emitter layer. TL is a layer of mesoporous PS, LI is the local interconnection); (3) equivalent circuit.*

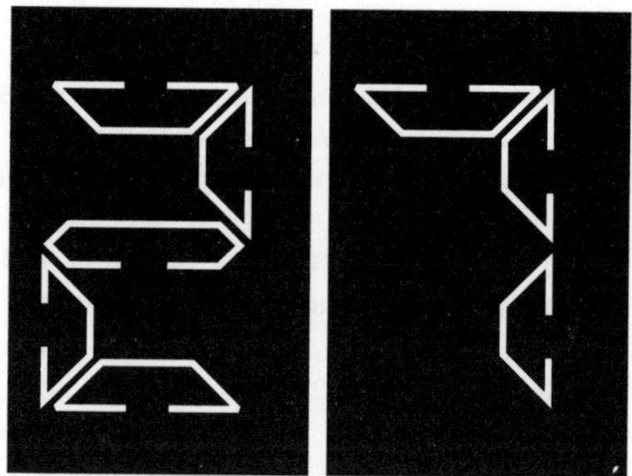

Fig. 4.9 *A simple alphanumeric display. Every seven segment display is made with LEDs in PS contacted in the superior part that emit only on the borders.*

Ge-SiO_2 Blue LEDs

Blue LEDs based on Ge-implanted SiO_2 have been fabricated. The Ge ions are implanted in SiO_2 layers whose thickness ranges from 130 to 500 nm. The peak Ge concentration ranges from 0.3% to 3%, and annealing is performed at 1273 K, for 5 to 30 s. An ITO layer is sputtered on top of the oxide layer which provides a transparent top contact. The LED structure is then coupled to a light detector as shown in Figure 4.10. Main emission is blue which is due to oxygen-deficiency defects. Electrical injection is obtained by tunneling of electrons from the Si substrate to the positive biased ITO electrode. The electrical excitation to light emitting states of Ge clusters is carried out by impact ionization by hot electrons, by direct field ionization of clusters. Power efficiency is 0.5%. An interesting feature of the device is the presence of an integrated detector, which suggests the possibility of introducing the devices in a CMOS line.

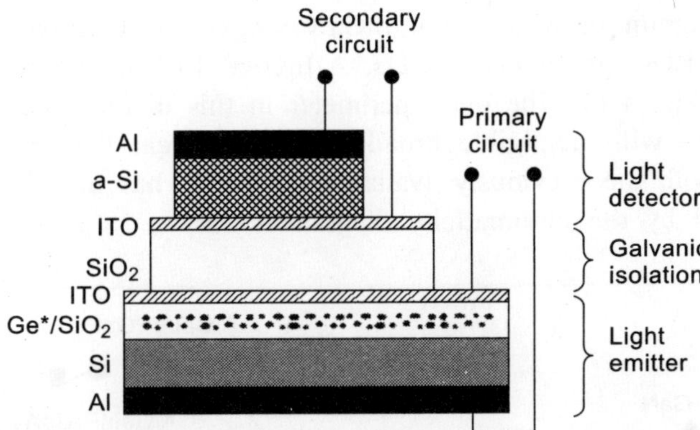

Fig. 4.10 *Schematic structure of an integrated optocoupler consisting of a light emitter based on Ge-implanted SiO_2 layers and a light detector made of amorphous Si.*

GaN and its Alloys for Blue and Blue-Green LEDs

By the advent of nitride semiconductors, LEDs in true green and blue are obtained. The band edge emissions in GaNa at 365 nm, which is in the UV. In order to cause a red shift in the band edge emission to the visible range, impurity assisted transitions and addition of InN to GaN, forming the alloy InGaN, is utilized.

Bright blue and blue-green LEDs are based on the InGaN/AlGaN double heterojunction structure. The solid state lighting (SSL) for first low level, followed by high level lighting is explored. The blue and green InGaN-based LEDs have produced power levels of the order of approximately 5(5.5 mW) and over 30 lmW^{-1} (4.9 mW) at a forward current level of 10 mA. Both color LEDs exhibit external quantum efficiencies of approximately 11%. A recently InGaN amber LED produce about 8 lm W^{-1}.

It relies on large mole fraction of InN and outperforms AlGaInP based LEDs. The spectral quality of the InGaN-based LED (about 30 meB) is not as good as that of the AlGaInP (about 10 meV). Details of the INGaN LED performance are shown in Table 1. It is clear that InGaN-based LEDs outperforms, others in both of the primary colors. The InGaN LED performance improves with time.

Table 4.1 *UV Blue and Green and Performance Fabrication in competing materials performance of Green and Blue LEDs at 20 mA and 298 K. Except II-VI which is at 10 mA*

Color	*Materials*	*Peak wavelength (mm)*	*Luminous Intensity (mcd) unless otherwise stated*	*Output power (μW)*	*External quantum efficiency (%)*
Green	AlInGaP	570	1000	400	1.0
	GaP	555	100	40	0.1
	ZnTeSe	512	4000	13000	5.3
	InGaN, 3.5 V	520		11,000	11.6
Blue	SiC	470	20	20	0.04
	ZnCdSe	489	700	327	1.3
	InGaN, 3.7 V	468		15,000	11.2
UV	FWHM: 8.6 nm	371		5000	7.5

The chromaticity diagram in which blue InGaN singal quantum well LEDs, green InGaN single quantum well LEDs, green GaP LEDs, AlInGAP LEDs, and red GaAlAs LEDs (are indicated) is shown in Fig. 4.11. The outer perimeter in this diagram indicates saturated colors, in other words emission with negligible broadening. It is clear that addition of nitride-based blue and green along with the previously available red LEDs has paved way to obtain 70% of the colors encompassed by the chromaticity diagram.

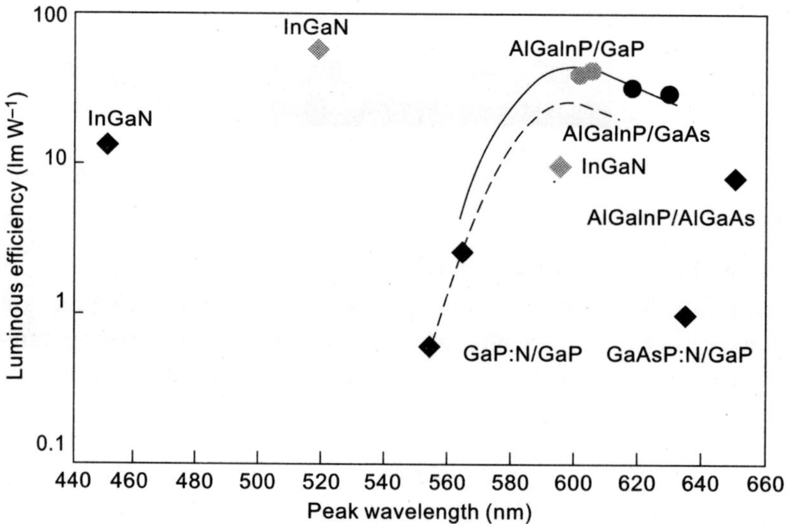

Fig. 4.11 *Luminous efficiency of blue InGaN single quantum well LEDs, green InGaN single quantum well LEDs, green GaPO LEDs, AlInGaP LEDs, and red GaAlAs LEDs.*

Also, green LEDs do not produce the saturated colors due to their relatively wide spectra. Soft white light is obtained by a combination of blue and yellow. This is accomplished by a combination of a blue LED, providing the blue color, and a solid state medium such as a Gd-doped YAG (yttrium aluminium garnet) providing the yellow component. White LEDs are constructed from a blue InGaN LED over-coated with a Ce-doped YAG inorganic phosphor. The InGaN LED generates blue light at a peak wavelength of about 460 nm, which excites the Ce^{3+}: YAG phosphor which emits pale-yellow light. The combination of the transmitted blue light from the LEd and the pale-yellow light from the Ce^{3+}: YAG medium leads to soft white light. It is possible to tune the emission spectrum of the YAG phosphor by substituting some or all the yttrium sites with other rare earths (REs) such as godalinium (Gd), terbium (Tb), etc. The RE^{3+}: YAG emission and absorption spectrum is fine tuned by substituting some or all of the aluminium sites by gallium.

The schematic representation of white length production from the combination of blue and yellow sources is depicted very clearly in the CIE diagram of Fig. 4.12.

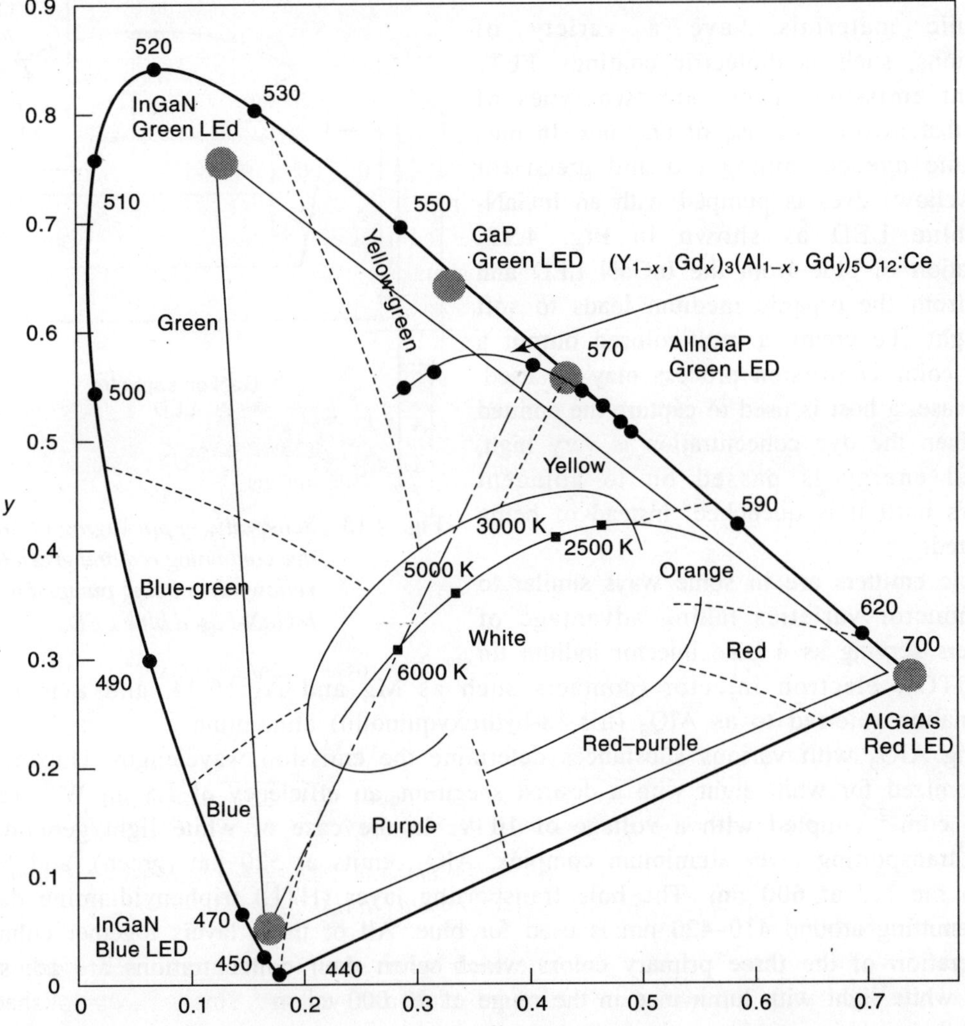

Fig. 4.12 *CIE diagram-superimposed with visible LEDs having three primary colors and soft white light obtained by pumping a YAG medium with a blue InGaN LED.*

The efficacy of the phosphor-white LEDs is as high as 15 lmW^{-1} with half-life times of 40000 h. When compared to multiple-chip LEDs for red, green, and blue color output, an advantage of phosphor-white or hybrid-white LED devices is that it only requires one blue or UV LED. White-light LEDs based on phosphors have shown to have relatively stable color with variations in temperature. The inorganic phosphors are replaced with organic polymers which are coated on the domed epoxy encapsulate of an InGaN LED lamp. They can also be used in the production of white light.

LEDs are used in indicator lights, flashlights, automotive, indicator signs including highway signs and traffic lights.

Organic LEDs (OLEDs)

Organic materials have a variety of applications, such as dielectric coatings, FETs and light emission. There are two types of devices that involve the use of organics. In one, an organic dye containing red and green (or simply yellow) dyes is pumped with an InGaN-based blue LED as shown in Fig. 4.13. Combination of blue from the InGaN LED and yellow from the organic medium leads to soft white light. To create a multicolored output a two-dye color conversion process may be used. In each case, a host is used to capture the emitted light. When the dye concentration is very high, absorbed energy is passed on to adjacent molecules until it is quenched, instead of being re-radiated.

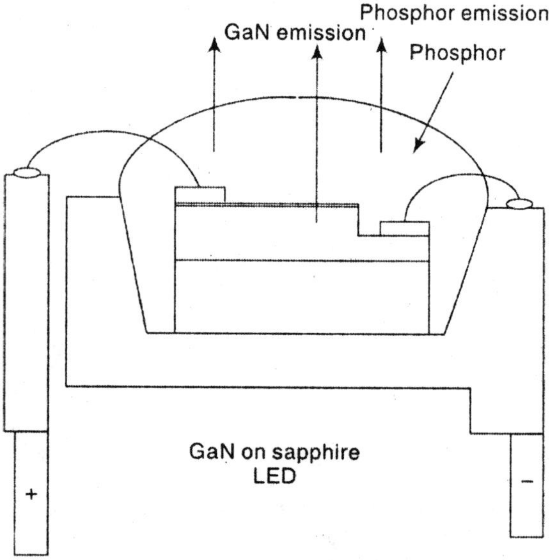

Fig. 4.13 *Schematic of packaging of an organic dye containing red and green (or simply yellow) dyes being pumped with a blue InGaN-based blue LED.*

Organic emitters are in some ways similar to semiconductor varieties taking advantage of multilayers serving as a hole injector indium tin oxide (ITO), electron injector (contacts such as Mg and Ag:10:1), and a medium for recombination referred to as AlQ_3 (tris (8-hydroxyquinolin) aluminium.

Doping AlQ_3 with various substances determine the emission wavelength. Dopant species are customized for white light with a desired spectrum. an efficiency of 1.5 lm W^{-1}, brightness of 1000 cdm^{-2} coupled with a voltage of 10 V. In the case of white light generation, the electron transporting layer aluminium complex, AlQ_3, emits at 520 nm (green), and Nile Red emits in the red at 600 nm. The hole transporting layer (HTL) triphenyldiamine derivative (TDP); emitting around 410–420 nm is used for blue. All of these layers together culminate in the generation of the three primary colors which when their concentrations are adjusted and result in white light with luminance in the range of 10 000 cd m^{-2}.

Here, the emission spectrum is wide as compared to semi-conductor emitters. Therefore, in applications where the hue is important, the spectrum is made narrower by placing the emitter region into a cavity. The concept is applied to plastic LEDs in cases where the light emission from the active medium cover a wide range, the cavity action is used to tune the wavelength.

The degradation is more severe as the wavelength of operation is extended toward blue. Higher luminance obtained with increased voltage shortens the lifetime by an equivalent amount. Real improvement in brightness must come from new material and efficient structures. The degradation mechanism is related to HTL wherein the radiative process is quenched with time of operation (mainly due to hole injection), as cationic AlQ_3 is not very robust and stable. Use of mixed emitting layer of hole and electron transporting molecules, placed between pure electron transport and hole transport layers improves lifetime. The device structure is shown in Fig. 4.14. A thin (5 nm) ALQ_3 layer is placed between two PNB (n, N'-di(napththalene-1-yl)-N, N'-diphenylbenzidine), the lower one of which facilitates hole injection from the ITO (anode) while the upper one transports holes to the cathode, but blocks electrons, from reaching the thin AlQ_3 layer. The preventation of electron injection into the thin (5 nm) AlQ_3 layer, reduces, the degradation of the thin AlQ_3 layer. The purpose of the upper AlQ_3 layer is to prevent shorting of the cathode to the 5 nm AlQ_3 layer, as the NPB layers are not perfect.

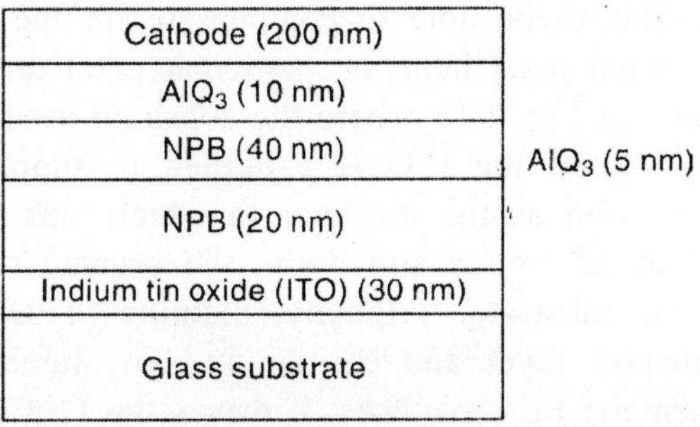

Fig. 4.14 *Schematic diagram of a modified plastic OLED structure where the holes are transport from ITO through a very think AlQ_3 layer. This thin layer is placed between two NPB (N, N'-di(naphthalene-1-yl)-N, N'-diphenylbenzidine), the lower one of which facilities hole injection from ITO (anode) while the upper one of which transports holes to cathode and blocks electrons, though not completely, from reaching the thin AlQ_3 layer. This approach, preventation of electron injection into the thin (5 nm) AlQ_3 layer, reduces the degradation of this thin AlQ_3 layer. The purpose of the upper AlQ_3 layer is to prevent shorting of the cathode to the 5 nm AlQ_3 layer as the NPB layer were not yet perfected.*

An all-organic LED is prepared in full and biased to produce light. The color of light is dependent on the types of dyes that are in the recombination medium. These plastic LED materials are attractive in that they have applications in large area displays for back lighting, active matrix displays, and even stimulated emission. In hybrid applications, these polymers can be conveniently produced to complement nitride based blue LED pumps to produce the desired color including white light. Large area, physical flexibility, and low cost are the attractive features afforded by the organic technology.

Light emitting polymers (LEPs) are a little more than a scientific curiosity. Bright future is seen for organic emitters for indoor displays, backgrounds panels, and nightlights by using large organic molecules.

LEDs in Displays

LEDs can also be used as pixels in flat panel displays such as in laptops. However, the mechanics of using semiconductor LEDs with TFTs are not combined for their manufacturing. But, organic LEDs and TFTs are integrated for active matrix displays.

The ease with which the organic materials are deposited allows very large surfaces to be patterned. What is more is that they are deposited to flexible substrates. The organic materials come in the form of monomers (individual molecules) or polymers (long-chain molecules that are made up of monomers that have bonded together). Monomers can be directed onto a substrate in a liquid both followed by polymerization as the solvent evaporates. For patterning purposes, the monomer solution can be deposited via screen and ink-jet printing, both of which are fast.

The development of high-performance LEDs is explored and found that it is less clear, how they are to be used and addressed in displays. Both passive and active addressings are demonstrated. In the passive case, a particular pixel only turns on when it is addressed. Since all pixel can not be addressed simultaneously, each pixel must be much brighter than would otherwise be necessary; it has to be able to compensate for the off time. In active addressing, FETs that are available in each pixel hold the particular pixel until it is updated. The schematic design of the pixel is shown in Fig. 4.15 where the substrate used is glass and is coated with the transparent conductor ITO. First, the ITO is patterned to supply the FET drain, and gold is used for the gate. Si_3N_4 is used as the insulator in which vias are etched for gate addressing. After depositing the gold metal source and drain, the organic semiconductor material PHT, is solution deposited onto the substrate. Triphenyl diamine (TPD) as a hole-transporting layer, AlQ_3 as the electron transport layer and emitter and an aluminum cathode are also used to create LED structure. When the FET switches, it drives the OLED on. Overall, the LED is able to supply 2300 cd m^{-2} of optical power.

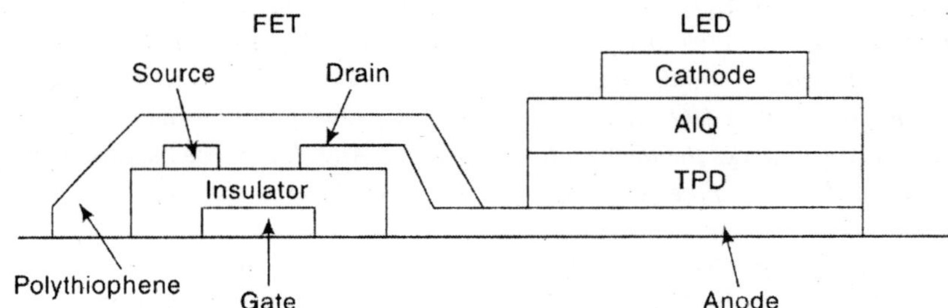

Fig. 4.15 *Schematic representation of an integrated smart pixel that contains an FET driving an OLED.*

p-n Solar cell

Figure 4.16 shows a schematic of a textured *p-n* solar cell. The device works as bulk-Si LED when forward biased. It also shows a schematic of the device and a room temperature emission spectrum. To increase the light extraction efficiency, the LED surface is texturized so that most of the internally generated light impinges on the external surface of the cell with an incident

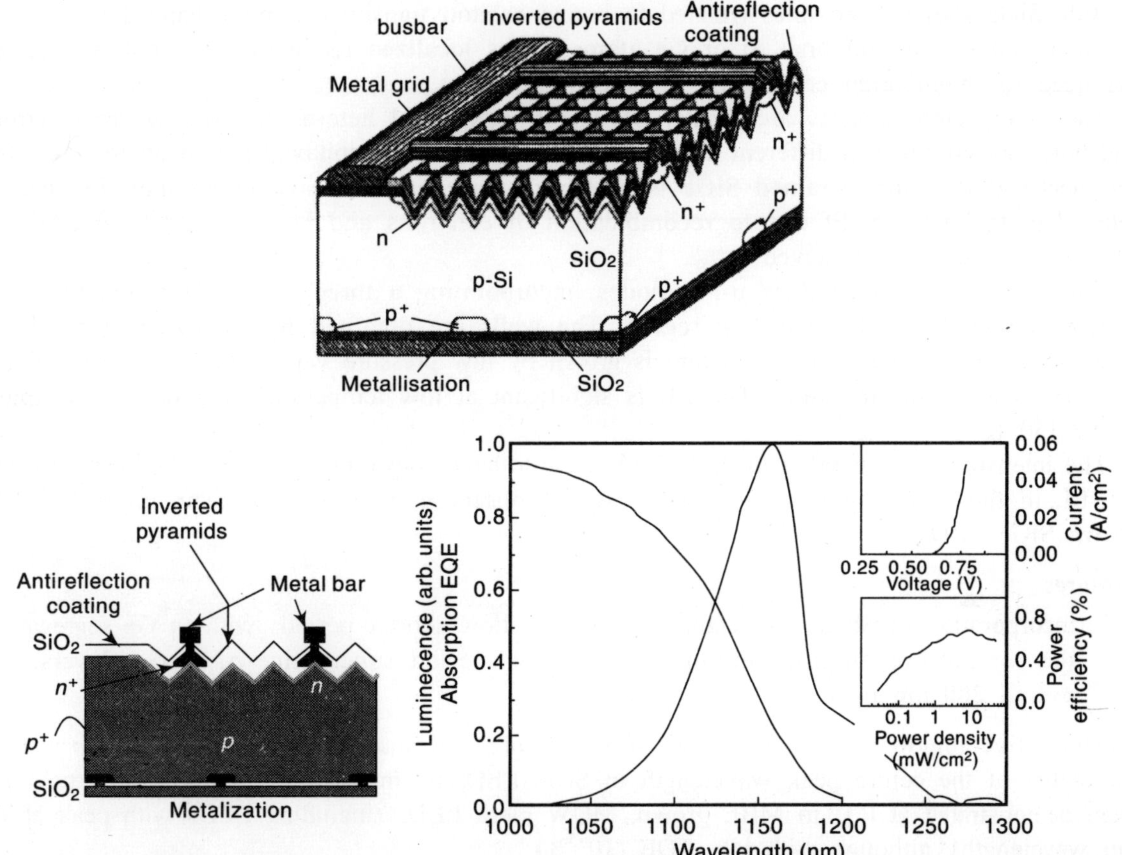

Fig. 4.16 *Bulk silicon LED Centre: sketch of the LED geometry. Right: Luminescence spectrum (red), absorption spectrum (green), power efficiency vs injected electrical power density (blu) and I-V characteristics (inset) at room temperature.*

angle lower than the critical angle for total internal refraction. Thus the light extraction efficiency is increased from a few % typical of a flat surface to almost 100% for the texturized LED. Finally to reduce free carrier absorption to a minimum, the electrodes, i.e. the heavily doped regions, are confined in very thin and small lines.

Thus, a plug-in efficiency (ratio of the optical power emitted form the LED to the electrical driving power) larger than 1% at 200 K is achieved. Most interesting the turn-on voltage of the device is the same as the forward bias of the solar cell, i.e. less than 1 V.

SiGe LED

SiGe has enhanced carrier mobility, bandgap and CMOS compatibility Si and Ge can be mixed to form alloys of type $Si_{(1-x)}Ge_x$ ($x \leq 1$). The SiGe material has the intriguing property that its bandgap can be continuously turned from 1.12 to 0.67 eV by adjusting x between and 0 and 1. This interval contains both 1.3 and 1.55 μm wavelengths, which are the most important for telecommunications.

Bulk SiGe shows more lines (related to bound exciton transitions) and enhanced intensity in the non-photon emission line: Si or Ge atoms act as localized recombination centres, relaxing the need for momentum conserving phonons.

The SiGe heterostructres are mostly of type II (staggered heterostructure), where electrons and holes are confined in different materials. SiGe are grown by epitaxial grown on Si substrates, for electrical transport, strained SiGe is preferred, since carriers have larger mobility than in relaxed epitaxial SiGe. El due to recombination of electrons and holes at the band edges of strained SiGe alloy is achieved.

The LEDs are fabricated as *p-i-n* diodes, incorporating a three-period multi-quantum well (MQW) structures as the emitting region. The wells are obtained by sandwiching $Si_{0.8}Ge_{0.2}$ layers between Si layers., The structure is grown by low pressure vapour phase epitaxy. Diodes are shaped as mesa structures. The EL is significant at low temperature but decreases rapidly above 130 K.

The intensity is comparable to the 1.1 eV band-to-band transition due to the bulk-Si substrate, RT EL in the wavelength region 1.3–1.7 μm is observed in molecular beam epitaxy (MBE) grown SiGe LEDs.

Features

1. incorporation of Sb as surfactant, in order to allow shorter periods without Ge segregation,
2. the composition of the emitting region, a 145-period superlattice of Si_6Ge_4 layers, and overall 200 nm thick.

The active region is included in a *p-i-n* structure and shaped as a mesa after fabrication. Tunability of the centre peak wavelength in SiGe LED, by modulation of MQW thickness has been demonstrated at RT, in MBE grown, MQW mesa LEDs (including LEDs with peak at 1.3 μm wavelength), although with low EQE (10^{-5}%).

SI Nanocrystals LED (on SiO_2)

Stimulated by the interesting results obtained in PS, many LEDs are based on nanostructured Si dispersed into a SiO_2 matrix. Si nanocrystal LEDs are obtained by capping the SiO_2 film with a conductive electrode of poly-Si, metal, or conductive and transparent Indium Thin Oxide (ITO). The main difficulty is in achieving efficient carrier injection Si nanocrystals are obtained by ion implantation and thermal treatments. Si and SiO_2 thickness are varied. Maximum external quantum efficiency is 3×10^{-5} for the LED with the thinnest SiO_2 thickness (10 nm).

Care is taken to the attribution of electroluminescence to recombination in nanocrystalline Si as SiO_2 on Si also emits photons under electrical injection. SiO_2 luminescence exhibits peaks at about 620–640 nm (1.9 eV), and is ascribed to non-bridging oxygen hole centers formed in silica.

In Figure 4.17 shows that, the Si nanocrystals are produced by plasma enhanced and chemical vapor deposition (PECVD). MOS structures are produced where the oxide is filled with Si nanocrystals. EL is excited by impact ionization of the nanocrystals. The largest EQE is obtained in films with a large dispersion in Si nanocrystals. The large Si nanocrystals contribute for the conductivity of the structures but internal quantum efficiency is largest in smallest Si nanocrystals. Turn-on voltages lower than 5 V are achieved in very thin SiO_2 films.

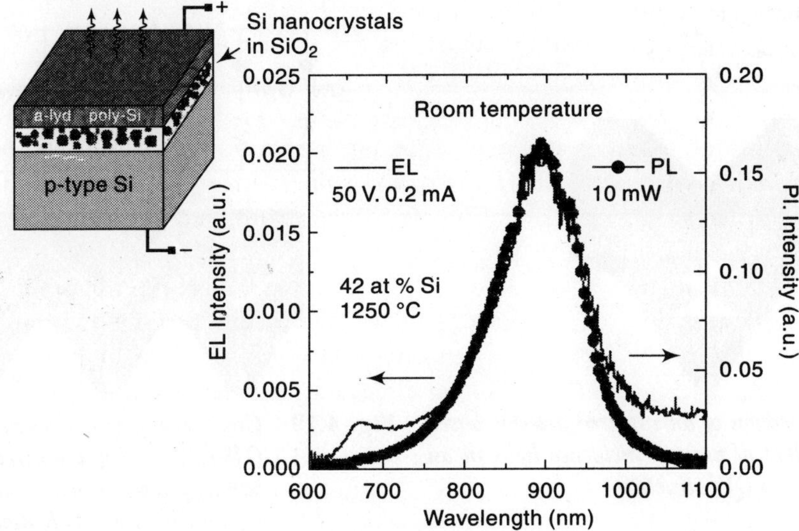

Fig. 4.17 *Comparison of the photoluminescence and electroluminescence spectra in MOS type structures with Si nanocrystals. The electroluminescence is provided by impact excitation of electron-hole pairs by tunneling current flowing in the vertical direction.*

The discovery of RT PL in Si/SiO_2 superlattices, with a size tunable emission energy, has motivated for fabrication of LEDs by Molecular Beam Epitaxy (MBE). The superlattice is formed by cycled deposition of a monolayer of absorbed oxygen and a thin Si MBE deposition. The growth procedure is obtained by alternating a step of exposure of the sample to oxygen with a step of epitaxial Si growth, at a deposition temperature of 823–873 K, and at a rate of 0.4 Å/s. The procedure allows the growth of single-crystal Si epitaxial layers over O-enriched SI layers. The latter layers form energy barriers of about 0.5 eV. Both PL and EL are observed.

V-GROOVE QWR LIGHT EMITTING DIODES

Quantum wire LEDs are fabricated by incorporating V-groove QWRs into *p-n* junctions. One of the major device issues in such a structure is related to the mechanism of carrier injection into QWRs. The QWRs are bounded by "parasitic" QW layers that grow on the sidewalls and the top (100) planes of the corrugated structure; thus, the QWRs occupy only a small fraction of the total device area even for the smallest-pitch arrays fabricated (see Figure 4.19). If the injection of electrons and holes proceeds uniformly across the QWR array, the injection efficiency into the QWR recombination region is poor. The carrier capture and relaxation to the ground states of the QWRs is important for obtaining efficient operation of such LEDs. The carrier injection and trapping efficiency is much simplified in QW-based LEDs due to their thin-film geometry. But configuration such as SCH are conceived and developed in order to maximize the carrier trapping efficiency into the extremely thin QW layers.

A useful approach for achieving efficient carrier injection into the V-groove QWRs is to utilize the connected AlGaAs VQWs (Fig. 4.18) as a preferential channel for channeling the

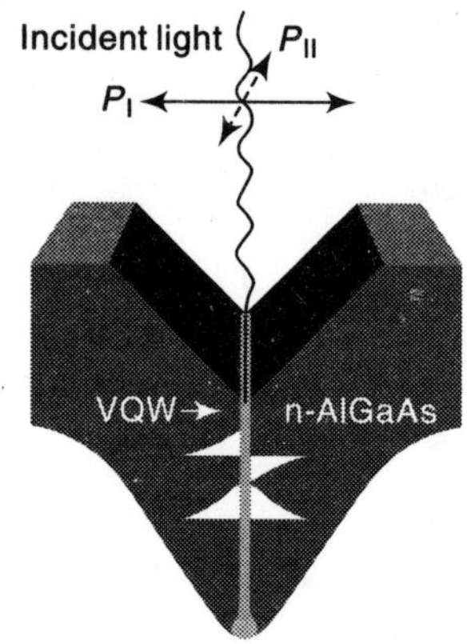

Fig. 4.18 *Cross section of an infrared detector based on ISBTs of normal-incidence light in an AlGaAs VQW structure*

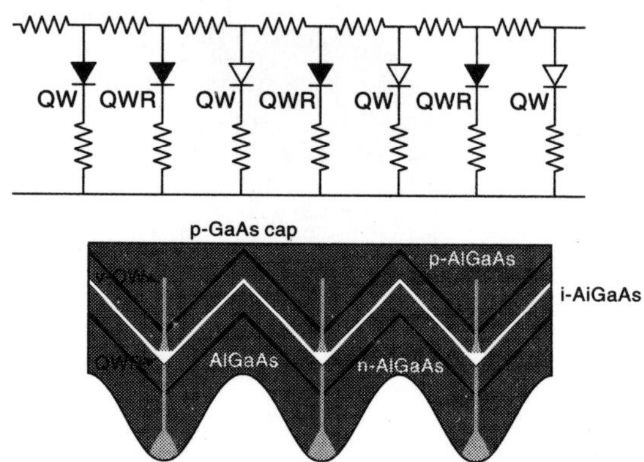

Fig. 4.19 *Cross section of a GaAs/AlGaAs V-groove QWR/QW light-emitting diode. The bandgap modulation at the AlGaAs barriers results in a QWR diode with a lowest barrier bandgap and therefore a lower turn on voltage. This allows the preferential injection of electrons and holes into the QWR via the AlGaAs VQW.*

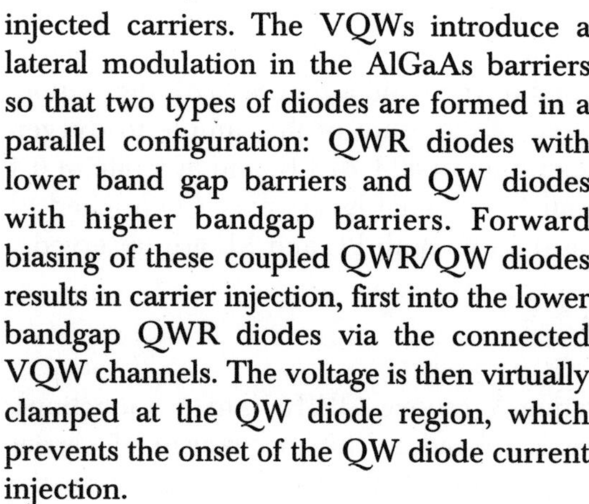

injected carriers. The VQWs introduce a lateral modulation in the AlGaAs barriers so that two types of diodes are formed in a parallel configuration: QWR diodes with lower band gap barriers and QW diodes with higher bandgap barriers. Forward biasing of these coupled QWR/QW diodes results in carrier injection, first into the lower bandgap QWR diodes via the connected VQW channels. The voltage is then virtually clamped at the QW diode region, which prevents the onset of the QW diode current injection.

Figure 4.20 show the measured light versus characteristics of a GaAs/AlGaAs QWR LED (x = 0.32, 0.5-μm pitch wire array), compared with the L-I curve of a planar GaAs/$Al_{0.32}Ga_{0.68}As$ QW LED grown side by side. The Al mole fraction in the QW LEDs similar to that near the QW regions in the QWR diodes. It is clearly seen that the turn-on voltage in the QWR LEDs is lower than that in the QW diodes.

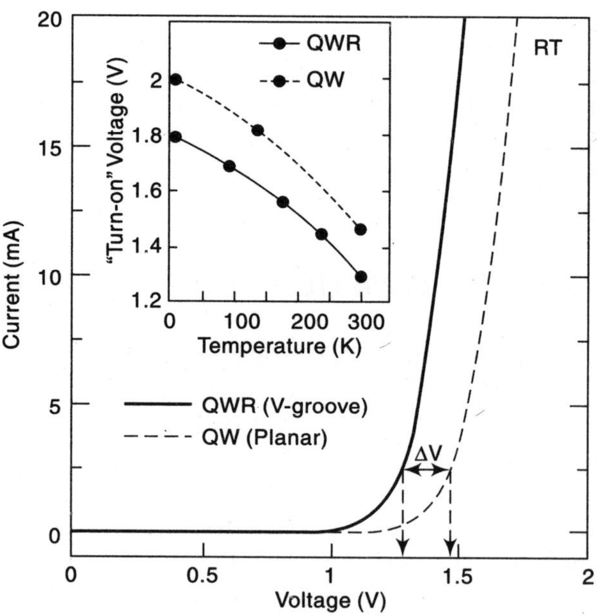

Fig. 4.20 *Light versus current characteristic of a V-groove QWR/QW and a planar QW GaAs/AlGaAs LEDs at room temperature. Inset shows the temperature dependence of the turn-on voltage of the QWR/QW and the QW diodes.*

The inset in Figure 4.20 shows that this difference in the turn-on voltage is maintained throughout the measurement range (4 to 300K). The difference in the turn-on voltage, $\Delta V = 170 \pm 20$ meV, is consistent with measured difference in the bandgap of the VQW and "background" AlGaAs barrier regions in the V-groove structure ($\Delta E = 190 \pm 40$ meV). This corroborates the conjecture that injection occurs via the narrow, lower bandgap VQW channels in the V-groove structure.

The preferential carrier injection made possible by the VQW channels results in preferential excitation of the QWR regions that are vertically sandwiched by the VQW barriers. This is demonstrated in Figure 4.21, which depicts the PL and electroluminescence (EL) spectra of the QWR LEDs, measured at low temperature (T = 10K). In the PL spectra, emission from both the QWR and the QW regions is evident, which results from an essentially uniform photo-excitation across the QWR array. In contrast, the EL spectra show only emission from the QWRs; at increased currents, band filling results in filling of several QWR subbands, which leads to a broadening and substructure in the EL line. Emission exclusively from on the QWR line is observed in these LEDs up to room temperature. The room temperature EL line width is

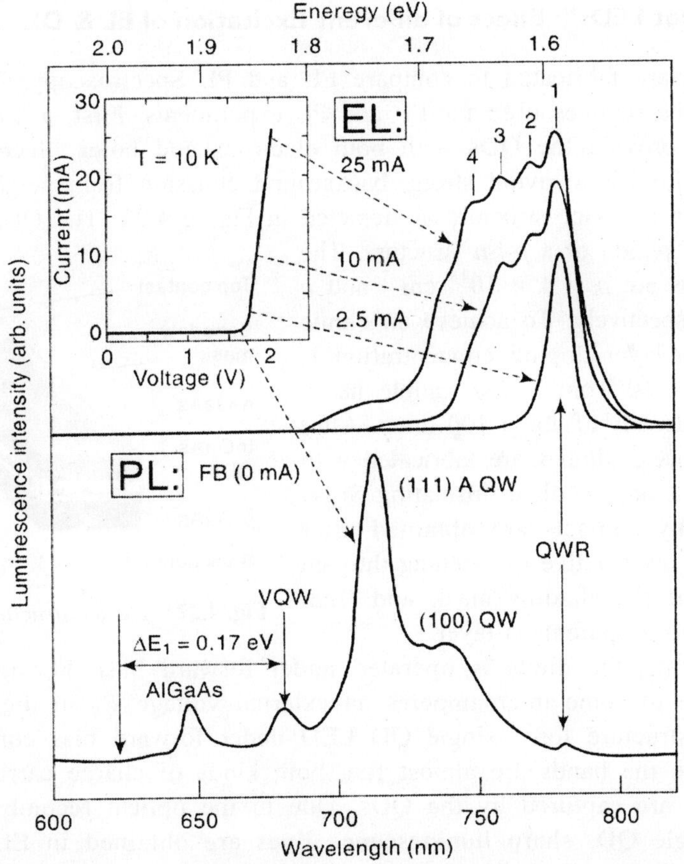

Fig. 4.21 *PL (bottom panel) and EL (upper panel) spectra of GaAs/AlGaAs V-groove QWR LEDs at 10 K. Inset shows the light versus current characteristics of the device.*

as narrow as 10 meV for an injection current of 10 μA. These linewidths are much smaller than the thermal broadening , which suggests the dominance of excitonic effects at elevated temperatures.

Enhanced (low temperature) QWR emission in V-groove diodes incorporate five vertically stacked GaAs/AlGaAs wires which grow by atmospheric pressure. QWR emission intensity that is comparable to that the surrounding QWs, is observed at low current levels. Exclusively QWR emission is not observed with these structures, due to larger (4 μm) groove pitch. Saturation of the emission from the ground QWR subbands and subsequently filling of the excited wire subbands is observed as the diode current is increased. The measured energy separations of the QWR transitions is 19 to 24 meV.

Several characteristics of the V-groove QWRs which make them suitable for applications in QWR diode lasers are the high interface quality, the minimization of nonradiative recombination processes and the achievement of high optical gain; the control of lateral quantum confinement, the large potential depth and lateral subband separation, operating at high temperatures, to bandgap-engineer the wire barriers, useful to efficient current injection.

Single Quantum Dot LEDs": Effect of different Excitation of EL & DL

Single QD LEDs are fabricated to compare EL and PL Spectroscopy. There are two main difference between the samples used for EL and PL experiments. First, a bipolar *p-i-n* device is required for EL to provide the QDs with both electrons and holes. Second, a local current injection is needed for EL to avoid strong background emission from neighboring QDs. These requirements result in a device structure as depicted in Figure 4.22. The QDs are centered in an 80-nm-thick intrinsic region of a *p-i-n* structure. The doping concentrations are $p = 4 \times 10^{18}$ cm^{-3} and $n = 5 \times 10^{17}$ cm^{-3}, respectively. To achieve an ohmic top contact, the cap layer doping concentration is increased to $n = 2 \times 10^{18}$ cm^{-3}. The sample has a InGaAs dot surface density of about 100 μm^{-2}. After the MBE growth process, diodes are fabricated with a metal shadow mask on top of an insulating Si_3N_4 layer. Local Schottky contacts are obtained after removal of the Si_3N_4 by reactive ion etching through the nanoapertures of the shadow mask and final deposition of a semitransparent Ti-layer.

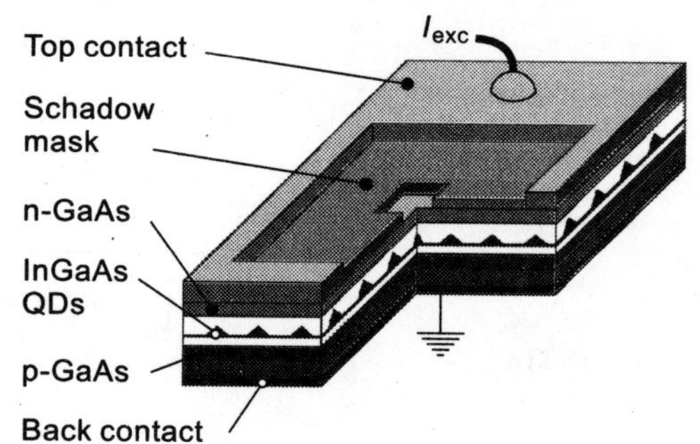

Fig. 4.22 *Device structure of a single QD-LED.*

For EL experiments, the diode is operated under forward bias. For injection of a direct current in the range of some microamperes an external voltage V_B in the range of +1.8V is applied. The band structure for a single QD LED under forward bias condition is shown in Figure 4.23(a). Since the bands are almost flat, both kinds of charge carriers diffuse into the intrinsic region and are captured by the QDs. Due to the optical recombination of electrons and holes in a single QD, sharp luminescence lines are obtained in EL. For $V_B = 0$, the corresponding band diagram is depicted in Figure 4.23(b). The QDs are empty, and no light is emitted by the QDs. An electric field of 165 kVcm^{-1} is calculated at $V_B = 0$V. Due to these high electric fields, a strong suppression of radiative recombination by tunneling is obtained in these diodes. Neglecting individual QD properties and assuming a threshold field of 35 kVcm^{-1}, radiative recombination is obtained only for $V_B > 1$V. For smaller values of V_B the carriers, tunnel efficiently out of the QD. This is in good agreement with V_B dependent PL measurements performed on these single QD LEDs. By increasing V_B, the first PL emission lines are found at about $V_B = 1$V. Without additional optical excitation EL starts to appear to about $I_{exc} = 3$ μA The interesting voltage region for comparisons between electrical and optical excitation is in the range between 1.7V and 2V. For, the influence of the excitation mechanism on the optical response of a QD, two kinds of experiments are done. In the first one, a constant bias voltage of 1.78 V is applied to the sample, resulting in a very weak but detectable EL signal. In addition, the optical excitation power is increased successively. This allows the observation of specific QD lines in both PL and EL. Moreover, it reveals the influence of increasing optical excitation power on the QD spectrum under comparable conditions to the second experiment. In this second experiment, the excitation current I_{exc} is simply increased without additional optical excitation.

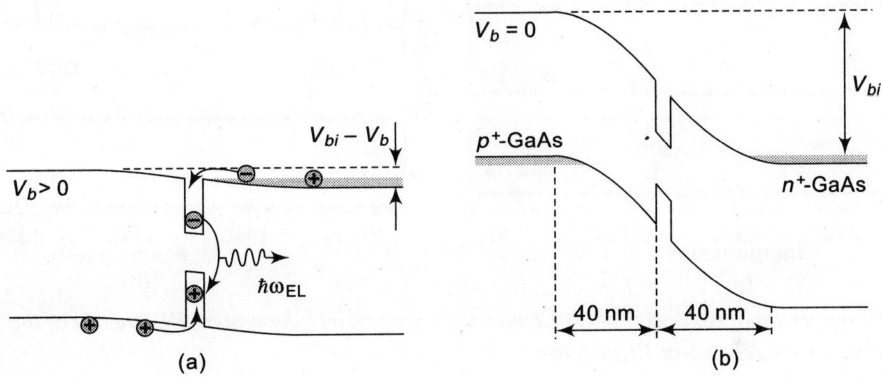

Fig. 4.23 *Schematic band diagram for a single Qd-LED (a) under forward bias condition and (b) at zero bias voltage.*

An EL spectrum recorded on a 300-nm aperture is presented in Figure 4.24(a). This spectrum is taken in the limit of low current injection with $I_{exc} = 4$ μA. Only two QD emission lines at 1,337.0 meV and 1,348.6 meV, marked as QD1 and QD2, contribute to the spectrum. In the upper part of Figure 4.24(a), PL spectra are presented as a function of the excitation power at fixed V_B. These spectra are recorded on the same aperture of the shadow mask as the EL spectrum shown below. Under high optical excitation, the low current EL contribution is negligible. In case of low optical excitation, the same lines are observed in OL ad in EL in the limit of low current injection. This is because V_B is the same in both measurements, and therefore the charging conditions of QD do not change. Increasing the optical excitation power by an order of magnitude, the line intensities of QD1 and QD2 rise; however, only one new line appears in the spectra 1.1 meV below the QD1 line. Since V_B is fixed in the experiment, the new line is tentatively assigned to the decay of a (charged) multiexciton state.

In Figure 4.24(b), EL spectra are shown as a function of I_{exc}. The EL spectrum changes dramatically increasing I_{exc} from 3 to 6 μA. The line intensities of the QD1 and QD2 lines rise strongly. Moreover, a number of new emission lines appear in the spectra. The new emission line that is observed in PL for high optical excitation power also appears in EL for increased electrical excitation. The other additional emission lines are not observed in PL. One possible explanation for these additional emission lines a is a change in the charging condition of the QDs for higher I_{exc}. The increase in I_{exc} is accompanies by a variation of V_B and new emission lines appear for changed charging conditions of the Qd. For higher I_{exc} multiexciton states are also contribute to the EL spectra.

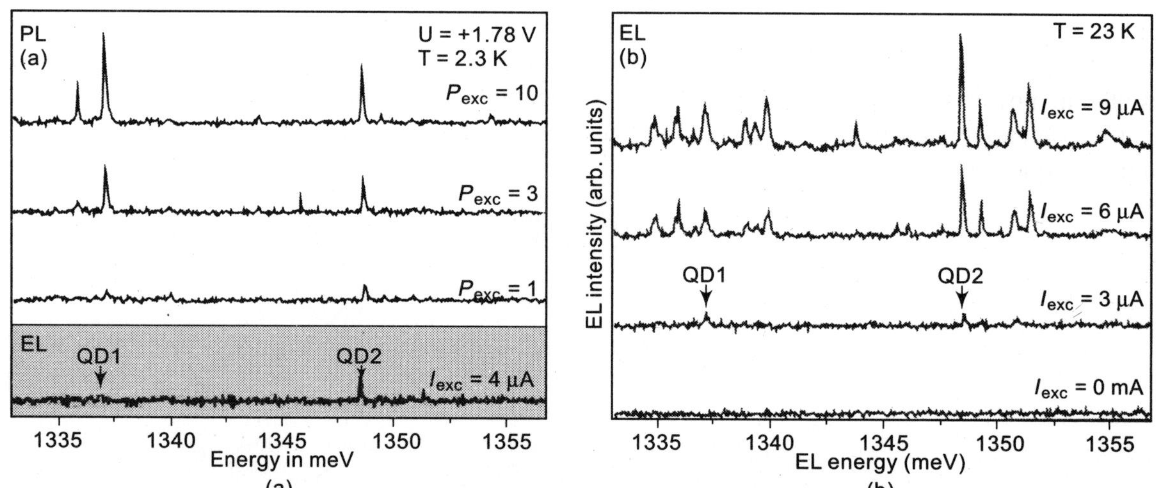

Fig. 4.24 *Comparison of (a) a excitation EL spectrum with power dependent PL spectra of the same QDs and (b) current dependent EL spectra.*

ADVANCED HETEROSTRUCTURE BLUE/VIOLET LIGHT EMITTERS

Vertical Cavity Emitters

The nitride optoelectronic materials have different properties as compared to III–V semiconductors. This fact extends to the device science and engineering of the vertical cavity blue and near-UV emitters. Firstly for the nitride vertical cavity emitters, the fabrication of a high-Q optical resonator is non-trivial. Secondly, due to the conductivity of the p-GaN and its alloys, the electrical injection incorporating lateral current injection into a nitride-based VCSEL is achieved.

Design and Fabrication: Optical Resolution

Microcavity resonators, in situ as-grown AlGaN/GaN distributed Bragg reflectors (DBR) are feasible for the fabrication of nitrides, but, due to the small index of refraction contrast, require a large number of layer pairs. The growth of such DBRs is made by epitaxial growth.

The vertical cavity, or "surface" lasing under optical intense pumping from GaN or InGaN MQW or thin-film heterostructures are encased by *in situ* grown AlGaN/GaN DBR reflectors. Stimulated emission employing a "hybrid" structure composed of one *in situ* grown AlGaN/GaN DBR and one dielectric DBR is achieved.

All-Dielectric DBR Resonator

A useful testbed for vertical cavity nitride light emitters is the all-dielectric mirror cavity. A fabrication technique is shown to produce microcavity resonators with quality factor Q approaching 1000. Here, the sapphire substrate is separated by pulsed laser ablation and a specific process sequence is used to sandwich the InGaN/GaN/AlGaN heterostructure between two DBR stacks of SiO_2/HfO_2.

VCSEL operation with a mean roughness of 2–3 nm over areas on the order of several hundred square micrometers is achieved. Although, Lasing is obtained but it is dominated by in-plane stimulated emission. Optically pumped quasi-CW VCSEL operation is achieved by photo-excitation at 355 nm, outside the reflectance band of the DBRs and slightly below the bandgap of the AlGaN cladding layer, ensuring a predominant creation of e-h pairs directly into the InGaN QWs. The upper trace of Fig. 4.25 shows spontaneous emission spectrum at temperature T = 258 K at an incident power of 17 mW. Several cavity modes are seen with a typical mode linewidth of approximately 0.6 nm. The bottom trace of Fig. 4.25 shows the onset of stimulated emission (at threshold input power of P_{th} = 32 mW).

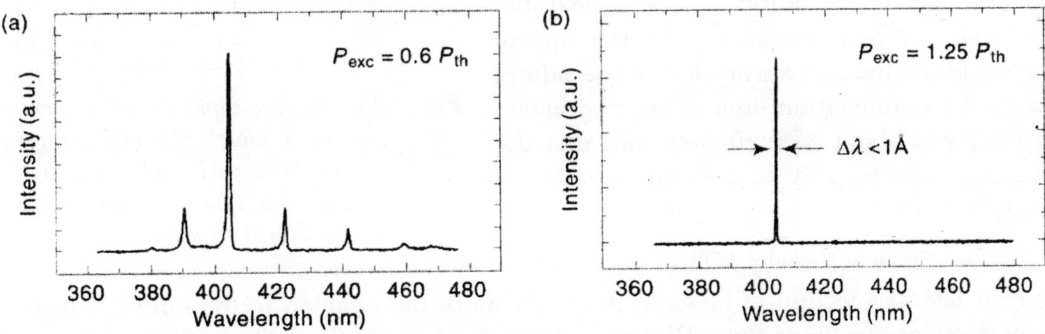

Fig. 4.25 *Upper trace spontaneous emission spectra of the optically pumped InGaN MQW VCSEL below threshold. Lower trace: stimulated emission spectra above the threshold under quasi-CW pumping conditions at T = 258 K.*

From the conditions of excitation, it is estimated that this type of optically pumped VCSEL corresponds to those in the CW edge-emitting diode lasers in terms of equivalent electrical injection.

AlGaN/GaN DBRs

To achieve the high reflectivities the large number of AlGaN/GaN layer pairs (>50) present sizable difficulties in the control of cracks and the DBR morphology.

AlGaN interlayers are effective in controlling mismatch-induced stress and suppresses the formation of cracks, occurred during growth of AlGaN directly upon GaN epilayers. Thus, stress evolution during growth of AlGaN/GaN DBR mirrors by metal–organic vapor phase epitaxy is controlled. The employment of an AlN interlayer at the beginning of a thick (5 μm) DBR growth leads to a the modification of the initial stress evolution. Tensile growth stress is brought under control and nearly eliminated through multiple insertions of AlN interlayers. So, crack-free growth of 60 pairs of $Al_{0.20}Ga_{0.20}N$/GaN DBR mirrors is achieved over the entire 5 cm wafer with a maximum reflectivity of at least 90%. Even the associated spectral bandwidths are relatively small (10–20 nm).

Such DBRs, with high-quality morphology and peak reflectivity R ~ 0.991, is applied to demonstrate quasi-CW operation, at room temperature, of an optically pumped $In_xGa_{1-x}N$ (x ~ 0.03) MQW VCSEL at near $\lambda = 383$ nm. The vertical cavity scheme combined a high reflectivity in situ grown multilayer GaN/$Al_{0.25}Ga_{0.75}N$ and post-growth dielectric SiO_2/HfO_2 DBR. A photograph of the side profile of the low divergence beam of circular cross section is shown in the inset of Fig. 4.26, which also displays the input/output power characteristics of a device with lasing threshold at pump power of 30 mW (average VCSEL output powers up to 3 mW were measured). However, finite thickness variation across the wafer led to spectral shifts of the cavity modes (relative to InGaN MQW gain spectrum) so that significant increases in threshold are encountered for device fabricated from near the edge of the wafer. When accounting for the optical excitation volume the fraction absorption of the pump, and using an e-h recombination time of approximately 0.5 ns, the device is about 25% efficient and that the threshold corresponds roughly to a carrier density of approximately 10^{19} cm^{-3}.

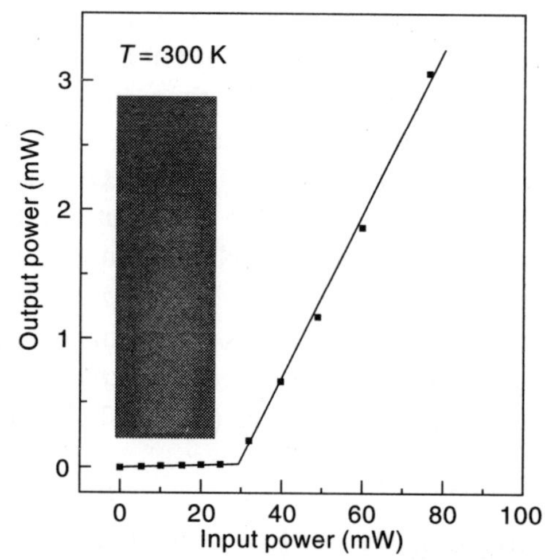

Fig. 4.26 *Average input vs output power of a violet optically pumped VCSEL device. The inset shows the beam far-field (side) profile captured on a screen.*

Electrical Injection: Resonant Cavity LEDs

The need for lateral injection of holes from the *p*-side of the junction to the optically active volume poses a major challenge for a III-nitride vertical cavity diode emitters. Due to the low conductivity of a MOCVD-grown p-GaN (on the order of 1 Ω^{-1} cm^{-1}), it requires that a high conductivity *intracavity* layer be inserted between the top p-GaN layer and the adjacent DBR. To avoid optical losses from the later, both its intrinsic absorption as well as morphology (in terms of optical scattering) must be considered. Two approaches are made in this direction. First, indium-tin oxide (ITO) is used. It is a thin film material both from the standpoint of low optical losses and in terms of forming an electrical contact to p-GaN resistivity of magnetron-sputtered and thermally annealed ITO approximately 4 × 10–4 Ω cm). Secondly nitride-based TH, provides a heterostructure building block which generally extends the design versatility for nitride optoelectronic devices. Useful demonstrations of blue and vertical cavity LEDs are made by employing ITO. It acts as an intracavity element for both approaches to optical resonator fabrication outlined above (all-dielectric DBRs and the hybrid configuration with one in situ grown AlGaN DBR, respectively).

The schematic diagram of a vertical cavity violet LED is shown in Fig. 4.27 which combines the hybrid resonators structure with a hole injection scheme that is based ion the use of p^{++}/n^{++} InGaN/GaN tunnel junction.

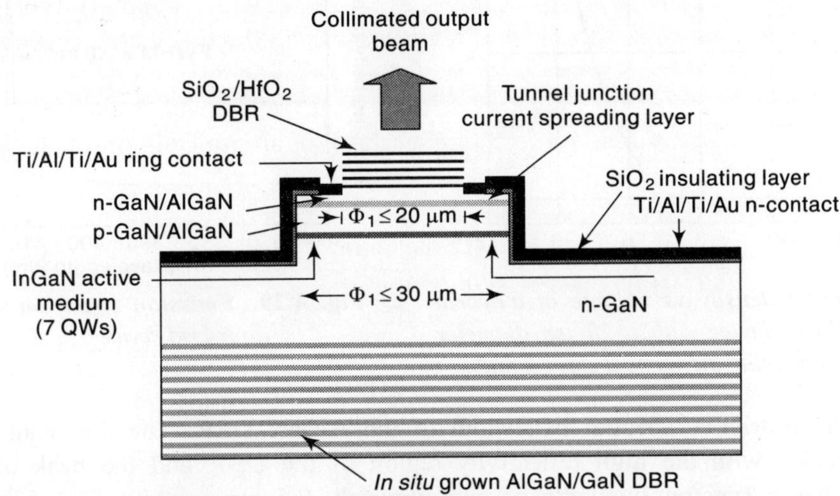

Fig. 4.27 *An RLED device with one as-grown GaN/AlGaN And one dielectric DBR mirrors, incorporating a nitride tunnel junction as the top current spreading layer.*

AlN strain-relief layers are used in the deposition of a 60-layer-pair quarter-wave GaN/$Al_{0.25}Ga_{0.75}N$ bottom DBR stack. Those growths that yielded a root-mean –square average roughness no words than 4 nm over an area of 1×1 mm^2 are deemed suitable for continuation of the epitaxy. The active p-n junction region is grown directly atop the GaN/(AlGaN) DBR, composed typically of seven $In_{0.08}Ga_{0.92}N$ QWs (L_w = 40 Å) with GaN barriers (L_B = 60 Å), and surrounded by approximately 1000 Å thick $Al_{0.07}Ga_{0.93}N$ current blocking barrier confinement layers. The tunnel junction is included in the superstructure of the nitride segment and capped by a n-GaN layer. The positive bias to the device is applied through contacts to this n-layer. Lateral current spreading ~100 μm givens nearly 100 times higher n-type conductivity of GaN. The TJ is grown atop the 1000 Å thick p-GaN layer as a p^{++}/n^{++} InGaN/GaN bilayer with the thicknesses of the layers approximately 150 and 300, respectively. The doping levels are approximately 1×10^{20} cm^{-3} Mg and 6×10^{19} cm^{-3} Si for the junction. The vertical cavity is completed by capping the structure with a multilayer $\lambda/4$ stack of SiO_2/HfO_2 (R > 0.995), deposited by reactive ion beam sputtering. The top dielectric DBR is patterned so that the device has an effective optical aperture varying from 10 to 30 μm.

The current density vs voltage of a typical device is shown in Fig. 4.28 upto a high continuous injection level (~1 kWcm^{-2}), which shows evidence of series resistance. Nonetheless, lateral current spreading was clearly accomplished so that the far-field light emission from the devices is uniform in its average intensity across the emitting aperture. Figure 4.29 shows the output spectrum of a typical device at operating current density of approximately 0.2 kAcm^{-2}. The emission is observed in the direction normal to the planar device, within an angular view of approximately 0°. While the optical resonator is thick (>10λ), only two vertical cavity modes

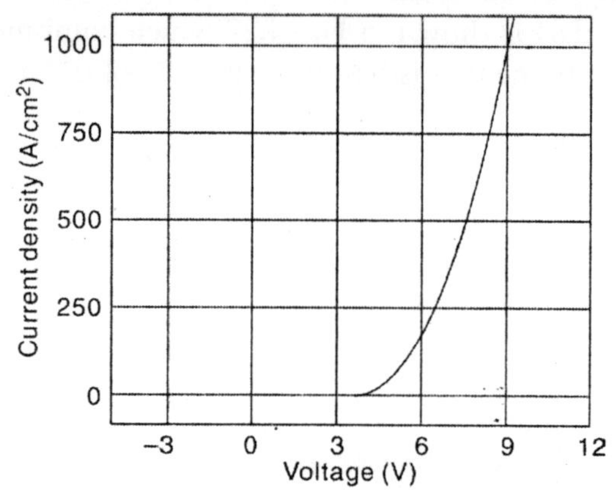

Fig. 4.28 *Current density vs voltage of a hybrid RCLED device with a 20 μm diameter mesa defining the vertical current path.*

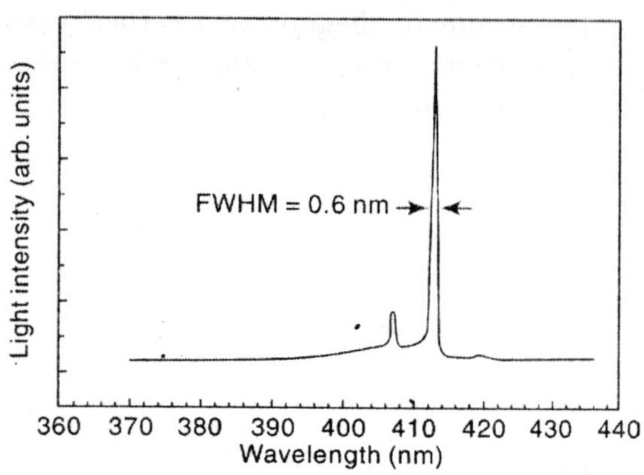

Fig. 4.29 *Emission spectrum of the RCELD hybrid device*

demonstrate the restrictive spectral bandwidth of the AlGaN DBR. The dominant mode at λ = 413 nm, coincides with the high reflectivity region of the DBR and the peak of the QW PL emission and has a spectral linewidth of approximately 0.6 nm.

Stimulated emission is obtained from these types of vertical cavity emitters, at high current injection levels (>10 kAm^{-2}). The top panel of Fig. 4.30 shows the *L–I* curve of a seven-QW device with ITO hole current spreading layer at two different temperatures. At 15 $kAcm^{-2}$, threshold-like *L–I* curve is observed, which suggests the existence of stimulated emission. Pulsed output powers in excess of 10 mW are measured. The far-field pattern of the dominant cavity mode is shown in the bottom panel of Fig. 4.30, showing a half angle of approximately 12°. This angle refers to the measured external angle of divergence after passing through the sapphire substrate. Spectral narrowing to less than 0.4 nm is observed. Although stimulated emission is achieved in these structures, several features remain unresolved.

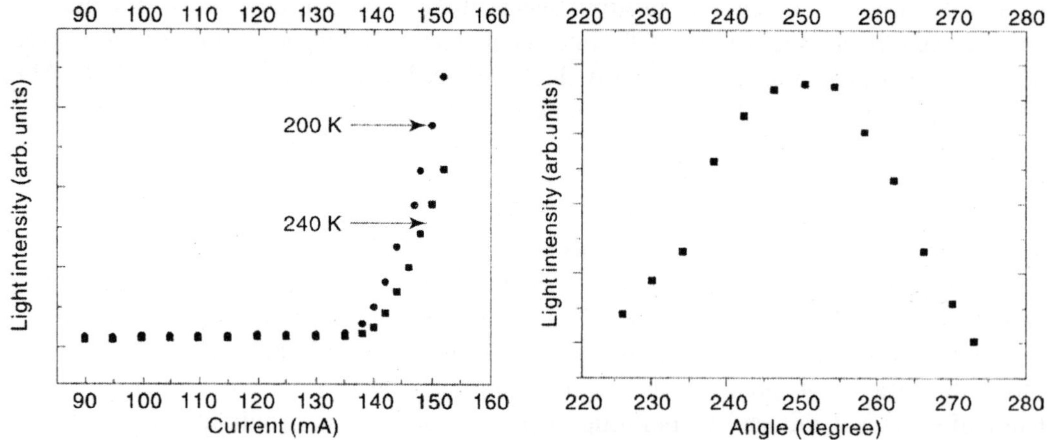

Fig. 4.30 *Top panel output vs current for vertical cavity device in high injection, showing a stimulated emission threshold. Bottom panel: corresponding far-field pattern for the dominant cavity mode.*

For example, the directionality of the emission is neither narrow nor polarized as with the optically pumped (albeit much simpler) structures. There is a finite, nearly isotropic back ground emission present. The presence of intracavity losses is a factor in raising the onset of threshold current, given the rather small total thickness of the InGaN MQW region (~300 A).

Thus, the vertical cavity LEDs provide important building blocks for the development of viable diode VCSELs in the violet. Th current device challenge is two-fold: high-quality epitaxy and relatively complex device processing. For example, though the layer thickness and composition (Al, In) control over a 5 to 8 cm wafer area is quite difficult, yet it is required that spectral overlap, between the high reflectivity band of the AlGaN DBR and the maximum gain of the InGaN MQW gain remain in spectral synchrony.

The application of the lateral epitaxial overgrowth techniques aid to create flexibility for designing and implementing blue/violet RCLEDs and VCSELs in two different ways. First, the patterned growth is adapted for creating a buried bottom dielectric DBR mirror. An illustration is shown in Fig. 4.31 where a patterned HfO_2/SiO_2 multilayer dielectric stack is deposited on a GaN buffer layer prior to subsequent regrowth of GaN. The particular dielectric stack is terminated with an SiO_2 layer so that the ELOG process occurs normally. The application of the lateral epitaxy pertains to building in a current aperturing scheme to alleviate the problem that arises from the competition between vertical and lateral transport in the nitride devices (especially on the p-side). Figure 4.32 shows the schematic of a possible arrangement where an SiO_2 defined current aperture is implemented in conjunction with lateral epitaxial growth.

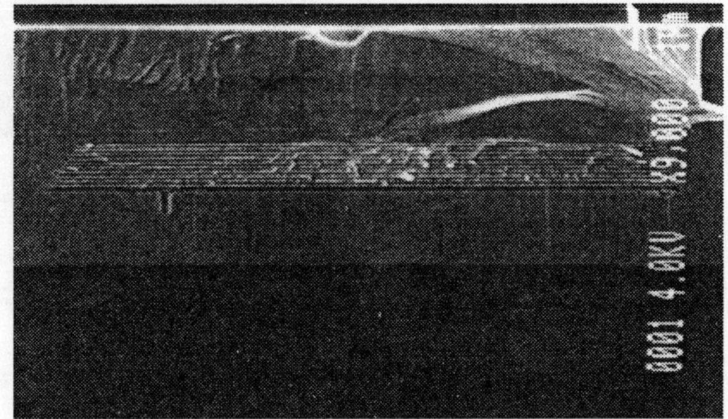

Fig. 4.31 *Use of lateral epitaxial GaN overgrowth to "bury" a dielectric DBR. The imperfections are due to breaking of the sample for access to an edge view in this cross-sectional SEM image.*

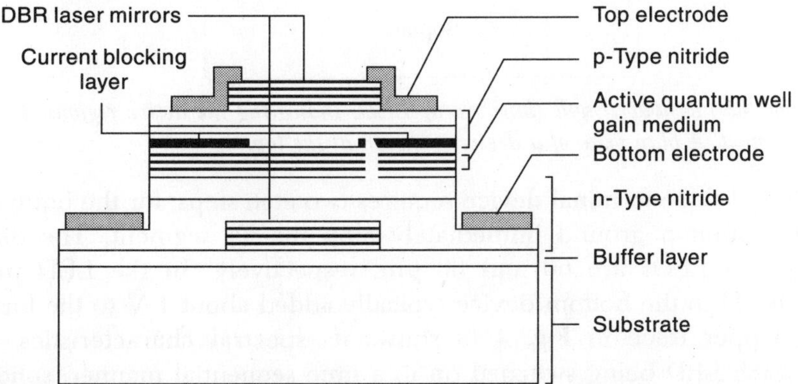

Fig. 4.32 *A blue VSEL device structure featuring a buried dielectric DBR and current confining aperture.*

Dual-Wavelength LED

A three-terminal, dual wavelength blue/green LED consisting of two electrically independent InGaN QWs of different indium composition within a single vertical heterostructure is demonstrated in the InGaN QW system. The device incorporating the p^{++}/n^{++} InGaN/GaN TJ a time-multiplexed two-color blue/green LED source (here at 470 and 535 nm). The nitride heterostructure used is shown in Fig. 4.33. The TJ segment is inserted between the two active $In_xGa_{1-x}N$ QW emitter segments (L_w = 30 Å). The TJ accomplishes two tasks: (a) electrically sectioning the nitride heterostructure into two independent LEDs; and (b) lateral current spreading for the bottom device.

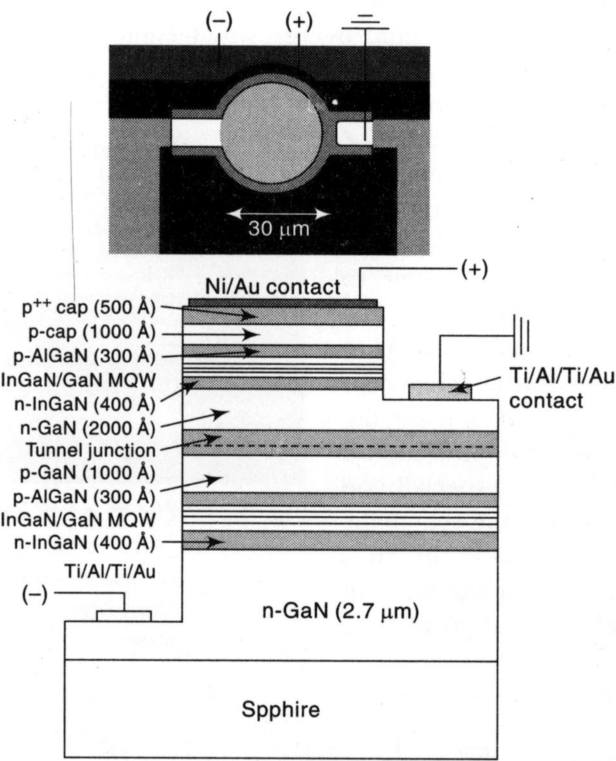

Fig. 4.33 *View of the two-wavelength (blue/green) LED, indicating the active regions, the TJ, and the bias arrangement. A plan view of a device is shown at the top.*

The fabrication of three-terminal device requires two etch steps, for the bottom LEDs n-GaN contact and the common ground immediately atop the TJ segment. The diameters of the top and the bottom LEDs are 60 and 80 μm, respectively. In the LED injection regime (~100 Acm^{-2}), the TJ in the bottom device typically added about 1 V to the forward "turn-on" characteristics. Upper trace in Fig. 4.33 shows its spectral characteristics. Dashed curve corresponds to each LED being switched on in a time sequential manner, solid line is for the LED operated as a simple two-terminal device with a constant voltage applied across the top p-GaN and lowest n-GaN). The electrical independence of the blue and green segments allows to "program: the 470 and 535 nm LEDs for any time sequence, up to speeds of 100 MHz. The bottom trace in Fig. 4.34 shows such time sequenced two-wavelength emission in the three-terminal configuration. However, as the current-voltage characteristics as well as their effective current apertures of the two LEDs are different, the applied voltage for each LED needed to be adjusted to obtain a comparable optical output from the devices. The device is applied to time-gated fluorescence fingerprinting of molecular dye labels because the fluorescent dyes are known in terms of their absorption and emission spectra. The emission wavelengths of the two LEDs are chosen to match their absorption spectra, respectively.

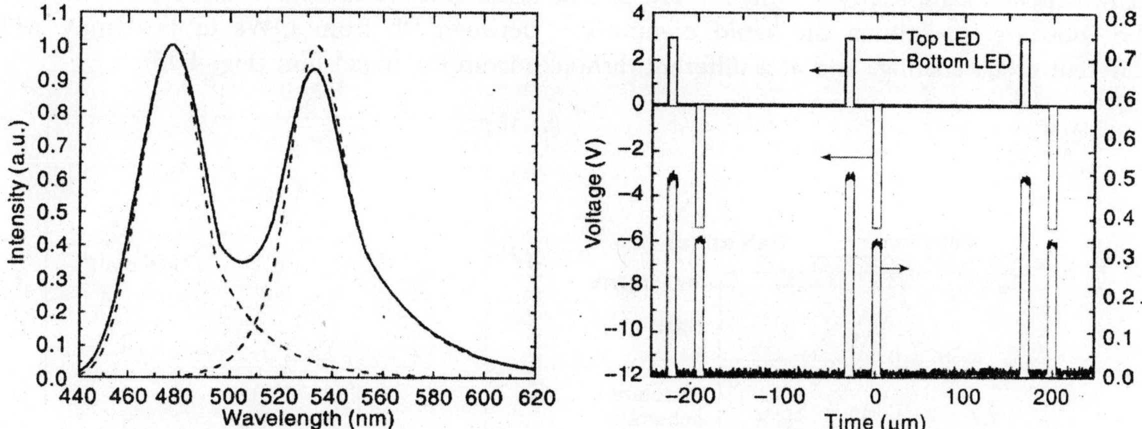

Fig. 4.34 *Top: Comparison of the emission spectrum from the blue and green LEDs when the devices are activated independently in a three-terminal case (weak solid trace), and their simultaneous activation as a two-terminal LED bold trace). Both: time-sequenced activation of the two LEDs (amplitudes and duty cycle chosen arbitrarily).*

The types of multiwave length emitters envisioned above, and single color devices, programmable 2D arrays are needed. For example, for fluorescence-based chipscale detection of chemical and biochemical species, such arrays add significantly to device throughput.

LIGHT EMITTERS BASED ON InGaN QUANTUM WELLS

InGaN light-emitting diodes (LEDs) have not fulfilled their original promise solid-state replacement for light bulbs because their light-emission efficiencies have been limited. Method to enhance the efficiency through the energy transfer between quantum wells (QWs) and surface plasmons (SPs) is given below. SPs can increase the density of states and the spontaneous emission rate in the semiconductor, and lead to the enhancement of light emission by SP–QW coupling. Large enhancements of the internal quantum efficiencies (η_{int}) are measured when silver or aluminum layers are deposited 10 nm above an InGaN light-emitting layer, whereas no such enhancements are obtained from gold-coated samples. The use of SPs lead to new class of very bright LEDs, and highly efficient solid-state light sources.

Surface plasmons, excited by the interaction between light and metal surfaces are known to enhance absorption of light in molecules and increase Raman scattering intensities, SPs are also used in LEDs. Although coupling of spontaneous emission from InGaN QWs into the SP modes of thin, silver films is indirectly observed through increased absorption at the SP frequency, and time-resolved measurements confirm that the recombination rate in QWs can be significantly increased, light has not yet been efficiently extracted from SP-enhanced InGaN emitters at visible wavelengths. Large photoluminescence (PL) increases are observed from InGaN/GasN QW material coated with metal layers. By polishing the bottom surface of sapphire-substrate-grown InGaN samples, QW emission are photoexcited and are measured through the back of the substrate, permitting the rapid comparison between PL from QWs in proximity with different metal coatings and at a different distances from the metal film (Fig. 4.35)

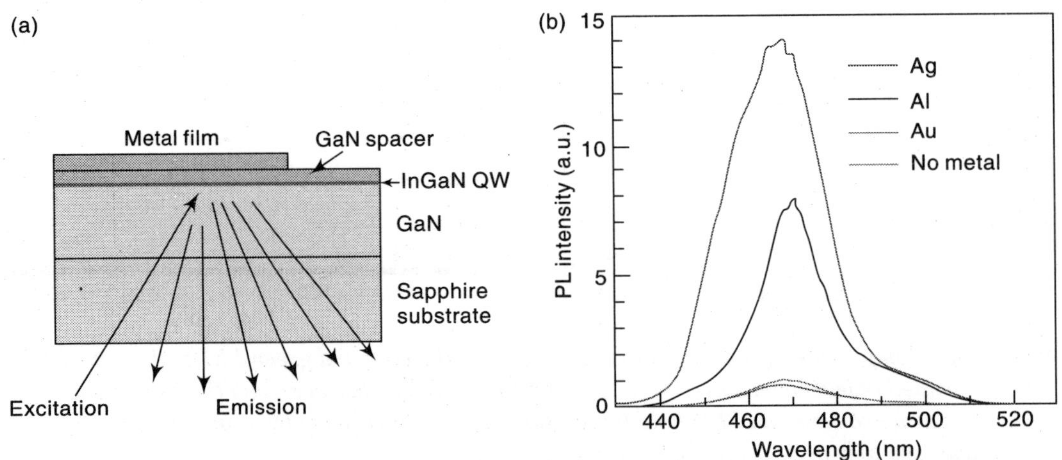

Fig. 4.35 *Photolumionescence measurements. (a) Sample structure and excvitation/emission configuration of PL measurements. (b) PL spectra of InGaN/GaN QWs coated with Ag (red line), Al(blue linw) and Au (green line). A.u.–arbitrary units. The distance between the metal layers and QWs is 10 nm. The peak intensity of uncoated InGaN/FaNQW (black line) at 460 nm is normalized to 1.*

Figure 4.35(b) shows typical luminescence spectra from an InGaN/GaN QWs separated from silver, aluminum and gold layers by 10 nm GaN spacers. For Ag coatings, the luminescence peak of the uncoated wafer at 470 nm is normalized to 1, and a 14-fold enhancement in peak PL intensity is observed from the Ag-coated emitter. The luminescence intensity integrated over the emission spectrum is increased by 17 times, whereas eightfold peak intensity and sixfold integrated intensity enhancements are obtained from AL-coated InGaN QW. The PL is not increased after Au coating. A small increase in the luminescence intensity is expected after metallization because the metal reflects pump light back through the QW, doubling the effective path of the incident light, but differences between AU and Ag reflectivities at 470 nm cannot explain the large difference in the measured enhancement alone.

Figure 4.36 a shows the enhancement ratios of PL intensities with metal layers separated from the QWs by 10 nm spacers as a function of wavelength. The enhancement ratio increase

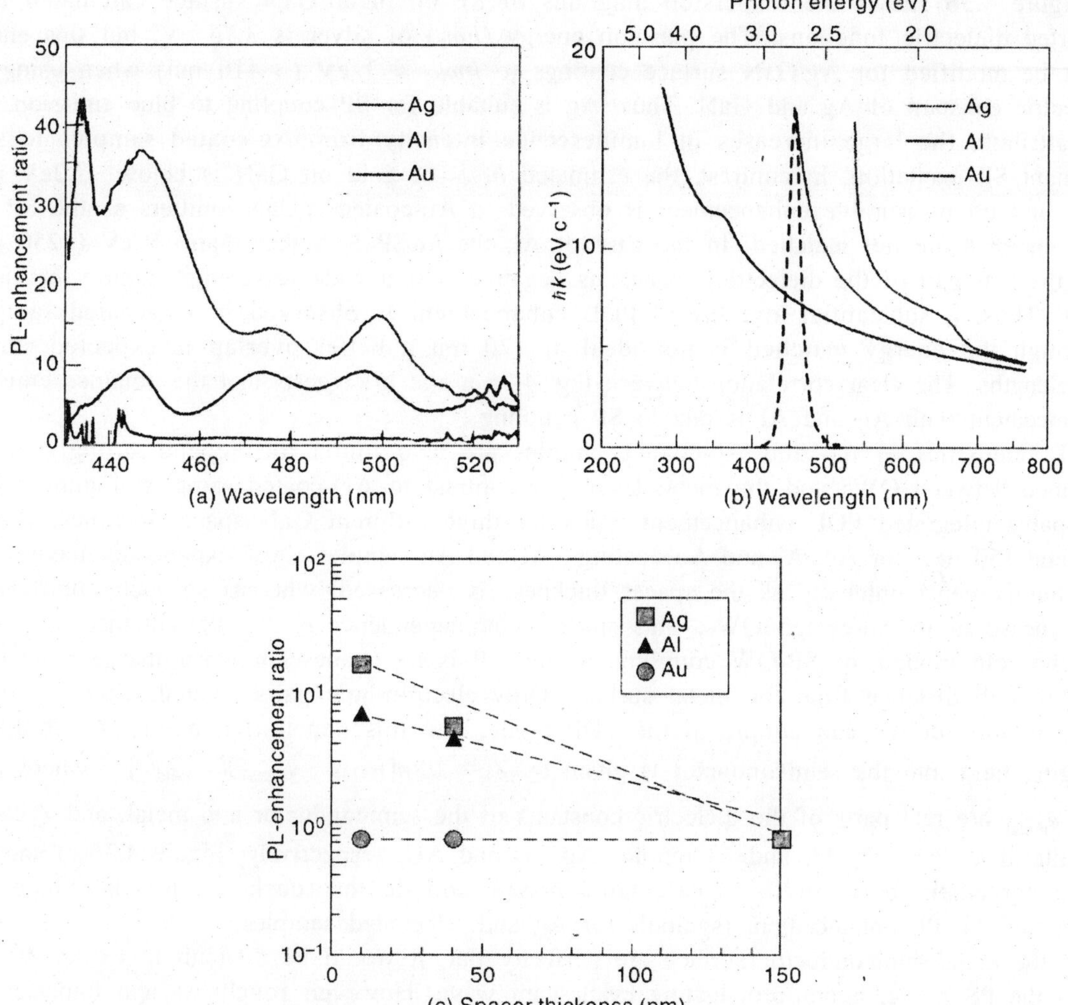

Fig. 4.36 *PL enhancement ratios. (a) PL enhancement ratios at several wavelengths for the same samples as Fig. 4.35(b). (b) Dispersion diagrams of surface plasmons generated on Ag/GaN (red line), Al/GaN (blue line) and Au/GaN (green line) surface. The dashed line is the PL spectrum of InGaN/GaN. (c) integrated PL enhancement ratios for samples with Ag (square), AL (triangle), and Au (circle) are plotted against the thicknesses of GaN spacers. The dashed lines are the calculated penetration depths.*

at shorter wavelength for Ag samples, whereas it is independent of wavelength for Al-coated samples. The PL enhancement after coating with Ag and Al is due to string interaction with SPs. Electron-hole pairs excited within the QW couple to electron vibration at the metal/semiconductor interface when the energies of electron-hole pairs in InGaN ($h\omega_{InGaN}$) and of the metal SP ($h\omega_{SP}$) are similar, where h and ω are the reduced Planck constant and the frequency, respectively. Then, electron-hole recombination produce SPs instead of photons, and this new recombination path increases the spontaneous recombination rate.

Figure 4.36 shows the dispersion diagrams of SP on metal/GaN surface calculated from reported dielectric functions. The plasmon energy ($h\omega_P$) of silver is 3.76 eV, but this energy must be modified for Ag/GaN surface coatings to $h\omega_{PS} \approx 3$ eV ($\approx$ 410 nm) when using the dielectric constant of Ag and GaN. Thus, Ag is suitable for SP coupling to blue emission, and we attribute the large increases in luminescence intensity from Ag-coated samples to such resonant SP excitation. In contrast, the estimated $h\omega_{SP}$ of gold on GaN is below ~2.2eV (~50 nm), and no measurable enhancement is observed in Au-coated InGaN emitters as the SP and QW energies are not matched. In the case of Al, the $h\omega$SP is higher than ~5 eV (~250 nm), and the real part of the dielectric constant is negative over a wide wavelength region for visible light. Thus, a substantial and useful POL enhancement is observed in AL-coated samples, although the energy matched is not ideal at 470 nm a better overlap is expected shorter wavelengths. The clear correlation between Fig. 4.36 a and b suggests that the obtained emission enhancement with Ag and Al is due to SP coupling.

PL intensities of Al- and Ag-coated samples are also found toe depend strongly on the distance between QWs and the metal layers, in contrast to AU-coated samples. Figure 4.36(c) compares integrated POL enhancement ratios for three different GaN spacer thickness (of 10, 40, and 150 nm) for Ag, Al and Au coatings. Al and Ag samples show exponential increases in the luminescence intensity as the spacer thickness is decreased, whereas so such improvement was measured in Au-coated QWs. This spacer-layer dependence of the PL enhancement ratios matches out models of SP-QW coupling, as the SP is an evanescent wave that exponentially decays with distance from the metal surface. Only electron-hole pairs located within the near-field of the surface can couple to the SDP mode, and this penetration depth (Z) of the SP fringing field into the semiconductor is given by $Z = \lambda/2\pi[(e'_{GaN} - e'_{metal})/ e'^{2}_{metal}]^{1/2}$ where e_{GaN} and e'_{metal} are real parts of the dielectric constants of the semiconductor and metal, and Z can be calculated as Z = 47, 77, and 33 nm for Ag, Al and AU, respectively. Figure 4.36(c) shows a good agreement between these calculated penetration depths (dashed lines) and measured values of the PL enhancement (symbols for Ag and Al coated samples.

If the metal/semiconductor surface are perfectly flat, it would be difficult to extract to light from the PS mode, a non-propagating evanescent wave. However, roughness and imperfections in evaporated metal coatings can efficiently scatter SPs as light. Such roughness in the metal layer is observed from topographic images obtained by shear-force microscopy of the original GaN surface (Fig. 4.37(a)) and the Ag-coated surface (Fig. 4.37(b)). Figure 4.37(c), d shows the depth profile along the dashed lines in Fig. 4.37(a), (b), respectively. Modulation depth on measuring of the Ag surface is approximately 30–40 nm whereas the GaN surface roughness on measured is below 10 nm. Higher-magnification scanning electron microscopy (SEM) images of the Ag and GaN surfaces are shown in Fig. 4.37(e), (f). The length scale of the roughness of the Ag surface was determined to be a few hundred nanometers. Figure 4.37(g) shows a fabricated metal granting, a geometry that is used to couple SP and photons. Microluminescence images of uncoated, coated and patterned grating structures of Ag on InGaN QWs with 10 nm spacers are shown in Fig. 4.37. There is a doubling of the emission from 133-nm-wide Ag stripes forming a 400 nm period grating, which are not observed from 200-nm-wide Ag stripes within a 600 nm period gratings. This measurement suggests that the size of the metal structure determines the SP–photon coupling and light extraction.

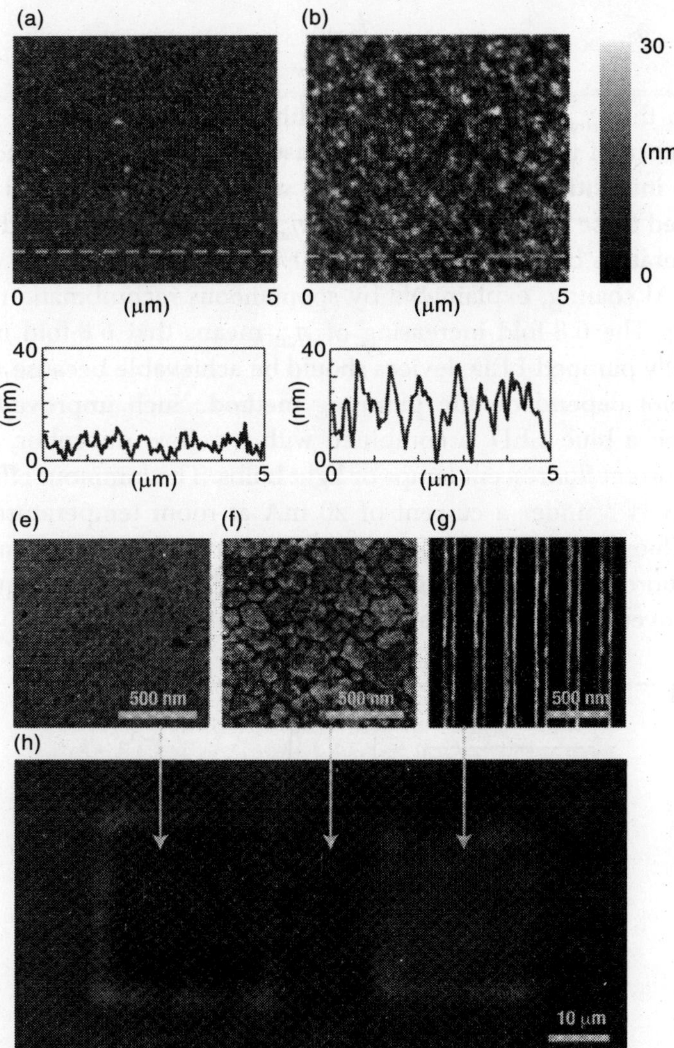

Fig. 4.37 *Topographic and luminescence images. (a) Topography image of the uncoated GaN surface. (b) Topographic image of a 50-nm-thick Ag film evaporated on GaN. (c) The depth profile along the dashed line in Fig. 4.37(a). (d) The depth profile along the dashed line in Fig. 4.37(b). (e) SEM image of the uncoated GaN surface. (f) SEM image of the 50 nm Ag film evaporated on GaN. (g) SEM image of the grating structure with 33% duty cycle fabricated within a 50-nm-thick Ag layer on GaN. (h) Microluminescence image including the areas of (e–g). Al samples used were coated onto the InGaN QWs with 10 nm GaN spacers.*

As spontaneous emission rates are increased, so are the internal quantum efficiencies η_{int}. The internal quantum efficiencies (η_{ext}) of light emission is given by the light-extraction efficiency (C_{ext}) and η_{int}, which is given by the ratio of the radiative (k_{rad}) and non-radiative (k_{non}) recombination rates of electron-hole pairs:

$$\eta_{ext} = C_{ext} \times \eta_{int} = C_{ext} \times \frac{k_{rad}}{k_{rad} + k_{non}}.$$

In order to obtain the η_{int} to separate the SP enhancement from other possible effects, the temperature dependence of the PL intensity is measured. Figure 4.38(a) shows Arrhenius plots of the integrated PL intensities from InGaN QWs separated from Ag and Al films by 10 nm spacers, and compared these to uncoated samples. η_{int} values from uncoated QWs are estimated as 6% at room temperature by assuming ηint ≈ 100% at 4.2 K. These ηint values increased 6.8 times (to 18%) after Al coating, explainable by spontaneous recombination rate enhancements through SP coupling. The 6.8-fold increasing of η_{int} means that 6.8-fold improvement of the efficiency of electrically pumped LED devices should be achievable because η_{int} is a fundamental property and does not depend on the pumping method. Such improved efficiencies of the white LEDs, in which a blue LED is combined with a yellow phosphor, are expected to be larger than those of current fluorescent lamps or light bulbs. The luminous efficacy of commercial white LEDs is 25 lm W^{-1} under a current of 20 mA at room temperature. This value is still lower than that of fluorescent tubes (75 lm W^{-1}). A threefold improvement is necessary to exceed the current fluorescent lamps or light bulbs. The SP coupling technique is very promising for even larger improvements of solid-state source.

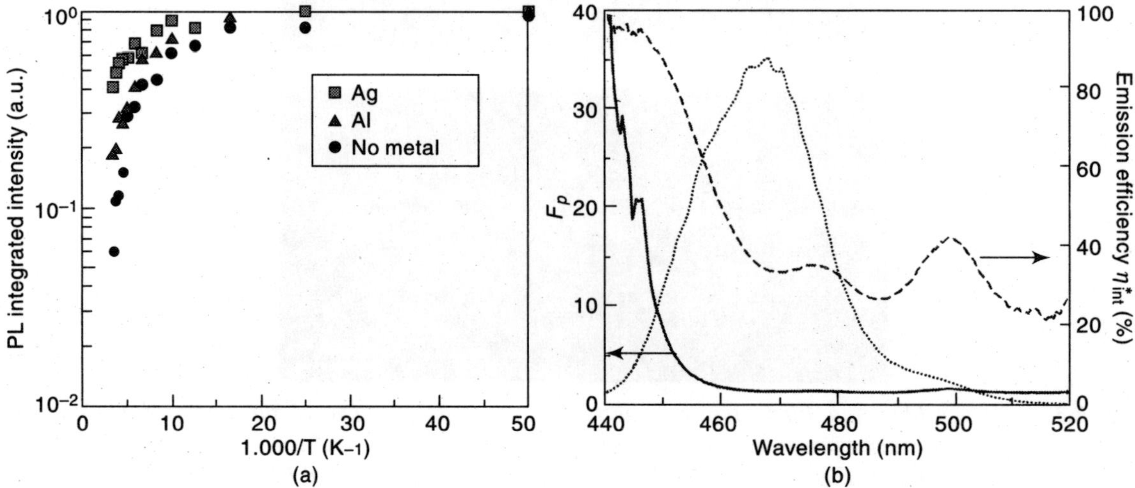

Fig. 4.38 *Temperature dependence and Purcell enhancement factors. (a) Arrhenius plots of the integrated PL intensities of InGaN/GaN with Ag (square), Al(triangle) and uncoated sample (circle) with 10 nm GaN spacers. PL integrated intensities at 4.2 K were normalized to 1. (b) Wavelength-dependent emission efficiencies (η^*_{int} (ω)) of the InGaN/GaN with Ag layer with 10 nm GaN spacers dashed line). The Purcell enhancement factor $F_p(\omega)$ estimated by η^*_{int} (ω) is also plotted (solid line). The dotted black line is the PL spectrum of the same sample.*

Wavelength –dependent enhanced efficiencies $\eta^*_{int}(\omega)$ are related to the coupling rate $k_{SP}(\omega)$ between QWs and SPs by the relationship:

$$\eta^*_{int}(\omega) = \frac{k_{rad}(\omega)+C'_{ext}(\omega)k_{SP}(\omega)}{k_{rad}(\omega)+k_{non}+k_{SP}(\omega)}$$

where $C'_{ext}(\omega)$ is the probability of photon extraction from the SP's energy, and is decided by the ratio of the light scattering and dumping of electron vibration. $C'_{ext}(\omega)$ depends on the roughness and structure of the metal surface (Fig. 4.37). Figure 4.38(b) shows the $\eta^*_{int}(\omega)$ of Ag-coated sample estimated from the OL-enhancement ratio (Fig. 4.36(a)) by normalizing the integrated η^*_{int} should be 41%. We find that $\eta^*_{int}(\omega)$ increased at shorter wavelengths where the plasmon resonance more closely matches the QW emission, and reaches almost 100% at 440 nm. The Purcell enhancement factor25 F_p quantifies the increase in spontaneous emission rate into a mode of interest, and is described by $\eta_{int}((\omega)$ and $\eta^*_{int}(\omega)$ when $C'_{ext} \approx 1$:

$$F_p(\omega) = \frac{k_{rad}(\omega)+k_{non}(\omega)+k_{SP}(\omega)}{k_{rad}(\omega)+k_{non}(\omega)} \approx \frac{1-\eta_{int}(\omega)}{1-\eta^*_{int}(\omega)}$$

Figure 4.38(b) also shows $F_p(\omega)$ estimated at each wavelength by assuming a constant $\eta_{int}(\omega)$ = 6%. $F_p(\omega)$ is significantly higher at wavelengths below 560 nm. The PL spectrum shape, plotted as a black dotted line, also indicated $F_p(\omega)$ values are higher for shorter wavelengths, a possible reason for the asymmetric in the luminescence peak of Fig. 4.38(b). Figure 4.38(b) suggests that InGaN QWs with peak positions around 440 nm matches with SPs from silver layers, yielding enhanced $\eta^*_{int}(\omega)$ approaching 100% throughout the luminescence spectrum. The SP is geometrically tuned to match out $\lambda \approx 470$ nm QW by fabricating nanostructures.

The SP enhancement of InGaN QW emission provides a promising methods for developing highly efficient solid-state light sources. The significant η_{int} increases are due to higher spontaneous recombination rate, and the distance and choice of patterned metal films can be used to optimize light emitters. SP coupling is one of the most interesting methods for developing efficient LEDs, as the metal can be used both as an electrical contact and for exciting plasmons.

The Method

$In_{0.3}Ga_{0.7}N$ QWs 3 nm thick are grown onto GaN (4 μm thick)/sapphire (0001) substrates by metal–organic chemical vapour deposition. To investigate the influence of the distance between the metal and the QW, 10, 40 and 150 nm thick GaN spacers are grown onto these QWs, and 50 nm thick Ag, Al or Au layers are then evaporated on top of the wafer surfaces (Fig. 4.35(a)). PL measurements are performed by exciting the QWs with a 406 nm diode laser from the back of the substrates, and collecting luminescence with a multichannel spectrometer. The excitation power is 4.5 mW. The temperature dependence of the luminescence intensity is determined by cooling the samples in a cryştal (Oxford Microstat) to ~4.2 K during such measurements. Topographic measurements are performed by a avin near-field optical microscope. Fluorescence microscopy is used with × 40 objective, a mercury lamp, and a colour charge-coupled device camera. Metal grating structures are fabricated by electron-beam lithography on a 50-nm-thick polymethylmethacrylate mask coated on the 50-nm-thick Ag surface. Electron-beam lithography is performed. The pattern is transferred through the top metal layer by argon-iron milling resulting in 400 nm and 600 nm period gratings on InGaN/GaN QWs with 10 nm GaN spacers. The duty cycle of these silver gratings is ~33%.

Nanoarrays

Carbon nanotubes (CNT) and semiconducting nanowires (NW) are built, with diameters measured in nanometers as well as thin molecular layers capable of forming non-volatile switches. Self-assembly techniques are used to produce sets of parallel NWs with nanometer spacings. One set is then placed above at right angles. The crosspoints in these arrays act as nanovolatile switching elements, allowing to control and differentiate the behavior of the assembled arrays at the nanoscale. Technology of this kind forms the basis for nanoscale memory devices and even programmable nanoscale logic arrays called **nanoarrays**, or a crossbar. (see Figure 5.1(a)). The dimensions of these nanoarrays (diameter of the wires, spacing between wires) are controlled to nanometer dimensions without using direct lithographic pattering Molecular seed catalysts control the diameter and physical forces between wires control spacing. Fig. 5.1 (a) A crossbar in which sites holding 1s are marked by dots and (b) (2,4)-hot addressing of six nanowires $\{nw_1, \ldots, aw_0)\}$ with four micro-level address wires $(aw_1, \ldots, aw_4\}$.

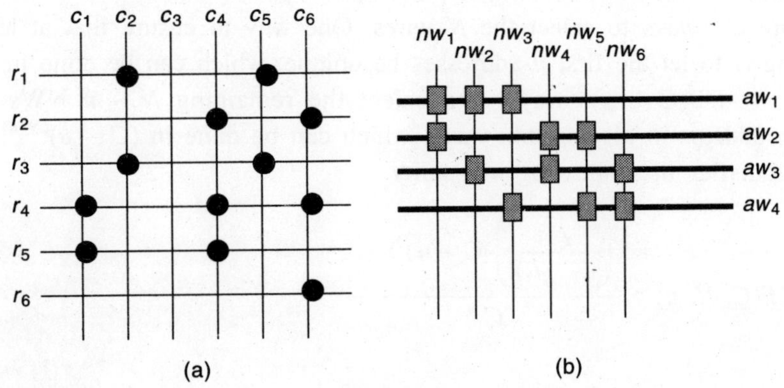

Fig. 5.1

Self-Assembly of NWs

Since NWs are too small to be manufactured lithographically and are too numerous to manipulate at the atomic level, some other means of assembling them is necessary. Directed self-assembly of undifferentiated NWs into sets of parallel wires. The assembly process mixes a set of modulation-doped NWs in a solution and flows the mixture into a trough.

NWs are then transferred to a chip. It is assumed that NW codes form a subset the (h, b)-hot addresses, and the each address is chosen at random with equal probability.

Stochastic Assembly

The stochastic selection of coded NWs from a sufficiently large ensemble of such NWs (the code space) ensures that all of almost all codes are unique. For the uses made of nanoarrays, it is not necessary that all codes be represented among the NWs.

Consider a large code space (e.g., 10^6 codes) and a very large number of wires of each code type (e.g., 10^6 of each, or 10^{12} total wires), and build a small array with ten wires in it. If each wire is selected randomly from the 10^{12} total wires, there is a very high likelihood that all ten wires are unique (in fact, over a 99.995% chance). There is an even higher likelihood that at least nine unique wires are obtained. From this example, it should be clear that one can randomly select the coded wires and obtain the independent nanoscale addressability that is desired.

It should be feasible to mix together a large number of NWs in solution in coded to achieve random coded mixing. Common techniques for aligning NWs are generally based on flow alignment is solution, so an additional mixing step is easy to accommodate.

Of course, large code space is not used as this does costs additional control wires. Consequently, the question becomes: How large does the size of the code space C need to be compared to the number N of NWs in an array in order to ensure that a large number of NWs have unique addresses?

So, a lower bound is achieved on the probability $P(C, N, u)$ that u unique codes in array of N wires randomly selected from a code space of size C by counting set sizes. Model the problem of code selection by assuming that each of the C codes appears equally frequently in the set of codes and that there are sufficiently many instances of each code that removing one does not change the probability $1/C$ of choosing a particular code.

Thus, there are C^N ways to select the N wires. One way to ensure that at least u NWs have unique addressing is to let the first u addresses be unique, which can be done in $C(C-1)(C-2) \ldots (C-u+1) = C!/(C-u)!$ ways, and select the remaining $N-u$ NWs from the set of remaining $C-u$ address in all possible ways, which can be done in $(C-u)^{(N-u)}$ ways. It follows that $P(C, N, u)$ satisfies the following inequality:

$$P(C, N, u) > \frac{\left(\dfrac{C!}{(C-u)!}\right)(C-u)^{(N-n)}}{C^N} \tag{3}$$

Table 1 *Probability that all N wires in a set are unique when selected from coded of size C*

Array Size (N = u)	*Code Size* (C)	*Probability* (*Eq. 4*)	*Estimates* (*Eq. 5*)
10	100	0.35	0.62
10	1,000	0.90	0.96
10	10,000	0.990	0.996
100	10,000	0.37	0.61
100	100,000	0.90	0.95
100	1,00,000	0.990	0.995
1000	1,000,000	0.37	0.61
1000	10,000,000	0.90	0.95
1000	100,000,000	0.990	0.995

A weaker and simpler bound is

$$P(C, N, u) \geq \left(\frac{C-u}{C}\right)^N = (1-x)^N \tag{4}$$

Here, $x = u/C$. It is straightforward to show that $(1 - x)^N \geq (1 - Nx)$ by induction where the base case is $N = 2$. It follows that if $Nu/C << 1$, $P(C, N, u)$ is close to 1. In fact, if $Nu/C \geq 10^{-2}$ then $P(C, N, u) > 0.99$.

Table I shows sample calculated lower bound probabilities for achieving unique sets of coded wires for 10, 100, and 1000 NW arrays using various size and code spaces. This data confirms that $C = 100 \times N^2$ is sufficient to yield almost all unique codes and $C = 10 \times N^2$ provides at most a 5% chance of not achieving unique codes.

For the even-weight codes described above, it would take a dense code with 14 bits to uniquely address over 1000 coded NWs. A code with 30 bits will support 155117520 unique codes, exceeding the $C = 100 \times N^2$ bound for $N = 1000$. In other words, this scheme requires a little over twice the number of control lines we would need of we could perfectly select and place coded NWs. This is true asymptotically since $\log(100N^2) = \log(100) + 2\log(N)$.

Alignment of NWs

By the self-assembly process the length-wise alignment of the doped regions on individual NWs with microscale address wires cannot be obtained.

So more potentially-doped regions to NWs are obtained than there are address wires. As a NW shifts relative to the address wires, different coded regions line up with the address wires, different coded regions line up with the address wires (Figure 5.2). This results in one of many different doping patterns being associated with each NW. For example, if multiple copies of some hot address are placed on an nanowire in repeating pattern, then when the nanowire is shifted, a new hot address is produced.

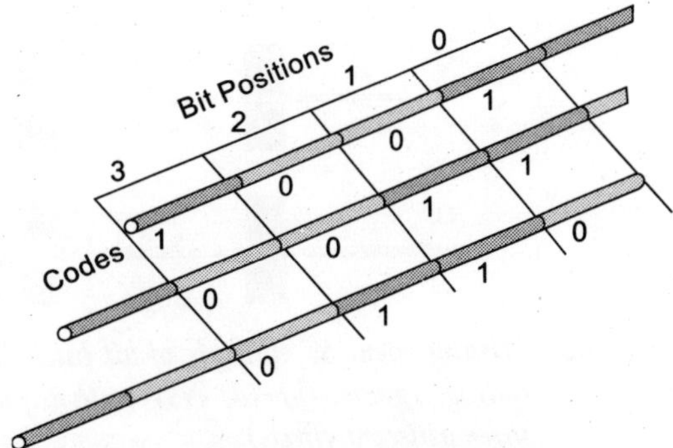

Fig. 5.2 *Three shifted copies of a nanowire with doping along its length. (NW coated with address)*

Alignment

NWs do not perfectly align to the control lines. The misalignment are due to: (1) multiples of the width of control wires i.e. the control of bit pitch; and (2) fractions of the bit pitch.

A. Multiples of Bit Pitch

Misalignments by multiple by pitches are done by repeating the code multiple times on the wire. For an n/2-hot code, if multiple copies of the code are [see Fig. 5.3(a)], any contiguous group of an coded bits will only be a rotation of the original code and, hence, is a valid code in this code space. As selecting codes is done randomly, random misalignment does not change the random code selection. In some applications (e.g., memories, when each nanowire layer does not see the field of the orthogonal nanowire layer), simply the code is repeated along the entire length of the wire, and the fact that the wire is coded along its length will not interfere with operation. For other applications, it is possible to mask off the address ends at a lithographic scale and bulk dope the non address sections of the nanowires, to extinguish the control regions outside of the addressing field. Alternatively, alignment within a few bit pitches is achieved by repeating the code (or a fraction thereof) for a distance equal to the alignment tolerance, (see Fig. 5.3). Here, the NW conducts across a coded region when there is no field applied; this way, the controllable bit code regions which ends upon either side the control wires all continue to allow signal conduction.

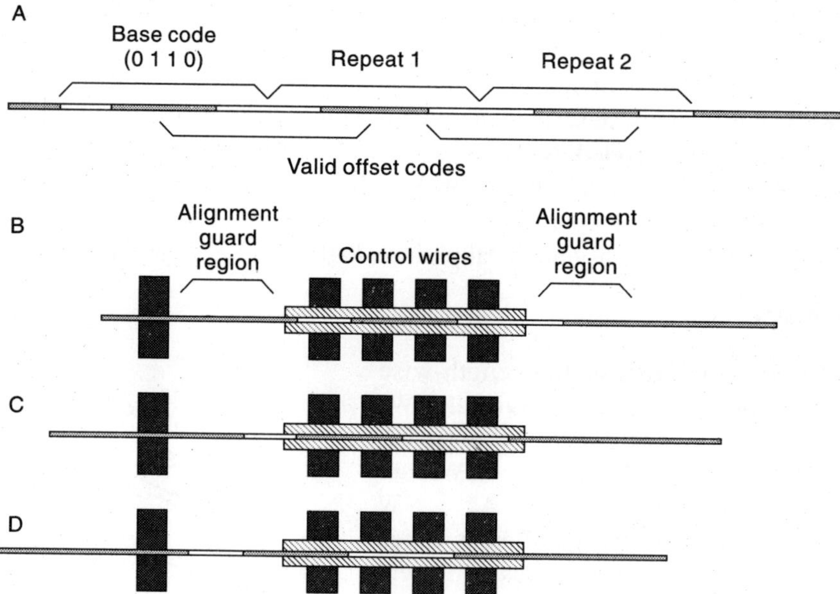

Fig. 5.3 *Misalignment by multiple of bit pitch (a) NW with three copies of code to tolerate misalignment by ± coding regions. (b)–(d) NW with a partial repeat of two bits to tolerate ±1 bit position (shown at three different effects).*

B. Fractions of Bit Pitch

In order to affect a controllable region, sufficient overlap between the field of one microwire and the doped controllab:e region is needed (see Fig. 5.4). Deplete carriers in a small region along the axis of the nanowire in order to stop conduction. Thus, the necessary overlap $W_{overlap}$ between the microwire field and the NW control region is likely to be small–on the order of a diameter or two of the NW (e.g., 5 nm). Overlaps between 0 and $W_{overlap}$ may only partially turn of conductions, resulting in intermediate current flow levels.

As shown in Fig. 5.4, a noncontrollable region between the fields of adjacent microscale control wires. Now, consider making the doped controllable region equal to the length of the noncontrollable region of nanowire plus $2 \times 3\ W_{overlap}$ as, there is always at least $W_{overlap}$ under once of the adjacent control fields, making every wire controllable and there is a window of alignments of alignments of length $2 \times W_{operlap}$ where the controllable region is affected by the fields of two microwires. For all other alignments, the controllable regions is under only one of the fields. Assuming all sub-bit-pitch misalignments are equally likely, the probability that a region is controlled by only one of the adjacent -microscale wire is

$$P_{control} = \frac{W_{bitpitch} - 2 \times W_{overlap}}{W_{bitpitch}}$$

$W_{bitpitch}$ is the bit pitch for the microscale wires; e.g., if the bit pitch is 210 nm and we conservatively assume a necessary overlap of 10 nm, $P_{control} = 190/210 \geq 0.90$. When controllabe region overlap multiple fields, it requires both fields to be zero to allow conduction. Since every n/2-hot code contains at least one transition between zeros and ones, the overlap case can end up with a least one more control region than any valid code. Consequently, no code in the standard *n*/2-hot code space will enable this small fraction of misaligned wires. Using codes outside of the code space, it is possible to still address some of these wires.

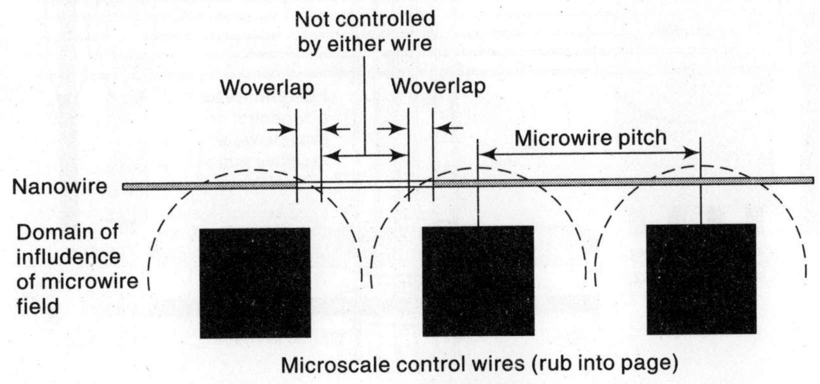

Fig. 5.4 *Design of modulation-doped control region length.*

It is not clear at this point far the microwire control field will extend beyond the width of each control microwire (Fig. 5.4). It is possible that the microwires are spaced wider than the minimum microwire pitch to prevent adjacent wire control fields from overlapping.

Forming Cross bars

A crossbar is formed by applying the self-assembly process twice, a second set of NWs is placed above and at right angles to first set (se Figure 5.5). This two-step process requires trimming of NWs that extend beyond the array, which is done lithographically.

Switches are defined at the crosspoints of NWs by either depositing a thin film of a supramolecule between the two parallel sets of wires or by coating one set of NWs with such a material (a form of radial doping). In both cases when large field is applied cross the switchable material it becomes conducting (represented by 1) or non-conducting (represented by 0) depending on the polarity of the field. The conductance of the material at each cross point is sensed by applying a field large enough to detect current flow but not large enough to change the state of the material.

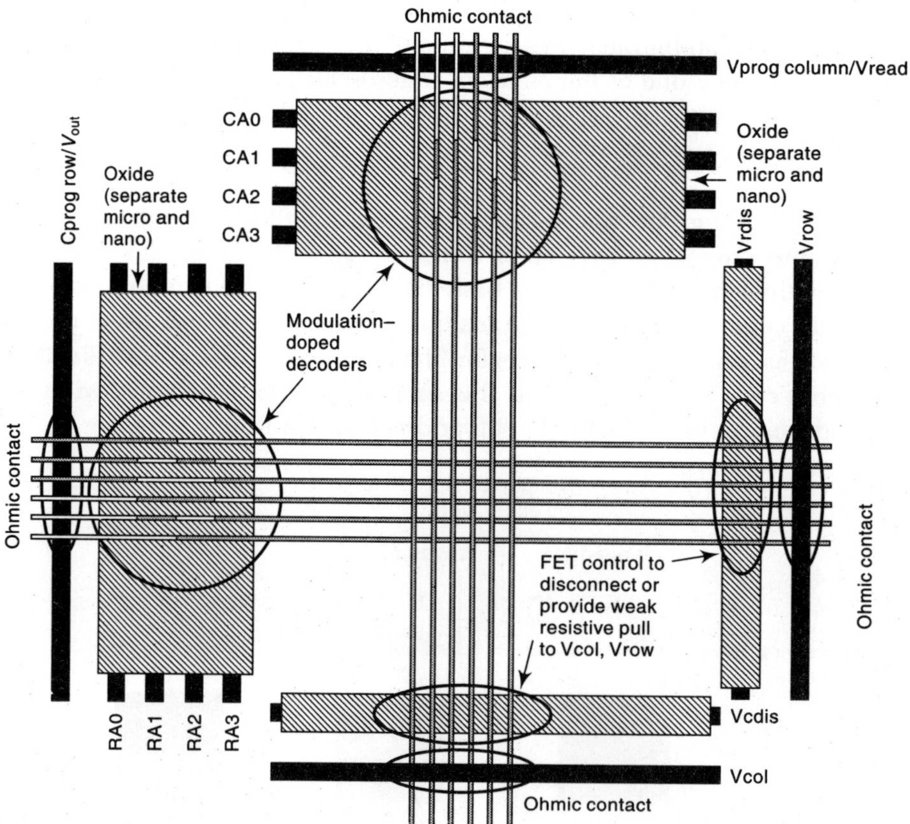

Fig. 5.5 *Nanoscale memory array interfaced using modulation-doped address decoders (shown with only a few nanoscale wires for clarity). A typical array size would have 100–1000 nanoscale wires addressed by only 24–30 microscale wires.*

Simple Memory Design

A programmable memory can be constructed by placing a decoder of the form described above on both sides of an NW array (see Fig. 5.5). There are multiple-scale technologies under consideration for placing nonvolatile memory bits at the crosspoints. These technologies are programmed by placing a large voltage across individual crosspoint junctions and are read by observing the current flowing through a junction, with programmed ON junctions acting as low-resistance paths, while programmed OFF junctions act as high-resistance paths.

The memory technology also supports writing blocks of 1s (**stores**) or 0s (**restores**) at the intersection of rows and columns of a crossbar. To execute such an operation, the rows and columns are selected and either a large positive or negative voltage is applied.

Note, that two NWs with the same encoding attached to a common ohmic region behave as a single NW with that encoding. When this happens the effective number of NWs is less than the actual number.

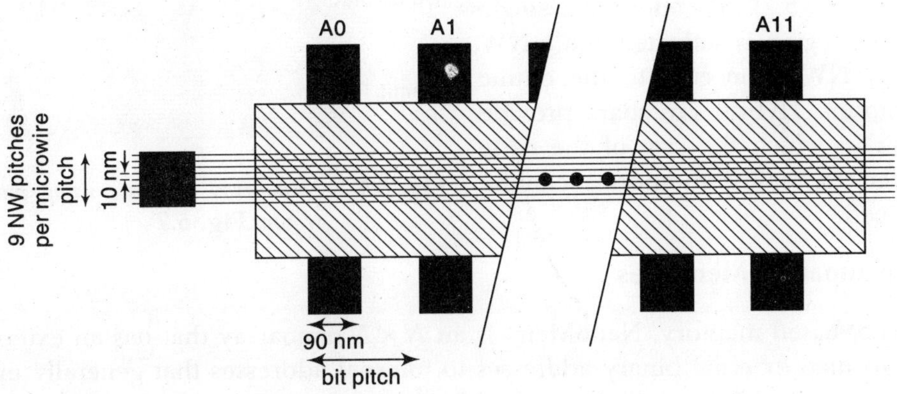

Fig. 5.6 *Small address space decoder.*

Using these addressable NWs, exactly one row and one column wire can be enabled so that we can apply a programming voltage across a single crosspoint. This will require care in the selection of voltage levels such that the crosspoints that are in the same row and column as the intended crosspoint are not also affected. These row and column neighbors will have one side pulled to the programming voltage, while the other is pulled to a nominal voltage, whereas the intended crosspoint is pulled to the programing voltage by both the row and column decoders and, hence, will see a greatly voltage differential. We also can generally arrange for the crosspoints to act as diodes to avoid parasitic paths in a partially programmed array.

Data bias are read from the array by again placing the appropriate control bits to enable only a single row and column. A high voltage is placed on the common column line, and the voltage on the common row line is observed. In this manner, only the intended crosspoint sees both a high input on its column line and a low-resistance path to the common row line. If the crosspoint is programmed ON, it will be possible to observe the current flowing out of the selected row line, perhaps raising the row line voltage. If the crosspoint is programmed OFF, there will be less current flow.

Crossbar Connection

To use a standard crossbar, an ohmic contact is needed at both ends of each set of parallel NWs, as shown in Figure 5.5, so that it is possible for current to flow through an individually addressed NW. It is also necessary to disconnect one of these two ohmic contacts so that current flows from a NW in one parallel set to an orthogonal NW in the other parallel set and be sensed. This is achieved by doping the ends of each NW just inside the "ground" ohmic contacts (associated with V_{ge} and V_{gr}) so that each NW is disconnected from its ohmic contact by a FET controllable region.

Hybrid crossbar use multiple ohmic regions, that is, the NWs in each parallel set are divided into sets of equal size and one ohmic region is attached to one end of each set and a common ohmic region to the other ends. (See Figure 5.7). To address a single NW, an ohmic region is activated and a NW address for a NW connected to the ohmic region is chosen. Hybrid crossbars provide several ways in which the size of the code space needed to realize memories can be greatly reduced.

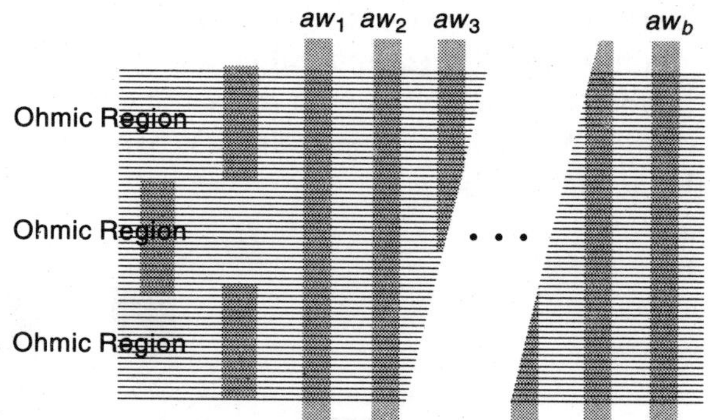

Fig. 5.7 *Nanoarray interface using multiple ohmic regions.***Fig. 5.7**

Designing Nanoarray Memories

A nanoarray-based memory, NanoMem, is an $N \times N$ nanoarray that has an external circuit, LithoMem, to map external binary addresses to internal addresses that generally unknown in advance. Each nanoarray has two sets of b address wires and two sets of m ohmic region.

NanoMems are constructed by a stochastic process.

Crossbar and Modulation-Doped Nanowires

Crystalline semiconducting wires with nanometer dimensions are grown by a vapor-liquid-solid growth process that controls the diameter of NWs. When the vapor level of reactants in this process is modulated over time, the semiconducting wires are doped along the axial dimension. The length of a doped region is controlled by the length of time during which doping reactants are in the vapor (**modulation-doping**); the transition between differently doped regions occurs over an axial length that is as small as fraction of a nanometer. The doped regions act as field-effect transistors (FETs). A NW is gated by a high electric field provided by a microwire at right angles to NW, Figure 5.1(b).

NWs are controlled by (h, b)-hot addressing. Each NW has b addressable regions exactly h of which are doped. Subsets of this set of codes C are used. It follows that a NW is conducting unless one of its doped region is adjacent to an address wire carrying an electric field of sufficient magnitude to reduce its conductance to near zero. Thus, exactly one NW in the set (h, b)-hot addressable NWs conduct if h of the address wires carry a large electric field. In order to create functioning memory, external binary addresses are mapped to the internal addresses of individual nanowires.

Doped NWs act as field-effect transistors (FETs) that is, conduction along the length of an NW is controlled by an applied voltage field. For the depletion-mode *p*-type devices a low voltage (or no applied voltage) allows good conduction, whereas a high applied voltage evacuates, carriers from the doped semiconductor, preventing conduction along the NW length. This allows to build a combining logic when several conductors cross a doped NW–if all the inputs are low, there is a conduction path from one side of the crossed wires to the other, if any of the inputs are high, there will be no conduction path (see Fig. 5.8).

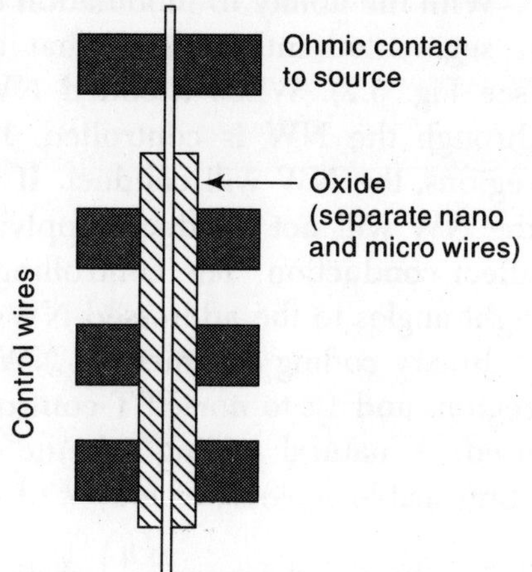

Fig. 5.8 *NWFETS with multiple gates wire crossings serve as an AND, allowing signal flow only when all control wires have suitable voltage.*

It is possible to control the doping profile or material composition along the axial dimension of an NW. By controlling the doping profile the threshold voltage for the FETs are controlled effectively.

That is, with high doping, it becomes very hard to deplete the carriers from the channel and stop conduction through wire; consequently, the threshold voltage is high. With low doping, there are fewer carriers, allowing a low voltage to deplete the channel and stop conduction. This allows to construct wires which are gateable in some regions but not gateable in others (see Fig. 5.9).

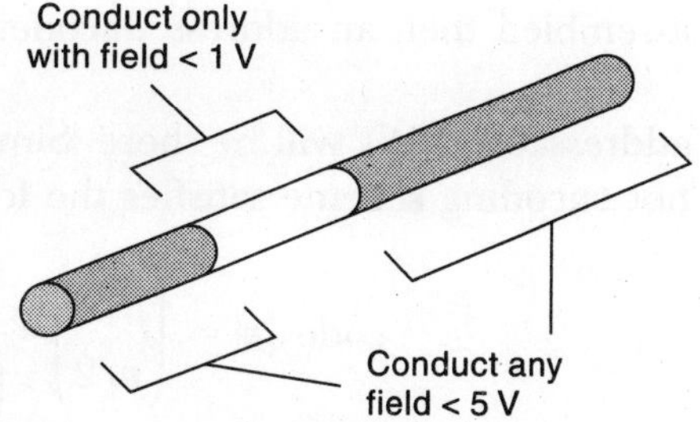

Fig. 5.9 *Modulation doping places selective gateable regions in an NW.*

The growth along the length of the NW is controlled by time. The NW crystal grows by incorporating new atoms into its lattice at one end. To control the dopant profile, simply the dopant concentration in the NW's growth environment over time is controlled. So, the width of each doping region is controlled by controlling the rate of the growth reaction and the introduction of dopants into the growth atmosphere at the appropriate times. The dimensions of the doping regions are thus defined completely without lithographic processing. It is observed that

1. the transition between materials occur over a 20-nm-length scale.
2. sharper transition (< 5 nm) are likely in smaller diameter NWs.
3. subnanometer transitions between materials.

The transition between a strongly doped (conducting) region and a weakly doped region of the same semiconducting material, is even easier using NWs that are just a few nanometers in diameter is of interest, but lithographic scale wires are tens of nanometers in width (e.g., 90 nm). Hence the transition region is small compared to the lithographic microwire pitch.

NW Coding

With the ability to modulation doped NWs, code words can be assigned to NWs. Each NW is segmented into regions that are doped as either FET-controllable or nanocontrollable (see Fig. 5.2). When a coded NW is aligned cross a set of microwires, the flow of current through the NW is controlled. If we apply a suitably low field on all the FET-controlled regions, the NW will conduct. If we apply a high field on any of the FET-controlled regions, the NW will not conduct. Applying a high field on the non-FET controlled regions will not affect conduction. The controlling voltages are provided by control microwires, which are at right angles to the addressed NWs (see Fig. 5.8).

Binary coding schemes for NWs are employed in which 0's correspond to FET-controllable regions and 1's to non-FET-controllable regions. There are many coding schemes that could be used. A natural coding scheme is the *k*-hot scheme in which each NW has *n* potentially controllable regions, exactly *k* of which are controllable (they must be "hot" to be controlled). This scheme allows for $\binom{n}{k}$ distinct codes.

Consider (*n*/2)-hot scheme. Place low voltages on the *n*/2 control lines the correspond to the 0's in a code word for an NW and high voltages on the rest, then this NW is the only one that can conduct. All other NWs will have a FET-controlled region where the code word has a 1 and will be disabled as a result. If exactly one of each type of coded NW into an array is assembled then an address decoder (see Fig. 5.10) with codes (*n*) = $\binom{n}{n/2}$ distinct codes for addressable NWs will be there. Simple calculations show that the number of codes in the *n*/2-hot encoding scheme satisfies the following relationships:

$$\text{codes}(n) = \binom{n}{n/2} = \frac{n!}{\left(\frac{n}{2}\right)!\left(\frac{n/2}{2}\right)!} \tag{1}$$

$$\frac{\text{codes}(n)}{\text{codes}(n-2)} = \frac{n\cdot(n-1)}{\left(\frac{n}{2}\right)\left(\frac{n}{2}\right)} = 4\left(\frac{n-1}{n}\right) \tag{2}$$

As the second calculation demonstrates, for large *n*, because codes(*n*)/codes(*n* – 2) approaches 4, the asymptotic growth of codes (*n*) approaches 2^n. Inverting this, to uniquely address *N* wires, we need no more than 1.1 $\log_2(N)$ + 3 address bits. Consequently, for large enough arrays, the overhead associated with control lines, even if built out of microscale wires, becoming small compared to the size of the nanoscale logic memory core which it addresses. The overhead remains modest even if *k*-hot addressing used with *k* much smaller than *n*/2.

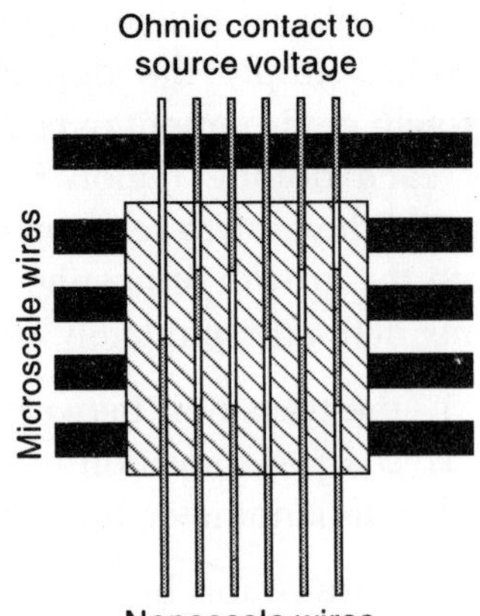

Fig. 5.10 *Decodes constructed from addressable NWs.*

Hybrid Control Memory

The simplest memory described above is easy to understand, but has the drawback that it requires a very large address space and, hence requires that first construct a very large collection of differently coded NWs (e.g., 25 million for a 500 × 500 array). Then, use a more modest number of NWs, if it is observe that only the stochastic addressing to distinguish among the number of wires that fit into one microscale wire width is needed. As shown in Fig. 5.6, selectively energize the endpoints of a collection of NWs at the lithographic scale; example, with a 10-nm pitch nanowires and a 90-nm-wide microwire nine nanowires at a time can be addressed. A 6-hot 12-bit code has 942 code words. With 942 code words, over a 96% probability is that all nine wires in a bundle will have unique codes.

By staggering adjacent microwave contacts, maintaining the tight NW pitch (see Fig. 5.11), perhaps losing one wire at the edge of each microwire group will be there. With a contact group length $W_{contact\_group\_len}$, uniquely address a wire group with N_g wires, as follows:

$$N_g = \frac{W_{contact\_group\_len}}{W_{nano\_pitch}}$$

A shown:

$$W_{contact\_group\_len} \geq 2 \times W_{litho\_pitch} - W_{metal\_width}$$

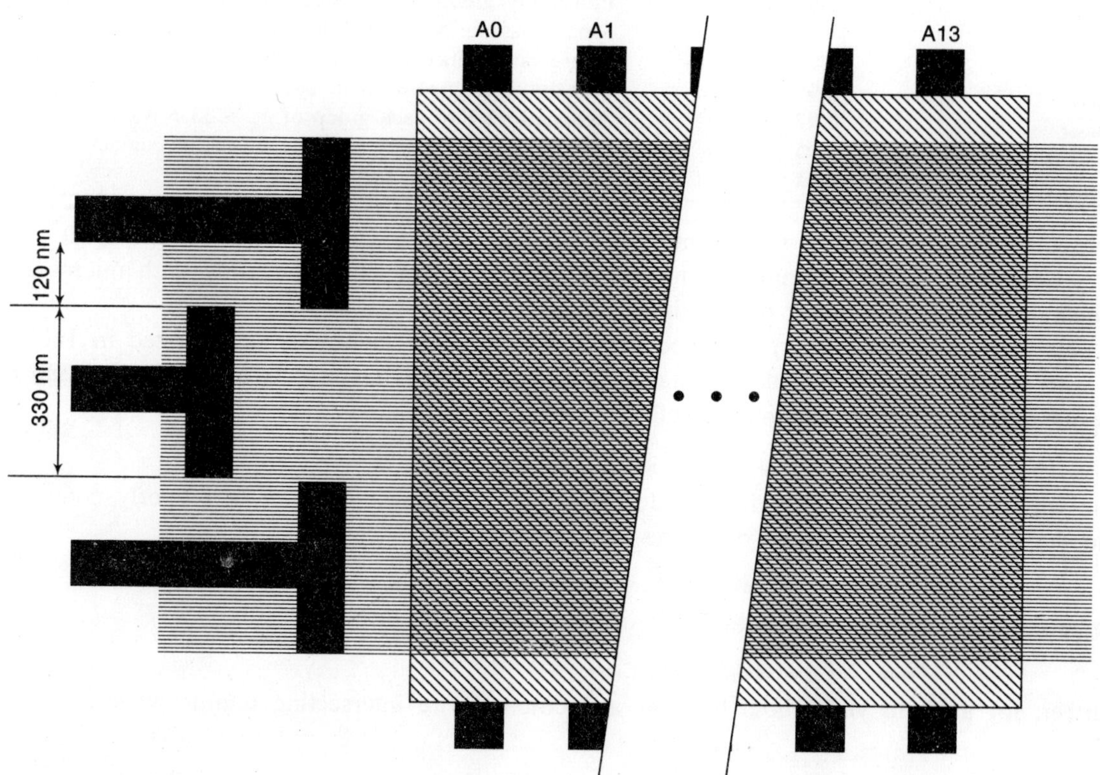

Fig. 5.11 *Small address space decoder with staggered microscale contacts.*

For a 90-nm process with $W_{\text{litho\_pitch}}$ = 210 nm, and $W_{\text{contact\_group\_len}}$ = 330 nm. With $W_{\text{nano\_pitch}}$ = 10 nm, N_g = 33. A 7-hot 14-bit code has 3432 code words thus, giving over an 85% chance of assembling a completely unique set of 33 coded NWs.

It is worthwhile to note that the microscale-to-nanoscale address area does not scale up with array size in this hybrid scheme. For each additional group of core wires add (e.g., N_g = 33 nanowires), so that an additional microscale wire for the contact is needed but the nanoscale addressing remains constant. To the 14 address bit pitches, address these 33 wires, add two bit pitches for the address contacts and two bit pitches for the load contact on the opposite side of the array. This allows to calculate the side length of the memory array including the microscale-to-nanoscale address translation $S_{\text{mem\_array}}$:

$$S_{\text{mem\_array}} = 18 \times W_{\text{litho\_pitch}} + N \times W_{\text{nano\_litch}}$$

For $W_{\text{nano\_pitch}}$ = 10 nm, N = 500, and $W_{\text{litho\_pitch}}$ = 210 nm (90-nm process), $S_{\text{mem\_array}}$ = 8780 nm. Including the decoder support, the memory area for the array is

$$A_{\text{mem\_array}} = S^2_{\text{mem\_array}}.$$

Table 5.2 *Memory Yield Factors*

Param	*Sample value*	*Description*
P_{unique}	0.85	Probability group of N_g wires contains unique members Sample assume N_g = 33, 7-hot, 14-bit code. (conservative since we yield a large fraction which are unique)
$P_{control}$	0.90	Probability aligned without overlap
$P_{wireend}$	0.97	Assume lose 1 of each group of N_g = 33 NWs
$P_{contact}$	0.95	Probability end makes adequate electrical contact
$P_{\text{nobreak (9 mm)}}$	0.91	(assume $P_{\text{break}}/W_{\text{nano-pitch}}$ = 0.0001)

Consequently, this gives a raw memory bit area of $A_{\text{mem-array}}/(N^2)$ = (8780 nm)^2/500^2 < 310 nm^2. Compared to the 100-nm^2 memory bit area in the NW core, the array with microscale-to-nanoscale decode is a factor of 3.1 times larger.

Less than (N^2) net bits are obtained due to a number of factors as summarized in Table 5.2. A wire is good only if it makes contacts at its ends where it connects to microwires and there are no breaks or shorts along its length:

$$P_{\text{goodwire}} = P^2_{\text{contact}} \times P_{\text{nobreak}} \times (P_{\text{nanowire}})$$

The wire is addressable only if the address group is unique, the wire is properly controllable, and the wire and connects to only one microwire group:

$$P_{\text{addressable}} = P_{\text{unique}} \times P_{\text{control}} \times P_{\text{wireend}}$$

For a wire to yield, it must both we good and addressable.

$$Y_{\text{wire}} = P_{\text{goodwire}} \times P_{\text{addressable.}}$$

Further, for a bit to yield, both the row and column wire intersecting it must yield:

$$Y_{\text{bit}} = Y^2_{\text{wire}}.$$

For the values in Table 5.2, Y_{bit} > 0.38. Combining with a raw area of 310 nm^2, this gives a yielded bit area of 800 nm^2.

Discovering Code Words

By design, the number of code words is large compared to the number of wires in any row, column, or microscale contact group ($C >> N$). Consequently, after fabrication discover which code words actually make up the set of live addresses.

A. Resettable Memory Technology

For the memory arrays based on resettable crosspoints initially, activate the row addresses with all zeros—this enables all of the row lines. Treat each column as a single bit rather than a collection of individual bits. This allows to attempt to program and read each possible column address. If the column address is programmable, then the address is present. If it is not programmable, it is not present. Once knowing which column addresses are present, each row address is tested using the known column addresses. Thus, both the column and row addresses are found and are operational in the array and crosspoint junction faults. For a $N \times N$ array, will takes stronger to test than an array with perfect, dense, predictable code. However, as long as $C = O(N^2)$, as derived above, it will still take only $O(N^2)$ total time.

With the hybrid address scheme, the testing overhead is only $O(N)$. For the 500×500 nanowire array above testing $(500/33) \times 3432 = 52000$ row addresses are needed and an equal number of column addresses to find which addresses are present. Note there are 250000 raw bits in the array, so the additional 104000 tests to fins valid addressed will not even double the number of test operations required. Compared to the $250000 \times 0.38 > 95000$ memory bits a yield from this array, the 534000 tests is less than four times the number of final yielded bits.

B. One-Time Programmable Crosspoints

Some crosspoints technologies set the junctions permanently during programming. These technologies, are tested for address presence without programming the crosspoints. As shown in Fig. 5.5, the row (and column) lines are connected to a common line (V_{row}, V_{col}). To test for presence of a row (column) address, weakly pull down V_{row} (V_{col}) and drive the row (column) address in question. If the address is present, it will be able to pull up V_{row} (V_{col}); if the address is not present, V_{row} (V_{col}) will be pulled down to a low voltage. By observing the voltage on V_{row} (V_{col}), detect the presence of absence of the address. This can be easily be done at voltage levels below the programming voltages so that no crosspoints are inadvertently programmed during address discovery. Again, as long as $C = O(N^2)$, we only $O(N^2)$ total row and column addresses to test to establish the set of row and column addresses present. This is reduced to $O(N)$ for the hybrid addressing scheme.

C. Mapping Present Address

Since the codes are sparse, it will also be necessary to keep tack of the live row and column addresses. There are N live row and column addresses, each of which is $O(\log(N))$ bits long; consequently, $O(N \log(N))$ bits of storage is needed to hold this address translation to build a monolithically addressed memory or about $14N$ for the hybrid addressing scheme. Since this is asymptotically smaller than the $O(N^2)$ bits in the memory, the memory to hold this translation is smaller than the memory that we are addressing. At the cost of multiple nanoscale reads to resolve an address, therefore apply this reduction trick repeatedly to reduce the number of bits needed to an arbitrarily small number which can then be store in a microscale memory.

Using the hybrid addressing scheme, 14 × 500 bits are needed to describe the nanoscale portion of the 500 row wires along with an equal number of bits for the column wires. Thus, ultimately 14000 bits of data are needed in order to retrieve the 95000+ vias in the memory. The roughly 7 : 1 reduction here is not adequate to reduce the information needed to address this memory to a sufficiently compact amount that it can be efficiently stored in a lithographic-scale memory; consequently, multiple stages of mapping will be necessary. An important area of future work will be the development of multistage nanoscale address mapping architectures where a nanoscale mapper can be programmed, using the same technique, to perform the address translation so as to provide a deterministic external set of memory addresses. Even if two programmed address mappers that are as large as the memory addressed are needed, still benefits from this are there.

FIELD EMITTER ARRAY

Flat Panel Displays for Head-mounted Applications

Field-emitter-array Flat-Panel Display (FED): It is a small, high-resolution, high-luminous-efficiency and high brightness display. It incorporates a high density, high-performance array of low-voltage field emitters, as shown in Figure 5.12.

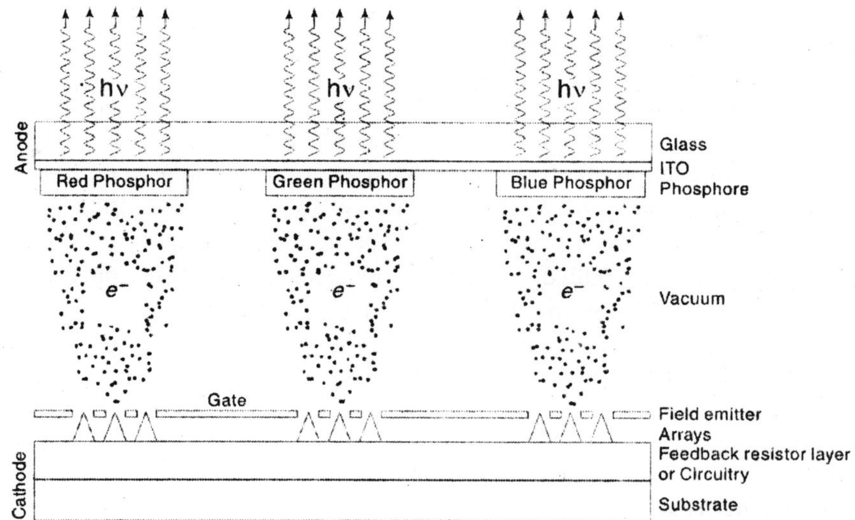

Fig. 5.12 *Concept of a field emission display pixel. The display uses a 2D matrix addressable electron source allowing temporal and spatial modulation.*

CMOS controlling electron emission from the tips impinges on a cathodoluminescent screen. It is thus possible to integrate the addressing and signal conditioning electronics on the same substrate as the Field Emitter Arrays (FEAs). The main advantage of this approach is the reduction of the number of wires and bond pads from about 2,000 to about 50. For example, it will be difficult to attach > 2,000 wires to bond pads in an area of 1.5″ × 1.5″ and obtain ultra-high vacuum in the display envelope. High resolution (> 1000 dpi) FEDs are only possible if the addressing/driver and other signal conditioning electronics are integrated on the same substrate as the field emitter arrays.

To demonstrate low-voltage field-emitter arrays fabricated using interferometric lithography for future integration with Si CMOS technology, interferometric lithography is used to define the emitter cone arrays that are spaced 200 nm tip-to-tip and <50 nm gate-to-emitter separation. Fabricated cone-field-emitter arrays with a 320-nm period have demonstrated emission currents 1 mA at a gate voltage of 20V from 900 cones in a 10 μm × 10 μm area. This current is more than adequate for a brightness of 1000 fL at a screen voltage of 500V.

To model the scaling behavior of FEA devices numerical simulation and computer models to predict FEA performance have been developed and continue to be refined. These models allow to explore the effects of device scaling on the emitter's output characteristics. Thus fabrication of devices whose performance will not only be better, but more dependent on geometries that can be well controlled in the manufacturing process is possible. Simulation results indicate that increase the current density and reduce the operating voltage, by decreasing the tip-to-tip separation to 200 mn is there.

FEAs of 200 nm period are fabricated by using interferometric lithography and standard processing techniques. Additional metallization layers and conventional lithography were used to create discrete Molybdenum Spind arrays for electrical characterization. The fabricated cone have similar size and structure to those simulated, Figure 5.13.

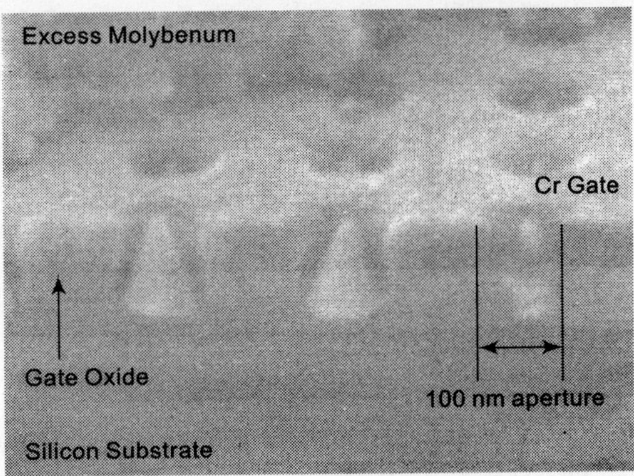

Fig. 5.13 *100 nm gate aperture molybdenum field-emitter cones with chromium gate formed using a vertical evaporation.*

Standard CMOS processing techniques are combined with the interferometric lithography to form 200 nm-period arrays of Si etched cones, Figure 5.14. TEM analysis of the oxidation sharpened emitter tips showed tips with radii as small as 2.5 nm, Figure 5.15.

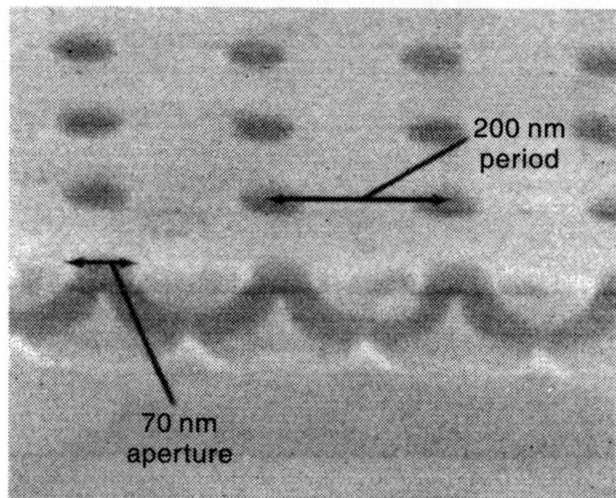

Fig. 5.14 *20 nm-period, 70 nm-aperture Silicon arrays with polysilicon gate. Array was formed by a rough cone formation, oxidation sharpening and CMP planarization.*

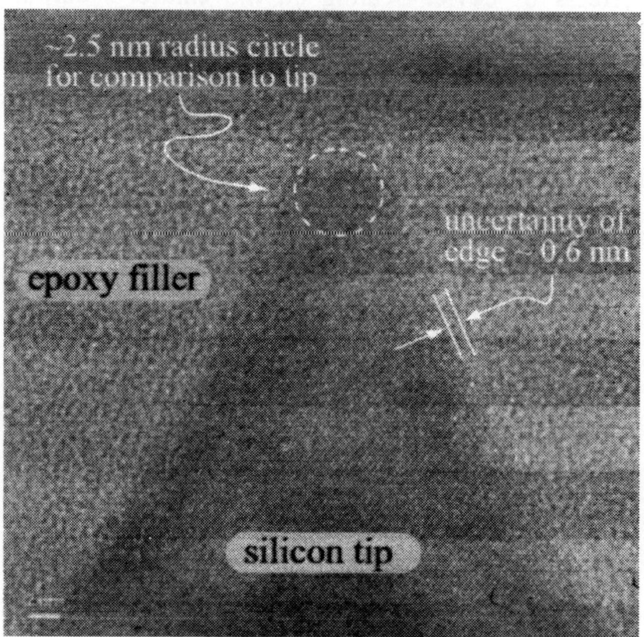

Fig. 5.15 *TEM of sharpened silicon tip.*

A semi-automated Ultra High Vacuum (UHV) probe chamber is developed for the electrical characterization of FEAs. The test bed allows the performance of the arrays to be evaluated without the lengthy overhead of vacuum packaging devices. Device performance showns it to be dependent on the device physical structure, and on surface contamination that may have resulted during fabrication and MEMS processing. The UHV probe chamber has the capability to do device conditioning including plasma surface cleans and wafer bake-out. The systems is combine with a UHV Scanning Maxwell Microscope, and allows the future expansion to include other surface analysis chambers including a Kelvin Probe and Auger.

Electrical characterization of the 100 nm-aperture Molybendum arrays and silicon arrays, (both 200 nm tip-to-tip spacing), has shown that arrays can operate at voltages as low as 16 volts and 13 volts respectively. Initial testing of low-gate-voltage FEAs with discrete solid state devices is demonstrated. The resistor that have been used to limit and control emission current with a MOSFET is replaced. Current control is critical to the uniformity of brightness across the display. It is possible to control the emitted current density and increase temporal stability using the gate voltage of the transistor load. This enables analog voltage gray scale or temporal gray scale. Figure 5.16.

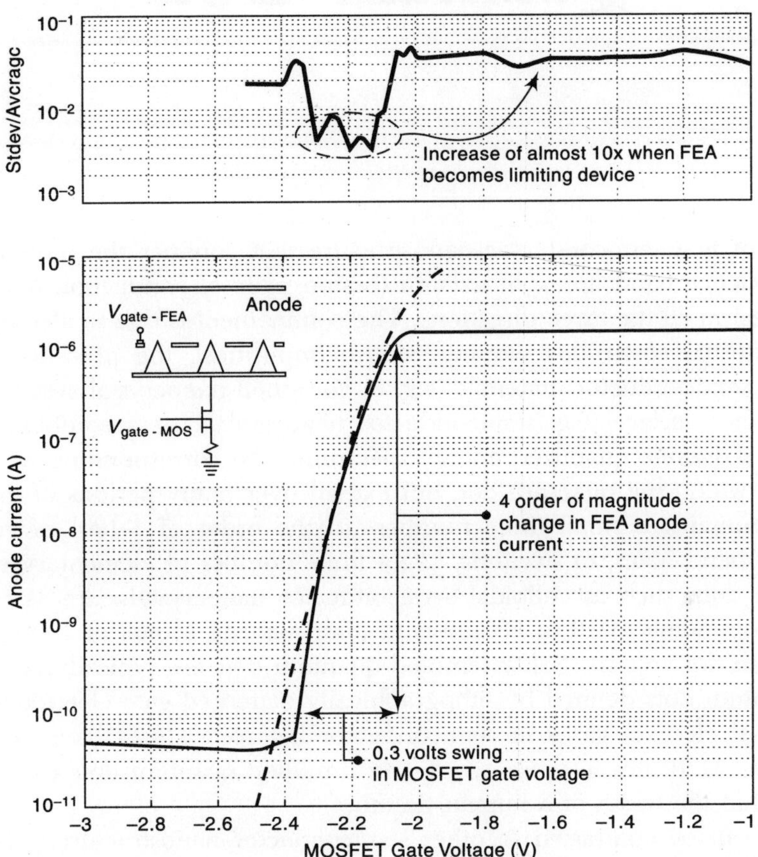

Fig. 5.16 *Silicon array anode current vs. MOSFET gate voltage showing control of over 3 orders of magnitude of the anode current with small variations in the MOSFET gate voltage. The temporal stability improves when in the MOSFET controlled regime.*

The above demonstration shows the feasibility of high brightness, high-resolution FEA image sources for head-mounted display (HMDs). HMDs are expected to have a variety of applications in military, medical, commercial and entertainment fields.

6

Quantum Dots

A **quantum dot** is a semiconductor nanostructure that confines the motion of conduction band electrons, valence band holes, of exciton (bound pairs of conduction band electrons and valence band holes) in all the three directions. The confinement is due to electrostatic potentials (generated by external electrodes, doping, strain, impurities), the presence of an interface between different semiconductor materials (e.g. in core-shell nanocrystal systems), the presence of the semiconductor surface (e.g. semiconductor nanocrystal), or a combination of these. A quantum dot has a discrete quantized energy spectrum. The corresponding wave functions are spatially localized within the quantum dot, but extend over many periods of the crystal lattice. A quantum dot contains a small finite number (of the order of 1-100) of conduction band electrons, valence band holes, or excitons, i.e., a finite number of elementary electric charges.

Small quantum dots, such as colloidal semiconductor nanocrystals, are as small as 2 to 10 nanometers, corresponding to 10 to 50 atoms in diameter and a total of 100 to 100,000 atoms within the quantum dot volume. Self-assembled quantum dots are typically between 10 and 50 mm in size. Quantum dots defined by lithographically patterned gate electrodes, or by etching on two-dimensional electron gases in semiconductor heterostructures have lateral dimensions exceeding 100 nm. At 10 nm in diameter, nearly 3 million quantum dots can be lined up end to end and fit within the width of a human thumb.

Quantum dots can be contrasted to other semiconductor nanostructures: (1) *quantum wires*, which confine the motion of electrons or holes in two spatial directions and allow free propagation in the third, (2) *quantum wells*, which confine the motion of electrons or holes in one direction and allow free propagation in two directions.

Quantum dots containing electrons can also be compared to atoms: both have a discrete energy spectrum and bind a small number of electrons. In contrast to atoms, the confinement potential in quantum dots does not show spherical symmetry. In addition, the confined electrons do not move in free space, but in the semiconductor host crystal. The band structure in a quantum dot plays an important role for all quantum dot properties. Typical energy scales, for example, are of the order of ten electron volts in atoms, but only 1 millielectron volt in quantum dots. Quantum dots with a nearly spherical symmetry, or flat quantum dots with nearly cylindrical symmetry show shell filling according to the equivalent of Hund's rules for atoms. Such dots are sometimes called "artificial atoms". In contrast to atoms, the energy spectrum of a quantum dot is engineered by controlling the geometrical size, and the strength of the confinement potential. Also in contrast to atoms it is relatively easy to connect quantum dots by tunnel barriers to conducting leads, which allows the application of the techniques of tunneling spectroscopy for their investigation.

Like in atoms, the energy levels of small quantum dots are probed by optical spectroscopy techniques. In quantum dots that confine electrons and holes, the interband absorption edge is blue; shifted due to the confinement compared to the bulk material of the host semiconductor material. As a consequence, quantum dots of the same material, but with different sizes, emit light of different colours.

Quantum dots are particularly significant for optical applications due to their high quantum yield. In electronic applications they operate like a single-transistor and show the Coulomb blockade effect. Quantum dots are also suggested for implementation of qubits for quantum information processing.

One of the optical features of small excitonic quantum dots noticeable to the unaided eye is coloration. While the material which make up a quantum dot defines its intrinsic energy signature, in terms of its coloration is its size. The larger the dot, the redder (the more towards the red end of the spectrum) the fluorescence. The smaller the dot, the bluer (the more towards the blue end) it is. The coloration is directly related to the energy levels of the quantum dot. Quantitatively speaking, the bandgap energy that determines the energy (and hence color) of the fluoresced light is inversely proportional to the square of the size of the quantum dots have more energy levels which are more closely spaced. This allows the quantum dot to absorb photons containing less energy, i.e. those closer to the red end of the spectrum. The shape of the quantum dot is also a factor in the colorization.

The ability to tune the size of quantum dots is advantageous for many applications; larger quantum dots have spectra shifted towards the red compared to smaller dots, and exhibit less pronounced quantum properties. Conversely the smaller particles allow one to take quantum properties.

Fabrication of Quantum Dots

Different techniques are used for the fabrication of quantum dots, namely: lithographic patterning and etching of quantum well structures growth on patterned substrates or by using stressors. QDs have also been obtained from monolayer (MLP) fluctuations of the well-barrier interface in quantum well structures. The major approaches to fabricate semiconductor quantum wire (QWR) and quantum dot (QD) structures include self-assembled growth, electron-beam lithography combined with dry etching, selective growth on patterned surfaces, and strain-induced structures. However, the most successful technique is the self-assembled growth on the Stranski-Krastanov (SK) mode, which allows the realization of quasi-0D semiconductor of excellent structural quality. In such a growth mode, the formation of 3D islands (QDs) is driven by the strain field induced by the deposition of a few MLs of a highly strained material on a buffer layer. When the deposited layer (wetting layer, WL) exceeds a critical layer thickness (CLT), a transition from the 2D to the 3D growth occurs. At the CLT, the accumulated strain energy due to the lattice mismatch makes the island surfaces energetically more favourable than the flat surface, resulting in a uniformly islanded surface. The islanding process uses a strain-induced roughened growth front model, in which a roughened surface is stable for wavelengths above a minimum value. Since the islands are capped; both the QDs, the WL and the barrier around the QDs are elastically strained thus influencing the carrier confinement properties.

To increase the optical density of QD structures several QD layers are grown in the structures. The local strain in the barrier above the dots is responsible for the vertical stacking of the dots, the top islands preferentially nucleate above the lower ones, forming vertical columns of dots. The distance among the layers is a very important parameter. When the distance is comparable to the dot height, tunnel coupling occurs and a sort of dot-molecule is formed. On the other hand, for dot distances larger than ~10 nm, vertically stacked dots turn out to be uncoupled.

Mass Production

In large numbers, quantum dots are also synthesized by means of a colloidal synthesis. Colloidal synthesis has the advantage of being able to occur at benchtop conditions. It is acknowledged to be the least toxic of all the different forms of synthesis.

Highly ordered arrays of quantum dot is self assembled by electrochemical techniques. A template is created by causing an ionic reaction at an electrolyte-metal interface which results in the spontaneous assembly of nanostructures, including quantum dots, on the metal which is then used as a mask for mesa-etching these nanostructrues on a chosen substrate.

Yet another method is pyrolytic synthesis, which produces large numbers of quantum dots that self-assemble into preferential crystal sizes.

Self-organised Growth of Quantum, Dots

Applications of QDs to lasers is possible when techniques of self-organized growth are developed to such a level that dense arrays of uniform QDs are produced and, defects of all types are reduced. Hundreds of billions of nanoscale islands, ordered in size and shape, are formed on a crystal surface per second and per square centimeter, without making any changes in growth equipment.

It is this islanding approach that has led to breakthrough in QD lasers. Photopumped lasing in QD heterostructures. InGaAs QDs are inserted in the central part of a GaAs cavity deposited on top of a thick AlGaAs layer and the structure is covered by a thin Al-GaAs layer to prevent surface recombination. Photopumped lasing via QDs is realized both at low and room temperature (see Fig. 6.1). It is close to the 2D-3D growth transition and has reduced the probability of defect formation, and uses low substrate temperature resulting in higher area density of islands and lower density of defects.

Low threshold injection lasing from InGaAs QDs occurrs at 1.24 eV and 1.31 eV, for two different structures at 77 K, and is within the range of the QD luminescence. Also, the threshold current density is low (< 100 A/cm^2) and is unaffected by temperature up to about 150 K. At high temperature the threshold current increases due to thermally activated escape of carriers from quantum dots to the surrounding GaAs matrix.

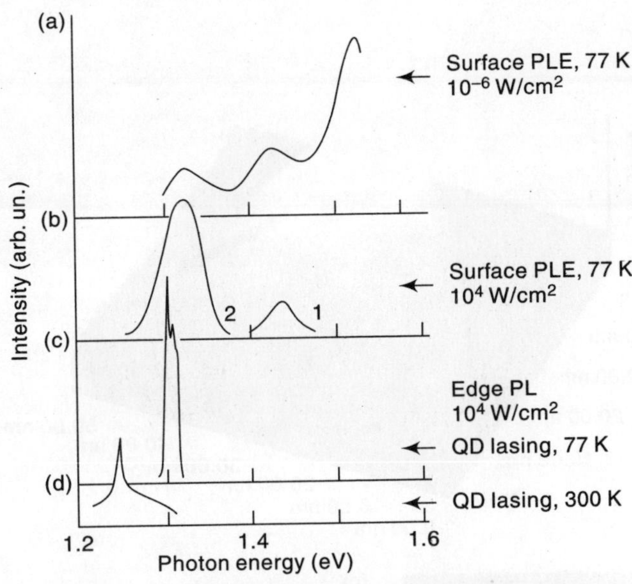

Fig. 6.1 *Photo-pumped lasing in quantum dots.*

Structural Analysis of InGaAs/GaAs

Consider the growth of InGaAs/GaAs single and six-fold stacked quantum dots, as samples, grown by metal-organic chemical vapour deposition (MOCVD) in the SK growth mode on (001) exactly oriented GaAs substrates.

The structural characterization of QD provides information about the dot shape and size. AFM and STM techniques are commonly used to investigate the structure of uncapped quantum dots. The first one gives information about dot density and size. Figure 6.2 displays a typical plan view of AFM image of a planar array of QDs. In this sample, the dot density is about 5×10^9 dots cm^{-2} and the size dispersion is reduced to some ±10%, resulting in a spectral broadening of the order of ±15 meV. Detailed information about the shape of uncapped QDs is obtained

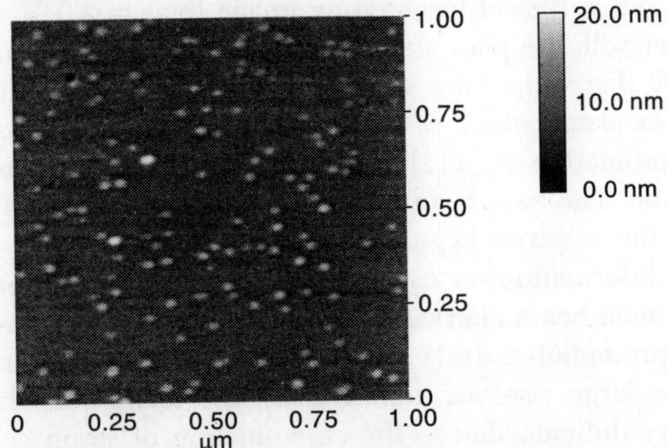

Fig. 6.2 *Topographic plan-view AFM image on uncapped MOCVD INGaAs/GaAs QDs.*

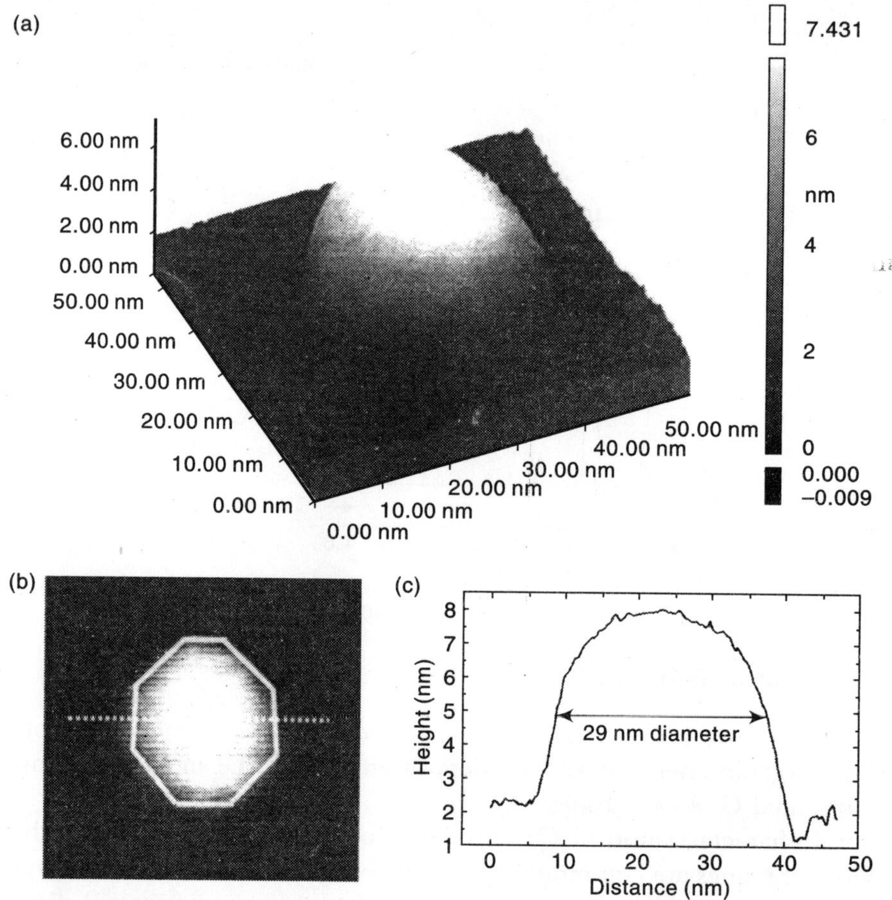

Fig. 6.3 *Topographic STM image (a) on a single uncapped QDs, (b) together with the plan-view of the STM image (c) and the line-scan profile across the QD.*

by high-resolution STM images which visualize the exact crystallogrpahic faceting of the nanostructures. In Fig. 6.3, a typical topography image [bias = 2.0 V, tunneling current = 0.5 nA; Fig. 6.3(a)], together with the plan view STM image [Fig. 6.3(b)] and a line-scan across the centre of the single QD [Fig. 6.3(c)] are shown. In the image displayed in Fig. 6.3(a), the QD shape exhibits a complex arrangement of crystal facets resulting in a slightly asymmetric shape. The side-wall are a combination of {111} and {110} planes resulting in steep inclinations with respect to the [100] plane. The base is an octogon with elongated sides aligned along the [001] and [010] directions of the substrate crystal.

To obtain structural information on capped QDs, TEM investigation is done; in particular, two beam and on zone multi-beam plan-view TEM images are used to investigate nanostructures. Figure 6.4 shows a representative (001) pan-view TEM image performed on a capped QD sample. In spite of the large contrast achieved by the technique, the interpretation of the diffraction image is very difficult, due to the convoluation of strain contrast and composition dependent contrast.

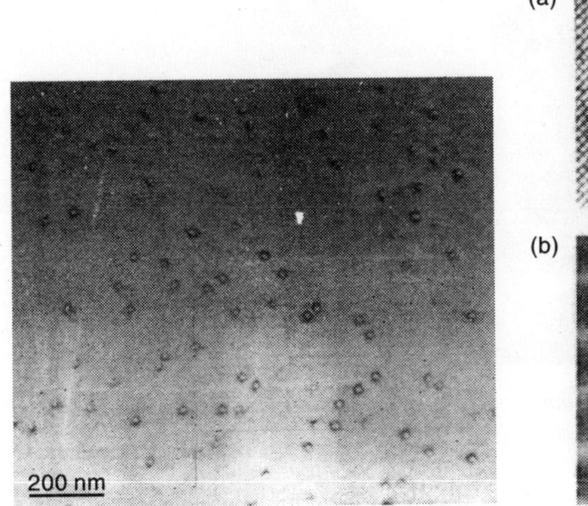
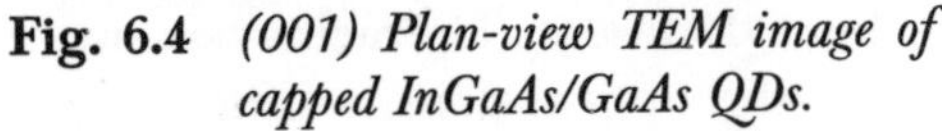

Fig. 6.4 *(001) Plan-view TEM image of capped InGaAs/GaAs QDs.*

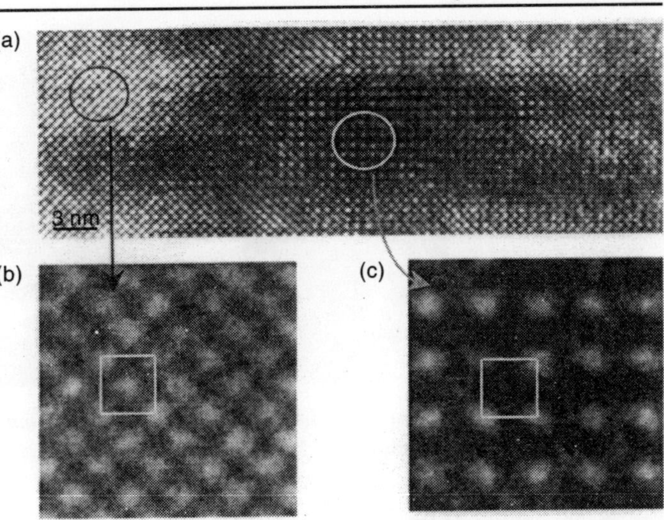

Fig. 6.5 *(a) High-resolution cross-section TEM image of a single dot along the ⟨001⟩ zone axis. (b) Diffraction pattern in the GaAs barrier region (c) the InGaAs dot region.*

Therefore, high-resolution TEM (HRTEM) measurements are performed along different crystallographic directions. In Fig. 6.5(a), and HRTEM image of one dot in the ⟨001⟩ zone axis is shown. Due to the chemical sensitivity of such zone axis, it is possible to distinguish from the phase contrast features the regions where In is located and, consequently, to get information about the real dot shape. In fact, the diffraction pattern of the GaAs lattice exhibits spots arranged in squares with a bright spot at the centre [Fig. 6.5(b)], whereas in the InGaAs lattice the central diffraction spot disappears. [Fig. 6.5(c)]. A detailed analysis of the HRTEM images shows that the capped MOCVD InGaAs/GaAs QDs have a cross-section similar to a truncated cone, with an angle between the dot base and the dot side equal to 54°. The sizes of the dots strongly depend on the growth conditions.

In vertically stacked QD samples, TEM measurements give additional information about the uniformity of the islands along the stack. Figure 6.6 shows the cross-sectional low-magnification TEM images obtained from three samples, consisting, respectively, of a single [Fig. 6.5(a)] and six-fold stacked dot layers [Fig. 6.6(b) and (c)]. The comparison between the images of the vertically stacked QD samples [Fig. 6.6(b) and (c)] shows that, although the dimensions of the bottom layers are nearly the same, the vertical size dispersion can be quite different. In particular, the sample in Fig. 6.6(b) shows dot which are vertically aligned without extended defects and with a rather uniform size (only the topmost dots exhibit a small enlargement). Conversely, the sample in Fig. 6.6(c) exhibits an increase of the dot size. These important differences are related to the different growth conditions that influence the quality of the stacked structures in terms of the vertical size uniformity. The QDs are formed after the deposition of only 4 ML of InGaAs in the sample showed in Fig. 6.6(b) and 6 ML in the sample of Fig. 6.6(c) resulting in a different total strain in the structure. The occurrence of size dispersion

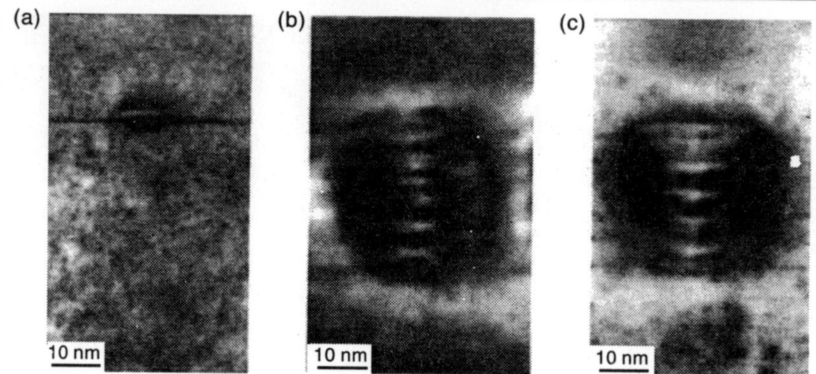

Fig. 6.6 *⟨001⟩0 cross-sectional low magnification (TEM images of a single dot sample (a), and of a uniform (b) and non-uniform (c) vertically stacked dot heterostructure.*

along the stack [Fig. 6.6(c)], in fact, causes considerable inhomogeneous broadening of the optical spectra thus preventing the use of these materials for QD lasers. The achievement of uniform stacked dots is thus a crucial prerequisite to increase the optical density of the active material without introducing large inhomogeneous broadening.

Quantum Wire: quantum confinement in two dimensions

In condensed matter physics, a quantum wire is an electrically conducting wire, in which quantum effects are affecting transport properties. Due to the confinement of conduction electrons in the transverse direction of the wire, their transverse energy is quantized into a series of discrete values E_0 ("ground state" energy, with lower value), E_1,... (quantum harmonic oscillator). Now, for calculating the electrical resistivity of a wire: $R = \rho l/a$ is not valid for quantum wire (where π ais the resistivity, l is the length, and A is the cross-sectional area of the wire).
A quantum wire consists of a strip of one semiconductor confined within a second, large band gap barrier semiconductor. Unrestricted carrier motion is now only possible along the length of the wire and is quantised along the remaining orthogonal directions. For simple wire shapes (square or rectangular cross sections) it is possible to calculate the quantisation energies for the two directions independently. These two quantisation energies are then added to the energy resulting from the unrestricted motion along the wire. Using the infinite-depth well approximation for the quantised energies, the total energy for a carrier in a quantum wire with z and y dimensions L_z and L_y respectively is

$$E_{n,m,k_x} = \frac{h^2 n^2}{8m^* L_z^2} + \frac{h^2 m^2}{8m^* L_y^1} + \frac{h^2 k_x^2}{2m^*} \qquad \text{(n, m = 1, 2, 3,...).}$$

The total energy depends on the two quantum numbers n and m the wave vector for free motion along the wire k_x. For each confined state, given by a particular combination of n and m, there is a subband of continuous states resulting from the unrestricted values of k_x. It is seen that real quantum wires have complex cross sections. This prevents the confined energies from being calculated by separating them into terms corresponding to the two directions perpendicular to the axis of the wire. Instead the confined energies of a quantum wire are obtained from a numerical solution of the Schrödinger equation.

Carbon Nanotubes as Quantum Wires

It is possible to make quantum wires out of metallic carbon nanotubes, at least in limited quantities. The advantages of making wires from carbon nanotubes include their high electrical conductivity (due to a high mobility), light weight, small diameter, low chemical reactivity, and high tensile strength. The major drawback is cost.

It has been claimed that ti is possible to create macroscopic quantum wires. With a rope of carbon nanotubes, it is not necessary for any single fiber to travel the entire length, since quantum tunneling allows electrons to jump from strand to strand.

Quantum Confinement in Three Dimensions

A quantum dot consists of a small region of one semiconductor totally surrounded by a second, larger band gap barrier semiconductor. Carrier motion is quantised along all three spatial directions and there remains no unrestricted carrier motion. For a simple shape such as a cube or cuboid, confinement for the three spatial directions is considered separately. IN the infinite-depth well approximation, the total energy for a carrier in a cuboid-shaped dot of dimensions L_z, L_y and L_x is a function of three quantum numbers n, m and l:

$$E_{n,m,l} = \frac{h^2n^2}{8m^*L_z^2} + \frac{h^2m^2}{8m^*L_y^2} + \frac{h^2l^2}{8m^*L_x^2} \qquad (n, m, l = 1, 2, 3, \ldots).$$

The energy is fully quantised and the states are discrete, in a manner similar to those of an atom. The shapes of real quantum dots are more complex than simple cuboids and a calculation of the confined energy levels requires a numerical solution of the relevant Schrödinger equation.

ENERGY STRUCTURE OF QD

Figure 6.7 shows the energy levels in InP QDs and InAs Qds as a function of height both for the conduction band and for the valence band. The hole level spacing in large InP QDs of the order of a few millielectronvolts, while for the smaller InAs QDs the level spacing is ten of millielectronvolts. The electron level spacing is much larger. Moreover, the sensitivity of the electron levels to small changes in dot size is comparable to the hole level spacing. This contributes to the need to use single dot spectroscopy to probe these dots. It is of interest to observe the photoluminescence from individual dots to see whether the theoretical predictions are observable and the distinctive differences are discernible.

Energy relaxation of excitons in self-organized QDs occurs extremely fast. Excitons excited in the wetting layer relax to the QD ground state in picoseconds or even sub-picoseconds Fig. 6.8.

The reason for the fast relaxation, and undoubtedly for the absence of a phonon bottleneck, is the close exciton energy spacings of the holes. The phonon bottleneck exists for QDs that contain only electrons as opposed to excitons since the electrons levels spacings exceed the maximum phonon energy.

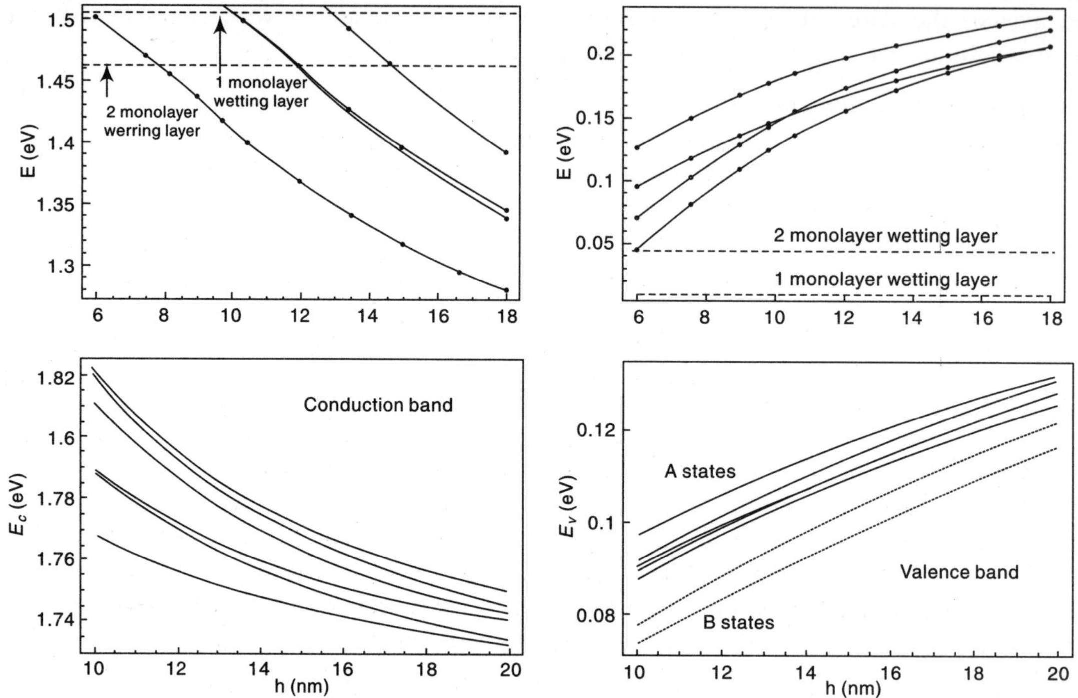

Fig. 6.7 *The single-particle energy levels for both InAs (top panels) and InP QDs (bottom panels). The level spacing in the InP QDs is considerably smaller than for the InAs QDs due to the larger size of the InAs QDs. In both cases the level spacing in the valence band is much less than in the conduction band.*

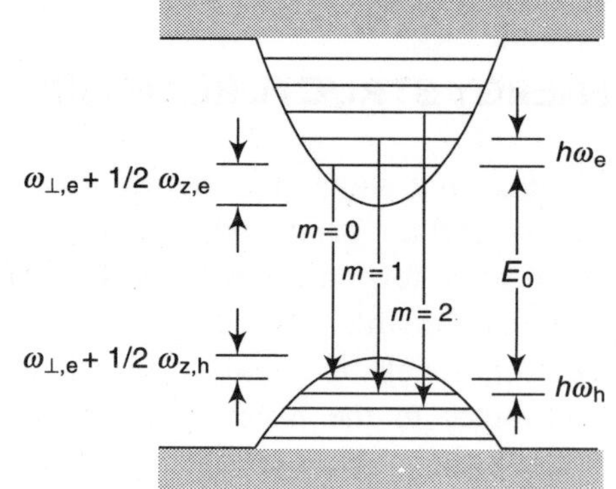

Fig. 6.8 *Illustration of the energy structure of QDs.*

The energy relaxation due to photons (light emission) follows dipole selection rules for the QD states that must be carefully calculated to determine the optically active transitions between electrons and holes. The Coulomb interaction should be included to find the complete set of energy states for the occupancy of the QD by a given number of electrons and holes. However, radiative transitions in the QDs occur between electron and hole quantum states with the same quantum number, so that a ground state electron recombines with a ground state hole. The resulting equations provide a working model for QD lasers in general, with the most important modification needed for real QD ensembles being that a more accurate description of the electronic states and transitions be used to improve agreement with experiment. The model below provides the insight necessary to predict important trends for QD lasers that arise due to their zero-dimensional energy spectrum.

Non-equilibrium Model

To simplify the description of the wetting layer, consider only lowest energy sub-band to provide a 2D electron reservoir to the QDs. Figure 6.8 illustrates the energy spectrum for this system. To relate the current and light emission to the voltage applied to the device, assume that the wetting layer electrons and holes reach quasi-equilibrium, and are thus described by quasi-Fermi functions set by the *n*- and *p*-sides of the laser diode. Therefore, the assumption becomes that the discrete levels of the QDs are driven by the wetting layer, which has different quasi Fermi energies for electrons and holes due to an external driving source.

Equilibrium Model (Doping)

Room temperature self-organized InAs for InGaAs QD lasers operate in regime of quasi-equilibrium. This allows to focus on the specific characteristics of the QD active material that is engineered to improve device performance. At higher temperatures, all thermally excited levels influence the ground state optical. Most important of these are the close energy spaced hole levels. The thermal hole distribution depletes the exciton ground state of some of its potential hole population, so a greater total number of electrons and holes are injected to obtain lasing, and this decreases the differential gain and increases the temperature dependence. *P*-type doping provides a means to overcome the limitation of the closely spaced hole levels.

The importance of the p-type doping is illustrated schematically in the energy diagram presented in Fig. 6.9. If the QDs are doped n-type, as in middle illustration of Fig. 6.9, the gain per injected carrier will not increase since the probability of the additional holes occupying the desired states is low. However, by doping the QDs p-type the critical ground state hole levels are occupied with holes, while additional electrons occupy the QD ground state. Therefore, the effects of closely spaced energy levels of one charge type is compensated by doping in charge carriers of that type. The result of the built-in carriers is an increase in the maximum optical gain and differential gain with increasing temperature. These effects strongly impact QD laser performance at room temperature.

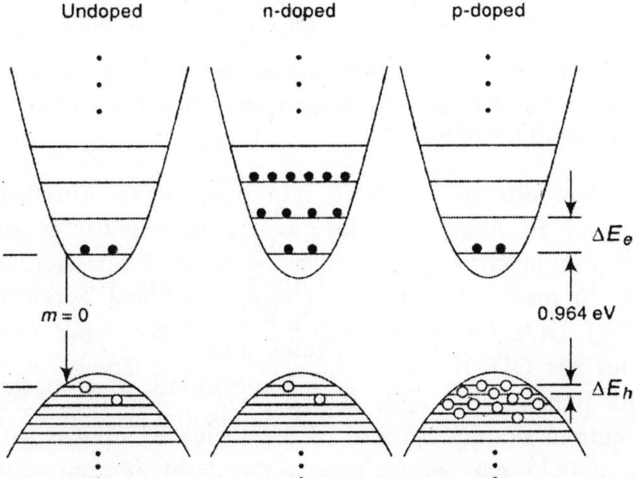

Fig. 6.9 *Schematic illustration of the energy diagram of p-type doping.*

Figure 6.10 shows the calculated 3 dB frequency response vs the square root of the difference between the input and threshold current density, the cavity lengths are chosen to optimize the frequency response for each doping level. Because the maximum gain is increased with *p*-type doping, a shorter cavity length is used to improve the modulation response. The modulation doping build in 50–100 holes per QD increases the modulation response of a QD laser from around 10 GHz maximum to over 20 GHz, and with a significantly better response at moderate current densities. If p-type modulation doping is used and the inhomogeneous linewidth is reduced to 10 meV; Fig 6.10 shows that the intrinsic bandwidth of the cavity and QD active region increases even more dramatically. The QD's room temperature homogeneous linewidth is ~5 meV (unless high injection levels are used), so that further improvement at room temperature is not possible for QD lasers by reducing the inhomogeneous broadening.

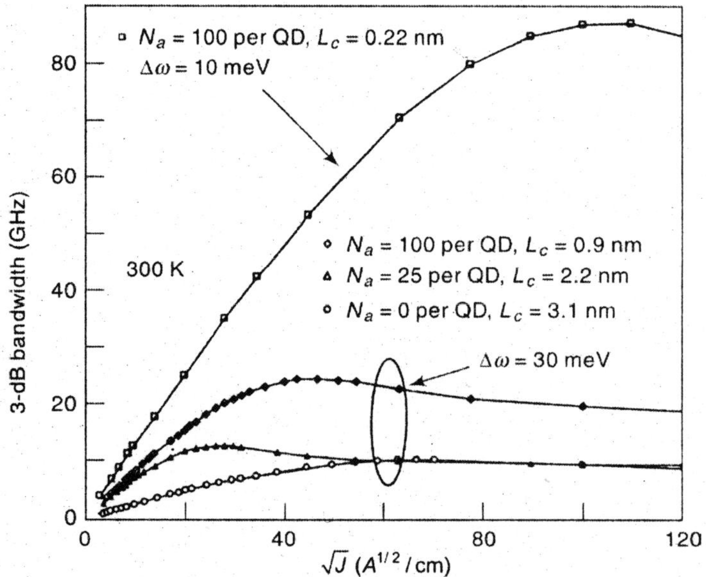

Fig. 6.10 *Calculated 3-dB frequency response vs the square root of the current density for different numbers of built-in holes. As in Fig. 6.10, the inhomogeneous linewidth is 30 meV, except for the upper curve which shows the result for 10 meV.*

To put the calculated results in Fig. 6.10 into perspective, suppose the QD laser has a threshold current density of 50 Acm^{-2}, and its modulation response is measured at 450 Acm^{-2}, or at none times the threshold current. The expected room temperature bandwidths intrinsic to the active regions with 30 meV inhomogeneous broadening and optimized cavities are ~4GHz for the undoped laser, 11 GHz for the laser doped to 25 holes per QD, and 15 GHz for the laser doped to 100 holes per QD. If the inhomogeneous broadening can be reduced to 10 meV the modulation response jumps to 25 GHz at 450 Acm^{-2}. And the higher speed comes with a much lower threshold current density because of the increased optical gain. A stripe width of 3 μm and cavity length of 0.22 mm would give a threshold of approximately 3 mA. The 450 Acm^{-2} current density then corresponding to ~27 mA. For these types of lasers increased reflectivity to further shorten the cavity lengths and increase the current density without increasing the total current reduces the operating current significantly.

Hence, that the QD laser's potential for high-speed modulation response at moderate power levels makes it an interesting source for fiber optics. Added to this is an additional benefit due to the discrete levels and inhomogeneous broadening that causes a symmetrical gain spectrum that is, relatively insensitive in shape to bias current. These unique features in the gain spectrum result in very low chirp in the modulation response. QD lasers therefore offer the possibility of low cost, high speed, high performance with temperature insensitive operation at the important fiber wavelengths.

Quantum Dots and Quantum Wells

Dots and wires are fabricated by lithographic techniques, by bottom up and top-down lithographic techniques. However, lithographic top-down has not provided the dot and wire structures needed. Fabrication by etching of quantum wells is also done. QDs were made from quantum wells by deep etching directly through GaAs/AlGaAs quantum wells, as shown in Figure 6.11. Implantation-enhanced interdiffusion is used to fabricate GaAs/AlGaAs QDs without exposing free surfaces. Here, diffusion of Al into the GaAs quantum well is enhanced in regions implanted with Ga. The interdiffusion of Al raises the local bandgap and provides a weak lateral confinement of regions not implanted with Ga. QDs are also fabricated in InGaAs/InP

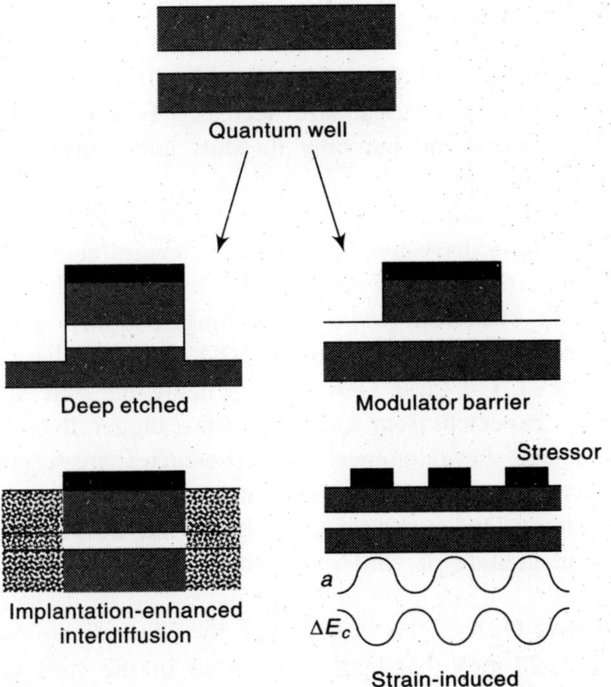

Fig. 6.11 *Top-down fabrication: etch/lithographic techniques for fabricating quantum dots and wires from quantum well structures.*

by deep etching because InGaAs/InP has stable surfaces and less surface recombination. Initial dot-array device-structures, made with GaInAsP/InP QDs by deep etching and buried by regrowth to make optical cladding layers, exhibited light emission by current injection. The smallest dots made are 30 nm wide. Most does are larger. Photoluminescence (PL) and cathodoluminescenec (CL) reveal small blue shifts of the emission that is attributed to lateral confinement. The emission peaks have a large inhomogeneous broadening, indicative of a large distribution of dot sizes.

The most direct way to fabricate QDs from quantum well structures is to etch directly through the quantum well. The lateral confinement is provided by the abrupt termination at the sidewalls. These deep-etched QDs have poor luminescence due to nonradiative recombination and trapping at etch-induced damage near the sidewalls. The details of the etching process are critical for determining the amount of damage the dot size, and the dot shape. Attempts to improve the optical quality of deep-etched dots are made by choosing the etch-process to minimize damage annealing the structures, or passivating the sidewalls by overgrowth. Instead of etching through the quantum well, the etch is stopped just above the quantum well to modulate the barrier above the well (see Figure 6.11). Regions of the well covered by the vacuum barrier have stronger vertical confinement, with higher confinement energies. The modulation of the confinement energy provides the lateral confinement. For barrier modulation, there is little etch damage to the well.

Deep-etched QDs made with a *dry-etch* technique, (such as $SiCl_4$ reactive ion etching), damage 100 nm in from the sidewalls. For these QDs, the effective dot size is much smaller than the geometrical QD size. A 250-nm-wide QD has 9-meV level shift due to lateral confinement because the active dot is about 30 nm wide. The luminescence of these deep-dry-etched dots drop rapidly as the geometrical size decreases below 500 nm. The luminescence is recovered by annealing or overgrowth, but only for dots larger than 250 nm. For smaller dots, there is little luminescence recovery.

Deep-etched QDs made with *low-damage etching*, (such as electron cyclotron or, *resonance etching* and *wet chemical etching*, have luminescence efficiencies comparable to the corresponding quantum well luminescene efficiencies. For example, by the fabrication of 25–60 nm GaAs/AlGaAs QDs using electron cyclotron resonance etching, 5% size uniformity in samples with 57-nm-wide QDs is obtained. The linewidth for the QD PL from these samples is same as for the quantum well PL (3 meV). For these QDs there is negligible blue shift between the QD and quantum well peaks, as is expected from QDs five time bigger than the exciton. These dots, with minimal sidewall damage, have luminescence efficiencies greater than for the unprocessed well. As the dot size decreases, the PL linewidth increases by a factor of two, indicating size fluctuations for smaller dots. The exciton energy in 25-nm dots shift by 7 meV due to the confinement. This is the magnitude of shift expected when the dot and exciton have the same size.

InGaAs/InP and InGaAs/GaAs QDs fabricated by *barrier modulation* also have high luminescence efficiency. A 24-meV bandgap modulation in the well is provided by removing the barrier above the well. The modulation provides energy blue shifts due to confinement in barrier-modulated dots that are less than level shifts in comparable deep-etch dots. Energy-level shifts of 10 meV due to confinement in 509-nm-wide, barrier-modulated QDs is observed.

These InGaAs/GaAs QDs are extensively characterized the PL from highly excited QDs exhibit transitions from excited states, broadening of the PL, thermalized state distributions, slower thermalization of the excited states, and a red shift of 2 to 6 meV of the main PL peak. Here the dots have 6 to 8 electron hole pairs. The results indicat significant carrier-carrier interactions. The bandgap red shift is typical of the energy expected for biexciton binding in a QD.

Strain in a semiconductor also modulates the bandgap. By tailoring the lateral distribution of strain in a well, a lateral bandgap modulation provides lateral confinement. As shown in Figure 6.11, a layer of stressed (compressed) material is grown on to of a quantum well structure. This stressor layer is etched, the compression relieved, and the strain transmitted locally into the well under the stressor. The areas dilates the well below the stressor and reduces the bandgap by red shift in the conduction band. Confinement is achieved without exposing free surfaces of damaging the well.

To produce stain-induced QDs, InGaAsP stressor layer 100-nm-thick under a strain of 0.5% is used. For 400-nm-wide QD stressors, the PL peak (red) shifts by 9 meV. The red shift is due to the strain-induced reduction in bandgap. For such a wide dot, the blue shift is only 0.5 meV. High QD luminescence efficiency is observed. Also Amorphous carbon is used as the stressor layer on GaAs/AlGaAs quantum well structures to achieve a larger bandgap modulation. For 800-nm stressors, the PL peak (red) shifts by 60meV. To eliminate the effects of homogeneous broadening and identify QD excited states. PL measurements on a single QD are taken. Line splittings of 2 meV are observed for dots defined by 300-nm stressors. This line splitting suggests that the effective dot diameter is 40 nm.

Similar strain-induced QDs are fabricated by growing compressed InGaAs layers on tip of GaAs/AlGaAs well structures. 14-meV red shifts in a well 22 nm below a 120-nm-wide stressors. By monitoring the red shifts of wells placed at different distances below the stressor layer, the strain-induced bandgap modulation propagates into the structure to a depth comparable to the stressor width and hence high QD luminescence efficiency. Only dots in the well closest to the stressor, and therefore closest to the surface, have lower luminescence efficiency than the corresponding well.

Techniques used to avoid exposing surfaces during QD fabrication is to interdiffuse AL from an AlGaAs barreir into a GaAs quantum well. Interdiffused regions, with a higher bandgap, provide lateral confinement. Interdiffusion is enhanced in regions implanted by Ga. An implantation mask, defined on the surface of the quantum well sample by use of high-resolution e-beam lithography, prevents implantation in the dot region. The confinement potential defined by this approach was determined by the mask size, by the straggling of implanted ions, and by the lateral diffusion of Al into the dot under the mask, which both raises the bandgap in the dot and broadens the confinement potential.

QDs with effective widths in the range 50 to 70 nm with lateral barriers of about 70 meV are made. CL and PL peaks (blue) shift by 5 to 7 meV, Linewidths for QD spectra are from 15 to 20 meV.

Implantation is achieved by direct writing with a Ga focused ion beam without the use of an implantation mask. Channeling of implanted Ga inhibits lateral diffusion, so that the small probe size of the focused ion beam is maintained 200 nm below the sample surface. As a result, 80-nm nanostructrues with good luminescence properties are possible by direct write techniques.

A focussed laser beam, instead of a focused ion beam, is used to thermally enhance interdiffusion and directly write a QD pattern. However, the pattern drawn by the focused laser beam and the effective dot size has the lowest PL peak as well as several higher energy PL peaks. Linewidths are 0.5 meV, much less than for a well, and clear evidence for discrete transition energies. As the nominal dot size is reduced to 450 nm, the lowest energy PL peak (blue) shifts by 10 meV, and the splitting to the next peak shift by 10 meV. As the dot size is reduced below 450 nm, the lowest energy PL peak continues to blue shift, with a 70 meV shift for a completely interdiffused well, but the splitting between peaks decreases. For dots wider than 450 nm, the potential barrier due to interdiffusion provides effective lateral confinement. The lowest level and the peak splitting increases with decreasing dot size. Smaller dots are interdiffused. The lowest peak shifts to higher energy because the bandgap is raised inside the dot, but the splitting decreases because the lateral barrier due to bandgap modulation is reduced. The estimated effective size of the 450-nm dot is about 25 nm.

Several techniques are developed to grow dots and wires from well structures. These techniques include cleaved edge overgrowth (CEO), self-assembled dots (SAD), and patterned growth, as shown in Figure 6.12. In CEO, a (100) quantum well is initially grown, example a GaAs/AlGaAs well. It is cleaved on its edge, leaving a free (110) surface. A layer of GaAs is overgrown on the surface and a barrier layer of AlGaAs is grown on to P of it. The two orthogonal quantum wells intersect at a T-shaped junction. The confinement energy is lower in the junction than in either of the two wells. This local modulation of the confinement energy

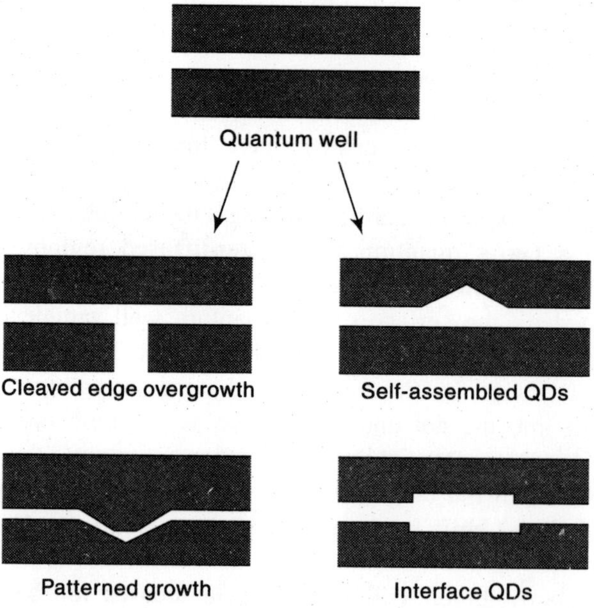

Fig. 6.12 *Bottom-up growth techniques for fabricating quantum dots and wires from quantum well structures.*

traps electrons and holes to the junction to form a T-shaped quantum wire. The difference between the CEO and barrier modulation techniques is that all of the dimensions in CEO are controlled with nanometer precision, by the growth of (100) and (110) wells. There are no damaged surfaces or junctions in CEO because all of the junctions are formed with precision MBE growth and cleave is along an atomic plane. Quantum dots can be formed from these wires by cleaving at a second, orthogonal plane and growing on the new exposed surface.

A second bottom-up approach is the growth of wires and dots on patterned substrates. For example, wires are grown in etched V-shaped grooves and dots in etched pyramids, as shown in Figure 6.12. The growth is done by first etching a AlGaAs substrate to form the V-shaped groove or pyramidal recess; and then a GaAs layer is grown in the groove or pyramidal recess. The well grows faster and thus thicker at the bottom of the groove or at the bottom of the pyramid because the atoms deposited during growth migrate away from the sidewalls. After the well is grown, a capping layer of AlGaAs is grown. The thicker well region acts as a quantum wire when grown in V-grooves and as a QD when grown in a pyramidal recess. The growth sequence is repeated to form a vertical stack of dots or wires in the same pyramid or groove. Patterned growth is a powerful fabrication technique, since the size and placement of the QDs and QWRs is controlled by the pattern etched in the substrate.

A example tetrahedral quantum dot structures are grown by using selective are MOCVD. By etching triangular-shaped holes, with the sides along {110)} directions, through a mask deposited on a (111)B GaAs substrate. MOCVD growth of AlGaAs inside the holes produced tetrahedral pyramids. GaAs is deposited at the apex of each pyramid to form a QD. The structures are then overgrown by AlGaAs. GaAs QDs, with a 0.1-μm-wide base and 20-nm height, are deposited on top of pyramids grown in 0.3-μm-wide holes. The PL from these QDs exhibits a 19-meV blue shift from the bulk GaAs PL.

PROPERTIES OF QUANTUM DOTS

There has been increasing interest in semiconductor quantum nanostructures, including wires and quantum dots (QDs), due to their unique optical properties. The additional quantum confinement in a wire or dot concentrates the density of states into a narrower energy range—at the subband edges of quantum wires and into a series of discrete states of quantum dots. This concentration of density of states greatly enhance the optical response of the nanosystem because more optical transitions are active at nearly the same energy, opening the possibility, for example, to engineer nanoscale lasers with greatly enhanced performance., Quantum dots, with their discrete electronic energy levels, are uniquely different from higher dimensional quantum structures, such as quantum wires and quantum wells.

These properties of the quantum dots are achieved by controlling the size, shape, and composition of the QDs. Moreover, QDs are fabricated directly in semiconductor devices structures. Understanding, identifying, and exploiting the atomic-like properties of QDs requires that individual dots be probed, one at a time, so that the effects of inhomogeneous broadening can be eliminated. It is possible to study individual quantum dots optically.

Electronic Structure

In order to understand the electronic structure of QDs, it is necessary to have realistic calculations of the single-particle levels that can be compared with experiment. This is a difficult problem since the structures are intrinsically three-dimensional. The number of atom sin a QD is of the order of 100,000, and if the matrix is included, millions of atoms are involved, precluding any type of routine ab initio calculations. Quantum dots of this type are usually compressively strained (i.e., the unstrained lattice parameter of the dot material is larger than that of the matrix) and the strain profile has to be calculated to correctly model the confining potential.

Figure 6.13 shows the potential energy diagram for the band-edges for InP and InAs quantum dots. Both types of dots are type I, localizing both electrons and holes. In the absence of strain, InP QDs would act as antidots for the hole bands. The strain profile is necessary to confine the holes in the InP QDs. As a consequence, the valence band well is shallower for InP QDs than for InAs QDs, and the hole states are less localized to the centre of the InP QDs.

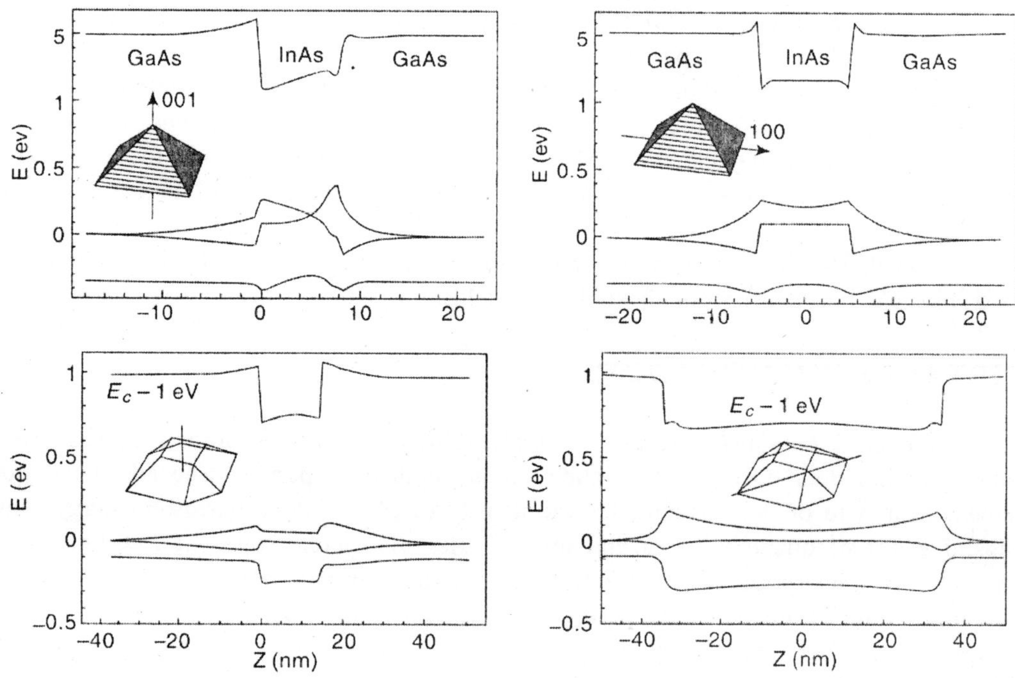

Fig. 6.13 *The potential profile of InAs (top panels) and InP QDs (bottom panels) along the directions indicated in the inserts. In both cases, the deepest well is in the conduction band. The strain also modifies the potential structure outside the dots.*

Photoluminescence

The calculation of the dot electronic structure is very useful not to study dependence of the dot properties on size, shape, and composition but also to interpret the photoluminescence (PL) experiments. In order to experimentally observe the photon emission from individual QDs, several techniques are developed. The easiest is to use a microscope and focus the emission from one dot on the entrance slit of a monochromator. This requires that the density of QDs be sufficiently low, which is the case for InP QDs if grown under suitable circumstances. For InAs QDs, a low-density sample is very difficult to grow. Due to the inverse relationship between size and density, small dots are formed under high density conditions. By growing lower density InAs QD samples simple microscopy can be used. However, for high-density samples emission from single QDs is observed, using a metal mask in which small holes are created.

Figure 6.14 shows the emission spectra from a single InP QD from a single InAs QD. These spectra are representative of most dots (>90%). A few unusual dots do occur, but these dots, in most cases, are actually pairs of dots.

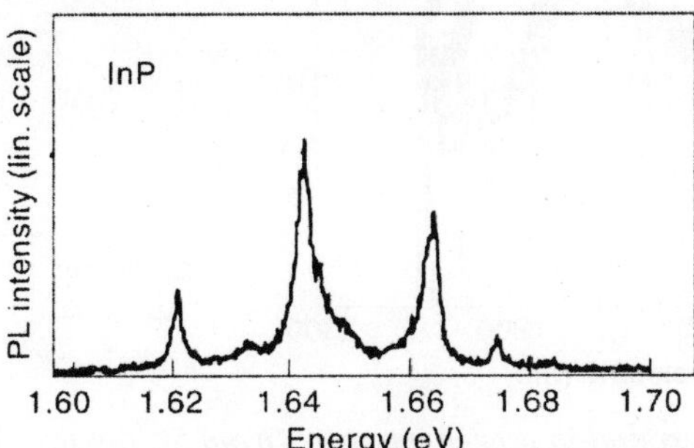

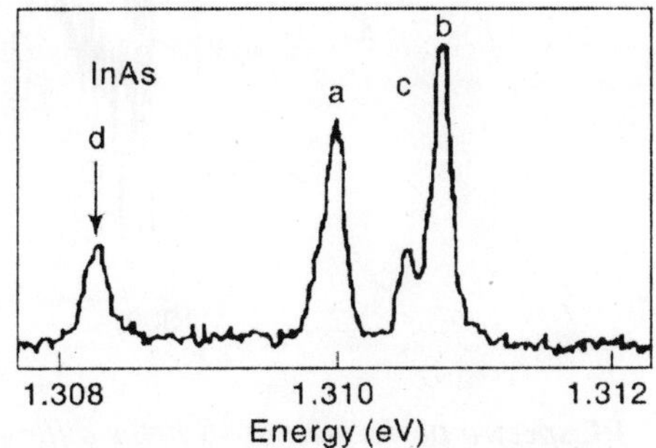

Fig. 6.14 *The photoluminescence of an InP QD consists of several lines having a linewidth of about 2 meV, while for a InAs only a few lines are seen having much smaller linewidths (about 0.11 meV).*

The spectra are very different from each other, despite being measured under essentially identical conditions. For instance, the linewidth is less than 0.1 meV for InAs QDs, while for InP QDs, the linewidth is about 2 meV (for fully developed InP QDs, whereas smaller InOP QDs have a narrower linewidth). Given an emission lifetime of about 1 ns, a line-broadening of 0.7 μeV is obtained. Each of the broad emission lines from the InP QD consists of many closely spaced lines, while from the InAs QD only a few lines are observed for sufficiently low excitation power density.

To decrease the number of emission lines from the InP QDs by decreasing the excitation power density is not possible. However, for InAs QDs in AlGaAs, multiple lines are also at very excitation power density.

There are several possibilities for the occurrence of multiple lines. For the InP QDs, the holes are localized either at the base or at the top of the pyramid, with very small overlap. The InP QDs are initially charged by electrons occupying several electrons levels. It is possible to change the number of electrons in the dot by applying bias to a Schottky gate on top of the sample. It is found that the number of lines decrease when the dot is depopulated.

PL spectra measured at 10 K for different excitation power intensities are shown in Fig. 6.15. With increasing the excitation intensity, the PL spectra show a clear band filling dynamics, which is due to QDs uniform size distribution. The monomodal size distribution is typical of samples having a low dot density ($\approx 10^9$ cm^{-2}). But, with dot density in excess of 10^{10} cm^{-2} shows a bimodal size distribution that results in emission spectra which exhibit structures of

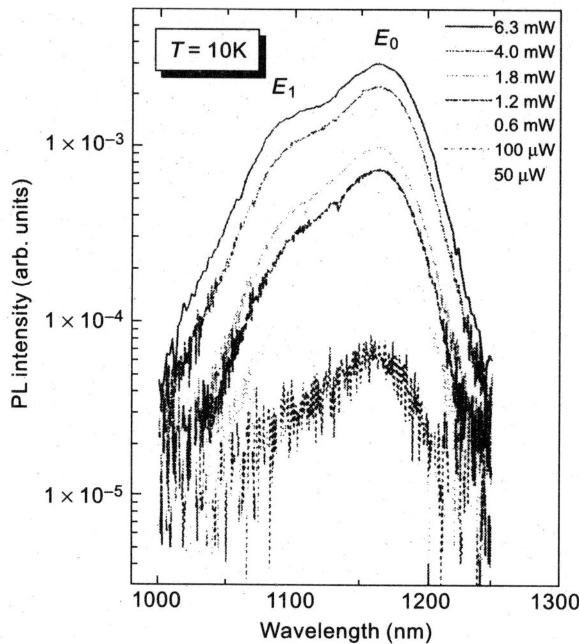

Fig. 6.15 *PL spectra performed at 10 K for different excitation power intensity (from 50 μW to 6.3 mW). E_0 and E_1 level the ground state and the first excited state peak transitions respectively*

constant intensity ratio. The ground level emission (E_0) peak is at 1167 nm with an FWHM of about 46 meV, whereas the first excited state (E_1) is observed at 1104 nm. The spectra shows a inhomogeneous broadening due to the non-uniformity of the QD size as well as the composition and shape. Even very high quality samples displays this type of broadening. The energy splitting between the ground level and the first excited level is about 60 meV.

The PL spectra of the QD structure for different temperatures are shown in Fig. 6.16. With increasing temperature, it is possible to observe two main effects: (i) a red-shift of the peaks; and (ii) a reduction of the luminescence intensity. The red-shift of the peaks us due to the energy-gap reduction with temperature whereas the intensity reduction is due to the thermal escape of the carriers from the dot levels and to the thermal activation of non-radiative recombination channels plotting the integrated emission intensity of the luminescence peaks vs 1/(kT) it is found that the depletion of the first excited state corresponds to a small increase of the integrated intensity of the ground level. This is due to the capture in the ground level of carriers escaped from the excited state. Above 180 K, the ground state also begins to empty due to the thermal escape effect.

Charging of Dots

The main difference between InAs dots in GaAs and InP dots in GaInP is the large number of emission lines for InP QDs. The reason for this is because the InP dots are highly charged charged with about 20 electrons. The matrix, GaInP, is *n*-type with a donor concentration of about 10^{16} cm^{-3}. By evaporation of a transparent gold film it is possible to perturb the dots by

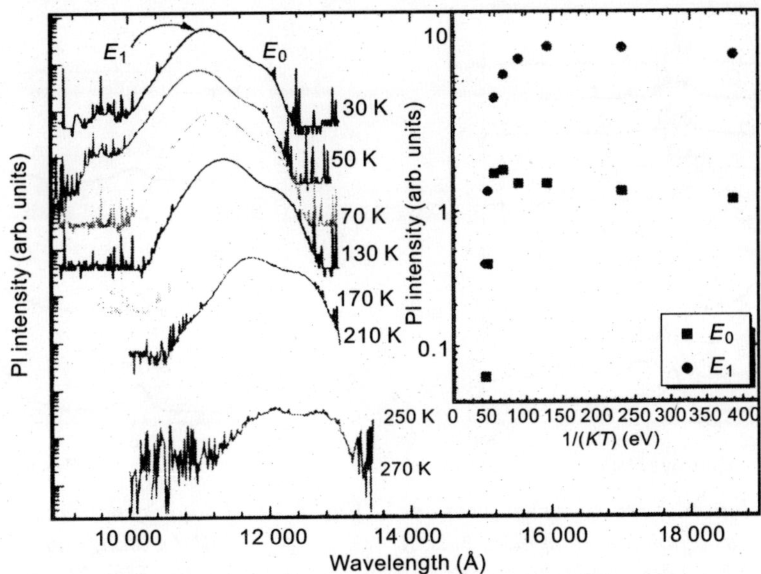

Fig. 6.16 *PL spectra performed at different temperatures. Inset: Arrhenius plot for the ground level and the first excited transition.*

an electric field and, to change the position of the Fermi-level in the vicinty of the dots as illustrated in figure 6.17. By using different sizes of the dots. eg: small dots (InAs) are expected to be less charged, or neutral. At an applied voltage of 1 V, the dots are filled with electrons. When changed to –1.0 V, the Fermi level is lowered in energy and the population of electrons in the dot is decreased. From Figure 6.18 it is observed that the number of emission lines decrease in the expected manner when the applied voltage is changed.

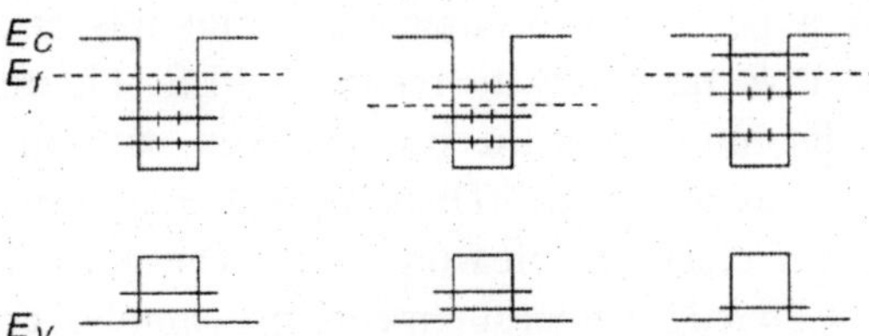

Fig. 6.17 *Influence of the Fermi level on the charging of quantum dots. If the Fermi level is above the ground state of the dot, the dot may be charged with electrons (left panel). By lowering the Fermi level (middle panel) or decreasing the size of the dot (right panel), it is possible to decrease the degree of charging. Each single particle level contains two electrons.*

When the dot becomes neutral (or weakly charged), the broad line at an energy of 1.65 eV splits into a set of sharp lines (100 μeV) due to charge fluctuations near the defect. The lines get narrowed at a temperature of 45 K but broad up at room temperature, meaning that such charge fluctuations are strongly dependent on sample.

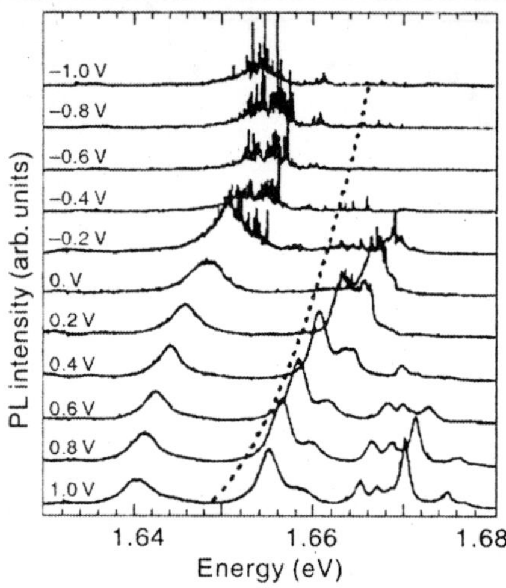

Fig. 6.18 *The evolution of the spectrum from one InP dots as a function of applied bias. A bias of 1.0 V corresponds to a dot filled with about 20 electrons, which give rise to a set of emission peaks. As the bias is reduced, the number of emission peaks becomes fewer due to a decrease of the charging. The lowest energy peaks splits into a large number of narrow peaks at a bias of –0.4 V. The dotted line shows the calculated energy shift.*

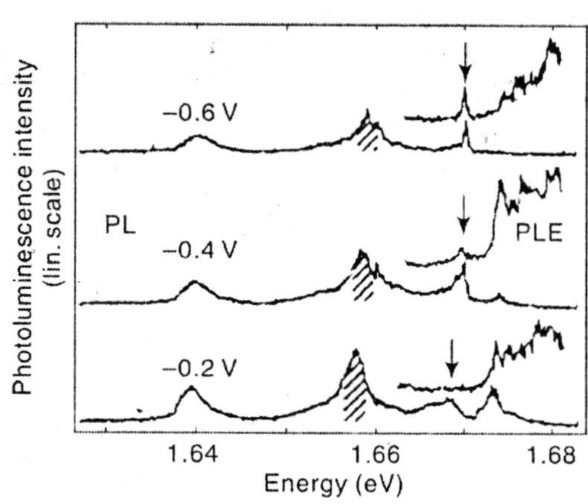

Fig. 6.19 *Photoluminescence exciton spectra of one InP dot along with PL spectra at different applied bias. The detection energy for the PLE is indicated by the shaded PL peak. One peak marked with an arrow becomes visible in PLE when the bias is decreased.*

If a states is filled with electrons, then absorption is forbidden into this state. However, if the state is empty, then there is a possibility of transitions into this state. Since absorption is very low in single quantum dots, it is necessary to resort to PLE. Figure 6.19 shows that when the bias is –0.2 V, there is no absorption into the peak marked with an arrow. At higher bias (e.g., –0.6 V), there is absorption into this peak. There is still emission from this peak in PL. The charging of the dots is seen in electrical measurements. These experiments measure the charge in the dots via capacitance. Reflectivity measurements on samples containing InP dots show the existence of charge close to the dots, which, is attributed to acceptors in the neighborhood of the dots.

As the dots become small they have fewer levels below the Fermi-level and the degree of charging becomes less, and finally the dots are neutral, as shown in Figure 6.17. This is shown in Figure 6.20 where dots emitting at an energy of 1.7 eV have many very sharp emission lines, and the dot emitting at an energy of 1.8 eV has only a few emission lines, in similarly to InAs dots in GaAs. The number of lines is about 16. For neutral dots, only a few lines correspond to excitons and biexcitons. The linewidth is about 100 μeV, which is typical for other types of SK dots such as InAs in GaAs.

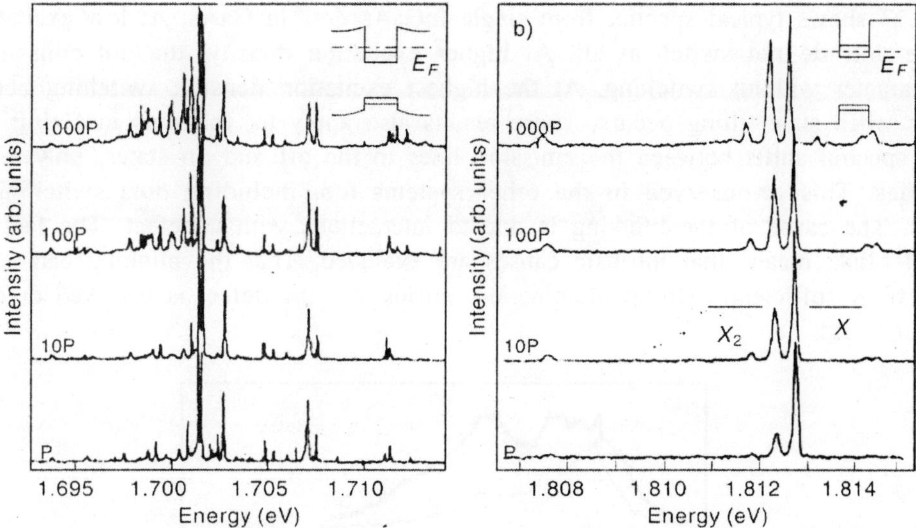

Fig. 6.20 *(a) Emission spectra from a small InP dot which still is charged. Many sharp emission lines are seen. (b) Emission spectra from a small InP dot which is not charged. The spectrum originates from the recombination of excitons, X_2 and biexcitons, X_2. The biexciton spectrum becomes more dominant at higher excitation power densities.*

Blinking Dots

Figure 6.21 shows blinking as in during emission, in InP dots in GaInP, when one dot has at least two PL intensity levels. Traces of the PL intensity, shown in Figure 6.22, show that the emission switches two levels on a very slow timescale. Most dots do not blink and the blinking increases in frequency when the excitation power density increases. Sometimes the blinking occurs between more than two intensity levels. The blinking is temperature activated, and is usually stable if the sample is heated to room temperature and then cooled again.

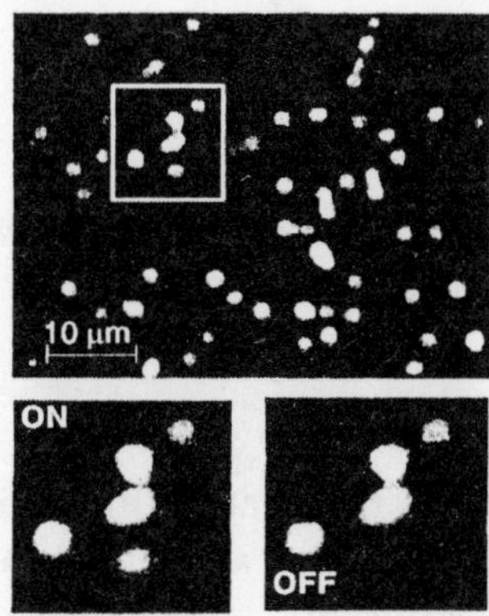

Fig. 6.21 *A top view of the emission from InP dots in GaInP. One dot changes intensity between two images.*

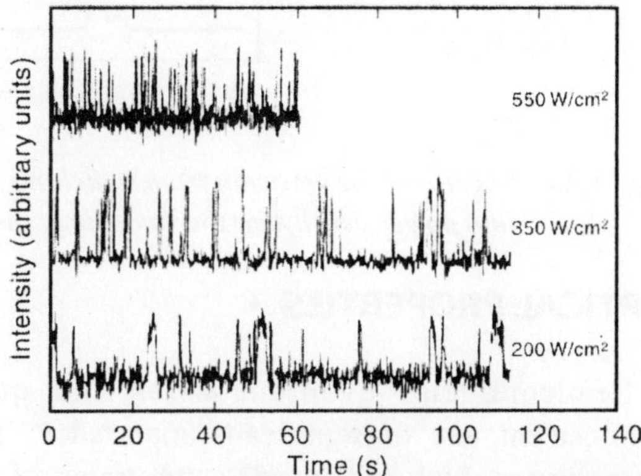

Fig. 6.22 *Intensity traces of the emission intensity from one switching InGaAs quantum dots in GaAs. The switching frequency increases with increasing excitation power density.*

Figure 6.23 shows typical spectra, from single InGaAs dots in GaAs. At low excitation power density, the dots do not switch at all. At higher excitation density, the dot emission slightly changes character without switching. At the highest excitation density, switching between two levels starts when state filling occurs. These results also apply to, InP dots in GaInP. There are hardly any spectral shifts between the emission lines in the off and on states, only a change of the intensities. This is observed in the other systems too, including dots switching between three levels. The cause of the blinking is due to interactions with a defect. The fact that most dots do not blink, means that intrinsic causes are excluded. That the blinking can permanently be turned off by sufficiently strong illumination, means that the defect is removed or deactivated by the strong light.

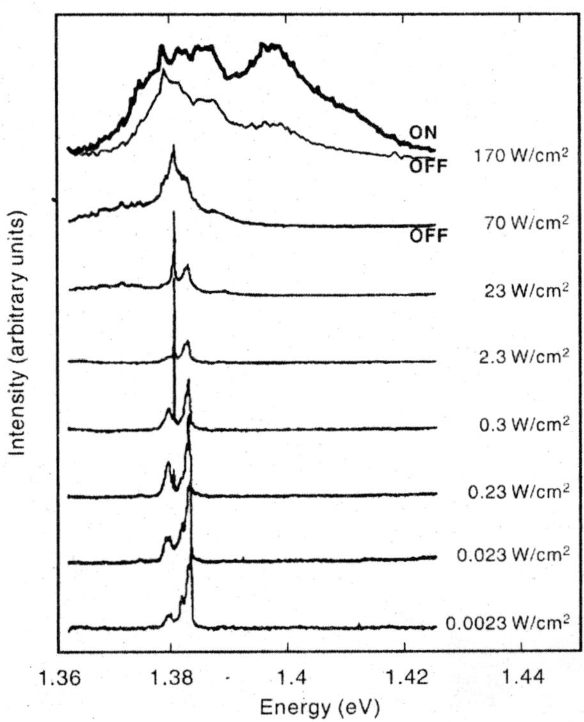

Fig. 6.23 *Spectra of the emission from switching InGaAs quantum dots. The switching occurs at high excitation power density. At the same time, the emission broadens due to state filling.*

OPTICAL PROPERTIES

Semiconductor quantum wires and quantum dots with two- and three-dimensional confinement, are of significant importance. To fabricate quantum wire and quantum dot arrays on patterned high-Miller-index substrates of unprecedented uniformity in size and composition is done with controlled positioning on the wafer. The one-and zero-dimensional nature of these quantum wire and quantum dot arrays manifests itself in the superior optical properties. The epitaxial growth on high-Miller-index substrates shows unique potential for creating high-quality quantum wires and quantum dots based on nanoscale lateral self-faceting and strain-induced organization in molecular-beam epitaxy (MBE) and metalorganic vapor-phase epitaxy (MOVPE). The self-organizing process are attributed to a lowering of the surface energy governed by the distinct asymmetry of band configuration and the related adatom migration and incorporation properties depending on the surface orientation.

The MBE growth on lithography patterned high-index substrates where surfaces of different orientation (i.e., index, surface energy, and adatom kinetics) are combined side by side establishing the driving force for selective growth and nanostructure formation on patterned substrates, example: GaAs (311)A. A new-phenomenon in the selectivity of growth for the fabrication of single and multiple GaAs/(Al,Ga)As sidewall quantum wires at shallow-patterned $[01\bar{1}]$ mesa stripes is found.

The formation of GaAs sidewall quantum wires on patterned GaAs (311)A substrates relies on a new type in the selectivity of growth. The surface energy ratios and adatom kinetics, are such that preferential migration of adatoms occurs away from the sidewalls to the mesa top and bottom. This leads to concave surface profile in corners, and provides the basis for the formation of V-groove and ridge-type quantum wires and quantum dots. In contrast, during MBE of (Al,Ga)As on patterned GaAs (311)A substrates, $[01\bar{1}]$ oriented mesa stripes develop a very smooth fast-growing convex sidewall on one side in the sector towards the next (100) plane, without any faceting, and a stable singular bottom corner, as shown schematically on the right-hand side in Figure 6.24(a) This new growth mechanism, being unique for the (311)A substrate, relies on the migration of adatoms from the mesa top and mesa bottom towards the sidewall. This is opposite in direction to that on patterned low-index substrates. The opposite sidewall [left-hand side in Figure 6.24(a)] as well as mesa stripes oriented along the perpendicular $[\bar{2}33]$ direction [Figure 6.24(b)] develops slow-growing side facets similar to patterned low-index substrates.

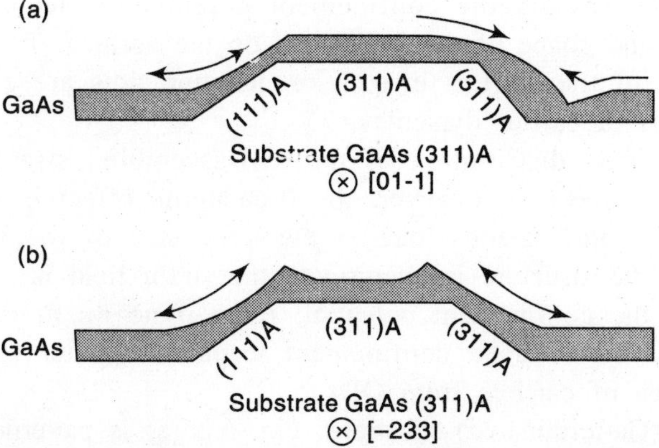

Fig. 6.24 *Growth mechanism on patterned GaAs (311)A substrates with mesa stripes oriented (a) along* $[01\bar{1}]$ *and (b) along the perpendicular* $[\bar{2}33]$ *direction. The arrows indicate the preferential migration of Ga adtoms resulting in the selectivity of growth across the mesa sidewall.*

Using mesa heights of 10 to 20 nm (i.e., I the quantum size regime), formation of the fast-growing sidewall allows the fabrication of quasi-planar lateral GaAs/(Al,Ga)As quantum wires of several tens of nanometers in width. Figure 6.65 shows the schematic illustration and TEM image of the 6-nm GaAs quantum-well layer embedded between 50-nm $Al_{0.5}Ga_{0.5}As$ lower and upper barriers; plus 50-nm GaAs buffer and 20-nm GaAs cap at the fast-growing sidewall of the 75-μm-wide and 15-nm-high $[01\bar{1}]$ mesa stripe.

Stressors are formed on QW structures by self assembled and strain induced structures.

Quantum dots induced by self-assembled stressor structures have high optical quality because the structure is fabricated in a single epitaxial growth run without any processing, and the active area of QD is located in nearly perfect lattice away from the surface. The strain field created by a self-assembled island is much larger than that of a stressor etched from a pseudomorphic layer. The structure is shown in the lower part of Figure 6.25. InP grows on GaAs in coherent Stranski-Krastanow growth mode, forming first a two-dimensional (2D) wetting layer having a thickness up to about two monolayers (ML). If the deposition thickness exceeds 2 ML, dislocation-free three-dimensional (3D) InP islands are assembled on top of the wetting layer. The lattice constant of InP is about 4% larger than that of GaAs, and the strain field is partly incorporated into the underlying substrate. The displacement of the atoms from their original positions results in a local modification of the band-edges of the materials under the stressor. Therefore, a lateral near-parabolic confinement potential in the QW plane for both electrons and holes is formed, as shown in Figure 6.25. The lateral stressor potential and the vertical near-surface QW potential form together the QD confinement potential. The electron and hole eigenstates are highly orthogonal, leading to a simple luminescence line spectrum corresponding to recombination of an electron and a hole in energy levels having the same level index, as seen in the upper part of Figure 6.25.

QD structures based on a near-surface QW and self-assembled stressors made of InP GaSb and Ge self-assembled InAs stressors are employed either below or above a QW.

With self-assembled stressors the confinement potential is accurately modeled, as it is determined by the size and shape of the InP island on the surface. Therefore, the energy levels and the wave-functions of the carriers in stressor quantum dots are calculated accurately, and used in detailed analysis of carrier dynamics.

The inhomogeneous size distribution of the self-assembled stressor islands broaden the transitions in the optical spectra. However, the broadening effect is less pronounced than in QDs based on self-assembled islands, due to the large size of the stressor islands. Another disadvantage is that, if the structure is overgrown, the strain field is redistributed, leading to a significant decrease in the confinement potential. Even if no misfit dislocations are generated during the overgrowth, the reduced confinement leads to smaller level quantization and to increased thermal escape of carriers from QDs.

The GaAs (311) A (Ga-terminated) substrates Fig. 6.26(a) is patterned by optical lithography and wet chemical etching in H_2SO_4:H_2O_2:H_2O (1 : 8 : 40) solution. The GaAs growth rate is 0.5 μm/h, the V-III-flux ratio 5, and the growth temperature 893 K which is optimum for GaAs (311) A.

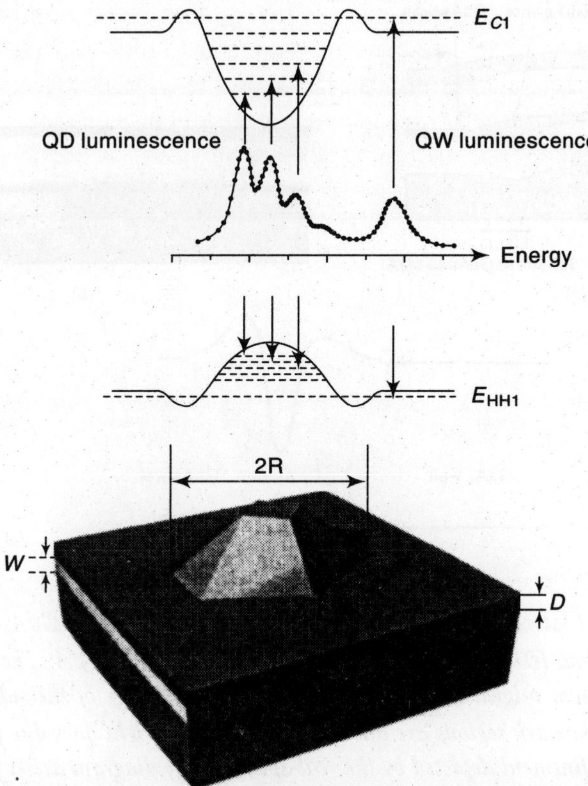

Fig. 6.25 *A model of the QD structure with a self-assembled InP island as a stressor. Shown above is the lateral confinement potential in the QW plane induced by the strain of the InP island together with a luminescence spectrum, and corresponding electronic transitions between the discrete QD levels.*

The one-dimensional confinement depicted by the real-space energy band diagram in Figure 6.26(c) arises form the strong lateral thickness variation along (233) of the narrow GaAs quantum well, in addition to the abrupt potential barrier in growth direction at the GaAs/(Al, Ga)As heterojunction. This lateral thickness variation translates directly into energy bandgap variations via the transverse quantum size effect. As a result, a lateral potential well for electrons and holes at the thicker part of the quantum well is obtained, which is of pure quantum mechanical origin.

The strong lateral carrier confinement and the structural perfection of sidewall quantum wires manifest themselves in narrow PL lines and high PL efficiency up to room temperature, as shown in the micro-PL spectra taken at 8 and 300 K in Figure 6.27 (the diameter of the optical probing area is 2 μm). On the flat area of the mesa, only the peaks originating from the 6-nm quantum well are observed, while the spectra excited at the sidewall split into two (three) lines at lower and higher energy due to the emission from the quantum wire (1.544 eV at 8 K) and from the adjacent quantum well (1.066 eV at 8 K). The PL peak energy position corresponds to a quantum wire thickness of 12 nm, (observe TEM data of Figure 6.26(b)). The blue shift of

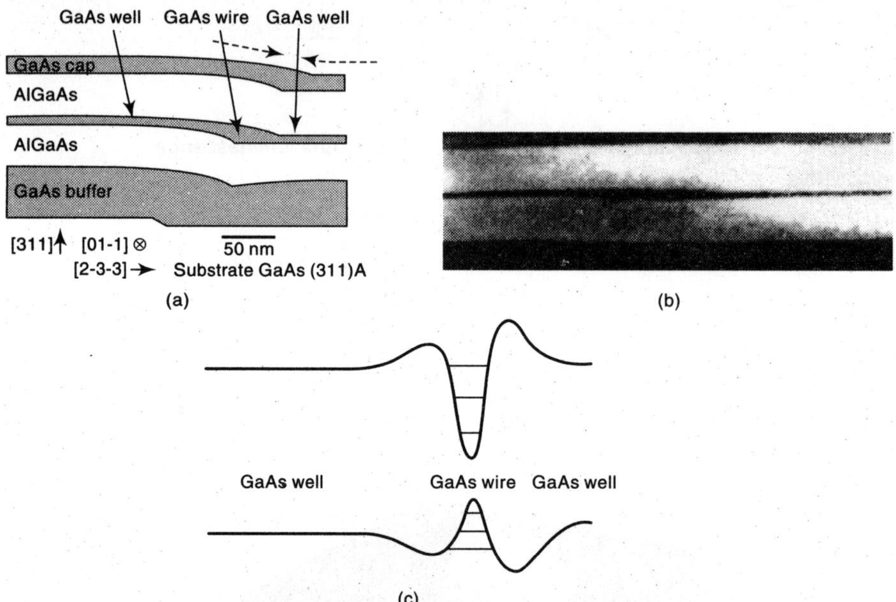

Fig. 6.26 *(a) Formation of lateral quasi-planar GaAs quantum wires in Al GaAs matrix at the fast-growing sidewall of shallow [01$\bar{1}$] mesa stripes on patterned GaAs (311) A. The dashed arrows indicate the direction of adatom migration. (b) Cross-sectional Tem image of the sidewall quantum wire taken along [01$\bar{1}$]. The dark regions are due to GaAs and the bright ones due to (Al, Ga)As. (c) The one-dimensional confinement depicted by the real-space energy diagram arises from the strong variation of the GaAs layer thickness.*

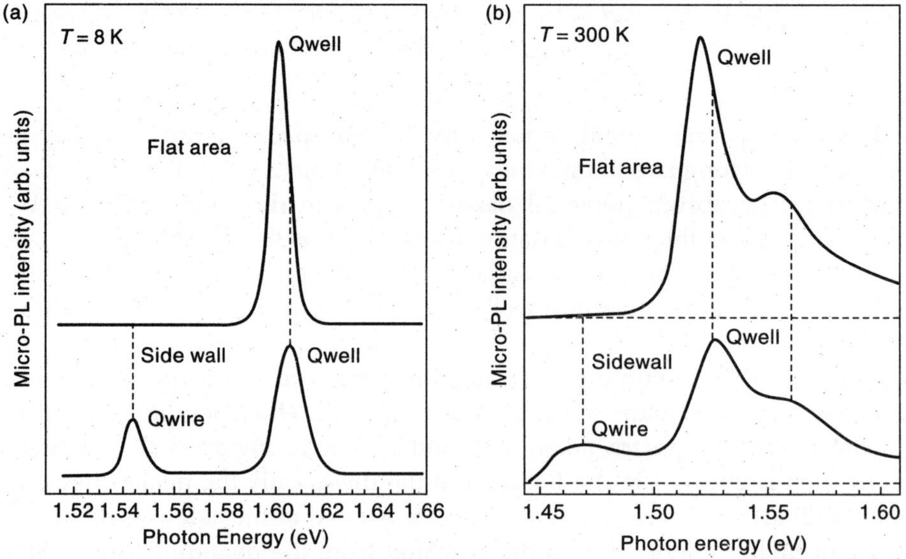

Fig. 6.27 *Micro-PL spectra taken (a) at 8 K and (b) at 300 K from the flat area of the mesa (upper curves) and from the quantum wire region at the fast-growing [01$\bar{1}$] mesa sidewall (lower curves). The diameter of the optical proving area is reduced to 2 μm by a confocal imaging system.*

the quantum well emission near the sidewall, to the flat part of the mesa, confirms the existence of narrower well regions in the vicinity of the wire, which form the lateral energy barriers. The PL intensity at room temperature is reduced by less than three orders of the magnitude compared to that at low temperature, and up to room temperature, miro-PL line scans reveal efficient lateral confinement of photogenerated carriers in the quantum wires.

SNOM measurements with a spatial resolution 250 m allows to map the lateral potential barriers in a single GaAs quantum wire due to the thinning of the wells. The confinement energies of electrons and holes with respect to the adjacent 6-nm quantum well are 60 and 15 meV, respectively, as indicated in the band diagram of Figure 6.26(c). In addition, the real space transfer and the trapping of photogenerated carriers from the adjacent quantum well, as well as the one-dimensional subband structure, are in good agreement with the model.

Due to the self-limiting growth mode of the convex surface profile, the sidewall quantum wires are stacked in growth direction with a perfect reproduction of shape and size. The periodicity of lateral quantum wire arrays is reduced to the submicrometer range. Figure 6.28 shows the schematic illustration and cross-sectional TEM image of a 500-nm pitch array of stacked sidewall quantum wires. The thickness of the GaAs quantum well is 3 nm and the Al

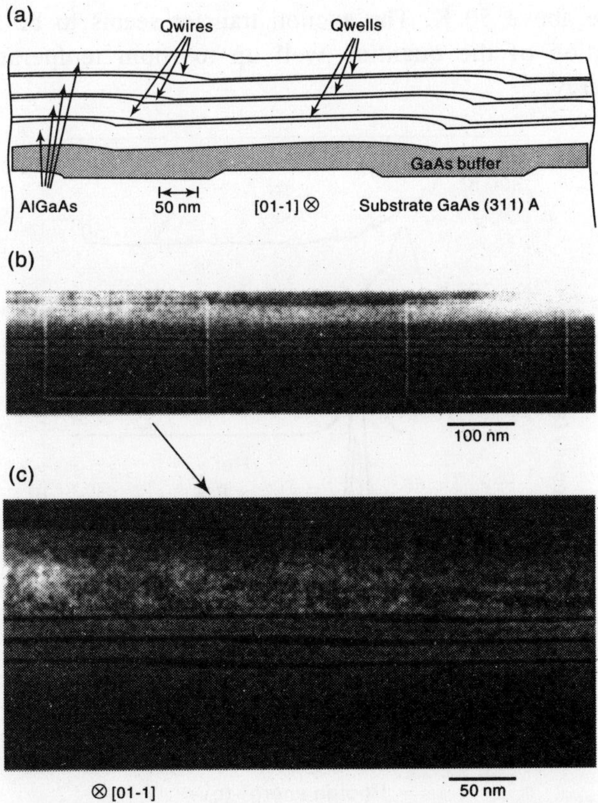

Fig. 6.28 *(a) Schematic illustration (b) cross-sectional TEM image of the 500-nm pitch array of vertically stacked sidewall quantum wires. In (c) the scale is enlarged.*

composition of the 10-nm (Al, Ga)As barrier layers between the wires is 0.5; the structure is completed by 50-nm (Al, Ga)As lower and upper barriers, 50-nm GaAs buffer, and 20-nm GaAs cap layer. The growth selectivity and overall uniformity of sidewall quantum wire arrays are even enhanced for submicrometer pitch arrays, where Ga adatoms migrate directly from the (slow-grown mesa sidewall to the fast-growing one. The slow-growing sidewall completely smears out to provide a uniform template for the quantum wells connecting the quantum wires, and the quantum well undergo stronger thinning compared to wide mesa stripes. This increases the lateral confinement energy considerably.

The low-temperature CL spectrum of the quantum wire array (Figure 6.29) consists of two lines at 1.814 and 1.606 eV, due to the emission from the quantum well regions and quantum wires. The large energy separation of the PL lines reflects the increased growth selectivity, resulting in lateral confinement energy of 210 meV. The strong thinning of the quantum well regions is evidenced by the substantial 75-meV blue shift with respect to the quantum well emission in unpatterned, planar areas (noted "Ref." In figure 6.29). Exciton localization in the quantum wells due to interface fluctuations, (giving rise to the quantum well luminescence) is very weak, even at low temperature. When the temperature is raised from 5 to 300K, the excitons in the quantum well become delocalized and the quantum well luminescence in Figure 6.29 is hardly detectable above 50 K. The exciton transfer seems to be almost perfect and there is no thermal repopulation of the quantum well up to room temperature owing to the large lateral confinement energy.

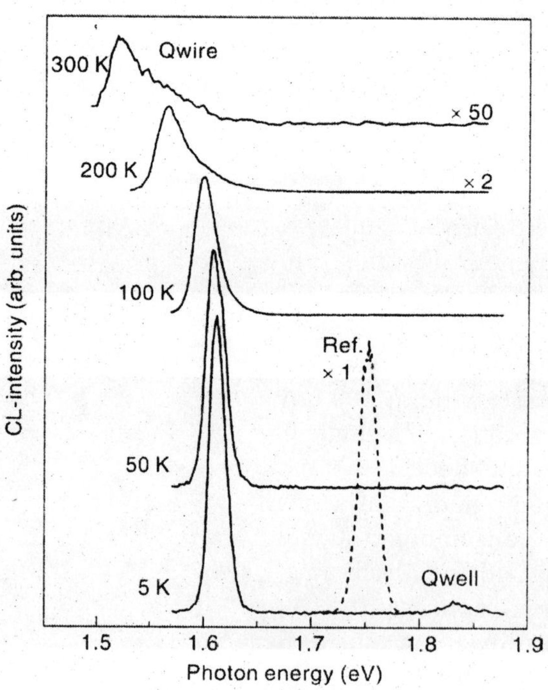

Fig. 6.29 *CLK spectra of the 500-nm pitch sidewall quantum wire array between 5K and room temperature (solid lines) and of the reference quantum well in unpatterned, planar areas. The CL spectra are averaged over an excitation area of 400 μm².*

For, electronically coupled quantum wires, the electronic coupling is confirmed by the distinct red shift of the PL when the (Al,Ga)As barrier layer thickness between the quantum wires is decreased from 10 to 1 nm. Moreover, surface acoustic wave (SAW)-driven one-dimensional carrier transport is demonstrated in these quasi-planar quantum wires. The device quality with enhanced functionality is achieved for this new type of sidewall quantum wire structure on patterned GaAs (311)A substrates.

High-Resolution Spectroscopy of Quantum Dots

Single dot resolution is obtained by placing a shadow mask from aluminium or gold on a quantum dot sample, as shown in the micrograph in Figure 6.30. The thickness of the mask is larger than approximately 100 nm to that it is not optically transparent and it blocks the incident laser light completely.

In it, holes are prepared with diameters as small as 50 nm, through which both optical excitation and detection of the single is done.

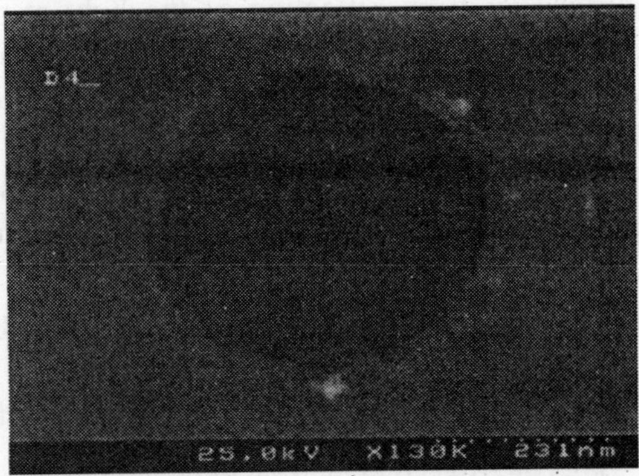

Fig. 6.30 *Scanning electron micrograph of a section of gold shadow mask with a hole. A diameter of approximately 300 nm isolates a single quantum dot underneath.*

For the occupancy of an optically addressed field by a single dot, photoluminescence spectroscopy at low excitation powers is used. Figure 6.31, shows spectra recorded $T = 2K$ on mesa structures with different lateral sizes. For the unstructured quantum dot sample, broad emission band with a width of about 30 meV reflecting the inhomogeneous broadening is observed. A mesa with a size of 300 nm, (the broad band splits into a set of sharp emission lines) shows the three-dimensional confinement of the carriers. With decreasing mesa size, the number of spectral lines decreases until a single sharp emission line for the mesa with a lateral size of *approximately* 100 nm is observed. Assume, the observation of a single line as an indication for a single dot in the mesa structure. In the spectroscopic finding of a single line at low excitation if there is another quantum dot in the structure, it may not be observable at low excitation (due to carrier capture at the close-by surfaces). This dot shows up, at higher excitation.

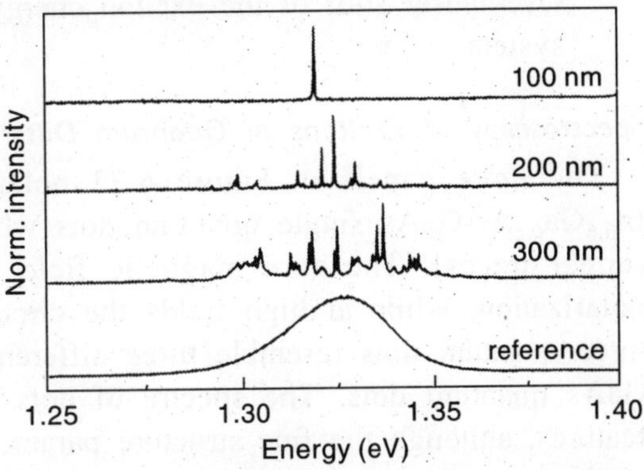

Fig. 6.31 *Photoluminescence spectra recorded at $T = 2K$ on a planar, unstructured quantum dot sample (bottom trace), and on mesa structures with varying lateral sizes, and indicated at each trace. Weak optical excitation powers were used.*

The structures are held in superfluid helium ($T < 2K$) in an optical cryostat with a split magnet coil for fields up to B = 8T. The orientation of the magnetic field is varied relative to the heterostructure growth direction by rotating the sample in the cryostat. Laser excitation is done by an Ar-ion laser. The quantum dot emission is dispersed by single or double grating monochromators depending on the required spectral resolution. The polarization of the emission is analyzed by suitable combinations of linear polarizers and quarter-wave retarders, Detection is done with a Si-charge coupled devices camera cooled by liquid nitrogen. The quantum dot is imaged into an intermediate plane, in which a small aperture that can move parallel to the plane is placed. Stray light might arise from scattering of incident laser light on the patterned surface, or from reflection of the light at the backside of the sample. Due to this scattering, mesa structure is excited. When the mesa is adjacent to big fields that are used for adjustment and contain a huge number of dots; the quantum dots emission is superimposed on a broad background.

From this spectrum, the background is obtained and subtracted from the emission spectrum recorded on the single dot mesa spectrum.

EXCITON COMPLEXES IN QUANTUM DOTS

Several interactions contribute to this fine structure

1. The first one is the exchange interaction between electron and hole, which couples their spins, or more precisely, their angular momenta.
2. In an external magnetic field, the Zeeman interaction of the electron and holes spins with B has to be added to the exchange interaction.
3. In general, the interaction of the carrier spins with the spins of the lattice nuclei needs to be included. Which is termed hyperfine interaction. This interaction results in the Overhauser shift of the exciton energy and is observed at interface or natural quantum dot system.

Spectroscopy of Excitons in Quantum Dots

The three panels in Figure 6.32 polarized photoluminescence spectra of three different $In_{0.6}Ga_{0.5}As$-GaAs single quantum dots which were recorded at B = 0 (lower traces) B = 8T (upper traces). The zero magnetic field spectra were analyzed with respect to their linear polarization, while at high fields the circular polarization of the emission was studied. These three quantum dots resemble three different classes of structures observed for the $In_{0.6}Ga_{0.5}As$-GaAs quantum dots. The spectra of dots belonging to one class all show the same principal features, although the fine structure parameters, exchange energies, and *g*-factors, vary from dot to dot.

The emission from dot 1 (left plane) shows no significant linear polarization at zero magnetic field. In contrast, the emission of the two other quantum dots is split into linearly polarized spectral lines at B = 0. The polarization direction coincides with the and crystal directions. The splitting between the lines is about 120 μeV for dot 2 (midpanel), and 150 μeV for dot 3 (right panel). In nonzero magnetic field, the emission split due to the Zeeman interaction of the

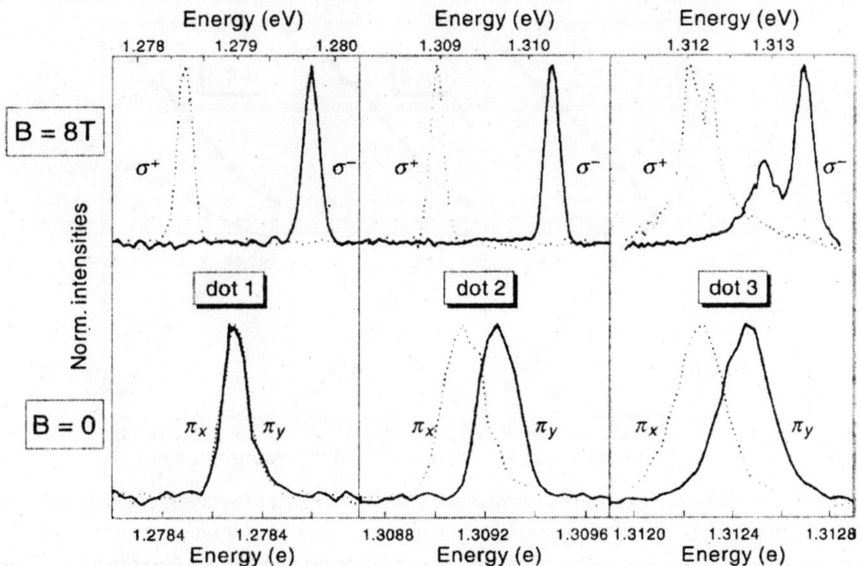

Fig. 6.32 *Polarized photoluminescence spectra of three different $In_{0.6}Ga_{0.5}As$-GaAs quantum dots recorded at zero magnetic field (lower half) and at B = 8T (upper half). At zero magnetic field, the circular polarization of the emission was analyzed; at high fields the linear polarization was studied.*

exiton spin with B. As a common feature at B = 8T, the spectra of all three dots show complete circular polarization. The low energy part of the spectrum is σ^+-polarized, and the high energy part is σ^--polarized. However, the number of spin-split lines varies from dot to dot. For the first two dots, a splitting into a doublet is observed, while the third dot exhibits a splitting into a quadruplet.

The theoretical analysis of the exciton fine structure for quantum dots of different symmetries given above permits to understand the observed features in the photoluminescence of the three dot classes. From the experimental data there are two characteristic quantities from which information about the quantum dot symmetry is derived. The first quantity is the magnetic field dependence of the energy splitting between the spectral lines. The second one is the polarization of the emission. Both quantities are discussed in the following.

The simplest situation for dots with D_{2d} symmetry, for which the exciton angular momentum M is a good quantum number is a single emission line observed at B = 0, due to recombination of the degenerate M = ±1 excitons, as observed for dot 1. Applying a magnetic field results in a spin-splitting. The symbols in Figure 6.33 show the observed exciton transition energies as functions of the magnetic field for the three dots of Figure 6.32. For, the fine structure effects, the energy of the center of the emission lines for each field strength is subtracted so that the center of emission shifts diamagnetically to higher energies with increasing B. The diamagnetic shift is a measure for the extension of the exciton wave function normal to the magnetic field. Due to the strong confinement of the carrier in the quantum dots, the shift remains considerably smaller than 1 meV, up to 8T. The emission of dot 1 splits linearly into a doublet with increasing B. ΔE equals 1.4 meV at 8T, corresponding to an exciton g-factor of –3.

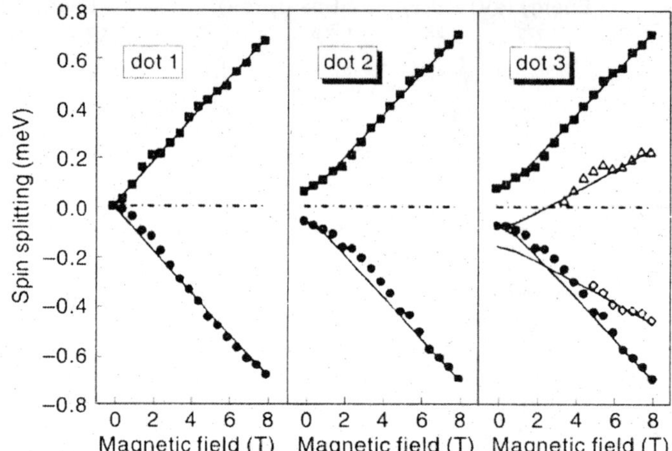

Fig. 6.33 *Magnetic field dependence of the exciton transition energies observed for three different $In_{0.6}Ga_{0.5}As$-GaAs quantum dots. Symbols give the experimental data, and lines give the results of fits to the data using forms obtained from diagonalizing the exciton fine structure Hamiltonian. For simplicity, the dimagnetic shift of the center of the emission has been subtracted in each case.*

For a dot a with a shape deformation, due to which the rotational symmetry is broken the diagonal elements in the subspace of the bright excitons in $H_{exchange}$ mix with the states $M = \pm 1$ and -1. The new coupled eigenstates repel each other, and the emission from these quantum dots is split by the exchange energy $\delta 1$, as observed for dot 2 (middle panel). From the transition energies, a δ_1 of about 120 μeV is obtained. In contrast to dot 1, at low B the energy splitting between the exciton transitions increase quadratically with B, and then transforms into a linear dependence. The transition into a linear dependence occurs at about 2T, because at these field strengths, the diagonal Zeeman interaction terms in the Hamiltonian are considerably larger than the off-diagonal δ_1. This means the rotational symmetry is moderately broken, and it is easily restored by a magnetic field.

Finally, for the behavior of dot 3 with a quadruplet splitting in magnetic field the observation of four exciton emission lines imply that the dark excitons become visible, and any possible dot symmetry is lifted. This symmetry breaking has different origins. First, the quantum dot symmetry is badly broken so that it exhibits no symmetry at all. In this case, all four band edge exciton states are observable at zero magnetic field. From the spectroscopic data, no clear proof for such a behavior is there, because at $B = 0$ only two emission lines are observed, which are rather broad. In addition, the symmetry breaking is magnetic-field induced, for if the dot structure (and therefore its internal [001] crystal axis) is slightly tilted with respect to the heterostructure growth direction, the spectroscopy is effectively no longer performed in FC, as there is a field component in the quantum dot plane. This component, described by the H_{Zeeman} for the VC causes a mixing of $|M| = 1$ and 2 excitons, making the dark excitons visible.

Similar to dot 2, for the bright excitons in dot 3, ΔE depends quadratically on B at low fields and shows a linear dependence at high B. The splitting ΔE between the "dark" excitons shows a linear B dependence in the field range in which these states can be resolved.

Correlation between Strain Field and TEM

The uniformity of vertically aligned dots is obtained directly from plan-view TEM images, once the correlation between TEM diffraction contrast and localized strain field of single and vertically stacked dots is known.

Figure 6.34(a)-(c) shows the on-zone plan-view bright field (BF) of single, uniform and nonuniform, vertically stacked dot. These images are obtained in the [100] zone axis, that is, with the electron beam propagating along the growth direction. Here, for the single dot [Fig. 6.34(a)], and for the uniform stack of dots, [Fig. 6.34(b)] the contrast is characterized by an external dark region, of nearly circular shape, with a bright spot at the centre. The stacked dots with a non-uniform size along the stack [Fig. 6.34(c)] show that the intensity modulation, results in a flower-like pattern. Figure 6.34(d)-(f) shows the contrast line-scan performed along the ⟨010⟩ directions, that is, the intensity modulation along the dashed lines in Fig. 6.34(a)-(c). A main central maximum is observed in the line-scan profiles of Fig. 6.34(d) and (e) whereas Fig. 6.34(f) shows three maxima of comparable intensity, the external one being lightly weaker and symmetric with respect to the central one.

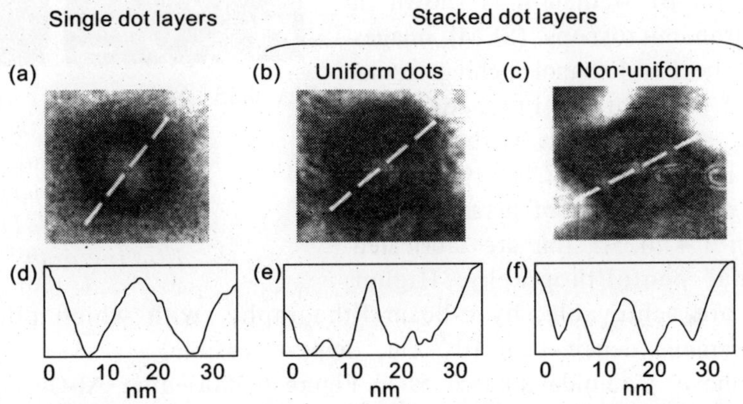

Fig. 6.34 *[100] Plan-view images obtained in the on-zone BF imaging conditions from the single (a), uniform (b), non-uniform (c), vertically stacked dot. The contrast line-scans, performed along the ⟨001⟩ directions on both images, are also displayed in (d) and (f).*

The white/black diffraction contrast in the TEM images is due to the inhomogeneous lattice strain associated to the 3D islands. This induces a local variation of the lattice planes orientation, resulting in a local modification of the electron diffraction conditions. The different contrast observed in the single and stacked dot samples is a general feature observed in many other samples, regardless of the growth conditions. Thus, the main parameter that affects the contrast pattern of plan-view images is the dot size uniformity along the stacking direction. The strain field associated with dots different sizes results in a modification of the electron diffraction condition. The correlation between the TEM diffraction contrast and the localized strain field in the single and vertically stacked dots, and, the effect of the linear combination of strain fields associated with dots of different dimensions exists.

INVERTED-PYRAMID QUANTUM DOTS

The self-limiting growth by OMCVD on V-grooved (100) GaAs substrates is useful for the preparation of high quality QWRs. Self-ordered QDs require the identification of a suitable set of crystalline facets that exhibit the characteristic self-limiting growth. The control of self-limiting profile via the adjustment of the surface diffusion length is necessary in order to control the size and shape of the resulting 0D structure. The formation of the QDs requires 3D self-limiting surface patterns. The simplest crystallographic configuration, is the corner formed by three {111} A planes which is etched into a GaAs (111)B-oriented wafer. Other 3D geometries utilized for the formation of nonplanar QD heterostructrues are tetrahedral mesas grown on GaAs (111)B substrates, and square-based inverted pyramids grown on GaAs (100) wafers using selective OMCVD growth on masked surfaces.

In the approach using pyramidal recesses etched into GaAs (111)B substrates; an array of inverted pyramids, is produced by optical lithography; pattern transfered into an SiO_2-mask and wet chemical etched with Br-Methanol; is shown in the scanning electron microscopy (SEM) images in Figure 6.35(a). The Br-Methanol etching results in pyramids with straight, {111}A oriented sidewalls forming sharp corners with radii of curvature in the 10-nm range. Closely packed, 2D hexagonal, or square lattices of pyramids with pitch ranging from 1.4 to 5.5 μm are fabricated using conventional photolithography, Higher pattern densities are achievable, by e-beam lithography, with which pitches of 100 nm corresponding to pyramid densities of 10^{10} cm^{-2} appear feasible.

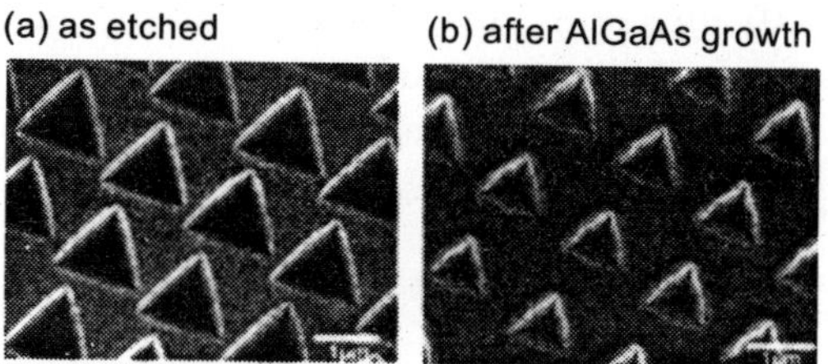

Fig. 6.35 *SEM top view images of sharp-bottom pyramids (a) before and (b) after AlGaAs deposition. In (b) as-etched {111} A pyramid sidewalls evolve into two vicinal {111} A facets during AlGaAs growth.*

Figure 6.35(b) shows pyramidal pattern as in Figure 6.35(a) after Al-GaAs deposition using optimized growth conditions (growth temperature 923k K), total pressure of 20 mbar, V/III ratio of 500, and a nominal (100) growth rate of approximately 0.1 nms^{-1}. Under these conditions, no growth takes place on the nonpatterned {111}B wafer surface. But the arriving growth species migrate and contribute to the growth inside the pyramidal holes.

In Figure 6.35(b), the originally planar {111}A pyramid sidewalls evolve into two vicinal {111} A planes during AlGaAs growth. The drawing in the inset of Figure 6.36(a) shows the topography of the pyramids after AlGaAs deposition.

This requires a method for 3D structural analysis with nanometer-size resolution. Cross-sectional AFM permits a 3D analysis using arrays of pyramids that are misaligned with respect to the {011} cleavage planes. Cross-sectional AFM images at positions through the pyramid are obtained by selecting a particular pyramid along the cleavage plane. Cross-sectional AFM images of a thick AlGaAs layer with thin GaAs marker layers are obtained at two such planes and are shown in Figure 6.36, a and a′ represent the slow growing, as-etched {111}A sidewalls while b, b′, and c correspond to the faster growing, vicinal {111}A planes. Two interfacet corners are also visible in the x-AFM image: the main pyramid corner between b and c and the less pronounced corner between the vicinal planes c and b′. The distance of these corners (e.g., the length of facet (c)is directly proportional to the horizontal distance of the cross sectional plane to the center of the pyramid.

Thus, in a cross section taken closer to the pyramid center [Figure 6.36(b)], facet *c* is smaller. The size of this facet can therefore be used to monitor the position of the cleaved surface with respect to the 3D hetrerostructure.

The inset of Figure 6.36(b) shows a magnified view of the corner region between the main facets. Right at the corner between the {111}A facets, an enhanced growth of GaAs is clearly visible. This corresponds to the formation of QWRs along the three primary corners between the {111}A sidewalls, similar to the case of the self-limiting V-groove structure made on (100) substrates. The intersection of these three QWRs at the tip of the pyramid along forms a QD at this point, due tc the superposition of the corresponding QWR potentials. A similar QD heterostructure is formed during a two-step cleaved edge overgrowth technique. Here, an additional thickening of the GaAs layer at the top of the pyramid leads to a much deeper QD potential. This additional thickening is brought about by the additional capilarity effects due to the further reduction of the surface chemical potential at the highly curved pyramid corner. This thickening is evidenced by measuring the layer thickness in a series of AFM cross sections taken at a range of positions around the center of the pyramid, as shown in Figure 6.37. The plot and also the x-AFM images on the left and right inset of Figure 6.37 show that the QD at the pyramid tip is thickner than the QWRs along the pyramid corners by a factor of approximately two. In addition, the measurements indicate a local reduction of QWR thickness (“necking”) just next to the QD.

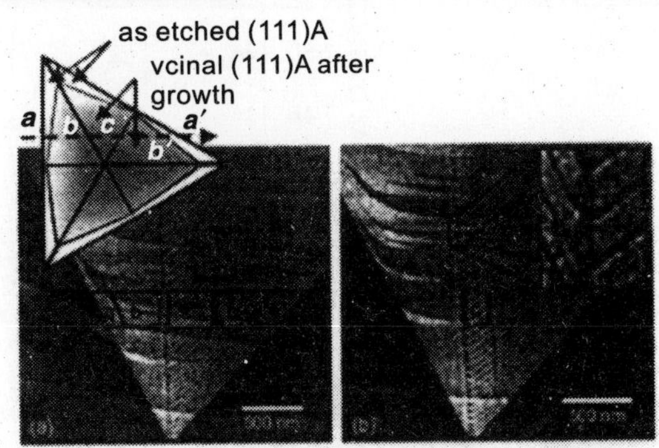

Fig. 6.36 *(a, b) Growth evolution in inverted pyramids as viewed by cross-sectional atomic force microscopy. The two cross sections correspond to two different planes cutting through the pyramid, as indicated in the schematic diagram in the inset.*

A lens-like QD structure with barriers of different dimensionality if formed by self-limiting growth in these inverted pyramids. Figure 6.38 shows the details of the resulting GaAs/AlGaAs QD heterostructure. Two classes of self-ordered nanostructures are formed, one due to the GaAs layer deposition, and the other as a result of the segregation effects in the AlGaAs layers. The self-limiting surface evolving during GaAs or AlGaAs growth is independent of the initial details of the shape of the etched recess. The growth of the GaAs layer on top of the self-limiting AlGaAs layer yields a lens-shaped GaAs QD laterally connected to three GaAs V-groove QWRs, which, in turn, are connected to the GaAs QRs grown on the vicinal {111}A facets. The thickness ratio of the GaAs QW:QWR:QD structure is approximately 1:3:6. Moreover, the resharpening of the pyramid tip during deposition of AlGaAs on top of a GaAs QD allows the vertical stacking of identical dots in a linear array at the center of the pyramid.

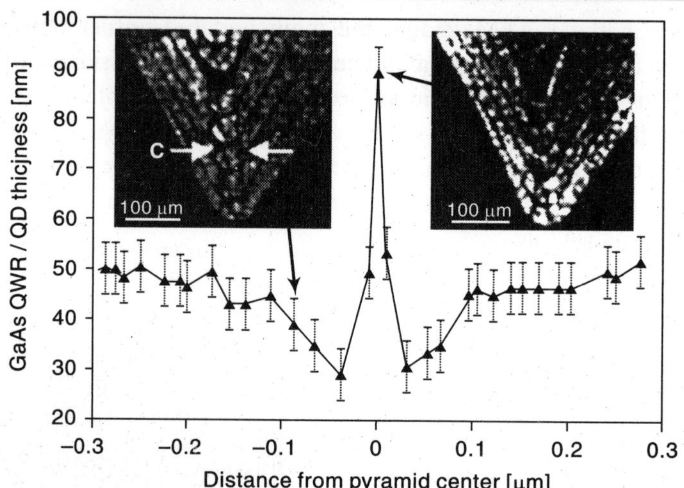

Fig. 6.37 *The QWR/QD thickness measured in cross-sectional AFM images plotted against the horizontal distance of the cross sectional plane from the pyramid center as deduced from the width of facet (c) (left inset, see also Figure 6.36). The left and right insets show cross-sectional AFM images taken close to and exactly at the pyramid center, respectively. At the enter, facet c is absent [see inset of Figure 6.36(a)].*

The darker stripes in the AFM images of Figures 6.37 and 6.38 represent the formation of self-ordered vertical AlGaAs QWs at the intersection of the {111}A planes, as for (100) V-groove structures. A new feature of the {111}B pyramids, however, is the formation of a self-ordered vertical, Ga-rich AlGaAs QWR along the line where the QWRs meet. The vertically stacked QDs are thus connected by vertical AlGaAs QWRs, which, unlike the corner GaAs QWRs, are interfaced to the dots without any intermediate potential barrier.

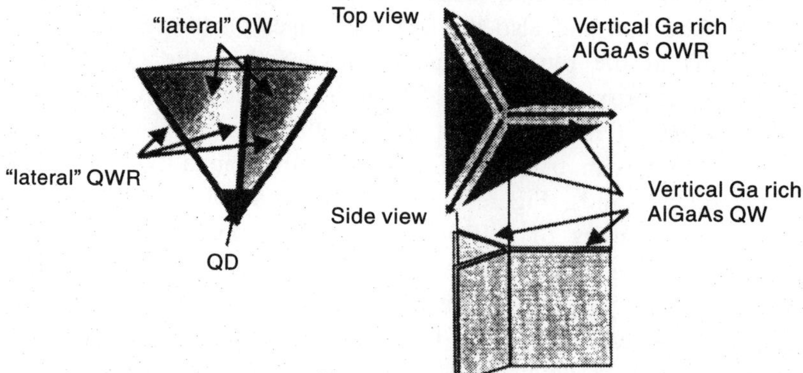

Fig. 6.38 *A system of connected nanostructures in the pyramids, as deduced from the structural investigation. (a) The GaAs QD at the tip of the pyramid is connected to lateral QWs on the pyramid sidewalls and "lateral" QWRs on the pyramid corners. (b) AlGaAs growth leads to the formation of vertical Ga-rich AlGaAs QWs on the pyramid corners, which meet at the center of the pyramid, forming a vertical Ga-rich QWR.*

The growth of GaAs in the inverted {111}B pyramids exhibits the same self-limiting growth behavior with the corresponding (larger) radius of curvature as determined by the effective surface diffusion length of the Ga adatoms. Such self-limiting GaAs pyramids are fromed the basis for InGaAs/GaAs QD heterostructures, similar to the formation of a Ga-rich AlGaAs VQWR in the case of AlGaAs deposition.

Optical Properties of Inverted-pyramid Quantum Dots

The luminescence properties of single-QD structures are investigated by employing inverted-pyramid arrays (1- to 5-μm pitch), using a single GaAs layer which is embedded in a self-limiting AlGaAs pyramidal structure. A single QD heterostructure is isolated using a low temperature micro-PL spectroscopy, set up with a lateral resolution of 0.7 μm. Figure 6.39 shows a typical low temperature (10K) micro-PL spectrum of a single pyramid, obtained using Ar-ion laser excitation at 514 nm with 350 nW excitation power. The three groups of peaks (observed in the spectrum) are due to recombination at the GaAs QW, QWR, and QD regions. The energy of the peaks is used to estimate the thickness of the corresponding GaAs quantum structures, assuming a uniform Al concentration of 45% in the barriers. The resulting thickness values, 1.5±0.3 nm, 3.6 ± 0.6 nm, and 7 ± 0.6 nm for the high, middle, and low energy peaks, correspond to the thickness ratios.

All three PL transitions in the single pyramid spectrum of Figure 6.39 show a distinct substructure composed of narrow lines, each 0.2 to 0.5 meV wide. For QWRs and QWs, they are attributed to the recombination of excitons localized at potential fluctuations. However, the small size of the pyramid, and hence the small area or length of the QW or QWR, effectively increases the spatial resolution, which resolves these very sharp lines. The single-QD line also shows several, very narrow peaks, with a typical energy separation of 1 to 5 meV. These single-QD spectral features are attributed to transitions with the participation of excited hole states

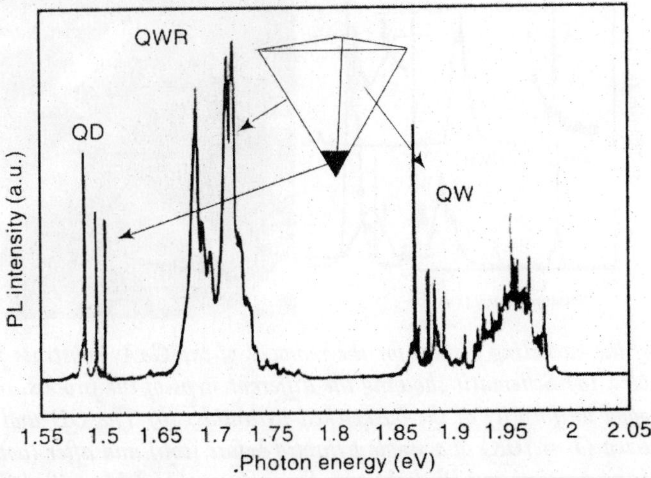

Fig. 6.39 *Micro-PL spectrum of a single 1-μm-wide pyramid measured at 10K. Inset shows the schematic drawing of the system of connected nanostructures that develops in the pyramid, indicating the origin of the different PL-transitions observed.*

(due to residual background doping and/or to multiexcitonic complexes formed in the QD. The fact that the QD PL fine structure and its excitation power dependence is similar for QDs in different pyramids is in contrast to the features associated with the QWRs and QWs.

For investigation of luminescence from single QDs at extremely low excitation power, it is important to maximize the excitation and light collection efficiency for the pyramidal dot structures. A higher excitation and light collection efficiency is achieved by exciting the QDs from the pyramid tip side [see Figure 6.40(b)], which, however, requires the removal of the substrate of these GaAs/AlGaAs structures by the backetching process illustrated in Figure 6.40(a). After depositing a gold layer on top of the sample and mechanical lapping, the sample is glued upside down onto a silicon support and the remaining GaAs substrate is selectively removed using an approximately 1:15 (by volume) NH_4OH:H_2O_2 solution. This backetching process results in upright, hollow AlGaAs pyramids, resting on a gold support, as shown in the SEM image of Figure 6.40(c). As shown in Figure 6.40(b), the back-etching leads to an approximately 5000-fold enhancement in the detected intensity of the QD PL. This strong enhancement of QD-PL efficiency permits PL measurements at extremely low excitation laser powers.

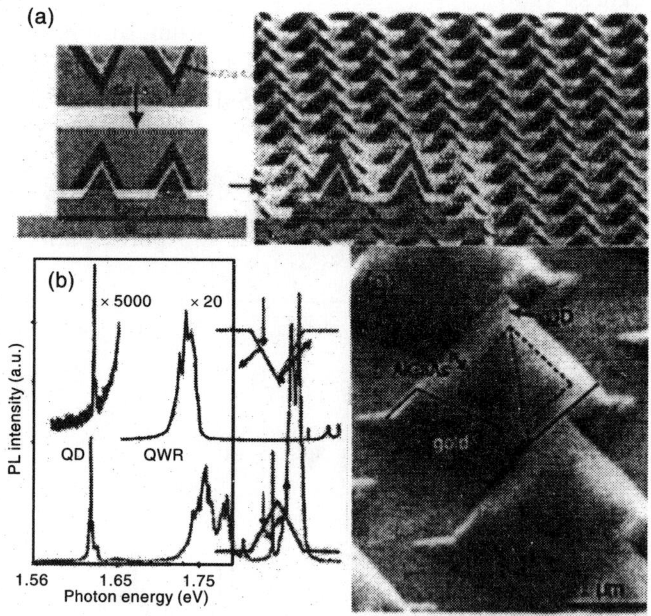

Fig. 6.40 *Illustration of the backeting process for the removal of the GaAs substrate from the epi-layers inside pyramidal holes. (a) Schematic showing the different steps of the process and a SEM image demonstrating the good uniformity of the backetched pyramids. (b) The QD and QWR spectral region of micro-PL spectra (T = 10K) of a single pyramid before (top) and after (bottom) backetching (shown is the path of the exciting and PL light rays for each case). (c) Magnified SEM view of a backetched sample showing the isolated pyramidal epi-payers resting on a gold support.*

For QD arrays cleaved at a small angle with respect to the array rows, the QD transition disappears abruptly when the excitation beam is moved past the point where the QD has been cleaved off. Another method for the peak-identification is illustrated in Figure 6.41. Here, micro-PL spectra (left) and spatial images of the PL-emission (right) are measured at different lateral positions along a line through the center of a (back-etched) pyramid. For position "1" (see inset), at the extreme corner of the pyramid, luminescence is only detected at the energy range of the QWR emission. As the laser spot is moved closer towards the pyramid center (position "2" and "3"), the QWR intensity increases, and, a very weak QD contribution becomes visible. When the laser spot is positioned exactly at the center of the pyramid (position "4," top spectrum/image), the QD-PL-intensity increases abruptly by two orders of magnitude. Thus, the fact is that the low energy PL-contribution is at .165 eV, stems from the QD situated at the center of the pyramid.

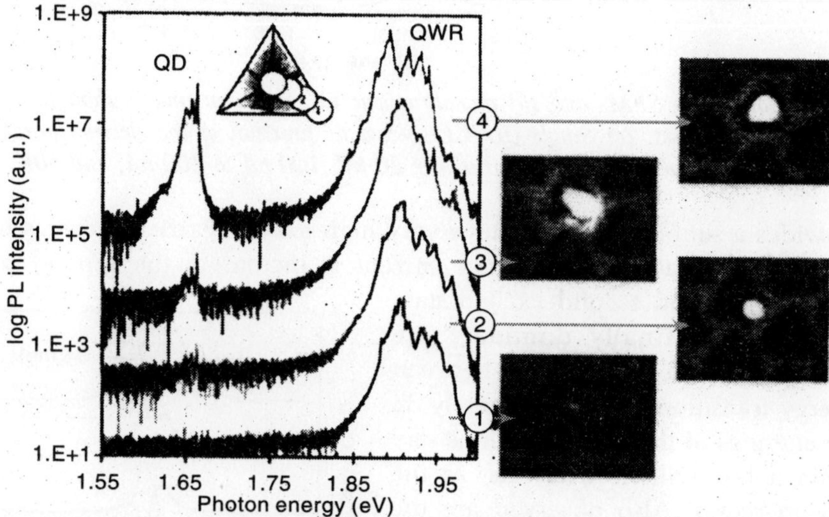

Fig. 6.41 *Micro-PL line scans as a means for identifying the origin of the different PL transitions. PL-spectra (left) and optical images of the PL-emission (right) have been collected for different spatial positions of the existing laser beam. The laser positions marked 1 to 4 in the inset correspond to the bottom to top spectrum/image.*

Consider, the spatial origin of the luminescence transitions in the QD heterostructure at low temperature (10K) CL spectroscopy; example, Figure 6.42 compares SEM and monochromatic CL images, measured at the same group of pyramids. The CL image is detected at 1.6 eV photon energy, corresponding to the spectral position of the QD emission and its intensity shows a maximum at the center of the pyramids, where the QD is situated. Three bright lines are visible along the pyramid corners connected to this central point. These lines originate from an efficient three-step capture process. First, carriers are captured from the AlGaAs barrier into the GaAs QWs and Ga-rich AlGaAs vertical QWs. Second, they transfer into the GaAs QWRs or into the Ga-rich vertical AlGaAs QWR in the vertical of the pyramid. Finally, carrier capture into the QD, via the vertical QWR, results in recombination at the dot and in the observed bright streaks in the monochromatic CL image.

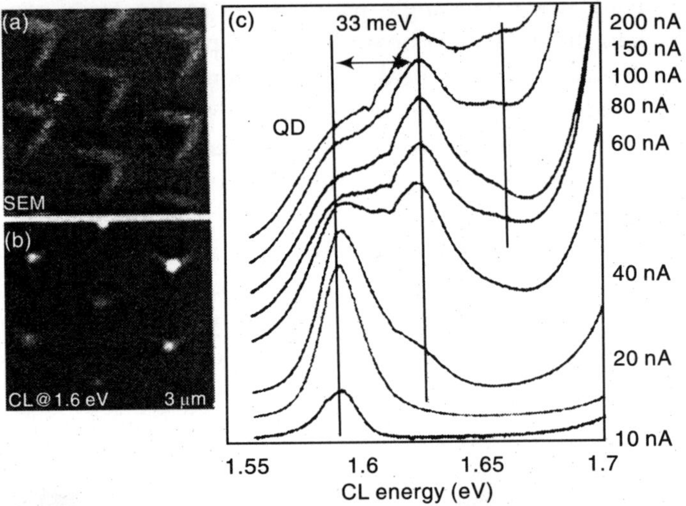

Fig. 6.42 *Comparison of (a) SEM- and (b) monohromatic CL top view images showing the same array of pyramidal structures. (c) Single-QD CL-spectra as function of the electron beam current. (Beam voltage, current and sample temperature are 20 kV, 100 pA to 200 nA, and 10K, respectively.)*

CL also provides a simple way to achieve very high excited-carrier densities in a given QD [see Figure 6.42(c)]. As the electron beam current is increased, the ground state transition saturates while the first and second excited state transitions appear and finally dominate the spectrum at high currents. The separation between the lowest energy transitions is approximately 33 meV, and the energies of the QD transitions stay constant (within a few millielectronvolts) as the carrier density increases. Also observed are the excited states of QDs in CL spectra of InGaAs/GaAs inverted-pyramid.

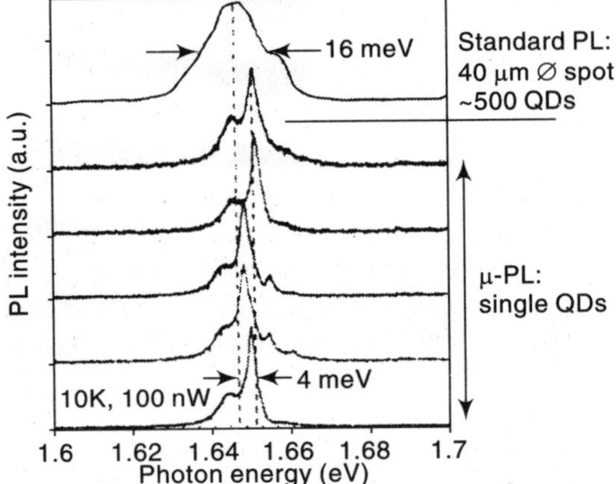

Fig. 6.43 *Comparison of micro-PL spectra (lower five sane) of several single-QDs separated by millimeter distances across the sample, and a large area PL spectrum over a large number of QDs (top scan). The PL peaks of the different single QDs fall within a narrow energy window of 4 meV, demonstrating the good QD size uniformity.*

Since a seeded self-ordering process is responsible for the formation of these QDs, the size and shape of each dot depends on the composition and the growth conditions. Thus, highly uniform dot arrays are formed by this method. The dot uniformity is probed by comparing the micro-PL spectra of dots at different pyramids, as shown in Figure 6.43. It seen that the energies of the emission peaks of dots separated by millimeter-long distances differ only as much as approximately 4 meV. This leads to typical values of 10 to 15 meV for the inhomogeneous broadening of standard PL spectra averaged over many QDs, as shown in the upper most spectrum of Figure 6.43. The corresponding size variation is thus considerably smaller than that for dots formed in a completely spontaneous self-ordering process For InGaAs/GaAs inverted-pyramid QDs, inhomogeneous linewidths of 24 meV at 77K

APPLICATION OF QUANTUM DOTS

Being zero dimensional, quantum dots have a sharper density of states than higher-dimensional structures. As a result, they have superior transport and optical properties, and are being researched for use in diode lasers, amplifiers, and biological sensors.

Quantum dot technology is one of the most promising candidates for use in solid-state quantum computation. By applying small voltages to the leads, one can control the flow of electrons through the quantum dot and thereby make precise measurements of the spin and other properties therein.

With several entangled quantum dots, or qubits, plus a way of performing operations, quantum calculations might be possible.

Another cutting edge application of quantum dots is also being researched as potential artificial fluorophore for intra-operative detection of tumors using fluorescence spectroscopy.

In modern biological analysis, various kinds of organic dyes are used. However, with each passing year, more flexibility is being required of these dyes, and the traditional dyes are simply unable to meet the necessary standards at time. To this end, quantum dots have quickly filled in the role, being found to be superior to traditional organic dyes on several counts, one of the most immediately obvious being brightness (owing to the high quantum yield) as well as their stability (much less photodestruction). For single particle tracking, the irregular blinking of quantum dots is a minor drawback.

Quantum dots have the potential to increase the efficiency of silicon photovoltaic cells. Quantum dots of lead selenide produce as many as seven excitons from one high energy photon of sunlight (7.8 times the bandgap energy). This compares favourably to photovoltaic cells which can only manage one exciton per high-energy photon, with high kinetic energy carriers losing their energy as heat. This would not result in a 7-fold increase in final output however, but could boost the maximum theoretical efficiency from 31% to 42%. Quantum dot photovoltaics can be made "using simple chemical reactions".

There are several inquiries into using quantum dots as lightemitting diodes to make displays and other light sources: "QD-LED" displays, and "QD-WLED" (White LED). Quantum dots are values for displays, because they emit light in very specific gaussian distributions. This can result in a displays that can more accurately reflect the colors that the human eye can perceive.

Quantum dots also require very little power since they are not color filtered. A LCD display, for example, is powered by a single fluorescent lamp that is color filtered to produce red, green, and blue pixels. Thus, when LCD display shows a fully white screen, two-thirds of the light is absorbed by the filters. Displays that intrinsically produce monochromatic light can for this reason be more efficient, since more of the light produced reaches the eye.

Self-assembled quantum dots

Self-assembled quantum dots find applications in wavelength domain *optical data storage medium*. Device concepts which allow for efficient electrical or optical detection of optically written information are developed. While optical writing has advantage of opening the wavelength domain, the resulting excitations (excitons) are relatively short-lived ($\approx$ 1 ns) and electrically neutral.

It is therefore promising to implement optical *write and read* from the underlying QD hardware. An optically gated FET, which incorporates a layer of InAs QDs gives preferential trapping of nonresonantly photogenerated holes in a QD layer.

Electric read out in smaller FET structures is demonstrated which allows to monitor discrete charging events of single QDs in the FET transport. The electric field-induced charge transfer is made, between a QD and a strain-induced potential minimum in an adjacent QW, in both directions. An advanced concept, based on a two-terminal diode structure is designed for optical write and read.

The working principle of the device is inferred from the band diagrams shown in Figure 6.44. A layer of self-assembled InGaAs QDs is embedded in the intrinsic region of a heterotype n-i-Schottky diode. The QDs again sandwiched between a GaAs layer and an $Al_{0.45}Ga_{0.55}As$

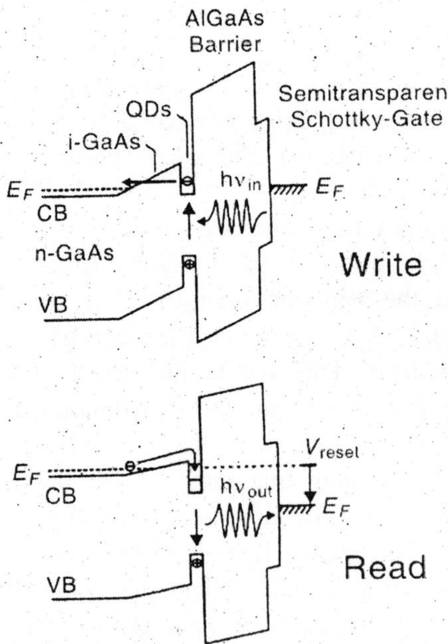

Fig. 6.44 *Band diagram and operating principle of an all-optical, wavelength domain memory structure. Optical excitation of e-h pairs in the QD-layer followed by electron tunneling allows wavelength-selective write operations in an in homogeneously broadened ensemble of dots (top).*

layer with substantially higher bandgap. In the "write" condition, the device is unbiased (V_{reset} = 0). The corresponding built-in potential leads to high electric fields in the intrinsic region of the device. Optical excitation within the QD layer leads, to the generation of an electron hole pair with subsequent electron tunneling through the GaAs barrier. The remaining holes are stored as there is exponentially smaller tunneling probability through the $Al_{0.45}Ga_{0.55}As$ barrier.

For resonant optical excitation in the region of the ground states of the QDs only a small subset of QDs is expected to be charged. For a given laser energy, $E_L = hv_{in}$, only QDs with matching ground state energies absorb and end up charged. Since excitonic ground state energies depend on the size and composition of a QD, subsets of different morphologic structures are selectively addressed by writing with different laser energies. Such manipulations are performed, either parallel in time or time-multiplexed, as long as the device remains biased in the hold condition (V_{reset} = 0).

In order to read out the previously written information in the wavelength domain, an energy-maintaining readout scheme is employed, by spectrally resolved optical detection. The memory device is switched to the "read" condition by applying a reset pulse (V_{reset} > 0), which brings the Schottky diode to a forward bias condition. At the rising edge of this pulse, electrons from the n-GaAs backcontact start to drift towards the QD layer, and lead to radiative recombination in QDs where stored holes are present (h_{vout}, see Figure 6.46). Light emission occurs, from previously "written" QDs. Since each QD represents a unique quantum mechanical nonenvironment so complete relaxation to the ground state is assumed for the relevant time scales, the photon energies of the originally employed storage procedure is reproduced. The present device therefore performs as wavelength domain memory. With typical QD surface densities in the regime of 300 μm^{-2}, even diffraction limited segmentation or imaging is possible.

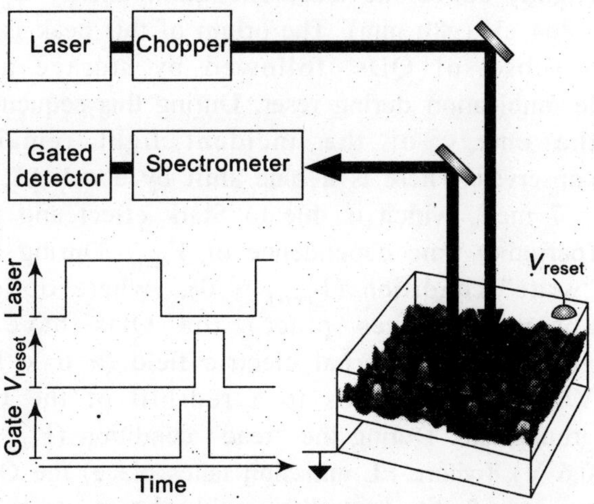

Fig. 6.45 *Experimental setup for the demonstration of a wavelength selective memory based on an inhomogeneously broadened ensemble of dots: After charging a subset of dots by resonant optical excitation in the ground state, the laser is switched off. Photon emission on readout occurs on the rising edge of V_{reset}. Prior to this, the gating of the detector is activated.*

The working principle described above is demonstrated in low temperature (T = 4.2 K) on InGaAs QDs. Samples are grown by MBE on a (100) oriented n-doped GaAs substrate. To reduce lattice defects a superlattice consisting of 25 periods of n-doped AlAs/GaAs (15/25Å) is grown, followed by 2,000Å n-doped GaAs layer. The intrinsic part of the structure starts with a 500Å GaAs layer, followed by a layer of InGaAs QDs, which are grown in the SK mode (7.5 ML of In0.5Ga0.6As at a growth temperature of 803K dot density about 300 μm^{-2}). The QDs are covered by 10Å of GaAs, followed by a 500Å thick $Al_{0.4}Ga_{0.6}As$ barrier, and a final 100Å GaAs cap layer. A Schottky junction is prepared by thermal evaporation of semitransparent 100Å Ti layer. The center of the ensemble ground state PL emission occurs at 1.264eV (980 nm).

By using set up as in Fig. 6.45, the wavelength domain memory effect, is verified. As laser sources, either a 980 nm diode laser or a cw Ti:sapphire laser (or both) are used. Those are either electrically or mechanically chopped, with a frequency of about 50 kHz. The resulting on-off ratio exceeds a factor 10^{10} (which is required for "monochromatic" EL). The diameter of the excitation spot on the sample is about 100 μm. The resulting EL emission is analyzed in a 0.64 m grating spectrometer, and detected by a gated photon counter (Si avalanche photodiode). Light emission from the QDs is only recorded while the laser beam is off. A corresponding pulse sequence is shown in the lower left of Figure 6.47. The gating of the detector is activated about 4 μs after laser turn-off, to allow for complete PL decay. The electric reset pulse is applied in the center of the gate pulse ($V_{reset} > 0$). EL photon emission appears at the leading edge of reset pulse. With such a timing, only recombination events of stored holes in QDs with electrons supplied during reset are recorded.

Data shown in Figure 6.46 is for diode laser excitation at 980 nm. The resulting EL spectrum is detected if readout pulses are applied, and it consists of two major contributions. The most intense peak appears at about E_{EL} = 1.27 eV, slightly above the laser excitation energy E_L = 1.264 eV (980 mm). The origin of this peak is in a subset of QDs, followed by storage and recombination during reset. During this sequence, the energy of the incident light remains conserved. There is a blue shift by about ΔE_{stark} = 7 meV, which is due to Stark effect and the (periodic) time dependence of V_{reset}. During the "write" condition (V_{reset} = 0), (where optical absorption takes place) the QDs have a perpendicular internal electric field (≈ 6 × 10^4 V/cm), which leads to a redshift of the QD groundstate. During the "read" condition (V_{reset} = 0.6 V), (where EL emission takes place) the QDs are in a flatband condition with almost vanishing electric field. The emission energy, therefore, blueshifts with respect to the absorption energy.

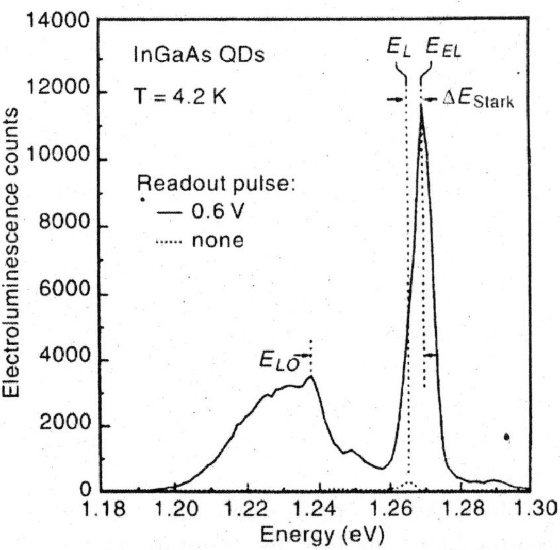

Fig. 6.46 *Spectrum of the EL output signal, which occurs on the rising edge of the readout pulse. The signal consists of a resonant part (E_{EL}), which is Stark-shifted with respect to the laser excitation energy E_L. The broad shoulder on the left corresponds to contributions from phonon-assisted absorption and excited states. The sharp feature labeled E_{LO} is separated by 32 meV from the main line, which corresponds to the InAs LO-phonon energy.*

In addition to the narrow main peak the spectrum (Figure 6.46) there is a broad emission feature, which appears redshifted by 30 to 60 meV. This emission feature contains some substructure with a main peak E_{LO}, which is about 32 meV below the resonant main peak.

For the wavelength domain storage, a two-color experiment (Figure 6.47) is performed. Here, two laser sources, a 980-nm diode laser and a tunable Ti:sapphire laser are used simultaneously for optical excitation. The resulting spectra exhibit the same features as the single sources spectrum. Tuning of the Ti:sapphire laser results in an EL emission peak with similar tuning behavior. The EL emissions corresponds to both excitation source and are clearly distinguished throughout the tuning range of the Ti:sapphire laser. The net emission is described in first order as linear combination of independent contributions form two different QD subensembles. This multiple energy storage in a QD ensemble is observed in low temperature experiments. Elevated temperatures lead to thermionic emission and spectral outdiffusion of the EL response in the current GaAs/InGaAs system.

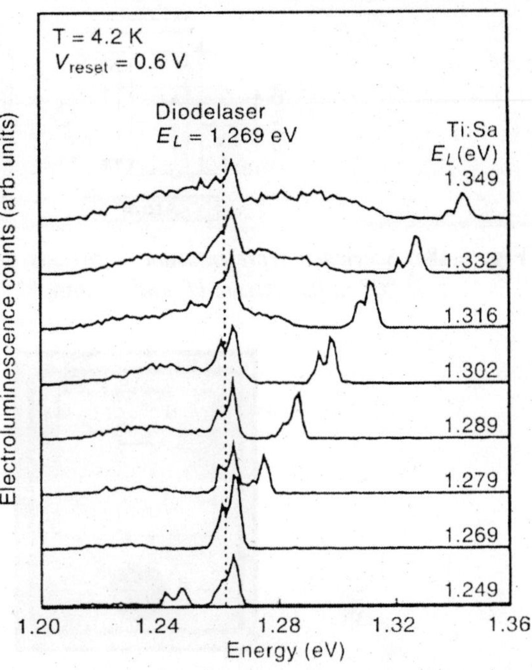

Fig. 6.47 *Demonstration of wavelength domain memory operation. Spectrally resolve readout signal for different two-color wire pulses (tunable Ti:sapphire laser ad fixed diode laser).*

SINGLE QUANTUM DOT PHOTODIODES

A quantum dot as an excitonic two-level system

Semiconductor QDs, are suitable entities to implement arrays of qubits of quantum information processing by using it in excitonic excitations in the ground state as basis for a two-level system (Figure 6.48). Rabi oscillations in the exciton population of single QDs are demonstrated. Low temperature dephasing times for excitons in self-assembled QDs exceeded several hundred picoseconds, basically allowing for high numbers of coherent manipulations with picosecond pulses from the ground states of single self-assembled $In_{0.5}Ga_{0.5}As$ quantum dots. Specifically, n-i-Schottky dot structures are grown by molecular beam epitaxy on a (100)-oriented n^+-GaAs substrate. (here a GaAs n-i-Schottky structure). The only optically active part is a single self-assembled $In_{0.5}Ga_{0.5}As$ QD contained in the intrinsic layer of the diode (see Figure 6.49). A semitransparent Schottky contact is provided by a 5-nm-thick titanium layer. The optical selection of a single QD is done by shadow masks with apertures from 100 to 500 nm, which are prepared by electron beam lithography from a 80-nm-thick aluminum layer.

Such single quantum dot photodiodes allow for resonant optical excitation in the ground state of a given quantum dot, which is the two-level system under investigation. At the same time, the diode arrangement allows for photocurrent (PC) detection, a very sensitive and quantitative way to determine the excitonic occupancy of the two-level system. PC experiments

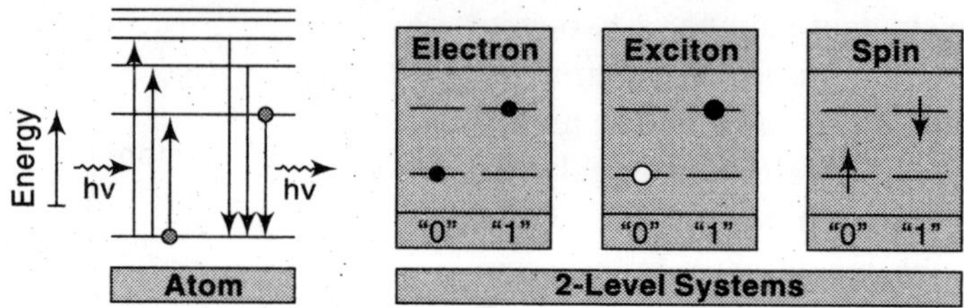

Fig. 6.48 *Schematic representation of various two-level systems. An excitonic two-level system is in the state "0" in the absence of, and in state "1" in the presence of an exciton, and otherwise in a mixed state.*

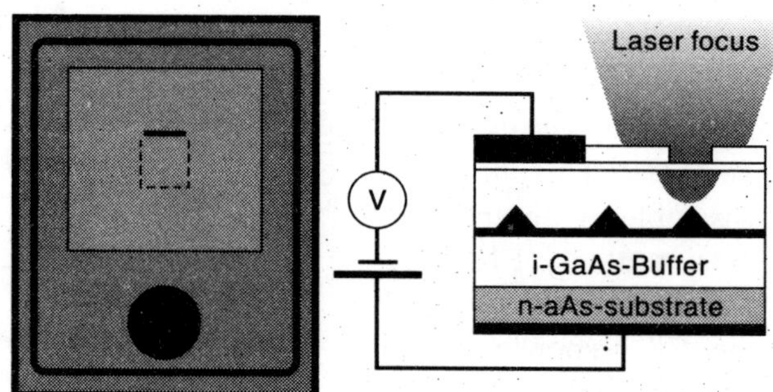

Fig. 6.49 *Micrograph and schematic view of a n-i-Schottky diode with e-beam written shadow masks for optical access to single self-assembled InGaAs quantum dots.*

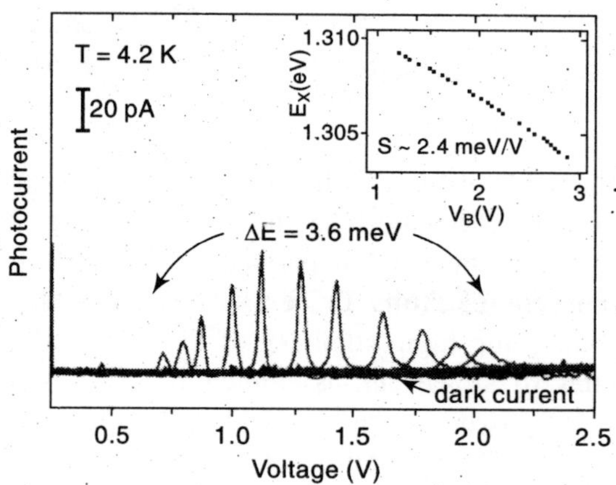

Fig. 6.50 *Stark effect tuning characteristics of a single quantum dots photodiode. The narrow ground state absorption line can be tuned by more than 3 meV.*

on single self-assembled QDs show the discrete absorption characteristics of single QDs thus resulting in sharp spectral features. By use of the quantum confined Stark effect, the quantum dot ground state, and hence the eigenenergy of the two-level system, is tuned in energy (see Figure 6.50).

Charging Q.D: (Photo diodes)

Single Electron Charging

In low dimensional semiconductors, charged excitons (artificial ions) are observed in quantum well structures. In QDs, charged excitons in inhomogeneously broadened ensembles, as well as in interband transmission and in single, optically tunable QDs, as well as electrically tunable quantum rings and quantum dots by PL are observed. The interaction of electrons and holes within a *charged QD*, introduced into an electric field tunable photodiode is given below.

For controlled charging of individual QDs, two different electric field tunable n-i-structures are grown by molecular beam epitaxy. Firstly, $In_{0.5}Ga_{0.5}As$ QDs are embedded in an I-GaAs layer 40 nm above an n-doped GaAs layer 5×10^{-18} cm^{-3}, which acts as a backcontact. The QDs are then overgrown by a 270-nm i-GaAs layer, a 40-nm-thick $Al_{0.3}Ga_{0.7}As$ blocking layer, and a 10-nm i-GaAs cap. A 50nm-thick Ti layer is used as a semitransparent Schottky gate. With respect to the distance between the n-GaAs layer and the QDs, this is the 40-nm-sample. Secondly, in a 20-nm-sample—the width of the intrinsic region and the distance between the n-GaAs layer and the QDs is only one-half as wide. Both samples are processed as photodiodes combined with electron-beam-structures shadow masks, using apertures ranging from 200 nm to 500 nm. The general arrangement of QD photodiodes is shown in Figure 6.51(a). The band diagrams of the samples are displayed in Figure 6.51 for the 20 nm- and the 40-nm sample.

The occupation of the QDs is controlled by varying the external bias voltage V_B between the Schottky gate and the backcontact. For increasing V_B the band flattens, and the QD electron levels shift below the Fermi energy E_F of the n-GaAs region. This results in a step-by-step occupation of the QD with electrons: first, the QD s-shell is brought below E_F. As a consequence, the s-shell is occupied with one electron via tunneling between the n-GaAs region and the QDs. Only a distinctly high voltage a second s-shell electron tunnels into the dot. This is due to the fact that a Coulomb charging energy of about 20 meV is taken into account for the second electron. Thus, due to the confinement energies of the higher shells, the voltages (for which the QD is occupied with three and more electrons) are shifted increasing to higher V_B values.

T-SHAPED QUANTUM WIRES AND DOTS

Cleaved Edge Overgrowth

It is another method for the fabrication of QWRs and QDs. It makes use of MBE grown on atomically cleavage planes of previously prepared multilayer structures is applied. This so-called cleaved edge overgrowth (CEO) technique enable us to prepare quantum-size structures with characteristics dimensions down to a few lattice constants.

The concept of CEO is illustrated in Figure 6.52. Two or three growth steps, separated by an cleave are performed in an ultrahigh vacuum environment. The highly reactive Al is accompanied by the formation of a stable oxide. To avoid contamination of the (110) cleavage face by impurities from the residual gas atmosphere, the time between the cleave and the initiation of overgrowth is reduced. The fabrication of a CEO structure is done as under:

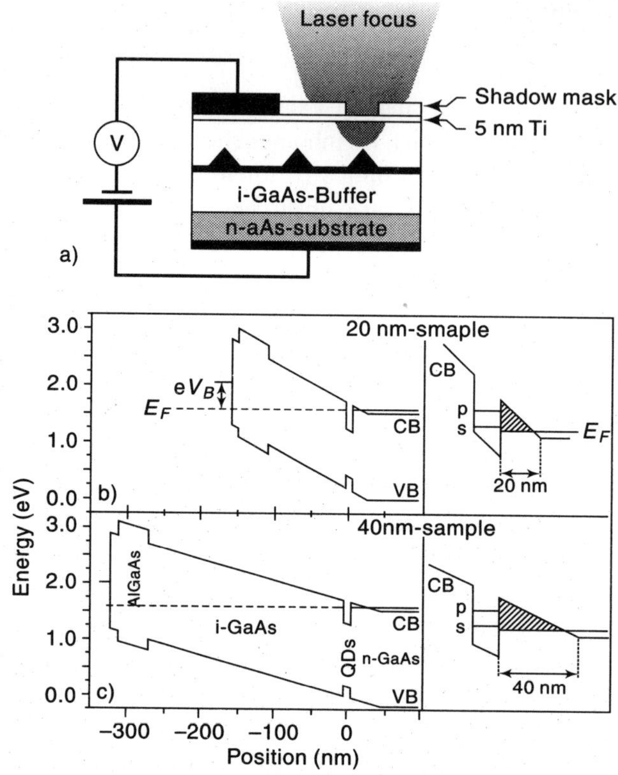

Fig. 6.51 *(a) QD photodiode combined with a near-field shadow mask. Energy-band structure: Overview and detail of the conduction band close to the QDs for (b) the 20-nm sample and (c) the 40-nm sample.*

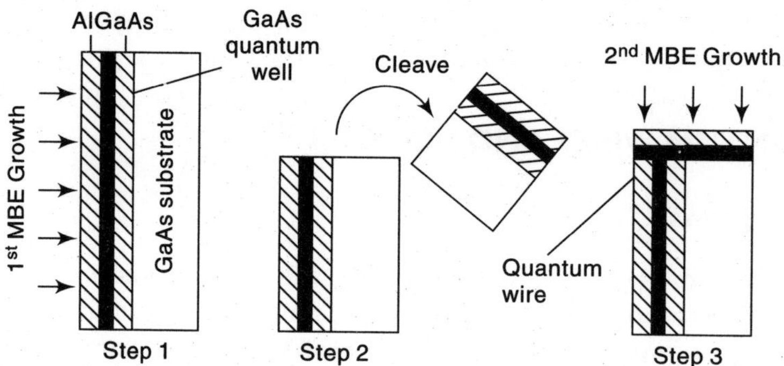

Fig. 6.52 *The concept of cleaved edge overgrowth: two MBE growth steps, separated by an in situ cleave.*

After the first MBE growth, wafer is removed from the machine and thinned from the backside, using Br_2CH_3OH, to a thickness of 100 to 150 μm. The thin wafer is then scribed and cleaved into rectangular pieces of approximately 5 by 10 mm. A scratch is made where the cleave is to occur. Up to eight of these pieces are mounted next to a GaAs(110) monitor wager, as shown in Figure 6.53. Thus, the (110) surface is exposed after the cleave is parallel to the surface of the monitor wafer. The monitor wafer provides a surface of proper emissivity for the measurement of the growth temperature, which turns out to be a critical parameter for GaAs (110) epitaxy, using a pyrometer. The CEO substrate holder assembly depicted in Figure 6.53 is loaded into the growth chamber. After the machine is brought into running condition, the oxide from the (110) monitor wafer is desorbed so that a buffer layer can be grown. Then the cleave is performed. Within seconds, MBE growth is initiated and proceeds both on the newly exposed cleave as well as on the adjacent monitor wafer.

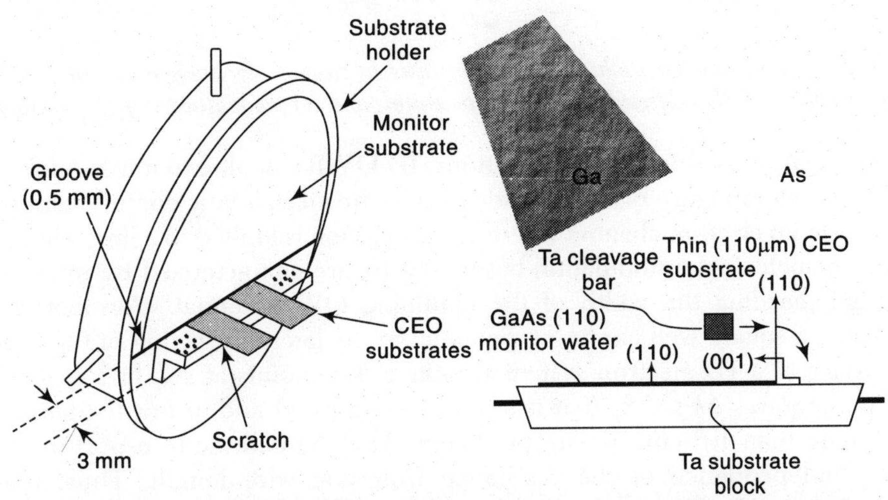

Fig. 6.53 *Experimental realization of the cleaved edge overgrowth method.*

Types of Low-dimensional Structures Achievable by Cleaved Edge Overgrowth

The precise control of layer thicknesses and layer compositions currently achievable by MBE allow the design of very sophisticated bandgap engineered multi-layer structures, with applications ranging from light emitters and detectors based on tailored transitions between quantum-confined states, to high electron mobility transistors (HEMTs) where Coulomb scattering is suppressed because of spatial separation of the charge carriers from the ionized donors. The possibility overgrow a heterostructure by another set of layers adds another degree of freedom in the design parameters. From the vast variety of structures which are prepared by CEO, Figure 6.54 illustrates three classes of QWRs. The intersection two QWs, as shown in Figure 6.54(a), confines both electrons and holes, and is used for interband transition experiments. While this structure results in the formation of a QWR, overgrowth by a third QW produces QDs, as described in the laser section of this chapter. Overgrowth of a periodic sequence of δ-doped layers embedded in barrier material by QW leads to a periodically modulated electron density in this QW, (Figure 6.54(b)). The excitation spectrum of such a lateral superlattice is characterized by particle properties of the 1D quantum-confined electron system. Since electrons as well as holes are confined in this structure, Raman scattering resonant to an interband

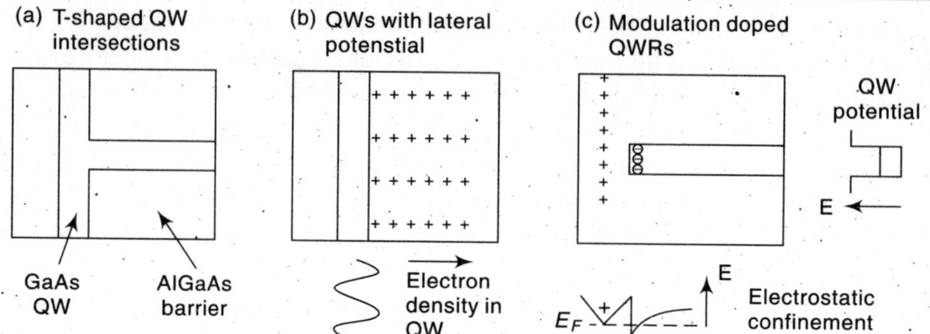

Fig. 6.54 *Classes of quantum wires which can be fabricated by cleaved edge overgrowth: (a) T-quantum wires, (b) quantum wells subjected to a lateral potential, and (c) modulation-doped quantum wires.*

transition in the QW is used for these excitations. ID QWRs, analog to a high-mobility electron layer in 2D, is shown in Figure 6.54(c). In addition to the electrostatic confinement the potential of the QW leads to an electron channel where carrier motion is limited to a line. Magnetotransport measurement on such single modulation-based QWRs are characterized by retarded levels for cycltron orbits exceeding the width of the confining QW potential. Consider, a theoretical model employing a square-well confining potential in the lateral direction of the quantum wire. The Fermi surface of a 1D electron system shrinks to two points at $\pm k_F$ in k space. So, a small angle scattering requires $\Delta k << k_F$ that is strongly suppressed and at low-temperature electron mobilities of more than 10^8 cm^2/Vs are predicted. Also the ballistic transport in high-mobility QWRs is the independence of the resistance from the wire length. Thus, the quantized conductance of these 1D electron systems is found to be smaller than multiplies of the universal value $2e^2/h$ by 25%, a behavior that cannot be understood within existing models.

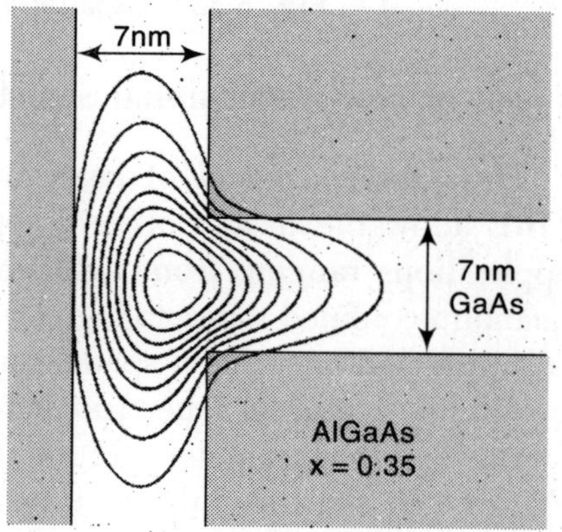

Fig. 6.55 *T-shaped intersection of two quantum wells in cross section. The contours are lines of constant probability ($|\psi^2|$ = 0.1, 0.2, ..., 0.9) for electron confined in the T-shaped potentials.*

Theory

The theoretical treatment of quantum-confined states at the intersection of two QWs requires solving of the 2D schrödinger equation for the T-shaped potential. Figure 6.56(a) shows the electron probability density $|\psi|^2$ of the electron ground state for the structure depicted in Figure 6.55 (i.e., for two 7-nm-wide GaAs QWs embedded in $Al_{0.35}Ga_{0.65}As$ barrier material obtained by the free-relaxation method), where an anitotropic effective mass and nonparabolicity can be included.

In order to optimize carrier binding to the 1D QWR state, the energy of lowest wire-like state is calculated as a function of the width b of the first-grown QW for a fixed width a of the overgrown QW (Figure 6.57). With the

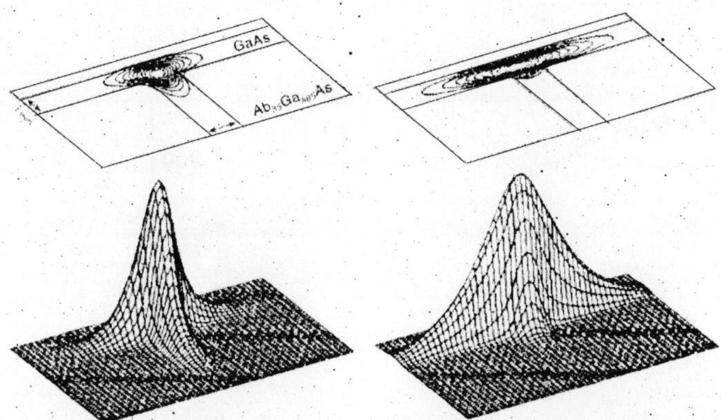

Fig. 6.56 *(a) The electron and (b) the hole probability density for the wire like state in the GaAs/$Al_{0.35}Ga_{0.65}As$ T-shaped structure with 7-nm-wide wells.*

change of b, the electron wavefunction rearranges between the two QWs. With b → 0, the QWR state asymptotically approaches that of a 2D electron in the overgrown well (indicated by a dotted horizontal line). But, with increasing *b*, the electron penetrates more deeply into the well of width *b*. At finite *b*, when the QWR energy reaches that of the unbound 2D QW state (given by the dotted line) is proportional to b^{-2} (Figure 6.57), the lowest electron state loses its wire-like character and becomes purely 2D. The separation of the 2D-1D states (i.e., the binding energy of electrons to the QWR) is represented by the dashed line in Figure 6.57. The QWR electron binding energy peaks sharply at the point $a = b$. This fact defines a criterion for an optimal choice of the parameters in the T-shaped structure: the quality of the energy values of 2D states in the two QWs. This criterion also holds for more complex CEO geometries. For the case $a = b$, the dependence of the QWR electron binding energy on the QW is plotted in Figure 6.58 (dashed line). The energetic positions of the QW and QWR states are indicated by the dotted by solid curves, respectively. The QWR binding energy decreases with decreasing well-widths, and reaches a maximum at $a = b \approx 2.5$ nm. This a consequence of the finite conduction band offset that allows penetration of the electron wavefunction into the barrier material. For thinner QWs, the electron is even more pushed out into the barrier regions, and electron quantum confinement reduces. In order to predict transition energies of interband QWR transitions, hole confinement and excitonic affects are included. The small heavy-holed binding energies of less than 2 meV are obtained by a simple one-band calculation. The masses are determined via the diagonal term in the Luttinger Hamiltonian with the angular momentum quantization axis parallel to the [110] overgrowth direction. The eigenvalues for the two QWs are different and the holes of equal well widths $a = b$ are much more spread out into the overgrown QWs, (Figure 6.56(b)). Thus maximum binding of carriers to the T-QWR, but a (110) QW width somewhat smaller than that of the (001) QW is required. Again the maximum energetic difference between the 2D QW states and the 1D QWR state is obtained when the QW transition energies, including electron and hole confinement, are matched.

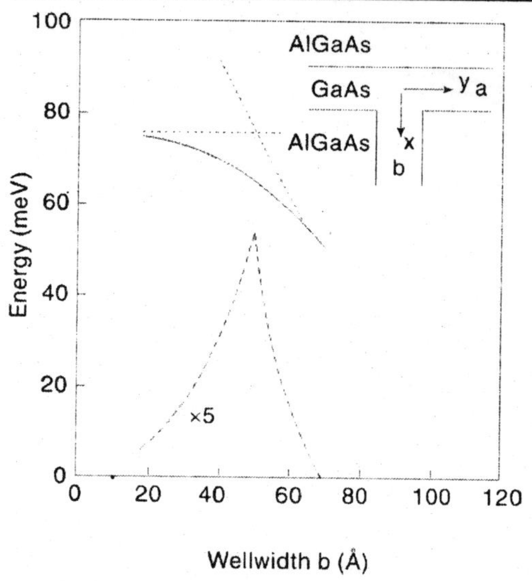

Fig. 6.57 *The dependence of the energy of the lowest electron state in the GaAs/$Al_{0.35}Ga_{0.65}As$ T-shaped structure on the width b of the (001) oriented QW (solid line) for fixed width of the overgrown QW of a = 5 nm. The energies of the lowest QW states are represented by the dotted lines. The 2D-1D energy separation multiplied by a factor of five is given the dashed line.*

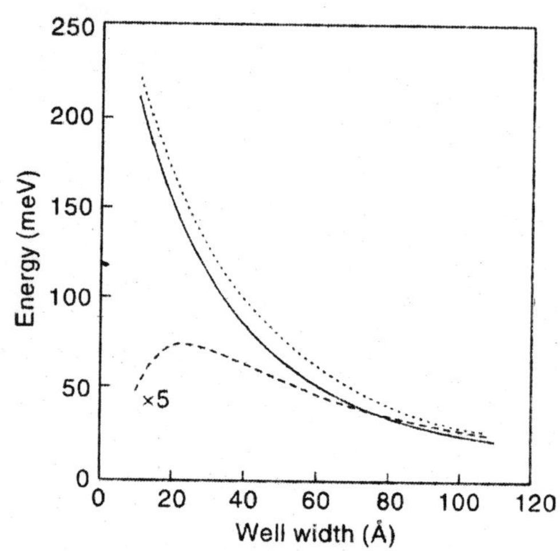

Fig. 6.58 *The dependence of the energy of the wire like electron state in the T-shaped structure on the width of two equally wide QWs (solid line). The lowest QW states are represented by the dotted line. The dashed curve shows 2D-1D energy separation magnified by a factor of five.*

The enlargement of exciton binding in such T-shaped 1D structures is estimated. Figure 6.59 shows the dependence of the exciton binding energy on the QW widths $a = b$ (solid line). For comparison, the exciton binding energy in QWs obtained is plotted in the figure (dotted line). Due to the T-shaped potential, the binding energy of the QWR exciton increases with decreasing well widths more rapidly than that of the 2D exciton. Their difference (dashed line) has a maximum at a = b ≈ 2 nm. The energy separation of the 2D and 1D exciton states is the sum of the confinement energy reduction of the free-electron-hole pair and the exciton binding energy enhancement. Combining the results (Figures 6.58 and 6.59), a maximum separation of approximately 33 meV is reached for the GaAs/$A_{0.35}Fa_{0.65}As$ heteropair. This value slightly underestimates the maximum achievable QWR binding energy since the calculation is based on a structure with identical well-widths that is only optimal if the effect of different hole confinement energies is neglected. For QW widths a = b = 7 nm, the 2D-1D exciton energy separation is 41 meV.

It is found that 1D hole confinement is not necessary for the formation of a 1D exciton, and that a 2D hole bound to a 1D electron is a good approximation for an exciton in a T-QWR.

Modification of the conventional T-shaped structure leads to an enhanced separation between the 2D and 1D single-particle states. One possibility is the doubled T-shaped structure, as indicated in the inset of Figure 6.60, where two individual intersections are close enough to

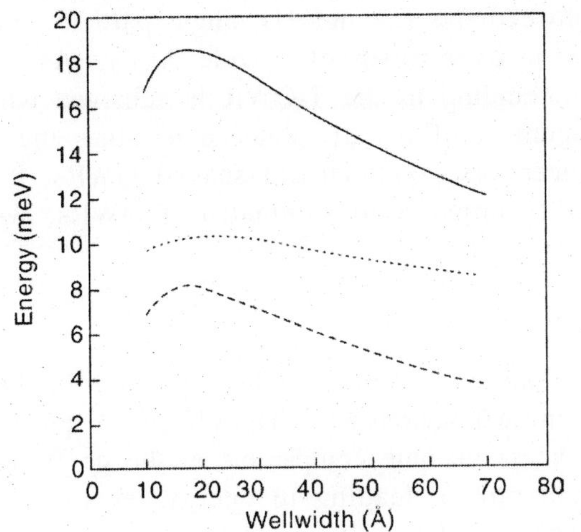

Fig. 6.59 *The binding energy of the 1D (solid line) and 2D excitons (dotted line) as a function of the well width a = b. Their difference is shown as the dashed curve.*

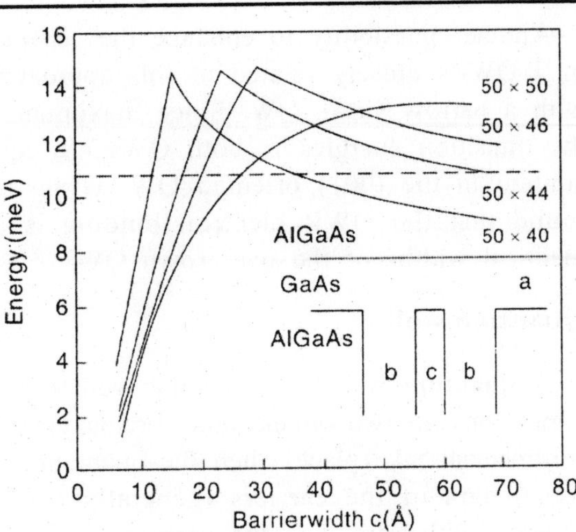

Fig. 6.60 *The 2D-1D separation of the electron states in the double T-shaped structure a × b (in Å) as a function of the barrier width c. The dashed line shows the same quantity for the single T-shaped intersection with a = b = 5 nm.*

each other to permit coupling and repulsion of the 1D states. The ground state wavefunction has a large extension into the well at the T-junction (see Figures 6.55 and 6.56), but negligible penetration into the barriers. Thus, there is a range of barrier widths where coupling of 2D states is negligible, and the interaction of the 1D states is large at the same time. The difference of the lowest 2D and 1D electron states as a function of the barrier width c in the double T-QWR structure is presented by the solid curves in Figure 6.60. For comparison, the corresponding energy of the single T-shaped interaction with $a = b = 5$ cm is also indicated in this figure (dashed line). When the QW widths are equal ($a = b = 5$ nm), the ground state in the double QW always represents the lowest 2D state in the system. For large barrier widths, the T-QWRs decouple and their electron binding energy asympotically approaches that of a single T-junction. For barrier widths in the range of 5 to 7 nm, however, the double QWR electron binding energy is indeed enhanced to more than 13 meV, compared to approximately 11 meV for the single T-QWR. For even narrow barriers, considerable lowering of the 2D state in the double QW with respect to the corresponding state in the overgrown QW occurs as a result of increased electron tunneling through the barrier. In this regime, the double T-QWR structure represents an effective single T-junction of a wide QW with a narrow one. The explains the drastic reduction of electron binding below that in a single T-QWR for barrier widths $c < 2.5$ nm. The set of well widths a = b considered so far, however, does not satisfy the criterion formulated above for maximum electron binding to the QWR. Consequently, by adjusting the width of the barrier separating two narrower QWs ($b < a$) so that the 2D state of the coupled QWs matches that of the overgrown QW, additional enhancement is seen.

Another possibility to enhance the separation between the 2D and 1D single particle states in T-QWRs closely related to this approach relies on overgrowth of a wide $Al_xFa_{1-x}As$ QW with a narrow GaAs QW. Since maximum electron binding to the T-QWR is achieved when the transition energies in both QWs are equal, alignment of the 2D states determines the Al content in the (001) oriented QW. Using this concept of asymmetric T-shaped QWRs, it is found that the QWR electron binding is doubled, compared to symmetric T-QWRs, with identical widths of the overgrown QW.

Quantum well

A **quantum well** is a potential well that confines particles, which are free to move in three dimensions, to two dimensions, forcing them to occupy a planar region. The effects of quantum confinement take place when the quantum well thickness become comparable at the de Brogile wavelength of the carriers (generally electrons and holes), leading to energy levels called "energy subbands", i.e., the carriers have discrete energy values.

Fabrication

Quantum wells are formed in semiconductor by having a material, like gallium arsenide sandwiched between two layers of a material with a wider bandgap, like aluminium arsenide. These structures can be grown by molecular beam epitaxy or chemical vapor deposition with control of the layer thickness down to monolayers.

In a bulk semiconductor, carrier motion is unrestricted along all the spatial directions. However, a nanostructure has one or more of its dimensions reduced to a nanometer length scale and this produces a quantisation of the carrier energy corresponding to motion along these directions.

Now, consider initially an isolated, thin semiconductor sheet of thickness. L. Carrier motion is unrestricted along the two orthogonal directions within the plane of the sheet, but is quantized perpendicular to the plane, forming a one-dimension quantum well. The resultant quantised energy levels are found by solving the one-dimensional form of the time-independent Schrödinger equation:

$$-\frac{\hbar^2}{2m^*}\frac{d^2\psi_n(x)}{dx^2} + V(x)\psi_n(x) = E_n\psi_n(x),$$

where $V(x)$ is the potential and $\psi_n(x)$ and E_n are the wavefunction and energy of the nth confined state. For the present case, $V(x)$ is zero within the semiconductor (which extends from $x = 0$ to $x =$ L) and is infinite elsewhere; this is the infinite-depth potential well model. Solving the Schrödinger equation and applying the boundary condition that the wavefunctions must be zero at the edges of the sheet, results in the following energies and wavefunctions:

$$E_n = \frac{h^2n^2}{8m^*L^2} \qquad \psi_n(x) = \sqrt{\frac{2}{L}}\sin\left(\frac{n\pi x}{L}\right) \qquad (n = 1, 2, 3, \ldots, \infty)$$

Figure 6.61 shows the energies and wavefunctions for the first three confined states (n, 1, 2 and 3) of an infinite-depth potential well.

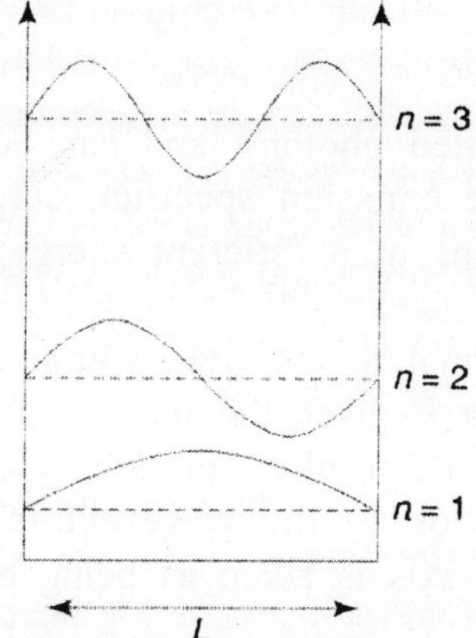

Fig. 6.61 *The energies and wavefunctions of the first three confined states of an infinite-depth quantum well.*

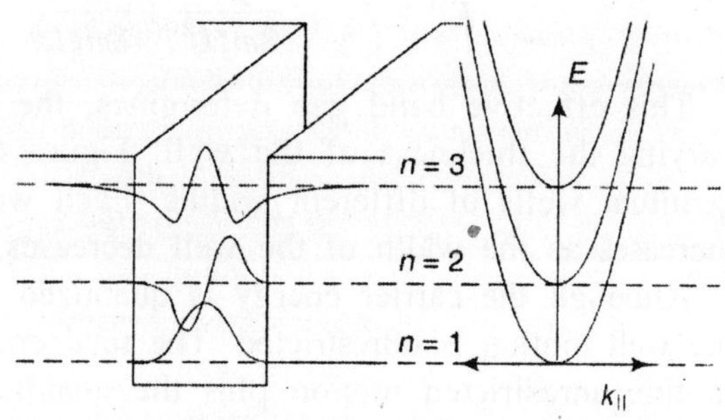

Fig. 6.62 *Energies and wavefunctions of the confined states in a finite-depth quantum well (left-hand side). The dispersion resulting from the unrestricted motion of the carriers in the plane of the well is shown to the right.*

A thin, free-standing semiconductor sheet possess negligible mechanical strength, and quantum wells are formed by sandwiching a thin layer of one semiconductor between two layers of a second, large band gap semiconductor, which form the barriers. This results in a finite-depth potential well as shown in Figure 6.62. The wavefunctions and energies of the confined states are again determined by the solution of the Schrödinger equation with the appropriate potential, which now remains finite outside of the well. The left-hand side of Figure 6.66 shows energies and wave-functions for a finite-depth well. In contrast to the infinite-depth well, there are not only a finite number of confined states and the wavefunctions penetrate out of the well and into the barriers. For a finite-depth well it is not possible to obtain analytical forms for the confined energies and the Schrödinger equation must be solved numerically. However, for many applications the energies and wavefunctions of an infinite-depth well can be used as reasonable approximations, particular for states that lie close to the bottom of the well.

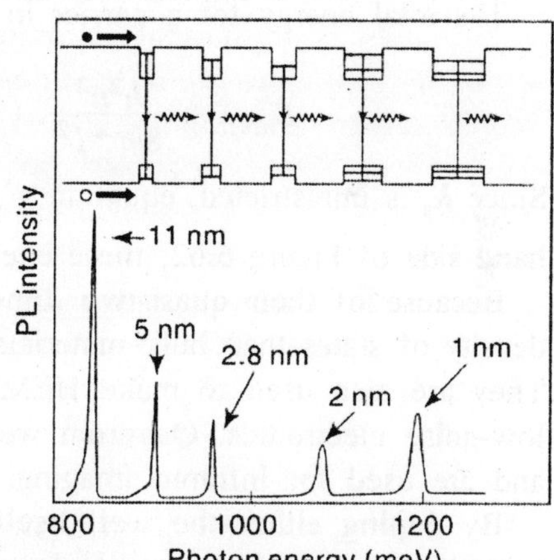

Fig. 6.63 *Emission spectrum of a quantum well structure containing five walls of different thicknesses. The wells are Ga0.47In0.53As and the barriers are InP. The inset shows the electronic structure and the nature of the optical transitions.*

For a semiconductor quantum well both the electron and hole motion normal to the plane is quantised resulting in a series of confined energy states in the conduction and valence bands (inset of Figure 6.63). One consequence of this quantum confinement is that the effective band gap of the semiconductor E_g^{ef} is increased from its bulk value by the addition of the electron and hole confinement energies corresponding to the states with $n = 1$:

$$E_g^{ef} = E_g + \frac{h^2}{8m_e^* L^2} + \frac{h^2}{8m_h^* L^2}$$

This effective band gap determines, the energy of emitted photons, and can be altered by varying the thickness of the well. Figure 6.63 shows the emission spectrum containing five quantum wells of different widths. Each well emits photons of a different energy; the energy increases as the width of the well decreases.

Although the carrier energy is quantized for motion normal to the well, within the plane of the well motion is unrestricted. The total energy of a carrier is given the sum of the energy due to this unrestricted motion plus the quantisation energy. The in-plane motion is characterized by a wavevector, $k_{\|}$, which corresponds to the combination of the wavevectors for motion along two mutually orthogonal in-plane directions. If the z-axis is taken as being perpendicular to the plane of the well, then the two in-plane directions are x and y and $k_{\|} = \sqrt{k_x^2 + k_y^2}$

From the relationship between momentum, $p = m^*v$, and wave sector, $p = \hbar k$, where $\hbar = h/2\pi$ and the definition of kinetic energy

$$E = \frac{1}{2} m^* v^2 = \frac{p^2}{2m^*},$$

The energy corresponding to in-plane motion can be written in the form $E = \dfrac{h^2 k_{\|}^2}{2m^*}$.

The total energy for a carrier in the nth confined state is therefore given by

$$E_{n,k_{\|}} = \frac{h^2 n^2}{8m^* L^2} \cdot + \frac{\hbar^2 k_{\|}^2}{2m^*}.$$

Since $K_{\|}$ is unrestricted, equation $E_{n,k_{\|}}$ gives energies for each value of n, as shown in the right-hand side of Figure 6.62; these energy bands are known as *subbands*.

Because of their quasi-two dimensional nature, electrons in quantum wells have a sharper density of states than bulk materials. As a result quantum wells are in wide use in diode lasers. They are also used to make HEMTs (High Electron Mobility Transistors), which are use din low-noise electronics. Quantum well infrared photodetectors are also based on quantum wells, and are used for infrared imaging.

By doping either the well itself, or preferably, the barrier of a quantum well with donor impurities, a two-dimensional electron gas (2DEG) is formed. This quasi-dimensional system as interesting properties at low temperature. One such property is the quantum Hall effect, seen at high magnetic fields. Acceptor dopants lead to a two-dimensional hole gas (2GHD).

Nanocrystal Solar Cell

Nanocrystal solar cell or quantum dot solar cells, are solar cells based on a silicon substrate with a coating of nanocrystals.

While previous methods of quantum dot creation relied on expensive molecular beam epitaxy processes, fabrication using colloidal synthesis allows for a more cost effective manufacture. A thin film of nanocrystals is obtained by a process known as "spin-coating". This involves placing an amount of the quantum dot solution onto a flat substrate, which is then rotated very quickly. The solution spreads out uniformly, and the substrate is spun until the required thickness is achieved.

Quantum dot based photovoltaic cells based around dye-sensitised colloidal TiO_2 films were investigated in 1991 and were found to exhibit promising efficiency of converting incident light energy to electrical energy, and were found to be incredibly encouraging.

The future quantum dot based photovoltaics offer advantages such as mechanical flexibility (quantum dot-polymer composite photobvoltaics) as well as low cost, clear power generation.

Lead selenide (PbSe) semiconductor, as well as cadmium telluride (CdTe), are well established in the production of "classic" solar cells.

Quantum Dot Lasers

I. QUANTUM DOT LASERS

A **quantum dot laser** is a semiconductor laser that uses quantum dots as the laser medium in its light emitting region. Due to the tight confinement of charge carriers in quantum dots, they exhibit an electronic structure similar to atoms. Lasers fabricated from such quantum dots exhibit device performance that's is closer to gas lasers. Improvements in modulation bandwidth, lasing threshold, relative intensity noise, linewidth enhancement factor and temperature insensitivity are all observed. The quantum dot active region is engineered to operate at different wavelengths by varying dot size and composition. This allows quantum dot lasers to be fabricated; to operate at wavelengths previously not possible using semiconductor laser technology.

A semiconductor laser is a compact device which emits coherent light. They, convert electrical energy into light with high efficiency, have very long operating lifetimes, and can be switched on and off (i.e., modulated), extremely rapidly. The optical radiation is produced by carrier injection into a medium where it recombines and produces light which is then amplified in an optical cavity formed by the semiconductor itself in edge emitters and combination of the semiconductor and reflector stack in vertical emitters. The cavity is made somewhat leaky so that part of the radiation is collected for use. Lasing occurs when the gain in the medium overcomes the losses. Generally, the tide in favor of the gain is accomplished by increasing the injected carrier concentrations.

A laser is an optical oscillator and therefore contains a gain mechanism by which light is amplified. Gain is the inverse of absorption and in a semiconductor it occurs when there area a large number of electrons in the conduction band and a large number of bolts in the valence band, a condition known as a population inversion. A population inversion is achieved in a forward-biased p-i-n structure where large numbers of electrons and holes are injected into the intrinsic region. For the system to oscillate or lase, the total gain must balance the total loss. This requires the creation of sufficient densities of electrons and holes, (by the injection of current) to achieve the required gain. The gain is proportional to the density of occupied states.

As the density of states in a bulk semiconductor increases with energy, the gain will increase with increasing carrier density because higher energy states are successively filled. The threshold current, the current that must be applied for the laser to operate, is relatively high and increases rapidly with device temperature. In a bulk system it is therefore necessary to 'fill' the density of states to a point where sufficient gain is reached, the carriers at lower energies effectively wasted. This process is shown in Figure 7.1(a). These wasted carriers result in a relatively large threshold current.

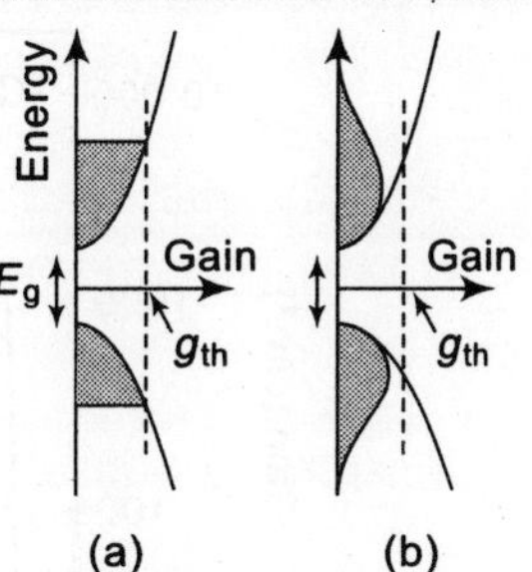

Fig. 7.1 *Electron and hole distributions in the conduction and valence bands of a bulk semiconductor laser at (a) absolute zero temperature and (b) non-zero temperature. In (b) thermal excitation of carriers to higher energy states has reduced the maximum gain below the value, g_{th}, required for laser action.*

Figure 7.1(a) shows the lowest energy states occupied at absolute zero temperature. At non-zero temperatures carriers are thermally excited to higher states (Figure 7.1(b)). As the temperature is increased, an increasing fraction of the carriers are thermally excited out of the lasing states (Figure 7.1(b)) and the maximum gain is now below that required for lasing action. Consequently, a higher current must be applied to achieve the required gain. Hence, the threshold current of a laser increases with increasing temperature.

The modification of the density of states in a nanostructure overcomes, the limitations of bulk semiconductor lasers. As the density of states is increased at low energies with respect to higher energies, it results in a decrease in the absolute threshold current and is also sensitivity to temperature; as in example: a quantum dot having only one confined electron state and one confined hole state, all of the carriers have the same energy at all temperatures so the laser exhibits a very low and temperature-insensitive threshold current.

Figure 7.2 shows the gain of systems having different dimensionalities developed with increasing current. It also shows the dependence of the threshold current on temperature. A given gain reaches successively lower currents as the dimensionality is decreased, with threshold current densities of 1050, 380, 140 and 45 A cm^{-2} calculated for bulk, quantum well, quantum wire and quantum dot lasers, respectively.

The temperature dependence of the threshold current density of a semiconductor laser is described by an equation of the form

$$J_{th} = J_{th}^{0} \exp (T/T_0),$$

where J_{th}^{0} is the threshold current density at 273 K and T_0 is the temperature sensitivity; a larger T_0 corresponds to a lower sensitivity. For the idealised lasers considered in Figure 7.2. T_0 values at 377 K, 558 K, 754 K and infinity demonstrate a reduced sensitivity with the progression bulk → well → wire → dot.

At room temperature, threshold current densities are approximately one-third the value of comparable quantum well lasers, and the stability of the threshold current is improved by approximately a factor of two as summarised in Figure 7.2. This is due to the departure of self-assembled dot from idealised systems. In particular, self-assembled dots have a number of

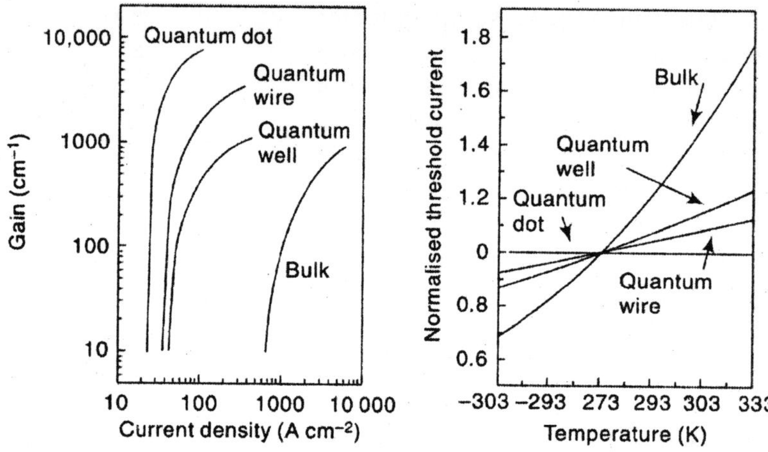

Fig. 7.2 *(a) Left-hand panel: calculated gain variation with current density for lasers having different dimensionalities. For a given current density, the gain increases as the dimensionality decreases. (b) Right-hand panel: calculated temperature variation of the threshold current density for bulk, quantum well, quantum wire and quantum dot lasers. Decreasing the dimensionality increases the temperature stability.*

confined levels into which carriers are thermally excited, and the surrounding barrier material results in a finite T_0. In addition, there is non-radiative process and the dot ensemble is inhomogeneously broadened. Figure 7.3 shows a comparison of the temperature dependence of the threshold current densities of a self-assembled quantum dot laser and a quantum laser. At low temperatures the expected temperature stability of the quantum dot laser is achieved. But below 200 K the quantum dot laser is significantly more stable than the quantum well laser. However, above 200 K the threshold current density of the quantum dot laser increases and at 300 K quantum well and dot lasers exhibit very similar temperature sensitivity.

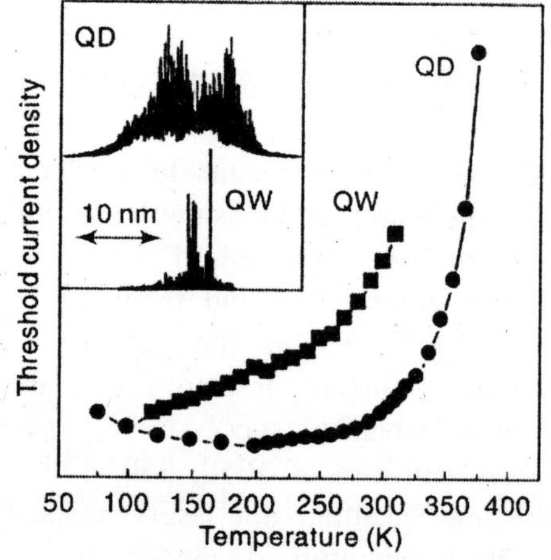

Fig. 7.3 *Comparison of the temperature dependence of the threshold current density of a self-assembled quantum dot laser (circles) and a quantum well laser (squares). Because of the different devices designs, the threshold current densities are normalised to their 100 K values. The inset compares emission spectra of a quantum well (QW) laser and a quantum dot (QD) laser.*

In addition to their low threshold current density and high temperature stability, quantum dot lasers offer a number of additional advantages over lasers of higher dimensionality. Their maximum modulation frequency is higher. This may be compromised if carriers are unable to relax rapidly to the states from which lasing occurs. Once carriers are captured into a quantum dot, their subsequent motion is restricted. But in a quantum well laser the carriers which are captured into the well are free to move within the plane of the well. This carrier localisation prevents the diffusion of carriers to non-radiative centres which are formed on the surface of the device or within the device. Quantum dots are therefore suitable for the fabrication of very small lasers, with a high area-to-volume ratio. Carrier localisation prevents carriers in different dots from directly interacting and subsets of dots with similar emission energies act as independent lasers. The dot ensemble therefore behaves as a collection of sub-lasers, with lasing occurring over a significant fraction of the inhomogeneous width. Figure 7.3 compares quantum dot and quantum well laser spectra; the increased width of the quantum dot spectra is clearly visible.

Quantum dots allow emission at new wavelengths to be obtained. Lasers with 0.8 μm wavelength are used for short haul and 1.3 μm one for long haul fiber optic communication systems with very large band-widths. The lasers having InGaAs strained active layers emitting at 0.98 μm are use to pump Er-doped optical fibers (distributed amplifier) used in long haul communication networks. GaInAs quantum well lasers grown on GaAs substrates are limited to <1.2 μm by the critical thickness of this strained system. However, InAs quantum dots allow the fabrication of lasers operating in the important 1.3 μm telecommunication band, with some prospects for devices operating in the related 1.55 μm band. Current 1.3 and 1.55 μm lasers require the use of quantum well lasers grown on less technologically convenient InP substrates.

Devices based on quantum dot are finding commercial application in medicine (laser scalpel, optical coherence tomography), display technologies (projection, laser TV), spectroscopy and telecommunications. A 10 Gbit/s quantum dot laser that is insensitive to temperature fluctuation is used in optical communications and optical networks. The laser is capable of high-speed operation at 1.3 μm wavelengths, at temperature from 293–343 K. It works in optical data transmission systems, optical LANs and metro-access systems. In comparison to the performance of conventional strained quantum-well lasers of the past, the quantum dot lasers achieve higher stability of temperature.

Semiconductor lasers have a respectable wavelength range of about 0.4–11 μm. Reliable lasers below orange color are due to the wide bandgap semiconductor, GaN and its alloys.

Lasers exhibit longevities of over a million hours with output powers near 200 mW. The quantum efficiencies are greater than 90%.

Semiconductor lasers, based on InGaP or AlGaAs are breaking the power barriers with power levels near 10 W. These high-power lasers are used as pumps for other solid state media for lasing such as in Ti-sapphire tunable lasers.

The most common type of semiconductor laser is based on a geometry where light propagates in the plane defined by the quantum well or dots. In this geometry the mirrors at the ends of the optical cavity are formed by cleaving the semiconductor along a particular crystal direction to give atomically flat surfaces. The refractive index difference between air and the semiconductor produces mirrors with a reflectivity of typically ~30%. Although the reflectivity is relatively low, and represents a large loss for photons within the cavity, the gain of the system is increased to compensate for this loss by increasing the length of the cavity. Typical cavity lengths are ~1 mm.

Geometry in which the light propagates normal to the plane, to give a vertical cavity surface-emitting laser (VCSEL) is used. In this geometry the available gain is much smaller than a transverse laser as the light passes only once through a layer of quantum dots or a quantum well, instead of along the layer. The losses must be reduced significantly and a mirror reflectivity of ~99.9% is required. As this value cannot be achieved with a simple cleaved semiconductor-air interface. So the mirrors are formed by depositing alternating layers of two semiconductors having different refractive indexes. The resultant is a reflectivity that increases with increasing layer number, allowing a reflectivity >99.9% to be obtained with typically ~20-30 layer pairs. Although the growth sequence of a VCSEL is more complex than that of a transverse laser, the subsequent fabrication is greatly simplified, with devices formed by lithography and etching to produce circular mesas. Dense two-dimensional arrays can be formed, with the circular cross section of the devices resulting in symmetrical output beams.

CHARACTERISTICS OF LASERS IN Q.D

Threshold Current Density

Threshold current density is an important characteristic of a laser. It gives the current density at which the modal gain overcomes overall (internal and external) losses, thereby enabling lasing. External losses are reduced by deposition of dielectric high-reflectivity (HR) coatings. However, as internal losses (scattering, free-carrier absorption, self-absorption, etc.) cannot be made negligibly small, the value of the threshold current density serves as an important quality indicator.

QD lasers have shown their potential for devices which operate at threshold current densities in the range 10–20 Acm^{-2} with HR facet coatings. For conventional devices, HR front facet coatings cannot be applied. Here, the measure of quality is the value of the threshold current density per single quantum well or quantum dot insertion. The best values of 7–10 Acm^{-2} per QD insertion are realized in devices with a large number QD stacks.

A record low transparency current, which is the extrapolated value of the threshold current density for infinite cavity length, 6 Acm^{-2} per dot layer, an internal quantum efficiency of 98%, and an internal loss below 1.5 cm^{-1} is observed.

Characteristic Temperature T_0

T_0 is the temperature at which the device is heated to increase the threshold current by a factor of e = 2.718. The QD injection lasers have infinite characteristic temperature range below 150–180 K. However, at high temperatures, the threshold current density is less temperature-stable and T_0 is inferior (50–80 K) to GaAs-based quantum devices.

T_0 values are improved by placing QDs in a quantum well and increasing the bandgap of the matrix. In some cases T_0 values are increased up to 350 K; the increase in threshold current density is also high. First, a high characteristic temperature (160 K); at ambient temperature operation (below 313 K) together with low threshold current density operation (70 Acm^{-2} for 3-fold-stacked QDs) is realized. As the devices operate in the range 1.26–1.28 μm, with T_0 values near 50 K, is a remarkable success. So it is possible to build the first high-power continuous wave (CW) QD lasers (3 W in CW operation. However, in the CW regime, the T_0 value is lower than 160 K.

Very high characteristic temperature in QD lasers (180 K and 230 K) is achieved by applying a QD *p*-modulation doping technique which reduces differential efficiency (<20%). The advantage is the increase in the modulation bandwidth beyond 20 GHz. By using undoped 5-fold-stacked QDs, increase in the T_0 value to 170 K (up to 338 K) without increased internal losses is possible Fig. 7.4. This has made possible to design 1.3 μm GaAs devices operating at 85% differential efficiency, low threshold current density (10 Acm^{-2}, 1.5 mm cavity length, uncoated) and high characteristic temperature, all in the same device.

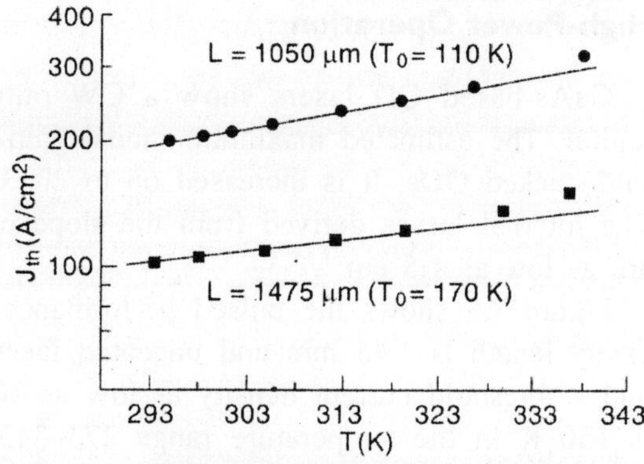

Fig. 7.4 *Temperature dependence of the threshold current density of a quantum dot (QD) laser with 5-fold QD stack and uncoated facets.*

Spectral Features of QD Lasers

It is found that the lasing spectrum of broad stripe QD lasers exhibits laser mode grouping. Intensity modulation of the lasing spectrum is observed and spectral shifts of the mode groups with temperature are found in devices with 40–60 μm stripe widths. This is due to the formation of a transverse resonator, affecting the gain spectrum in the longitudinal direction.

The spacing between the mode groups is defined by homogeneous broadening. The mode grouping effect is performed on QD lasers with a range of stripe widths and in an extended temperature range. The spacing between the mode groups is not a function of temperature and is defined by the stripe width. By introducing a proper period etching profile at the edges of the stripe, it is possible to suppress or enhance the hole-burning effect in the gain spectra due to transverse cavity modes at particular wavelengths and, potentially, realize that wavelength is stabilized.

Time-Response of QD Lasers

It is known that electron relaxation occurs at a very slow rate in high-purity bulk materials. In GaAs exciton photoluminescence (PL) evolution the time is up to 4 ns. This so-called phonon bottleneck effect for electron energy relaxation is reduced in QDs, where photoluminescence (PL) evolution has time of the order of 10–40 ps, depending on the particular QD geometry, even at low temperatures and excitation densities. This opens the way to devices operating in the 20–100 Gb/s range. Relaxation oscillations in practical devices also indicated a potential for modulation band-widths larger than 10 GHz. Under conditions of high excitation density and lattice temperature, the time response is even faster.

At high temperature, population of matrix states have an adverse effect on time response. Using the idea of resonant tunneling; carrier injection into QDs devices are made, which operate in the 1 μm wavelength range at 15 GHz and at room temperature. Furthermore, *p*-doping QDs increase the high-frequency response of QD lasers up to 20 GHz.

High-Power Operation

GaAs-based QD lasers show a CW output power of ≈ 3–6 W depending on the spectral region. The estimated maximum modal gain for QD ground state lasing is about 14 cm^{-1} for 3-fold-stacked QDs. It is increased up to 20–35 cm–1 for 5- to 10-fold-stacked QD active regions. The internal losses derived from the slope of the inverse differential efficiency vs. Cavity length are as low as 1.5 cm^{-1}.

Figure 7.5 shows the pulsed performance of a QD laser based on 10-fold stacked QDs. The cavity length is 1.45 mm and uncoated facets are used. A differential efficiency as high as 85% and a threshold current density as low as 100 A/m^2 are achieved. The characteristic temperature is 150 K in the temperature range 273–343 K. All the key values are significantly better than those for commercial InP-based 1.3 μm devices. This QD laser is used as a standard transmitter, since the ground state modal gain is sufficient to maintain in 300-μm-long devices (with HR-coated rear facet). It is also used in high-power applications, for example in Raman pumps, where high efficiency is essential at long cavity lengths. Raman pumps in the 1.2–1.3 μm wavelength range are used in new types of fiber, enabling low losses over the whole 1.25–1.65 μm spectral range. Narrow 7-μm-wide stripes exhibit single-transverse-mode kink-free operation up to 300 mW for uncoated facets (see Fig. 7.6)

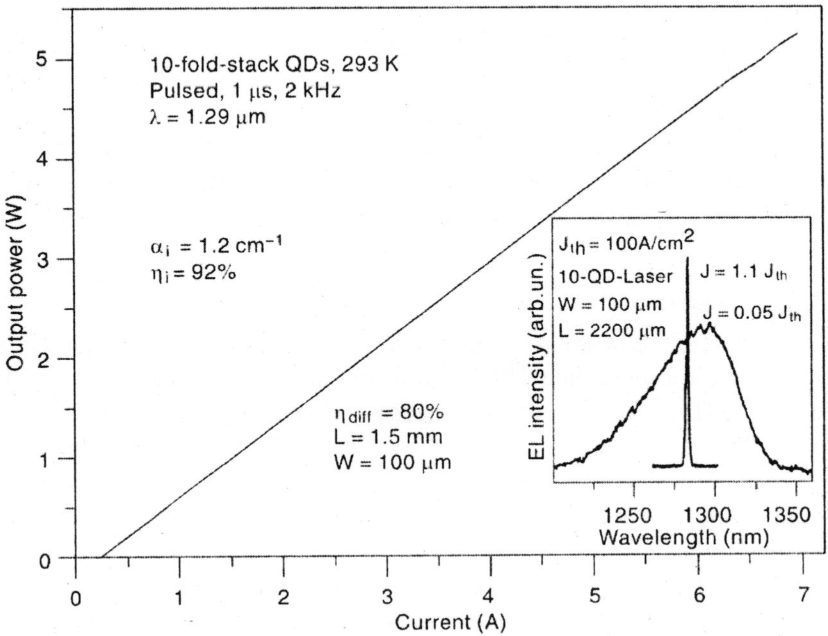

Fig. 7.5 *High-power operation of a long-wavelength laser. A 10-fold stack of QDs is used. The inset shows electroluminescence (EL) and lasing spectra.*

The laser beam quality are performed for three different types of InGaAs/GaAs laser:

- MOCVD-grown 1.1 μm quantum well lasers,
- MOCVD-grown 1.1 μm lasers,
- MBE-grown 1.3 μm QD lasers.

An intrinsic filamentation suppression is observed for narrow stripe gain guided QD lasers compared to QW lasers with identical waveguide structure and similar lasing wavelength. Filamentation suppression is even stronger for 1.3 μm QD lasers due to the longer emission wavelength. This effect is attributed to reduced in-plane carrier diffusion and the reduced α-factor in QD lasers compared to QW lasers.

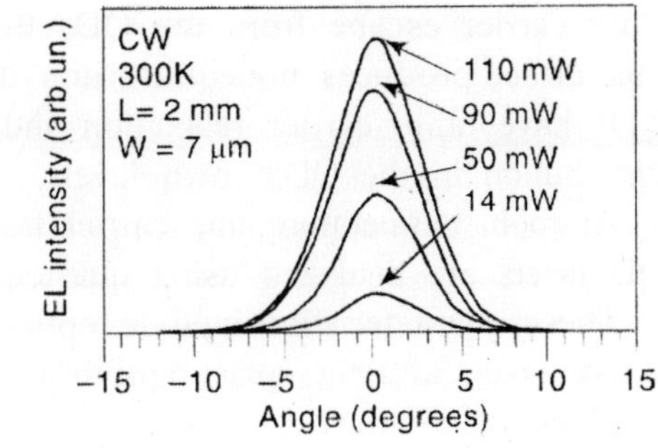

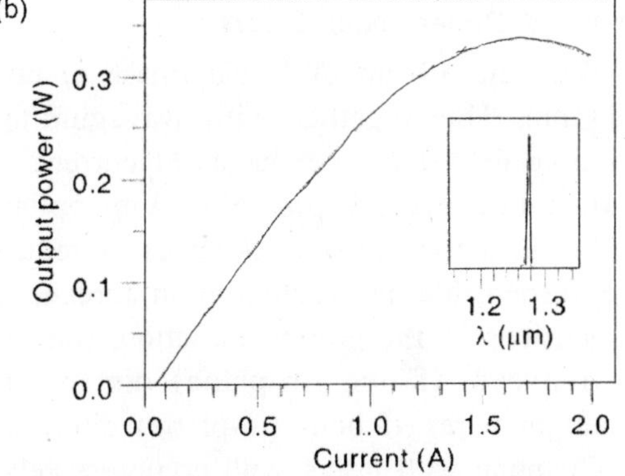

Fig. 7.6 *High-power kinds-free single-mode operation of a long-wavelength QD laser. (a) Far-field pattern in the lateral direction. (b) Output power. The lasing spectrum is shown in the inset.*

Laser C.D

Semiconductor lasers based on quantum wells are now used extensively for many applications, including CD and DVD data storage and fibre-optic data transmission. Quantum wire lasers have been fabricated, although their performance has been limited by the lack of systems with suitable energy spacings between the confined subbands. Greater progress has been made with quantum dot lasers based on self-assembled dots.

Many of the ideal features of QDs are realized by self-organized III-V. The three-dimensional (3D) confinement of charge to dimensions of its wavelength leads to discrete energy levels for electrons and holes. Lasers operating on these discrete levels exhibit new phenomena when compared to planar quantum well lasers. QD lasers are temperature insensitive.

For small inhomogeneous linewidths a QD active layer's optical gain exceeds that of a planar quantum well, at a lower current density. This combination gives rise to very high differential gain and speed. Self-organized QD lasers are strongly limited in their performance by their large inhomogenous linewidths that typically exceed ~30 meV.

Self-organized QDs in non-ground state levels become thermally occupied at room temperature. For InAs and InGaAs QDs, the thermal smearing of the hole population due to their closely spaced energy levels degrades laser performance. However, doping is used to build in excess holes and limit the thermal effect of the close hole spacings. The p-type doping brings more ideal temperature and speed performance to QD lasers. Non-equilibrium effects need attention for QD devices.

For, energy relaxation in QD ensembles and its impact on QD laser performance. Non-equilibrium effects QD lasers for low temperature due to their inhomogenous broadening and slow carrier escape from the QDs that thermalizes carrier distributions are known. However, this effect produces nonequilibrium distributions between different QDs, while the individual QDs have rapid carrier relaxation and appear to satisfy quasiequilibrium predictions for charge distribution in the QDs' own levels.

At room temperature, the carrier transport is so fast that the most important characteristics of QD lasers are analyzed using quasiequilibrium solutions to the rate equations.

These characteristics include optical gain and the modulation response intrinsic to the QD active layer, and the quasi-equilibrium analysis is effective in analyzed and understanding QD laser characteristics.

Reduced Dimensional Lasers

With the advent of liquid phase epitaxy (LPE), active layer thicknesses are reduced to near 100 nm. This together with waveguiding structures imbedded in the heterostructure made it possible to reduce the threshold current to the point where continuous wave operation at room temperature became possible. With epitaxial deposition deposition methods such as MBE and OMVPE, active layer thicknesses is reduced to the electron wave confinement. These structures are responsible for reduction in threshold current by battery operation. Also, reduction of the dimensions in the growth direction, (one of both dimensions) in the plane of the layers can also be reduced. If one (in-plane) dimension is reduced down to nanometer scale, one obtains quantum wires if both (in-plane) dimensions are reduced one obtained quantum dots.

Quantum well lasers with compressively strained InGaAs active layers have led to considerable reduction of the threshold current density. Both compressive and tensile string lead to smaller threshold currents due to alterations in the valence band structure. With coherently strained active layers based on GaAS, lasers with longevities superior to those with lattice matched channels are obtained. With strained channels, 0.98 μm wavelength radiation are produced with important applications to Er-doped fiber amplifiers. The strained layer concept is used in yellow and green lasers made in InGaAlP and ZndSe/ZnCSSe, respectively. The latter has short longevity whereas InGaN lasers have sufficient longevity.

Semiconductor lasers have application in communications, pumping sources, and application of GaN-based lasers in digital versatile disks. DVD for short. The DVD has the spot size and the storage density is diffraction limited. The present CD players utilize GaAsIR lasers produced by MBE. With red lasers pit dimensions of about 0.4 μm can be read. Using a two-layer scheme in a DVD, the density can be increased from 1 Gb to about 17 Gb per compact disk. Also, the violet or the blue laser can be implemented.

The nitride-based lasers with their inherently short wavelength, when adopted, offer much increased data storage capacity possibly in excess of 40 Gb per compact disk.

The DVD and CD formats share the same basic optical storage technology and information as represented by microscopic pits, formed on the surface of the plastic disk when the material is injected into a mold. The pitted side of the disk is then coated with a thin layer of aluminum which, in the case of a CD, is followed by a layer of protective lacquer and a label Top read the data, the player shines a small spot of laser light through the disk substrate onto the data layer as the disk rotates. The intensity of the light reflected from the disk's surface varies according to the presence (or absence) of pits along the information track. When a pits lies directly underneath the "read-out" spot, much less light is reflected from the disk than when the spot is over a flat part of the track. A pohotodetector and other electronics inside the player translate this variation into the 0s and 1s of the digital code representing the stored information. A schematic representation of a CD player is shown in Fig. 7.7.

The smallest DVD pits intended for red lasers are only 4.0 μm in diameter, whereas the equivalent CD pits are nearly twice as large, or 0.83 μm wide. Also, DVD data tracks are only 0.74 μm apart, whereas 1. 6μm separate CD data tracks. To read the smaller pits, a DVD player's readout beam must achieve a finer focus than a CD player by using shorter wavelength lasers. Also, DVD players employ a more powerful focusing lens that has a higher numerical aperture than the tens in a CD player. Added density is in part due to better error correction and control (EC) techniques that require special algorithms to compute additional data bits stored along with the user data. These additional bits reduce the fraction of the total disk capacity available.

Fig. 7.7 *A schematic representation of the main components of a CD-player using the DVD technology.*

InGaN/GaN injection lasers operating room temperature are there. Also developed are continuous wave lasers at room temperature having over 10 000hours of operating time. Small quantities of InGaN-based lasers operating near 410 nm are developed. The lasers utilize a complex heterostructure and mirrors that are etched since the substrate on which it is grown is sapphire and does not cleave along the GaN cleavage planes. A schematic of the epitaxial layer structure and one indicating the device structure are shown in Fig. 7.8.

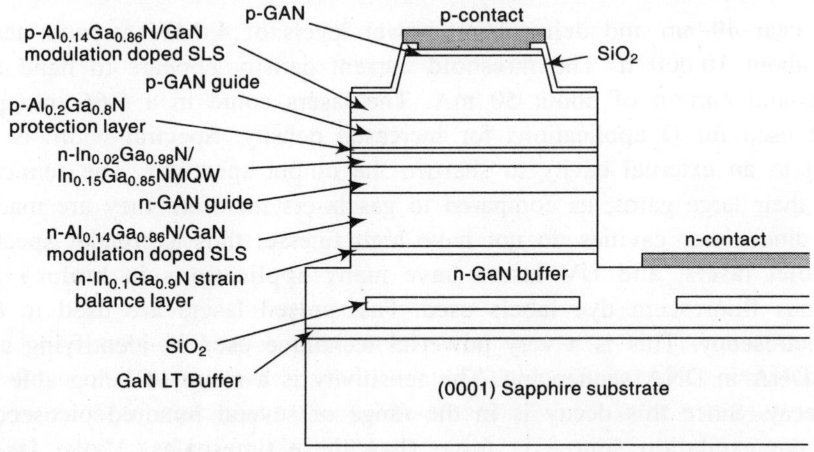

Fig. 7.8 *A schematic of the epitaxial layer structure of InGaN laser structure.*

For a three period (35 Å)/$In_{0.02}Ga_{0.98}N$ (70 Å) quantum well structure at room temperature, CW operation is done for some 10 000 h. CW light and voltages vs current characteristics of an InGaN MQW laser having three $In_{0.15}Ga_{0.85}N$ quantum well gain media are shown in Fig. 7.9. Threshold current densities as low as 16 mA, and voltages across the device at threshold as low as 4V is reported. The lowest current density at threshold is 0.5 kA cm^{-2}. Maximum operating temperature is 273 K and the maximum pulsed power output is about 500 mW. These data are tabulated in Table I. For read only DVD applications a CW power of a few milliwatts would suffice while higher power levels would be required for writing. The emission wavelength range covered with InGaN lasers is between 380 and 450 nm with about 410 nm as being the best. The operation of the InGaN laser, in light of the fact that the GaN active layered devices do not lase, is attributed to compositional inhomogeneous in InGaN such as self-formed clusters or data.

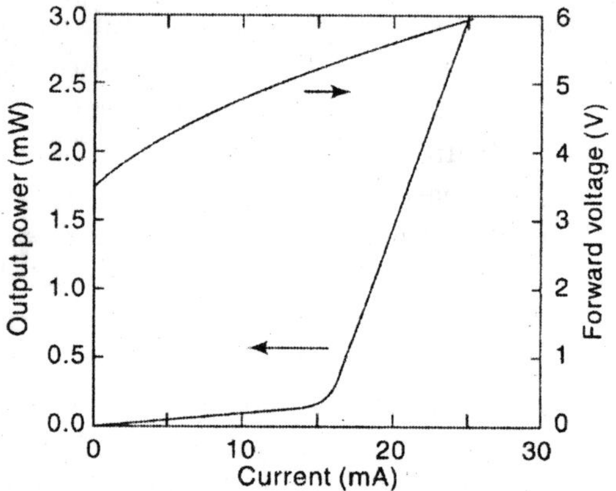

Fig. 7.9 *I-L characteristics (current vs light) and I-V (current-voltage) characteristics of a InGaN laser.*

Table I *Summary of InGaN Laser Performance*

Lowest threshold current	16 mA
Lowest threshold current density	1.2 kA cm^{-2}
Emission wavelength	380–450 nm
Highest output power	420 mW
Highest operation temperature	100°C
Operation voltage	4.5 V
Longest lifetime	10 000 h

Lasers emit near 40 nm and deliver CW power levels of 4 MW at room temperature with longevities of about 10,000 h. The threshold current density appears to hand around 3.9 kA cm^{-2} with threshold current of about 50 mA. The lasers come in a 5.66 mm package. These devices can be used for D applications for increased density, Spectral width is important. The devices are put in an external cavity to sharpen the output spectrum. The semiconductor lasers are known for their large gains, as compared to gas lasers therefore they are made smaller. The semiconductor diode laser cavities do not have high finesse, thus a broader spectra is obtained.

Compact violet lasers, and UV lasers have many applications in biology as sources for activating various fluorescent dye labels used. Fast pulsed lasers are used in time correlated fluoresence spectroscopy. This is a very powerful technique used in identifying a single antigen or a strand of DNA in DNA sequencing. The sensitivity is a result of being able to measure the fluorescence decay. Since this decay is in the range of several hundred picoseconds to several nano-seconds, the excitation source is faster than these timescales. Violet lasers emitting at about 400 nm are used to produce some 50 ps pulses.

II. QUANTUM DOT HETERSTRUCTURE LASERS

The lasers are divided into two types. 1. In *edge-emitting* lasers (Fig. 7.10), the active medium, (a thin layer) is placed in waveguide region having a larger refractive index than the surrounding cladding layers. The laser light is confined in this narrow waveguide. The advantage of this laser is that a compact output aperture is realized simultaneously with a high light output power. High reflection and antireflection dielectric coatings are deposited on the rear and front facets. 2. In a *vertical cavity surface-emitting laser* (VCSEL) the photons are bounced in the vertical direction in a high finesse cavity (Fig. 7.10). The cavity is very short and the gain per cycle is very low. There are losses at each reflection, therefore to provide very high reflectivity at the cavity edges, these losses should be low, otherwise lasing will not be possible.

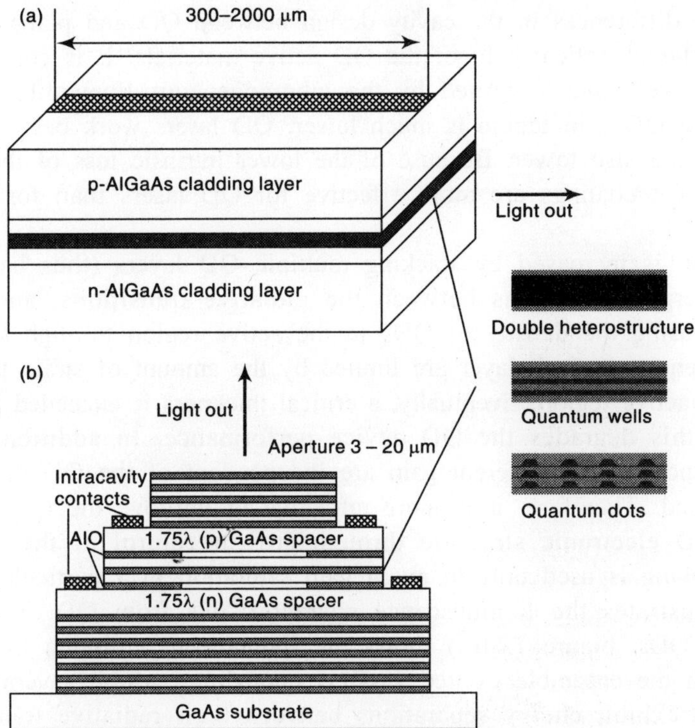

Fig. 7.10 *Comparison of (a) edge- and (b) surface-emitting lasers. In vertical cavity surface-emitting lasers (VCSELs), all-semiconductor (e.g., AlAs-GaAs) or selectively-oxidized AlO-GaAs distributed Bragg reflectors (DBRs) are generally used to achieve high reflectivity. The active region is typically made up of ultrathin layers or, more recently, stacks of coherent nano-inclusions (quantum dots), as shown in the figure. For single-mode operation, the thickness of the waveguide region in an edge-emitting structure should not generally exceed 0.6–0.8 μm, and for VCSELs the lateral dimensions of the microcavity should lie within 3–5 μm, depending on the specific design.*

Edge-Emitting Lasers are a new types of laser based on self-organized QDs. Although the laser performance is far inferior to planar quantum well lasers, InAs or InGaAs QD lasers that operate at ~1.3 μm are based on GaAs substrates.

The first 1.3 μm QD laser generated a new effort in QD lasers because of its wavelength. Since then threshold currents in the 1 mA range, and threshold current densities less than 20 Acm^{-2} are achieved. Improvements in p-type and modulation doping has increased the T_0 of the 1.3 μm QD lasers to over 200 K between 0 and 353 K. These superior characteristics are a direct consequence of the discrete electronic states of the QDs. However, QD lasers have two major hurdles; the low modulation response and low internal efficiency. Most QD lasers so far have been grown using MBE. The QD formation is sensitive to the total deposition thickness to even 1/10 of a monolayer (ML) which is achieved using MBE: It is at present the crystal growth approach for high quality QD lasers.

There are some differences in the cavity design between QD and planar quantum well lasers, because of the different optical gain of the QD active materials. It is considerably less than for a planar quantum well and is limited by the inhomogeneous linewidth. However, the optical loss intrinsic to the QD gain region is much lower. QD lasers work best with low loss cavities. The doping levels are also lower. Because of the lower intrinsic loss of the QD active material, high reflectivity facet coatings are more effective for QD lasers than for planar quantum well devices.

The optical gain is increased by stacking multiple QD layers (thus have high density) and creating wide energy separations between the radiative transitions, and modulation doping these layers. Increasing the density of QDs in the active region through stacking and trying to obtain high QD density in each layer are limited by the amount of strain that is incorporated in the volume of the active region. Eventually, a critical thickness is exceeded for which dislocations are formed, and this degrades the QD device performance. In addition, certain critical QD laser parameters such as the different gain are independent of the QD density as long as gain saturation is avoided. Therefore, it is more attractive to improve the QD laser performance by optimizing the QD electronic structure through growth control of the QDs and modulation doping, while stacking is used only to avoid gain saturation in a particular cavity design.

Figure 7.11 illustrates the luminescence characteristics from QD. There are different for different types of QDs. Figure 7.11(a) shows the spontaneous emission from a 1.3 μM InGaAs QD ensemble with the ensemble excited strongly enough to exhibit several radiative transitions. The InGaAs QDs exhibit energy separations between their radiative transitions of ~66 meV. This energy separation (≈ 300 Å) is set by the lateral dimensions of the InGaAs QDs, after covering with GaAs epitaxial. Figure 7.11(b) shows the spontaneous emission characteristics from an InAsQD ensemble that emit at close to the same wavelength. However, the InAs QD ensemble exhibits energy separations between its radiative transitions of ~104 meV, considerably larger than for the InGaAs QDs.

Egde-emitting lasers are fabricated from both types of QDs, and their threshold vs temperature is characterized (Figure 7.12). The InGaAs QD laser exhibits an abrupt increase in threshold above ~280 K. The InAs QD laser, has a less increase at the highest measurement temperatures. The results show the dramatic impact that the electronic structure plays in setting the high temperature performance of QD lasers.

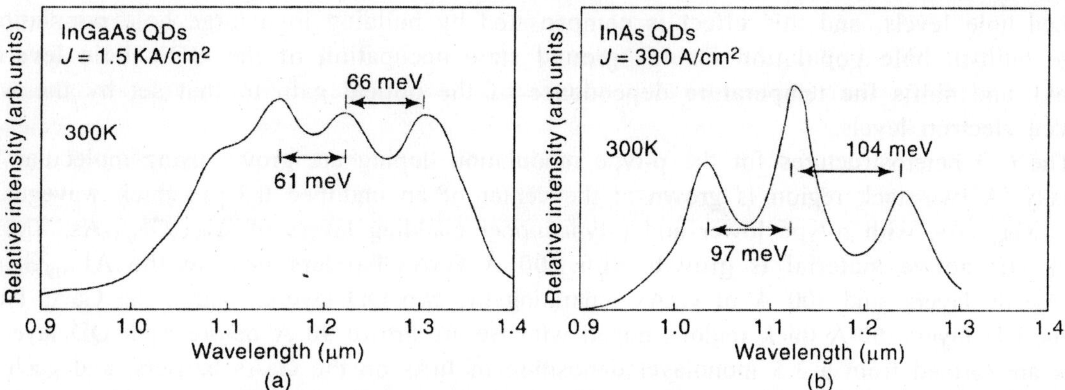

Fig. 7.11 *Comparison of energy separations for the radiative transitions between large InGaAs QDs and smaller InAs QDs that emit at nearly the same wavelength.*

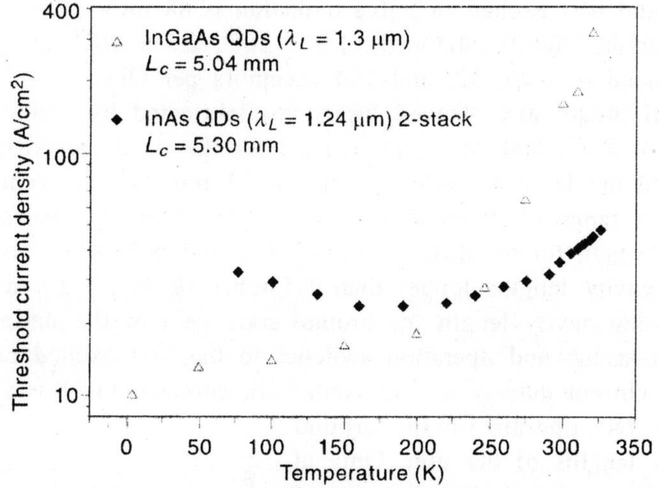

Fig. 7.12 *Comparison of threshold current vs temperature for lasers made from large InGaAs QDs with an energy separation of ~65 meV between the ground and first excited radiative transition, and InAs QDs with an energy separations of 104 meV between the ground and first excited radiative transition.*

It is also important to note that both the InGaAs and InAs QDs have wells with numerous discrete electronic states. These wells insure that the electrons and holes are confined in the QDs. In shorter wavelength QDs electrons and holes are thermally excited to the QD wetting layer. As the energy of separation between the QD ground state and wetting layer are too small. As the wetting layer has a very high density of states, it holds a large electron-hole population in thermal equilibrium.

The low T_0 of the QD lasers (Fig. 7.12) provide little advantage over InP-based devices. These lasers have closely spaced hole levels. The InAs QDs have hole levels spaced ~10 meV apart while the electron levels are separated by ~80 meV. The T_0s are mainly set by the closely spaced hole levels, and this effect is compensated by building in a large hole population. A large built-in hole population insures ground state occupation of the QD's hole levels (by holes), and shifts the temperature dependence of the optical gain to that set by the widely spaced electron levels.

The QD heterostructures for the p-type modulation doping are grown using molecular beam epitaxy. A two-stack region is grown at the center of an undoped 0.2-μm thick waveguide of $AL_{0.05}Ga_{0.95}As$, with n-type lower and p-type upper cladding layers of $Al_{0.85}Ga_{0.15}As$. The two-stack QD active material is growth with 200 Å GaAs barriers next to the $Al_{0.05}Ga_{0.95}As$ waveguide layers, and 300 Å of GaAs separating the two QD layers. Within the GaAs barriers of the QD layers, 60-Å thick regions doped with Be are grown 90 Å before each QD layer. The QDs are formed from a 3.5 monolayer deposition of InAs on the GaAs barriers, and each layer is covered with 50 Å $In_{0.15}Ga_{0.85}As$. The QD density in a single layer is measured on calibration growth to be 3×10^{10} cm^{-3}. The room temperature ground state energy separation between electrons and holes in the QD active material is 0.958 eV. The measured energy separation between the ground and first excited radiative transition is 88 meV. The Be doping levels in the 60-Å regions are calibrated for 0 (undoped), 1.3×10^{18}, 2.6×10^{18}, and $5.2 \times 10_{18}$ cm^{-3}. The doping levels correspond to 0.26, 52, and 104 acceptors per QD.

Various lengths of broad area cleaved lasers are fabricated by etching 27-μM wide ridges which are metalized with Cr and Au, with In used for the n-side metalization. The laser facets are left uncoated, and the laser operation occurs at 1.3 μm. Device testing is performed under pulsed operation for a range of temperature. Figure 7.13 shows the measured threshold current density versus cavity length for the different lasers. The undoped lasers have the lowest threshold current density (for cavity lengths longer than 1.5 mm) 38 Acm^{-2} for a cavity lengths of 2.9 mm. However, at 1-mm cavity length the ground state gain of the undoped active material is insufficient to obtain lasing, and operation switches to the first excited radiative transition with a jump in threshold current density to 343 Acm^{-2}. In contrast, the lasers doped with either 26 or 52 acceptors per QD operate on the ground state even for cavity lengths of 0.9 mm. Only at 0.64 mm does the lasing switch to the first excited transition. The threshold current densities for the p-doped lasers are somewhat higher than the undopped laser for cavity lengths greater than 1.5 mm, but become lower with either 26 or 52 acceptors per QD for shorter cavity lengths. The threshold current densities are 64 Acm^{-2} for 26 acceptors per QD and L_c = 2.3 mm, 70 Acm^{-2} for 52 acceptors per QD and L_c = 3.0 mm, and 96 Acm^{-2} for 104 acceptors per QD and L_c = 3.1 mm. For shorter cavities, the threshold current densities are 108 Acm^{-2} for 26 acceptors per QD and L_c = 1.5 mm, 99 Acm^{-2} for 52 acceptors per QD and L_c = 1.6 mm, and 141 Acm^{-2} for 104 acceptors per QD and L_c = 1.4 mm.

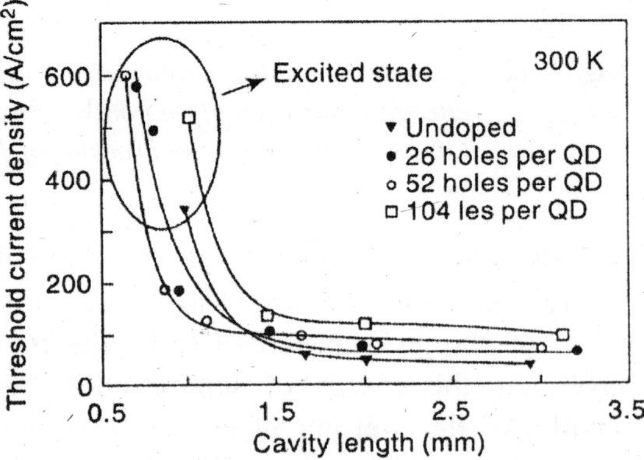

Fig. 7.13 *Threshold current density vas cavity length for differently doped, 1.3 μm QD lasers. Doping levels correspond to 0.26, 52, and 104 acceptors per QD.*

Figure 7.14 shows *the threshold versus temperature for the lasers doped with 0, 26, 52 acceptors per QD*. A significant improvements is obtained in the temperature sensitivity of the threshold for the p-doped QDs. The T_0 is sensitive to the cavity length, since gain saturation due to thermal excitation causes the take-off in threshold the optimum doping level appears to be ~52 acceptors per QD to maximize T_0 from 273–373 K C for these lasers. The T_0 of 171 K from measured for the 3-mm laser doped with 26 acceptors per QD is the highest T_0 for a 1.3-μm laser. This value has now been exceeded through use of a p-type modulation doped triple stacked QD laser.

Two-stack laser with 52 acceptors per QD show that the ground state laser operation is obtained upto 430 K. The lasing wavelength shifts from 1.295 μM at 297 K to 1.380 μm at 430 K. The threshold current density remains reasonably low at 347 Acm^{-2} at 430 K. That the QD laser doped with 52 acceptors per QD performs better at 430 K than the laser doped with 26 acceptors per QD. Therefore, higher doping becomes more important at higher temperatures.

Increasing the number of QD stacks to three, combined with the *p*-type doping, yields further improvement in the T_0. The QD laser structure is schematically illustrated in Fig. 7.15.

A *three-stack QD* active region is grown at the center of an undoped 0.2-μm thick waveguide of $Al_{0.05}Ga_{0.95}As$, with n-type lower and p-type upper cladding layers of $Al_{0.85}Ga_{0.15}As$. The QD active material is grown with 200 Å GaAs barriers next to the $Al_{0.05}Ga_{0.95}As$ waveguide layers, and 300 Å of GaAs separating each of the three QD layers. Within the GaAs barriers of the QD layers, 60-A thick regions doped with Be are grown 90 Å before each QD layer. The QDs are again formed from a 2.5 monolayer deposition of the InAs on the GaAs barrier, and each layer is converted with 50 Å of $In_{0.15}Ga_{0.85}As$.

The lasers are fabricated by etching the ridges through the p-side to within ~1 μm of the active region, and then oxidizing ~0.5 μm of the upper $Al_{0.85}Ga_{0.15}As$ cladding layer. The oxidized region provides electrical isolation and optical waveguiding. The p-side of the device is metalized with non-alloyed Cr/Au and In is used for n-side metalization. A high reflectivity facet coating, consisting of four

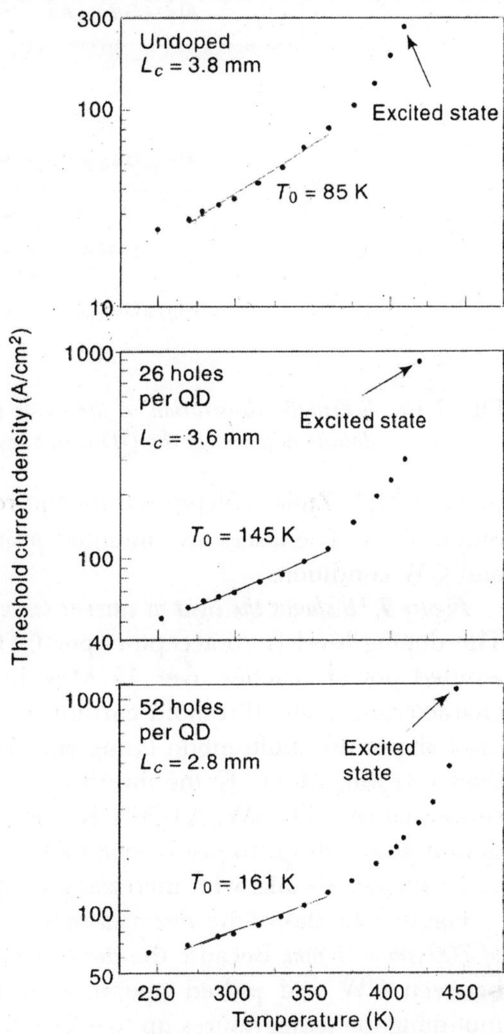

Fig. 7.14 *Threshold current density vs cavity lengths for the lasers doped with 0.26, and 52 acceptors for QD, for different cavity lengths. The maximum T_0 = 160 K between 273 K and 353 K is achieved for the laser doped with 52 holes per QD, and L_c = 3.2 mm.*

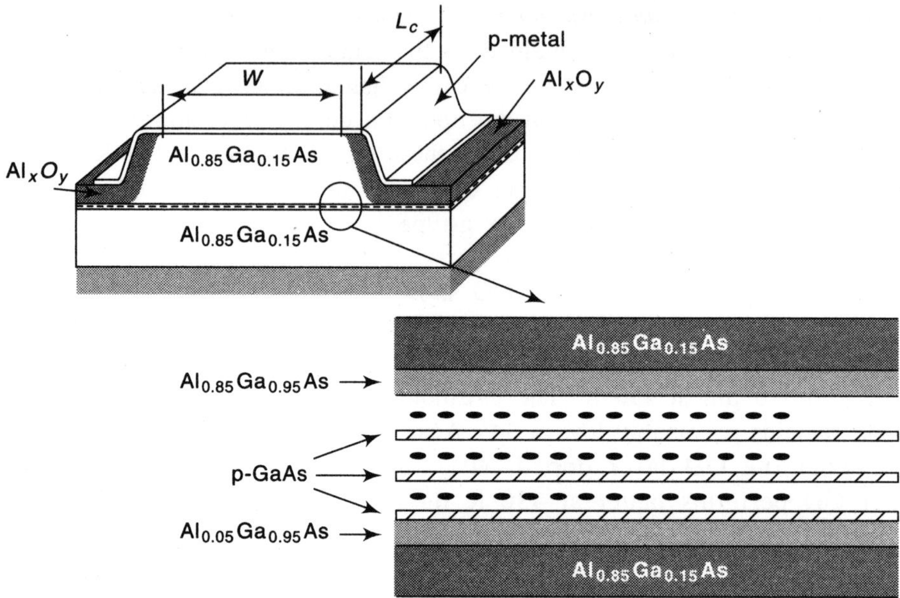

Fig. 7.15 *Schematic illustration of the oxide stripe QDF laser design. Be is used for the p-type modulation doping adjacent to the QD active layers.*

pairs of MgF/ZnSe is deposited on the rear facet, and a single MgF/ZnSe pair is used on the output facet. The lasers are mounted *p*-side up on a copper heat sink and tested under pulsed and CW conditions.

Figure 7.16 shows the light vs current curves for a 5-μm wide, 700-μm long device operated CW. The doping level is 26 acceptors per QD. The threshold current at 299 K is 4.4 mA, and the emitted power reaches over 15 MW before the onset of significant roll-over in the output characteristics. The threshold current density is 127 Acm^{-2}, or ~42 Acm^{-2} per QD layer. The inset shows the multi-mode lasing spectrum at 7.6 mA with lasing obtained at slightly longer than 1.31 μm. At 354 K, the threshold current increases to 6.4 mA, the maximum output drops somewhat to ~11 mW. AT 374 K, the threshold current increases to 8.4 mA, the maximum output power drops to just over 8 mW. The lasing wavelength at 374 K increases to 1.345 μm, and bandgap shrinks with increasing temperature.

Figure 7.17 shows the *threshold current versus temperature measured for two different cavity lengths of 700 μm or 1 mm.* Because the devices operate at a low threshold power density, the difference between CW and pulsed operation is not too significant for either device even for p-up mounting for temperatures up to ~273 K. The T_0s under pulsed operation are 213 K for either cavity length between 273 K and 273 K. The T_0 decreases to 189 K for the 1-mm long cavity, and 196 K for the 700-μm long cavity for CW operation. The higher T_0 for the shorter cavity (700 μm) suggests that the total power, as opposed to the power density, is also impacting the T_0. In addition, the threshold current reduces proportionally to the cavity length. The room temperatures threshold current density of the 1 mm long laser is 122 Acm^{-2}, vs 127 Acm^{-2} for the 700-μm long laser. So, even shorter cavities have better power conversion and temperature

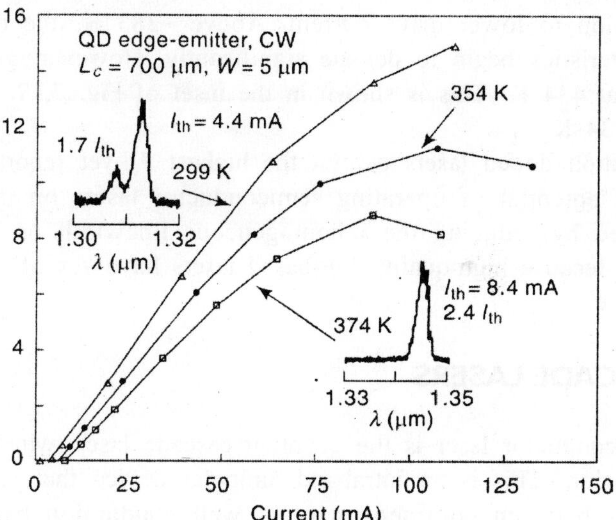

Fig. 7.16 *Light vs current characteristics measured under VW operation (p-up mounting) for three different temperatures. The threshold currents are 4.4 mA at 299 K 6.4 mA at 354 K, and 8.4 mA at 374 K. The insets show the lasing spectra at different temperatures.*

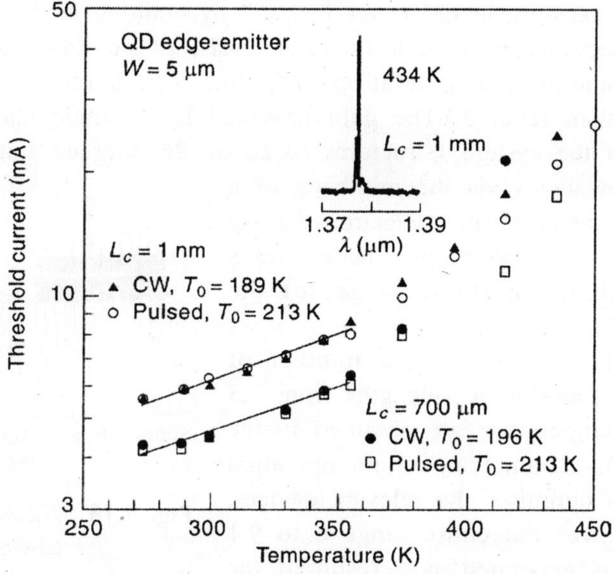

Fig. 7.17 *Threshold current density vs temperature for either pulsed or CW operation (p-up mounting) and two different cavity lengths of 1 mm or 700 μm. Because of the low power densities, pulsed and CW operation result in nearly the same T_0s for either cavity length between 273 and 353 K. Inset shows ground state CW operation can be achieved even at 434 K.*

characteristics, in addition to lower drive current. Above ~383 K, the difference between the CW and pulsed characteristics begin to deviate significantly. However, ground state CW lasing operation are obtained at 434 K. This is shown in the inset of Fig. 7.17. The lasing wavelength shifts to ~1.38 μm at 434 K.

These *p*-type modulation doped lasers exhibit the highest T_0 yet reported for a QD laser, or any 1.3 μm laser. The potential of operating semiconductor lasers on discrete levels for high performance is achieved by reducing the inhomogeneous linewidth or increasing the number of QD layers, or both. Because high-quality InP-based lasers have T_0s of 60–70 K or less due to their electronic structure.

III. QUANTUM CASCADE LASERS

A new type of semiconductor laser is the quantum cascade laser, which provides emission in the infrared spectral region. This is an intraband, unipolar device that relies only on electrons, which make transitions between confined quantum well conduction band states. Figure 7.18 shows a typical band structure of a quantum cascade laser. The lasing transition occurs between quantum well levels 3 and 2, the separation of which is typically in the range of ~12–350 meV, corresponding to wavelengths ~100–3.5 μm. The lasing wavelength is varied by selecting a suitable well width. Electrons are injected into the upper levels 3 by tunneling through the left-hand barrier, B, before relaxing to the lower levels 2 by emitting a photon. Level 1 is required to efficiently remove electrons from the lower laser levels, as a build-up of electrons in this level can prevent the attainment of a population inversion, which requires a greater number of electrons in levels 3 than level 2. The gain provided by a single stage is not sufficient to overcome the losses of the system. Structures based on 25 coupled stages are used. Electrons are transported between stages via the miniband of a superlattice, which is designed to inject electrons having the correct energy into the next stage. The electrons cascade down through the multiple stages of the structure.

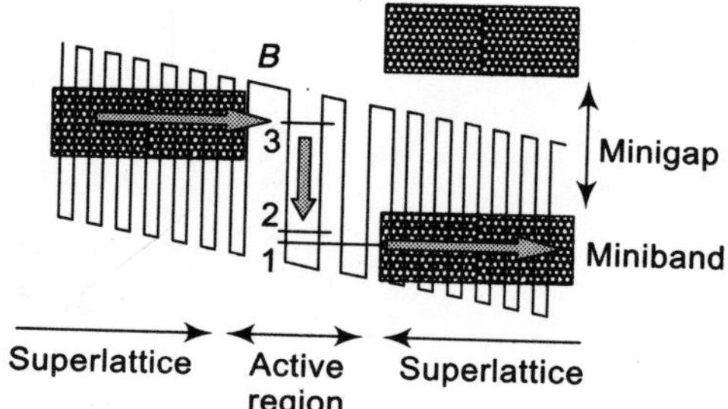

Fig. 7.18 *The band structure of a quantum cascade laser. Electrons are transported between stages via the miniband of a superlattice, before tunnelling through barrier B into the upper lasing state, state 3. The minigap formed between the minibands of the superlattice prevents electrons tunnelling directly out of state 3.*

Quantum cascade lasers based on a number of different designs, with emission wavelengths from 3.5 to 106 μm, at room temperature are obtained in the range of 4.5 to 16 μm. Room temperature operation at the two wavelengths optimises the relevant devices, and similar operation over the entire range 6 to 9.1 μm is possible. Heat is generated as a result of the large threshold current densities, which are typically up to two orders of magnitude higher than those of interband lasers. This is because electrons are able to relax between the lasing levels by emitting a phonon instead of a photon, resulting in a large fraction of the electrons being wasted. In addition, electrons are thermally excited out of the quantum well from the upper lasing state. However, despite their high threshold current densities, quantum cascade lasers are investigated for gas monitoring, free space communication systems and medical imaging.

Operating Principle and Structure

The quantum cascade laser (QCL) is the equivalent to the quantum well infrared photo detector (QWIP) with regard to the optical emitter. The emission of light quanta, is not based on *inter*band transitions but on *intra*band transitions. Radiating transitions between different energy levels within individual neighbouring quantum wells are used in QCL. Therefore, optoelectronic devices operating in the far infrared (3.8–200 μm) with material systems based on semiconductors with a relatively large band gap are obtained, example; use of GaAs/GaAlAs and InP/InAlAs is made. The energy difference between the energy levels in the quantum well not only depends on the barrier height of the quantum well but also on its width. This method of *band gap engineering* enables the implementation of emitters with very different emission wavelengths.

Moreover the quantum cascade laser is a unipolar device, which means that the emission is based on electronic transitions in the conduction band. Since the charge carriers do not recombine during radiative transition, they are used several times for the light emission. So, quantum efficiencies >1 are achieved. In state-of-the-art quantum cascade lasers, the basic units are repeated between 20 and 40 times.

Despite the operating principle being similar, the quantum cascade laser exhibits a substantially more complicated structure in comparison to the quantum well infrared photodetector (QWIP). This difference is due to the population inversion between the two energy levels required for the operation of the laser. Thus, the higher energy level must have a high electron concentration than the lower level. In principle, the operation of the laser can be described with the help of the conduction band profile of a basic unit of this laser (Fig. 7.19). In the case illustrated, the

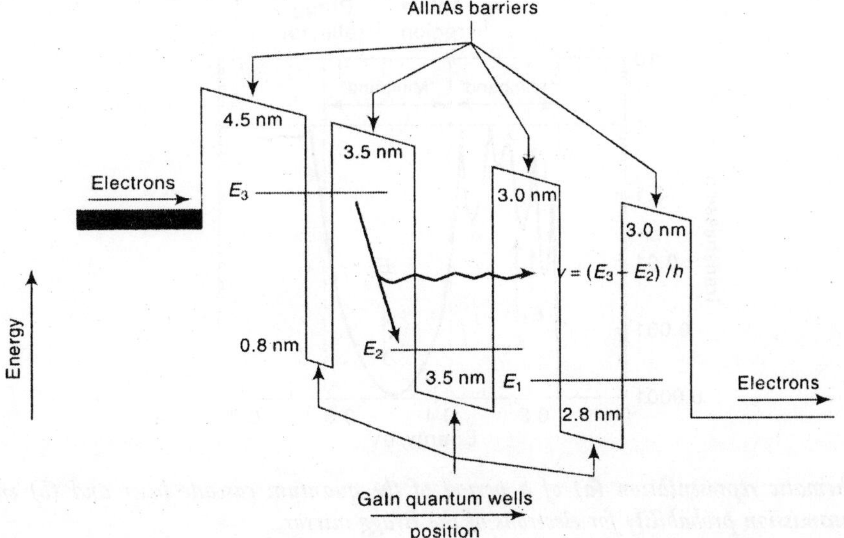

Fig. 7.19 *Conduction band profile in the area of the active layer of a quantum cascade laser*

barriers are made of AlInAs layers, and the quantum wells consist of GaInAs. Electrons tunnel from the injector into the electronic level E_3 and fall on level E_2 by sending out light quanta with $v = (E_3 - E_2)/h$, followed by a non-radiative transition from energy level E_2 to level E_1.

To achieve population inversion and laser operation, the electron tunneling rate from the two levels E_2 and E_1 into the neighbouring conduction band must be higher than the tunneling rate from the level E_3. Therefore, an electronic Bragg reflector consisting of a semiconductor superlattice is inserted into the QCL structure behind the active layer. This structure is depicted in Fig. 7.20(a). The probability of transmission of the Bragg reflector's electrons as a function of the energy position relative to the conduction band minimum is shown in Fig. 7.20(b). Also shown is a forbidden band (mini-gap) which develops and takes effect within the energy region of the level E_3. The tunneling probability from the level E_3 is two orders of magnitude lower than from the levels E_2 and E_1, whose energies correspond to the energy region of the mini-band in the Bragg reflector selection of the laser. Thus, the necessary condition for population inversion is ensured.

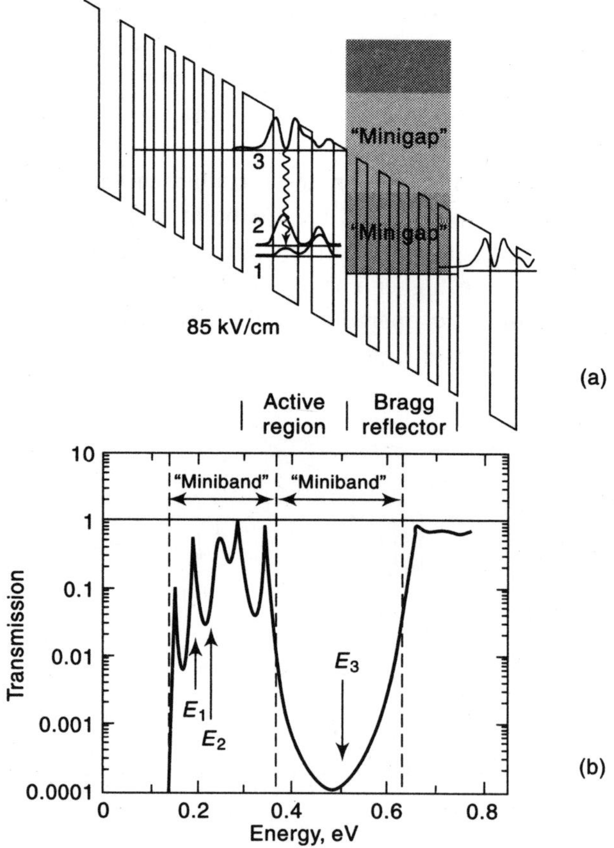

Fig. 7.20 *Schematic representation (a) of a period of the quantum cascade laser and (b) energy dependent transmission probability for electrons of the Bragg mirror.*

The first quantum cascade laser ever devised achieved a few mW optical power at cryogenic temperatures under pulsed operation conditions. As an example of these early QLs, the emission spectrum and the emitted power as a function of the laser current of an InGAs/InAlAs quantum cascade lasers with an emission wavelength around 4.5 μm are depicted in Fig. 7.21.

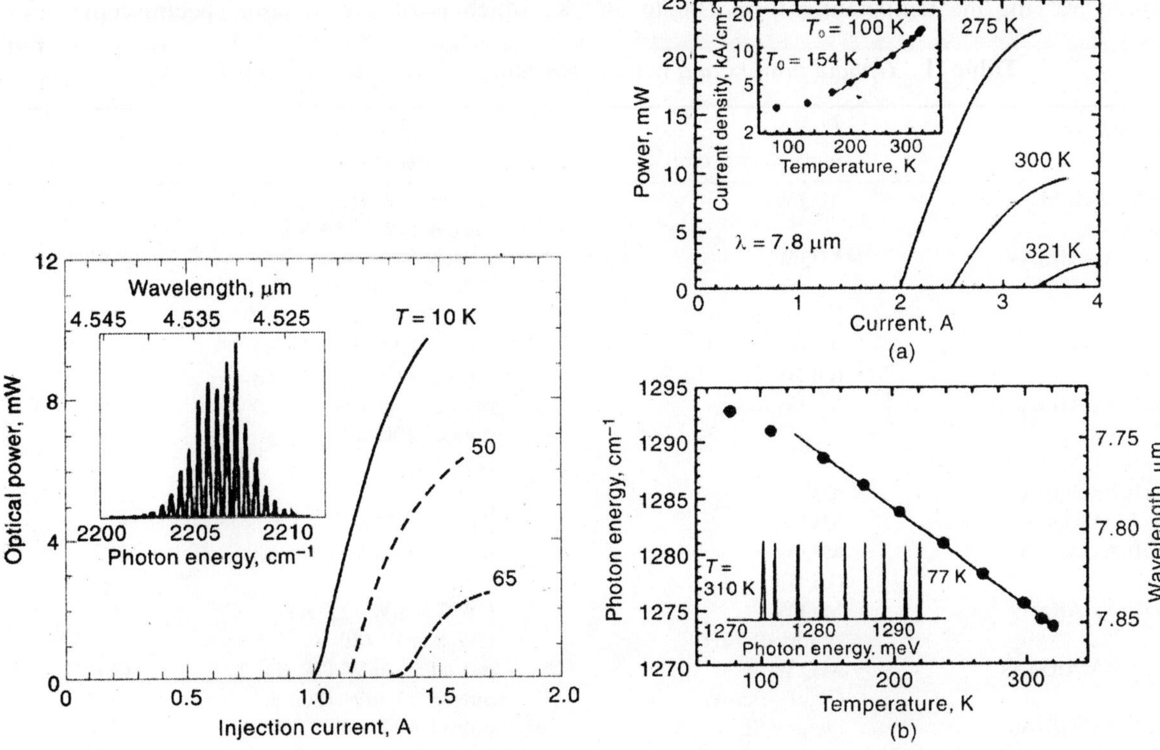

Fig. 7.21 *Optical power as a function of laser current and optical emission spectrum of a Fabry-Perot InGaS/AlInAs quantum cascade laser.*

Fig. 7.22 *(a) optical power as a function of laser current. Inset: temperature dependence of the lasing threshold current density. (b) Temperature dependence of the emission wavelength of a distributed feedback InGaAs/AlnAs quantum cascade laser. Inset: optical emission spectra for different temperatures between 77 and 300 K.*

The optical power and the wavelength ranges of quantum cascade lasers is given in Table II. An emitted optical power of more than 1 W near room temperature and of up to 6 W at a temperature of 80 K is achieved under pulsed operation. Also, a wavelength range from 3.8 up to 24 μm is covered with QCLs using only three different combinations of active layer materials; example, this wavelength range is used for optical data transmission within the earth's atmosphere, in two spectral range windows from 3.5 to 4 μm and from 8 to 10 μm. The full infrared spectral range is covered with III-V compound semiconductor lasers. Conventional semiconductor lasers based on electron-hole recombination cover the wavelength range up to 3.25 μm, while QCL lasers cover the longer wavelength range. In Fig. 7.22, the optical power-laser current characteristics and the optical emission spectrum of a DFB InGaAs/AlInAs quantum cascade laser are depicted. At room temperature, this laser emits around 7.8 μm with an optical power of up to 10 mW. Furthermore, the emission wavelength can be tuned between 7.75 and 7.85 μm by varying the temperature form 100 to 300 K, which is of use in laser spectroscopy.

Table II *Optical powers and wavelength ranges of quantum cascade lasers*

Active layer material	*Electric (optical) structure*	*Max. optical power (at temperature)*	*Emission wavelength, μm*
InAs/GaInSB/ AlSb	MQW	pulsed 6 W (80 K) pulsed 1 W (150 K)	3.8
GaInAs/AlInAs	MQW	pulsed 900 mW (770 K), pulsed 240 mW (300 K)	5.0
GaInAs/AlInAs	undotietes SL (DFB)	pulsed 1,15 W (273 K) pulsed 92 mW (393 K)	5.3
GaInAs/AlInAs	MQW	pulsed 750 mW (80 K) pulsed 200 mW (210 K)	8.0
GaInAs/AlInAs	nipi SL		8.8
GaAs/AlGaAs	MQW	pulsed > 1 W (77 K)	9.7
GaInAs/AlInA	MQW (DFB)	pulsed 80 mW (300 K)	10.2
GaInAs/AlInAs	MQW (Fabry-Perot)	CW 75 mW (25 K) CW 8 mW (80 K)	11.1
GaAs/AlAlGaAs	MQW (Fabry-perot)	pulsed 75 mW (78 K) pulsed 23 mW (300 K)	12.6
GaInAs/AlInAs	*chirped* SL (Fabry-Perot)	pulsed 400 mW (210 K) pulsed 150 mW (300 K)	16.0
GaInAs/AlInAs	undoped SL (Fabry-Perot)	pulsed 14 mW (70 K)	19.0
GaInAs/AlInAs	undoped SL (Fabry-Perot)	pulsed 3 mW (70 K)	24.0

For quantum cascade lasers with emission wavelengths below 4 μm, InAs/GaInSb/AlSb is preferred as active layer material. With regard to the type of the electronic structure, nipi-superlattices (layer sequence of n-type, intrinsic, p-type, intrinsic) or intrinsic superlattices are used as the active layer of QCLs, besides the widespread multi-quantum-well (MQW). Thus, there are lasers with a high optical power in the spectral range around 5 μm and lasers with emission wavelengths up to 24 μm.

GaAs/GaAlAs quantum cascade lasers with emission at wavelengths between 100 and 200 μm have been produced. These monochromatic Tetrahertz emitters close the gap between electronic and optical oscillators in this wavelength range. Terahertz emission has interesting applications in biomedical imaging.

IV. GROOVE QWR LASERS

Different types of diode laser configurations are possible with V-groove QWR heterostructures (see Figure 7.23). The tight V-shaped optical waveguides are utilized in order to construct a single- or multiple-QWR laser structure in which the optical beam propagates parallel to the axis of the wires [Figure 7.23(a)]. The FP optical cavity is obtained by cleaving two opposite and parallel (011) planes that are perpendicular to the axis of the single V-groove. Large optical confinement factors are achieved and lasing is done with a single QWR at room temperature. The optical gain is reduced. Higher optical gain is obtained with light propagating perpendicular to the wire and polarized along their axis, with transitions involving heavy hole states but very high optical confinement due to the tight 2D waveguide helps to compensate for this shortcoming. Extremely low threshold currents are possible with longitudinal cavity V-groove QWR lasers.

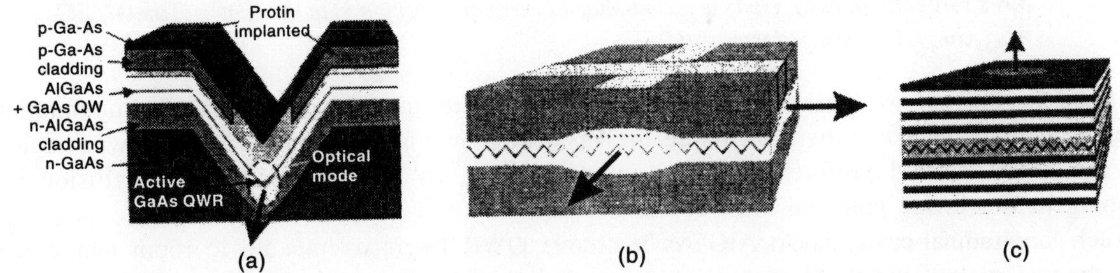

Fig. 7.23 *Different configurations of V-groove QWR lasers. (a) Longitudinal cavity laser, (b) slab waveguide V-groove QWR array laser; (c) vertical cavity surface emitting V-groove QWR laser.*

Another cavity configuration [Figure 7.23(b)] involves V-groove wires incorporated in a slab optical waveguide. Light propagation is either parallel or perpendicular to the wires. In the latter case, DFB optical cavities are obtained using resonant Bragg reflection, which is obtained with submicon pitch V-groove gratings. Lasers made on GaAs substrates, the shallow gratings necessitate a two-step epitaxial growth process, in order to achieve a sufficiently thick high refractive index buffer layer for confining the optical mode. Thus, InGaAs/GaAs V-groove QWR structures are useful, as they involve regrowth on GaAs, rather than AlGaAs, corrugated substrates.

Vertical cavity V-groove QWR lasers are obtained by sandwiching the active (QWR array between two multilayer Bragg mirrors, as shown in Figure 7.15(c). Here, the polarization anisotropy of the optical matrix elements of the QWRs is utilized in order to control the polarization of the surface emitting laser.

A TEM cross section of the core of a multiple-QWR GaAs/AlGaAs V-groove laser heterostructure is shown in Figure 7.24. In this case, the wires are placed at they centre of the $Al_{0.25}Ga_{0.75}As$ waveguide layer, which is sandwiched between $AL_{0.55}Ga_{0.45}As$ cladding layers. The minimum separation between the wires is limited by the minimum barrier layer required for recovering the curvature at the bottom of the groove after deposition of each GaAs layer. The nominally undoped wavaguide layer and the p- and n-doped waveguide cladding layers form the p-i-n structure for providing carrier injection in to the reactive region. The current is

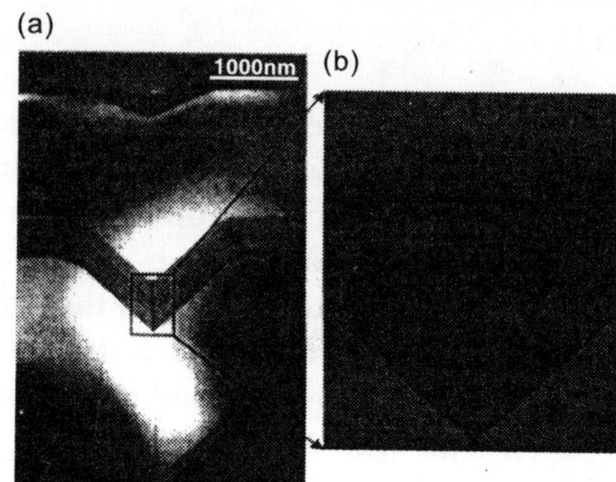

Fig. 7.24 *TEM cross section of a V-groove 3-QWR GaAs/$Al_{0.25}Ga_{0.75}As$ laser grown by low-pressure OMCVD. (a) Overview of the entire QWR laser heterostructure; (b) magnified view of the active region. The three vertically stacked, GaAs crescent-shaped wires are positioned at the centre of the $Al_{0.25}Ga_{0.75}As$/ $Al_{0.55}Ga_{0.45}As$ V-shaped waveguide.*

confined to the narrow wire region using different techniques, such as proton implantation, reserve pn junctions, or growth on oxide masks. Carrier are further confined to the wire regions via the VQW-assisted preferential injection mechanism, as well as because of diffusion and trapping in the lower potential well at the QWRs.

Such longitudinal-cavity GaAs/AlGaAs V-groove QWR lasers operate up to room temperature and show very low threshold currents, due to their small active volume. Threshold currents as low as 3.5 mA for uncoated-facets, single-QWr devices, and 0.6 mA facet-coated 3-QWR devices, are achieved with OMCVD V-groove structures in which proton implantation is employed for current confinement. Similar, MBE grown, V-groove QWR lasers, incorporating strained InGaAs/AlGaAs QWRs and employing growth on an oxide mask, have shown threshold currents as low as 200 μA for uncoated facets. These latter devices also exhibit very weak dependence of the threshold current on temperature (T_0 = 263 K).

Properties

In spite of the extremely small cross-sectional area of the crescent-shaped QWRs, *optical confinement factors* comparable to those in high performance QW lasers are achievable with the longitudinal cavity geometry. This is due to the very tight 2D optical waveguiding introduced by the V-shaped waveguide. With Waveguiding formation of a fundamental optical mode with a heart-shaped intensity distribution is there and is with a typical full-width at half-maximum of less than 0.5 μm. The optical confinement factors are increased by using several wires in the optical waveguide; however, an optimum number of wires exists, as a finite separation between wires in necessary for forming identical QWRs. The optical confinement factor in such V-structures is maximized by shifting the wires upwards with respect to the geometrical centre of the waveguide, towards the point where the optical mode peaks.

The *subwavelength* features in the near-field patterns of the modes of a V-shaped waveguide make far-field optics techniques of little use in their characterization. In fact, near-field images of the V-groove lasers yield circular mode shapes, approximately 1 μm in diameter, but is limited by the spatial resolution of the imaging system. Scanning near-field optical microscopy

(SNOM) is used in the near-field patterns of these lasers. Here, the image is acquired by scanning a tapered, metal coated optical fiber tip over the cleaved laser facet. The fiber aperture (approximately 100 nm) determines the spatial resolution. Spectrally resolved SNOM images are obtained using a monochromator at the output of the fiber. SNOM images of a single-QWR GaAs/$Al_{0.25}Ga_{0.75}As$ V-groove laser measured at two different wavelengths (805 and 805.8 nm) above threshold are shown in Figure 7.25(a). Two different lateral modes are resolved at the different wavelength. In this case, the modal gain provided by the wire is not sufficient to yield lasing, due to recombination in the wire, and gain is provided by the sidewall GaAs QWs. The SNOM image of a similar V-groove laser containing three identical GaAs/ $Al_{0.25}Ga_{0.75}As$ QWRs in a similar V-shaped waveguide structure as for the single-QWR device of Figure 7.25(a) is shown in Figure 7.25(b). The lasing occurs at a longer wavelength (816 nm), corresponding to transitions in the wires. The heart-shaped intensity distribution, with a full width at half maximum of 0.47 μm, corresponds to the fundamental mode of the V-shaped waveguide. The calculated pattern of this mode, shown in Figure 7.25(c), closely resembles the measured pattern.

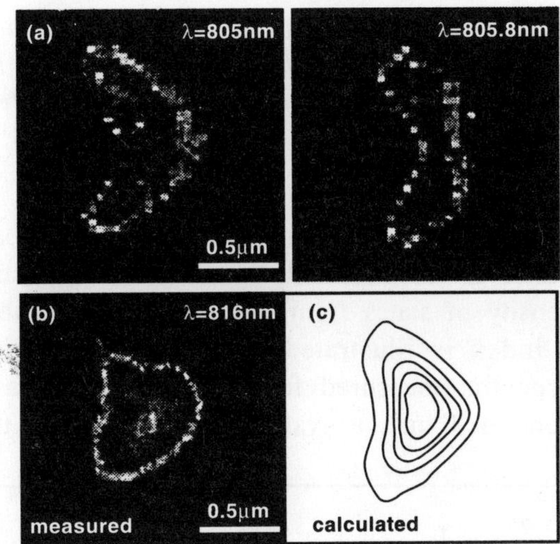

Fig. 7.25 *(a) SNOM patterns of a single-QWR GaAs/AlGaAs V-groove laser measured the two different lasing wavelength indicated. (b) SNOM pattern of a 3-QWR GaAs/AlGaAs V-groove laser measured at the lasing wavelength. (c) Calculated near field pattern of the laser in part (b).*

Evidence for lateral confinement effects in the V-groove QWR lasers is provided by the *emission spectra* of these lasers. An example is shown in Figure 7.26, which presents the amplified spontaneous emission (ASE) and lasing spectra of two 4-QWR GaAs/$Al_{0.3}Ga_{0.7}As$ V-groove lasers. The wires in structure A [Figure 7.26(a)] are thicker than those in structure B [Figure 7.26(b)]. Peaks in the ASE are evident in both cases, indicating local maxima in the optical modal gain at the corresponding energies. These peaks are due to transitions between the 1D wire states in conduction and valence bands. The measured peak separations and the calculated energy difference for the 1D transitions, obtained with the aid of the TEM is same. Above threshold, lasing is observed at transitions between excited 1D states.

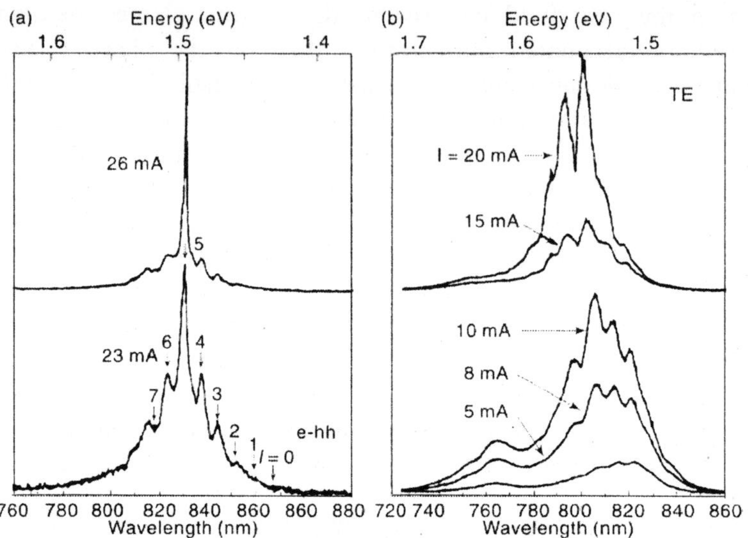

Fig. 7.26 *(a, b) Emission spectra of two-4-WER GaAs/AL0.3Ga0.7As V-groove QWR lasers grown by atmospheric pressure OMCVD. The wires in structure A are thicker than for structure B. The arrows indicate the calculated transition energies for the corresponding wire structures as determined by the wire shape observed in TEM cross sections.*

The QWR lasers of Figure 7.26 do not lase at the 1D ground states. This is because the corresponding modal gain is insufficient to compensate for the optical cavity losses. The modal gain, increases with the index of the subbands involved in the transition. This is due to the finite contribution of the density of states from lower energy 1D subbands. This increase in the modal gain with subband index is illustrated by the spectral gain measurements shown in Figure 7.27(a). The gain spectra, measured for an asymmetric-waveguide, 4-QWR V-groove laser with a threshold current of 4 mA, a evaluated by measuring the cavity finesse at several

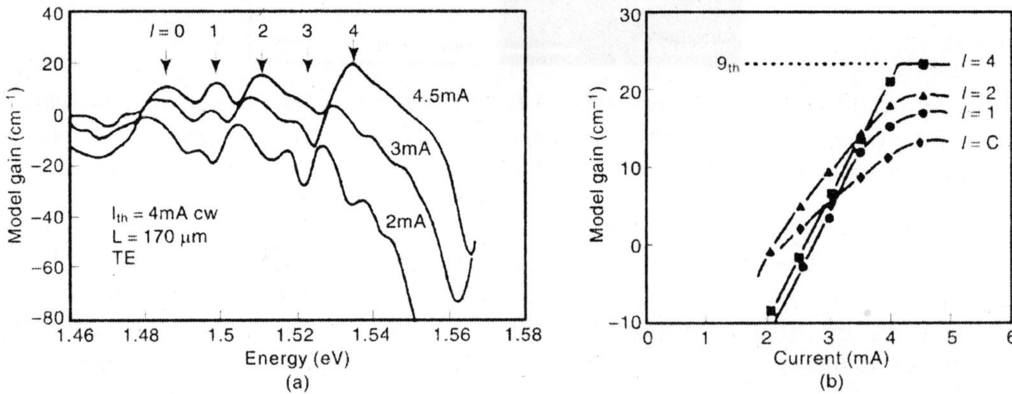

Fig. 7.27 *(a) Modal gain spectra measured at room temperature for a 4-QWR V-groove diode laser. The calculated transition energies for the different 1D subbands are also indicated. (b) Measured modal gain at the peaks of the gain spectra of (a) as function of the diode current.*

currents. The spectra exhibit peaks with energy separation that agree with the calculated energy difference between the 1D transitions. The peak values of the modal gain at each transition are depicted in Figure 7.27(b) as a function of the diode current. It is seen that at low currents the lowest energy transition exhibits the largest modal gain. However, this gain saturates at a value of approximately 13 cm^{-1}, which is lower than the optical cavity losses. The gain saturation is due to the finite density of states at the 1D subbands. At higher currents, the carrier density at the higher energy subbands increases, and owing to their higher density of states, higher modal gains result at the corresponding photon energies. Finally, filling of the fourth excited subbands allows for achieving the threshold modal gain, which is 23 cm^{-1} for this particular device.

Lasing at the QWR ground subbands would be possible by increasing the modal gain for that transition, or else by decreasing the optical cavity losses. One way to increase the modal gain is by minimizing the thermal population of the excited 1D states. In fact, consecutive decrease in the index of the lasing subband has been achieved at reduced temperatures, finally resulting in lasing from the ground state at temperature below 70 K.

In an attempt to increase the power output of QWR lasers, GaAs/AlGaAs V-groove QWR laser arrays are fabricated. The devices were grown at 973 K using low pressure OMCVD. It consist of four 5-nm-thick GaAs wells sandwiched in a graded-index (GRIN) SCH waveguide structure, grown on a 6-μm pitch grating. Arrays of 19 grooves are operated in parallel, with the current confined to the QWRs using proton implantation. Threshold currents of 0.7A are obtained at room temperature under pulsed operation, with output power greater than 100 mW.

V-groove QWR laser arrays incorporating pnp junctions for blocking the currents between the wires are also reported. Each V-groove contained there vertically stacked, 5-nm-thick GaAs/GaAs crescents, and the conducting channels through which the current is injected into the V-grooves are 1.5 μm wide. Room temperature threshold current of 276 mA is obtained for 10-groove devices with uncoated facets.

A TEM cross section of the core of an InGaAs/GaAs V-groove QWR laser grown on a submicron pitch grating is shown in Figure 7.28. The device is fabricated by first growing a "half" GaAs/AlGaAs laser structure consisting of the lower $Al_{0.7}Ga_{0.3}As$ layer and half of the

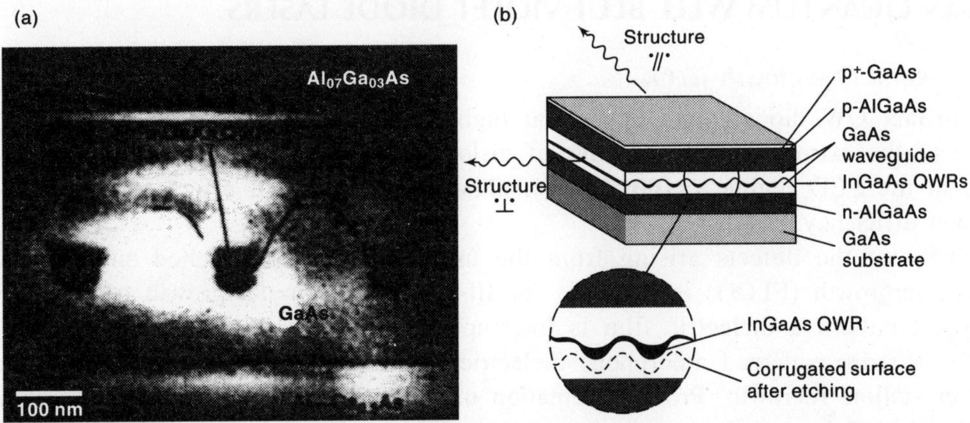

Fig. 7.28 *(a) Schematic and (b) TEM cross sections of an InGaAs/GaAs V-groove QWR laser grown on a submicron pitch grating. The crescent-shaped QWRs inserted in a GaAs slab wave-guide, which is sandwiched between $Al_{0.7}Ga_{0.3}As$ classing layers.*

GaAs waveguide layer. The upper surface of this wafer is then corrugated with 0.25-μm pitch periodic corrugations aligned in the $[00\bar{1}]$ direction, and the second "half" of the laser, including the InGaAs active layer, is regrown. This results in the formation of crescent-shaped InGaAs/GaAs QWRs at the center of the GaAs core of the slab waveguide.

Depending on the orientation of the contact stripe, two cavity configurations are possible, corresponding to either light propagating along the grooves or perpendicular to them. The latter configuration allows for taking advantage of the transitions involving the heavy-like holes, and exhibit a higher gain. The coherent Bragg reflection is induced which generat a DFB optical cavity. The pitch of the grating in this particular device corresponds to second-order Bragg DFB configuration. Wavelength filtering due to this resonant Bragg reflection is observed in the emission spectra of these QWR devices.

Optically pumped vertical-microcavity surface emitting V-groove lasers are fabricated by first growing a lower Bragg reflector stack consisting of λ/AlAs/AlGaAs layers, on which $In_{0.3}Ga_{0.7}As$/GaAs V-groove wires are grown using a dielectric mask pattern. Then an upper epitaxial Bragg mirror is grown. Lasing at the cavity mode is observed at 77 K. Weak coupling effects are observed in InGaAs/GaAs V-groove QWRs integrated in a Bragg-mirrow microcavity. Drastic reduction in QWR emission linewidth (to about 1 meV) and polarization anisotropy associated with the 2D confinement is observed in low temperature PL spectra of these structures.

One optical application of QWR active regions is their incorporation in III-V compound lasers grown on Si substrates. With this material efficient lasers on Si substrates for optical interconnect applications are fabricated. The lattice mismatch between Si and the GaAs/AlGaAs system, results in a high density of defects. The advantage of a QWR laser is to reduce the probability of introducing a defect into the extremely narrow active region. V-groove QWR GaAs/AlGaAs diode lasers are also grown by atmospheric pressure OMCVD on GaAs/Si substrates. They have threshold currents significantly lower than those for GaAs/AlGaAs QW lasers grown on planar Si substrates (16 to 20 mA versus 24 to 64 mA, respectively).

V. InGaN QUANTUM WELL BLUE/VIOLET DIODE LASERS

Epitaxial Lateral Overgrowth (ELOG)

The InGaN QW diode lasers operate at high current densities (>1 $kAcm^{-2}$) even with the presence of the extremely high densities of dislocations. These defects do not effect the device lifetime in the LED regime (<100 Acm^{-2}), although they have a effect on electrical-to-optical conversion efficiency.

The density of the defects arising from the use of lattice mismatched substrates an lateral epitaxial overgrowth (ELOG) is reduced. For III-nitrides, the initial growth of the GaN layer to deposit and pattern a dielectric film is interrupted before resuming the growth. As shown in Fig. 7.29, a stripe-patterned amorphous dielectric film (SiO_2) defines windows of exposed GaN for the crystalline regrowth. Proper orientation of the stripes relative to crystallographic axes of the underlying GaN promotes rapid lateral growth over the dielectric and planarization of the growing GaN layer. By this process truncation at the dielectric of the threading dislocations that originate at the sapphire GaN interface is done, and also the near absence of these extended

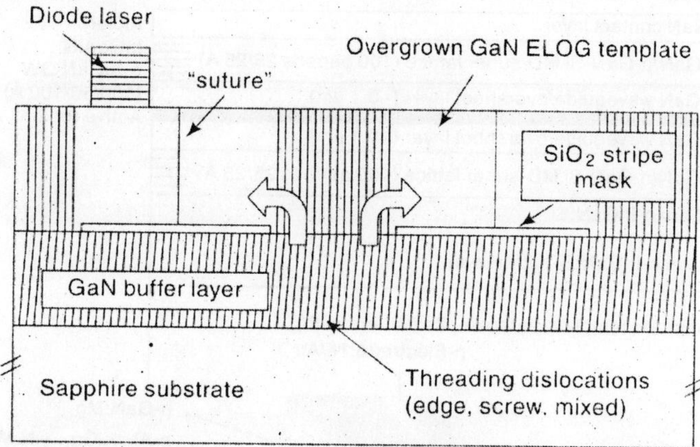

Fig. 7.29 *Schematic of epitaxial lateral overgrowth (upper panel). The lateral growth fronts, emanating from spaces between the SiO_2 stripes, coalesce near the center of each mesa to for a "suture" region which commonly still contains defects. The electron micrograph images show how the "wing" regions between the seed areas and the sutures have significantly fewer dislocations.*

defects in the laterally grown GaN directly above the dielectric stripes is achieved. Above the dielectric, the two laterally propagating growth fronts from the opposite sides of a stripe coalesce to form an interfacial layer (Fig. 7.39), the structural quality of which depends on the degree of the tilt misalignment between the coalescing fronts. The region above the dielectric between the coalescence interface and the edge of the window defines the "wing" of the ELOG material where the density of threading dislocations is reduced by several orders of magnitude, on either sapphire of SiC substrates; to values as low as 10^5 cm^{-2} and below. One measure of the impact of reduced dislocation density of ELOG is the greatly reduced reverse leakage current obtained in GaN p-n junctions. In connection with ultraviolet AlGaN photodetectors, a reduction of the reverse-bias leakage current by up to six orders of magnitude is achieved.

The use of ELOG "substrate template" has improved the performance of laser diode, extended the device lifetime. Lasers with threshold current densities below ~ 3 kA cm^{-2} are fabricated from the portions of the ELOG wafer where dislocations are minimized. The characteristics of lasers fabricated from such ELOG wafers have in terms of variation in threshold current density for diodes fabricated above the wing, window, and coalescence regions been investigated. The threshold current density remains high even in the best lasers.

Design and Performance of the InGaN Diode Laser—Blue Violet

The active region of the violet edge-emitting nitride diode lasers and high power blue LEDs are based on the InGaN/AlGaN/AlGaN MQW separate confinement heterostructure (SCH). They are grown either on the (nonconducting) sapphire or conducting SiC substrate. In Fig. 7.30, the layering details are shown for InGaN SDCH/MQW diode laser, which combines the features of an edge-emitting diode laser, within the constraints imposed by the growth of the AlGaN, GaN, and InGaN cladding, waveguide, and QW active layers, respectively. A wide range of temperatures is required when growing the three different materials, anywhere from about T = 1073 K for InGaN to nearly T = 1373 K for AlGaN.

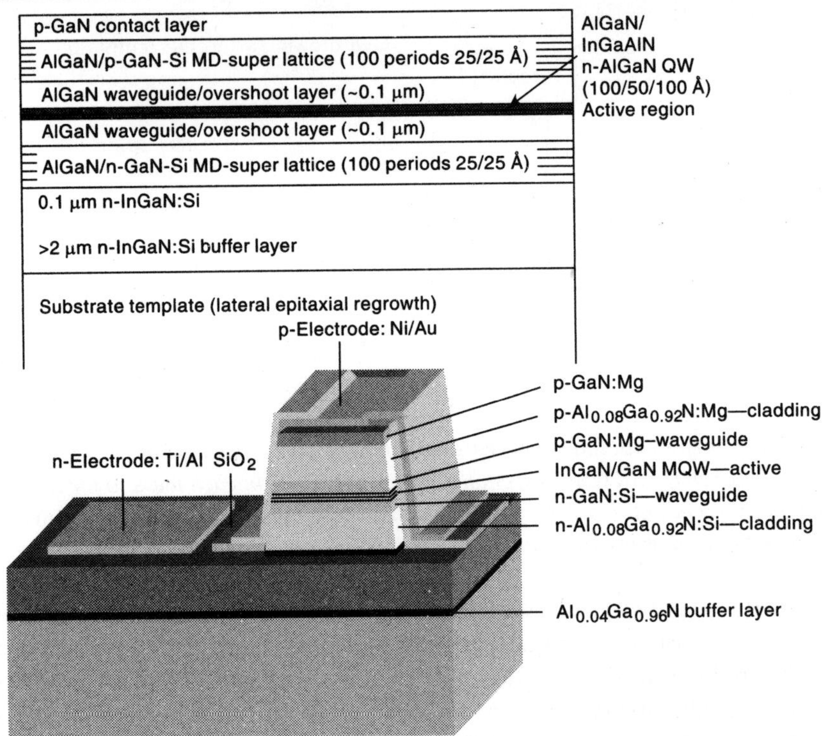

Fig. 7.30 *Schematic of layer structure for a SCH MQW violet laser AlGaInN heterostructure.*

In the heterostructure diode laser structural schematic of Fig. 7.30. (upper trace), a low temperature grown GaN nucleation layer (~250 Å, not shown), a pair of modulation doped superlattices on both n- and p-side of the junction, and the active MQW regions sandwiched between AlGaN waveguide layers are observed. The super lattice approach to modulation doping is useful for the *p*-doping in terms of lateral and vertical transport, and is shown as a 'dislocation'. The AlGaN waveguide layers have additional thin ALGaN layers (not shown) for blocking the electron/hole overshoot of the active QW region, strategically positioned near the gain medium. The blocking layers prevent the InGaN decomposition. Whereas In concentrations for LEDs ranges from about x ~ 100–0.4, the range for InGaN diode lasers at practical current threshold is (x ~ 0.10–0.15), with typical wavelength emission near $\lambda = 410$ nm. The threshold current density increases rapidly with the indium concentration, by more than a factor of 5 over the range from about 410 to 450 nm.

For blue and UV diode lasers (lifetime > 10^4 h), the high dislocation density for example, in InGaN diode lasers is very high. This is due to the high threshold current density in the edge-emitting devices. The ELOG technique is increasingly becoming a standard building block for laser design/ fabrication. It is patterned GaN growth is implemented by employing a few micrometer apertured SiO_2 mask on the GaN buffer layer. The epitaxial growth nucleates on the exposed GaN within the stripes of the SiO_2 template and, under a sufficiently high temperature (T> 1323 K) and constituent flow in the MOCVD reactor, acquires a large lateral growth rate, once the aperture area is filled with GaN. By the standards of the nitrides, there are very few (<10^5 cm^{-2}) dislocations in the overgrown areas above the SiO_2 mask, specifically "wing area". These tend to be of the edge dislocation type, parallel to the (001) plane after a 90° bend in the regrown region. After a thick (say > 20 μm GaN) layer is overgrown, this "substrate" is ready for the subsequent deposition of the device heterostructure. With SiO_2 the device lifetime increases by more than 1000 hours under high power (~50 mW) CW operation. Thus ELOG techniques, while being very desirable for the advancement of the violet lasers, are likely to be essential for nitride-based short wavelength photodiode detectors. This follows from the necessity of reducing the reverse bias leakage current, dark current, which is otherwise greatly increased by contributions from the dislocations as well as related extended state and points defects.

The relationship between the dislocation density (measured as etch density) and PL efficiency of the approximately 5-μm thick GaN substrate is shown in Fig. 7.31, whereas Fig. 7.32 shows the spatially resolved PL emission, indicating that the "best quality" material is found under the wing region. Only the mixed dislocation type is found in this region, at concentrations less than 10^5 cm^{-2}. To prepare the optical resonator required for the complete diode laser structure, the sapphire substrate is thinned to about 100 μm and the laser diode facets are subsequently fabricated by cleaving or by dry etching.

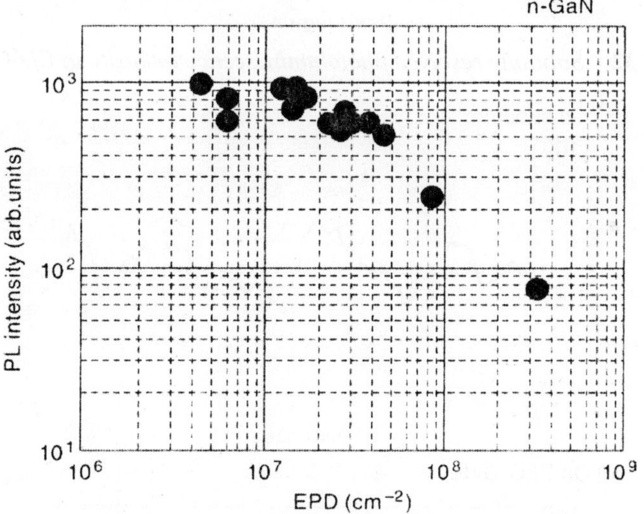

Fig. 7.31 *Correlation between photoluminescence intensity and etch pit density (dislocation density) in GaN substrate, demonstrating the impact of ELOG techniques.*

The surface flatness and morphology are superior for cleaved facets that coincided with the wing region of the ELOG substrate. Figure 7.33 shows the results of AFM profiling where the smoothness of the laser facet reaches values better than 1 nm, in contrast with roughness measured on cleaved laser devices grown directly on sapphire (without the ELOG process) beyond 10 nm. Lifetime tests under 30 mW CW output power indicate good performance beyond 1000 h.

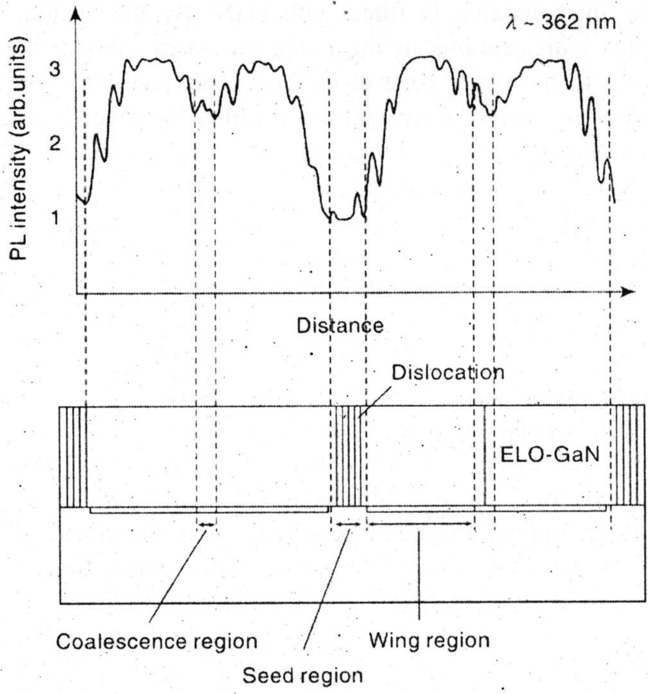

Fig. 7.32 *Spatially resolved photoluminescence intensity in ELOG-GaN*

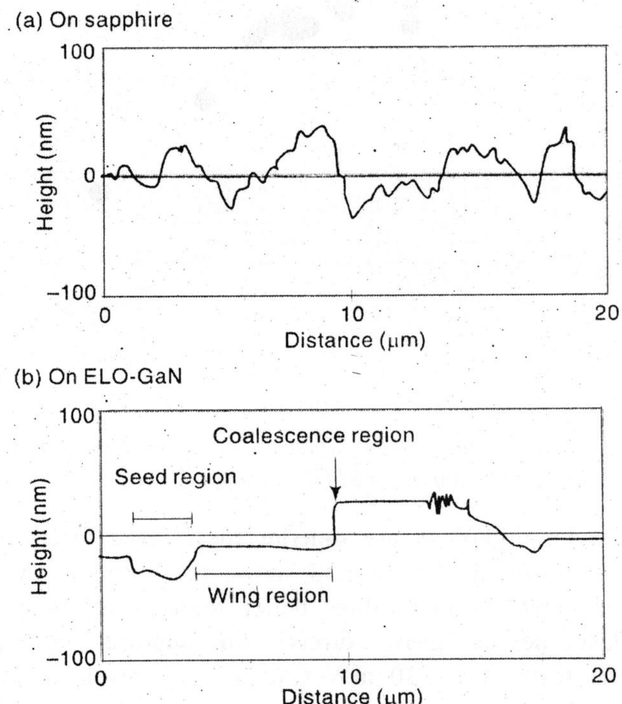

Fig. 7.33 *Effect of GaN ELOG substrate/template on facet roughness of cleaved diode laser*

Figure 7.34 shows comparison of three different substrate/template cases where the ELOG idea is implemented at levels of successively increasing complexity and physical thickness. The first case corresponds to the usual ELOG. The second involves the use of hydride vapor phase epitaxy (HVPE) to synthesize a very thick (200 μm) GaN atop the ELOG undercarriage; thus most of the substrate and HVPE-grown thick GaN are removed by polishing leaving an about 150 μm thick free-standing GaN substrate. The HVPE layer is assumed to direct residual dislocations towards the edge of the wafer. In the third case, the substrate/template growth continues atop the HVPE GaN with another ELOG step, which is then ready for the hetero-epitaxy of the active device. The backside of the device is polished to remove the substrate. The dislocations are reduced with the increased complexity of the ELOG-based substrate/templates, specifically in terms of their density in GaN above the SiO_2 mask window region. With the laser devices grown on the three substrate/templates, the lifetimes at 30mW average power are 700, 300, and 15000h, respectively, at a temperature of 333 K.

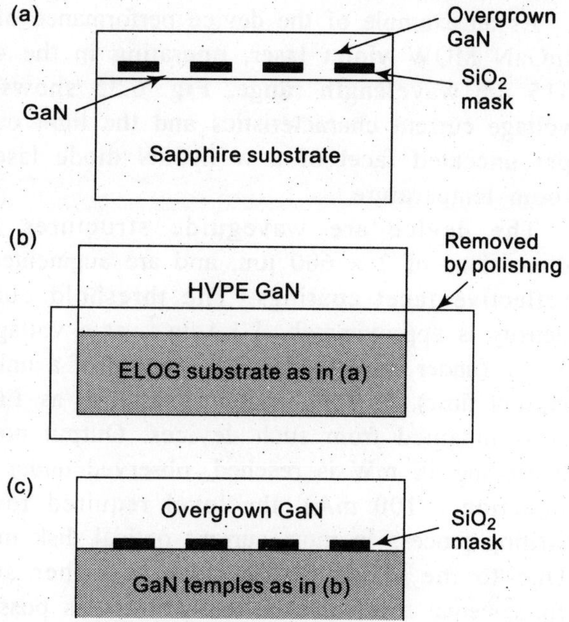

Fig. 7.34 *Schematic of ELOG substrate/templates for InGaN MQW diode layer, at varying levels of complexity.*

Thus dislocations are reduced to achieve long-lived violet diode lasers. The ELOG intensive approach leaves a residual dislocation density which is unrealistic for a viable conventional III-V laser.

Edge Emitter Diode Laser Performance

The SCH/MQW InGaN/AlGaN heterostructure material is fabricated into index-guided mesa structures by employing dry-etching techniques (Fig. 7.30). The electron cyclotron resonance plasma (ECR), chemically assisted ion beam etching (CAIBE), or standard reactive-ion etching, produce etch for the vertical walls as well the resonator end facets. The etched n-GaN surface onto which metallization is applied for the n-type contact has an acceptable electronic quality. In case of the HVPE grown thick buffer layers, or the SiC substrate, the optical resonator is formed by cleaving of free standing ~100 μm films.

There is variation in cavity losses in the InGaN MQW diode lasers, due to the uneven pace of device development. This loss is from threshold current density variation as a function of resonator length and is 20–50 cm^{-1} (for the optical losses). Some of this is due to the quality of the end mirror facets. for the devices grown on sapphire, where dry etching is required for mirror fabrication.

As an example of the device performance of the InGaN MQW violet laser, operating in the 405–415 nm wavelength range, Fig. 6.35 shows the voltage current characteristics and the light output per uncoated facet of a w InGaN diode laser at room temperature.

The device are waveguide structures with dimensions of 2 × 600 μm, and are augmented by reflective facet coatings. The threshold current density is approximately 4 $kAcm^{-2}$, at a voltage of 5.5 V (under the uncertain assumption of a uniform current flow). A slope efficiency as high as 1.0 W/A is obtained from such devices. Output powers exceeding 50 mW is reached, observed under high injection (>100 mA), the level required for the writing process in most current optical disk media. Due to the small ridge width, a rather stable fundamental transverse mode operation is possible, however, in order to avoid the dependence of the far field pattern on the injection level (so-called "kinks") the ridge waveguide design requires high accuracy in the index of refraction data as well as high precision in the ridge etch depth.

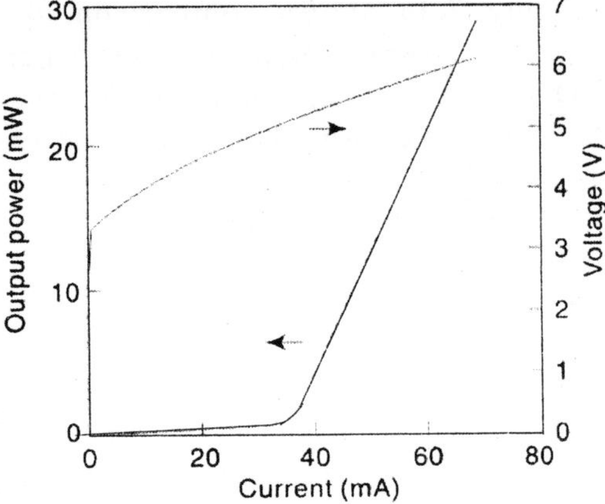

Fig. 7.35 *I-V characteristics and the optical power output per uncoated facet of a CW InGaN diode laser at room temperature.*

These constraints are stabilized by using a spin-on-glass (SOG) insulator in place of SiO_2 as the dielectric insulator at the mesa edges (Fig. 7.30). The light absorptive characteristics of this process material constrain and stabilize the lowest transverse mode. Figure 7.36 shows the *L–I* characteristics and the "kink-free" far-field pattern, the latter showing FWHM angles parallel and perpendicular to the junction plane of 9.1 and 22.4°, respectively.

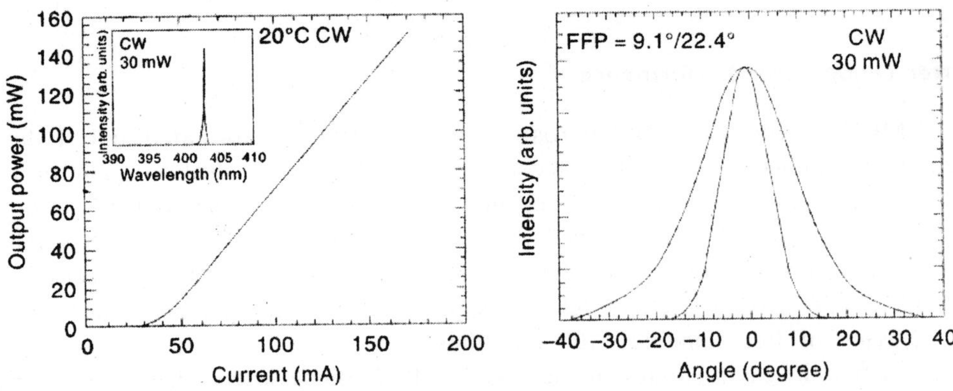

Fig. 7.36 *L–I characteristics of a diode laser using SOG current blocking layer (left panel); far-field emission pattern from the transverse-mode stabilized device at 30 mW output power.*

An example of the emission spectrum below and above lasing threshold is shown in Fig. 7.37 for an InGaN MQW diode laser grown on an SiC substrate and displaying a single longitudinal mode.

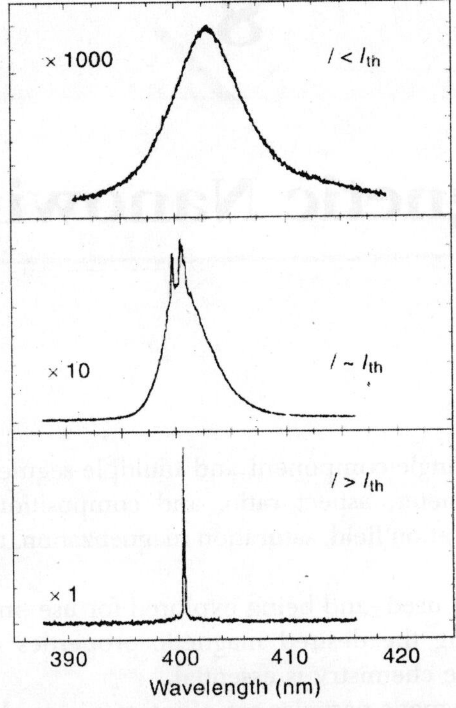

Fig. 7.37 *Emission spectrum below and above lasing threshold for an InGaN MQW diode laser grown on SiC substrate*

InGaN MQW diode laser efficiency has improved by employing the ELOG-based substrate/template. Threshold current densities are reduced to the range of 2–3 $kAcm^{-2}$ and the slope efficiency increased up to about 1.5 W/A. The threshold voltages, have dropped to the 4–5 V range.

Magnetic Nanowires

The magnetic property of single-component and multiple-segment magnetic nanowires and the influence of particle diameter, aspect ratio, and composition on many their magnetic properties: the orientation saturation field, saturation magnetization, and remanent magnetization are given below:

Magnetic particles are being used–and being explored for use–in fields as diverse as biology and data storage. To obtaining the desired magnetic properties controlling of particle size, shape, composition and surface chemistry is essential.

Suspensions of superparamegnetic particles are of interest in application such as ferrofluidics. In biomagnetics, supermagnetic particles are used commercially for cell sorting and are being explored for radiation treatment. Magnetic particles are used in magnetic tweezers where the force required to displace magnetic particles bound to cells or proteins is used to probe the micromechanics of cells and the torsion of DNA molecules. Such particles are also being explored for use in drug delivery and gene therapy. The use of magnetic field gradients to transport magnetic particles is termed magnetofection.

In contrast to spherical particles, nanorods or nanowires exhibit degrees of freedom associated with their inherent shape anisotropy and their length, is shown in Figure 8.1. For example, ferromagnetic nanowires exhibit unique and tunable magnetic properties that are different from those of bulk ferromagnetic materials, thin films, and spherical particles. The introduction of multiple segments along the length of a nanowire leads to further degrees of freedom associated with the shape of each segment and the coupling between the layers.

The diverse range of applications has resulted in interest in nanoparticles with a wide range of magnetic properties. For example, the magnetic beads used in cell sorting are superparamagnetic in order to avoid aggregation; however, it is desirable that they have a relatively high magnetic moment so that they can be separated from the suspension at relatively low fields. In contrast, magnetic particles for magnetic recording should be ferromagnetic, with high saturation magnetization and high coercivity.

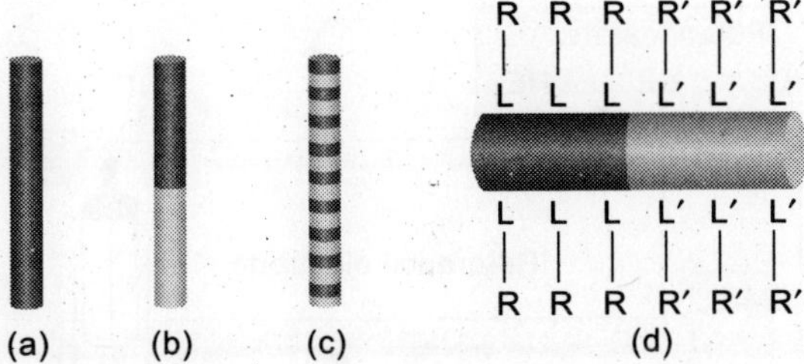

Fig. 8.1 *The inherent shape anisotropuy associated with nanowires: (A) single-component nanowire; (b) two-component nanowire with segment aspect ratio > 1; (c) two-component multilayer nanowires with segment aspect ratio < 1; (d) functionalization of a two-component nanowires for which L and L′ represent ligands that bind selectivity to the two components, and R and R′, represent spatially separated functional groups.*

Both single-component and multiple-segment magnetic nanowires are of interest in these emerging fields, since magnetic properties such as the orientation of the magnetic easy axis, Curie temperature, coercivity, saturation field, saturation magnetization, and remanent magnetization can be tailored by modifying the diameter, composition, and layer thicknesses in multiple-segment FM/NM (FM = ferromagnetic, NM = nanomagnetic material) nanowires.

SYNTHESIS OF NANOWIRES

The deposition of structures, such as nanowires, with a high aspect ratio is difficult to achieve by vapor-phase techniques but is relatively straightforward using electrochemical template synthesis. Nanowires with diameters from 5 nm to more than 1 μm and with lengths up to 100 μm are produced by electrochemical deposition into a nonconducting membrane having an array of parallel-sided pores. The electrochemical deposition of nanowire arrays involves the use of a metal film (sputtered into one side of the membrane template) that serves as the working electrode (WE) in a three-electrode electrodeposition cell, as shown in Figure 8.2. The nanowires form as the pores in the template are filled by the deposition material. This is used to fabricate nanowires of metals, alloys, semiconductors, and electronically conducting polymers and also to synthesize multiple-segment nanowires, as illustrated in Figure 8.1. Another advantage of electrochemical template synthesis is that large quantities of monodisperse nanowires can be produced.

Templates for electrochemical deposition of nanowire arrays include porous anodized aluminum films, etched nuclear particle tracks in various materials and block copolymer films. The properties of the nanoporous template, such as the relative pore orientations in the assembly, the pore size distribution, and the surface roughness of the pores, have significant influence on the intrinsic properties of the nanowires.

Nanoporous alumina templates are formed by anodic oxidation of aluminum. Pore sizes typically range from 5 to 500 nm and up to about 200 μm in thickness, permitting the fabrication of nanowire with aspect ratios in excess of 1,000. Pore densities range from 10^8 to 10^{10} cm^{-2}. Large domains of hexagonally ordered pores are achieved. Longer-range ordering is obtained by using stamping.

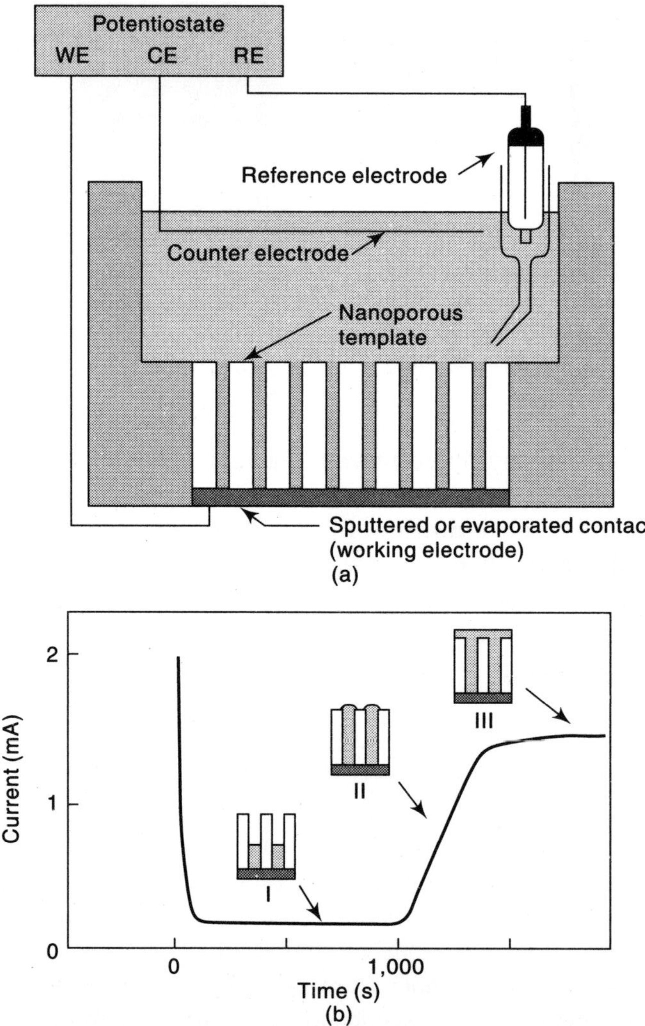

Fig. 8.2 *(a) A cell for electrochemical deposition of nanowires. (b) deposition current-time transient for deposition of 60-nm-diameter Ni nanowires in a 6-μm-thick polycarbonate template (pore density* $= 10^8\ cm^{-2}$*) from nickel sulfamate solution having a pH of 3.4. (WE: working electrode; CE: counter electrode; RE: reference electrode.)*

By nuclear track etching the choice of material, pore size, and template thickness is made. Damage tracks are created in polymeric (e.g., polycarbonate) or inorganic (e.g., mica) insulting films by the passage of high-energy particles. With a suitable etchant, the etch rate of the damage tracks is several times higher than that of the surrounding bulk material, so that uniform, parallel-sided pores are created. Further etching results in enlargement of the pores at a rate determined by the bulk etch rate. Pore diameters are typically 10–1,000 nm, and pore densities of $1–10^9\ cm^{-2}$ are achieved. The pore distribution is random. Figure 8.3 shows a plan view scanning electron microscope (SEM) image of etched particle tracks in single-crystal mica.

Block copolymer templates have pore sizes ranging from 10 to 100 nm and pore densities ranging from 10^8 to 10^9 cm^{-2}. The template thickness is limited. It can be increased by annealing in the presence of an electrical field, resulting in aspect ratios in excess of 100.

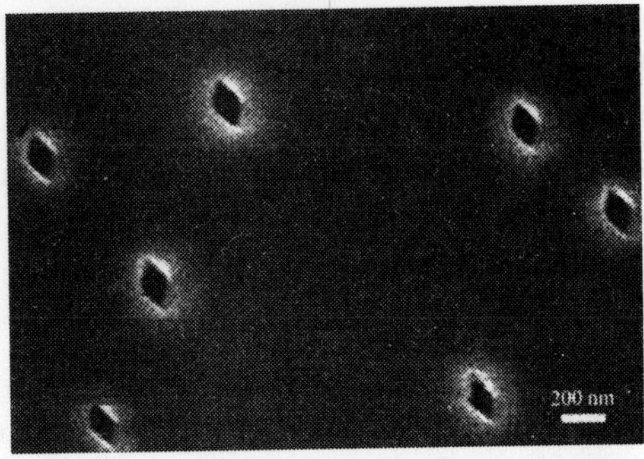

Fig. 8.3 *Plan view SEM image of etched particle tracks formed in single-crystal mica.*

Ferromagnetic properties of nanowires

1. Hysteresis Loops

Magnetization hysteresis loops, which display the magnetic response of a magnetic sample to an external field, are used to characterize the behavior of nanostructured magnetic materials. Figure 8.4 shows typical magnetization hysteresis loops with the applied

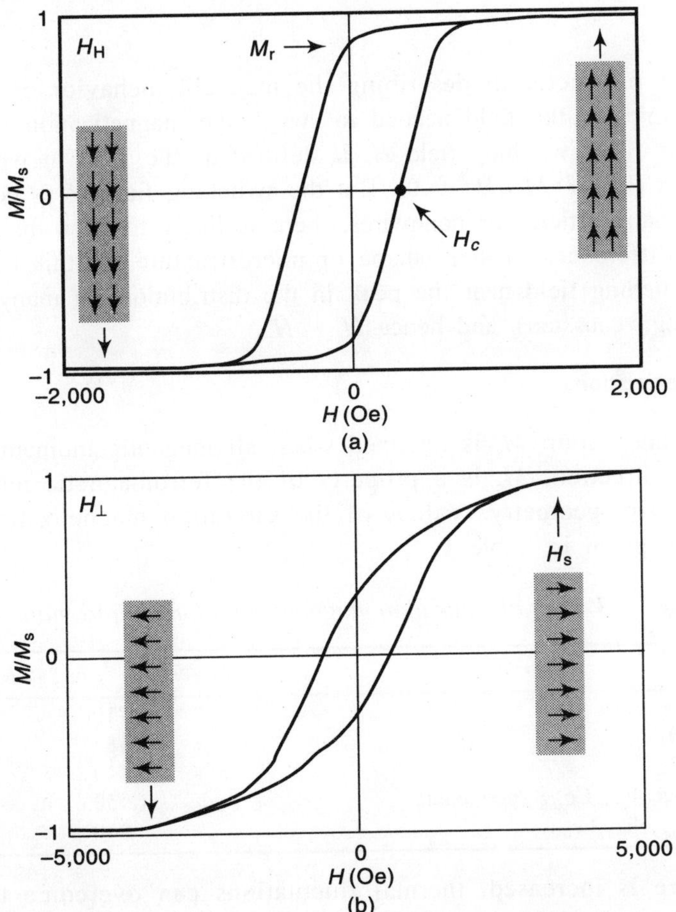

Fig. 8.4 *Typical hysteresis loops for an array of nickel nanowires 100 nm in diameter and 1 mm long with the applied field H parallel (a) and perpendicular (b) to the wire axis.*

magnetic field parallel and perpendicular to the nanowire axis for an array of single component ferromagnetic nanowires. The characteristic features of the hysteresis loop are dependent on the material, the size and shape of the entity, the microstructure the orientation of the applied magnetic field with respect to the sample, and the magnetization history of the sample. For arrays of nanoparticles, the hysteresis loop depends on interactions between individual particles.

2. Magnetization

The common parameters used to describe the magnetic properties are the saturation M_s, the remanent magnetization M_t, the coercivity H_c, and the saturation field H_{sat}.

The coercivity H_c is the applied field at which the magnetization M becomes zero; the saturation field H_{sat} is the field needed to reach the saturation magnetization; and the remanent magnetization M_r is the magnetization at $H = 0$. For the case of magnetic nanowires, H_c, H_{sat}, and M_r are strongly dependent on the size and shape of the sample and the orientation of the applied magnetic field.

3. Switching Field

Another important parameter in describing the magnetic behavior of nanoparticles is the switching field H_s, which is the field needed to switch the magnetization from one direction to the opposite direction. The switching field H_s is defined as the field in which the slope of the M–H lop is maximum (i.e., $d^2M/dH^2 = 0$). But the switching field H_s is equal to the coercivity H_c. For an array of nanoparticles or nanowires, there is likely to be a distribution of switching fields due to small differences in size, shape or microstructure. In this case, H_s characterizes nanowires with a switching field near the peak in the distribution. In many cases the switching field is symmetric (e.g., gaussian), and hence $H_s = H_c$.

4. Saturation Magnetizations

The saturation magnetization M_s is obtained when all magnetic moment in the material are aligned in the same direction. M_s is a property of the ferromagnetic material and hence is independent of nanowire geometry. Values of the saturation magnetization in emu cm^{-3} for different materials are given in Table 1.

Table 1 *Values of saturation magnetization for various materials.*

Material	M_s *(emucm^{-3}) at 290 K*
Fe	1,710
Ni	485
Co	1,440
bcc $Fe_{0.5}Co_{0.5}$ (permedur)	1,950
f_{cc} $Ni_{0.78}Fe_{0.22}$	800

As the temperature is increased, thermal fluctuations can overcome the ordering of the magnetic moments in a ferromagnetic material. The transition between ferromagnetic and paramagnetic behavior occurs at the Curie temperature T_c, which is a property of the material and is dependent on sample dimensions. For magnetic nanowires, all of these characteristic magnetic properties can be controlled by choice of material and dimensions.

In principle, magnetization hysteresis loops for an arbitrary entity can be calculated by minimizing the total free energy in the presence of an external field. The state of magnetization is determined from the eigenvalue of the magnetization vector configuration that minimizes the total system energy. The total energy E, can be expressed as

$$E = E_{ex} + E_H + E_{EA} + E_{ca} + E_D. \tag{1}$$

where E_{ex} is the exchange energy, E_H is the Zeeman energy, E_{EA} is the magnetoelastic energy, E_{ca} is the crystalline anisotropy energy, and E_D is the magnetostatic energy (demagnetization energy).

The exchange energy E_{ex} is related to the interaction between splits that introduces magnetic ordering and is written as

$$E_{ex} = -2\sum_{i>j} J_{ij} S^2 \cos\theta_{ij} \tag{2}$$

where J is the exchange integral, S is the spin associated with each atom, and θ_{ij} is the angle between adjacent spin orientations. For ferromagnetic materials, J is positive, and the minimum-energy state occurs when the spins are parallel to each other ($\theta_{ij} = 0$). Exchange interactions are inherently short-range and are described in terms of the exchange stiffness constant A. The exchange stiffness is a measure of the force acting to keep the spins aligned and is given by $A = JS^2\, c/a$, where J is the exchange integral, S is the magnitude of the individual spins, c is a geometric factor associated with the crystal structure of the material ($c = 1$ for a simple cubic structure, $c = 2$ for a body-centered cubic structure, $c = 4$ for a face-centered cubic structure, and $c = 2\sqrt{2}$ for a hexagonal close-packed structure), and a is the lattice parameter. For most ferromagnetic materials, A is between 1×10^{-11} and 2×10^{-11} Jm^{-1}.

The Zeeman energy E_H is often referred to as the magnetic potential energy, which is simply the energy of the magnetization in an externally applied magnetic field. The Zeeman energy is always minimized when the magnetization is aligned with the applied field; it can be expressed as

$$E_H = -\mathrm{M} \cdot \mathrm{H}, \tag{3}$$

where H is the external field vector and M is the magnetization vector.

The magnetoelastic energy E_{EA} is used to described the magnetostriction effect, which related the influence of stress, of strain, on the magnetization of a material. For particles in suspension this effect can be neglected; however, in other cases it may be important.

The last two terms in equation (1), the crystalline anisotropy energy E_{ca} and the magnetostatic energy (or demagnetization energy) E_D, describe the dependence of the total free energy on the direction of the magnetization with respect to lattice orientation and the shape of the sample, respectively. These two terms give rise to the size-dependent magnetic properties in nanostructures; they are discussed in more detail below.

MAGNETOCRYSTALLINE ANISOTROPY

In a magnetic material, the electron spin is coupled to the electronic orbital (spin-orbital coupling) and influenced by the local environment (crystalline electric field). Because of the arrangement of atoms in crystalline materials, magnetization along certain orientations is energetically preferred. The magnetocrystalline anisotropy is closely related to the structure and symmetry of the material.

For a cubic crystal, the anisotropy energy is often expressed as

$$E_{ca} = K_0 + K_1(\cos^2\theta_1 \cos^2\theta_2 + \cos^2\theta_2 \cos^2\theta_3 + \cos^2\theta_3 \cos^2\theta_1) + K_2 \cos^2\theta_1 \cos^2\theta_2 \cos^2\theta_3 + \cdots, \quad (4)$$

where K_0, K_1, K_2, ... are constants (ergcm^{-3}) and θ_1, θ_2, and θ_3 are the angles between the magnetization direction and the three crystal axes, respectively. K_0 is independent of angle and can be ignored since it is the difference in energy between different crystal orientations that is of interest. In many cases, terms involving K_2 are small and can also be neglected. If $K_1 > 0$ E_{ca} is minimum in the $\langle 100 \rangle$ directions; hence these directions are the easy axes. Conversely, if $K_1 < 0$, the axes correspond to the $\langle 111 \rangle$ directions. The difference in magnetocrystalline energy between the [111] direction and the [100] direction, $\Delta K_{[111]-[100]}$ is equal to $K_1/3$. Similarly, the difference between the [110] direction and the [100] direction $\Delta K_{[110]-[100]}$ is equal to $K_1/4$.

The magnetocrystalline anisotropy energy associated with a hexagonal close-packed crystal is often expressed as

$$E_{ca} = K_0 + K_1 \sin^2\theta + K_2 \sin^4\theta \quad (5)$$

where K_0, K_1, and K_2 are constants (ergcm^{-3}) and θ is the angle between the magnetization direction and the c-axis. As described above, K_0 is independent of angle. For most cases in which K_2 can be neglected, if $K_1 > 0$, the energy is smallest when $\theta = 0$ i.e., along the c-axis, so that this axis is the easy axis. If $K_1 < 0$, the basal plane is the easy axis. As a result of the symmetry of the hexagonal close-packed lattice, the magnetocrystalline anisotropy is a uniaxial anisotropy. Examples of K_1 for various materials are given Table 2.

Table 2 *Values of the magnetocrystalline anisotropy energy constant K_1 for various materials.*

Material	*K_1 (ergcm^{-3})*
fct $Co_{0.5}Pt_{0.5}$	5×10^7
fcc Ni (easy axis [111])	-4.5×10^4
bcc Fe (easy axis [100])	4.8×10^5
fcc $Ni_{0.78}Fe_{0.22}$	0
hcp Co (easy axis [0001])	4.5×10^6

Shape Anisotropy

The magnetization of a spherical object in an applied magnetic field is independent of the orientation of the applied field. However, it is easier to magnetize a nonspherical object along its long axis than along its short axis. If a rod-shaped object is magnetized with a north pole at one end and a south pole at the other, the field lines emanate from the north pole and end at the south pole. Inside the material the field lines are oriented from the north pole to the south pole and hence are opposed to the magnetization of the material, since the magnetic moment points from the south pole to the north pole. Thus, the magnetic field inside the material tends to demagnetize the material and is known as the demagnetizing field H_d. Stated formally, the demagnetizing field acts in the opposite direction from the magnetization M which creates it, and is proportional to it, namely

$$H_d = -N_d M, \tag{6}$$

Where N_d is the demagnetizing factor. N_d is dependent on the shape of the body, but an only be calculated exactly for an ellipsoid where the magnetization is uniform throughout the sample. The magnetostatic energy E_D (ergcm^{-3}) associated with a particular magnetization direction can be expressed as

$$E_d = \frac{1}{2} N_d M_s^2 \tag{7}$$

where M_s is the saturation magnetization of the material (emucm^{-3}), and N_d is the demagnetization factor along the magnetization direction.

For a general ellipsoid with $c \geq b \geq a$, where a, b, and c are the ellipsoid semi-axes, the demagnetization factors along the semi-axes are N_a, N_b, N_c, respectively. The demagnetization factors are related by the expression $N_a + N_b + N_c = 4\pi$.

Figure 8.5 shows the ellipsoids of particular interest in the study of nanowires: the prolate spheroid (or ellipsoid of revolution), slender ellipsoid, and oblate spheroid:

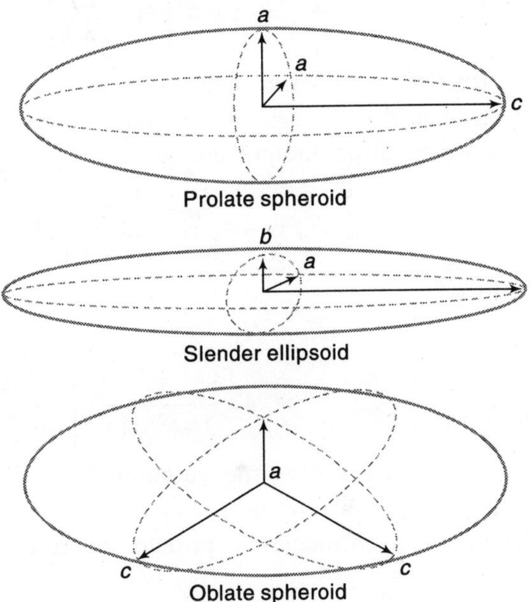

Fig. 8.5 *Illustration of ellipsoids for the study of nanowires.*

(a) *Prolate spheroid* (where $c > a = b$). This ellipsoid is of interest as an approximation for a single-component nanowire with circular cross section. When the aspect ratio is defined as $m = c/a$, its demagnetization factors are given by

$$N_a = N_b$$

$$= 4\pi \frac{m}{2(m^2-1)} \times \left[m - \frac{1}{2(m^2-1)^{1/2}} \times \ln\left(\frac{m+(m^2-1)^{1/2}}{m-(m^2-1)^{1/2}} \right) \right] \tag{8}$$

and

$$N_c = 4\pi \frac{1}{m^2-1} \times \left[\frac{m}{2(m^2-1)^{1/2}} \times \ln\left(\frac{m+(m^2-1)^{1/2}}{m-(m^2-1)^{1/2}} \right) - 1 \right] \tag{9}$$

(b) *Slender ellipsoid* (where $c >> a > b$). This ellipsoid is of interest as an approximation for nanowires deposited into templates with noncircular cross sections, such as nanowires deposited in pores in single-crystal mica that have a diamond-shaped cross section. Its demagnetization factors are given by

$$N_a = 4\pi \frac{b}{a+b} - \frac{1}{2}\frac{ab}{c^2} \ln\left(\frac{4c}{a+b}\right) + \frac{ab(3a+b)}{4c^2(a+b)} \tag{10}$$

$$N_b = 4\pi \frac{b}{a+b} - \frac{1}{2}\frac{ab}{c^2} \ln\left(\frac{4c}{a+b}\right) + \frac{ab(a+3b)}{4c^2(a+b)} \tag{11}$$

and

$$N_c = 4\pi \frac{ab}{c^2}\left[\ln\left(\frac{4c}{a+b}\right) - 1\right] \tag{12}$$

(c) *Oblate spheroid* (where $c = b > a$). The ellipsoid is of interest as an approximation or disk-shaped magnetic segments in multiple-segment nanowires. Defining the aspect ratio $m = c/a$, its demagnetization factors are given by

$$N_a = 4\pi \frac{m^2}{m^2-1}\left[1 - \frac{1}{(m^2-1)^{1/2}} \times \arcsin \frac{(m^2-1)^{1/2}}{m}\right] \tag{13}$$

and

$$N_b = N_c$$

$$= 4\pi \frac{1}{2(m^2-1)} \times \left[\frac{m^2}{(m^2-1)^{1/2}} \times \arcsin \frac{(m^2-1)^{1/2}}{m} - 1\right] \tag{14}$$

Figure 8.6 shows the calculated dependence of the demagnetization factors ($N_d/4\pi$) of a prolate spheroid on its aspect ratio (c/a). A nanowire with a high aspect ratio (infinitely long cylinder) is considered a prolate spheroid with a high aspect ratio (large c/a). In this case, the

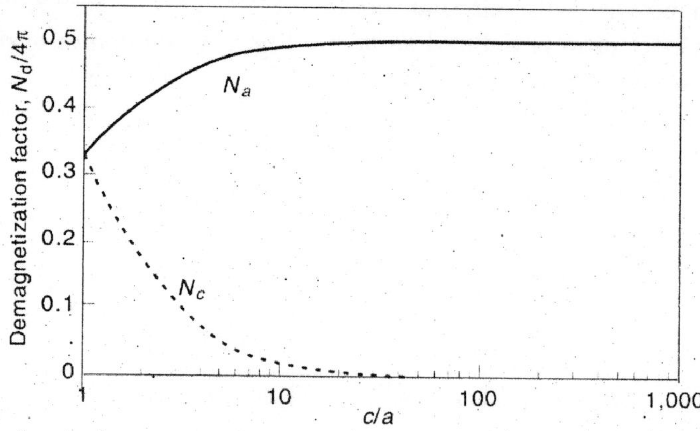

Fig. 8.6 *Calculated dependence of the demagnetization factor ($N_d/4p$) of a prolate spheroid on its aspect ratio c/a. For c/a > 10, $N_d/4p$ = 0.5 or $N_a \approx 2p$.*

demagnetization factor along the hard axis, perpendicular to the wire axis, is equal tot 2π, and the demagnetization factor along the easy axis, parallel to the wire axis, is 0. Thus, the shape anisotropy energy difference along the two axes obtained from equation (7) is $K_u = \Delta E_D = E_{Da} - E_{Dc} = \pi M_s^2$. Note that the infinitely long cylinder approximation is used for nanowires having an aspect ratio greater than about 10.

A disk-shaped ferromagnetic segment in a multiple-segment nanowire can be considered to be an oblate spheroid with a very low aspect ratio (small c/a). In this case, the demagnetization factor N_a along the hard axis, parallel to the wire axis, is 4π, and the demagnetization factor N_c along the easy axis, perpendicular to the wire axis, is 0. Thus, the shape anisotropy energy difference along the two axes is $K_u\ n = \Delta E_D = E_{Da} - E_{Dc} = -2\pi M_s^2$.

Saturation Field

The saturation field H_{sat} is the external field needed to overcome the anisotropy energy and align the magnetic moments along the field direction. If we consider only shape anisotropy, from Equation (6) it can be seen that for a nanowire with a high aspect ratio ($N_a = 2\pi$ and $N_c = 0$), the saturation field parallel to the wire direction is 0, whereas the saturation field perpendicular to the wire axis is $2\pi M_s$. Referring to the *M–H* loops of Figure 8.4 (for 100-nm-diameter nickel nanowires with the applied field parallel and perpendicular to the wire axis), with the field parallel to the wire axis, the loop is relatively square, indicating that the magnetizations are easily aligned in this direction. The slight shearing of the curve is due to the fact that the pores in the polyarbonate templates are not all aligned perpendicular to the film plane. With the field perpendicular to the wire axis, the loop is sheared significantly. The field required to align the magnetizations perpendicular to the wire axis is about 3500 Oe, close to the value of $2\pi M_s = 3050$ Oe.

Switching in Single-domain Particles

In principle, the magnetization configuration in a magnetic nanowire can be determined from the Brown equation by minimizing the total free energy. To simplify the problem, it is common to ignore the magnetoelastic energy (E_{EA}) and the crystalline anisotropy energy (E_{ca}). The magnetoelectric energy is usually very small for nanostructures. The crystalline anisotropy energy is important for single crystals or highly textured structures but is usually much smaller than the shape anisotropy. Other effects related to the microstructure and surface effects are also ignored. The remaining terms are analyzed by considering the collective spin motions. In bulk ferromagnetic materials, the energy of the system can be minimized by forming multiple magnetic domains in which the atomic magnetic moments are aligned. However, there is a critical size below which a particle remains in a single-domain state during switching.

For a prolate spheroid, the critical radius r_{sd} (in the short axis) for a single-domain particle is expressed as

$$r_{sd} = \sqrt{\frac{6A}{N_c M_s^2}\left[\ln\left(\frac{2r_{sd}}{a_1} - 1\right)\right]} \tag{15}$$

where A is the stiffness constant (ergcm^{-1}), N_c is the demagnetization factor, M_s is the saturation magnetization (emu/cm3), and a_1 is the near-neighbor spacing (cm). The critical radius for a single-domain particle is dependent on the materials parameters A, a_1, and M_s as well as demagnetization factor N. Note that we use the near-neighbor spacing a_1 and that A contains the influence of the crystal structure. From Figure 8.7 it is seen that the critical radius increases with increasing aspect ratio ($m = c/a$) because of the decrease in the demagnetizing factor N_c. Thus, nickel nanowires with an aspect ratio of 10 ($N_c = 0.255$) is expected to be single-domain nanowires for diameters less than about 600 nm ($r_{sd} \approx 300$ nm), whereas cobalt and iron nanowires with the same aspect ratio are expected to be single-domain for diameters less than about 140 nm ($r_{sd} \approx 70$ nm).

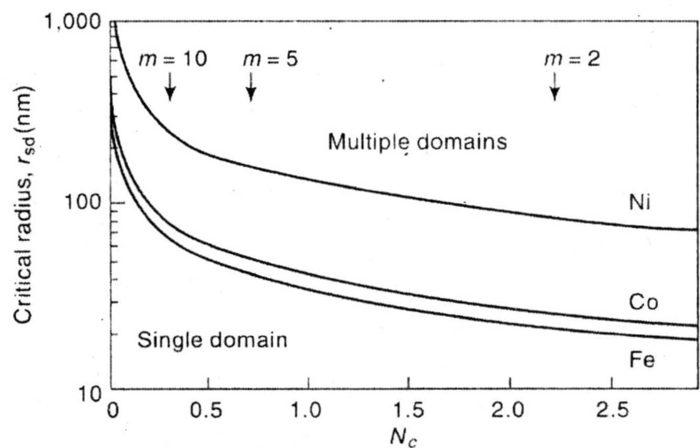

Fig. 8.7 *Critical radius for a single-domain prolate spheroid vs. demagnetization factor along the c-axis N_c from Equations (9) and (15). $M_s(Ni) = 485$, $M_s(Co) = 1{,}440$, $M_s(Fe) = 1{,}710$, $a_l(Ni) = 0.2942$ nm, $a_l(Fe) = 0.2482$ nm, and $a_l(Co) = 0.2507$ nm. In all cases $A = 1 \times 10^{-6}$ erg cm^{-1}.*

For single-domain particles, the most common magnetization reversal modes are modeled by *coherent rotation or curling.* In the absence of an applied field, the magnetic moments align along the easy axis, which is the longest major axis in an ellipsoid. In the coherent rotation model, all magnetic moments remain parallel to one another and rotate away from the easy axis during the reversal process, thus minimizing the exchange energy in the system. However, when there is a magnetization component along the hard axis, the demagnetization energy increases. In the curling model, neighbouring magnetic moments are not constrained to be parallel to permitting the formation of configuration with no net magnetization along the hard axis and hence minimizing the demagnetization energy. However, if the magnetic moments are not parallel to one another, the exchange energy increases. Coherent rotation and curling are illustrated in Figure 8.8.

The switching mode is determined by the competition between the exchange energy and the demagnetization energy. In the curling model, the exchange energy density increases with decreasing size because of the increase in relative angle between neighbouring moments; hence, coherent rotation is favorable. But, the demagnetization energy density increases with increasing aspect ratio, favoring curling. Thus, there is a critical size between the two magnetization reversal processes. When the magnetic easy axis is aligned with the applied field, the critical radius r_c (cm) for the transition is defined by

$$r_c = q\left(\frac{2}{N_a}\right)^{1/2} \frac{A^{1/2}}{M_s} \tag{16}$$

where q is the smallest solution of the Bessel functions and is related to the aspect ratio of the prolate spheroid, A is the exchange stiffness constant (ergcm^{-1}), Ms is the saturation magnetization (emucm^{-3}), and N_a is the demagnetizing factor along the minor axis. The value of q varies between the limits of 1.8412 for a cylinder with an infinite aspect ratio and 2.0816 for a sphere with an aspect ratio of 1. For an infinitely long cylinder, $N_a = 2\pi$; hence, $r_c = (q/\sqrt{\pi})\ (A^{1/2}/M_s)$.

There are three terms, with different physical meanings, commonly used to describe the magnetization reversal process: the nucleation field H_n, which describes the field at which a change in magnetization just starts in a saturated single-domain state; the switching field H_s, which is the field at which an abrupt change in magnetization occurs; and the coercivity H_c, which is the field at which the magnetization changes sign. The nucleation field is a theoretical concept. The switching mechanism is usually determined from the angular dependence of the coercivity. Analytical solutions for two reversal models with the external field applied at an angle θ_0 to the magnetic easy axis are given below.

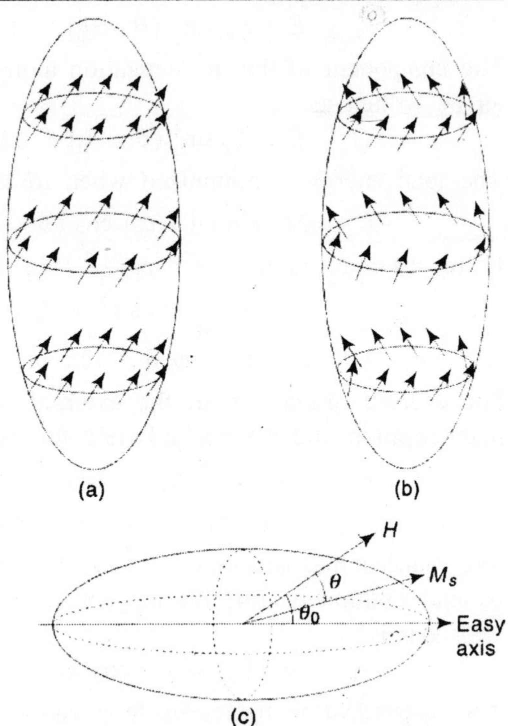

Fig. 8.8 *Illustrations of magnetization reversal in a single-domain prolate spheriod: (a) Coherent rotation; (b) curling; (c) coordinate system used for analysis of coherent rotation.*

(A) Coherent Rotation

In the coherent rotation, the total free energy consists of the magnetostatic energy (or demagnetization energy) and the Zeeman energy due to the external magnetic field, viz.,

$$E = E_D + E_H \tag{17}$$

The coordinate system used for the analysis of coherent rotation is shown in Figure 8.8. The magnetostatic energy for a prolate spheroid is given by the expression

$$E_D = K_u \sin^2(\theta - \theta_0) \tag{18}$$

where θ_0 is the angle between the external field and the magnetization easy axis, θ is the angle between the magnetization and the external field, and Ku is the uniaxial shape anisotropy constant. [Note that for a prolate spheroid with high aspect ratio ($N_a \approx 2\pi$ and $N_c \approx 0$), $K_u = \pi M_s^2$.] As, $E_H = -\text{H} \cdot \text{M}$ (from equation (3))

$\therefore$ $$E = K_u \sin^2 (\theta - \theta_0) - \mathrm{H} \cdot \mathrm{M_s}. \quad (19)$$

The component of the magnetization along the field axis is $M = M_s \cos \theta$; hence, Equation (19) an be written as

$$E = K_u \sin^2 (\theta - \theta_0) - HM_s \cos \theta. \quad (20)$$

The total energy is minimized when $dE/d\theta = 0$; hence,

$$2K_u \sin (\theta - \theta_0) \cos (\theta - \theta_0) + HM_s \sin \theta = 0. \quad (21)$$

Using the reduced field $h = H/(2K_u/M_s)$, Equation (21) is written as

$$\sin (\theta - \theta_0) \cos (\theta - \theta_0) = -h \sin \theta, \quad (22)$$

or $$\sin [2(\theta - \theta_0)] = -2h \sin \theta \quad (23)$$

For a given orientation of the external field θ_0, the equation defines the angle θ between the magnetization and the applied field for each value of h. equation (23) is written as:

$$\sin 2\theta \cos 2\theta_0 - \cos 2\theta \sin 2\theta_0 = -2h \sin \theta \quad (24)$$

or $$2 \cos \theta (1 - \cos^2 \theta)^{1/2} \cos 2\, \theta_0 + (1 - 2 \cos^2 \theta) \sin 2\theta_0 = -2h \sin \theta. \quad (25)$$

The reduced magnetization $m = M/M_s$, and since $M = M_s \cos \theta$, therefore, $m = \cos \theta$. Thus, the general solution relating the magnetization m top the applied field h at a specific angle θ_0 to the easy axis is

$$2m (1 - m^2)^{1/2} \cos 2\theta_0 + (1 - 2m^2) \sin 2\theta_0 = \pm 2h(1 - m^2)^{1/2}. \quad (26)$$

The magnetization hysteresis loop can be calculated from this equation by solving for m as a function of h. The switching field can be obtained when

$$\frac{\partial h}{\partial m} = 0 \quad \text{and} \quad \frac{\partial^2 E}{\partial \theta^2} > 0 \quad (27)$$

and is written as

$$h_s = (\cos^{2/3} \theta_0 + \sin^{2/3} \theta_0)^{-3/2}, \quad (28)$$

where $h_s = H_s/(2K_u/M_s)$.

The switching angle, i.e., the angle at which the magnetization flips, is obtained from

$$\tan^3(\theta_0 - \theta_s) = -\tan \theta_0. \quad (29)$$

Figure 8.9 shows m vs. h curves for the coherent rotation model. For a given value of θ_0, the m–h loop is calculated in the following way. First h_s and θ_s are determined from equations (28) and (29). equations (23) is then used to calculate θ at each value of h. Finally, the relation $m = \cos \theta$ is used to calculate the m vs. h curve. The calculation is divided into two regions corresponding to the positive-going branch and the negative-going branch of the m–h loop, starting and $h = 1.5$ to $h = -1.5$; (1) $1.5 > h > -h_s$, with θ restricted to values between 0° and θ_s, and (2) $-h_s > h > 1.5$, with θ restricted to values between 180° and 180° + θ_0.

Here, two regimes are identified on the basis of the dependence of the switching field (h_s) and coercive field (h_c) on the angle (θ_0) between the external field and the magnetization easy (nanowire axis). When the applied field close to the magnetic easy axis, $0° < \theta_0 < 45°$, the hysteresis loop is relatively square; hence, the change in sign of the magnetization, corresponding to the coercivity, occurs at the switching field. In this regime the coercivity H_c is equal to H_s and is determined from equation (28).

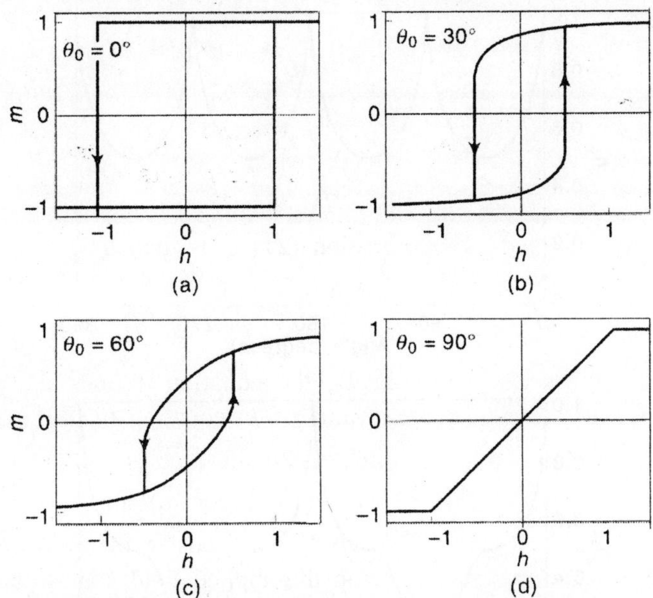

Fig. 8.9 *Calculated m–h curves for the coherent rotation model assuming an applied field at an angle θ_0 with respect to the magnetization easy axis (nanowire axis). The predicted behavior is shown for $\theta_0 = 0°$, 30°, 60°, 90° and 90°.*

For 45° < θ_0 < 90°, the applied field is oriented closer to the magnetic hard axis, and the hysteresis loop is sheared such that switching occurs after the magnetization changes sign. In this case, $H_c \neq H_s$, and the coercive field is determined from equation (26) by setting $m = 0$, leading to the expression

$$h_c = \sin\theta_0 \cos\theta_0, \tag{30}$$

where $h_c = H_c/2(K_u/M_s)$.

Figure 8.10 shows the angular dependence of the reduced switching field h_s and reduced coercivity h_c, according to equations (28) and (30), respectively. Comparison of the two figures shows that $h_s = h_c$ when 0 $\theta_0 < 45°$, whereas for $45° < \theta_0 < 90°$, $h_s \neq h_c$.

A key feature of the coherent rotation model, as seen from equations (28) and (30), is that the switching field and coercivity are independent of particle size.

(B) Curling

For particle sizes larger than the critical size but still in the single-domain regime, magnetization reversal occurs by curling. Here, the magnetization switching is an abrupt process, and the switching field is very close to the nucleation field; hence, $H_c = H_s$ for all angles. Furthermore, H_c and H_s are dependent on both the aspect ratio and the size of the ellipsoid.

Analytical solutions for m vs. h curves in the curling model are not available. The angular dependence of the normalized nucleation field for a prolate spheroid based on the curling model is given by

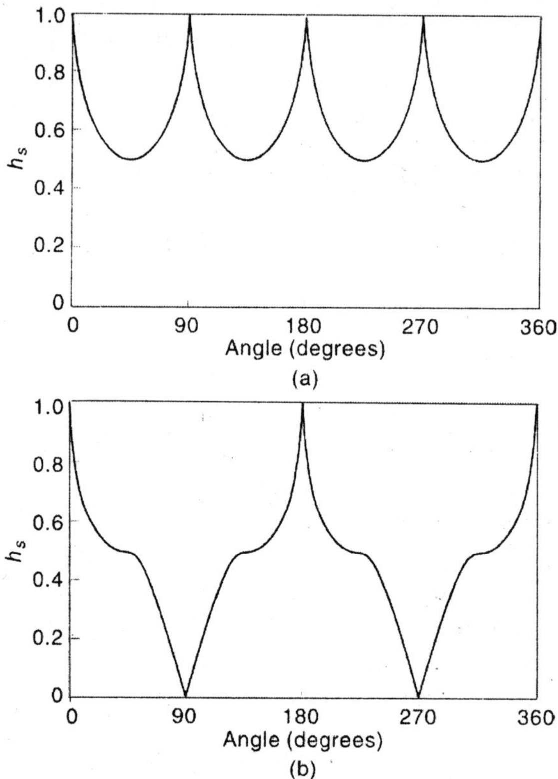

Fig. 8.10 *Calculated angular dependence of (a) the reduced switching field h_s and (b) the reduced coercivity h_c for the coherent rotation model, from equations (28) and (30). In both cases the fields are normalized by the factor $2K_u/M_s$.*

$$h_c(\theta) = \frac{\left(2N_c - \frac{k}{S^2}\right)\left(2N_a - \frac{k}{S^2}\right)}{\sqrt{\left(2N_c - \frac{k}{S^2}\right)^2 \sin^2\theta + \left(2N_a - \frac{k}{S^2}\right)^2 \cos^2\theta}} \tag{31}$$

where $h_c = H_c/2\pi M_s$, N_c and N_a are the demagnetizing factors of the spheroid along the major and minor axes; S is the reduced radius r/r_0, where $r_0 = A^{1/2}/M_s$; $k = q^2/\pi$, where θ is the same geometrical factor used in equation (16) ($q = 1.8412$ for a cylinder and 2.0816 for a sphere); and θ is the angle between the external magnetic field and magnetic easy axis (c-axis). Note that r_0 is independent of the size and shape of the spheroid and is related to the critical size for the transition between curling and coherent rotation via $r_c/r_0 = r/\sqrt{2}$.

Figure 8.11 shows the angular dependence of the coercivity of a slender spheroid with high aspect ratio ($q = 1.8412$, $N_c = 0$, and $N_a = 2\pi$) predicted by the curling model [calculated using equation (31)] as a function of diameter and aspect ratio. The corresponding angular dependence of the coercivity from the coherent rotation is shown for comparison. The figure illustrates that the angular dependence of the normalized coercivity predicted by the curling rotation mode.

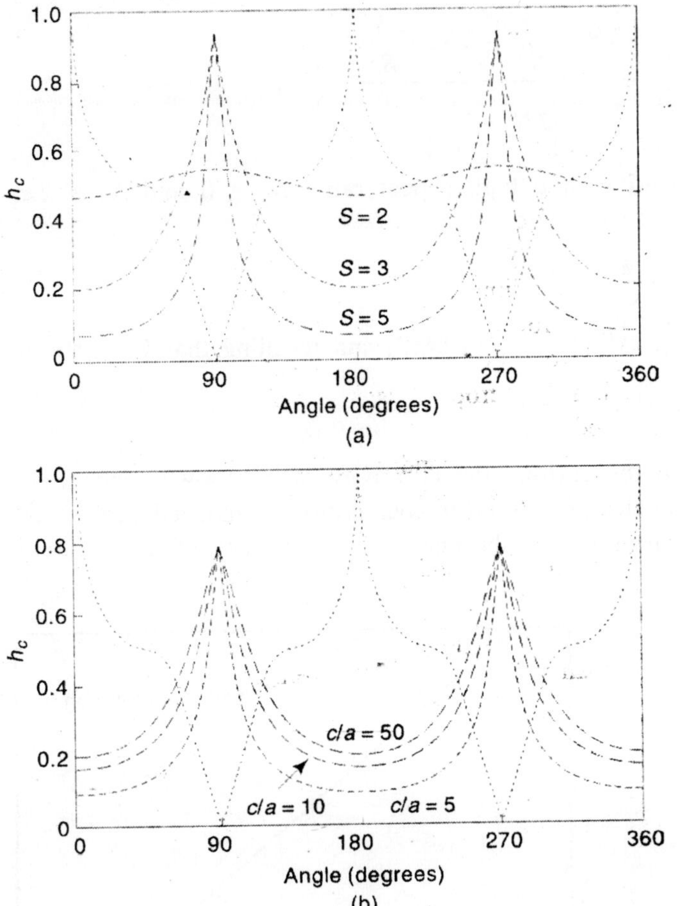

Fig. 8.11 *Calculated angular dependence of the reduced coercivity $h_c = H_c/2pM_s$ based on the curling model, from equation (31): (a) Dependence on nanowire diameter $S = r/r_c$, assuming $c/a = 10$; (b) dependence on aspect ratio c/a assuming $S = 3$. The angular dependence of the coercive field calculated via the coherent rotation model for an infinitely long cylinder is also shown (dotted curves).*

Just as the magnetization configuration adopts the lowest-energy state, switching occurs by the process that gives the lowest coercivity. At lower angles, for which the applied field is aligned closely with the easy axis (along the wire axis), switching will occurs by curling if the nanowire diameter is larger than the critical size [$r > r_c(\theta)$]. Note that equation (16) describes the critical radius only when the applied field is aligned along the wire axis ($\theta = 0$). However, at higher angles, switching is predicted to occur by coherent rotation. The angle at which the switching mode changes from curling to coherent rotation increases for larger wire diameter and lower aspect ratio.

For an applied field oriented along the magnetic easy axis (*c*-axis), for which $\theta = 0$, equation (31) is expressed as

$$h_c = \frac{\left(2N_c - \frac{k}{S^2}\right)\left(2N_a - \frac{k}{S^2}\right)}{\left|2N_a - \frac{k}{S^2}\right|} \tag{32}$$

For a prolate spheroid with a high aspect ratio, $N_c = 0$ and $N_a = 2\pi$; hence, equation (32) reduces to

$$h_c = \frac{k}{S^2} \tag{33}$$

For an infinitely long cylinder, $q = 1.8412$; and recalling that $k = q^2/\pi$ and $S = r/r_0$,

$$h_c = 1.08\frac{1}{(r/r_0)^2} \tag{34}$$

Thus, according to the curling, the coercivity of a nanowire increases with $1/r^2$. Figure 8.12 shows the reduced nucleation field (or coercivity) vs. reduced radius, as predicted by equation (34). This figure suggests that in the curling the coercivity of nanowires increases, on decreasing the wire diameter.

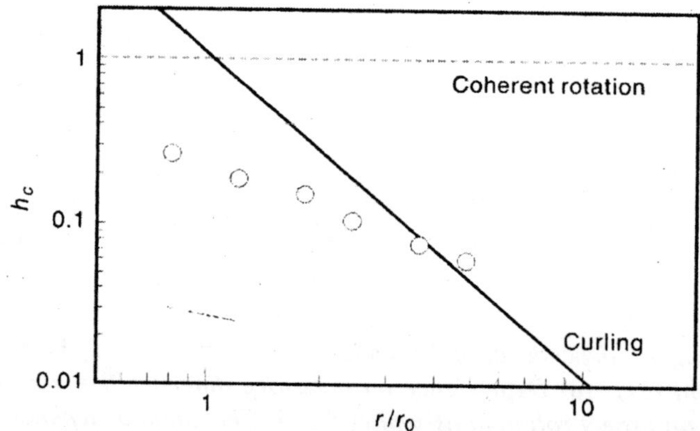

Fig. 8.12 *Calculated dependence of the reduced coercivity ($h_c = H_n/2pM_s$) on reduced radius r/r_0 for a slender ellipsoid ($r_0 = A^{1/2}/M_s$) according to the curling model with the applied field parallel to the c axis (solid curve). In the coherent rotation model, the coercivity is independent of the reduced radius. Also shown are measured values for nickel nanowires (M_s = 485 emucm^{-2} and A = 1 × 10^6 ergcm^{-1}).*

(C) Superparamagnetism

At finite temperature, the process of magentization reversal is also viewed as overcoming a single energy barrier. Thermal fluctuations allow the magentization to surmount the energy barrier and switch from one stable direction to the other. The switching probability per unit time $P(s^{-1})$ is obtained from the Arrhenius relation,

$$P = \nu_0 \exp\left(-\frac{\Delta E}{kT}\right), \tag{35}$$

where v_0 is the thermal attempt frequency, which is usually assumed to be 10^9 s^{-1}; ΔE is the energy barrier; k is the Boltzmann constant; and T is the temperature. A reversing field aligned in the opposite direction from the magnetization direction acts to lower the energy barrier, thereby increasing the probability of switching. The dependence of the applied field on the energy barrier is described by the expression

$$\Delta E = U\left(1 - \frac{H}{H_0}\right)^{1/n}, \tag{36}$$

where Y is the energy barrier at zero applied field, H_s is the applied field, and H_c is the field needed to overcome the barrier at zero temperature. The parameter n reflects different switching mechanisms.

Assume that the switching field H_s is equal to H_c and take the coercivity as the applied field; thus, by combining equations (35) and (36)

$$H_c(t, T) = H_0\left[1 - \left(\frac{kT}{U}\ln(2f_0 t)\right)^n\right]. \tag{37}$$

This equation indicates that the parameter n and energy barrier U is obtained by measuring the time dependence and temperature dependence of the coercivity.

For single-domain particles having a uniaxial shape anisotropy, the energy barrier is just the energy required to switch by coherent rotation. Thus, the barrier is equal to K_uV, and K_u is the uniaxial shape anisotropy constant and is equal to $M_s(N_a - N_b)$, and V is the particle volume. H_0 and ΔE can be rigorously derived as follows:

$$H_0 = \frac{2k_u}{M_s}; \tag{38}$$

$$\Delta E = K_uV\left(1 - \frac{H}{H_0}\right)^2; \tag{39}$$

For more complex switching mechanisms, such as curling, the following equations are adopted for characterizing an assembly of randomly oriented single-domain magnetic particles:

$$H_0 = \frac{2K_{eff}}{M_s} \tag{40}$$

$$\Delta E = K_{eff}V\left(1 - \frac{H}{H_0}\right)^{3/2}; \tag{41}$$

where the switching volume V is a fraction of the particle volume which rotates in order to initiate the reversal process of the entire particle. K_{eff} is called the effective anisotropy constant; it reflects the combined effect of all of the applicable energy terms on the magnetization rotation process.

Since in general the energy barrier for magnetization switching in a particle is proportional to the particle volume, a particle with volume below a critical value does not have an energy barrier high enough to prevent spontaneous switching at room temperature. The particle behaves like a paramagnetic material, but with a much higher susceptibility. This phenomenon, called *superparamagnetism*, has become of considerable interest in high-density magnetic recording media. For a particle to retain its magnetization at room temperature (298 K) on average, for $t = 1$ s, $U \approx 21kT$; hence, if U < 21kT, we regard the particle as being superparamagnetic at room temperature.

For magnetic nanowires, an estimate for the energy barrier in the remanent state ($H = 0$) is determined by assuming $K_{eff} = K_u$ (shape anisotropy). For a prolate spheroid, $Ku = N_a - N_c) M_s^2$, provided the aspect ratio $c/a > 10$. In that case $N_a = 2\pi$ and $N_c = 0$, so that $K_u = 2\pi M_s^2$ (see Figure 8.6). On the basis of this assumption, Figure 8.13 shows the dependence of the energy barrier on radius for nicker nanowires as a function of their length. For magnetic recording media, stability over a time scale of ten years is usually required, giving a criterion of $U/kT > 40$. The energy barrier for nickel nanowires with radii larger than about 2 nm and more than 100 nm in length is sufficiently large to satisfy this criterion. For particles in suspension that have been magnetized, magnetostatic interactions may lead to aggregation. The figure shows that short nickel nanorods 10 nm in length would be expected to be supermagnetic for radii less than about 10 nm, since the switching probability is close to 1 for a time period one second. Consequently, aggregation induced by magnetic interactions should not be significant, resulting in a stable suspension.

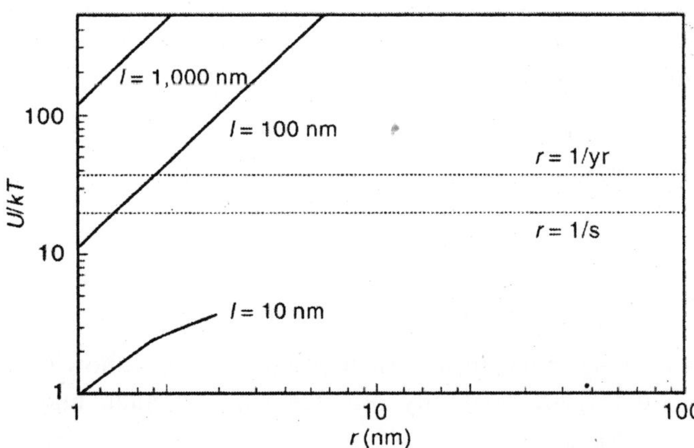

Fig. 8.13 *Calculated dependence of the energy barrier for spontaneous switching on nanowire radius calculated from $U = K_u V$, assuming $K_u = (N_a - N_c) M_s^2$ and $M_s = 485$ emucm^{-3}. The dotted lines correspond to switching probabilities of one per second and one per year, respectively.*

Magnetostatic Interactions between Nanowires

In the discussions above it is assumed that the magnetostatic interaction between nanowires is negligible. The magnetic field H_x created by a dipole with moment m and length l, at distance x in the direction perpendicular to the dipole, is given by

$$H_x = \frac{m}{\left(x^2 + \frac{1^2}{4}\right)^{3/2}} \tag{42}$$

where the magnetic moment $m = M_s V$, and l and r are respectively the length and radius of the dipole. This dipole approximation gives a valuable estimation of the magnitude of the magnetostatic interaction between nanowires. Figure 8.14 shows a plot of the field H_x at a distance x from a cylindrical nickel nanowire (M_s = 485 emucm^{-3}). For a nanowire having a radius of 100 nm, the field is calculated to be about 110 Oe within 100 nm but to decrease to small values at about 1 μm. For longer nanowires, the maximum field is calculated to decrease significantly. The field for smaller-diameter nanowires is calculated to be significantly smaller and to drop off at shorter distances. Since the coercivity of nickel nanowires is typically 100–1000 Oe, magnetostatic interaction are expected to be important only for longer-diameter nanowires that are closely spaced, as would be typical in alumina templates.

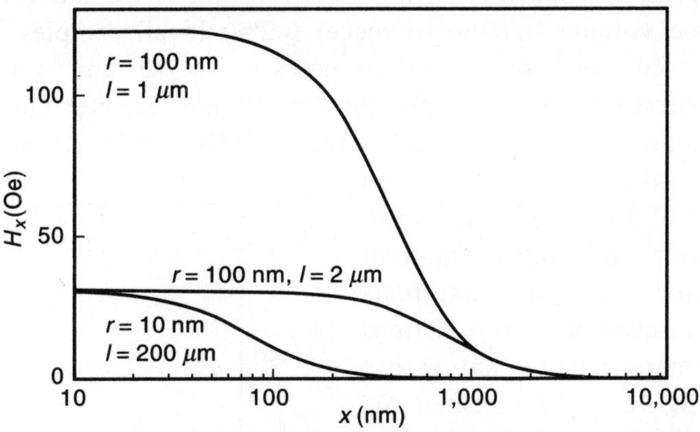

Fig. 8.14 *Magnetic field at a distance x from an nickel; nanowire (M_s = 485 emucm^{-3}) of radius r and length l.*

Results

For the magnetic properties of a wide range of single-component ferromagnetic nanowires, Figure 8.15 shows *M–H* loops for 12-nm-diameter nickel nanowires in a single-crystal mica template. A plan view of a single-crystal mica template formed by nuclear track etching is shown in Figure 8.3. The pores have a diamond-shaped cross section with angles of 60° and 120°. The pore density is controlled by the flux of high-energy particles passing through the membrane, and the pore size is controlled by the etching time. The etching rate along the particle track is more than three orders faster than the lateral etching rate, resulting in uniform, parallel-sided nanopores perpendicular to the film plane. The axes of the pores are aligned.

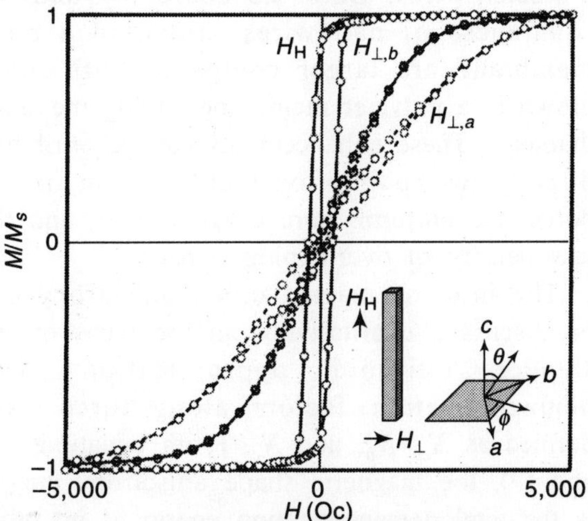

Fig. 8.15 *Typical M–H loops for nickel nanowires with an effective pore diameter of 120 nm; magnetic field applied parallel ($H_{\|}$) and perpendicular ($H_{\perp}$) to the wire axes.*

The crystal structure of mica determines the etching anisotropy. For convenience, define an effective diameter of a pore as the diameter of a circle with same area.

The magnetization hysteresis loop obtained with the applied field parallel to the wire axis is relatively square, characteristic of the easy axis, with a remanence M_r/M_s of 0.89. Since the pores have a diamond-shaped cross section, the nanowires exhibit magnetic anisotropy in the plane perpendicular to the wire axis, as shown in Figure 8.15. Rotating the field from along the short diagonal a to along the long diagonal b results in a decrease in the saturation field from 4650 Oe to 3380 Oe, as shown in the figure. Concurrently, the coercivity decreases from 220 Oe to 80 Oe, and the squareness of the loop decreases from 0.066 to 0.05.

Figure 8.16 shows the dependence of the coercivity (H_c) and loop squareness on the effective nanowire diameter with the applied field parallel to the wire axis. In order to maintain approximately the same volume fraction of nickel (≈2%) in all samples, the nanowire density was increased from 5×10^7 cm^{-2} for the 200-nm pores to 2×10^9 cm^{-2} for the 30-nm pores. The average wire spacing decreased from 1.4 μm for the 200-nm-diameter nanowires to 220 nm for the 30-nm diameter nanowires. The measured coercivity [Figure 8.16(a)] increases with decreasing diameter, reaching a value of 800 Oe at an effective wire diameter of 30 nm. Figure 8.16(b) shows the magnitude of the remanent magnetization obtained from the hysteresis loops and plotted as the squareness (SQ), defined as the ratio of remanence to saturation magnetization ($SQ = M_r/M_s$). The value of SQ is as high as 0.96 for the 30-nm-diameter wires, decreasing gradually to 0.83 for th e200-nm-diameter wires. Both the coercivity and the squareness of nanowires grown in a mica membrane are larger compared with those grown in a polymer membrane having the same diameter. These enhancements can be attributed directly to the improved collimation of the pores, the uniform, pore cross section, and the low density of overlapping pores.

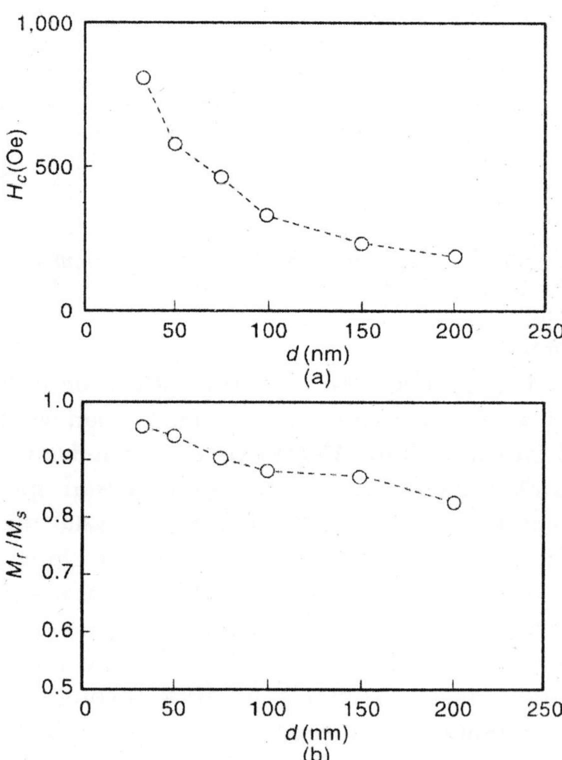

Fig. 8.16 *(a) Coercivity and (b) squareness of hysteresis loop for nickel nanowire arrays in single-crystal mica films as a function of the effective wire diameter measured along the wire axis.*

The in-plane magnetization anisotropy can be discussed qualitatively on the basis of the slender ellipsoid approximation with demagnetization factors along three axes defined as N_a, N_b, and N_c. From equations (6) and (7), the magnetic shape anisotropy energy is the total demagnetization energy if no other anisotropy is considered. The magnetization energy can be written as the sum of the demagnetization energy along the three axes, viz.,

$$E = \frac{1}{2}(N_a M_a^2 + N_b M_b^2 + N_c M_c^2)$$

$$= \frac{1}{2} M_a^2 (N_c \cos^2\theta + N_a \sin^2\theta \cos^2\phi + N_b \sin^2\theta \sin^2\phi), \tag{43}$$

where θ is the angle between the magnetization and the wire axis (c-axis), and ϕ is the angle between the in-plane projection of the magnetization and the short diagonal of the cross section (a-axis).

When the magnetization is aligned in the *ab* plane ($\theta = 90°$), equation (43) reduces to

$$E = \frac{1}{2}(N_a \cos^2\phi + N_b \sin^2\phi)$$

$$= K_a \cos^2\phi + K_b \sin^2\phi, \tag{44}$$

where K_a and K_b are the uniaxial anisotropy constants along the *a* and *b* directions. From the expression, we can write the saturation field as

$$H_{sat} = M_s(N_a \cos^2\phi + N_b \sin^2\phi)$$

$$= \frac{2K_a}{M_s}\cos^2\phi + \frac{2K_b}{M_s}\sin^2\phi \tag{45}$$

Figure 8.17 shows the agreement obtained between the measured angular dependence of the saturation field and equation (45). From the fig, we obtain $K_a = 1.13 \times 10^6$ ergcm^{-3} and $K_b = 8.2 \times 10^5$ ergcm^{-3}. The corresponding demagnetizing factors are $N_a = 0.764$ and $N_b = 0.556$. These values can be compared to the calculated demagnetizing factors for a slender ellipsoid with $c/a = 43.8$ and $c/b = 25.3$ [equations (10) and (11)], of $N_a = 0.33$ and $N_b = 0.365$. These values are in reasonable agreement given the difference in cross section between a diamond and an ellipse. Thus, the in-plane shape anisotropy caused by the diamond-shaped cross section can be described by a simple model based on two mutually perpendicular anisotropic axes.

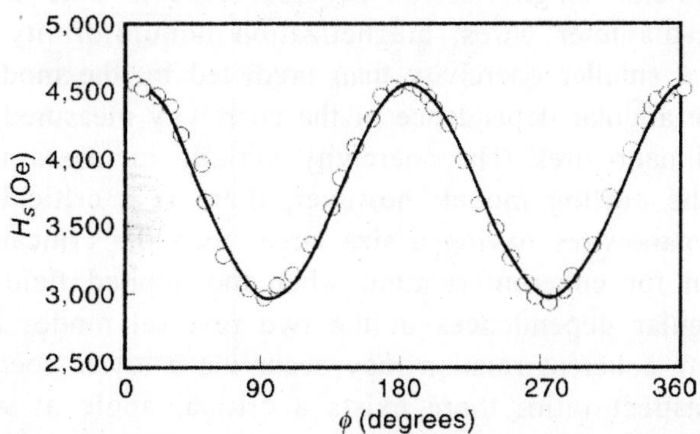

Fig. 8.17 *Angular dependence of the saturation for 120-nm-diameter nickel nanowires with the external field perpendicular to the wire axis. $\phi = 0°$ corresponds to the short diagonal. The solid curve depicts equation (45).*

Now, consider, the magnetization when the external field is rotated from the direction parallel to the wire axis ($\theta = 0°$) to being perpendicular to the wire axis ($\theta = 90$) in the *ac* or *bc* plane. When $H = 0$, the magnetizations in the nanowires are aligned along the c-axis (the magnetic easy axis). However, the magnetization is measure at an angle θ with respect to the c-axis. Thus, the component of the remanent magnetization $M_r(\theta)$ at an angle θ to the c-axis is cos θ. Therefore,

$$M_r(\theta) = M_r(\theta = 0)|\cos\ \theta|, \tag{46}$$

where $M_r(\theta = 0)$ is the remanent magnetization when the applied field is along the wire axis ($\theta = 0$). Figure 8.18 shows that the remanent magnetization of the hysteresis loops decreases with increasing θ according to equation (4).

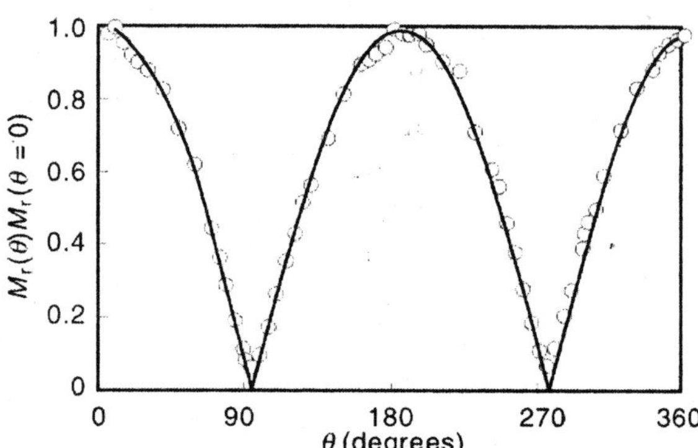

Fig. 8.18 *Dependence of remanent magnetization on the angle θ of the applied magnetic field respect to the wire axis and the short diagonal in the ac plane. The solid curve corresponds to $SQ = SQ_0\ |\cos\ \theta|$.*

The critical radius r_c between coherent rotation and curling for an infinitely long nickel cylinder obtained from equation (16) to 20 nm. Thus, switching in nickel nanowires is expected to occur by curling. As was shown in Figure 8.16, the easy-axis coercivity increases with decreasing diameter, as is consistent with the curling model. Measured values for the normalized coercivity vs. the normalized nanowire radius using $A = 1 \times 10^6$ ergcm^{-3} and $M_s = 485$ ergcm^{-3} are shown in Figure 8.12. For larger-diameter nanowires, the coercivity shows good agreement with the curling model; however, at smaller diameters the coercivity does not increase as rapidly as predicted by the mode. Similar results are reported for cylindrical Ni nanowires in porous alumina templates. Note, the earlier magnetization reversal models take into account only shape anisotropy. For smaller-diameter wires, magnetization nonuniformity can cause noncoherent switching and result in a smaller coercivity than predicted by the model.

Figure 8.19 shows the angular dependence of the coercivity measured in the *ac* and *bc* planes for 120-nm-diameter Ni nanowires. The coercivity initially increases with increasing angle, in good agreement with the curling model; however, there is a critical angle above which H_c decreases abruptly. For nanowires having a size larger then the critical size, the coercivity due to curling is lower than for coherent rotation when the applied field is along the wire axis. However, since the angular dependences of the two reversal modes have opposite trends, at higher angles, reversal by coherent rotation may occur via a lower coercivity. Thus, for a given particle diameter and aspect ratio, there exists a critical angle at which the magnetization reversal modes change from curling to coherent rotation. For a constant diameter, the critical angle increases with increasing aspect ratio. For a constant aspect ratio, the critical angle decreases with decreasing particle diameter. As out measurements on the prismatic Ni nanowires indicate, the transition from curling to coherent rotation occurs at about 78° when the field is aligned in the *ac* plane but increase to about 85° when the field is aligned in the *bc* plane.

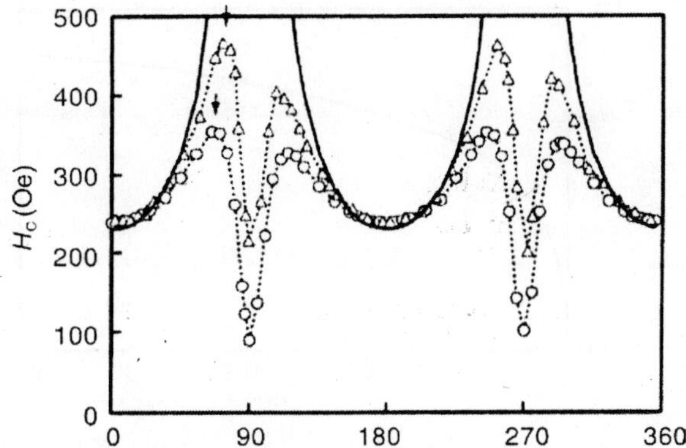

Fig. 8.19 *Dependence of coercivity on the angle θ of the applied magnetic field with respect to (○) the wire axis and the short diagonal in the ac plane and (Δ) the wire axis and the long diagonal in the bc plane for 120-nm-diameter nickel nanowires. They solid curve corresponds to* $H_c = H_c(\theta + 0°)/\cos\theta$.

Curie Temperature

Supermagnetic behavior describes the influences of thermal fluctuation on small magnetic particles that randomize the magnetization as a function of time. In superparamegnatic particles, the thermal energy overcomes the anisotropy energy; however, thermal energy is still smaller than the exchange energy, and the magnetic moments are still aligned parallel to one another in the particles. The temperature at which the exchange energy between magnetic moments becomes smaller than the thermal energy and ferromagnetic order disappears is defined as the Curie temperature.

At 0K, all magnetic moments are aligned along the same direction. When the temperature increases, spin-orientation fluctuations begin to occur. The distance over which fluctuations in the system are correlated is known as the correlation length ξ. In bulk magnetic systems, it increases with temperature and diverges at the bulk transition temperature $T_c(\infty)$, at which the magnetic disorder at one point randomizes the system and a ferromagnetic-to-paramagnetic phase transition occurs. When the physical dimensions of a system are reduced, finite size effects occur, changing the materials properties. In systems with reduced dimensions, the growth of ξ with temperature is limited by the smallest dimension *d*, and the system displays a reduced transition temperature T_c(d) because of the finite-size effects.

Figure 8.20(a) shows the resulting decrease in the Curie temperature for nickel nanowires in mica templates. The Curie temperature is reduced by about 50 K for a nanowire diameter of 30 nm. Owing to the relatively large diameters, the nanowires are expected to behave as a constrained three-dimensional system.

The correlation length $\xi(T)$ of a magnetic system increases with temperature, and at temperatures close to the bulk transition temperature $T_c(\infty)$, it exhibits asymptotic behavior described by

$$\xi(T) = \xi_0 \left| 1 - \frac{T}{T_c(\infty)} \right|^{-\nu}, \tag{47}$$

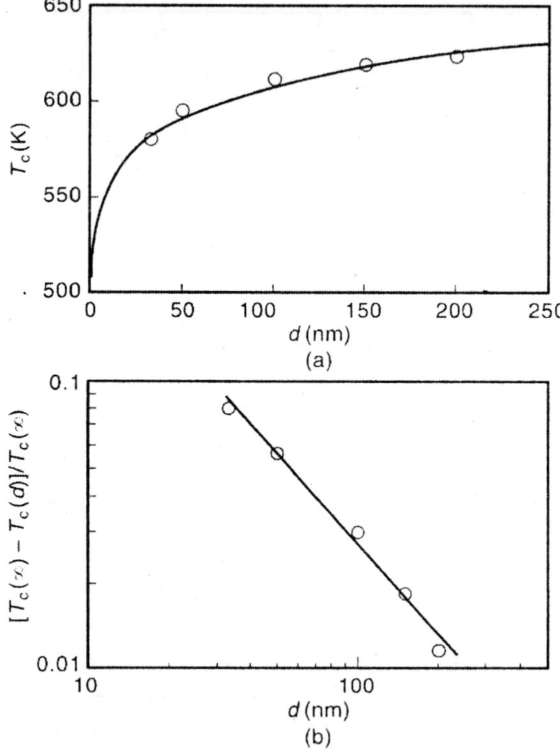

Fig. 8.20 *(a) Curie temperature of nickel nanowire arrays vs. wire diameter d, illustrating the decrease in T_c with decreasing diameter. (b) Log–log plot showing $[T_c(\infty) - T_c(d)]/T_c(\infty)$ normalized to the Curie temperature for bulk Ni $[T_c(\infty) = 631$ K] vs. wire diameter. The solid curve corresponds to $\lambda = 0.94$ and $\xi_0 = 22$ Å.*

where ξ_0 is the correlation length extrapolated to $T = 0$, and v is the critical exponent for correlation. ξ_0 is typically of the order of the near-neighbor spacing, although for some ferromagnetic materials, such as nickel, it can be somewhat larger. For nanowires having a small diameter, the growth of $\xi(T)$ with increasing temperature is constrained by the wire diameter d, resulting in a reduced Curie temperature defined by

$$T_c(d) = T_c(\infty)\left[1 - \left(\frac{\xi_0}{d}\right)^{\lambda}\right] \tag{48}$$

or

$$\frac{T_c(\infty) - T_c(d)}{T_c(\infty)} = \left(\frac{\xi_0}{d}\right)^{\lambda} \tag{49}$$

where T_c(d) is the Curie temperature for nanowires with diameter d, and $\lambda = 1/\nu$ is the shift exponent.

Figure 8.20(b) shows a log-log plot of the reduced temperature $[T_c(\infty) - T_c(d)]/T_c(\infty)$ vs. wire diameter d, illustrating that the measured values for $T_c(d)$ follow the finite-size scaling relation. From the figure we obtain $\lambda = 0.94$ and an extrapolated value of $\xi_0 = 22$ Å. The observed exponent of $\lambda = -0.94$ is lower than the theoretical values predicted by the three-dimensional Heisenberg model ($\lambda = 1.4$) and the three-dimensional Ising model ($\lambda = 1.58$); for both models, however, it is assumed that near-neighbor interactions occur, whereas nickel is a ferromagnet that exhibits longer-range interactions. The extrapolated correlation length $\xi_0 = 22$ Å is close to the value of 20 Å reported for polycrystalline nickel thin films, although it is somewhat larger than the value for 4–10 Å obtained for epitaxial single-crystal nickel films.

MULTIPLE-SEGMENT NANOWIRES

The magnetic properties of nanowires are further modified in multiple-layer or multiple-segment nanowires consisting of ferromagnetic (FM) and nonmagnetic segments (NM). The magnetic properties of FM/NM multiple-layer (or multilayer) nanowires are tuned by varying the size, aspect ration, and spacing of the FM segments.

Electrodeposition of multilayer films from a solution containing the ions of two components is achieved by modulation of the potential or current and usually results in the deposition of a compositionally modulated multilayer with a bilayer repeat unit of the form A_xB_{1-x}/A_yB_{1-y}. In the systems $Cu_xBi_{1-x}/Cu_yBi_{1-y}$, $Cu_xZn_{1-x}/Cu_yZn_{1-y}$, $Ni_xMo_{1-x}/Ni_yMo_{1-y}$, $Ag_xPd_{1-x}/Ag_yPd_{1-y}$, and $Au_xAg_{1-x}/Au_yAg_{1-y}$, the compositional modulation is small, typically less than 30 at.%.

A major breakthrough in multilayer deposition is associated with the recognition that multilayers of the from A_xB_{1-x}/A with $x \approx 0.01$, deposited for systems for which the difference in equilibrium potentials between A and B is sufficiently large (typically > 0.4 V) and the concentration of the component is very low (i.e., $A^{n+}] << [B^{m+}]$). This approach is used in the deposition of films with alternating FM and NM layers that exhibit giant magnetoresistance (GMR). Examples of electrodeposited FM/NM multilayers include Ni/Cu, Co/Cu, CoNi/Cu, and NiFe/Cu.

Figure 8.21 shows a current vs. voltage curve for the electrodeposition of copper and nickel into the pores of a 6-μm-thick nanoporous polycarbonate membrane. Electrodeposition is from a solution containing 0.5 mol L^{-1} $NiSO_4$, 0.005 mol L^{-1} $CuSO_4$, and 0.6 mol L^{-1} H_3BO_3. The equilibrium potentials for copper and nickel in this solution are $U_{eq}(Cu^{2+}/Cu) = 0.15$ V (vs. Ag/AgCl) and $Ueq(Ni^{2+}/Ni) = -0.468$ V (vs. Ag/AgCl). On scanning the potential negative from the open-circuit potential, the current onset at about +0.05 V is followed by a peak associated with the nucleation and diffusion-limited growth of copper. Because of the low Cu(II) concentration in solution, the deposition is diffusion-limited over a wide potential range. At more negative potentials, the onset of nickel deposition is seen at about −0.7 V. On the reverse scan the diffusion-limited deposition of copper is seen up to about 0 V, followed by a peak associated with the stripping of copper. Note that nickel dissolution is negligible in this solution.

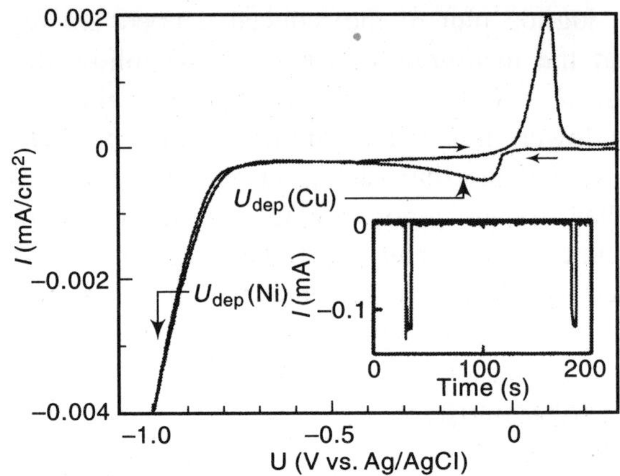

Fig. 8.21 *Current-voltage curve during electrodeposition onto a gold film at 10 mV/s through the pores of a 6-μm-thick nanoporous poly-carbonate membrane. Electrodeposition was from a solution containing 0.5 mol/L $NiSO_4$, 0.005 mol/L $CuSO_4$, and 0.6 mol/L H_3BO_3. The inset shows part of a current-time curve for the deposition of 50-nm-diameter [Ni(125 nm)/(Cu(125 nm)]10 multilayer nanowires. The two short (≈ 4 s) sections correspond to the deposition of Ni at −1./0 V, whereas the longer (≈140 s) section corresponds to the deposition of Cu at −0.16 V.*

Figure 8.22 shows a transmission electron miscroscopy (TEM) bright-field image of part of a 120-nm-diameter $[Ni(20\ nm)/Cu(10\ nm)]_m$ multilayer nanowire. Figure 8.22(b) is an electron energy loss spectroscopy (FELS) image of part of a 30-nm-diameter $[Ni(4\ nm)/Cu(1\ nm)]_n$ multilayer nanowires. The layered structure and the reproducibility of the layer thickness are clearly seen.

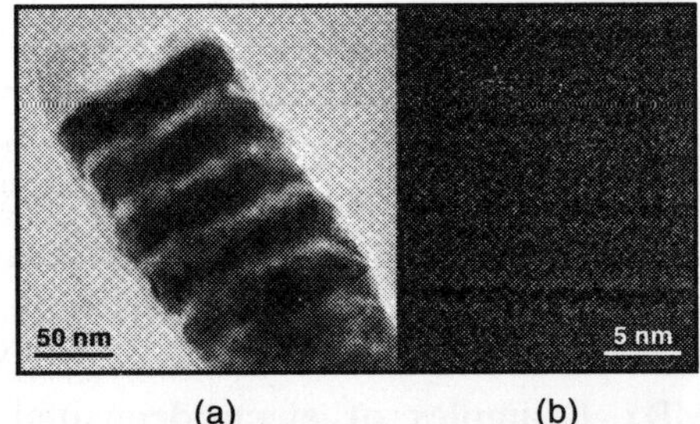

Fig. 8.22 *(a) TEM image of a 120-nm-diameter nanowire with 20-nm-thick Ni layers and 10-nm-thick Cu layers. (b) FELS image of a 30-nm-diameter nanowire with 1-nm-thick Ni layers and 4-nm-thick Cu layers.*

For the multilayer nanowire shown in Figure 22(a), the Ni segments have an aspect ratio of 0.17, and from equation (15) are expected to be single-domain. From the pore density, the average spacing between the nanowires in the templates is estimated to be 305–400 nm. These dimensions are sufficiently large to permit wire-wire interactions to be neglected. Furthermore the crystalline anisotropy of nickel is relatively small; hence the magnetic response is dominated by the shape of the FM segments.

Figure 23 shows typical *M–H* shows magnetization curves for 50-nm-diameyer $[Ni(120\ nm)/Cu(125\ nm)]_{10}$ multilayer nanowires. For rod-shaped magnetic segments (aspect ratio > 1), the easy axis is parallel to the wire axis, as can be seen from the large remanence and coercivity. Perpendicular to the wire axis, the magentization curves are characterized by very small remanences and coercivities. The measured saturation field along the hard axis is about 5,000

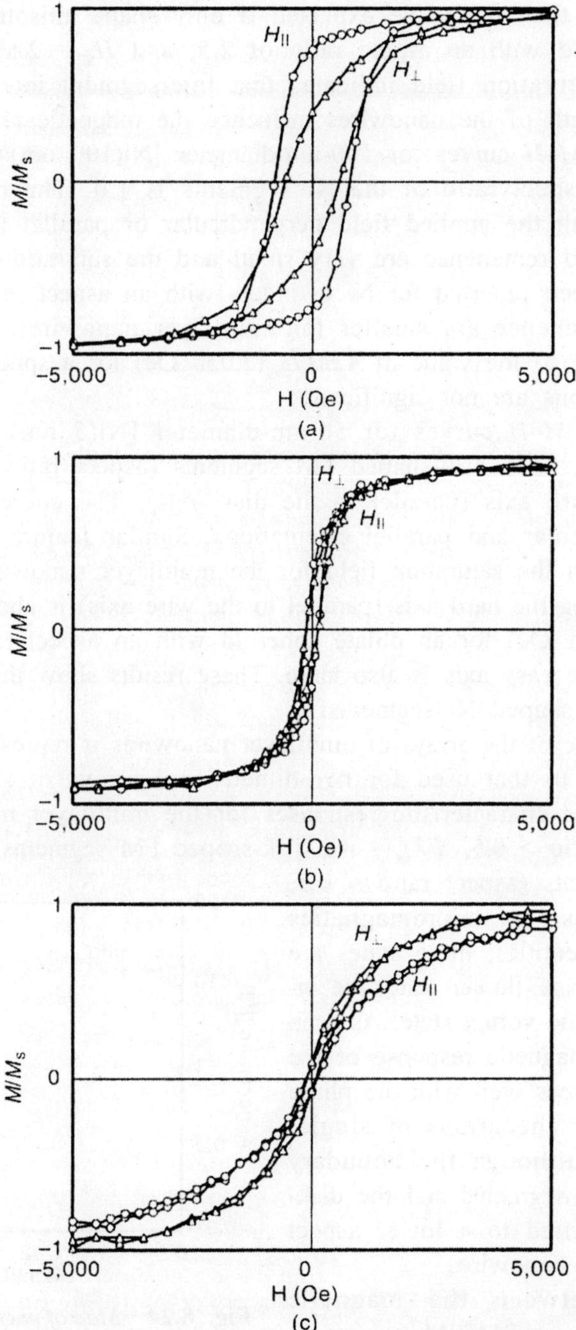

Fig. 8.23 *Magnetic hysteresis loops of electrodeposited Ni/Cu multilayer nanowires: (a) [Ni(125 nm)/Cu(125 nm)]10 having a diameter of 50 nm (rod-shaped Ni segments; aspect ratio = 2.5). (b) [Ni(100 nm)/Cu(100 nm)]$_{30}$ having a diameter of 100 nm (intermediate-shaped Ni segments; aspect ratio = 1.0). (c) [Ni (5 nm)/Cu(5 nm)]250 having a diameter of 50 nm (disk-shaped Ni segments; aspect ratio = 0.1).*

Oe, significantly larger than the value expected if only shape anisotropy is considered. $H_s = 1.73\pi M_s$ for an ellipsoid with an aspect ratio of 2.5, and $H_s = 2\pi M_s$ for an infinitely long cylinder. The larger saturation field indicates that intersegment interactions between the Ni segments along the length of the nanowires influence the magnetic response.

Figure 23(b) shows *M–H* curves for 100-nm-diameter $[Ni(100\ nm)/Cu(100\ nm)]_{30}$ multilayer nanowires. Here, the aspect ratio of the Ni segments is 1.0. The magnetization curves are essentially identical, with the applied field perpendicular or parallel to the wire axis. In both cases, the coercivity and remanence are very small and the saturation field is relatively large. Similar features have been reported for Ni cylinders with an aspect ratio close to 1.0; however the coercivity and remanence are smaller for multilayer nanowires. The saturation fields in both directions are close to the value of $4\pi M_s/3$ (2,030 Oe) for a sphere, indicating that in this case interlayer interactions are not significant.

Figure 23(c) shows *M–H* curves for 50-nm-diameter $[Ni(5\ nm)/Cu(5\ nm)]_{250}$ multilayer nanowires. In this case, with disk-shaped FM segments (aspect ratio = 0.1), the easy axis is perpendicular to the wire axis (parallel to the disk axis). The coercivity and remanence are small in both perpendicular and parallel orientations. Similar features are observed for arrays of nickel disks, although the saturation field for the multilayer nanowires is significantly larger. The saturation field along the hard axis (parallel to the wire axis) is about 5,300 Oe, close to the value of $3.44\pi M_s$ (5,241 Oe) for an oblate spheroid with an aspect ratio of 0.1; however, the saturation field along the easy axis is also large. These results show that dipolar interactions are also important for disk-shaped Ni segments.

The magnetic response of the arrays of multilayer nanowires is represented on a micromagnetic phase diagram, similar to that used for two-dimensional arrays of single-component entities. Figure 8.24 shows three characteristic responses for the multilayer nanowires with rodshaped FM segments (aspect ratio > 0.5, $d/\lambda_{ex} < 4$), disk-shaped FM segments (aspect ratio < 0.5), and intermediate FM segments (aspect ratio < 0.5, $d/\lambda_{ex} > 4$). On the basis of micromagnetics simulations of isolated entities, these states are referred as the out-of-plane flower state, the in-plane flower state, and the vortex state. As seen from Figure 8.24, the magnetic response of the multilayer nanowires agrees well with the phase boundaries defined for the arrays of single-component elements, although the boundary between the rod-shaped segments and the disk-shaped segments is shifted to a lower aspect ratio for the multilayer nanowires.

The difference between the magnetic response of the multilayer FM/NM nanowires and two-dimensional arrays of FM entities is due to the dipolar magnetostatic interactions between the FM layers in the multilayer

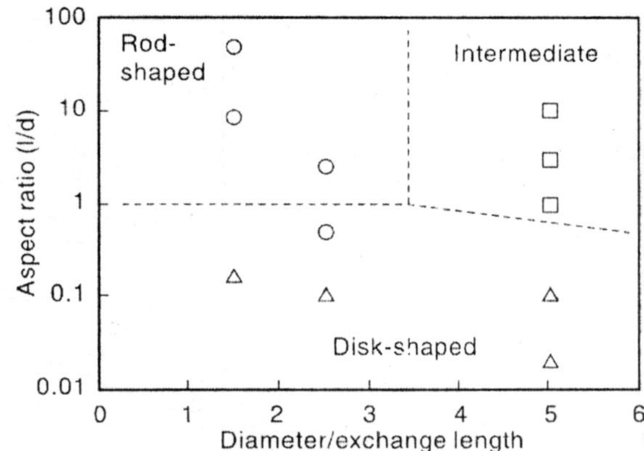

Fig. 8.24 *Map of micromagnetic states for the electrodeposited Ni/Cu multilayer nanowires. The dotted lines represent the boundaries of micromagnetic states obtained for arrays of two-dimensional FM entities.*

structures. For the multilayer nanowires with rod-shaped FM segments, magnetostatic interactions between the segments favor head-to-tail alignment of the magnetic moments along the wire axis. Thus, the remanence and coercivity along the easy axis (parallel to the wire axis) are larger than expected for a single FM rod of the same directions. In contrast, for multilayer nanowires with disk-shaped segments, magnetostatic interactions favor antiparallel alignment perpendicular to the wire axis. Thus, the remanence and coercivity along the easy axis (perpendicular to the wire axis) are smaller than obtained for a single disk. The shift in the boundary between the rod-shaped segments and the disk-shaped segments to a lower aspect ratio suggests that the head-to-tail alignment parallel to the wire axis is favored over the antiparallel alignment down to an aspect ratio of ~0.5.

For, the magnetostatic interlayer interaction, fabricate a series of 50-nm-diameter [Ni(5 nm)/Cu(t nm)]$_n$ nanowires with Cu layer thickness ranging from 5 to 100 nm. The Ni segments are 5 nm thick, corresponding to an aspect ratio of 0.1. Figure 8.25 shows the remanence measured with the field along the easy axis (perpendicular to the wire axis) vs. Cu layer thickness. The remanence increases from 0.09 to 0.26 as the Cu layer thickness increases from 5 to 100 nm, showing that the magnetostatic interaction between nickel segments decreases with increasing NM layer thickness.

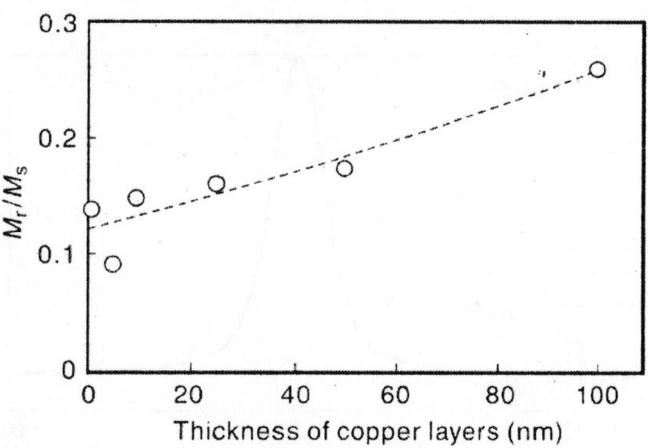

Fig. 8.25 *Remanent magnetization vs. copper layer thickness for 50-nm-diameter [Ni(5 nm)/Cu(x nm)]$_n$ multilayer nanowires.*

GIANT MAGNETORESISTANCE

Figure 8.26(a) shows the M–H loop for 30-nm-diameter [Co(5 nm)/Cu(0.8 nm)]$_{1000}$ multilayer nanowires at 293 K with the field perpendicular to the wire axis. The aspect ratio of the Co segments is 0.125; hence, the direction perpendicular to the wire axis is the magnetic easy axis. The [Co(5 nm)/Cu(x nm)]$_{200}$ multilayer nanowires are deposited from 50 gL^{-1} $CuSO_4 \cdot H_2O$, 0.5 gL^{-1} $CuSO_4 \cdot 5H_2O$ and 40 gL^{-1} boric acid. Cu and Co are deposited at –0.16 V (Ag/AgCl) and –1.0 V (Ag/AgCl), respectively, In order to minimize Co dissolution during the copper deposition cycle, a two-seconds segment at the open circuit potential is introduced into the deposition cycle after a Co deposition segment At –0.16 V, only Cu is deposited, whereas at –1.0 V, the two metals are co-deposited. The concentration of copper in the cobalt layers is 7 at. %. Diffraction peaks up to seventh order are observe in low-angle X-ray diffraction patterns, indicating good multilayer quality.

Figure 8.26 shows the corresponding GMR effect plotted as the change is resistance-normalized with respect to saturation resistance at 50 kOe. At room temperature, 11% GMR is observed with a zero field resistance of 38 Ω, with no discernible hysteretic characteristics.

Figure 8.27 shows a plot of the GMR effect at 5 K and 293 K for 400-nm-diameter [Co(5 nm)/Cu(x nm)]$_n$ multilayer nanowires vs. Cu layer thickness. As the Cu layer thickness increases,

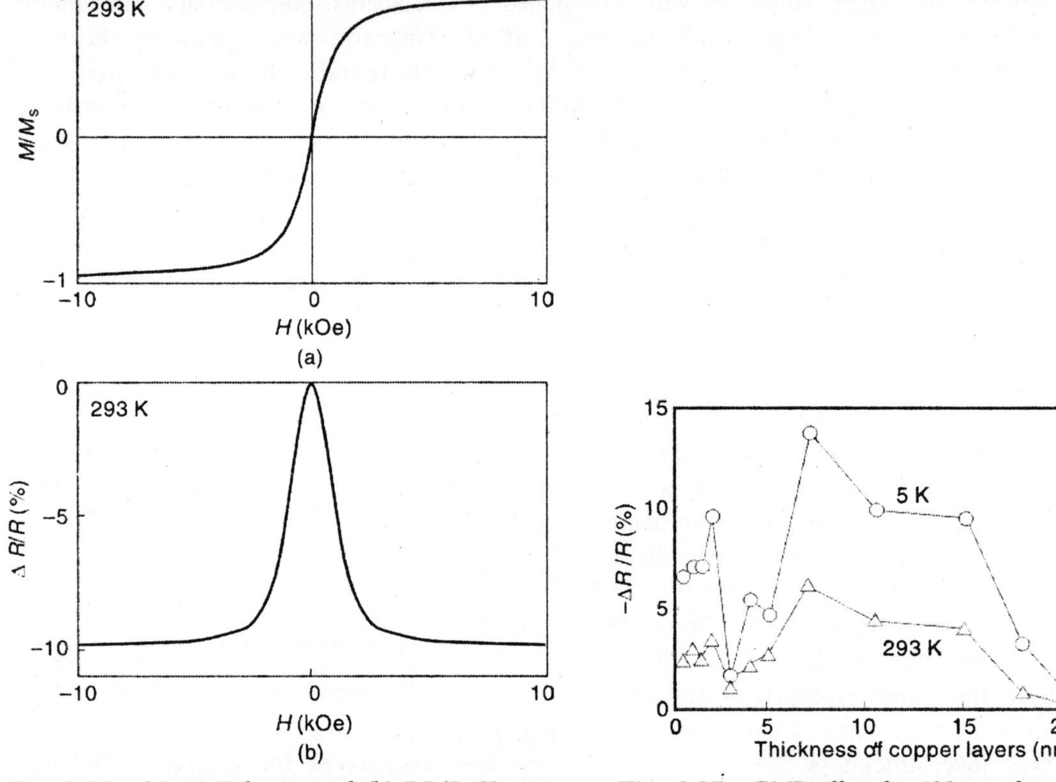

Fig. 8.26 *(a) M–H loops and (b) DR/R–H curve for 30-nm-diameter [Co(5 nm)/Cu(0.8 nm)]1,000 nanowires with the magnetic field perpendicular to the wire axis.*

Fig. 8.27 *GMR effect for 400-nm-diameter [Co(4 nm)/Cu(x mm)]$_n$ multilayer nanowires at 5 K and 293 K vs. Cu layer thickness.*

the GMR effect decreases; it approaches zero when the thickness approaches 21 nm. In this current-perpendicular-to-the-plane (CPP) geometry, the GMR effect is no longer observed when the copper layer thickness becomes comparable to the spin diffusion length (the length over which the conduction electron spin is relaxed). When the Cu layer thickness exceeds the spin diffusion length, the spin of the conduction electrons is randomized as they pass through adjacent Co layers. Thus, from this figure, the spin diffusion length in copper is estimated to be about 21 nm. As can also be seen, for smaller Cu thicknesses (<10 nm), there appear to be clear oscillations of the GMR effect, a phenomenon well established in sputtered and MBE-grown Co/Cu multilayers.

NANOWIRE SUSPENSIONS

Magnetic interactions between particles in suspension influences the stability of the suspension and is helpful in the assembly of larger-scale structures. For example, rod-shaped magnetic nanowires with high remanences act as small bar magnets in suspension and form one-dimensional chains because of the attractive wire-wire interactions. Nanowire suspensions are formed by dissolving the template in a suitable solvent. For example, polycarbonate is dissolved in dichloromethane. To ensure that the surface of each nanowire is clean, residual polycarbonate is removed by sequential centrifugation and the addition of clean solvent, or by solvent exchange. The nanowires is then suspended in water or an organic solvent.

The multilayer FM/NM nanowires in suspension to a small magnetic field is controlled by the aspect ratio and thickness of the FM and NM layers. Figure 8.28 shows suspensions of Ni/Cu multilayer nanowires in a small (≈10 Oe or 0.8 kAm^{-1}) external magnetic field. Figure 28(a) shows $[Ni(1,000\ cm)/Cu(1,000\ nm)]_3$ nanowires having a diameter of 100 and rod-shaped FM segments (aspect ratio = 10). The nanowires aligned parallel to the applied magnetic field, since their easy axis was parallel to their physical axis. Figure 8.28(b) shows $[Ni(10\ nm)/Cu(10\ nm)]_{300}$ nanowires of the same diameter and having disk-shaped FM segments (aspect ratio = 0.1). The nanowires have the same overall length and the same amount of nickel as those of part (a), but are aligned perpendicular of the applied field, since their easy axis is perpendicular to their physical axis. Thus, the shape anisotropy is exploited in tuning the response of magnetic nanoparticles in suspension.

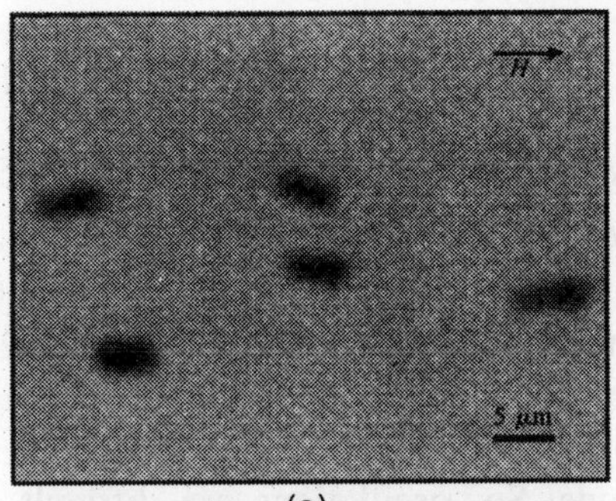

(a)

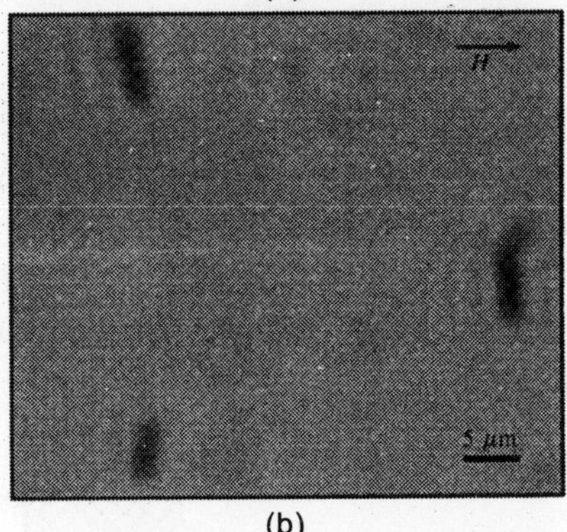

(b)

Fig. 8.28 *Optical microscope images of Ni/Cu multilayer nanowires suspended in 1:1 octadecane and hexadecane mixture under an externally applied magnetic field. In both cases the nanowires are 100 nm in diameter and about 6 mm long, with an average composition of 50 at.% nickel, (a)* $[Ni(1,000\ nm)/Cu(1,000\ nm)]_3$ *with a diameter of 100 nm and (b)* $[Ni(10\ nm)/Cu(10\ nm)]_{300}$ *with a diameter of 100 nm.*

MAGNETIC NANOSTRUCTURE ELECTRONIC STATES

As the structures become smaller the energy spacings of the states increase eventually becoming greater than the thermal energy k_T. If that happens for states at the Fermi level E_F, only a single quantum level is accessible thermally. Consequently, the dimensionality of electrons is reduced in a nanostructure because they cannot propagate along the directions of the confinement. These effects are analysed for the electronic states (in thin metal overlayers) by photoemission and inverse photoemission. The two techniques are useful since they make it possible to determine the complete set of quantum numbers that characterize an electron in a crystal, i.e., energy E, momentum p (or wave vector $k = p/h$), angular symmetry, spin photoemission measures occupied states and inverse photoemission unoccupied states. Electronic states near the fermi level are essential transport properties, such as magnetoresistance, and drive magnetic transitions.

The density of unoccupied states at E_F with parallel momentum $k_{\|} = 0$ is shown in Figure 8.29 for Cu films grown epitaxially on an fccFe which in turn is grown epitaxially on a Cu(100) single crystal. As the thickness of the Cu overlayer is increased, the density of states oscillates with a period of about six atomic layers (about 1 nm), which brings up into the nanometer regime. The amplitude of the oscillations is significant—there is no zero offset in Figure 8.29. The amplitude is limited by the smoothness of the interfaces. The smoother the films, the larger amplitude. So, oscillations are difficult to be observed in thin films.

The density-of-states oscillations can be understood by a simple interferometer model (Figure 8.29, top). Reflection of electrons at the interfaces builds up standing waves whenever the round-trip phase is a multiple of 2π. The resulting interference fringes appear with a period of $\lambda/2$, where λ is the wavelength of the electrons. The wavelength of electrons is measured using interferometers. In the case of Cu(100) at $E = E_F$ and $\mathbf{k}_{\|} = 0$, a wavelength of 2 nm is obtained. This value is much larger than the Fermi wavelength, which is comparable to a lattice constant. The connection between density of states and quantum well states is obtained from the whole

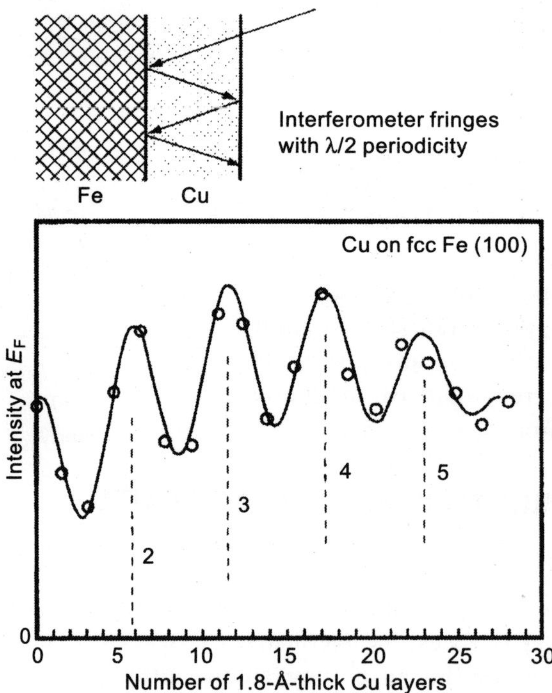

Fig. 8.29 *Oscillations in tne density of states versus thickness for Cu grown epitaxially on an fcc Fe film which was grown epitaxially in a Cu(100) single crystal. A simple interferometer model explains the oscillations as interface fringes of electron weaves reflected back and forth at the interfaces. This allows a direct measurement of the wavelength of the envelope wave function.*

spectrum of electronic states (Figure 8.30). For a bulk Cu(100) crystal (top spectrum), a continuous spectrum is observed which corresponds to the *s*, *p* band with Δ_1 symmetry at the Fermi level and has an upper cutoff at 1.8 eV about E_F, which corresponds to the p_z-like X_4' point of the Cu band structure. In the thin-film spectra, the continuous spectrum breaks up into discrete quantum well states. These states change their energy with decreasing Cu thickness and cross the Fermi level at regular thickness intervals. These Fermi-level crossings give rise to the density-of-states maxima in Figure 8.29. The fact that quantum well states change their energy with thickness is easy to understand by electron counting. In a finite slab of *N* atomic layers, one has *N* states at each $\mathbf{k}_{\|}$ point. Therefore, the energy interval between adjacent states shrinks as $1/N$ with increasing film thickness and approaches zero for the bulk. For the very thin films in Figure 2, the spacings are still large compared to kT = 0.026 eV (at room temperature): e.g., five unoccupied states within 2 eV for a 20-monolayer film, giving an average spacing of 0.4 eV films 15 times thicker (about 50 nm) are made so as to reach the point at which the energy spacing becomes comparable to kT.

Oscillatory Magnetic Coupling and GMR

The quantum well states seen in Figures 8.29 and 8.30 are closely correlated with magnetic properties such as oscillatory magnetic and giant magnetoresistance (GMR). Figure 8.31 compares the density of states, spin-polarizations, and magnetic coupling for the fcc Cu/Co(100) system. All three quantities are found to oscillate with the same period of about six atomic layers (1 nm). Giant magnetoresistance in trilayers exhibit the same period, since it is tied to antiparallel coupling. Such an agreement is found wherever comparable data exist, e.g., for Ag/bcc Fe(100), Au/bcc Fe(100), Cu/fcc Co(100), and Cu/fcc Fe(100). For the short-period oscillations in Cu/ fcc Co(100), density-of-states oscillations are also observed for transition-metal spacers: i.e., Cr/ bcc Fe(100). The physical connection between quantum well states and oscillatory magnetic coupling is done by calculating the total energy of all occupied quantum well states and observing it jump periodically with increasing thickness whenever a new state crosses the Fermi level.

That antiparallel coupling correlates with a density-of-states, maximum at the Fermi level E_F is shown in Figure 8.31. A high density of states at E_F is energetically unfavourable. The magnetic arrangement undergoes a restructuring to minimize the density of states at E_F. In a trilayer, confined quantum well states occur only for parallel orientation of the magnetic layers as shown in Figure 8.32(a). For an antiparallel orientation, the wave functions extend into one of the magnetic layers. Therefore, a transition to antiparallel coupling circumvents the high density of states created by a quantum well state. Quantum well states are observed in Figures 8.30 and 8.31 for spacer thicknesses characteristic of antiparallel coupling. The density-of-states for Cu/Co(100) bilayers, where one of the ferromagnetic layers is replaced by vacuum, confines the wave function.

Quantum well states are spin-polarized for transmitting magnetic information, even in a noble-metal spacer such as Cu, as confirmed by spin-polarized photoemission (Figure 8.31, middle) and by magnetic circular dichroism. Since they are formed by reflection of electrons at the interfaces, any spin-dependence in the reflectivity affects their confinement. In particular,

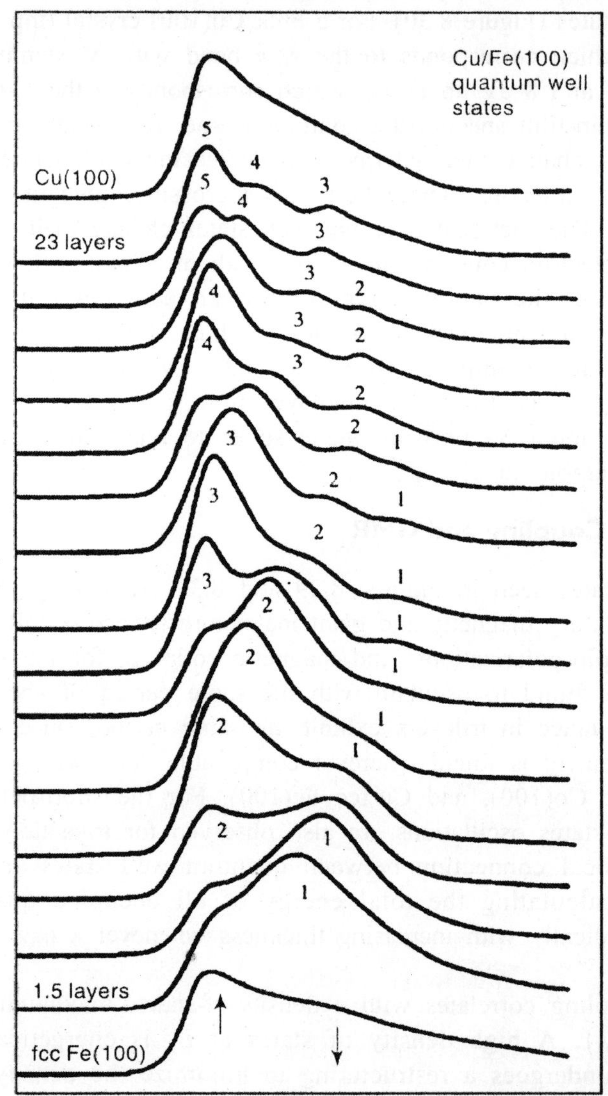

Fig. 8.30 *Inverse photoemission spectra for Cu films of varying thickness on fcc Fe(100). The s, p-band continuum of bulk Cu (top spectrum) splits into discrete quantum well states (numbered).*

the reflectivity at a ferromagnet differs between majority and minority spins, because of the magnetic exchange splitting of the bands. Figure 8.32(b) shows the interface between fcc Ag(100) and bcc Fe(100) at $\mathbf{k}_{\|} = 0$. Minority spin states at E_F in the Ag layer are totally Bragg-reflected. But, Majority states, exist on both sides of the interface and are able to propagate. This leads to quantum well states with minority spin character, which is observed by spin-polarized photoemission. They are superimposed on a continuum of majority spin states. In general, there is no clear distinction between the reflectivity of majority and minority spins. But, there is always a spin-dependent potential step at the interface that makes the reflectivity

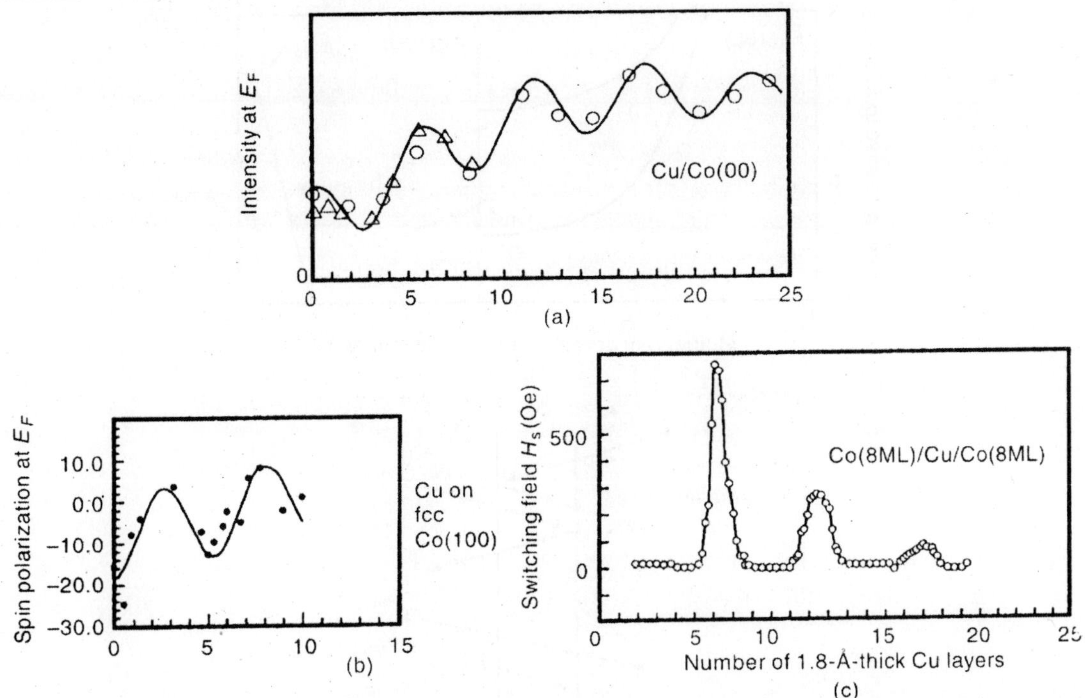

Fig. 8.31 *Comparison of oscillations in density of states (a), spin-polarization (b), and magnetic coupling (c) for fcc Cu/Co(100) sandwich structures. Identical periods of six monolayers (ML) suggest a connection between quantum well states at the Fermi level and oscillatory magnetic coupling.*

spin-dependent. For spacer layers to the right of the ferromagnets in the periodic table, the majority bands are more closely aligned than the minority bands. This explains the observation of quantum well states with minority spin in hole metals. For spacers to the left of the ferromagnets, e.g., Cr, the situation is reversed, the spin-polarization in Cr has not yet been measured.

Spin-dependent reflectivity of electrons at interfaces is important for quantum well states and magnetic coupling. It has an effect on giant magnetoresistance as well. Models of the magnetoresistance for "conductivity in-plane" (CIP) and "conductivity perpendicular to the plane" (CPP) provide spin asymmetries for interface and bulk scattering in magnetic multilayers. Typically, interface spin asymmetries are four times larger than in the bulk, which leads to the fact that the CPP geometry exhibits 3–10 times higher magnetoresistance than the CIP configuration. These asymmetries are due to diffuse scattering and other parameters for magnetoresistance. Concentrating on interface scattering in the CPP geometry, one can construct a simple polarizer-analyzer model for giant magnetoresistance: Each interface with a ferromagnet acts as a polarization filter for electrons. If two magnetic layers are aligned antiparallel, one has the analog of crossed optical polarization filters, i.e., little transmission. If they are parallel, they both transmit electrons. The smaller magentoresistance of the CIP geometry has a shunting

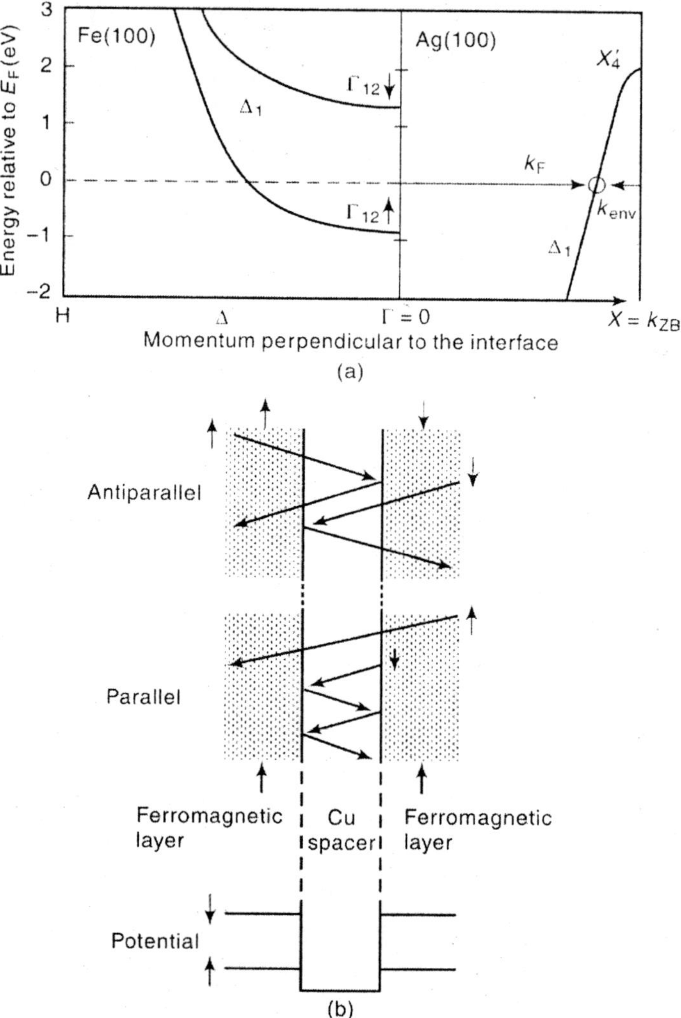

Fig. 8.32 *(a) Connection between the spin-polarization of quantum well states and spin-dependent reflectivity. Minority spin states cannot propagate from fcc Ag(100) into bcc Fe(100) at the Fermi level E_F, since there are no minority states with the same D_1 symmetry in Fe(100). Instead, they form standing waves, i.e., discrete quantum well states. Majority spin states are able to propagate and remain a band-like continuum. (b) possible wave functions in trilayers, where two ferromagnetic layers are separated by a nonmagnetic Cu spacer.*

effect whereby electrons are able to avoid spin-dependent scattering at the interfaces by travelling inside the layers. In order to achieve the high magnetoresistance of the CPP geometry several attempts have been made to produce one-dimensional wire or thin-film stripe structure, in which the current runs perpendicular to the interfaces but is confined to the wire or the thin film of stripes. Lithographic patterning has been used to produce sawtooth-shaped Si substrates for the deposition of magnetic stripes. Magnetic wires perpendicular to the surface have been obtained by filling pores in a polymer film with electroplated metals. Designing such structures opens up a fascinating world of low-dimensional growth phenomena. Below is a new approach for producing stripes with nanometer width at stepped surfaces.

Since electron reflectivity at interfaces plays and important role in the magnetic phenomena the phenomena is modified by coating the interfaces with a different material. Using one or two monolayers of a ferromagnet with large magnetic band splitting, such as Co, it is possible to enhance the magnetoresistance of permalloy/Cu/permalloy structures by more than a factor of 2. Electronic states are found at such "doped" interfaces by inverse photoemission, e.g. in Cu/Co/Ni(100). A Co3*d*-like interface state builds up as the same fast rate at the enhanced magnetoresistance. There is a magnetic band splitting in Co, which is three to four times as large as in Ni. (The spin splitting is about 2 eV in bcc Fe, 1 eV in fcc Fe, 1 eV in fcc and hcp Co, and 0.3 eV in fcc Ni.) Quantum well states in the subsequent Cu layer are enhanced near the interface, indicating stronger confinement by higher, spin-dependent reflectivity. Going to thicker ferromagnetic overlayers, quantum well states act as the noble-metal spacer and as ferromagnet. In the optical interferometer; reflective or antireflective coatings on the electron mirrors, and the reflectivity oscillates with the thickness of the ferromagnetic overlayer. This affects the strength of the magnetic coupling, but not its periodicity, which is determined by the noble-metal spacer layer.

Stripes and Dots at Stepped Surfaces

For growing one-dimensional nanostructures, consider growth at stepped surfaces, where the steps provide a template for attaching magnetic wires or stripes. The first monolayer growth on a stepped surface in two dimensions, growth modes are observed, layer-by-layer. Same is observed for growth in three dimensions. As shown in Figure 8.33, Cu grows on a stepped Mo(110) surface in parallel stripes that correspond to a row-by-row growth mode. On W(110), only the first row of Cu atoms are observed at the step edge, and additional Cu grows in monolayer-height islands that are attached to the step edges. An additional phenomenon at stepped surfaces is the presence of an energy barrier for crossing steps. This is found at an uphill crossing, since the atom incorporated at the step edge is less coordinated on the terrace. Even for a downhill crossing, the atom gives up neighbours when crossing the step edge. An atom exchange mechanism eliminates this barrier on certain metals during homoepitaxy. In a heteroepitaxial system, such as Cu/Mo(110), the Cu atom crossing a Mo/Cu boundary trades Mo neighbours for Cu neighbours, which provides less binding energy. Such a barrier causes the inhomogeneous width of the Cu stripes on Mo(110) in Figure 8.33(b). The wider the terrace on which a Cu stripe resides, the wider the stripe. Apparently, the Cu atoms migrate toward the uphill edge of an individual terrace. It requires higher annealing temperatures to let Cu atoms cross step edges and equilibrate their stripe width, independent of the terrace width. Stripes with a uniform width of 3 nm have been obtained for Cu on stepped Mo(110) after annealing to 873 K. The existence of uniform stripes in spite of nonuniform terraces indicates that the binding energy of Cu increases monotonically toward the step edge.

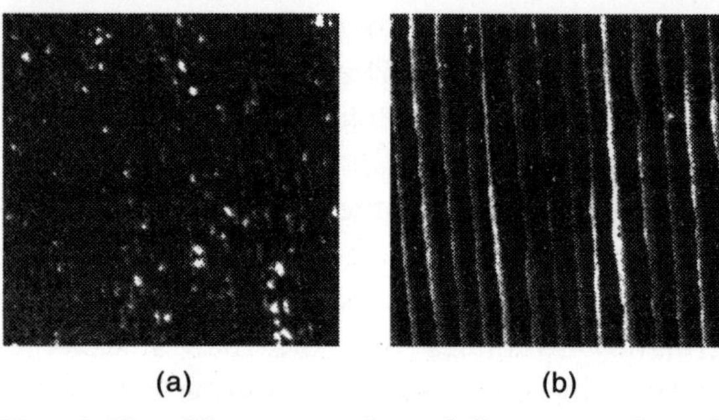

Fig. 8.33 *Observation of two different growth modes in two dimensions. Cu grown in stepped W(110) (a) and Mo(b) respectively displays the analog of Stranski–Krastanov and layer-by-layer growth. The average step spacing is 50 nm.*

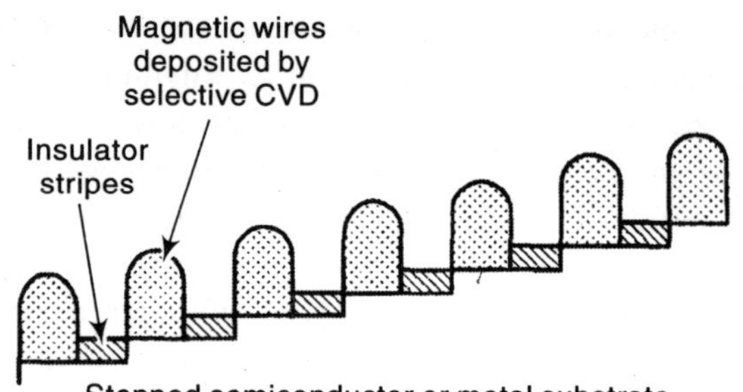

Fig. 8.34 *Illustration of a proposed "universal" deposition method for producing striped structures on a stepped surface. Initially, a set of inert striped structures on a stepped surface. Initially, a set of inert stripes, e.g., an insulator, is deposited by step-flow growth. These act as templates for selective deposition of the magnetic wires on the remaining substrate.*

To make step decoration a two-step process is used (Figure 8.34): In the first step, a template of inert stripes is grown by step flow. In the second step, the desired material is deposited selectivity on the remaining stripes of the reactive substrate using chemical vapour deposition (CVD). For example, one could consider a stepped Si(111) surface as substrate; passivating stripes of lattice-matched CaF_2 are grown along the step edges, and deposit metals are grown by selective CVD on the remaining Si. Compounds for substrate-selective CVD are developed for contacting Si devices through via holes in SiO_2 without shorting out the SiO_2 insulation. Selective deposition of W or Ta via WF_6 or TaF_5 is useful to create a reaction barrier on the Si stripes against the formation of silicides with Fe,Co, or Ni. For the deposition of noble-metal spacers, such as Au, organometallic precursors are used. The spacers are evaporated after selective CVD of ferromagnetic wires.

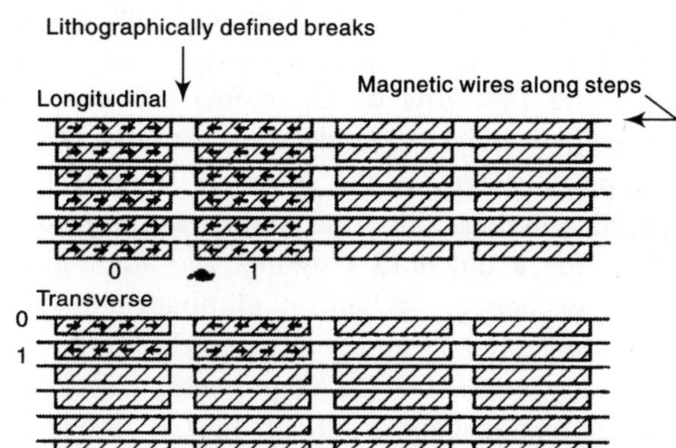

Fig. 8.35 *Proposed array of anisotropic magnetic dots produced by patterning a set of stripes lithographically (typical size 20 × 200 nm^2). Such a structure might be useful as a storage medium with uniform longitudinal recording; pairs of dots could be use in a transverse recording scheme.*

If it becomes feasible to grow magnetic nanowires by step decoration, an array of regular dots by patterning the wires lithographically can be produced (Figure 8.35). Using a UV laser, a standing-wave pattern with a periodicity of 200 nm is produced on a photoresist that allows elongated dots with typical dimensions of 20 × 200 nm^2. Such a regular array of highly anisotropic magnetic particles make a good medium for the ultimate goal in magnetic recording, single-particle-per-bit storage. Current media with randomly segregated particles in a CoPtCr alloy typically use 10^3 particles per bit to reduce the readout noise introduced by irregular particles and domains. Working with a small subarray of magnetic dots per bit (say 1×100) reduces this large number significantly without deviating too far from the longitudinal recording geometry. When the number of dots per bit becomes too small, there are tracking problems. Other geometries have a reasonably wide track, e.g., the two-particle-per-bit transverse recording scheme shown in Figure 8.35. The stray field between antiparallel-magnetized pairs of elongated dots becomes the readout signal. The two bit levels corresponds to magnetizations pointing toward each other or away from each other.

Element-sensitive STM via Resonant Tunneling

For determining how stripes grow at steps, it is essential to have a high-resolution microscope that not only resolves the topography but is also able to distinguish between the substrate and the new material incorporated at the step edges. For perfect step-flow growth of lattice-matched materials, there is no topographic contrast. To achieve chemical contrast, an element-specific version of scanning tunneling microscopy (STM) is developed. It is based on resonant tunneling into empty surface states that are specific to each metal. Thus, the electronic structure of nanostructures again becomes important. Figure 8.36 shows two STM images of the same three Cu stripes (about 3 nm wide) on a stepped Mo(110) surface. At specific bias voltage, it is possible to make the Cu appear either brighter than Mo (at +5 V) or darker (at +0.06 V). Such chemical contrast results from resonant tunneling of electrons from the tip into unoccupied surface states of the Cu or Mo, respectively. With most other bias voltages there is little contrast between the Cu and Mo (not shown). This reflects the topography, since Cu has the same size as Mo and grows in registry with the substrate lattice.

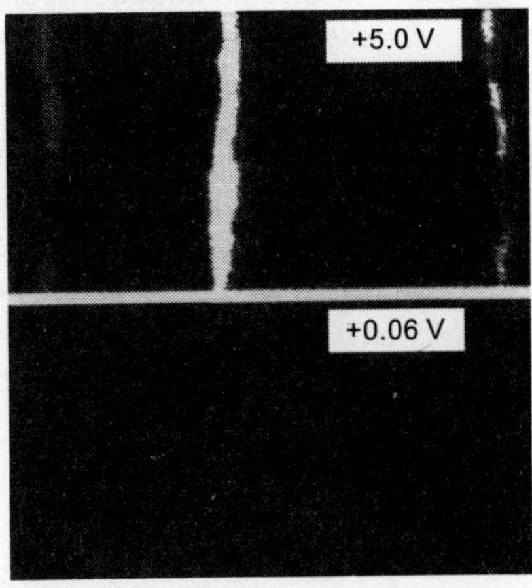

Fig. 8.36 *STM images of 3-nm-wide Cu stripes on stepped Mo(110) with a step spacing of about 50 nm. Specific sample bias voltages can be used to enhance the brightness of either the Cu or Mo via resonant tunneling into surface states and image states.*

The states available for resonant tunneling are found with inverse photoemission (Figure 8.37, shaded states). There are two types of element-specific surface states for Cu on W(11) and Mo(110). One is a p_z-like surface state near the Fermi level, the other an image state near the vacuum level. Mo appears bright at a bias of +0.06 V because of tunneling into its p_z-like surface state; Cu appears bright in Figure 8 at +5 V bias because of tunneling into its image state. While the upward shift of the p_z-like surface state from W to Cu is nontrivial, the downward shift of the image state is easily understood. It tracks the work function reduction $\Delta\Phi$ from W to Cu, since image states are bound to the vacuum level E_v and the energy zero is the Fermi level E_F in inverse photoemission and STM ($\Phi = E_v - E_F$). The work functions of metal surfaces are well known and

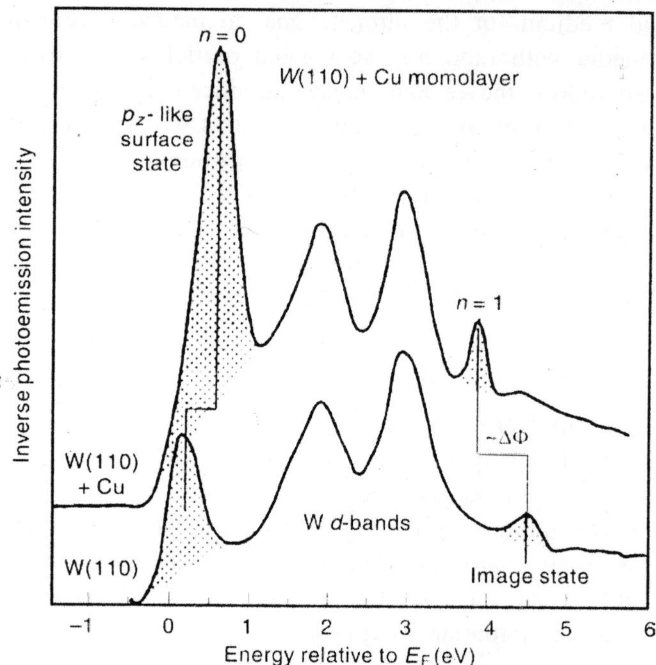

Fig. 8.37 *Inverse photoemission spectra from surface and image states of clean W(110) and from a Cu monolayer on W(110). Resonant tunneling into the shaded states provided contrast between the Cu and W in STM and allows their identification.*

are tabulated, allowing identification of different metals at a surface. Such tunneling resonances into image states are modeled with realistic tip geometries.

Resonant tunneling, is observed at we display the enhancement of the Cu stripes in the constant-current mode in Figure 8.38. The apparent height of the Mo–Cu step is normalized to Mo–Mo step height of 0.2 nm. At bias voltages corresponding to the n = 1, 2, 3 image-state resonances, sharp resonances with enhancements of up to 100% are observed. The tunneling

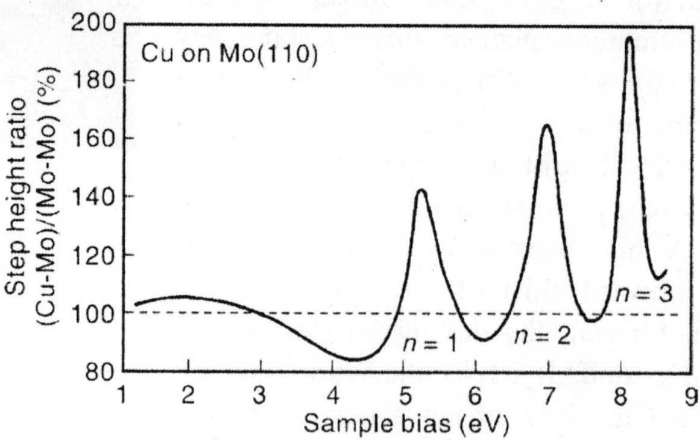

Fig. 8.38 *Resonant tunneling via image states for Cu stripes on stepped Mo(110). The apparent step height from a Mo terrace to a CU step is plotted versus sample bias voltage. Enhancements of up to 100% are seen at the n = 1, 2, 3 image-state resonances.*

current increases typically by two orders of magnitude over such a distance in STM, so a effect on current images at constant height is expected. This enhancement is due to narrow intrinsic width of the image states, which can be as low as 0.02 eV in Cu. The p_z-like surface state is broader and generally does not produce a large effect, but it allows higher spatial resolution, since the tip comes closer to the surface at low bias voltages. For example, the narrow parts of the Cu stripes in Figure 8.36 are resolved in the high resolution image at +0.06 V but are lost at +5-V bias. Thus, the low-work-function materials are firt identified by its $n = 1$ image state and then, by tuning to a low-bias surface state, proceed to a higher resolution.

SPIN ANGULAR MOMENTUM TRANSFER

Spin angular momentum transfer, or spin-transfer, is the transfer of spin angular momentum between a spin-polarized current and a ferromagnetic conductor. The angular momentum transfer exerts a torque spin-current induced torque, (or spin-torque) on the ferromagnetic conductor. When its dimensions are reduced to less than 100 nm, the spin-torque becomes comparable to the magnetic damping torque at a spin-polarized current of high current density (above 10 Acm^{-2}), giving rise to a new of current-induced excitation and magnetic switching phenomena. This is observed in sub-100-nm current perpendicular spin-valves and magnetic tunnel junctions, and appears promising as a basis for direct write-address of a nanomagnetic bit when the lateral bit size is reduced to well below 100 nm.

The configuration of magnetization orientation in a metallic ferromagnetic system affects the electron transport properties of the system, example, in a multilayered magnetic and nonmagnetic metal thin-film stack, the resistance of the stack depends on the relative magnetic orientation of the individual magnetic layers. The change in resistance resulting from this is known as the "giant magnetoresistance".

Spin angular momentum transfer, or spin-transfer, pertains to reverse effect: the influence of a spin-polarized current on its host magnetic conductor, as shown in Figure 8.39. Such spin-transfer-induced magnetization reversal is observed only in magnetic structures smaller than ~0.1 μm in size.

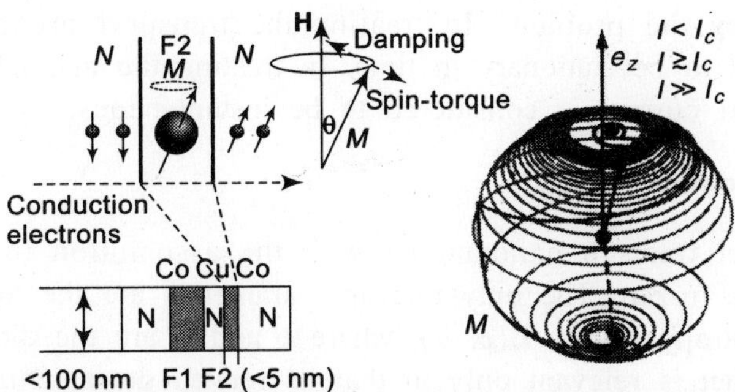

Fig. 8.39 *Illustration of spin-transfer and associated macro-spin dynamics. A uniaxial anisotropy is assumed to exist with its easy axis. The magnetic field H is applied in the same direction M designates the magnetic moment of the ferromagnetic layer.*

Spin-transfer and its related macro-spin dynamics is shown in Figure 8.39. At the lower left in the figure is a two-ferromagnet layered spin-value structure. The current passes through the left ferromagnet (F1) and becomes spin-polarized. When it passes through the second, thinner ferromagnet on the right (F2), the polarization direction of the current changes depending on the relative orientation of F2 and F1. This is shown at the upper left in the figure, where N designates a nonmagnetic conductor. The second ferromagnet experiences an effective torque. This spin-current-induced torque, or spin-torque, is in a direction that opposes the magnetic damping torque for F2, as shown in the figure. For a large enough current, the spin-torque overcomes magnetic damping. This causes an instability to develop, and the precession cone angle increases with time. When the cone angle increases past the equator, both the damping torque and the spin-torque point toward the south pole, which becomes a stable point for F2, thus completing the magnetic reversal, as depicted at the right in the figure. The situation for reversed current direction is a bit more complex, but the net spin-torque on F2 remains proportional to the current. The reversal process remains the same as described above.

The presence of a spin-torque on a ferromagnet due to spin-transport, affects magneto-transport on many levels. First, the process is microscopic and quantum-mechanical in nature and involves spin-polarized transport. The collective effect of the spin-transport presents a new torque that affect the magneto-dynamics of the ferromagnet. The magnetic response of the ferromagnet, in turn, affects the electronic transport. But, electronic transport process and the magneto-dynamics operate on two very different time scales. So it is possible to treat the transport and the magneto-dynamics processes separately.

The collective response of the magentization in a ferromagnet is governed by the effective magnetic field through a dynamic equation at a time scale that is greater than 100 picoseconds. For an isolated macro-spin, the time scale τ is related to an effective magnetic field H, with the relation $2\pi/\tau \sim 2\mu_B H/\hbar$, or about 0.5 nanosecond for every kilo-oersted of magnetic field. In the expression, μ_B is the Bohr magneton, and $\hbar = h/2\pi$, with h being Planck's constant. The spin-polarized electronic transport, tends to have a response time faster than or of the order of the spin-flip-scattering lifetime, which is of the order of several tens of picoseconds for materials with relatively little spin-flip scattering, such as Cu, and shorter for materials with strong spin-flip scattering, such as Pt. The difference in these two time scales make it possible to mathematically simplify the problem. In treating the transport process, the moment of the ferromagnet is assumed to be stationary in time. In treating the magnetic dynamics, the related adjustment for transport current is considered to be instantaneous.

Basic Macro-spin Dynamics

A macro-spin model treats a nanomagnet with the assumption that its internal magnetic degrees of freedom are frozen. The only relevant parameters are the total magnetic moment **m** and the magnetic anisotropy energy $U(\theta, \varphi)$, where θ and φ are the direction angles of **m**. The shape of the nanomagnet is relevant only in that its related demagnetization energy contributes to the total anisotropy energy function $U(\theta, \varphi)$.

Dynamics of a Nanomagnet under Spin-current-induced Torque

When a spin-polarized current passes through a ferromagnetic electrode, the ferromagnetic repolarizes the current in the direction of its magnetization. In the process, some of the angular momentum from the electron spins is absorbed by the ferromagnet, resulting in the exertion of a net torque (spin-torque) on the ferromagnet.

For a nanomagnet macro-spin within which the magnetization is uniform, the transverse component of the spin-torque is

$$\Gamma = -g(\mathrm{n_m}, \mathrm{n_s})[\hbar(2e)](\eta I/m^2)(\mathrm{n_s} \times \mathrm{m}) \times \mathrm{m}, \tag{1}$$

where m is the magnetization vector, m is its magnitude, $\mathrm{n_m}$ is its unit vector direction, $\mathrm{n_s}$ is the direction of spin-polarization of the incoming current, and $\eta = (I_\uparrow - I_\downarrow)/(I_\uparrow + I_\downarrow)$ is the spin-polarization factor, where $I_\uparrow$ and $I_\downarrow$ are the majority and minority spin-polarized currents with their polarization axis defined by the polarizing magnet (F1 in Figure 1). The term $g(\mathrm{n_m}, \mathrm{n_s})$ is a numerical prefactor that describes the angular dependence of the efficiency of spin-angular momentum transfer, originating from the quantum-mechanical nature of the interaction between spin-polarized current and the macro-spin; it also depends on the global spin-current and the boundary condition of the spin-density. The case of a constant $g(\mathrm{n_m} \cdot \mathrm{n_s}) \equiv 1$ within the macrospin-based model describes a simple redirection of the spin-current polarization direction and complete absorption of its transverse angular momentum by the macrospin.

For simplicity assume a constant $g(\mathrm{n_m}, \mathrm{n_s})$ and use Equation (1) as the basic interaction that enters the magneto-dynamics equation for the motion of the macro-spin. The macrospin dynamics equation is:

$$(1/\gamma)\frac{d\mathrm{m}}{dt} = \mathrm{m} \times \left[\mathrm{H} - (\alpha/m)\mathrm{m} \times \left(\mathrm{H} + \frac{\eta\hbar I}{2em\alpha}\mathrm{n}_s\right)\right], \tag{2}$$

where γ is the gyromagnetic ratio $g\mu_B/\hbar \approx 2\mu_B/\hbar$, and α is the LLG damping coefficient.

Threshold Current for Magnetic Amplification

For simple geometries and under a macro-spin approximation, Equation (2) can be linearized and solved for its stability boundary. For a thin free-layer nanomagnet in a collinear geometry with the easy axis of its uniaxial anisotropy field aligned with that of the applied field and the easy-plane anisotropy sharing its easy plane with the film plane, a stability threshold current I_c of

$$I_c = \left(\frac{2e}{\hbar}\right)\left(\frac{\alpha}{\eta}\right) m\,(H + H_k + 2\pi M_s) \tag{3}$$

is obtained. Here M_s is the saturation magnetization of the free layer (F2), and $m = (abt)M_s$ is the total magnetic moment of the free layer, with a, b as its lateral dimensions and t as its thickness.

Equation (3) gives a current threshold above which the *linearized* LLG equation becomes unstable over time, and a net gain of precession cone-angle results. In comparing with experimental results, however, effects of large cone-angle precession must often be carefully taken into account, since the development of an initial cone-angle increase dictated by the linear stability threshold may not necessarily lead to complete magnetic reversal. However, in many simple systems such as those with uniaxial-only anisotropy or thin-film nanomagnets with a strong easy-plane anisotropy (due to demagnetization and a moderate in-plane uniaxial anisotropy), the linear stability threshold often leads to the reversal of the magnetic moment.

Several factors affect the switching current example the simple stability threshold expression [equation (3)] does not account for them; the finite temperature effect; the actual LLG damping coefficient for a particular device structure and of the exact spin-polarization value η; details of the spin-transport, whether it is ballistic or diffusive, or whether the interface contributes to spin-flip scattering. For simplicity, equation (3) is derived using spin-torque expression of equation (1) with $g = 1$. Detailed calculations for each of these specific spin-transport possibilities gives rise to additional angular dependences of the spin-torque as a function of the relative orientation between F1 and F2.

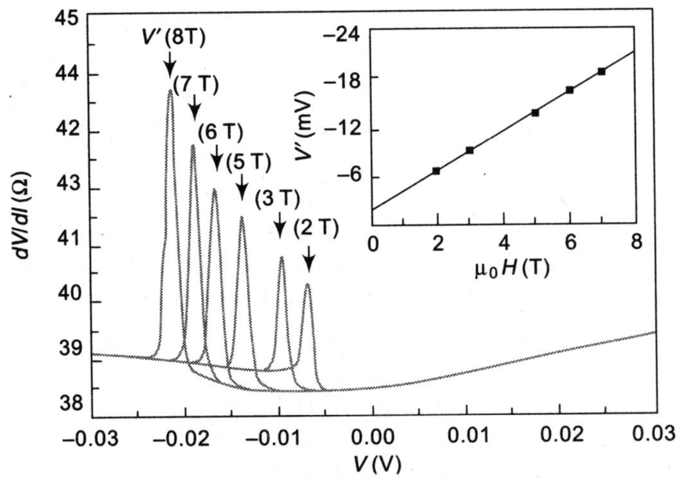

Fig. 8.40 *Differential resistance of a point-contact junction formed by a silver tip and a Cu|Co|Cu|Co multi-layer thin film. Inset: The threshold voltage (or current) depends linearly on the magnitude of the external magnetic field applied perpendicularly to the film surface.*

The intercept-to-slope ratio of the threshold boundary $I_c(H)$ is defined as $R_{IS} = I_c(0)/(dI_c/dH)$. For equation (3), $R_{IS} = H_k + 2\pi M_s$. A lower value of R_{IS}, by almost an order of magnitude is found experimentally and this decrease in R_{IS} well below $H_k + 2\pi M_s$ is due to the finite temperature effect. Equation (3) represents a zero-temperature stability threshold. At finite temperature, additional thermal agitation is present, making the apparent threshold current lower and yielding a smaller R_{IS} value.

Energy Flow during Precession and the Berger Voltage

When the threshold current I_c is exceeded, precession at a large cone angle usually follows. Such large-cone-angle precession cause an additional voltage rise across the spin-valve structure because of energy conservation as the dissipation associated with magnetic damping is proportional to the square of the sin of the precession cone-angle. Hence, to maintain a large-cone-angle precession, energy must be supplied from the transport system, resulting in an additional voltage rise.

For large-amplitude magnetic excitation in a highly asymmetric magnetic stack (where the free magnetic layer is much thinner than the thick layer), the asymptotic limit of the voltage rise due to precession is predicted to be $\Delta V(\omega) = (\hbar\omega/2e)\,(\sigma_1 - \sigma_2)/(\sigma_1 + \sigma_2)$, where $\sigma_{1,2}$ are the majority and minority channel conductivities defined by the thick ferromagnetic layer, corresponding to a spin-polarization factor of $(\sigma_1 - \sigma_2)/(\sigma_1 + \sigma_2) = (I_\uparrow - I_\downarrow)/(I_\uparrow + I_\downarrow) = \eta$. Replacing $\omega = 2\mu_B H/\hbar$ as the precession frequency gives

$$\Delta V = \eta\left(\frac{\mu_B}{e}\right)H \tag{4}$$

Equation (4) follows from energy conservation. To maintain the nanomagnetic precession at a cone angle θ, energy is delivered from the transport current to the magnetic system. This dissipation power, when supplied by a transport current of $I = I_c$, must give rise to a voltage difference ΔV, which gives $I_c\Delta V = -dU(\theta)/dt = \alpha\gamma m\mathrm{H}^2\sin^2\theta$. For maximum spin-wave excitation, $\theta = \pi/2$, and hence $e(\Delta V) = \eta\mu_B H$. Here, $U(\theta) = -mH\cos\theta$.

Size

There are two mechanisms that cause interaction between a magnetic moment and a transport current: current-induced magnetic field (the oersted field) and spin-polarized current-induced spin-torque. A current-induced magnetic field for a wire of radius r is related to the maximum field (usually around the surface of the wire) and the current passing through the wire I. From Maxwell's equations, the relation is $I = (c/2)rH$ (in gaussian units, c is the speed of the light). A spin-valve of similar lateral size ($2r$) has a spin-torque threshold current [from equation (3)] of the order of $I_c \approx (H + H_k + 2\pi M_s)(4r^2 t)M_s(\alpha/\eta)(2e/\hbar)$.

The spin-torque threshold is proportional to r^2, and the oersted-field-related current (for a given threshold field, such as the anisotropy field H_k) is proportional to r. Thus, at large dimensions the threshold from the oersted field is the lower threshold. The crossover point for high-moment thin films such as cobalt. With $H \sim H_k << 2\pi M_s$, is roughly

$$r_c = \left(\frac{c\hbar}{4e}\right)\left(\frac{\eta}{\alpha}\right)\left(\frac{1}{M_s t}\right)\left(\frac{H_k}{2\pi M_s}\right) \tag{5}$$

which gives, for 30-Å-thick cobalt, an $r_c \approx 0.04$ μm, assuming a spin-polarization factor of $\eta \approx 0.1$, an LLG damping coefficient $\alpha \approx 0.01$, and $H_k \sim 100$ Qe. Thus, a practical crossover dimension for a pillar-structured spin-valve is of the order $2r_c \approx 0.1$ μm, below which the spin-torque effect is more significant.

Spin-current-induced Magnetic Excitation

Spin-current-induced magnetic excitation is observed in a point-contact junction on giant-magnetoresistance (GMR) multilayers and magnetic switching in highly spin-polarized manganite junctions. It also reveals the effect of the spin-torque on spin-transfer-induced magnetic reversals. The effect of the spin-angular momentum of a carrier on abrupt magnetic domain walls and the presence of a spin-torque from a spin-polarized current in a magnetic multilayer geometry assuming ballistic transport and using WKB wave function is also predicted. It is observed that magnetic excitation results from, bringing a point-contact tip made of silver into contact with a multilayered Cu|Co|Cu|Co ... thin film. The current density under the point contact is high enough to exceed the spin-torque excitation threshold. So, a local excitation and reversal of magnetic moment results—probably only for the first cobalt layer. This manifests itself as a step in the current–voltage (*I–V*) characteristics of the point-contact junction, as shown in Figure 8.40. The threshold voltage varies linearly with the magnetic field *H*, applied perpendicular to the film surface, and is large enough to overcome the easy-plane demagnetization field of cobalt (which is about $4\pi M_s \approx 17.6$ kOe). This linear dependence of threshold current vs. applied field is consistent with equation (3). The presence of a voltage threshold for spin-wave (mangon) emission at a certain energy, with the voltage threshold determined by the Zeeman splitting, $V_c \approx (g\mu_B/e)H$ is there.

The point contact onto magnetic layers is also obtained from a special wafer with a SiN membrane. It is used, with a nanometer-size hole which is lithographically fabricated into the SiN membrane. The magnetic multilayers are then deposited on one side of the wafer, and the metal point contact (copper in this case) is deposited on the other side, forming a metallic contact, through the SiN hole and thus a point-contact junction. Not only signatures of magnetic excitation, but current, induced hysteretic magnetic reversal is observed.

The current-induced magnetic reversal is also observed in manganite-based all-oxide tri-layer magnetic junctions such as the one shown in Figure 8.41. The junction is 1 μm × 1 mm in size, although the actual current path is much smaller.

The junction-switching behavior is consistent with a spin-transfer-induced magnetic reversal process. This observation stimulate a renewed interest in the spinangular moment transfer process in spin polarized transport systems.

The transfer occurs in only a small fraction of the junction prepared. It originates from interface-inhomogeneity-related current paths, and the switching occurs for only those junctions in which the interface inhomogeneity is at the right place with the right size. These junctions, are, switched with a well-defined threshold current whose value showes a dependence on applied magnetic field.

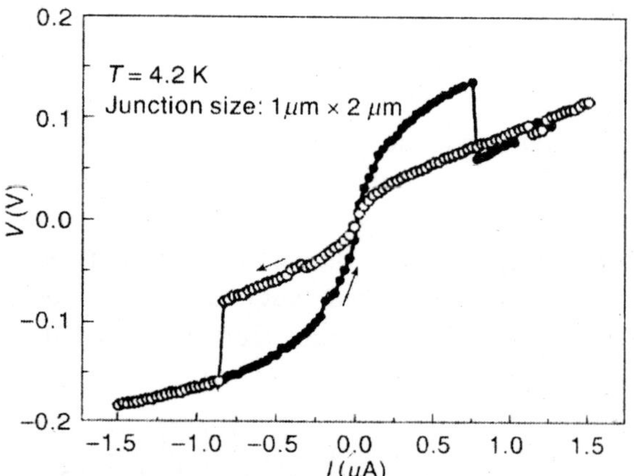

Fig. 8.41 *The bias-current-induced switching of resistance in a ||LSMO|LSMO|| trilayer junction. (LSMO is the Mn perovskite* $La_{0.7}Sr_{0.3}MnO_3$; *STO is the insulator* $SrTiO_3$.*) The* R_{high} *and* R_{low} *values of the current-induced states correspond to those of the magnetic-field-induced* R_{high} *and* R_{low} *values.*

The interplay between spin-polarized current and the magnetic-field-driven reversal in nanomagnetic electrode is observed; by using an electroplated Ni wire, 80 nm in diameter and about 500 nm in length. A shift in the threshold magnetic field for a resistance-field hysteresis loop when a 10^7-Acm^{-2} pulsed current is observed. The change of threshold field is about 100 Oe at a current pulse of 0.15 mA, larger than any induced magnetic field the current could generate because the spin-polarized current is affected by the magnetic reversal threshold field.

The quantitative proof of a spin-transfer-induced magnetic reversal is shown for metal current-perpendicular (CPP) spin-value nanomagnets by using electron-beam lithography to define a PP spin-value nano-pillar, about 100 nm in lateral dimension, in Figure 8.42. A clear signature of magnetic reversal in observed at a threshold current density in the mid-10^7-Acm^{-2} range. The threshold current demonstrates the characteristic linear dependence on applied magnetic field, with a slope.

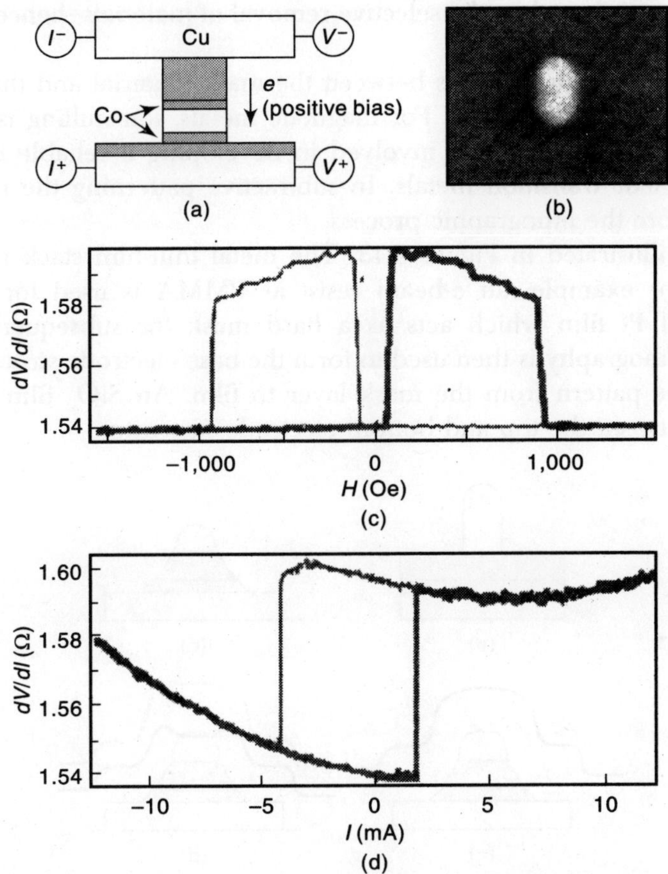

Fig. 8.42 *(a) Current-induced reversal of magnetization in a current-perpendicular spin-valve. (b) The lateral size of the junction ranged from 120 nm in diameter to 70 nm by 120 nm, as shown in the scanning electron micrograph (SEM), (c) The resistance vs. magnetic field sweep (d) the resistance vs. bias current sweep observed is shown. Resistance corresponds to dV/dI, which is measured using a locking detection method with an ac-bias current superimposed on the dc-bias current.*

Lithographic Fabrication

Quantitative experimental investigation of the spin-transfer effect requires access to well-defined magnetic nanostructures of the order of 100 nm or less in lateral size. This is usually done using electron-beam lithography. The strategies are successfully used for the fabrication of spin-transfer devices with CPP structures by a subtractive process and an additive process.

In the subtractive process a blanket multilayer film on a wafer with CPP spin-valve layering structure is formed as multilayers. Usually such multilayer are formed *in situ* in order to preserve the integrity of the thin-film interfaces. The interfaces are critical to spin-polarized transport and to the magnetic properties of the layers. The required CPP pillar shaped structure is then formed by masked etching steps to remove certain parts of the films while preserving other areas. The process is based on the selective removal of materials–hence the term *subtractive patterning.*

The selectivity of the etching process between the mask material and that of the multilayer films forming the device is considered. For magnetic metals, ion milling is the most practical method. This is due to the difficulties involved in developing a reliable dry plasma etching process for ferromagnetic transition metals. In subtractive patterning the magnetic multilayer stack is prepared before the lithographic process.

Such a process is illustrated in Figure 8.43. The metal thin-film stack is formed first–with sputter deposition, for example, an e-beam resist as PMMA is used for the lift-off pattern transfer of 500 Å of Pt film which acts as a hard mask for subsequent ion-mill etching. Conventional photolithography is then used to form the base electrode structure. Ar ion milling is used to transfer the pattern from the mask layer to film. An SiO_2 film is usually used for electrical isolation between the top and bottom electrode structures.

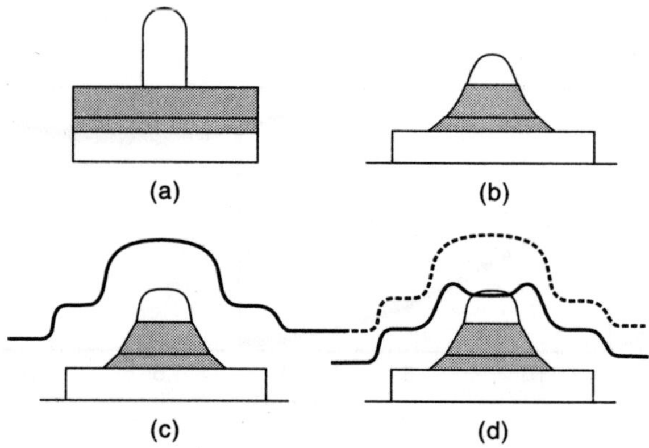

Fig. 8.43 *A subtractive process: (a) Pt mask fabricated by lift-off of an e-beam-defined bilayer PMMA resist pattern. (b) The result of ion milling in order to transfer the metal mask pattern onto the junction stack. An optical lithography step is then used to define the base-electrode via (c) blanket coating of the entire structure with SiO_2 and "etch-back" in CF_4 of a (d) photoresist planarized surface. The etching stops on the Pt, exposing the top of the junction for cross-wire contacting.*

The additive process uses a predefined structure on a substrate (such as a lift-off photoresist mask or a stencil mask of some other type) to define the necessary device structures. An example of a batch-fabricatable additive process is illustrated in Figure 8.44. In this process the cycle time between magnetic multilayer film deposition and final device testing is shortened

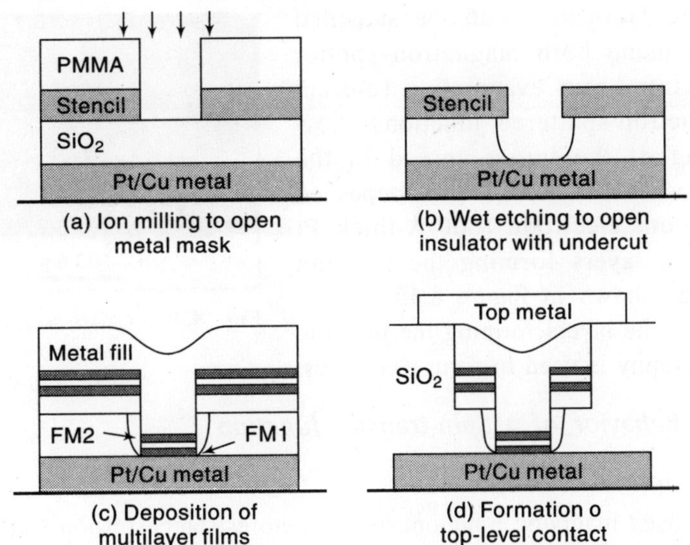

Fig. 8.44 *The stencil process: (a) E-beam lithography and pattern transfer onto the stencil; (b) wet etching to open the insulator spacer and create a controlled amount of undercut; (c) deposition of magnetic stack followed by metallic filling to from top electrode contact; (d) optical lithography to define the wiring.*

and because novel magnetic materials are difficult to etch, a process allowing for controlled shape definition without having to etch magnetic thin films is also flexible.

Three types of thin films, Pt, Si and Ge, are used as the masking layer, each having its own advantages and disadvantages. A Pt stencil is chemically more resilient, yet it is more difficult to etch. Ar ion milling often leaves uneven edges because the surface morphology of the Pt film reflects the underlying grain structure of the bottom electrode film. The use of a Ge and Si stencils results in much better shape definition because there are well-developed dry-etch processes for them. However, their chemical stability is relatively poor against selective undercut SiO_2 etching, and could lead to long-term-storage-related stencil degradation and substrate contamination.

Examples of stencil structures formed by Pt and Ge are shown in Figure 8.45. Both are grown on a relatively thick (100–150-nm) copper base electrode.

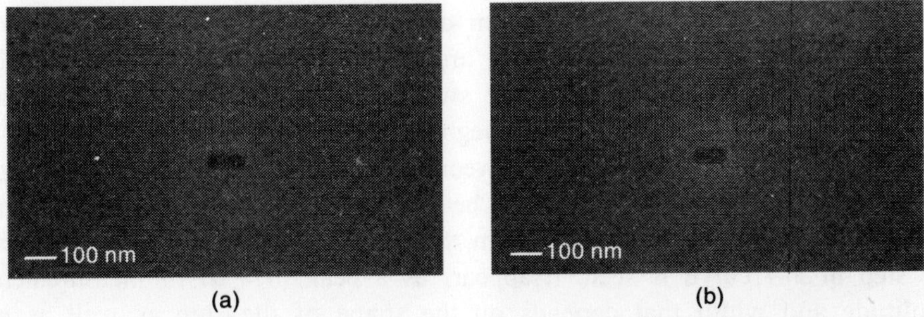

Fig. 8.45 *Scanning electron micrograph of illustrative stencils formed using (a) Pt and (b) Ge.*

CPP junctions are fabricated with the stenciled substrate approach using both magnetron sputter deposition and electron-beam evaporation (e-beam evaporation). Magnetron-sputtered junctions show more edge tapering of the larger spread in the incoming angle of the atomic beam. When deposited on a smoother bottom electrode (600-Å-thick Pt layer), the individual layers forming the junction device is resolved, as shown in Figure 8.46.

After deposition of the layers forming the junction device, optical lithography is used to form electrodes.

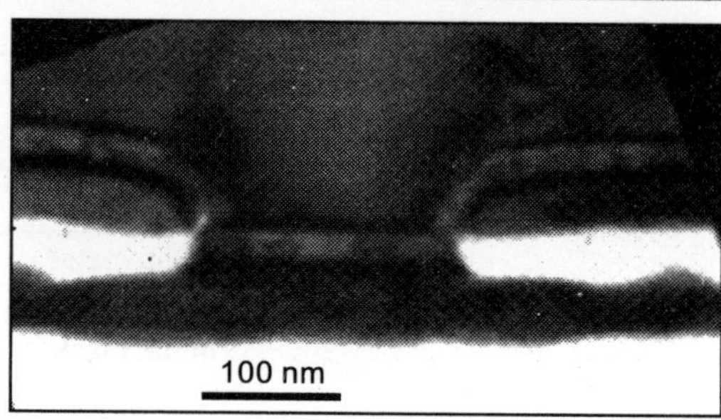

Fig. 8.46 *Cross-section transmission electron microscopy view of a sputtered magnetic tunnel junction.*

Magneto-transport Behavior of a Spin-transfer Junction

Quasi-static Magnetoresistance Properties

For a spin-valve-based magnetic junction, its magnetoresistance response to the combined effect of an applied magnetic field H and a bias current I shows three distinctively different response regions. They are identified in the (I, H) parameter space.

Such magnetoresistance measurements is taken in a quasi-static setup (with measurement response time at 1 ms or less). A dc bias current is applied to the junction, and the junction resistance is measured. The method is useful when the junction MR is only a small percentage of the resistance. In quasi-static measurement the bias current is stepped at a rate of approximately 0.2 to 2 mA/min, while the magnetic field is of the order of 500–1,000 Oe/min. The resistance response is hysteretic with respect to both applied field sweep and bias-current sweep. The values of the switching threshold current I^+ and I^-, corresponding to the resistance high-to-low-switching threshold and low-to-high threshold, are functions of the applied magnetic field.

Figure 8.47 shows the hysteretic current and field dependence of the junction resistance. Parts (a) to (d) correspond to the sweeping of the bias current in both directions while the magnetic field is stepped, either up or down, between each bias-current sweep. The directions of the current sweeping and field steeping are indicated, respectively, by the vertical and horizontal arrows at the upper right corner of each part. Light color in the contour represents high resistance, dark color, low resistance. The magnetic spin-valve layer stack for this device is ||3Co|10Cu|12|Co|200Cu|10Pt||. The numbers before the elements indicate the layer thicknesses in nm. This switching-boundary phase diagram depends on the direction of the variables being swept and the history of the junction. There are three regions in these plots, as illustrated in Figure 8.48. The first region shows hysteretic switching between parallel and antiparallel states. The second region shows a large amount of telegraph noise, often involving two-level fluctuations, signaling thermally activated transitions between two metsatable states. These two states are two orbits of persistent precession. They are between two stable points, or between one stable point and one orbit of persistent precession. In the third region (high-field, high-current-density), a reversible step in V–I curve is seen. It appears as a peak in a dV/dI measurements; the peak has an amplitude and width that depends on the shape of the step as well as measurement conditions.

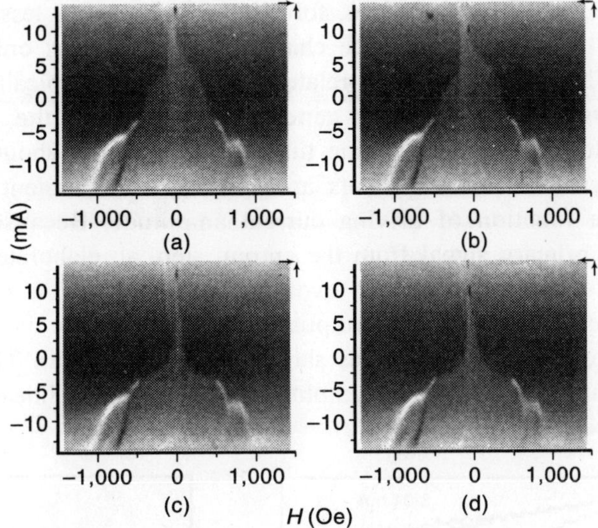

Fig. 8.47 *Contour plots of switching boundaries of a 0.05-μm × 0.10-mm junction at ambient temperature. The magnetic field is applied along the easy-axis direction and the current is swept one full circle at a constant bias field. The bias field is then stepped to the next value. The vertical and horizontal narrows at the upper right corners of parts (a)–(d) respectively indicate the direction of current sweeping and field stepping.*

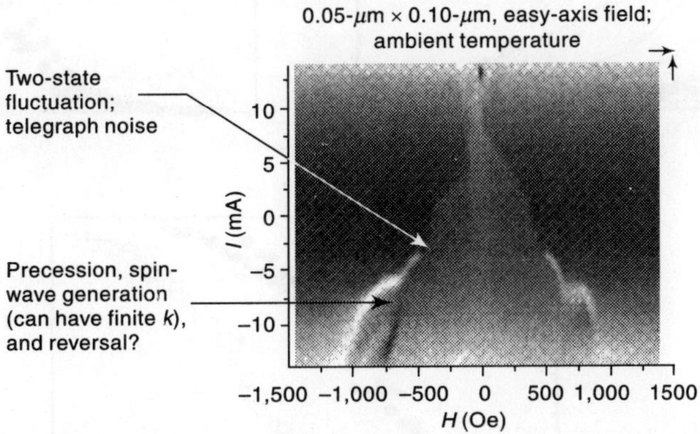

Fig. 8.48 *Three regions commonly seen for the spin-torque excitation of a CPP GMR nanojunction under current sweeping and field stepping.*

Time-dependent Magnetoresisting during Magnetic Reversal

The response of a nanomagnet to spin-torque is dynamic as is investigated by time-resolved transport measurements. Such measurements shed light on spin-torque switching time and is dependent on the conditions of the driving current.

Spin-transfer-induced magnetic reversal follows a different type of dynamics than those involved in magnetic-field-driven reversal. For present-day spin-valve devices, direct measurement of the switching speed of spin-transfer junctions is nontrivial because of the relatively small signal level involved. For CPP spin-valves, for sizes of 100 nm or less, junction resistance is still less than 10 Ω, and the magnetoresistance change is even smaller; only about 3–5% of the total junction resistance. This results in a MR-related voltage signal typically of the order of 0.1 mV. Dynamic calculations indicate that the generic time scale of the reversal is approximately $(2\pi M_s)\gamma$. This estimate places the switching time in the range of about 1 to ns.

Most switching dynamics measurements are performed at ambient temperature example; the switching speed as a function of driving current amplitude. Because of the small MR signal level above the large primary signal from the current step, an elaborate signal-averaging sequence is devised to extract the time-dependent evolution of the junction voltage related to magnetic reversal. With proper averaging, the output voltage difference is normalized to reflect the ensembble-averaged reversal probability, as shown in Figure 8.49(a). The corresponding switching speed as a function of the drive current amplitude is shown on a linearly scale in Figure 8.49(b), and on a log-linear scale in Figure 8.49(c).

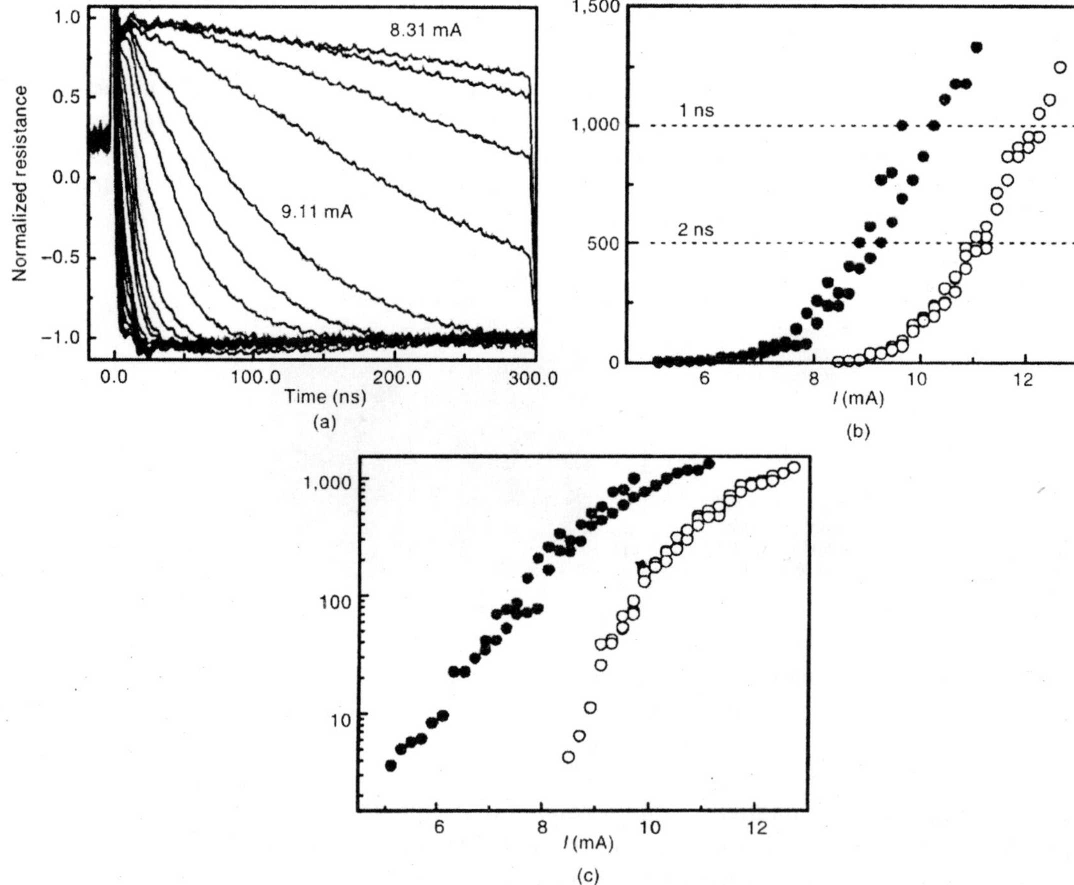

Fig. 8.49 *(a) Time-dependent switching probability. (b) Switching speed τ^{-1} extracted from (a). Horizontal dashed lines with labels indicate the switching speed corresponding to reversal time of 1 and 2 ns, respectively. (c) Switching speed plotted on a log-linear scale. The open and closed circles respectively represent the switching threshold on the positive and the negative current step of junction.*

Two aspect of the data are shown in Figure 8.49(b) and (c). First, at high speed limit, the dependence of τ^{-1} on bias current I is linear. Second, in the subthreshold, large-τ regime, the linearity gives way to a curved pnset which is exponentially dependent on the bias current, as shown in Figure 8.49(c). The linear τ vs. I dependence stems from spin-transfer angular momentum conservation, and the curved onset related to thermal activation. Both are adequately described by spin-transfer dynamics in the presence of thermal noise. In addition, threshold current observed in Figure 8.49(b) is more than a factor of 2 large than the corresponding threshold currents measured quasi-statically. The difference between these two values lines in the vastly different time scales over which they are measured.

This experiment reveals the switching junction voltage response in time. The detailed oscillation in the voltage related to magnetic precession is sensitive to the initial condition, which is thermally randomized. A more recent experiment has revealed the dynamic voltage output of the switching junction and also the actual oscillations that reflects the dynamic precession accompanying the reversal. It is achieved by using a junction with a non-collinear magnetic moment arrangement between its fixed and free layers, introducing a distinctive initial condition for the precession dynamics upon the presence of a step-function driving current and thus preserving the phase information of the oscillations upon multi-trace averaging. Thus, the effect of spin-current on the damping characteristics of the nanomagnet by relating the oscillation envelope to the spin-current amplitude is observed.

Spin-wave Excitation and Microwave Emission

Spin-transfer-induced magnetic excitation are observed in transient magnetic precession and reversal, and in magnetic precession as well. Hence, in the emission of spin-wave excitation in the presence of a nanometer-scale patterned boundary confinement results in discrete modes. This is responsible for the complex behavior seen in the reversible (I, H) region, Figure 8.47. Spin-wave excitations occur in magnetically confined nano-pillar spin-value geometries or magnetically extended structures in which the magnetic film under excitation is extended in nature and the transport charge current is confined by the arrangement of electrodes. These types of structure are successfully prepared and associated spin-wave emission from spin-current is demonstrated, through the direct observation of microwave output spectra as a function of junction current and magnetic field bias. Figure 8.50 gives an example of one such measurement. The junction used in the measurement is fabricated via a magnetic stack of $||2.5Ta|50Cu|20Co_{90}Fe_{10}|5Cu|5Ni_{80}Fe_{20}|1.5Cu|2.5Au||$. The magnetic layers are continuous, whereas the contact is about 40 nm in diameter at its upper surface.

The process of spin-wave excitation in isolated nanomagnetic pillar structures is well modeled by a macro-spin model, although magnetic excitation of finite wavelength comparable to the lateral dimension of the structure are also likely to be present.

Spin-wave excitation of an extended magnetic film under a localized point-contact spin-current excitation is examined theoretically, resulting in an analytical solution. The spin-wave produced is shown to have a half-wavelength about the size of the point contact, resulting in a threshold current that is essentially independence of the contact. Experimental work using mechanical point contacts to bilayers of magnetic thin film is found to be in semi-quantitative agreement with predictions.

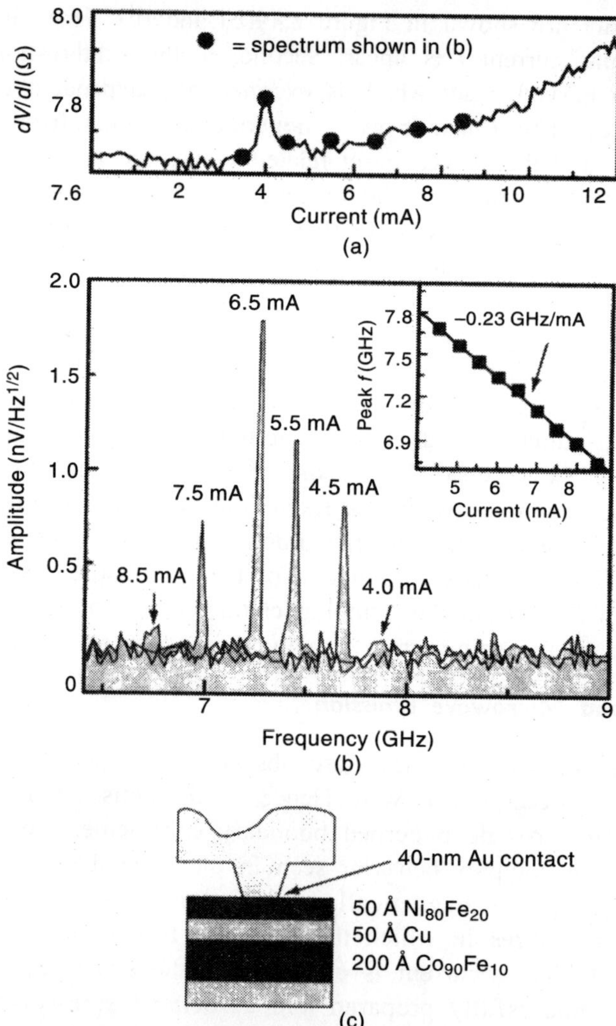

Fig. 8.50 *Microwave emission from spin-current excitation in nano-contact junctions: (a) Section of the I–V curve of a junction from which such emission is measured. (b) Microwave power spectra as a function of bias current, with inset showing the dependence of frequency on the bias current. (c) Geometry of the junction.*

At a high current density, the internal magnetic degrees of freedom of the nanomagnet become very important because energy is directed into spin-wave modes of different wavelengths. Full micromagnetic modeling in necessary to understand the large-amplitude behavior in this region. The basic features between experiment and full micromagnetic modeling agree with each other. The excited spin-wave modes depend not only on device design but on other subtle features of the magnetic boundary conditions of the nanomagnet, such as shape and materials irregularities formed during fabrication.

For spin-wave excitation measurements, the magnetic-field dependence of the threshold current, dI_c/dH, is found. The junction sample is placed in a magnetic field which is applied perpendicularly to the film surface that is large enough to overcome the thin-film-shape-related demagnetization field 4πMs (which for cobalt is about 17 kOe). The use of such a large field makes finite-temperature fluctuation within the approximation of a macrospin mode. The large amount of spin-polarized current is necessary to excite the appropriate magnetic procession and/or reversal because the applied magnetic field involved is relatively large.

High-field Response and Constraints on Signal Voltage Amplitude

Experimentally, a spin-wave nanomagnet junction under a strong, perpendicular applied magnetic field exhibits two main characteristics. First is the presence of a (often non-hysteresis) voltage step in a certain bias current, as shown in Figure 8.51. Second, the voltage step height is seen to be similar to both the Zeeman-energy related voltage ($\hbar\omega/2e$), (where $\hbar\omega \approx g\mu_B H$ corresponds to the ferromagnetic precession frequency,) and the giant-magnetoresistance-related voltage associated with a magnetic reversal, $V_c \sim I_c\Delta R$, (where δR is the resistance change of the junction between the magnetically parallel and anti-parallel states and I_c is the position of the current step) as shown in Figure 8.52. These observations are either as a consequence of magnetic reversal of the free layer , or as a result of magnon-emission-related magnetic excitation.

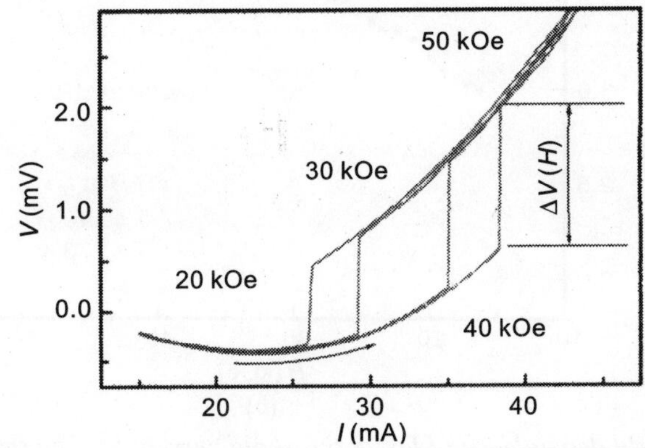

Fig. 8.51 *I–V characteristics measured for a 0.05-μm × 0.20-mm CPP junction. A linear background of 0.935 W is subtracted from the data to make the threshold behavior clearly visible. The characteristics shown are for applied values of 20, 30, 40, and 50 kOe. The curvature is commonly seen for CPP junctions. The arrow shows the direction of current sweep. The data is obtained at 5 K.*

This similarity has been seen over many samples and from results obtained by many different groups. A summary of the experimental observations is given in Table 3. It indicates that the spin-values are within the limit of spin-pumping-dominated dissipation. Spin-pumping-induced dissipation provides an additional relationship between the MR of a spin-valve and that of the spin-torque threshold current I_c, leading to the relationship $I_c\delta R \sim (2\mu_B/e)H$.

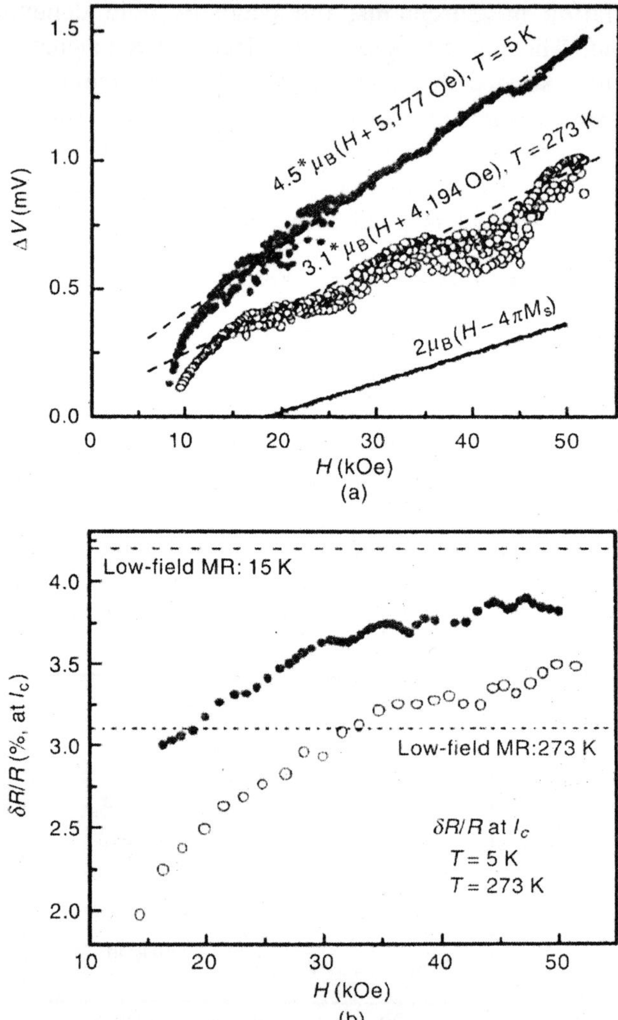

Fig. 8.52 *(a) Magnetic field dependence of DV at threshold current I_c. The threshold current and DV shows that current increases with the applied magnetic field. The solid line on the bottom is the voltage corresponding to Zeeman energy $(2\mu_B/e)\,(H - 4\pi M_s)$. (b) The dashed lines indicated low-field R(H) measurement-based MR at 15 K and 273 K.*

Table 3 *Measured field-dependent slopes $V_c = I_c\delta R$, from results in the literature. The numbers preceding the element designations indicate layer thickness in nm.*

Junction stack	*Lateral size (nm × nm)*	*δR (Ω)*	*dI_c/dH (A/Oe)*	*$(dI_c/dH)\ \delta R$*	*T (K)*
‖120Cu|10Co|6Cu|2.5Co15|Cu|3Pt|60Au‖	100 × 100	0.075	2.90×10^{-7}	3.76	300
‖10Cu|3Co|10Cu|12Co|300Cu|10Pt‖	90 × 140	0.03	4.51×10^{-7}	2.34	4.2
‖80Cu|40Co|10Cu|3Ni80Fe20|2Cu|30Pt‖	70 × 130	0.129	2.16×10^{-7}	4.82	4.2
‖150Cu|20Pt|10Cu|2.5Co|10Cu|12Co|250Cu|10Pt‖	50 × 200	0.039	4.80×10^{-7}	2.23	5

Potential Applications

Spin-transfer-induced magnetic excitation and magnetic reversal is a relatively new phenomenon that begins to dominate magnetic behavior for junction devices below about 0.1 to 0.2 μm in lateral size. As the critical dimensions of current electronic devices shrink below this length scale, the spin-transfer mechanism becomes important in several aspects. It is used for localized write-addressing of magnetic random memory element, or for on-chip generation of tunable microwave radiation. The mechanism has significant impact on the design and operation of magnetic disk read heads.

A two-terminal spin-vales or magnetic tunnel junction that can be current-switched between two stable resistance states constitutes a memory element. For such a memory element to be integrated into the existing CMOS circuit technology, some basic device requirements like device impedance, device voltage swing between the two stable states, and threshold current required to switch the device must be met. Switching speed and its relationship and tradeoff with switching current are also important.

For effective integration, the current density required for device switching must be comparable to that supplied by a typical CMOS circuit of comparable density. If this were to be supplied by a MOSFET transistor, it would usually be of order of 0.5 to 1 mA/μm of channel width. This determines the upper limit of the switching current. Diode selection could in principle allow for higher current density, although there are additional concerns regarding associated impedance (mis)match and uniformity of device characteristics over large numbers of junctions and diodes.

The junction switching current must be large enough to switch a nanomagnet that has sufficient stability to retain its remanent state at room temperature. This requires a magnetic anisotropy energy $U_k = (1/2)mH_k$ of the order of 40 to 60kBT. The high-speed switching threshold current I_c in Equations [(16) (Page 293)] and [(3) (Page 325)] are directly related to this uniaxial anisotropy energy in the form of $I_c \approx (2r/**)(\alpha/\eta)\ U_k$. Depending on the values of the damping α and spin-polarization factor η, this gives $I_c \approx 10$–100 μA for $U_k \approx 60\ k_BT$.

Present-day spin-transfer switching junction devices involve a quasi-static switching current of the order of 0.1–1 mA for a device cross section size of approximately 50–100 nm, resulting in a quasi-static switching threshold current density of the order of 10^6Acm^{-2}. For CMOS integration, the high-speed switching threshold, can be significantly higher (by perhaps a factor of 2–5 depending on device structure details). Existing low-impedance (1 to Ω/μ^2) magnetic tunnel junctions support a transport current of the order of 10^7 Acm^{-2} before inducting is sufficient to demonstrate spin-transfer effect. A large part of the threshold current of spin-transfer device comes from the easy-plane demagnetization field due to its thin-film geometry. This type of anisotropy does not contribute to thermal stability, but significantly increases the spin-transfer switching current. Therefore, for reducing the switching current of a spin-transfer device; reduce or eliminate this easy-plane anisotropy from the system. This is achieved by careful engineering of the interface magnetism of the free layer or control of its stress field. Although these possibilities are theoretically feasible, significant materials and fabrication challenges have to be overcome before they are successfully implemented for manufacturing.

The threshold current, as expressed in equation (3), provides estimate for the switching current necessary for memory circuit operation. To achieve sufficiently rapid switching, the drive current has to be greater than the threshold current, by a significant amount. The switching-speed vs. switching time is shown by the curves presented in Figure 8.49. It is the linear slope in front of the second equation in Equation (16). For the 50-nm × 10-nm × 3-nm cobalt nanomagnet of Figure 8.49, the figure of merit for $(I - I_c)\langle\tau\rangle$ is about 8 pC.

For finite-temperature operation as a memory element, the collinear switching geometry is unlikely to be the best, because for that geometry the initial condition of the switch depends sensitivitely on the thermal distribution. A noncollinear arrangement between the orientation of the "free" layer of the nanomagnet and that of the spin-polarized current is more desirable. At the same time it is helpful through layout design to add a current induced magnetic field as a transit "tipping" field in order to create a non-equilibrium initial state when switching the nanomagnet.

Spin-transfer devices are used as compact, on chip-microwave oscillators. Tunable microwave output from spin-transfer-based magnetic junction structures at frequencies ranging from 1 to 20 GHz and at full-width-half-maximum power linewidth at least 10,000 times below the center frequency is achieved. Phase-locking between magnetic precession-induced microwave oscillation and additional input tune has also been demonstrated.

Spin-transfer also affects the performance of magnetic read heads in modern hard drives, acting as a negative influence by amplifying the thermal and other magnetic noise of a read head, decreasing the signal-to-noise ratio, and changing the dynamic characteristics of the read-head performance.

Spin-transfer-related magnetic excitation is observed to move magnetic domain walls in narrow ferromagnetic wires. The interaction between spin-polarized current and a ferromagnetic domain wall was one of the first phenomena for which a spin-angular-momentum transfer process was considered. The unambiguous presence of a spin-torque term is the cause of domain wall motion under applied current. The mechanism, at a low enough current density, has significant implications for memory devices.

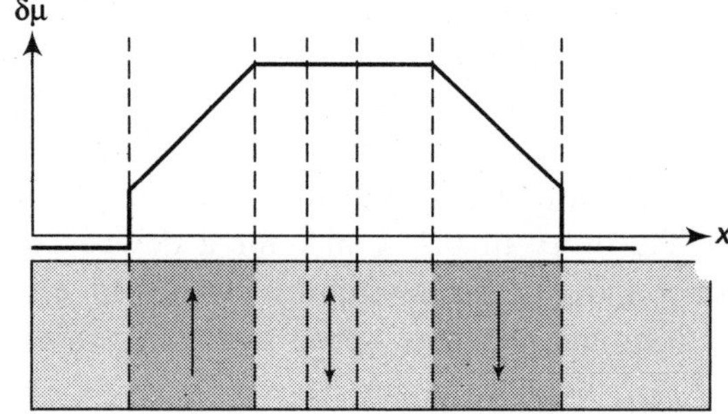

Fig. 8.53 *Conceptual use of a three-magnetic-layer structure to increase the efficiency of a spin-current-induced magnetic switch. The vertical axis of the upper part indicates the amount of Fermi-level splitting (δμ) between the spin-up and spin-down electrons as a function positions x. The corresponding layer structure for the ferromagnetic films is illustrated in the lower part.*

The applications in integrated circuits, the spin-transfer device must have a lower threshold current density and it must have much larger voltage output than the demonstrated value of several hundred μV.

One proposal for reducing the current required to switch a nanomagnet is shown in (Figure 8.53). For a nanomagnet sandwiched between two oppositely fixed magnetic polarizer layers, a sizable enhancement of the spin-transfer effect, and sixfold net reduction of the threshold current is achieved.

The presence of a spin-transfer-induced magnetic reversal process in magnetic tunnel junction structures; is achieved (see Figure 8.54). One challenge here is to quantitatively separate the spin-polarized tunneling current from other parallel channels e.g., pinhole conduction channels that are present in such low-resistance tunnel devices.

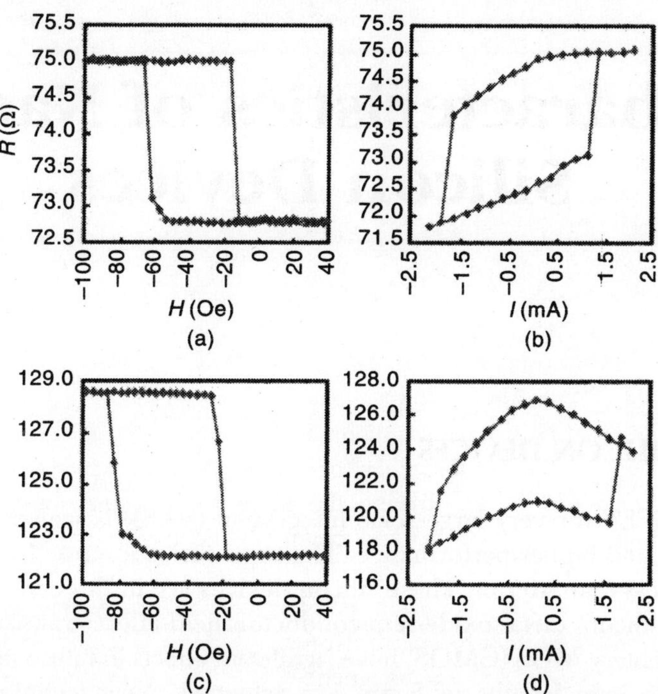

Fig. 8.54 *A magnetic tunnel junction switching under a current bias. (a) and (c), magnetoresistance change when the junctions are swept by an applied magnetic field. (b) and (d), behavior of the corresponding junctions when swept by their bias current. The junctions are approximately 100 nm × 200 nm in size, comprising a stack structure of || Ta 2|NiFeCr 3.5|PtMn 14|CoFe 2|Ru 0.8|CoFe 2.2|AlO$_x$|CoFe 1|NiFe 2|Ta 5|| (numbers are layer thicknesses in nanometers).*

Characteristics of Nano Silicon Devices

I. NANOSCALE SILICON DEVICES

The silicon MOSFET for very large scale integration (VLSI) is scaled down to attain higher levels of integration and higher performance. The miniaturization rate has accelerated, and the gate length is now less than 40 nm. These silicon devices are in the nanometer regime. Further miniaturization of silicon metal-oxide-semiconductor field-effect transistors (MOSFETs) into nanoscale complementary MOS (CMOS) have significant effects in future information technology. CMOS devices with gate lengths of 5 nm are achieved. New techniques to overcome the scaling and performance limits of conventional CMOS devices are urgently needed. One of the most promising techniques is the utilization of new physical phenomena that appear in silicon nanostructures but have not yet been utilized in nanoscale devices.

Stages in Nanoscale Silicon Devices

Quantum effects and single-electron charging effects, occur even at room temperature in these devices. New functionalities due to these phenomena have a huge potential in information processing or data storage. These new effects in nanoscale structures are utilized for future integrated devices. There are three stages in silicon nanodevices that are used in nanoscale structures depending on how the devices are applied to integrated circuits.

First stage: Enhancing Performance

The basic operating principle of the nanoscale device is conventional CMOS, but new physical phenomena like quantum effects and ballastic transport in nanodevices enhance the performance of nano-CMOS.

In the first stage of developments, the performance of CMOS is enhanced by these physical phenomena. Since the present CMOS platform need not be changed for system design and manufacturing, the first stage is achieved. In the middle of the first stage, paradigm shift takes place from top-down (fabricated by lithography and etching) to bottom-up-type nanodevices (formed by a self-assembly process). If the operating principle of a device fabricated by the bottom-up process is the same as for conventional CMOS, this device is classified into the second half of the first stage. For example, carbon nanotube (CNT) FETs, formed by the bottom-up process, are classified into the second half of the first stage if the integrated CNT-FET form CNOS circuits.

If the performance of CMOS is enhanced, it will exceed the performance limit and scaling limit of CMOS. For example, the performance of nanoscale CMOS devices is controlled and improved by the quantum confinement effect. In the characteristics of nanoscale narrow-channel MOSFETs and nanoscale thin-channel MOSFETs on silicon-on-insulator (SOI) substrates it is found that the carriers are respectively confined into a one-dimensional narrow channel and a two-dimensional thin channel.

Figure 9.1(a) shows a SEM image of a fabricated nanoscale *narrow channel*. The channel width is less than 10 nm. In the nanoscale narrow channel, the carries are confined in a one-dimensional channel, and the confinement is stronger than in a thin, planar channel. Therefore, more evidence of quantum effects is expected. The threshold voltage of narrow MOSFETs is varied and controlled depending on the channel width because the ground energy of carriers is raised by quantum confinement.

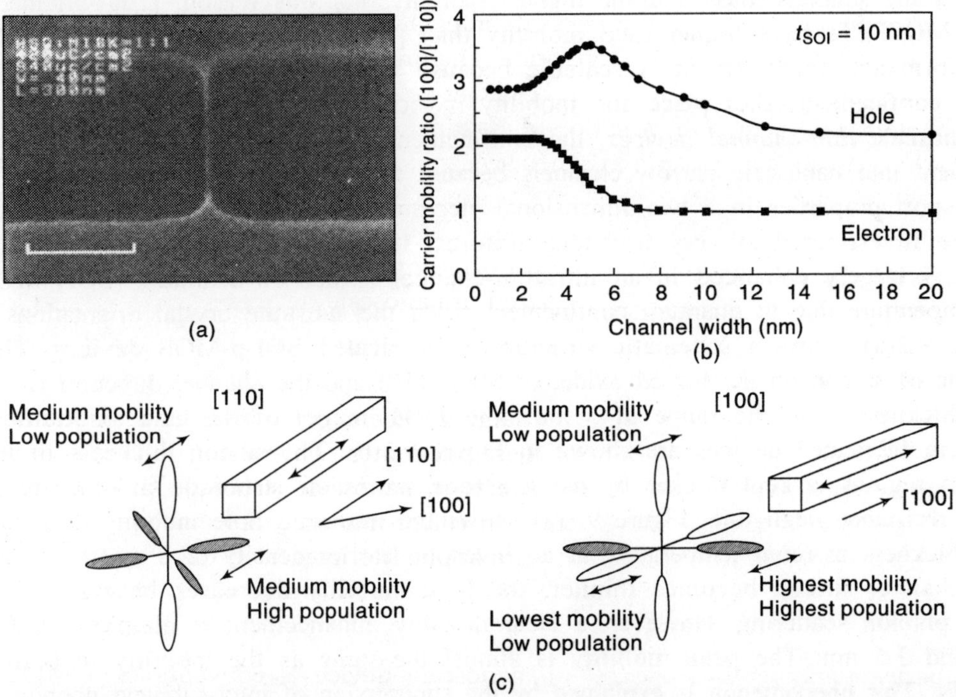

Fig. 9.1 *(a) SEM image of nanoscale narrow channel. The narrow channel with widths of less than 10 nm ions formed between source and drain. (b) Stimulated mobility ratio of [100]-oriented channel to [110]-oriented channel as a function of channel width. (c) Conduction band structures in [110]-oriented and [100]-oriented nanoscale narrow channels.*

Also, by simulation the mobility is modified in nanoscale narrow-channel MOSFETs. Figure 9.1(b) shows the ratio of mobility of [100]-oriented narrow-channel MOSFETs to that of [110]-oriented devices as a function of channel width. Conventional MOSFETs have the channel in the [110] direction, and the [100] direction is rotated form the [110] direction by 45 degrees. This larger mobility in [100]-oriented devices is due to the anisotropic effective mass of silicon. In an n-type MOSFET, this mobility enhancement is due to the electron population of valleys in the conduction band.

Figure 9.1(c) shows the conduction-band structure in [110]- and [100]-oriented channels. Owing to the anisotropy of the electron mass, the rise of valley energy is different depending on the direction of the ultranarrow channel. In the [110]-oriented ultranarrow channel, six equivalent valleys split into two sets of valleys; twofold degenerate valleys and fourfold degenerate valleys. When the channel width becomes narrower, the energy of the twofold degenerate valleys increases more than that of the fourfold degenerate valleys; therefore, more electrons are populated in the latter. However, fourfold degenerate valleys have larger mass and smaller mobility along the [110] direction than twofold degenerate valleys, as shown in Figure 9.1(c). Accordingly, the total electron mobility of [110]-oriented narrow-channel MOSFETs is smaller.

In the [100]-orientation narrow channel, on the other hand, six equivalent valleys split into three sets of twofold degenerate valleys, as shown Figure 9.1(c). Most of the electrons are populated in the valley shown (in red) because the energy of this valley is the smallest, and this valley has the smallest mass and the highest mobility. For this reason, [100]-oriented narrow-channel MOSFETs have higher total mobility than [110]-oriented narrow-channel MOSFETs. The performance enhancement is scalable because when the device becomes smaller, more quantum confinement takes place and mobility is increased.

In nanoscale *thin-channel devices*, the confinement of carriers is weak compared with the confinement into nanoscale narrow channels because the carriers have two degrees of freedom. The transport properties in a two-dimensional electron gas of a quantum well in silicon or in GaAs are investigated at very low temperatures. It is found experimentally that the hole mobility is largely enhanced in an ultrathin-channel silicon-on-insulator (SOI) MOSFET at room temperature due to quantum confinement when the substrate crystal orientations is (110).

Figure 9.2(a) shows a schematic structure of fabricated SOI-p-MOS devices. The crystal orientation of silicon on the buried oxide, or SOI, (110) and the channel direction is [110]. The silicon thickness is in the range of 3 nm. The TEM images of the gate oxide/silicon/buried oxide form fabricated devices are shown in Figure 9.2(b). The silicon thickness of the source and drain regions is kept thicker by using a local oxidation technique in order to make the parasitic resistance negligible. Figure 9.3(a) shows the measured hole mobility as a function of silicon thickness at room temperature at an inversion carrier density of 3×10^{12} cm^{-2}. As the narrow-channel silicon becomes thinner, the hole mobility decreases because of increased acoustic phonon scattering. However, a clear mobility enhancement is observed at thicknesses of 3.4 and 3.6 nm. The peak mobility is almost the same as the mobility in bulk (110) p-MOSFETs. This phenomenon is explained by the suppression of inter-subband phonon scattering assisted by optical phonon absorption that is the transition between the two lowest-lying heavy-hole subbands. Note that the increase in hole mobility is observed only in (110) ultrathin-body p-MOS and not in (100) because of the high degree of degeneracy of heavy and light holes.

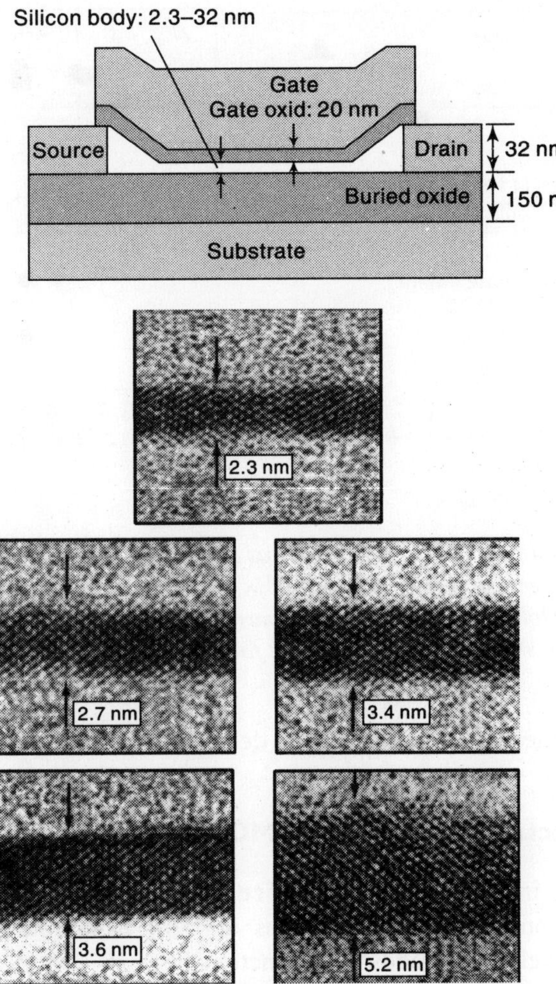

Fig. 9.2 *(a) Structure of (110)-oriented ultrathin-body SOI p-MOS devices. (b) TEM images of measured ultrathin-body p-MOS devices.*

Figure 9.3(b) shows measured hole mobility as a function of SOI thickness at room temperature at an inversion carrier density of 1×10^{13} cm^{-2}, where the density is higher than in Figure 9.3(a). The (110) p-MOSFETs retain high mobility even when the body thickness is thinned down to 3 nm, and the mobility is higher than that of other devices, including Si (110) bulk, Ge-rich strained SiGe on insulator (SGOI) and strained SOI-p-MOSFETs at a thickness of less than 6 nm. In this thin-body regime, the dominant scattering mechanism is the scattering induced SOI thickness fluctuation. The high mobility in (110) p-MOSFEt in the extremely thin-body regime is due to the suppression of the SOI-thickness fluctuation-induced scattering compared with that in conventional (100) p-MOSFETs. This is because the (110) p-MOSFET is less sensitive to SOI-thickness-fluctuation-induced scattering due to heavier hole effective mass normal to the channel surface, as predicted in. In the nanoscale CMOS, the crystal orientation, channel

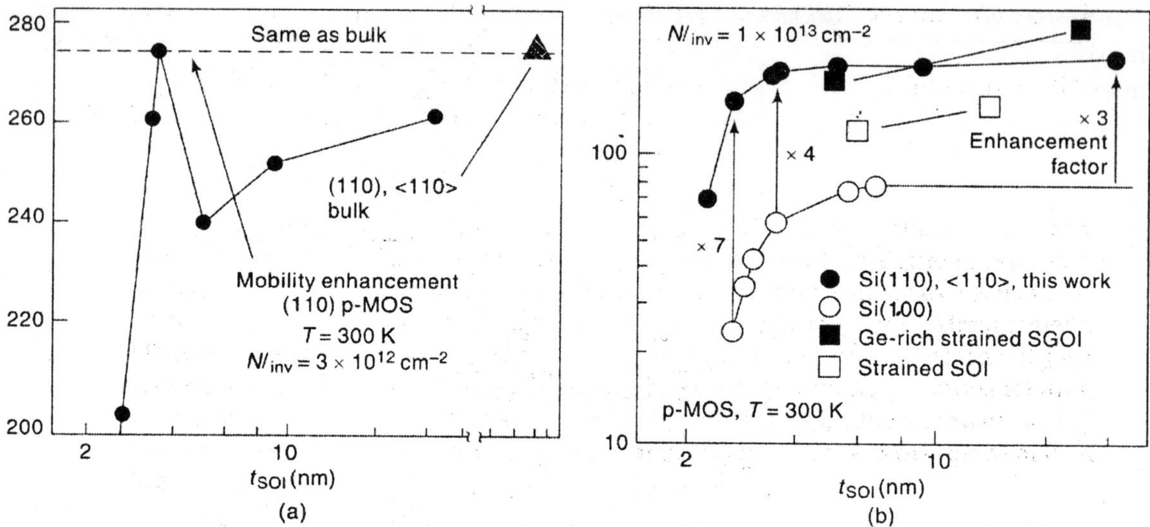

Fig. 9.3 *Measured hole mobility of (110) p-MOSFETs as a function of SOI thickness at room temperature. (a) Hole mobility at inversion carrier density of 3×10^{12} cm^{-2}. The mobility in a bulk (110) p-MOSFET is also shown. (b) Hole mobility at inversion carrier density of 1×10^{13} cm^{-2}. Mobility data for other p-MOSFETs is also shown. The enhancement factors compared with those for Si(100) p-MOSFETs are shown.*

direction, and device dimension are carefully determined in order to maximize the device performance.

Second Stage: New Functions Merged into CMOS

New function appears in the nanoscale device, and the devices are merged into CMOS circuits to add new functionalities. The operations are still based on CMOS.

The second stage of development adds new functionalities in CMOS. The last half of the first stage and the second stage will complete for the development of future integrated devices. At both the first and second stages, the life of CMOS is prolonged, and these stages will have a great impact on the future development of CMOS and all information technologies.

The best example of the second stage is *memory* chip, which is composed of memory cell arrays that store digital data and peripheral circuits that write and read the data. In a new memory, nanostructures are adopted and new functions are utilized only in the memory cells, while conventional CMOS devices are utilized in the peripheral circuits. Therefore, all of the new and nanoscale memory devices are classified into the second stage.

In a new function in silicon nanocrystal memories silicon nanocrystals are embedded in gate oxide and it act as site for charge storages. Physical separation of nanocrystal improves the retention time by limiting the lateral flow of charges. The new function that appears in a silicon nanocrystal memory cell is the two-bit-per-cell operation. The electrons are locally injected only near drain and/or source by hot-carrier injection, and there are four states depending on where the electrons are injected. Distinct four-threshold voltages are observed that can be read out.

A *single-electron transistor* is one of the best-known nanoscale devices. The single-electron transistor has a unique feature of Coulomb blockade oscillations in *I–V* characteristics, and therefore has great potential to add new functionalities to future VLSI.

Let a single-electron/hole transistor be in the form of a point-contact MOSFET. The silicon quantum dot is self-formed in the very narrow channel, and the device acts as a single-electron/hole transistor. Some single-electron/hole transistors are in the form of an ultranarrow-channel MOSFET, as shown in the inset of Figure 9.4(a). Single-electron/hole transistors operate only at very low temperature. Efforts are made to raise the operation temperature by making the silicon quantum dot smaller. Figure 9.4(a) shows the *I–V* characteristics of a single-hole transistor at room temperature. It shows the largest Coulomb blockade oscillations in a single-dot system at room temperature. The peak-to-valley current ratio (PCVR) is 395; the estimated dot size is 2 nm. Since the dot is extremely small, the quantum-level spacing in the dot is not negligible, and negative differential conductance (NDC), is observed at room temperature. The PVCR of NDC is 106 at room temperature.

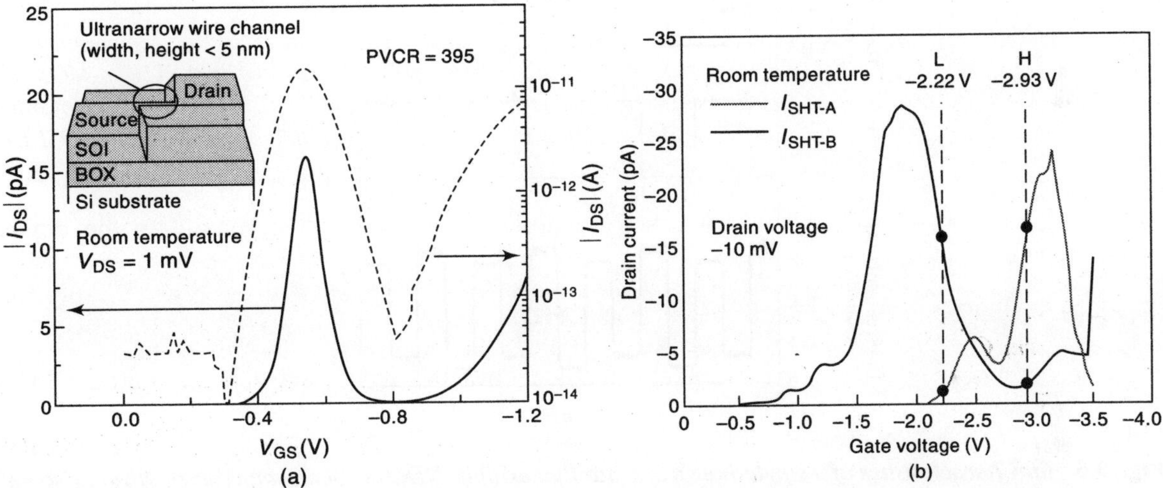

Fig. 9.4 *(a) Coulomb blockade oscillations of a single-hole transistor at room temperature. The device structure is shown in the inset. (b) Characteristics of two integrated room-temperature-operating single-hole transistors.*

Efforts are also made to integrate room-temperature-operating single-electron/hole transistors. Figure 9.4(b) shows Coulomb blockade oscillations of two integrated single-hole transistors that form a directional current switch at room temperature. This is the first integration of the room-temperature-operating single-electron/hole transistors. Moreover, each single-hole transistor has silicon nanocrystals embedded it the gate oxide and acts as a non volatile memory. Therefore, the peak position of coulomb blockade oscillations are controlled by applying gate pulse voltage that injects electron into silicon nanocrystals.

Consider application of single-electron/hole transistors in digital circuits. If a part of conventional CMOS circuits is replaced by single-electron/hole transistors, new functions are added with extremely low per consumption. Figure 9.5(a) shows characteristics of an exclusive-

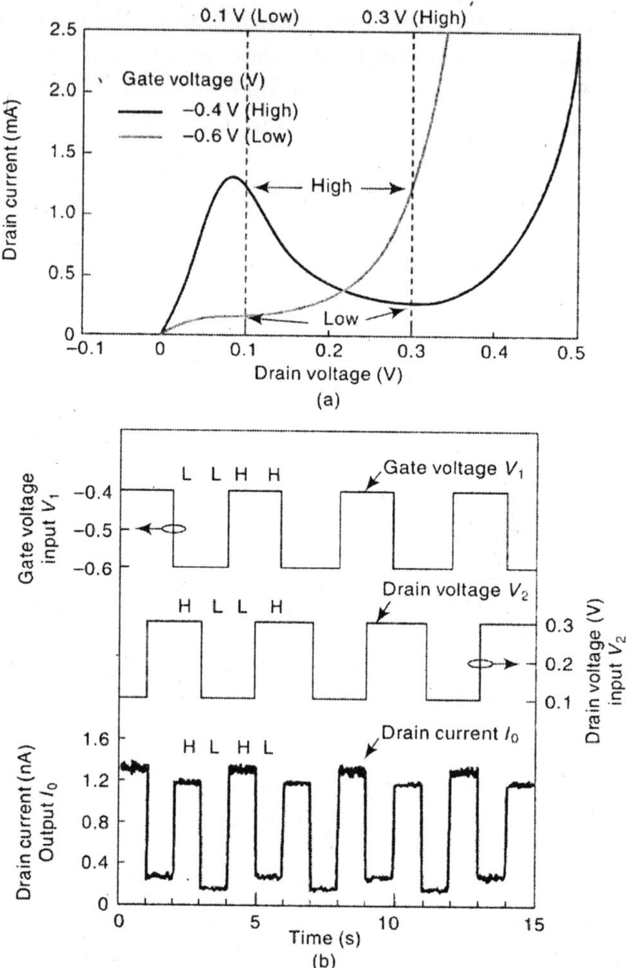

Fig. 9.5 *(a) Characteristics of a single-hole transistor that exhibits NDC at room temperature. (the horizontal axis is drain voltage, not gate voltage.) (b) Operation of exclusive-OR function by only one single-hole transistor at room temperature.*

OR circuit. The circuit is composed of only one single-hole transistor that exhibits NDC at room temperature. The modulation of NDC characteristics of by gate voltage is utilized. When the gate voltage and drain voltage are inputs of the circuits, the output current shows the exclusive-OR function, as shown in Figure 9.5(b). Compared with the conventional exclusive-OR circuit by CMOS, the number of devices is greatly reduced when the single-electron/hole transistor is used because of its high functionality.

Another application of single-electron/hole transistors in the second stage is analog circuit. The bell-shaped *I–V* characteristics is utilized for the analog pattern matching circuit. Since the Coulomb blockade oscillations have bell-shaped *I–V* characteristics, they are applied to analog pattern matching, the matching is performed by the single-hole transistors, and other calculations are done by conventional CMOS circuits. Therefore, this new circuit-scheme utilizing single-electron/hole transistors and adding new function to CMOS is in the second stage.

Third stage: The nanoscale devices which operate by new principles are integrated to form new circuits. The circuits are no longer CMOS.

These are completely new types of nanoscale devices, such as spin transistors, that are integrated. The system architecture is not as the conventional one. Therefore, the third stage is realized as a mainstream device technology only in the distant future; the first two stages are of much more important for current developments.

II. AFM-TIP-INDUCED AND CURRENT INDUCED LOCAL OXIDATION OF SILICON AND METALS

(a) AFM-Tip Induced Local Oxidation

There are two different processes that can be used to locally oxidize silicon or metals and are promising for the fabrication of model nanoelectronic devices. The first involves oxidation induced by a negatively biased conducting atomic force microscope (AFM) tip. The second approach involves local oxidation induced by high current densities generated by forming constrictions in the current-carrying sample. The novel local oxidation process is used to generate oxide tunneling barriers of 10–50 nm. Thirdly, is the use of proximal probes as tools for the fabrication of nanoelectronic devices.

Among the different approaches, the most promising is the atomic force microscope (AFM) induced oxidation of silicon and metals. Tip-induced oxidation of silicon is done under a scanning tunneling microscope (STM). Here, an H-passivated Si surface is scanned in air by a *positively* biased tip to generate surface oxide features. By a different approach, an AFM with a conducting tip biased *negative* with respect to the sample is used to induce oxidation, so that thicker, non-conducting oxides can be produced. This approach is used to fabricate simple device structures.

The AFM-tip induced oxidation process is based on negatively biasing the tip with respect to the substrate under ambient conditions, which is either a semiconductor or a metal (Fig. 9.6). The substrate locally oxidizes upon moving the tip in contact mode across the surface. The oxidant for the chemical reaction is provided by OH^- ions in the water droplet that is formed between the tip and the sample. Thus, the lateral resolution of the AFM oxidation process depends strongly on the humidity in the air.

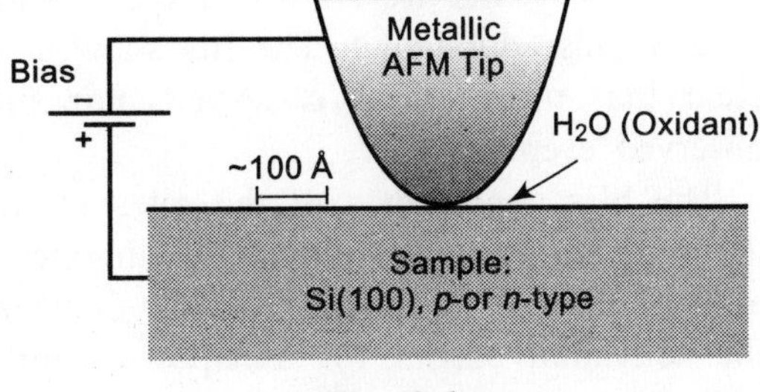

Fig. 9.6

The tip-induced oxidation on tip bias shows that the process is affected by the generated electric field. The initial density of surface OH groups is rate limiting, the fact that water from the ambient is necessary for the oxidation and that the process is analogous to electrochemical anodization. No current flow is found during the reaction.

To optimally use this oxidation process in nanofabrication requires understanding, the factors that control its characteristics.

The mechanism of the process involves the rate of the reaction, its dependence on electric field strength and oxide thickness, and the factors that control the thickness and lateral extent of the oxide, (i.e., the lithographic resolution).

The *n*-type (8–12 Ω cm) Si(100) samples are cleaned by removal of the native oxide in aqueous 10% HF solution. Local oxidation is performed in the ambient using conducting p^{++}-Si tips (radius <100 Å) and AFM microscope. The relative humidity is kept constant (during experiments) at values ranging from 10% to 95%.

In Fig. 9.7(a) and (b), show two grids of oxide lines written at a rate of 0.3 μm/s using –10 V tip bias. In Fig. 9.7(a), the relative humidity is 61% and the linewidth ~90 nm. Lowering the humidity to 14% reduces the width by a factor of ~4. The height of the oxide barely changes.

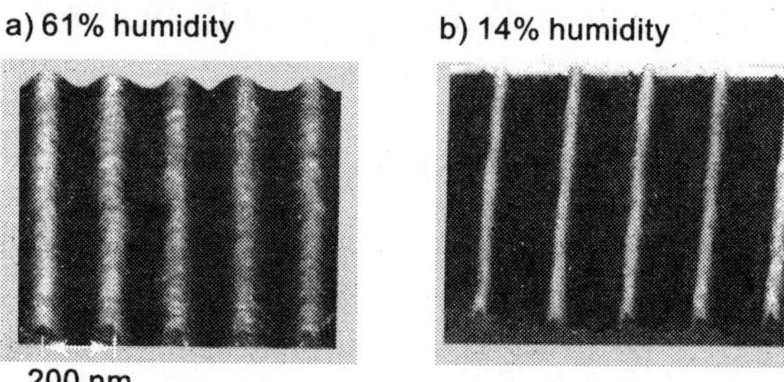

Fig. 9.7 *The aspect ratio (height/width) of oxide lines improves significantly when the relative humidity is lowered from 61% to 14%.*

These results show the strong influence of external factors on the characteristics of the oxidation process for example the information needed in considering the use of the process in fabrication is its intrinsic rate. To determine the reaction kinetics, voltage pulses with the tip stationary over a surface site applied which is then varied between –2 and –2 V and the pulse duration is increased from 10 ms to 1000 s while the tip is move to a new position before each pulse. The width and height of the resulting oxide dots is obtained from AFM images. To obtain the total amount of Si oxidized, the indentations left after the oxide is selectively etched away by aqueous HF and imaged. In this way, apparent volume expansion upon oxidation of 3.0 ± 0.4 is found. This is higher than the expected increase by a factor of 2.27 anticipated for formation of amorphous SiO_2.

Kinetics results are shown in Fig. 9.8(a) where the height of the oxide dots is plotted versus voltage pulse duration t. The fits show a rapid decrease of the growth rate with time as $1/t$. No clear bias threshold is observed. For sufficiently long pulses, growth of shallow oxides is observed even at –2 V.

The bias dependence of the rates indicates that the electric field plays an important role in the process. The high initial growth rates occur at extreme electric field strengths near the tip axes of up to ~10^8 V·cm^{-2} in the field enhanced thin film oxidation. The electric field lowers the activation barrier for transport of ionic species across the oxide. The decrease of growth rate observed here is due to the reduction of the electric field strength as the oxide thickness increases. To confirm the involvement of ionic species in the tip-induced oxidation process, the weak Faraday currents are searched. The measured current in Fig. 9.8(b) shows the same behavior as the Faraday current searched for anodic oxidation, if calculated from the measured volume growth according to the electrochemical reaction: $Si + 4h^+ + 2OH^- \rightarrow SiO_2 + 2H^+$. Here the charge efficiency is ~50%. The efficiency, varies due to oxide formed at the tip apex and differences in surface passivation.

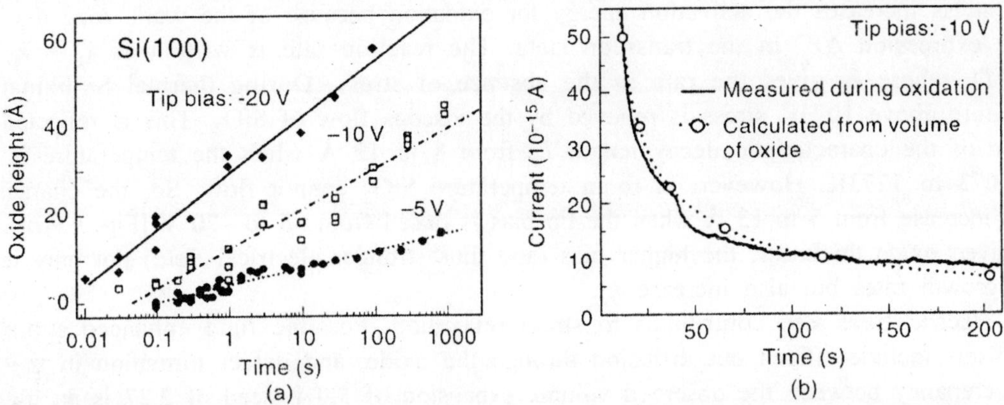

Fig. 9.8 *(a) Kinetic of oxide dot growth for different tip bias at ~50% humidity. (b) Current measured during the oxidation process.*

From the straight line fits the Fig. 9.8(a), the growth rate as a function of electric field strength is obtained [Fig. 9.9(a)]. The high initial rates (~10^3 Å/s for –20 V) decrease fast with decreasing field strength and the oxide practically ceases to grow at field strengths $<1 \times 10^7$ V cm^{-1}. That the rate of oxidation is not only a function of electric field strength but also depends on the applied bias—in particular for high fields.

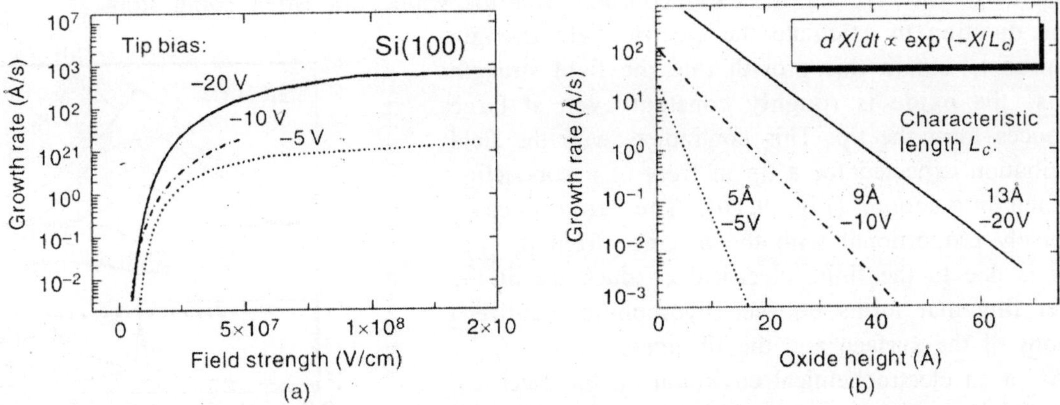

Fig. 9.9 *(a) The growth rate versus electric field strength at the tip apex. (b) Growth rate versus oxide height at three tip biases.*

When the oxidation rate is plotted versus oxide thickness it shows that the rate constant is proportional to exp ($qa\Delta\Phi/2XkT$). Here, q is the charge of moving ions, a gives the distance between interstitial sites of ions, $\Delta\Phi$ is the potential drop across the oxide, and X the oxide thickness. Deviations are also observed in thermal oxide growth. This is partly attributed to the influence of stress, σ, on the oxidation rates. Large stresses build up during oxidation due to the large volume mismatch V_{Si}/V_{SiO_2} = 20 Å^3/45 Å^3 between Si and SiO_2.

The stress increases the activation energy for oxidation because of the work done due to the volume expression ΔV^+ in the transition state. The reaction rate is written as $k_S = k_0 \exp(-\sigma\Delta V^+/kT)$, where k_0 gives the rate in the absence of stress. During thermal Si oxidation at temperature above 1073K stress is relieved by the viscous flow of SiO_2. This is reflected in an increase of the characteristic decay length L_c from 8 to 12 Å when the temperature is raised from 1073 to 1273K. However, at room temperature SiO_2 cannot flow. So, the characteristic lengths increase from 5 to 13 Å when the tip bias is raised from –5 to –20 V [Fig. 9.9(b)]. Thus, for a given oxide thickness, the higher bias (and thus stronger electrical field) not only leads to higher growth rates but also increase L_c.

The electric field also contributes to strain relaxation. Possible field enhanced strain relief mechanisms include silicon out diffusion through the oxide, and defect formation in the oxide. The discrepancy between the observed volume expansion of 3.0 instead of 2.27 is an indication that the high strains are relieved by the formation of a rather open and defect-rich oxide.

The reaction kinetics are further influenced by the field dependence of the concentration of reactant species and their solubility in the oxide. The self-limiting oxide growth puts limitations on the use of this process as a high speed lithographic tool, used for multiple writing tips.

To determine the parameters that affect the lateral resolution of the oxidation process are investigated by the lateral growth kinetics using oxide dot profiles (Fig. 9.10). The strong field dependence of the oxidation kinetics, together with the anticipated focusing of the electric field at the tip apex (radius <100 Å), allows to pattern very narrow oxide structures. The increase of oxide height in Fig. 9.9 is nearly constant over the whole dot when going from 10 to 100 s pulse duration to calculate the electric field strength required to obtain this growth rate the field strength across the oxide is roughly constant even at large distances form the tip. This contradicts with the field distribution expected for a tip in front of a conducting surface in vacuum (Fig. 9.10). The latter decays, inversely proportional with the distance from the tip. This is due to the finite electrical conductance of the water film that forms between hydrophilic (oxidized) regions of the surface and the tip apex.

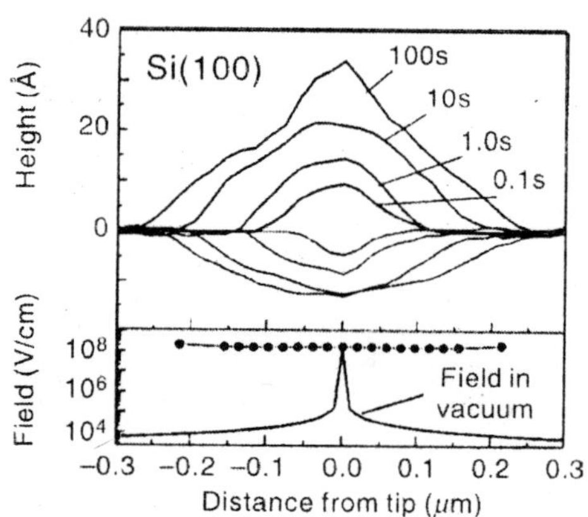

Fig. 9.10 *Profiles of oxide dots grown at 0 V tip bias. Lower panel: The calculated electric field for a tip 30 Å in front of a conducting surface in vacuum, (solid line) and the field calculated form the observed growth rate (solid circles). High growth rates at the edges of the dots show that the field is strongly defocussed by the water film on the oxide.*

As in an electrochemical environment, the electrical potential is nearly constant within the water film and drops in the proximity of the electrode solvent interface, i.e., in the electrochemical double layer. The extent of the water film which "defocuses" the electric field thus plays a decisive role in the resolution of the process. The formation and extent of this film is governed by a number of factors including capillary forces, high electric field gradients at the tip apex, the wetting behavior of the substrate and the tip and the relative humidity. Writing with *hydrophobic* carbon nanotubes appears promising. The effect of the ambient humidity is shown in Fig. 9.7 where the resolution is significantly enhanced at lower humidity. The stress which develops during the oxidation also favours oxides with low curvature, (i.e., shallow oxides.) By increasing the field strength, thicker oxides (e.g., >100 Å) are grown but with a decreasing aspect ratio.

(b) Current-induced Local Oxidation (CILO)

A novel route is introduced for oxidizing thin metal films with nanometer-scale resolution in air. By locally subjecting Ti films to high in-plane current densities, metal-oxide tunneling barriers are formed. The oxidation is triggered by current-induced atomic rearrangements and local heating.

At the final stages of the barrier formation, when only atomic-scale channels remain unoxidized, the oxidation rate decreases drastically while the conductance drops in steps of about the conductance quantum. This behavior gives evidence of ballistic transport and a superior stability of such metallic nanowires against current-induced forces compared with the bulk metal. The current-induced local oxidation process is used to fabricate a single-electron transistor, which exhibits a Coulomb staircase at room temperature.

The CILO process is based on thin Ti films as illustrated in the sketch and the AFM picture above. Subsequently, oxide notches are via AFM oxidation such that a narrow metallic gap is formed. Upon applying a constant bias voltage across the gap under ambient conditions (in air),, the measured current starts to decrease rapidly as a function of time, stabilizing seconds later at a two orders of magnitude lower value. In the meantime, the gap has been closed by a current-induced nanometer-scale oxide barrier. After the barrier formation the current-voltage curve became strongly nonlinear (top right), while it was linear before. The CILO process occurs at current densities of 10^7 Acm^{-2} and is triggered by current induced atomic rearrangements and local heating.

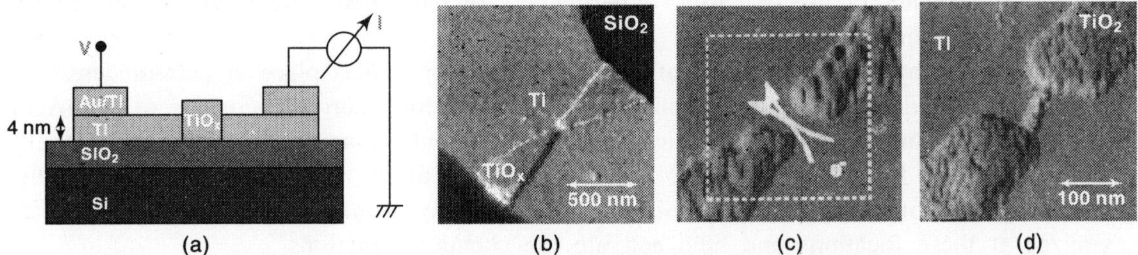

Fig. 9.11 *The Current-induced Local Oxidation (CILO) Process*

The conductance-time curve provides a measure for the oxidation rate during the CILO process. At the final stages of the barrier formation, when only nanometer-scale channels within the gap remain metallic, the oxidation rate decreases drastically and conductance drop sin steps of about $2e^2/h$ (top left and right). These steps give evidence of conductance quantization as a consequence of ballistic transport. The reduced oxidation rate indicates a superior stability of such metallic nanowires against current-induced forces compared with the bulk metal. Thus, the CILO process provides a striking example, where quantum transport influences a chemical reaction.

The high current densities that can be generated with the STM are shown to lead to the breaking of chemical bonds via a multiple-vibrational excitation mechanism. The high electric current densities that are generated inside nanowires and nanoconstrictions is achieved. The reaction rate is a strong function of the current density and this leads to nanometer scale spatial resolution. Now, an electric current flowing through a conductor exerts forces on defect atoms in the conductor and two types of forces are there. One, is the electron wind force, results from the momentum transfer in electron–atom collisions. The direction of this force is the same as that of the electron flow, and its magnitude is $F_{\mathrm{EW}}\ \alpha\ J_e\ \alpha\ Q_E$, where J_e is the electron current density and σ of the conductivity. The other force is the result of the electric field acting directly on the nominal charge (ze) of the atom, i.e. $F_{\mathrm{EF}} = zeE$, and the opposite direction to F_{EW}. In most cases $F_{\mathrm{EW}} > F_{\mathrm{EF}}$ and the atoms move in the same direction as the electrons. The resulting net atomic current can be written as

$$J_{\mathrm{A}}\ \alpha\ \frac{NJ_e}{kT}[D_0\ \exp(W_{\mathrm{D}}/kT)]$$

where N is the number of the atoms, and D_0 and W_{D} are their diffusion coefficient and diffusion activation energy, respectively. From the above equation, the electric-current-induced atomic motion, so-called electromigration, becomes important at high electron current densities or electric fields. Along with the atomic current a vacancy current J_{V} is formed which moves in the opposite direction. Joule and local heating effects further increase the atom/vacancy currents. Permanent structural changes occur at places where the divergence of the current is not zero, i.e. $\nabla J_{\mathrm{A,V}} \neq 0$. Such locations act as sources or sinks of atoms. The atom/vacancy current is a function of several other variables as the local composition or the temperature, and the continuity equation $\nabla J_{\mathrm{A,V}} = -\partial N/\partial t$ suggests that the sites of local modification will be at the places where the gradients of these variables are large. Grain boundaries where the mobility of the atoms and the above gradients are highest, form the preferred sites for structural modification. The thin metal films used in the oxidation experiments are nanocrystalline with an average grain size of only a few nanometers.

The destruction by electromigration of the thin metal films takes place at current densities of $\sim 10^8$ Acm^{-2}. To induce local nanochemistry we must generate current densities of 10^7 Acm^{-2}, vacancy aggregation at grain boundaries creates nanocracks that act as channels that allow oxidants (O_2, H_2O) from the ambient to reach the interior of the film. The large field gradients present at grain boundaries help in this transport. Finally, Joule and local heating are also maximized at these locations and help activate the chemical reactions.

The use of high local current densities to induce nanochemistry is shown in Fig. 9.12. The structure used is the side arm of a Ti star-shaped film. AFM oxidation is done at the place by making its use so as to generate two oxide notches that constrict the current path. The constriction shown in Fig. 9.12(a) has a width of ~140 nm. By applying a voltage bias of 2 V ("stress voltage") a current density of $\sim 5 \times 10^7$ Acm^{-2} is obtained. Under these conditions, in a short period of time, a barrier is formed that bridges the constriction as shown in Fig. 9.12(b). In this case, the width of the barrier convolved with the tip width is 28 nm. Widths as low as 10 nm in this manner is produced. For a given constriction, the barrier width is a function of the applied stress voltage. An example is shown in Fig. 9.12(c). The strongly non-linear I–V obtained after

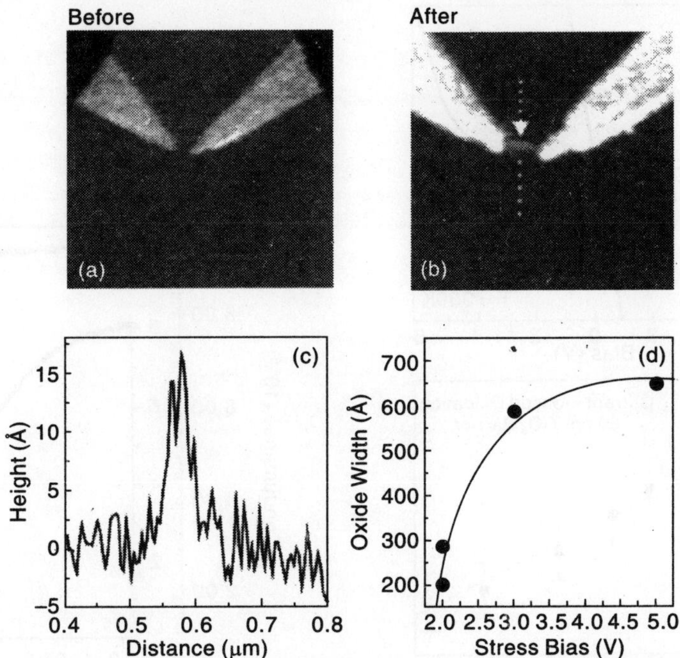

Fig. 9.12 *A AFM fabricated oxide constriction on a 30-nm-thick Ti film. (b) After stressing by applying a 2-V bias a 28-nm oxide barrier if formed. (c) Line scan through the current-grown oxide. (d) Dependence of the oxide barrier width on the stress voltage.*

stressing a 150 nm constriction at a voltage of 5 V is shown in Fig. 9.13(a). By plotting the current as a function of $1/T$, the effective tunneling barrier height Φ_B^{eff} is obtained. This height is a function of the electric field and is given by $\Phi_B^{\text{eff}} = \Phi_B^0 - e(eE/4\pi\varepsilon_0\varepsilon)^{1/2}$, where the second term accounts for the reduction of the barrier by the electric field and by image effects. To obtain the field free value Φ_B^0; Φ_B^{eff} as a function of $\sqrt{E}$ is plotted and is extrapolated to $E = 0$ (See Figure 9.13(b)). This field-free value of the barrier height is 0.24 eV, which is very close to the value of 0.28 eV reported for the Ti oxide produced by tip oxidation. Thus, the barrier is formed by the oxidation of Ti. A measure of the oxidation rate is provided by the time evolution of the conductance. Figure 9.14 shows the behavior of the conductance of a 140-nm constriction during current-induced oxidation. Initially the rate of change of the conductance is slow but as the cross section of unoxidized part of the constriction decrease, the current density increases and the oxidation rate becomes very high. Hence, the rate is a string function of the current density. This non-linear dependence of the rate on current density is analogous to the behavior observed for hydrogen desorption induced by STM tip-emitted electrons. In both cases the strong dependence of the rate on current density tends to localize the reaction, to area where Je is the highest and this leads to high spatial resolution. Conductance curves such as that shown in Fig. 11 also show kinks.

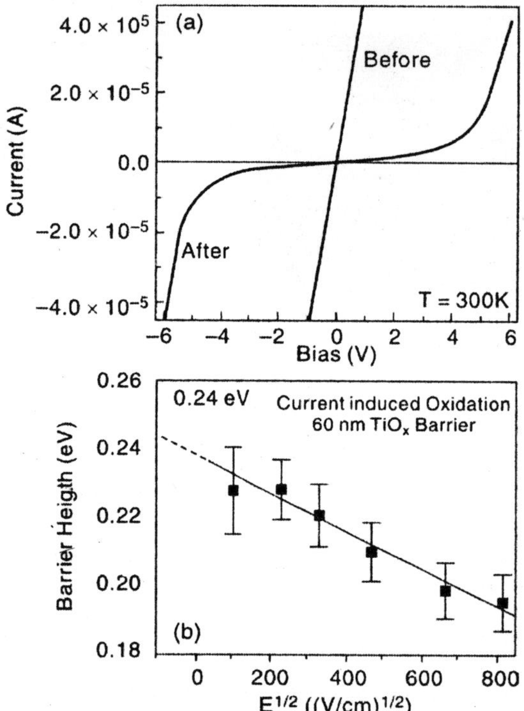

Fig. 9.13 *Top: I–V curves before and after a 5-V stress of a 150-nm constriction on a 4 nm-thick Ti film. Bottom: Determination of the field-free value of the tunneling barrier generated by the 5-V stress.*

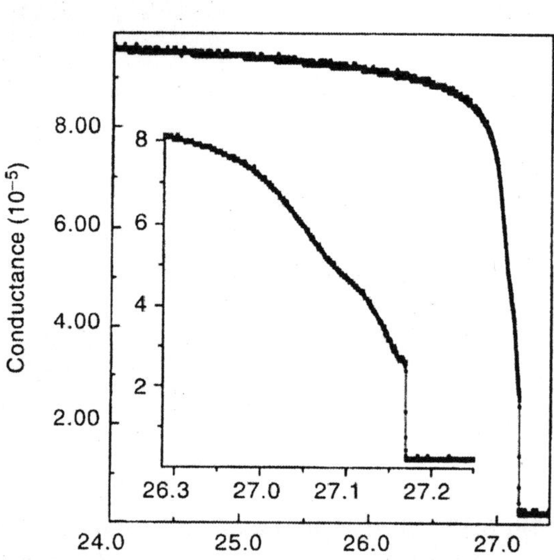

Fig. 9.14 *Time dependence of the conductance through a 140-nm constriction following the application of a 2.5V stress voltage.*

III. HIGH-κ DIELECTRIC STACKS WITH POLY Si AND METAL GATES

(a) Poly Si/high-κ Gate Stacks and Hf-based Gate Dielectrics

PolySi-based devices, are usually annealed at high temperatures (>1273K) in order to activate dopants in the gate and source/drain regions. The requirements of thermal stability in contact with the poly Si gate electrode and negligible metal diffusion into the Si channel have ruled out successful integration of high-κ materials such as ZrO_2 and Al_2O_3. The thermal stability of the dielectric, electrical thickness scaling, carrier mobility, p-FET threshold voltage, and long-term stability/reliability are needed for the development of Hf-based gate stack materials for polySi-gated devices.

Thermal Stability

In contrast to ZrO_2, no detrimental silicide formation occurs with HfO_2 in contact with polySi gates. However, polySi/HfO_2 gate stacks undergo substantial changes during thermal processing. Dopant activation requires annealing to temperatures of ~1273 K or more, much higher that crystallization temperature of amorphous HfO_2. Depending on HfO_2 thickness, crystallization into predominantly monoclinic polycrystalline films occurs at 573 K – 773 K. Also, the formation of additional interfacial SiO_2 is often observed, degrading gate stack capacitance.

For HfO_2, the grain boundaries in polycrystalline films constitute electrical leakage paths, giving rise to increased gate leakage currents. Experimentally, only minor, increase in leakage with polycrystalline HfO_2 is observed. However, amorphous high-κ layers are preferred.

It is found that heterogeneous grain orientations in the dielectric layer may give rise to spatially varying electric fields and cause carrier scattering, thereby degrading mobility. Also, with continuing scaling, the gate length becomes comparable to the HfO_2 grain diameter.

The grain boundaries in poly- or nano-crystalline material are found to be responsible for unoccupied states below the metal d-state-derived conduction band edge in ZrO_2. Such defect states are observed by optical and X-ray absorption spectroscopy as well as by photoconductivity measurements. The band-gap defect states occur in HsO_2 if the dielectrics exhibit crystallinity as detected by infrared spectroscopy, X-ray diffraction (XRD), and vacuum ultraviolet spectroscopic ellipsometry (VUV–SE). This is exemplified by the imaginary part ε_2 of the dielectric function for HfO2 films grown by atomic layer deposition (ALD), as shown displayed in Figure 9.15. When HfO_2 thickness is increased, an absorption feature emerges at ~5.8 eV. The correlation between crystallinity and electronic defects holds also for HfO_2. These observations are significant from a device perspective, as HfO_2 gate dielectrics occurs via trapping sites located a few tenths of an eV below the HfO_2 conduction band edge. It is unlikely that such states are a limiting factor in high-κ-based CMOS technologies, since they line up close to the insulator band edge and are therefore not accessible at the low gate voltages in high-performance and low-power technologies. As additional grain-boundary-induced defect states exist deeper in the bandgap. It is therefore preferable to employ amorphous high-κ materials.

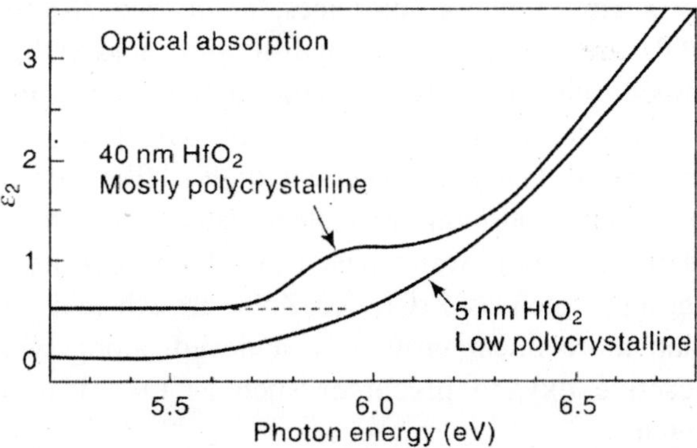

Fig. 9.15 *Imaginary part ε_2 of the dielectric function, determined from VUV–SE data, for 5- and 40-nm-thick ALD-grown HfO_2 films.*

In order to prevent gate dielectric crystallization and to minimize interfacial SiO_2 formation, the thermal stability of the HfO_2 dielectric must be increased. This is achieved by addition of Al, Si, and/or N. HfAlO gate dielectric is found to reduce carrier mobility, due to fixed charge near the high-κ/channel interface.

Increased thermal stability is achieved at a comparatively low Si content; (Si(Hf + Si) = 20%). After rapid thermal anneals to 1273 K for 5 s, the films do not exhibit any infrared phonon mode characteristics of monoclinic HfO_2, HfSiO remains mostly amorphous, although partial crystallization into the tetragonal or orthorhombic phase is not excluded. However, longer 1173–1273K anneals, HsSiO crystallizes and decomposes into HsO_2 and SiO_2. Further increased thermal stability is achieved by additionally introducing N. The tendency to crystallize under 1273 K dopant activation anneals is completely suppressed in HfSiON with a N content of N/(O + N) > 10%. In addition, boron penetration is more effectively prevented by HfSiON than by HfSiO.

Electrical Thickness Scaling

The electrical thickness of Hf(Si)O(N)/SiO(N) gate dielectrics is determined by the sum of the electrical thickness of the high-κ and the interfacial SiO(N) layer (if present). Therefore, a combination of strategies are pursued to minimize total electrical thickness, which are:

(i) Minimize high-k thickness, while maintaining (a) a closed high-k layer and (b) sufficient Hf constant of the gate stack, ensuring a gate leakage advantage over pure SiON gate dielectrics.

(ii) *Minimize the interfacial SiON thickness*, by maintaining (a) appropriate Si surface functionalization to ensure homogeneously high nucleation, and (b) high carrier mobility.

(iii) *Increase N concentration* in the interfacial SiON and high-k layers so as to increase the dielectric constant and reduce interfacial layer growth during thermal processing, while maintaining low charge trapping and high carrier mobility.

Surface Preparation

To fabricate atomically sharp Si/high-κ interfaces, oxide-free *H-terminated Si(H/Si)* substrates are utilized. Such H/Si(100) are prepared by a hydrofluoric acid (HF) wet etch of SiO_2/Si with subsequent water rinse. Such substrates are resistant to oxidation, and even in O_2^- or H_2O- at 573 K. While this low surface reactivity is in terms of oxidation resistance, it also causes the poor nucleation characteristics of many ALD-grown high-κ films, resulting in nonlinear growth kinetics and the formation of discontinuous and electrically leaky gate stacks. Examples are; the $HfCl_4/H_2O$ process for HfO_2 growth, water-based ALD process having metal precursors that react with surface—OH groups, such as $ZrCl_4$ for ZrO_2 growth and $Al(CH_3)_3$ for Al_2O_3 growth. Nucleation is enhanced and more linear growth is achieved, along with the activation of the H/Si surface by a more reactive oxygen precursor such as O_3. However, substantial interfacial SIO_2 formed during growth.

To overcome nucleation on H/Si without having reactive O is *via initial extended H/Si exposure to $Al(CH_3)_3$*. During exposures ~1,000 times larger metal–organic functional groups are introduced onto the Si surface. Such groups are reactive toward the water precursor. On Al–

organic functionalized Si, HfO_2 and Al_2O_3 nucleation is achieved. Note that, large H_2O exposures of H/Si leave the H termination unaffected and therefore do not lead to enhanced high-κ growth. Thus, a possible shortcoming of Al-organic functionalization is the excessive $Al(CH_3)_3$ exposure time.

Another approach to optimize nucleation and minimize interfacial SiO_2 formation of H/Si is based on *ALD growth at reduced temperatures* (e.g., 323–373 K). Interfacial SiO_2 formation is prevented using both the $Al(CH_3)_3/H_2O$ process for Al_2O_3 growth and the tetrakis(ethylmethylamino)hafnium/water process for HfO_2 growth.

To passivate the surface and prevent oxidation, hydrogen is replaced with other is surface passivant; for example chlorine. The monolayer *chlorine pasivation* is achieved by a simple Cl_2 gas treatment of H/Si, where reaction rates are enhanced by ultraviolet (UV) light. It gives a minor thickness advantages over SiON interfaces, but nucleation is poor.

The nitride-based interfaces such as high-nitrogen-concentration SiON or pure silicon nitride often are good nucleation layers. The nitridation increases interface permittivity (e.g., k_{SiO_2} = 3.9; $k_{Si_3N_4}$ = 7–8) and thermal stability. This scaling is realized, for example, with interfacial SI(O)N layers fabricated by H/Si, anneal in NH_3 at 923 K to form thin Si_3N_4. PolySi/HfSiO stacks on such high-nitrogen-content films exhibit lower electrical thickness than on 1.1-nm low-N-content SiON control substrates. But with high-nitrogen-content interfaces there is carrier mobility loss.

Carrier Mobility

High-κ materials are observed to degrade carrier mobility in the transistor channel. A number of mechanisms are responsible for remote phonon scattering through emission of absorption of low-energy phonons in the high-κ material and remote Coulomb scattering off fixed or trapped charges in the gate dielectric. But, mobilities with nominally similar gate stacks have improved. Thus certain defects in the high-κ materials such as electrical trap sites and impurities give rise to fixed charges which are minimized by process engineering.

However, remote phonon scattering limits the performance of HfO_2-based devices, which is improved by incorporating Si into the HfO_2. Since Si–O bonds are stiffer than Hf–O bonds, soft phonon modes are reduced in intensity. Also a drop in the remote phonon scattering cross section is observed to result in near-complete carrier mobility recovery even when ZrO_2 is replaced by $ZrSiO_4$. The same is true for halfnium silicates. The mobility advantage comes at the expense of a reduced dielectric constant (20–25 for HfO_2 compared to 10–15 for HfSiO.

Carrier mobility and thermal stability are the main characteristics for shifting the focus from polySi/HfO_2 to polySi/HfSiO(N) gate stacks. Indeed, even Si concentrations in HfSiO as low as Si/(Hf + Si) = 20% give excellent mobilities. For example, polySi/HfSiO/SiON n-FETs fabricated with EOT =- 1.6 nm exhibits electron peak mobility identical to that for low-N-content SiON control devices. High-field electron mobility is degraded by 10%.

Mobility improvements are also achieved by reduced charge trapping as reduces the Coulomb scattering rate. The charge-trapping behavior of HfSiO is better than that of HfO_2.

Nitrogen is often introduced into high-κ gate stack to enhance thermal stability and reduce electrical thickness. Here, the carrier mobility is reduced, for HfSiO on Si_3N_4 interface layers. This is illustrated in Figure 9.16 which shows n-FET electron mobility for various HfSiO/

Si(O)N gate stacks. Mobility at high field (black symbols) is extracted from full mobility curves (inset). It is measured using the split *C–V* technique. With interfacial Si_3N_4 formed by an NH_3 anneal of H/Si at 923 K (N areal density ~2×10^{15} Ncm^{-2}), high-field mobility is degraded by 20–25% compared with now-N-content interfacial SiON layers (~7×10^{14} Ncm^{-2}). Even upon introduction of O into the nitride using NO gas anneals at 973K–1073 K mobility recovers. High nitrogen concentrations reduce mobility even with SiON gate dielectrics. For Coulomb scattering; fixed charges is SION creted by nitrogen are responsible for the mobility loss both in SiON and high-κ stacks. However, other physical causes are also observed in N-induced mobility degradation in high-κ gate stacks. Three scenarios are observed in mobility reduction: (a) Slow interface states (areal density N_{it}) or (b) fixed charges (areas density N_{ox}) causes Coulomb scattering of channel electrons; and (c) charge trapping causes Coulomb scattering or induced hysteresis which distorts the inversion charge and mobility measurement.

A combination of electrical measurement techniques aid in assessing which of the above causes is dominant in mobility degradation. In (a), N_{it} is measured by amplitude-sweep charge pumping. Independently of O content, all nitride-based interfaces exhibited 3–5 times higher $N_{it}(1.3–1.9 \times 10^{11}$ $cm^{-2})$ than low-nitrogen-content control SiON interfaces. Whether thin N_{it} difference is sufficient to explain the mobility loss, a corrected mobility is calculated that is measured if N_{it} is reduced to zero without modifying the gate stacks. After this correction, the mobility trend with N and O content remain virtually unchanged Figure 9.16, (white symbols), show that mechanisms other than slow interface states are responsible for N-induced mobility loss. In (b), by contrast, is supported by *C–V* measurements: threshold voltage V_t is ~0.1 V lower with all Si_3N_4-based interfaces than with the control SiON interface. The shift corresponds to an areal density of positive fixed charge of N_{ox} ~8×10^{11} cm^{-2}, independent of O content.

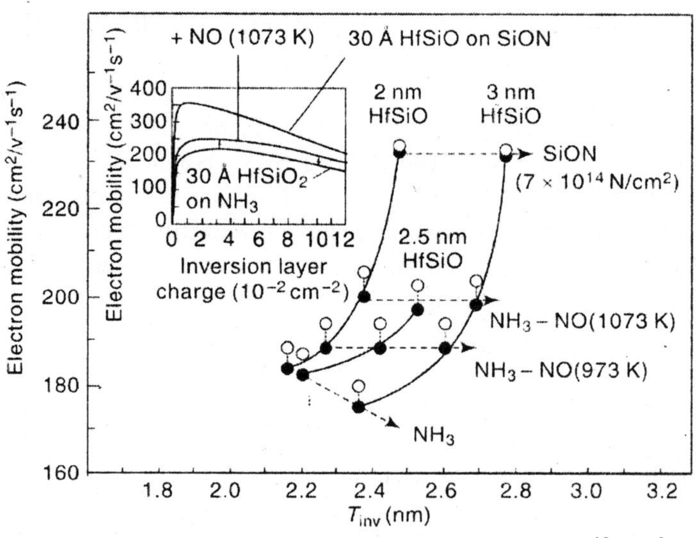

Fig. 9.16 *High-field electron mobility (at inversion charge density $N_{inv} = 10^{13}$ cm^{-2}) as a function of inversion thickness (T_{inv}) values for various Si(O)N processes and HfSiO thicknesses. Black as-measured. White: after N_{it} correction. Inset: Electron mobility as a function of N_{inv} for various SiO(N) processes at a HfSiO thicknesses of 3 nm.*

Thus, N_{ox} is higher than N_{it} and explains the observed mobility loss with reoxidized nitride interface. With pure nitride interface, distortion of the inversion charge measurement due to transient charging (c) occurs in addition. However, fixed charge likely is the main case of carrier mobility loss with interfacial N.

Nitridation of HfSiO layers degrades mobility. However, mobility impact is greatest if nitridation conditions allow N to permeate the entire HfSiO film, while near-surface nitridation preserves mobility. This indicates that N in or close to the interfacial layer has by far the greatest impact on mobility. This is due to (a) the rapidly decaying electrical field strength around a fixed charge, giving rise to simultaneously dropping Coulomb scattering cross section; and (b) a lower fixed charge per N atom in HfSiON than in SiON.

Good electron mobility with HfSiON/SiO_2 is shown in Figure 9.17. Appropriate a high-temperature plasma nitridation conditions ensures a high proportion of near-surface N. Under such conditions, gate stacks incorporating such HfSiON show N-induced T_{inv} reduction by up to 0.1 nm, confirming the scaling benefit of N. At N concentrations as high as[N/(N + O)] ~21%, the N-induced V_{fb}/V_t shift to more negative values is smaller than 0.02 V, showing that little positive fixed charge is created for from the gate electrode. Trap density remains low. Thus by the mobility degradation mechanisms with interfacial N, mobility is nearly identical to that of a low-N-content SiON control (Figure 9.17).

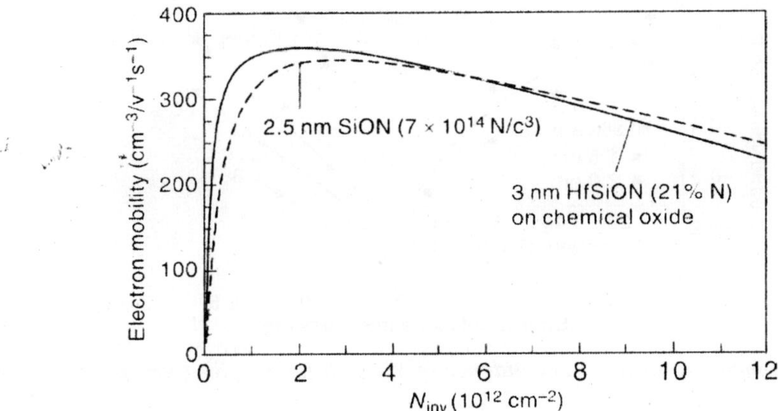

Fig. 9.17 *Electron mobility with 3 nm top-nitrided HfSiON with [N/N + O)] ~21%, compared with low-N-content SiON.*

Threshold Voltage

The threshold voltages of polySi-gated high-κ n-FETs and p-FETs are different from those achieved with corresponding SiO(N) devies. Using Hf-based high-κ materials, n-FET V_t is excess by ~0.2 V, while p-FET V_t is less by ~0.6 V. While threshold voltage their optimum values the device performance degrades with excessive tuning. So, n-FET devices designed to offset the materials-induced shift of ~0.2 V. But with ~0.6-V shift for p-FET devices, one cannot fabricate good-performance Hf-based polySi/high-κ devices. So, the gate stack itself must be understood and modified.

The choice of dopant, method of doping (implant vs. *in situ* doping with CVD precursors), and thermal processing on testing helped to control V_t. The V_{fb}/V_t shifts a measured after gate-stack fabrication steps. Measurements with undoped and unactivated polySi gates are achieved by recording electrical data at elevated device temperatures (up to 473 K). The results indicate that V_{fb}/V_t ratios are largely set during polySi deposition and remain unchanged during gate implantation and thermal activation, independent of the p-type dopant (B, Al, Ga). The p-FET V_t shift is thus not easily prevented by employing modified polySi/Hf(Si)O(N). A reaction of Si with the Hf-based material, occurring during polySi deposition, appears to be the root cause for the poor V_{fb}/V_t control.

The introduction of Si or N into the Hf-based layer has a limited impact on V_t/V_{fb}. When utilizing HfSiO with increasing Si content, V_{fb} gradually approaches the value observed with SiO_2 (Figure 9.18, inset). However, in order to bring Bt to within less than 0.3 V from the target value, Hf contents below ~210% are required. At such compositions, the dielectric constant is only marginally higher than for SiON, making implementation unattractive. The fixed charge from N incorporated into the gate stacks in the bottom interface is another means of controlling V_t. But, only a limited degree of V_t improvement (by up ~0.1 V) is achieved, at the expense of mobility loss.

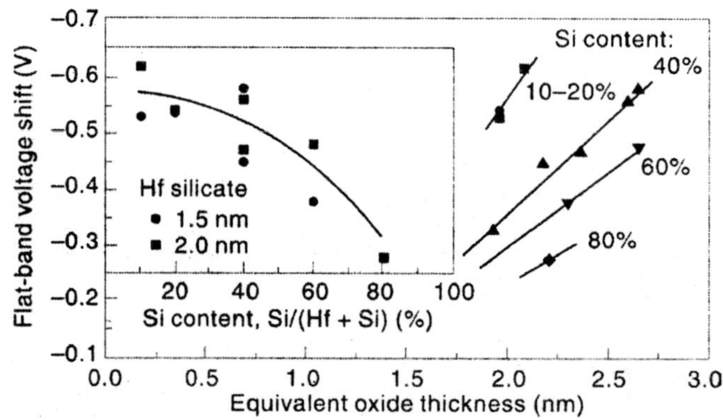

Fig. 9.18 *EOT and composition (inset) dependence of V_{fb} shift for polySi-gated p-FETs with HfSiO gate dielectric.*

Experimental evidence indicates that oxygen plays a prominent role in the p-FET V_t shift phenomenon. It is shown that oxidation of the polySi/high-κ stack by lateral indiffusion of oxygen alleviates the p-FET V_t shift of transistor devices with channel lengths below ~1 μm at the expense of EOT. Also, optical spectroscopy is used to trap levels in HfO_2 to oxygen deficiencies.

Fermi-level pinning causes the V_t shift, by a high areal density of interface states whose occupation charges (as the gate voltage is swept) from accumulation to inversion. The interface states partially screen the electric field from the gate electrode, preventing in from reaching the channel. The extent of gate-induced turning of the channel carrier occupancy is thus greatly reduced. Also, fermi-level pinning just below the polySi conduction band is caused by Hf-Si bonds at the high-κ/polySi interface.

However, defect levels and fixed charge in the Hf-based gate dielectric also causes V_t shifts for example, O vacancy formation in HfO_2 is favorable when the HfO_2 is in contact with a p-doped polySi gate, since such defect states are stabilized by the transfer of two electrons to the gate elctrode; this transfer cannot occur in contact with n-doped polySi gate. Positive fixed charge is thus created inside the HfO_2, shifting the p-FET V_t to more negative values.

Generally, physical causes for fixed charge are vacancies or interstitials, foreign atoms such as Si, N or gate dopants diffused into the high-κ layer. Si and N do not cause the p-FET V_t problem.

The importance of O, and a better understanding of the electronic structure and formation enthalpy of O vacancies in HfO_2, such vacancies create V_t shifts. However, in order to distinguish such defects from interfacial Hf-Si bonds more physical characterization is required.

It is critical to determine whether the p-FET V_t can be shifted closer to the target value by choosing appropriate processing conditions. Lateral oxidation of the high-κ layer brings relief for short-channel devices. However, the growth of SiO_2 at the gate electrode interface increases the EOT; O indiffusion. Hence V_t are dependent on channel length, and it is unclear whether the O content of the gate stack is maintained during the entire device fabrication process.

Thin dielectric cap layers inserted between the Hf-based dielectric and the polySi electrode, weaken the Hf–Si bond. For example, Si_3N_4, SiC:4, and HfON cap layers lead to only very small V_t. With SiO_2 cap layers, V_t improvemt (by 0.3 V) is achieved at a cap thickness of 1 nm. However, SiO_2 capping severely limits thickness and effectively defeats the purpose on introducing high-κ materials.

Consider Al_2O_3 cap layers growth by ALD. Improvements range from 0.1 to 0.3 V on HfSiO to 0.6–0.7 V on HfSiON. These variations indicate the process control. Here, V_t improvement is through negative fixed charge introduced into the Hf-based material by the AL. But Al_2O_3 cap layers there is charge trapping under operation conditions. In Al_2O_3/HfSiO stacks, the degree of trapping inside the HfSiO is independent of cap thickness. It is be beneficial to reduce cap thickness to a minimum; since with increasing cap thickness the distance of the trap sites from the gate electrode increases and in turn, the V_{fb} shift induced by trapped charges increases.

Now, aluminium nitride (AlN) is used as a material that ensures sufficient V_t at very low cap thickness and high effective permittivity. Hf-based stacks are thus engineered such that the n-FET and p-FET V_t are sufficiently low, with excellent device characteristics. The AlN cap is deposited onto the HfSiO on both p-FETs and n-FETs, and subsequently etched of the n-FETs. Selective capping of p-FETs is achieved. Separate wafers a employed, but full CMOS integration is possible through masking/etching. Figure 9.19(a) shows *C–V* curves for optimized p-FET and n-FET polySi/AlN)/HfSiO gate stacks. The physical thickness of the AlN cap is only 0.4 nm; and contributes only 0.1 nm to the total EOT, ensuring scalability. The p-FET V_t shift is reduced by –0.22 V to –0.31 V compared with SiON, depending on the Si/high-κ interface layer. For n-FETs, ΔV_t = 0.21 V, similar to polySi/Hf(Si)O stacks. Thus, nearly symmetric *C–V* characteristics with low V_t is obtained. These findings are confirmed by I_d–V_g data [Figure 9.19(b)]. A small subthreshold swing of 71 mV/dec indicates that the interface state density is low. Thus it is shown that n-FET D_{it} ~ 10^{10} eV^{-1}-cm^{-2} and p-FET D_{it} ~ 7×10^{10} eV^{-1}-cm^{-1}. The p-FET V_t improves slightly with decresaing thickness, indicating further scalability. Also, the

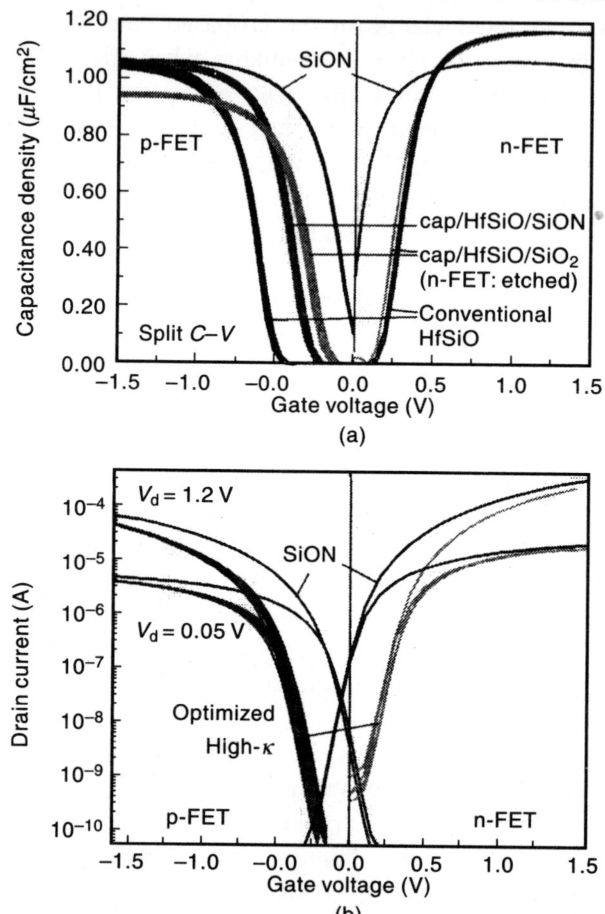

Fig. 9.19 *(a) Split C-V characteristics for AIN-capped p-FET and AIN-capped/etched n-FET high-κ stacks on two interfaces compared with conventional HfSiO and SiON. (b) I_d–V_g characteristics for AIN-capped p-FET (EOT = 1.9 nm) and AIN-capped/etched n-FET (EOT = 1.9 nm) and AIN-capped/ etched n-FET (EOT = 2.0 nm) high-κ stacks compared with SiON.*

V_t -optimized high-κ based FETs shows good performance. Mobilities and drain currents for p-Fets and n-Fets range between and 90 and 110% of those for a SiON control. A narrow distribution of breakdown voltage indicates the uniform quality of the dielectric. Stress induced n-Fet V_t shifts as the charge trapping are sufficiently low. By combining this capped gate stack with modertate implant engineering for final V_t adjustment, short-channel polySi(AlN)/HfSiO devices with acceptable performance are achieved.

(b) Metal Gates

While high-κ dieletrics are clearly required to scale beyond the 45-nm node, the interaction of Hf-based dielectrics with polySi electrodes suffers from a number of drawbacks, including high p-FET V_t and difficulty in scaling below T_{inv} of 2 nm. The use of metal gates helps to overcome some of these hurdles. In this section, we summarize the advances and challenges remaining for metal/high-κ stacks. We show that aggressively scaled metal/high-κ stacks (T_{inv} = 1.4 nm) with high electron mobility can be achieved in a conventional self-aligned process by careful process optimization, including the use of non-nitrogen interface layers, high-temperature processing, and appropriate electrode structures to prevent regrowth. However, V_{fb}/V_t instability after high-temperature processing remains the biggest challenge to overcome, with oxygen vacancies in the high-κ resulting in large V_{fb}/V_t shifts for high-workfunction (φm) metal gates.

Thermal Stability

An electrode is considered unstable if the XRD analysis shows a deviation from the linear decrease in diffraction angle (2ω) as a function of temperature. The reaction and/or formation of a new phase with a different crystal structure, give possible stable electrode choices as shown in Figure 9.20. Most of the low-φ_m elemental metal gates (φ_m = 4.1 to 4.3 eV), indicated by

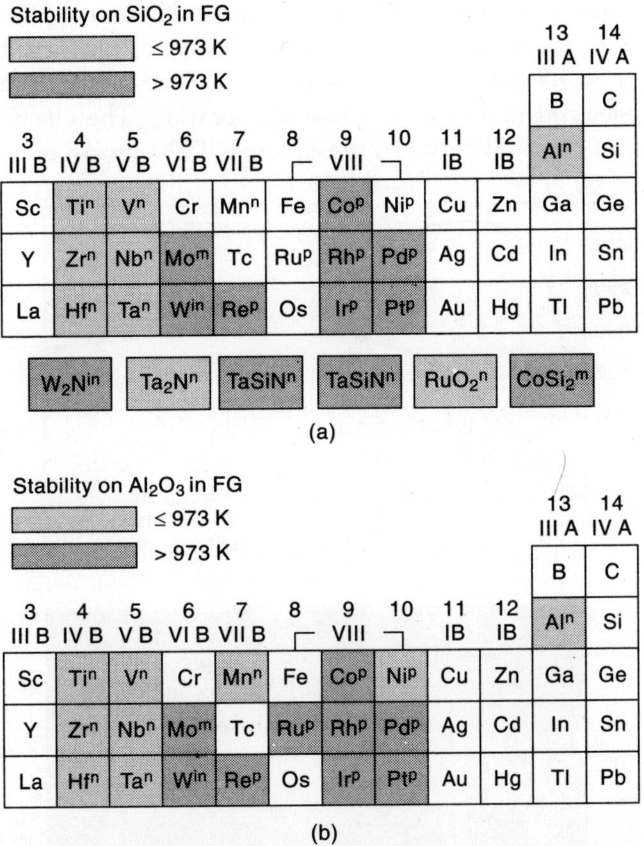

Fig. 9.20 *(a) Periodic table indicating the thermal stability of different electrode materials on SiO_2 evaluated using in situ XRD, resistance, and optical scattering analysis techniques. Superscripts following the chemical symbols indicate the type of work function: ^{n} = n-FET, ^{m} = midgap, and ^{p} = p-FET. The shading indicates whether the thermal stability is less then 973 K (light gray) or greater than 972 K (darker gray). (b) Thermal stability of different electrode materials on Al_2O_3.*

light shading, are reactive and do not withstand conventional CMOS annealing temperature. The exceptions, TaN and TaSiN, have to lower, n-FET φ_m but remain stable to high temperature. Most of the midgap (including TiN) and high-φ_m metal gates (γ_m = 4.9 to 5.2 eV), indicated by darker shading, remain stable to high temperatures (1073–1273 K). While most of the p-FET gate metals and alloys structurally stable at high temperatures, but, CMOS processing requires temperature greater than 1223 K.

These thermal stability constraints act as a catalyst for the development of a gate-last or replacement-gate process. The process requires that after a source/drain (S/D) activation and anneal and silicide scheme, nitride and oxide are deposited, and this is followed by planarization using chemical–mechanical polishing (CMP). The polySi gate and SiON dielectrics are removed, and the new SiON (or high-κ) is grown (or deposited), followed by deposition of the metal gate. After deposition, at highest temperature to which the gate stacks are exposed (shown in the back end) is < 773 K. Using the replacement-gate integration and CVD W as a metal gate, CMOS transistors down to 0.1 μm a successfully fabricated. It is shown that while the hole mobility of p-FETs remains better than that of polySi/SiON controls, the electron mobility for W/SiO_2, W/SiON, and W/HfO_2/SiON [Figure 9.21(a)] are degraded by more than 20% compared with polySi/SiON gate stacks of similar T_{inv}. The presence of N in the gate stack further degrades the electron mobility for a low temperature. The electron mobility for a low-temperatures integration Fig. 9.21(b) is a representative TEM image of center of a 1-μm trench

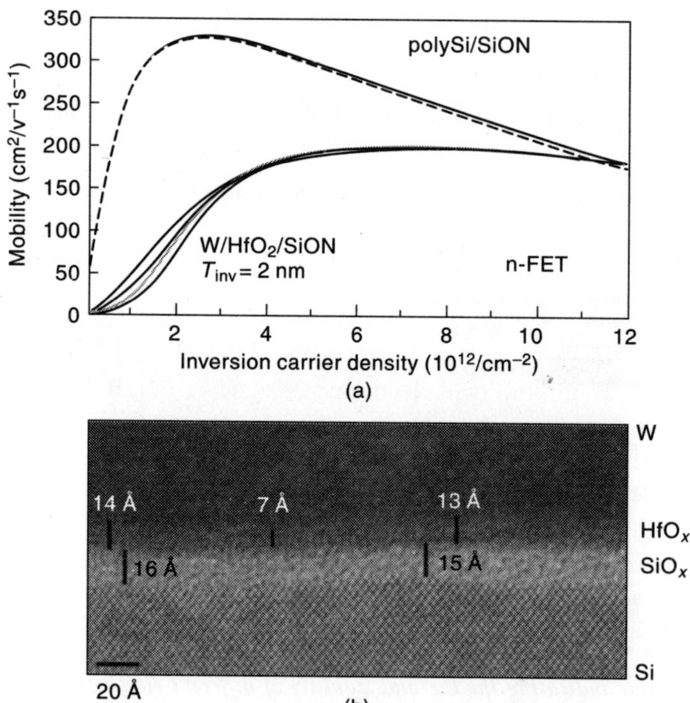

Fig. 9.21 *Electron mobility of a CVD W/HfO$_2$/SiON replacement-gate device compared with a polySi/SiON control device. (b) TEM image of the center of a replacement-gate trench.*

for a W/HfO_2/SiON stack. The thickness variation of the HfO_2 (≈ 2.5 nm) shows confirmation for gate lengths of less than 45 nm. Advances are made with the replacement-gate process, in which high electron mobility (250 $cm^2V^{-1}s^{-1}$ peak and 190 $cm^2V^{-1}s^{-1}$ at 1 $MVcm^{-1}$) at T_{inv} of 1.5 nm A_{is} observed for a non-nitrogen-based electrode and dielectric process. However, the viability of the replacement gate process for gate lengths of less than 25 nm depends on the development of extremely conformal dielectric and electrode deposition process.

Metal/SiON vs. Metal/High-k stacks

Metal/SiON gate stacks are not suited for high-performance logic applications because they do not provide an improvement in leakage and deposition processes. Low-damage deposition processes such as ALD or CVD are preferred, as physical vapor deposition (PVD) processes results in damage to the thin oxynitride dielectric layer. Such requirements on deposition processes are not required for the integration of metal/HfO_2 gate stacks, since these stacks have a physically thicker high-κ dielectric (compared with SiON) that results in lower gate leakage. These stacks are more thermodynamically stable at elevated temperatures than metal SiO_2 stacks. This allows for a "gate-first" processes integration scheme. Thus, introducing a metal gate for high-performance, CMOS requires that both metal and high-κ be introduced at the same time to achieve scaling and leakage. Unfortunately, aggressively scaled metal/high-κ stacks suffer from electron mobility degradation and V_{fb}/V_t instabilities.

Electron Mobility

The low-temperature-processed metal./high-κ devices have degraded electron mobility. The effect of high-temperature processing on the n-FET mobility of W/HfO_2/SiO_2 stacks is evaluated using a simple non-self-aligned integration flow, with devices processed between 873 K and 1273 K. It is shown that even with low interface-state densities (N_{it}), low-temperature (<873 K) processing resulted in extremely low electron mobilities. Increase in the thermal, results in improved mobilities, but at the expense of T_{iv} due to interlayer (IL) regrowth. Using a non-self-aligned for, n-FET-like CVD TaSiN/HfO_2 gate stacks a fabricated, and the mobility compared with CVD W/HfO_2 after high-temperature processing (Figure 9.22). It is observed that the presence of N at the interface for both gate electrode stacks *degrades the electron mobility*. Nitrogen is also the potential cause of the reduction of mobilities for aggressively scaled polySi/SiON devices. For the non-nitrogen IL, it is suggested that the Hf intermixes with *non-nitrogen* IL to form a high-κ Hf-siliacte IL which results in higher mobility than HfO_2, as it has a weaker coupling of the SO phonons compared with HfO_2. But, chemically analyzed IL, using low-loss electron energy loss spectroscopy and medium-energy ion scattering show no evidence for Hf-silicate formation upon annealing. It is suggested that the IL is Si-rich and results in a dielectric constant greater than that of SiO_2.

The effect of processing temperature on electron mobility and IL thickening from mobility improvement of metal/HfO_2 stacks, PVD TaSiN is used, which minimizes IL thickening. For, of dopant activation at low temperature, the solid phase epitaxial regrowth (SPER) process is used. Here, high-energy As implants for S/D amorphization is used followed by a 873 K anneal, in combination with NiSi S/D and gate contacts to fabricate self-aligned n-FETs at low

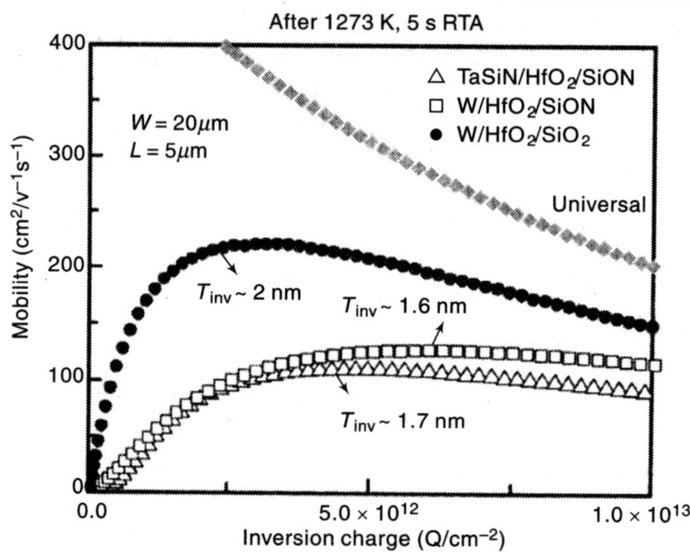

Fig. 9.22 *Electron mobility comparison of TaSiHfO$_2$/SiON, W/HfO$_2$/SiON, and W/HfO$_2$/SiO$_2$ stacks after high-temperature annealing.*

temperatures. Some wafers at 1073 K, 5 s and 1273 K, 5 s are annealed after SPER and prior to NiSi formation to observe the impact of high-temperature activation. Figure 9.23(a) shows that substantial improvement in mobility (25%, peak) is observed for both TaSi/HfO$_2$/SiON and a control TaSiN/SiON stack only after 1273 K anneals with little change in T_{inv}. The mobility increase is neither related to IL regrowth nor affected by T_{it} variations, as the mobility curves are corrected for T_{it}. Thus, the high thermal modifies the dielectric stack without interfacial regrowth to enhance the mobility. The mobility enhancement is also related to formation of a relaxed IL/Si interface at T > 1273 K or, in addition, especially for the high-κ gate stacks, to structural relaxation and modification of the HfO$_2$/IL interface.

High-mobility devices are obtained at aggressive T_{inv}, for self-aligned metal-gated high-κ transistors with oxide starting surfaces by capping different thin metal gate stacks such as PVD TiN, ALD TaN, and CVD W with polySi [Figure 9.23(b), and (c)]. To prevent reactions between W and polySi at T> 1073 K, a TiN barrier layer is inserted between the W and polySi layers. PolySi/TiN/HfO$_2$ gate stacks show electron mobilities at a T_{inv} to be 1.4 nm better. By the use of non-nitrogen interface layers, high-temperature processing, low T_{it} (<3×10^{10} cm^{-2}-eV), and appropriate electrode and electrode structures; Coulomb scattering. This results in high mobility in aggressively scaled metal/high-κ stacks that are competitive or better than aggressive polySi/SiON stacks [Figure 9.24(a)]. These high-mobility stacks still maintain more than 4–5 orders of leakage reduction compared with polySi/SiON [Figure 9.24(b)].

Metal gate screening of the soft optical phonon modes in the high-κ (the primary reason for reduced mobility of polySi/high-κ) has been proposed for improvement in mobility. However, with low-temperature mobility measurements of aggressively scaled metal-gated high-κ stacks; electron mobility is still limited by HfO$_2$ SO-phonon scattering.

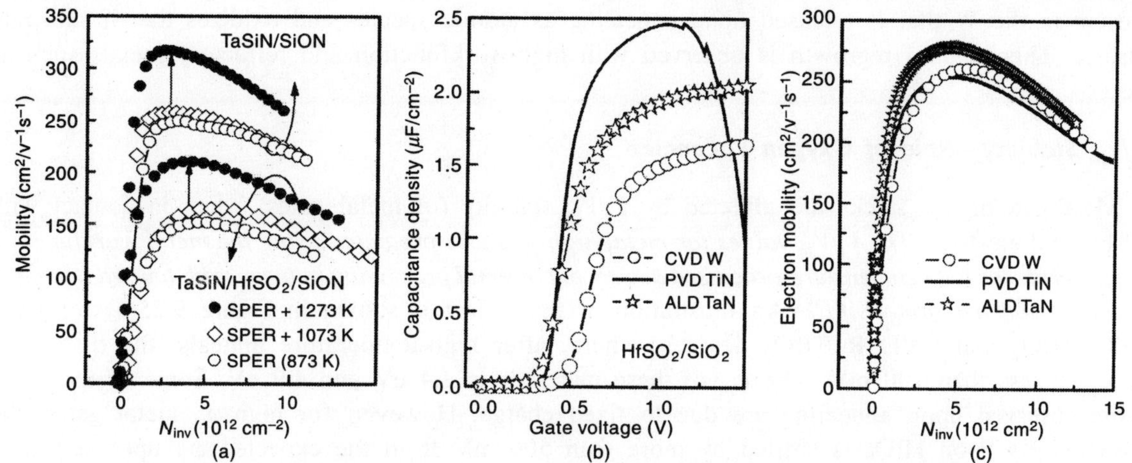

Fig. 9.23 *(a) Comparison of electron mobilities for TaSiN/HfO$_2$/SiON and TaSiN/SiON gate stacks, for three different process temperatures. Inversion thickness (T_{inv}) values corresponding to these process temperatures are respectively 1.7 nm, 1.5 nm, and 1.6 nm for the HfO$_2$/SiON stacks and 2.3 nm, 2.3 nm, and 2.4 nm for the SiON stacks. (b) Inversion split-CV characteristics and (c) electron mobility curves of self-aligned metal/HfO$_2$/SiO$_2$ n-FET gate stacks after a 5-s RTA. T_{inv} values for CVD, PVD TiN, and ALD TaN are 2.05 nm, 1.4 nm, and 1.7 nm, respectively. Channel doping = 1×10^{17} cm^{-3} for all stacks.*

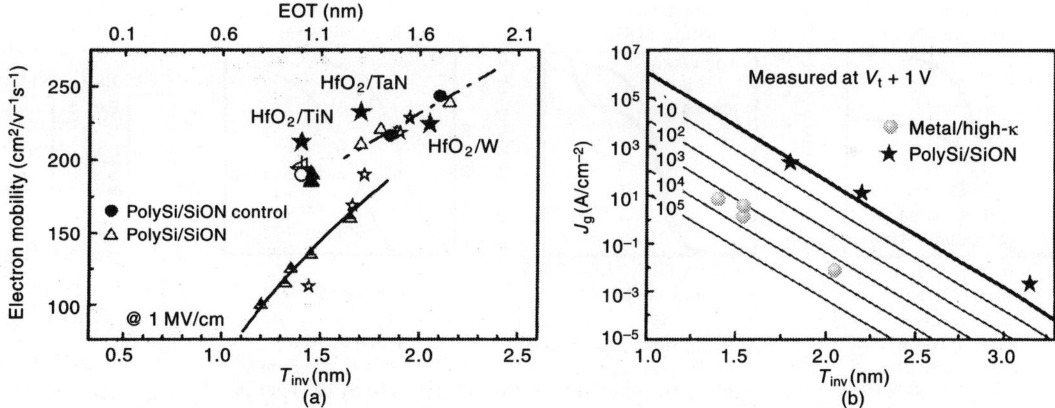

Fig. 9.24 *(a) Comparison of electron mobility (at 1 MV/cm) vs. inversion thickness (T_{inv}) for different metal-gated HfO$_2$ gate stacks. (b) Comparison of gate leakage as a function of T_{inv} for metal/high-k and polySi/SiON gate stacks.*

T_{inv} Scaling

Figure 9.24(a) and (b) shows difference in T_{inv} for different electrode stacks. At a high-temperature W-gated devices capped by TiN and polySi are at least 0.5 nm thicker in T_{inv} than an equivalent polySi/TiN device. Since the W is completely encapsulated during the S/D activation by TiN/polySi and nitride spacers, the increased T_{inv} is due to residual oxygen present in the W that is released upon annealing as atomic species and oxidizes the Si substrate surface. This kind of regrowth is observed with high-workfunction and refractory metals such as Re.

V_t/V_{fb} Stability—Role of Oxygen Vacancies

Metal-gate/high-κ stacks are affected by V_t/V_{fb} stability (of metal gates) when in contact with Hf-based dielectrics. *The V_t/V_{fb} values for metal/high-κ stacks predicted using the metal workfunction are accurate for low-temperature-processed device, but thermal processing induces drift, toward a midgap effective workfunction (EWF).* An illustration of this effect is shown in Figure 9.25(a) for PVD TaSiN/HfO_2 and CVD Re/HfO_2 devices, where, after high-temperature anneals, the difference in V_{fb} is less than 100 mV. The φ_m of these materials is 4.4 eV and 4.9 eV. Some of these V_{fb} shifts observed upon annealing are due to fixed charge. However, for high-φ_m metal gates, the observed EWF on HfO_2 is shifted by more than 500 mV from the expected φ_m upon exposure to moderately high temperatures and/or reducing ambients. The shift in V_{fb} is qualitatively similar to the high V_{fb} shift observed for p$^+$polySi/HfO_2 gate stacks, where the shift is attributed to Fermi-level pinning. Re/HfO_2 gate stacks, show that for room-temperature-deposited e-beam Re, reducing ambients at moderate temperature shift the V_{fb} of MOS capacitors by ~7600 mV [Figure 9.25(b)]. The V_{fb} for the forming-gas-annealed e-beam Re/HfO_2, stacks is very similar to as-deposited CVD Re/HfO_2 films that are grown at 773 K under reducing conditions. These kinds of shifts are observed for Ru/HfO_2 and Pt/HfO_2. However, by using appropriate low-

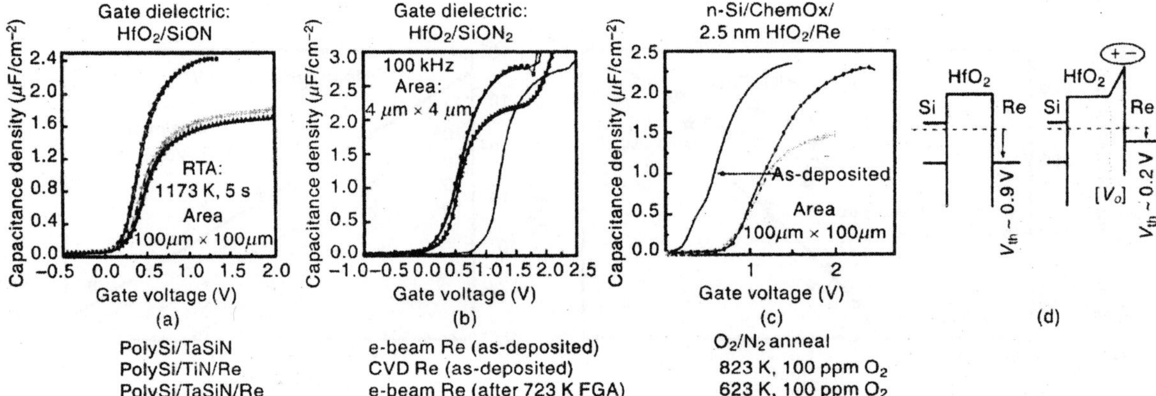

Fig. 9.25 *(a) High-frequency (100-kHz) C–V characteristics of different polySi-capped metal gated stacks clearly showing the midgap-like V_{fb} for both TaSiN and Re gate stacks after 1173 K, 5-s RTA, (b) Comparison C–V characteristics of Re/HfO_2/SiO_2 gate stacks with CVD Re (grown at 773 K) and e-beam-evaporated Re 298 K, respectively. (c) High-frequency (100-kHz) C–V characteristics of SiO_2/HfO_2/CVD Re showing that flatband voltage shifts are induced in oxidizing ambient without incurring interfacial oxide regrowth, if low temperatures and low O_2 partial pressures are used. The impact of oxygen vacancies, [V_O], and of "dipole" formation due to electron transfer from the HfO_2 to the Re. The magnitude of the V_{fb} shift depends on the oxygen vacancy concentration and the distribution in the HfO_2 layer.*

temperature oxidizing ambients, some of this V_{fb} shift is recoverable without interfacial regrowth [Figure 9.25(c)]. Thus, V_{fb} modulation is related to the oxygen vacancy concentration [Vo] in the HfO_2 near the Re contact.

The Fermi-level pinning effect for p^+polySi/SfO_2 is due to the generation of an interfacial dipole formed by the evolution of charged oxygen vacancies. Also, the introduction of a high-φ_m metal gate adjacent to the HfO_2 allows for the following reaction: $O_0 \rightarrow V_o^{++} + 2e^- + \frac{1}{2}O_2$. This reaction is thermally activated. The oxygen vacancy defect level in the HfO_2 is aligned close to the silicon conduction band. At moderately high temperatures, the presence of a high-φ_m metal with its Fermi level aligned close to the valence band of Si provides the necessary driving force to generate charged oxygen vacancies and lose $2e^-$ to the metal. This results in a dipole layer that changes the effective gate workfunction and the corresponding V_{fb} by pulling it toward midgap, as shown in Figure 9.25(d). By introducing oxygen to the system, the oxygen vacancies near the metal/high-κ interface are neutralized thereby recovering the high φ_m of the metal gate, as shown in Figure 9.25(c).

These shifts are attributed to metal-induced gap states (MIGS), an intrinsic effect in which the EWF is modulated by the charge neutrality level and pinning parameter. It is shown that for as-deposited high-φ_m gate metals, the EWF is predicted by the MIGS; however, upon annealing the realized EWF is explained *only* by the vacancy model. Thus, oxygen movement and its role in modulating oxygen vacancy (V_o) formation in the high-κ, strongly coupled with the gate electrode is responsible for the low EMFs that is observed for materials that have high φ_m.

Hence, most of the n-FET metals or alloys are either unstable at high temperatures or have EWFs that are more than 200 mV from the Si conduction band edge (for example TaSiN, TasSi2.5, or TaC). ON the other hand, p-FET metals and alloys, though stable at high temperatures, have unusually high V_{fb} shifts that are related to the oxygen vacancy concentrations in the HfO_2 or HfSiO gate dielectric.

Charge Trapping and NBTI

Unlike polySi/HfO_2 stacks which suffer from significant charge-trapping concerns, metal/high-κ gate stacks have very good V_t stability under constant stress conditions. This is shown in Figure 9.26. Compared with W/HfO_2, both FUSI/HfO_2 and polySi/HfO_2 suffer from significant charge trapping. These observations combined with the metal gate data strongly indicate that reactions(s) between polySi gates and high-κ dielectrics is responsible for defect creation that leads to enhanced charge trapping. Most of these trapping effects are eliminated by the use metal gates. Degradation related to NBTI (negative biased temperature instability) in scaled W/HfO_2 replacement-gate p-FET is comparable to polySi/SiON.

Thus, substantial mobility improvements in metal-gated high-κ systems at an aggressive T_{inv} of 1.4 nm achieved. High-temperature processing and nitrogen in the interface layer improved mobility. Workfunction stability overcomes, with oxygen vacancies in the high-κ which result in large V_{fb}/V_t shifts for high-workfunction metal gates. Low-workfunction metal gates are either unstable at high temperatures or are shifted form the Si conduction band edge. Therefore changes are required to obtain high-mobility and band-edge workfunction metal/high-κ stacks.

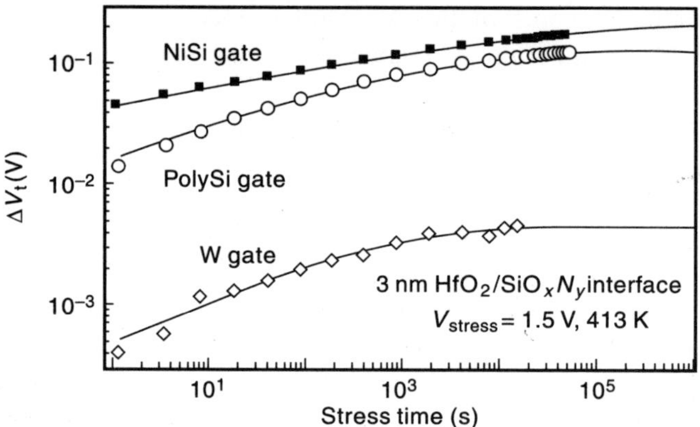

Fig. 9.26 *Comparison of V_t instability in polySi, fully silicided (FUSI), and metal gate stacks with the same high-k stressed under identical conditions.*

(c) Gate Stacks with FUSI metal Gates

Metal gates are used for CMOS scaling, as it lowers T_{inv}. Metallic material deposit directly on gate dielectric regardless of the "gate-first" or "gate-last". For fabricating metallic gates polySi gate is converted into a silicide material which, after silicidation transformation, is in direct contact with dielectric film. Most metal silicide materials have low electrical resistivities, 10–100 μΩ-cm. Low resistivity and selectivity to form silicides in areas where metal is in direct contact with silicon are used in ULSI transitors.

Silicide gates are fabricated by CMP Figure 9.27 shows a front-end-of-line (FEOL) process flow that includes polysilicon gate definition and patterning, ion implantation into extension regions, spacer formation, source/drain ion implantation and silicide contacts, and an oxide passivation layer. After that, FUSI-specific steps include 1) CMP planarization of the passivation

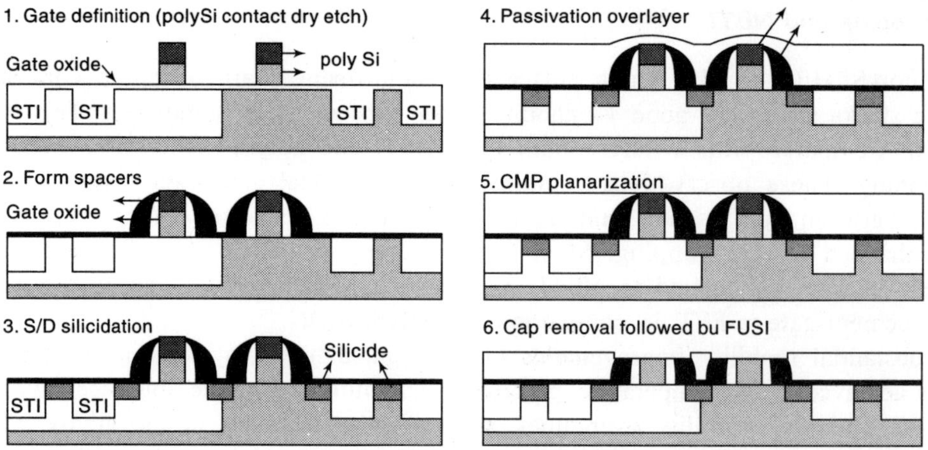

Fig. 9.27 *Integration scheme for fully silicided (FUSI) gate dielectrics utilizing the CMP approach. In this approach, source/drain and gate are silicided separately.*

overlayer; (2) removal of the cap layer on the top of the polySi gate; and (3) metal deposition at the thickness sufficient to fully silicide the polySi gate after moderate annealing, at 673–873 K. The source/drain and gate silicidation are performed separately. FUSI gate integration is similar to the conventional CMOS process flow, and therefore has several advantages over the more complex standard metal gates. Short-channel FUSI devices are demonstrated for 650-nm- and 45-nm- nodes.

Silicide materials used for FUSI gates in source/drain contacts or other microelectronic processes, are molybendum silicides, tungsten silicides, titanium silicides, hafnium silicides, platinum silicides, cobalt silicides and nickel silicides, germmanides, and alloy. Nickel-based silicide materials are also used for FUSI gates for several reasons; (1) low resistivity (~15–25 μΩ-cm; 2) low volume expansion (less than 20%); and (3) the fact that this material is used in Si FEOL processing for sub-90-nm-technology nodes. Nickel silicide is formed by Ni indiffusion into the polySi gate, thereby allowing complete silication without forming voids. A comparison of cobalt silicide and nickel silicide gates is shown in Figure 9.28. In contrast to nickel silicidation, silicon atoms are the main diffusing species during cobalt monbosilicide formation, resulting in void formation at gate dielectric interface. Silicidation is a complex multistep process involving diffusion and phase transformation. Depending on the ratio of nickel to silicon, different phases are formed (e.g., Ni_3Si, Ni_3Si_2, Ni_2Si, N_3Si_2, NiSi, $NiSi_2$), with different workfunctions.

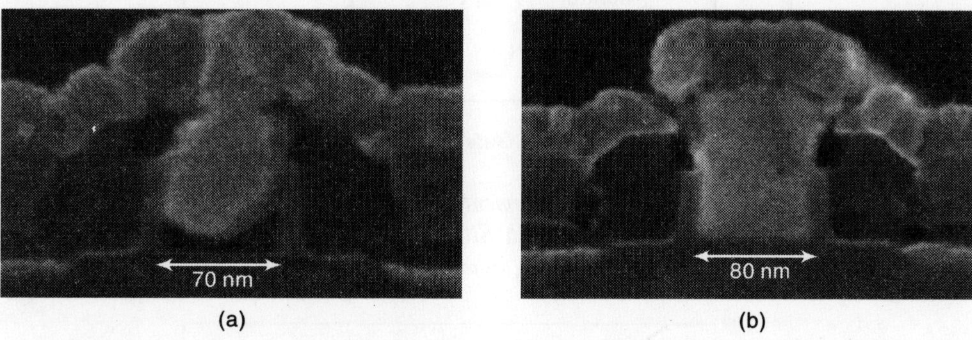

Fig. 9.28 *Cross-sectional SEM image of short-channel fully silicided (FUSI) devices with (a) $CoSi_2$ and (b) NiSi gates.*

FUSI gates show a metallic behavior (due to silicidation) with no signature of polySi depletion for both high-κ and SiO_2 gate dielectrics (Figure 9.29). Accumulation and inversion capacitances are equal, and this is true for both n-FET and p-FET devices. Slight increase of the capacitance in accumulation is observed. The gain of T_{inv} due to FUSI is 0.3–0.5 nm, especially over the polySi/high-κ devices without polySi-pre-doping (Figure 9.29). The combination of polySi-depletion elimination and the high permittivity of the high-κ layers results in very significant gate leakage current reduction, plotted against T_{inv} (Figure 9.30). A high-κ layer (with polySI gates) contributes to a gate leakage reduction of approximately 10^3–10^5, while FUSI gates offer additional reduction by a factor of ~100.

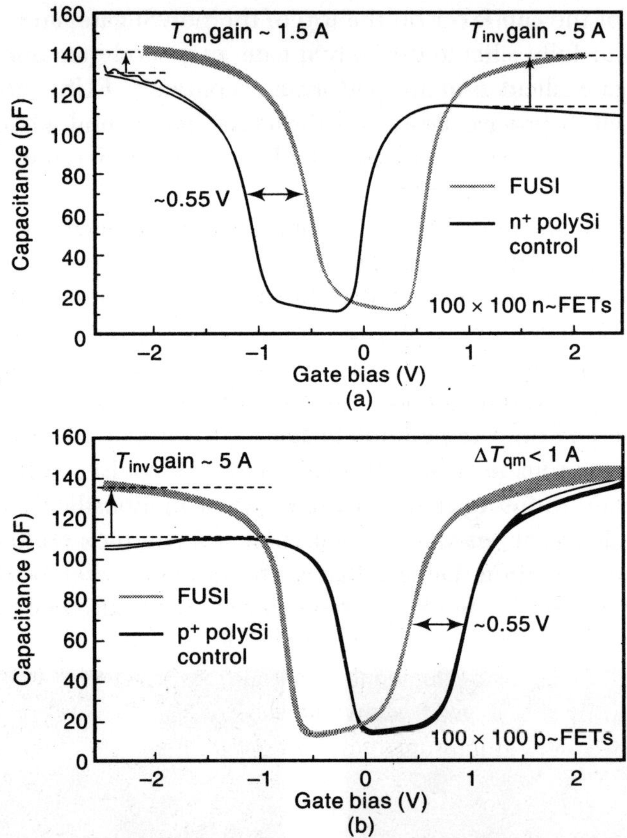

Fig. 9.29 *High-frequency (100-kHz) CV measurements on (a) n-FET and (b) p-FET devices with fully silicided (FUSI) and polySi devices with SiON gate dielectrics. T_{inv} gain due to polySi-depletion elimination and V_t shifts are shown. T_{qm} = oxide thickness metric.*

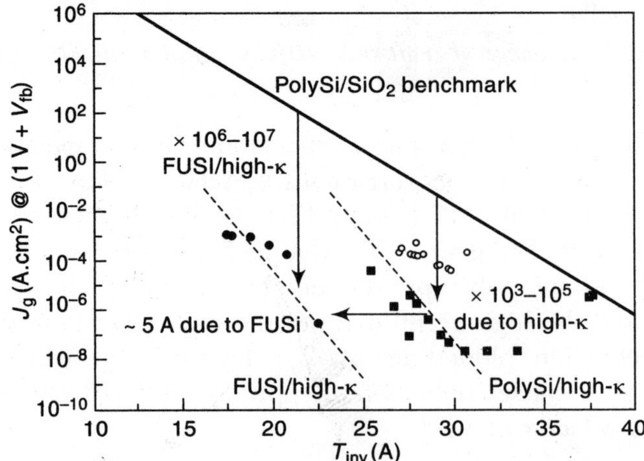

Fig. 9.30 *Gate leakage current density as a function of inversion thickness for polySi-gated devices with SiO_2 and high-k dielectrics and fully silicided (FUSI) devices with high-k dielectrics.*

As the threshold voltage control is a challenge for both metal-gate (band-edge metals) and polySi/high-κ devices, achieving band-edge workfunctions for CMOS is a key issues with FUSI gates. Undoped NiSi gates show a mid-gap workfunction, for example, V_t shift by ~0.5 V from n^+ Si and p^+Si controls is shown in Figure 9.29. Several techniques are used to adjust the workfunction of FUSI gates toward band edge; (1) pre-doping of polySi gates with common n^+ and p^+ dopants before gate silicidation; (2) changing the composition of FUSI gates, alloying Ni with other elements (for example, Pt or Ge for p-FET shifts and Al for n-FETs); (3) using different silicide phases (4) utilizing ultrathin cap materials between the gate dielectric and the FUSI gate; (5) bottom interface engineering; and (6) channel pre-doping.

With the help of polysilicon pre-doping of FUSI gates on SiO_2-based gate dielectrics, V_t is adjusted within ~150 meV (for p-FETs) and 300 meV (for n-FETs) from the mid-gap value of the undoped NiSi (Figure 9.31). The dopant is optimized because some EOT loss and adhesion is observed at high ion implant. There is a tradeoff between the value of the V_t shift and the degree of delamination (for n-type dopants), and also EOT loss. PolySi pre-doping becomes less efficient in the case of FUSI gates on high-κ dielectrics due to Fermi-level pinning. For FUSI gates use is made of; (1) metal-rich phases of nickel silicides; (2) platinum silicides or platinum alloys; and/or (3) more stable silicate and nitride silicate materials. Now, different phases of nickel silicides exhibit workfunctions ranging from ~4.3 eV (for $NiSi_2$) to ~4.7 eV (for Ni_2Si). This silicidation tunes the workfunctions of FUSI gates. Another factor in adjusting V_t is to alloy nickel silicides with elemnents that help to move the workfunction toward band edges. Devices with NiPtSi FUSI gates show threshold voltages close to a "quarter-gap" p-FET value, whereas alloying with aluminum shifts the workfunction almost to the n^+ band edge (Figure 9.32).

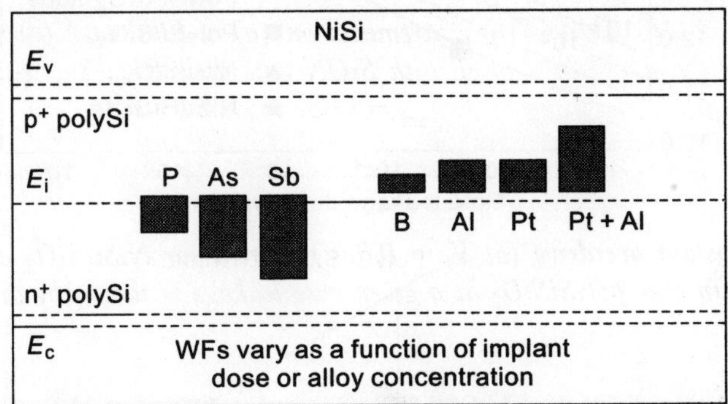

Fig. 9.31 *Workfunction control for fully silicided (FUSI) gates by polySi pre-doping with typical n-type and p-type dopants.*

In terms of device performance, long-channel FUSI-gated $HfSi_xO_y$ devices show carrier mobilities close to that of the SiO_2. This fact combined with reduced T_{inv} (Figure 9.29 and 9.30) results in significant drive current improvements. Figure 9.33 shows drive current in the linear regime as a function of gate leakage. The upper x-axis shows gate oxide thickness extracted from gate current density assuming SiO_2 tunneling behavior. At a given gate leakage, the

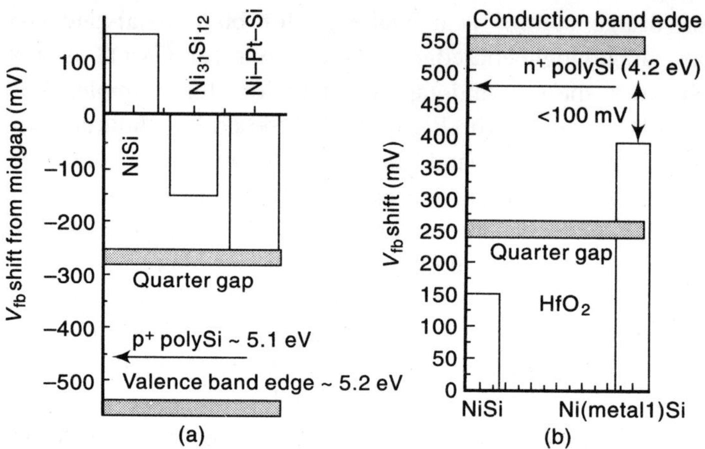

Fig. 9.32 *Workfunction adjustments for fully silicided (FUSI) gates by alloying polySi (a) with Pt and Ni-rich silicides for p-FETs and (b) with Al for N-FETs.*

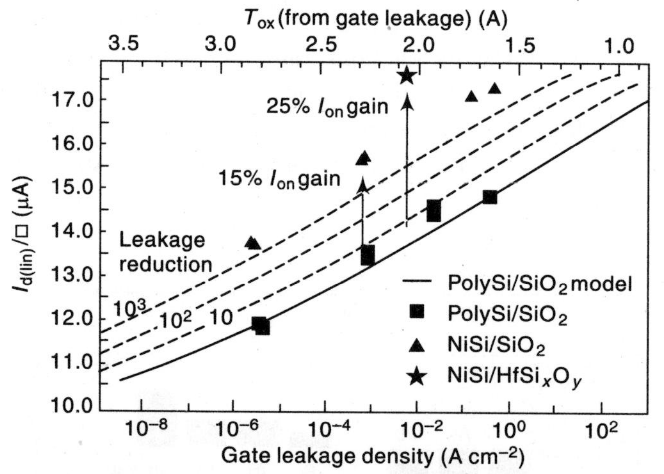

Fig. 9.33 *Normalized constant overdrive (at V_t + 0.8 V) current for NiSi/SiO$_2$ and NiSi/HfSiO n-FETs. Performance gain over polySiSiO$_2$ at a given gate leakage is shown by arrows. ($I_{d(lin)}$: linear drive current.)*

n-FET performance gain is ~25% for NiSi/HfSi$_x$O$_y$ and ~15% for NiSi/SiO$_2$. Also, figure 9.33 shows that, for a given driven current NiSi/HfSi$_x$O$_y$ has six orders of magnitude lower gate leakage.

Charge (electron) trapping causes V_t instabilities and drive current degradation. FUSI-gated devices exhibit charge-trapping behavior similar to that of the polySi/high-κ stacks. FUSI on HfO$_2$ show significantly V_t instability, whereas charge trapping in both doped and undoped FUSI ion HfSI$_x$O$_y$ is negligible.

"Gate-last" FUSI devices are subjected to processing (starting from polysilicon deposition) and a thermal budget similar to that of polySi/high-κ stacks. One conventional way to scale down the electrical equivalent thickness of the stack is to combine an optimized thin SiO_2-like interface and a reduced high-κ layer thickness. Electrical thicknesses in inversion (T_{inv}) as thin as 1.6 nm is achieved for NiSi/HfSiO devices.

In summary, the FUSI device is an attractive metal-gate integration option that offers a number of device benefits such as sub-2-nm T_{inv}; performance gain over polySi/SiO_2 at a given gate leakage; six to seven orders of gate leakage reduction (at a given T_{inv}); V_t control for both n-FET and p-FETs, and negligible charge trapping.

IV. LOW TEMPERATURE Si AND Si: Ge EPITAXY

The ultrahigh-vacuum/chemical vapor deposition (UHV/CVD) process fig. 9.34 is a chemical vapor deposition process for which operating temperature is 773 K. At this temperature the rates of such diffusive processes become negligible and the thickness of a strained Si:Ge epitaxial layer is relaxed. This relaxation produces structures and devices of Si:Ge (and dopant composition) with accurate control of composition on the atomic length scale. The ability to customize the bandgap and doping of bipolar or field-effect Si offers for device performance, based on III-V materials such as GaAs and GaAlAs.

Molecular beam epitaxy (MBE) is the primary technique employed to prepare metastable materials in the Si system.

Surface segregation of species such as Ge, Sb, B, is reduced or absent in the chemical vapor deposition of silicon (at processing temperatures) thereby improving dopant and alloy transitions.

The Chemical Process

To deposit silicon epitaxially onto a substrate, two fundamental conditions must be met. (1) The initial growth interface must be atomically perfect and chemically pure. (2) For, the growth process the controlled introduction of dopant species species for the purpose of alloy formation is the sole process of determining and controlling the final composition of a film (on the scale of atomic length). For, reduced film growth rates, the reactor geometry is shown in part (a) Figure 9.34. It is selected for implementing the UHV/CVD process. The placement of wafers is coaxial within a hot-walled furnace, as shown in Figure 9.34(b). A high wafer-packing density is achieved. So the product of the number of wafers processed is high.

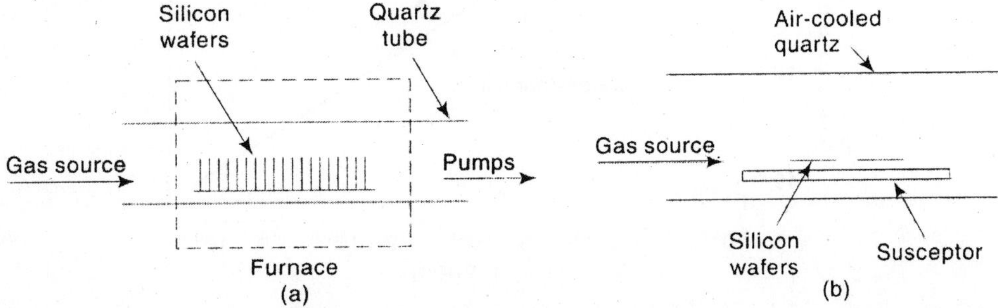

Fig. 9.34 *(a) UHV/CVD reactor, showing high-density coaxial wafer placement; (b) A conventional cold-walled epitaxy reactor.*

The interface, in an epitaxial growth, serves as a template for the layers to follow. The interfacial defects are replicated in the resultant layers. It contaminates during the growth, thus leads to the introduction of impurities. In silicon epitaxy, the primary contaminating species encountered are oxides of silicon. So, an oxide-free silicon surface is devised and maintained.

Consider, the oxidation reaction Si undergoes in the presence of two contaminants, water vapor and oxygen.

At high partial pressures of H_2O, a stable oxide is formed in the silicon surface by the reaction

$$Si(s) + 2H_2O \rightarrow SiO_2 + 2H_2 \uparrow.$$

But, at low partial pressures of H_2O, the silicon surface is etched because of the formation of volatile suboxide SiO, viz.,

$$Si(s) + 2H_2O \rightarrow SiO_2 \uparrow + 2H_2 \uparrow.$$

So, AT equilibrium (1173–1423K) a crossover occurs from oxidation to etching is shown in Figure 9.35. The extrapolation indicates that to conduct epitaxy at temperatures lower that 973 K, the partial pressure of H_2O at ultrahigh-vacuum levels should be maintained. Similar requirements are observed for oxygen. So, the epitaxial growth of silicon, with operating conditions, within which an oxide-free silicon surface is maintained. Note that these conditions are applicable only to a static silicon surface in an oxidizing ambient. Under dynamic conditions, there is a continuous flux of additional silicon to the growth surface; thus greatly reducing its sensitivity to oxidation. Thus, oxygen is incorporated as a contaminant in the resultant layers, but it also disrupts film growth. Therefore, to prepare a silicon wafer with such an optimal surface and subsequently transfer it to a reactor without altering its state; it is preferable to start by producing an air-stable yet oxide-free silicon surface. This is achieved by employing a hydrogen-peroxide-based chemical cleaning process, followed by a ten-second dip in 10:1 H_2O/HF. Terminating this sequence with a dip in HF results in the formation of a hydrophobic,

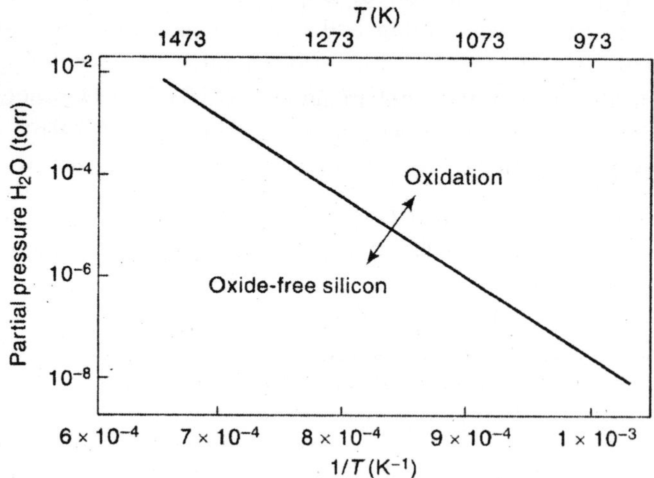

Fig. 9.35 *Extrapolation of the equilibrium data for the formation of SiO_2 versus Si etching by SiO evaporation, shown as a function of H_2O partial pressure.*

air-stable, hydrogen-passivated silicon surface. The degree to which the surface is thus stabilized shown in Figure 9.36. Spectrum (a) is composed of photoemission data taken from a wafer. The dominance of the SiO_2 peak indicates that the formation of its native oxide is complete.

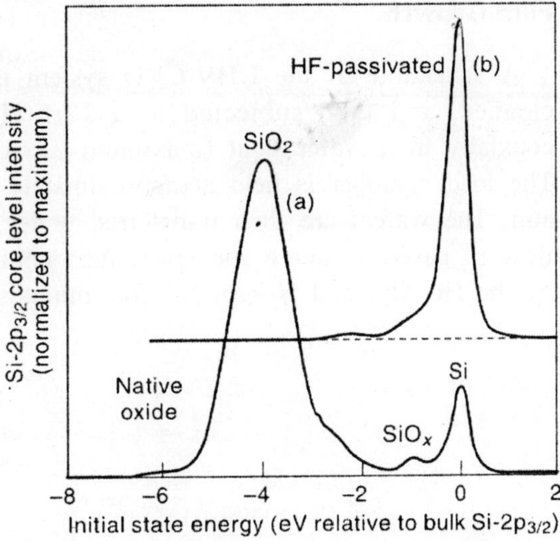

Fig. 9.36 *Photoemission spectra from the (111) surface of a Si wafer (a) as received and (b) after HF-passivation and subsequent exposure to room ambient for ten minutes.*

Spectrum (b) is composed of photoemission data taken after the above treatment i.e. H-passivated by an HF dip, and then allowed to remain room ambient for ten minutes. No evidence of reoxidation is found, with the hydrogen adlayer acting as a chemical barrier to recontamination. So, a 13-order-of-magnitude reduction in the reactivity of the silicon surface with respect to oxidation by such hydrogen passivation is achieved. Silicon wafers prepared are loaded into the UHV/CVD reactor through the ambient without oxidation and/or contamination of their surfaces. In contrast to other forms of low-temperature epitaxy by CVD no *in situ* cleaning step is performed, as no native oxide exists as the inception of film growth. Secondary-ion mass spectroscopy (SIMS) show that there is less than about one hundredth of a monolayer oxygen trapped at the initial growth interface at the completion of film deposition.

In order to have uniform and particle-free epitaxial silicon films in a high-density multiwafer processing geometry as that in Figure 9.34(a), homogeneous silane must be suppressed. The silicon hydride system leads to the formation of the highly reactive intermediate silylene (SiH_2) which inserts into the parent silicon hydride, forming higher silanes i.e.

$$SiH_4 \rightarrow SiH_2 + H_2.$$

The reaction forms the highly reactive silylene radical, which reinserts into the parent species by the reaction

$$SiH_4 + SiH_2 \rightarrow Si_2H_6.$$

or it inserts into Si-H bonds at a growth surface with near-unity efficiency. Species Si_2H_6 and SiH_2 react with surfaces but do not transport into the high-aspect-ratio cavities existing between closely packed wafers, because, the gaseous source is rapidly depleted of these species when they collide with surfaces. This leads to an excess of film growth at wafer edges. This is observed for films grown directly from disilane and for films grown in environments where disilane is formed *in situ*. With high total pressure and high temperature, it is possible to reinsert the silylene radical into parent hydrides. This leads to gas-phase nucleation of polymeric silicon hybride $(SiH_2)_x$.

Film Growth

A schematic of the UHV/CVD system is shown in Figure 9.37. The wafers are first RCA-cleaned, and then subjected to a 10:1 H_2OHF dip for ten seconds. They are then placed coaxially in a wafer boat (maximum 35 wafers), and the boat is placed within the load-lock. The load chamber is held at approximately 273 K while being evacuated to below 1.33×10^{-9} atm. The wafers are then transferred under hydrogen into the UHV portion of the system. The flow of gaseous source species is then commenced. The hydrogen surface passivation is created by the HF dip; and is kept for five minutes.

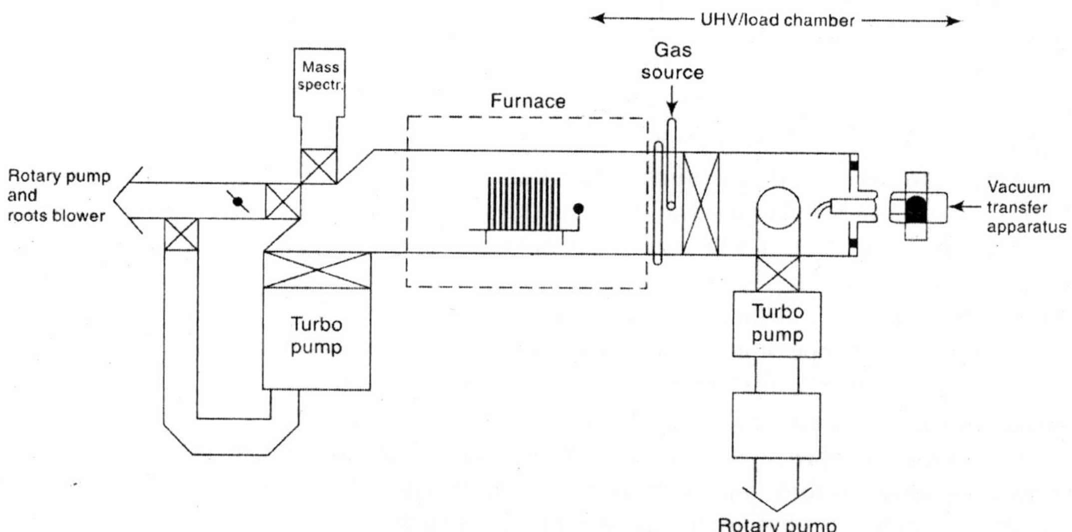

Fig. 9.37 *A UHV/CVD system.*

At nominal growth conditions, 1.33×10^{-6} atm, 823 K chemical processes such as dopant and germanium incorporation, are linear in gaseous content and are readily controlled. An example of boron profile control that is possible is seen in Figure 9.38. Discrete concentration steps over a narrow dopant range are readily achieved because dopant content depends linearly on dopant flow. This is in marked contrasts to beam deposition technique such as MBE, where reactant fluxes to surfaces are exponential in source temperature, as is dopant incorporation, making precision concentration control difficult. In the present process, dopant incorporation and Ge alloy formation are all found to be strong functions of the deposition temperature. However, it is straightforward to control reactor temperatures to within several tenths of a degree, so that film composition is controlled with precision over a wide dynamic range solely by the composition of the gaseous source employed. Film thickness is linear in time, and is controlled at the chosen growth rate. The growth conditions yield a growth rate of approximately 3 Å/min for intrinsic or boron-doped silicon epilayers. This value varies for the growth of thicker layers. The values widely vary as the growth rates are thermally, activated with an activation energy ~1.6 eV. The growth of Si:Ge alloy films reveals a more complex behavior

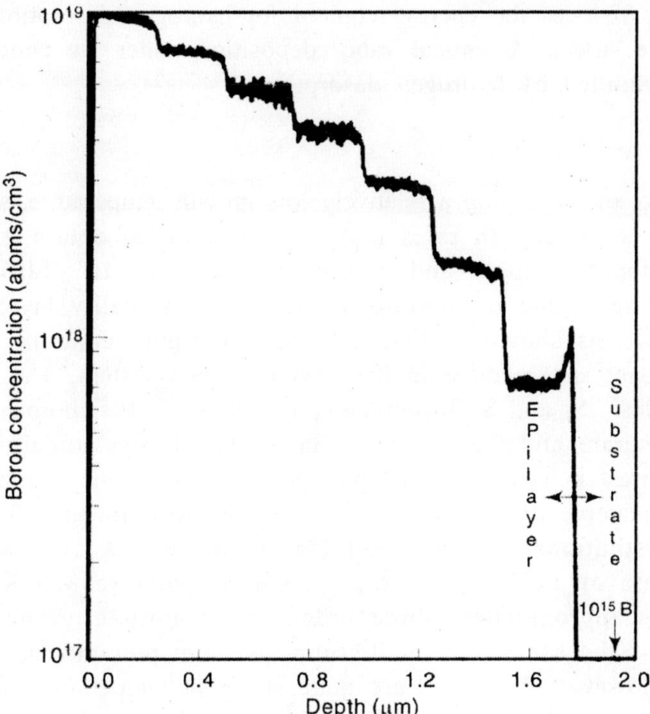

Fig. 9.38 *SIMS data for a stepwise profile of boron in silicon.*

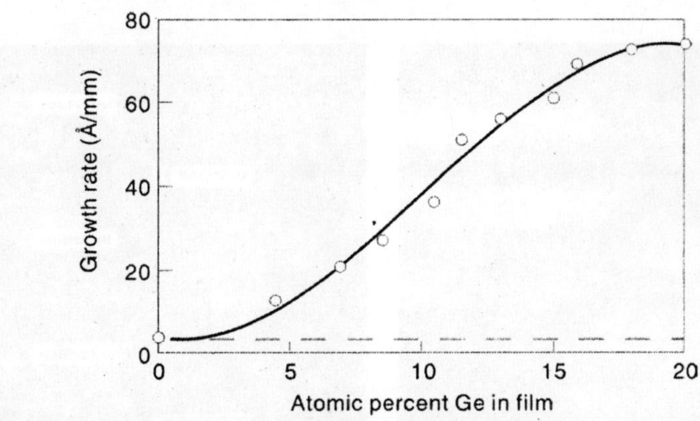

Fig. 9.39 *Film grown rate as a function Ge content in a Si:Ge alloy film.*

as, a cooperative growth phenomenon is encountered. The addition of germane to the silane results in the behavior seen in Figure 9.39, where growth rates are strongly enhanced by the presence of germane. It is seen theta germane greatly enhances the efficiency of the growth, accelerating it by a factor of about 25 for films containing about 20 percent germanium. The reaction efficiency of silane is enhanced considerably. It is found that the presence of germanium on the growth surface reduces the energy required for hydrogen desorption relative to its value for desorption in pure silicon. Chemical vapor deposition under the conditions used in UHV/CVD process is rate-limited by hydrogen desorption.

Materials

The combined effect of employing a relatively low growth temperature, short system residence times, and relatively slow growth rates makes it possible to obtain extraordinarily abrupt transitions in both dopant content and/or alloy composition. In addition, the use of such growth temperatures has made it possible to deposit high-quality layers of nonequilibrium materials, for example, as shown in Figure 9.40. The figure contains cross-sectional TEM micrographs of a dopant super lattice in the Si:B materials system. The superlattice comprise altering layers of intrinsic Si and Si:B containing 10% B (5×10^{21} boron atoms cm^{-3}), with all layers fully commensurate and free of precipitates. The layers containing boron are highly strained, and the contrast is visible in the micrographs. Each layer is roughly 40 atoms thick. This structure is an example of metastable material fabrication by the UHV/CVD process. The upper bound on substitutional boron content in silicon is 2×10^{30} boron atoms cm^{-3}, at temperatures in excess of 1473 K. AT the growth temperature 823 K, boron exceeds the equilibrium value by approximately three orders of magnitude. Annealing at 1173 K for several minutes brings the sample to equilibrium, as large precipitates form, and the silicon lattice is disrupted. However, the layers are quite stable at temperature below 973 K, and are employed in devices as highly degenerate contacts or single-crystal emitters. Such layers have carrier densities of 1.5×10^{20} cm^{-1} and a higher degree of activation (×2) with a degraded mobility.

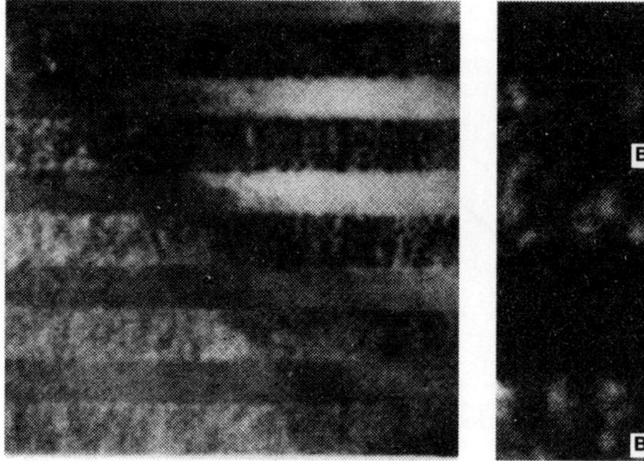

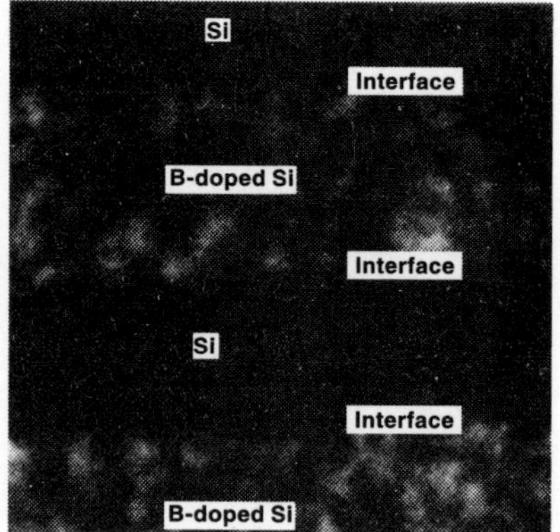

Fig. 9.40 *Cross-sectional TEM micrgrsphs of a Si/Si:B dopant superlattie. The superlattice in (a) has been lattice-imaged in (b), with a delineated "interface," added for identification purposes.*

The control of film concentration is absolute from the perspective of concentrationn and location. Transitions between regions of varying alloy composition and dopant content, are atomically abrupt.

A two-dimensional hole-gas (2DHG) structure, designated as *Sigmob1*, is deposited by UHV/CVD. The structrue and associated Hall-effect data are shown in Figure 9.41. Also shown are data for *Sigmob2*, which is uniformly doped across all of its layers, thus fails to exhibit enhanced mobility at low temperatures. The band structure of Sigmob1 is shown in Figure 9.42. The hole gases are formed (shown for T = 77 K), one each at the Si/Si:Ge and Si:Ge/Si transitions, each centered at a depth of 10–20 Å into the Si:Ge well. These gases serve as extraordinarily sensitive probes of the abruptness of the boron-dopant and Si:Ge alloy transitions because they are less than ten atomic layers removed from the boron-doped Si layers. The uncontrolled incorporation of residual boron into the Si:Ge well, greatly degrades the mobility of holes in the well. As growth proceeds, the first interface is formed moments after the cessation of intentional boron incorporation, while the second interfaces formed later. The hole gases formed at the interfaces are identical and are a highly sensitive measure of the abruptness of dopant transitions. If there is a chemical memory in the system, the hole formed at the first interface will be of higher density and lower mobility than that at the second. In the data shown in Figure 9.43. A single set of oscillations is seen in the magnetoresistance of the layers. The onset of oscillations in a SdH measurement occurs when μH > 1. Of the two hole gases formed each produce an independence set of oscillations. The exact alignment of these oscillations,

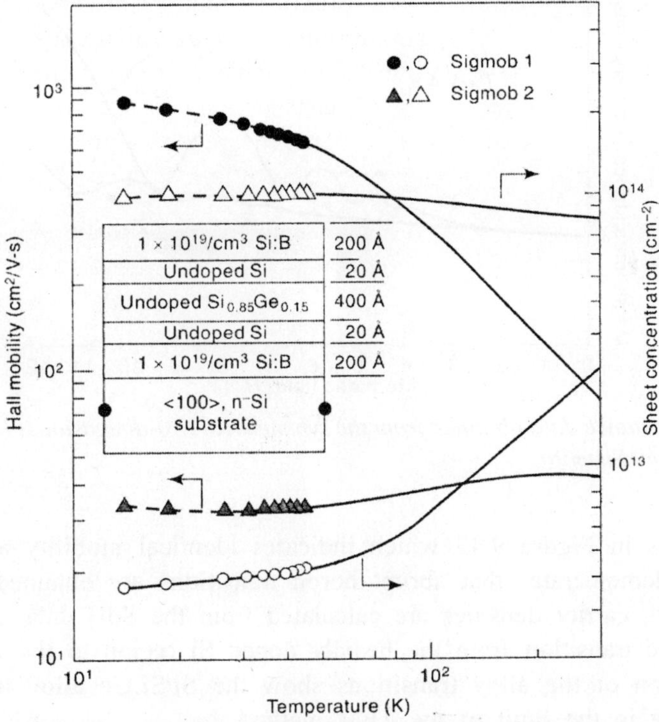

Fig. 9.41 *Hall-mobility and carrier-density data for two symmetrical two-dimensional hole-gas (2DHG) heterostrictures that a fabricated. A schematic of the cross section of one of the (Sigmob1) is included. The other heterostructure (Sigmob2) has the same cross section but is boron-doped throughout.*

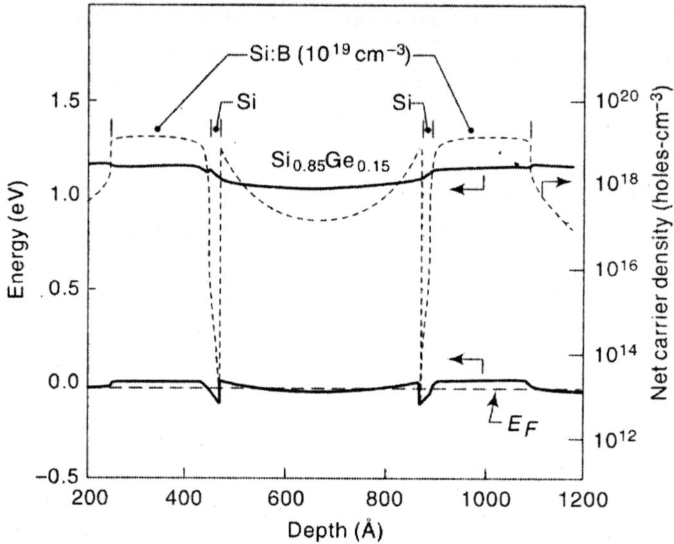

Fig.9.42 *Energy-band diagram for the heterostructure Sigmob1.*

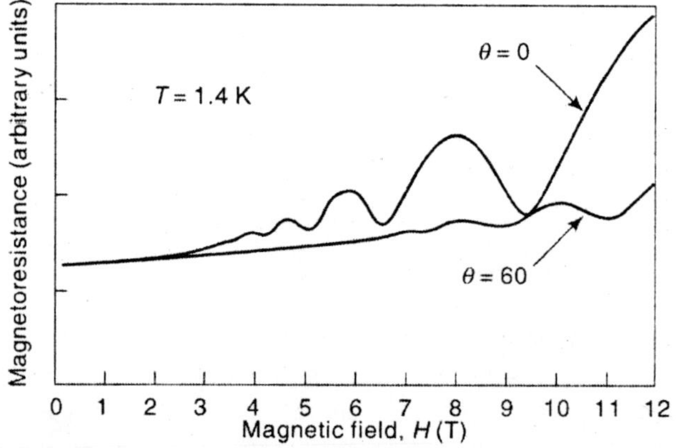

Fig. 9.43 *Magnetoresistance data obtained from the symmetrical two-dimensional hole gas generated in the heterostructure Sigmob1.*

produces single peaks in Figure 9.43 which indicates identical mobility and carrier density at two interfaces, and demonstrates that abrupt boron transitions are obtained. To verify that two hole gases are formed, carrier densities are calculated from the SdH data. A buffer width of 20 Å is sufficient for the transition from the heavily doped Si region to the intrinsic Si:Ge carrier well. TEM observation of the alloy transitions show the Si/Si:Ge alloy transitions to be both abrupt and defect-free in the limit of the TEM method.

V. SILICON CMOS DEVICE SCALING

Scaling

This section briefly reviews some of the basic principles of scaling. It shows the benefit when scaling works well, and when such benefits are greatly diminished. Table 1 shows the scaling rules for electric field. In one case the voltage is scaled down in direct proportion to physical dimensions, and in other case the electric field is allowed to be an independent variable E, defined as V divided by the dimensional scaling factor α.

Table 1 *Relationships for constant-field scaling and for generalized field scaling, a is the scaling factor for dimensions, and E = V/a is the normalized electric field.*

Parameter	*Constant-field Scaling*	*Generalized field scalingf*
Physical dimensions		
L, W, Tox, wire pitch	$1/a$	$1/a$
Body doping concentration	a	E/a
Voltage	$1/a$	E/a
Circuit density	$1/a^2$	$1/a^2$
Capacitance per circuit	$1/a$	$1/a$
Circuit speed	a	a (goal)
Circuit power	$1/a^2$	E^2/a^2
Power density	1	E^2
Power–delay product (energy per operation)	$1/a^2$	E^2/a^3

The simple concept of scaling for MOS transistors is to reduce all of the physical dimensions by the same amount α, while increasing the body doping and reducing the applied voltage to cause the depletion regions within the devices to scale as much as the other dimensions. Progress in microelectronics is also linked to scaling of the wiring dimensions, particularly the wiring pitch. For simplicity it is assumed that wiring pitch is scaled by the same factor α used in the device, (It has been shown that when devices and wires are scaled independently by different factors α_d and α_w, the speed of predominantly determined by α_d and the circuit density by α_w.

A first benefit of scaling is the increased *circuit density*, and the reduction of capacitance per circuit. It is due to the reduction of transistor widths and write lengths, with the capacitance per unit dimension (e.g., C/μm) remaining essentially unchanged by ideal scaling. (This ignores the trends towards thicker wires and low-*k* insulators between the wires, which tend to offset each other.)

Another benefit of scaling is *higher speed*. In constant-field scaling, it is shown that circuit speed increases directly with the amount of scaling α. In CMOS technology it is impossible to scale V. The maintained circuit speed increases due to constraints on the threshold voltage, to avoid rising standby power in the "off" transistors. The electric field E steadily increases by scaling V by less than α in order to meet the increasing circuit speed by α, (Table 1). That constant-field scaling provides much lower power per circuit, constant power density, thus the power–delay product (energy per operation) improves by α^3. As shown in Table 1, all of these are multiplied in generalized field scaling by E^2.

As the voltage is not scaled for a given application, the parameter E rises directly with α, and circuit power constant with scaling, power density rises as α^2, and the power–delay is improved only by α. With respect to the power and power density, it is assumed that the circuit speed actually increases with scaling, which is very difficult to achieve.

Power-constrained Device Scaling

The rapid growth in subthreshold leakage in CMDS. VLSI has altered the direction of power/performance improvements to CMOS technology. Figure 9.44(a) shows the improvements in intrinsic transistor delay to the power of –1; Figure 9.44(b) illustrates the growth of active and passive power density with scaling from 1-μm CMOS to 65-nm CMOS technologies. A significant transition occurs in the 130–65-nm regime, where passive power density moves from a minor part of the total to becoming dominant.

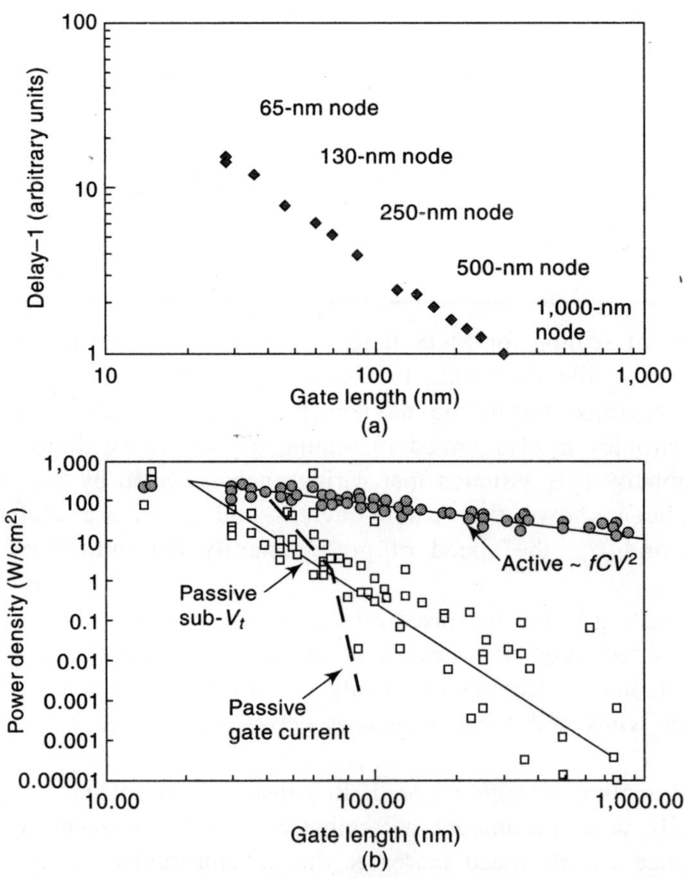

Fig. 9.44 *(a) MOSFET performance vs. gate length; normalized MOSFET intrinsic device delay (CV/I_{eff}) vs. gate length. (b) Power density vs. gate length.*

Power vs. Power Density

Two views are encompassed while examining power-related issues in CMOS scaling; (1) total power, or power per circuit, and (2) power density, or power per unit area. As classic scaling leaves power density fixed so the quadratically decreasing area per circuit, decreases the power per circuit quadratically. Thus, two metrics i.e. performance per power per circuit (F/P/Ckt) and performance per power density (F/Pd) are examined for scaling. In Figure 9.45(a) and (b), these two metrices are plotted vs. gate length L_g, from the same data that is used to construct Figures 9.44(a) and (b). While the performance (F) is continued beyond 130 nm ($L_g \sim$ 70 nm), the growth rate in F/P/Ckt dropped from a (classic scaling) $\sim L^3$ to below $\sim L^2$ on entering the 65-nm node ($L_g \sim$ 35 nm). Thus, the benefit to CMOS VLSI in terms of speed per power-function is drops from a cubic dependence to a much slower improvement rate. The improvement in F/Pd is not only slowed, but reverses to become degraded; that is, for power-density-limited functions, design innovation is required in order to avoid an increase in power density, even with no increase in circuit speed.

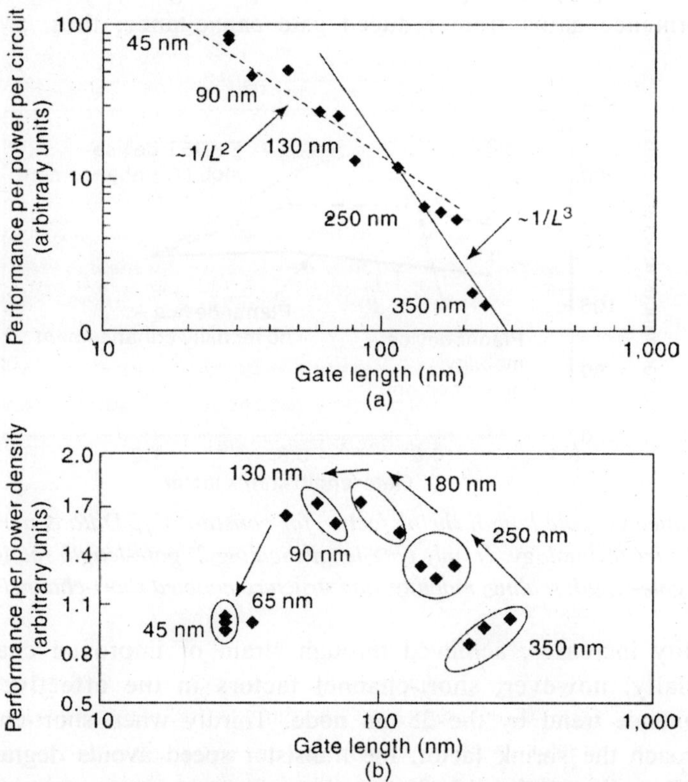

Fig. 9.45 *(a) Performance per power per circuit vs. Gate length. Data from Figure 9.45(b) used to calculate performance (frequency) normalized to power and circuit count with respect to scaled density. (b) Performance per power density vs. gate length. Slowing of voltage scaling causes a reversal of the trend beyond 130-nm-node technology.*

Transport vs. Electrostatics

In the 90-nm node grains in short-channel scaling shown are, aided by improvements in doping profile advances and limited to decreases in effective dielectric thickness T_{ox}. Most of the performance gain for the 90-nm node comes from electron and hole mobility enhancements. In Figure 9.46 three scaling cases are shown. In the first, no new structural innovations are introduced, and the mobility of transistors is held fixed at the values achieved in 65-nm CMOS. It shows no significant improvement in short-channel scaling. The second case assumes the same structural assumptions but the mobilities of electrons and holes are presumed to increase by 1.5x. In the third case, in addition to the mobility assumptions, structural innovations are introduced at each node to improve control of short-channel effects, effectively keeping DIBL and swing constant at each node while gate length is reduced. (These structural innovations are used in double-gate transistors with reduced body thickness.) In all cases the power-supply voltage is kept fixed at 1 V, as in the effective gate dielectric thickness. In the first scenario, minor improvements in junction technology are overwhelmed by intrinsic short-channel effects associated with shrinking gate lengths, and the degrading effects shown in Figure 9.45(a), outpace the performance gains from reduced gate capacitance; thus, a net loss in transistor speed results.

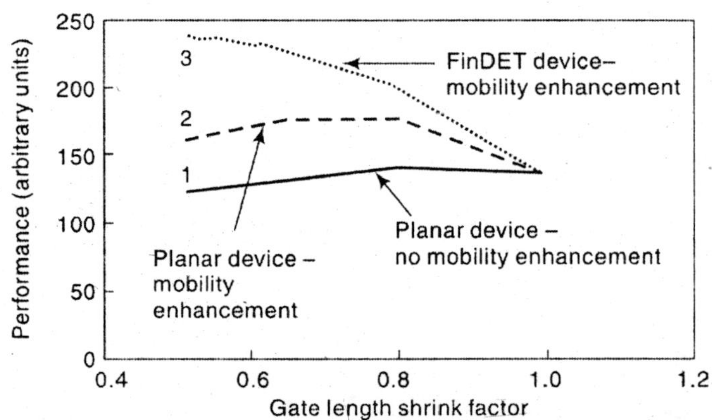

Fig. 9.46 *Performance vs. gate length shrink factor, for constant V_{dd}. Data is normalized to 650-nm node planar device technology. 1: only gate-length scaling 2; gate-length scaling and mobility improvement; 3; gate-length scaling, mobility, and structure-enabled short-channel-effect improvement.*

Secondly mobility increases, achieved through strain or improved channel materials, boost current drive initially; however, short-channel factors in the effective drive dominate and reverse the performance trend by the 25-nm node. Thirdly when short-channel effects improve at a rate that approach the shrink factor, the transistor speed avoids degradation until it reaches the 25-nm node. Thus, innovations in both enhancements to transport and short-channel effect suppression are necessary for continued advancement of CMOS speed when power is constrained.

Finally, it is important to consider the choice of power supply voltage V_{dd} in the power-constrained era. In Figure 9.47 transistor speed per power is plotted vs. transistor speed for a fixed L_g = 35 nm, 65-nm CMOS technology. Choices of V_{dd} from 0.8 V to 1.1 V are shown,

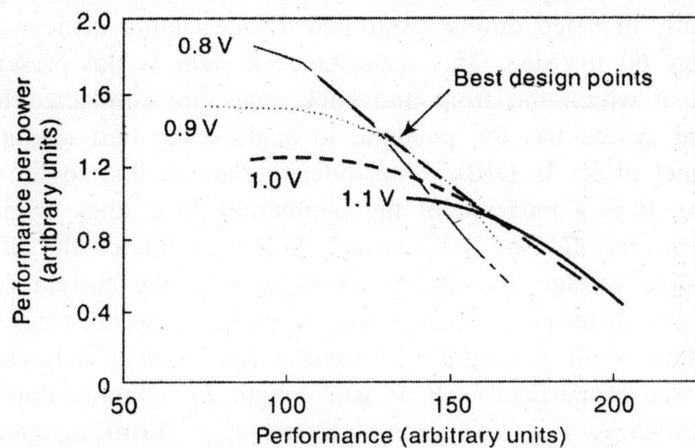

Fig. 9.47 *Performance vs. performance per power. For a given V_{dd}, I_{off} (off current) is varied (V_t setting from 1 nA/mm to 1 mA/mm. Solid curve: 1.1 V, dashed curve: 1.0 V, dotted curve: 0.9 V, dashed-dotted curve: 0.8 V.*

and for each the off-current is varied from 1 nA/µm to 1 µA/µm. An envelope of best design points is sketched on the basis of these curves. It shows that there is a direct tradeoff of transistor speed for power efficiency.

For greater speed V_t either decrease V_t or increase V_{dd}. When V_t is decreased the power eventually becomes dominated by subthreshold leakage; hence, the speed/power efficiency eventually degrades exponentially. So, increasing V_{dd} gives a favorable return of speed per power.

Device Scaling

Short-channel Effects

A successful device design delivers maximum on-current (I_{on}) at an acceptable device off-current (I_{off}). Given a constant supply voltage, high drive current requires a low threshold voltage, which is contrary to the desire for a low I_{off}. In addition to the threshold voltage (V_t), the subthreshold swing (S) also determines the device off-current:

$$I_{off} \propto 10^{-(V_t/S)}$$

Threshold-voltage reduction with gate length and subthreshold swing are related to the device structure. They are weakly dependent on the transport properties and are related to the electrostatic behavior of the device. The subthreshold swing is determined by the gate modulation of the potential barrier height ($\theta_{barrier}$) that separates the source from the drain. For a partially depleted doping-controlled device, this acts as a capacitance divider between gate dielectric C_{ins} and channel depletion and C_{depl}:

$$S = 2.3\frac{kT}{q}\left(\frac{\partial\theta_{barrierg}}{\partial V_g}\right)^{-1}$$

Consequently, a partially depleted doping-controlled device cannot achieve the ideal subthreshold swing of approximately 60 mV/dec. This capacitance divider is not present in a fully depleted doubled-gated device in which the front and back gates are connected to the same potential. Therefore, this type of device has the potential to achieve the best subthreshold swing.

A third short-channel effect is DIBL. This phenomenon is due to the modulation of θ_{barrier} with the drain voltage. It is a measure of the number of field lines originating from the drain that terminate at the source side of the channel. DIBL modulates the threshold voltages with respect to drain-to-source voltage and affects the effective drive current I_{eff}. Figure 9.48 shows the scaling behavior for different device architectures as obtained from a generalized scaling theory. The electrostatic scaling length Λ measures the scaling behavior of the device and related to various device properties, such as gate length L_g, channel doping N, body thickness T_{si}, gate dielectric thickness T_{ins}, and gate dielectric ε_{ins}. Shrinking gate length for partially depleted doping-controlled devices requires an ever-increasing doping level in the device, which may aggravate subthreshold, swing, DIBL, and junction leakage. The threshold voltage of a fully depleted (FD) device is set by the body thickness of the device. Owing to a better shielding of the drain and source fields, the double-gate device, at a given minimum gate length, requires a less stringent body thickness than a fully depleted SOI device. But, if $_{\text{si}}$ is not significantly smaller than the minimum gate length, doping in the body is needed to control the short-channel behavior.

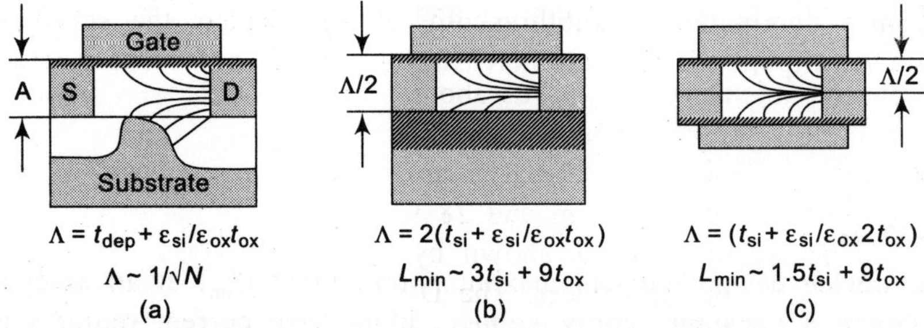

Fig. 9.48 *Scaling potential of different devices types; (a) bulk; (b) FDSOI; (c) FD double-gate device.* L_{thin} *estimates are done assuming a gate oxide dielectric and Si substrate and* $L_{min} = 1.5L$.

Single-gate Partially Depleted, Doping-controlled Devices

Consider a 2D process and device simulation, which includes mixed-mode simulations of delay chains. If not stated otherwise, these are fan-out-of-1 delay chains, with or without appropriate wire loads. Firstly, the transition, of a doping-controlled partially depleted (PD) SOI device, to a geometry-controlled fully depleted (FD) SOI device is investigated. The partially depleted SOI device behaves as a bulk device, with respect to short-channel scaling, except that charge is accumulated in the body. The impact of body thickness scaling, is shown in Figure 9.49; the saturation threshold voltage for a polySi-gate n-FET device vs. body thickness at 46-nm and 28-nm nominal gate lengths. Also shown is DIBL for the same devices at corresponding minimum channel lengths of 38 nm and 25 nm, respectively. For thick body scaling, the halo implants are adjusted to meet the leakage targets for the device so that the

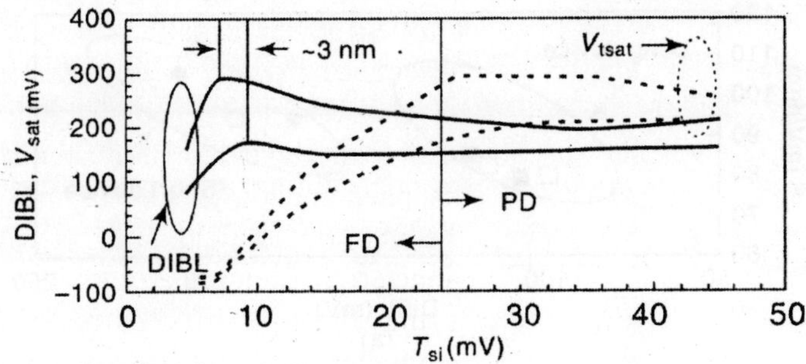

Fig. 9.49 *n-FET T_{si} scaling with fixed channel doping. Solid curves: V_{tsat} at L_{nom} and V_{ds} = 1 V; dashed curves: DIBL at L_{min}, Blue curves: short-device L_{nom} = 28 nm (L_{min} = 25 nm); red curves: long-device L_{nom} = 46 nm (L_{min} = 38 nm). (FD = fully depleted; PD = partially depleted.)*

bodies are thinned. Equivalent oxide thickness is fixed at 1 nm. There are three distinct regions: the partially depleted device for $T_{si} > 25$ nm, the fully depleted device for $T_{si} < 25$ nm, and the body-thickness-controlled device of $T_{si} < 7$ nm. Owing to the high halo for the shorter device, there is a higher V_{tsat} and DIBL for the partially depleted, PD region, which is largely independent of T_{si}. The fully depleted, FD device shows a decrease in V_{tsat} which is proportional to the loss of doping due to thinner body ($\Delta V_{tsat} \sim \Delta T_{si} \cdot N_{channel}$) and an increase in S. This for the shorter device is related to an increase in drain-to-source coupling for the FD body. For $T_{si} < 7$ nm, the device is controlled by the T_{si} and is largely independent of the doping. It is observed that for the shorter device, body control comes at a somewhat thinner body thickness. From the general scaling theory L_{min}/T_{si} of about 5 is required for a T_{si}-body-controlled SOI device [Figure 9.48(b)]. The presence of doping in the channel reduces this ratio. In Figure 9.50(a) shows subthreshold swing vs. DIBL for an FD device at T_{si} = 15 nm compared with its PD counterpart for nominal (L_g = 45 m and 28 nm) and minimum (L_g = 42 nm and 25 nm) gate length. The increase in DIBL is shown by migrating to shorted channel length. Also observed is a higher DIBL in FD devices. The DIBL increase has a direct consequence on the performance level.

DIBL for the POD device usually adds a constant contribution. The performance impact of T_{si} body scaling, is calculated from ring oscillator delays which also give simultaneously competing ac and dc effects. As, an FD device shows slightly degraded short-channel effects (if it is doping-controlled), the thinner body reduces the junction capacitance. In Figure 9.50(b) compares the ring oscillator delay for PD (T_{si} = 48 nm) and FD (T_{si} = 15 nm) vs. Device leakage. The devices are leakage-matched at minimum gate length L_g = 42 nm and L_g = 25 nm, respectively. For both cases the nominal gate length is 3 nm longer than the minimum gate length (allowing for presumed manufacturing tolerances). The graph compares performance gain through gate length with performance gain through body thickness reduction. Thinning the body results in 5% performance gain due to reduced junction capacitance and reducing the gate length from nominal 45 nm to 28 nm gives 14% performance gain for the both ED and PD devices. The limited performance increase is because the effective drive current is degraded by 16% due to short-channel effects for the shorter device, as shown in Figure 9.44. Short-channel scaling must be improved to further improve the effective drive current. One possible solution is to back off

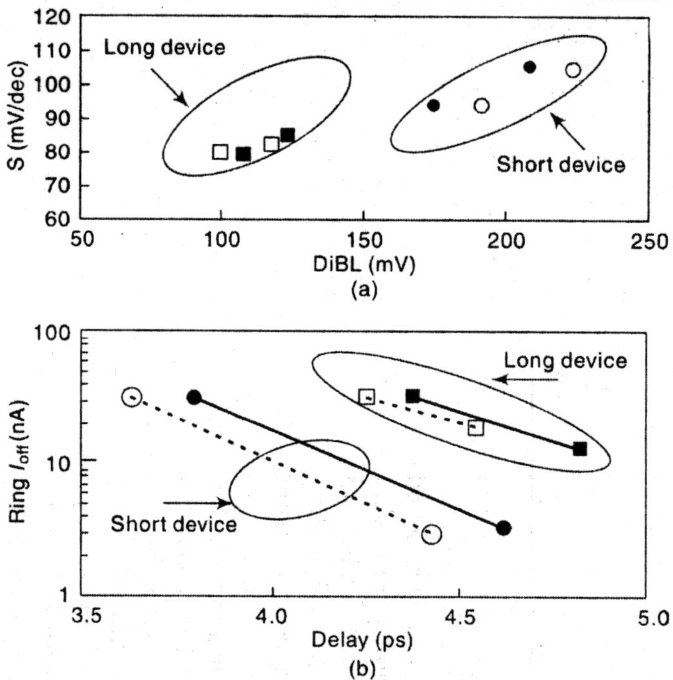

Fig. 9.50 *(a) DIBL vs. subthreshold swing for PDSOI and FDSOI devices; long devices L_{nom} = 45 nm (L_{min} = 42 nm); short devices L_{non} = 28 nm (L_{min} = 25 nm). Solid symbols—PDSOI devices with T_{si} = 45 nm. Open symbols—FD devices with T_{si} = 15 nm. All devices have the same I_{off} at L_{min}. (b) Ring oscillator delay vs. ring oscillator leakage current. Devices are leakage-matched at L_{min} = 42 nm (boxes) and 25 nm (cirles). Solid symbols—PDSOI devices with T_{si} = 45 nm. Open symbols—FDSOI with T_{si} = 15 nm.*

from channel-length reduction and find and optimal short-channel/delay design point. Another solution is to improve the gate coupling to channel by using metal gates and high-*k* gate dielectric material. The meal gate eliminates polkysilicon depletion and thus increases gate control of the channel potential. The device off-current, together with the metal-gate workfunction, determines the short-channel behavior of the device at a given dielectric thickness. In Figure 9.51 shows the modulation subthreshold swing and DIBL by the coinciding metal-gate workfunction with I_{off} at a minimum channel length of 17 nm. Halo doping is adjusted to meet the off-current at the minimum devices length, and gate length is varied through the trajectory. For both the PDSOI device with T_{si} = 48 nm and FDSOI device with T_{si} = 10 nm, a significant modulation of short-channel effects is observed with varying workfunction. Only for the FDSOI device with extremely thin body, T_{si} = 5 nm, the dependence of the short-channel effect on workfunction is weak. This is due to the confinement of minority carriers in the channel by the physical thickness of the body, rather than the fields. The lower doping levels required for more mid-gap workfunctions reduce the vertical field and spread the carriers into the body region during subthreshold operation. There is no shielding from drain to source and the subthreshold swing is degraded. Figure 9.52 shows the carrier distribution at zero gate voltage

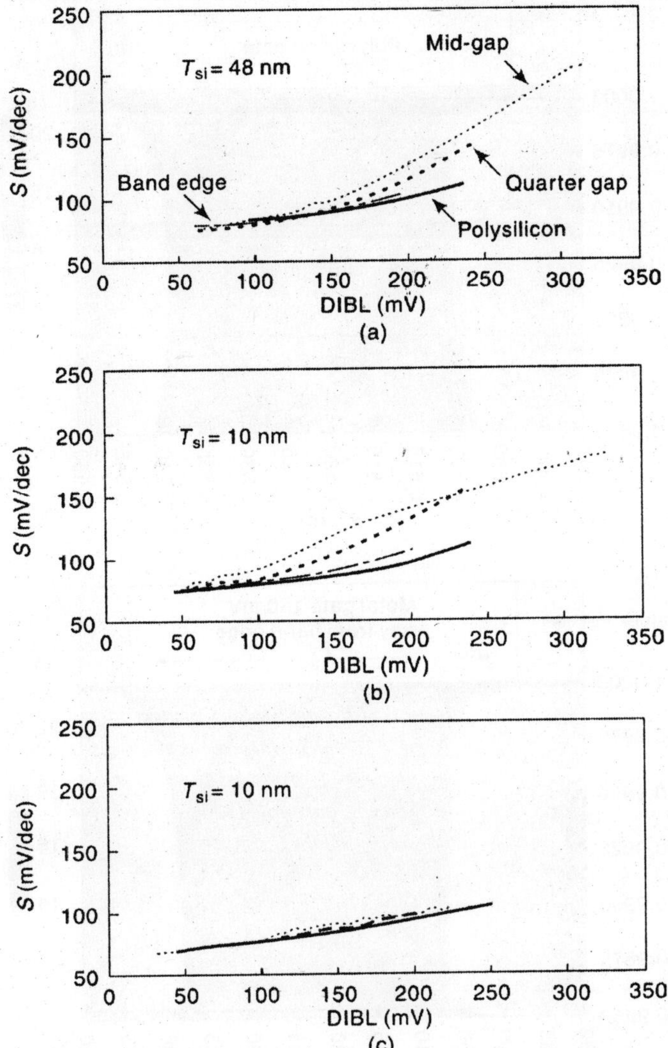

Fig. 9.51 *Subthreshold swing vbs. DIBL for metal gate with different work-functions and SOI devices with different body thicknesses. Gate length is varies: 17 nm, 22 nm, 30 nm, and 50 nm.* I_{off} *fixed at* L_{min} *= 17 nm by halo adjustment.*

for a device with a polysilicon gate and a device with a metal gate with a workfunction of 12 mV away from the band edge. Both devices are designed to have the same I_{off} and channel length. Figure 9.51 also suggests that an aspect ratio L_{min}/T_{si} > 3–4 provides short-channel control that is independence of gate workfunction for a doping-controlled FD device.

The performance advantage of a metal-gate device, is known from the delay of an inverter chain, the workfunction, and its sensitivity to high-*k* gate dielectric. The effect of a high-*k* gate dielectric is minimised by increase in the dielectric constant of the gate dielectric by a factor of 2, (from 3.9 to 7.8) and the insulator thickness constant at 1 nm. Figure 9.53(a) shows the delay

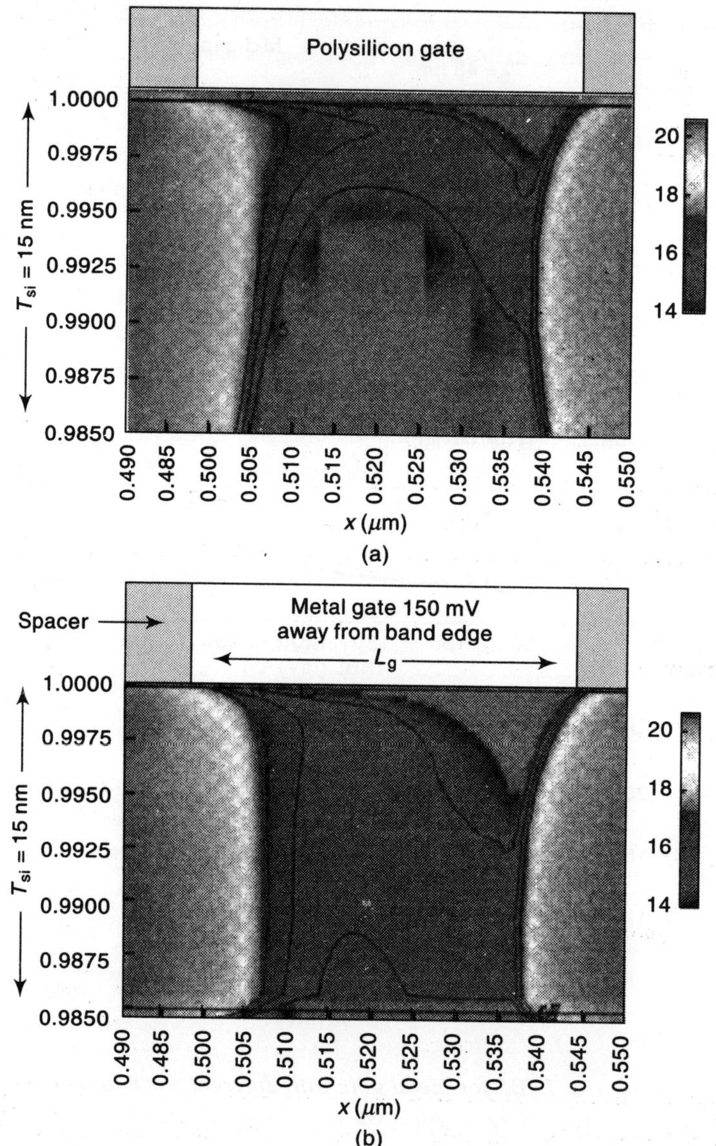

Fig. 9.52 *Carrier distribution in polysilicon-gate and metal-gate devices (150 mV away from band edge). Both devices have the same off-leakage at 42-nm gate length. The device shown has 45-nm gate length.*

for an unloaded ring oscillator of fan-out 1 for two different device lengths (L_{min} = 35 nm and L_{min} = 25 nm) and three metal workfunctions (quarter gap, band edge, and 110 mV away from band edge) over the range of gate dielectric scaling. For selected cases the effect of an additional wire load to mitigate the effect of increased gate capacitance on ac performance is shown. The delay is normalized to a polysilicon-gate device with the same gate length, and off-leakage current. The polysilicon-gate device has T_{inv} of 17.5 A for a 1-nm equivalent oxide thickness.

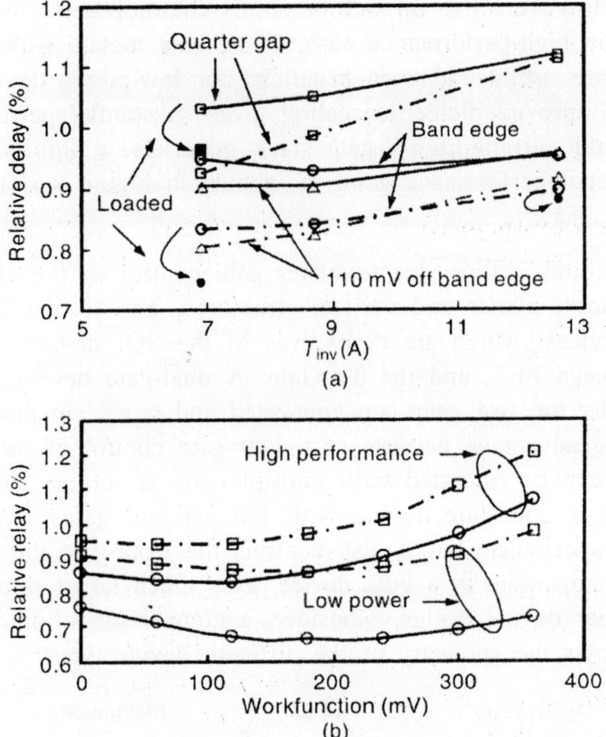

Fig. 9.53 *Ring oscillator delay vs. dielectric scaling and metal gate workfunction, for L_{min} = 35 nm (solid curve) and L_{min} = 25 nm (dashed-dotted curve) for PDSOI devices with T_{si} = 48 nm. Work-functions: open squares—quarter gap ±250 mV of mid-gap; open triangles—±110 mV off-band edge; open circles—band-edge metal. Solid symbols account for the effect of wire load for the short device. (b) Loaded ring delay vs. workfunction for high-performance and low-proper devices. Band-edge workfunction corresponds to 0 V. Open-boxes—high-performance devices: L_g = 37 nm, T_{inv} = 21 A; open circles—low-power devices: L_g = 42 nm, T_{inv} = 28 A (low gate leakage!); dashed-dotted curves—polysilicon gate replaced by metal gate; solid curves—high-k dielectric.*

Figure 9.53(a) shows that for a band edge and a quarter-gap workfunction, the metal-gate performance improves with respect to polysilicon-gate devices (for a scaled dielectric). But the gate length, off-leakage, and dielectric scaling space, the band-edge workfucntion are not optimal. This is due to the off-current constraint, since the band-edge metal requires higher doping in the channel, and therefore results in a mobility degradation which in turn degrades the drive current. The figure also shows that for high-*k* dielectrics (T_{inv} ~ 14–15 A), as performance gain for gate-loaded circuits is of the order of 5% to 10%. The gain is higher for partially wire-loaded circuits. Figure 9.53(b) compares the workfunction scaling behavior of high-performance (PDSOI) and low-power devices (bulk). The off-current leakage specification of the low-power devices is typically three orders of magnitude lower than that of high-performance devices. Therefore, the channel doping concentration (halo) is significant higher for the low-power device. To adjust for an off-band-edge workfunction, the halo dose is reduced to maintain the off-current leakage for maximum drive. However, the impact on short-channel behavior is less sensitive to workfunction than in the high-performance case. Thus, gate metals with workfunctions around ±250 mV off mid-gap are useful and even beneficial for low-power devices, primarily because of improved mobility. Improved dielectric scaling gives substantial performance gain for low-power devices. Assuming current high-*k* gate stack materials, a significant reduction in gate length results in substantial performance gains, as shown in Figure 9.53(b).

Multi-gate Devices

Multi-gate devices exhibit scaling due to better gate control of the channel. There are two types of multigate devices; planar and vertical structures. The former group contains ground plane and back-gate devices, which are derivatives of the SOI device. The vertical structures contain the FinFET Omega FET, and the Tri-Gate. A dual-gate devices are also used in two modes. In the first mode, the two gates are connected and move simultaneously. This provides the short-channel scaling advantage because of tighter gate control of the channel and superior subthreshold swing. It can be operated with multiple-gate structures such as the Tri-Gate or Omega FET. Operating a dual-gate device with independent gates requires introduction an additional gate-to-gate capacitance. but It also controls the front-gate threshold voltage with the back gate. Thus, it is comparable to a bulk device, as a much larger threshold-voltage swing is accomplished with proper device design. Consider, a comparison of multiple-gate and FDSOI devices. Figure 9.54 shows the geometry of the different device types.

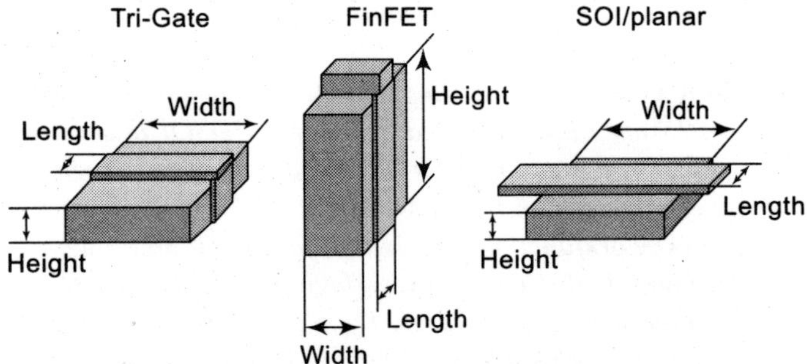

Fig. 9.54 *Geometry definition of multi-gate devices: Tri-Gate—three conducting surface; FinFET—two conducting surfaces (top surface not active).*

The Tri-gate is essentially a mesa-isolated SOI device in which the gate wraps around the active silicon area. All surfaces, sides of height H, and top surface of width *W*, contribute to channel conduction. The conduction in the corner, where vertical and horizontal surfaces, meet, is an essential part of this device. The FinFET has a higher aspect ratio than the Tri-Gate device; the top surface is usually covered by a thick oxide and does not contribute to the channel conduction. It height and width define the Fin. The height of the planar device is defined by the thickness of the active silicon. All devices are constrained to the same I_{off} leakage at a given channel length, and their short-channel effect is doping-controlled. Figure 9.55 shows the relative change in ring delay compared with that of a PDSOI device (with the same constraints) and is obtained by 2D calculations. There are small performance advantages

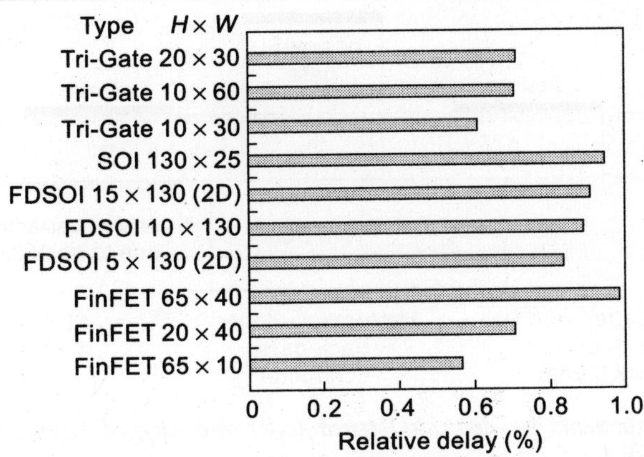

Fig. 9.55 *Relative delay for an unloaded fan-out-of-1 ring oscillator, normalized to a PDSOI device of same length. Devices have the same I_{off} equivalent oxide thickness of 1 nm, and are operated at 1 V. L_g = 23 nm.*

of FDSOI devices with respect to silicon body thickness as shown in Figure 9.50. The improved performance of FinFET for thinner width (body thickness) is directly related to the better short-channel scaling behavior of the device, which allows lower threshold voltage (as Figure 9.46). Many of these advantages carry over to the Tri-Gate structure, even for a relatively open aspect ratio, as may be seen by comparing the 10 × 30 (H × W) Twi-Gate, the 65 × 10 FinFET, and the 10 × 130 SOI structures in Figure 9.55. Much of this advantage is derived from the ability of the overlapping gates to partially screen the bottom of the SOI island from the drain field, which easily penetrate through the thick buried oxide to affect the short-channel behavior of the planar SOI device. There is a slight advantage from increased mobility due to low surface field, if the Fin is fully depleted. A advantage of the FinFET is that it can be operated with two independent gates, which offers a number of interesting benefits.

Although the FinFET offers improved scaling behavior it is a disruptive element with respect to circuit design, since the effective device width, and therefore the current, comes only in multiplies of Fins. A planar double-gate device has continuous width but suffers other drawbacks. One advantage for the FinFET is, that both gates are self-aligned with the junctions so that the impact of overlap capacitance is engineered by a proper junction design. A planar back gate, needs to align itself with the front gate. The simplest back gate is built on a SOI substrate with a thin buried oxide and a buried back gate. To operate the back gate in a reasonable voltage range, the body must be fully depleted and the buried oxide must be reasonably thin. Figure 9.56 shows the capacitance for an unpatterned (or non-self-aligned) back-gated device. For an unpatterned back gate, additional capacitance adds up through junction-to-back-gate coupling, that scales with the junction length. This capacitance mitigates by using a (non-self) aligned back gate which, also increases process complexity. For the choice of an unpatterned back gate, the proper performance design space is now an optimization of buried-oxide (back-gate dielctric) thickness, body thickness, and gate length, and the front-gate dielectric. Figure 9.57 shows a result of such an optimization. In this figure the delay impact to the SOI deviceis normalized (100 nm burined oxide) with the same body thickness and off-leakage for all devices is constrained

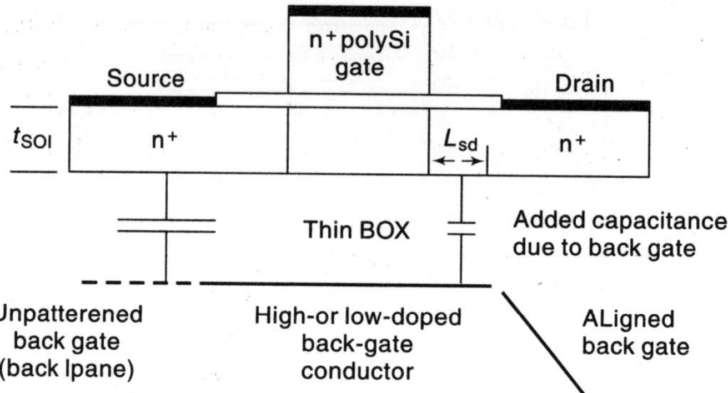

Fig. 9.56 *capacitance components for an unpatterned (left) and aligned (right) back gate with diffusion-to-back-gate overlap L_{sd}.*

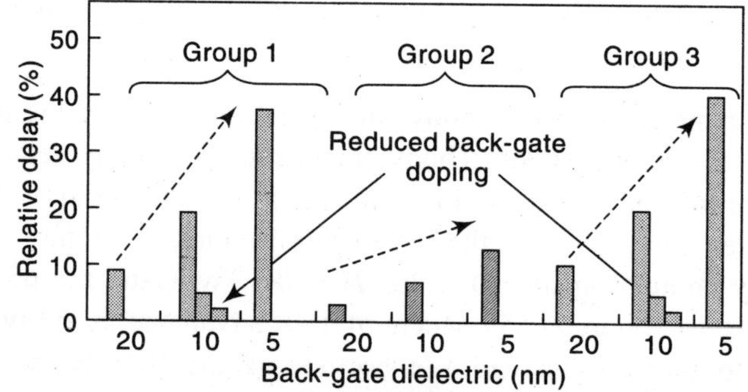

Fig. 9.57 *Performance impact of unpatterned and non-self aligned back gate normalzied to an FDSOI device of the same body thickness, channel length Lg = 25 nm, and I_{off}, Group 1: T_{si} = 10 nm, contact length 100 nm; Group 2: T_{si} = 10 nm, contact length 22 nm (aligned back gate);' Group 3: T_{si} = 20 nm, contact length 100 nm. Dashed arrow–performance degradation with thinner back-gate dielectric; solid arrows–reduced junction-to-back-gate capacitance due to more lightly doped back gate.*

to the same value. The front-gate dielectric is 1 nm equivalent oxide thickness. For the back-gated devices; a grounded back gate for the n-FET and the back gate at V_{dd} for the p-FET is chosen. Halo doping is adjusted to meet the device off-current for the minimum device. The performance impact due to the back-gate dielectrics scales linearly with its thickness, reaching 40% degradation for a 5-nm back-gate dielectric. The impact of the back-gate thickness does not depend significantly on the body thickness of the device. It allows the design point to be determined by other constraints. A reduciton of junction-to-back-gate capacitance, as obtained by a patterned aligned back gate, reduces its capacitance penalty linearly with back-gate-to-junction overlap. Reduction of the undesired back-gate capacitance for the unpatterned case is accomplished by reducing the doping levels in the back gate. For the device with a 10-nm and 20-nm body thickness and 10-nm back gate, the gate doping level is reduced from 10^{20} cm^{-3} to 10^{17} m^{-3} and further to 2×10^{16} cm^{-3}. This affects the effectiveness with which the back gate is operated. Regions where back-gate action is required receive a highly doped back gate, and those where the performance impact is detrimental is doped at a much lower level. So, performance devices are designed with halo doping and proper junction engineering. It provides an option to create SRAM device without channel doping and set thresholds by back-gate bias. This eliminates one component in the threshold variation and in addition, allows device V_t values to be set with appropriate control circuits. Figure 9.58 shows how two completely different devices that coexist on one wafer with the same device structure. The logic device is optimized to have an I_{off} ratio of 10 between the nominal and the 6σ short device at a fixed leakage for the 3σ short device. Its short-channel behavior is adjusted with a proper halo implant.

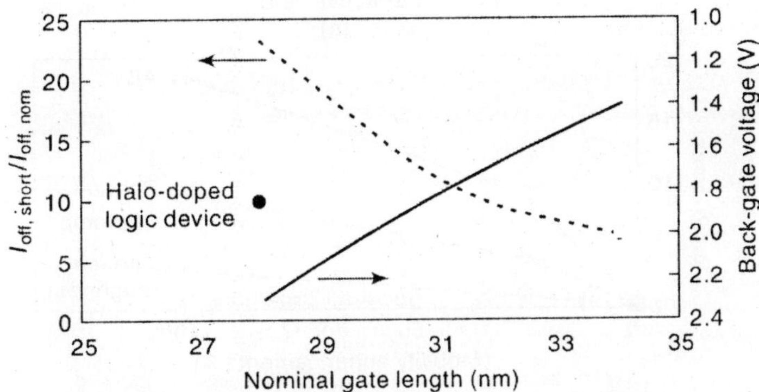

Fig. 9.58 *Leakage-limited design point for undoped body back-gate-controlled device, with 10-nm body thickness and 10-nm back dielectric thickness. Front gate is n^+ polySi and back gate is p^+ polySi. The nominal gate length is varied, and the ratio I_{off} short (L_{nom} – 6 nm) to I_{off} at L_{nom} is plotted (left side), with back-gate voltage plotted (right side) for constant leakage at a shorter (L_{nom} – 3 nm) device.*

High-mobility Channels

The channel mobility in a FET has three distinctive regions [Figure 9.59(a)]. For low vertical fields or weak inversion, mobility is limited by Coulomb scattering due to doping atoms or charges at the gate dielectric/silicon channel interface. At higher fields, phonon scattering dominates, and in still higher fields, surface roughness scattering becomes the limiting scattering mechanism for the channel mobility.

In the circuit operation the device switching trajectory passes through various regions of the V_{ds}–V_{gs} space, and it is of interest to understand how these different cattering components perform. Figure 9.59(b) shows the response of an unloaded ring oscillator to mobility change. The effect on performance of each scattering mechanism (Coulombic, phonon, interface roughness) is examined separately by increasing the corresponding mobility component by a factor of up to 2. For an inverter similar dependencies are obtained as that for other circuit elements. It is found that the phonon and Coulomb parts are similar in impact, whereas surface scattering has less effect. This is because the device spends less time in the high-gate field region (high V_{gs} and V_{ds}) than in the other regions of the V_{ds}–V_{gs} space. Figure 5.59(c) shows the performance impact over a wider range of total mobility variation, for two different device lengths, constrained to the same off-leakage. It is observed that relative performance impact depends weakly on the gate length, with a strong tendency to saturate at higher mobility

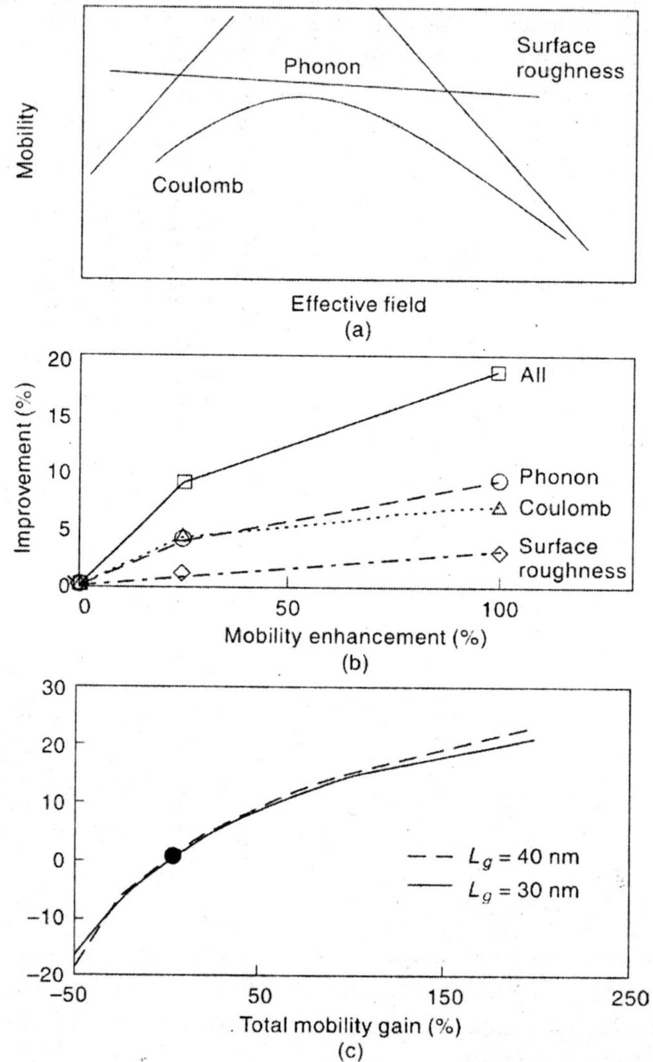

Fig. 9.59 *(a) Channel mobility components in a FET, (b) Impact of mobility component on ring oscillator delay. Dashed curve—phonon scattering; dotted curve—Coulomb scattering; dashed-dotted curve—surface roughness scattering. Calculation was done for a 38-nm PDSOI device with T_{si} = 48 nm and equivalent oxide thickness of 1 nm. (c) Relative delay impact of ran inverter delay chain with fan-out-of-3 vs. mobility enhancement. Model is calibrated to early 65-nm node (red dot). Devices with L_g = 30 nm and L_g = 40 nm have same I_{off}*

enhancements. The curve is calibrated to typical data for high-performance 65-nm-node-devices. For performance, mobility related to Coulomb and phonon scattering should be the focus of improvement; So, the Coulomb component is improved by reducing the doping in the channel, by short-channel degradation. This requires a 28-nm gate length and an equivalent oxide thickness of 1 nm, a silicon body thickness of approximately 5–6 nm for an undoped channel, and 7–8 nm for a doped channel. Mobility degradation due to quantum confinement becomes significant at body thicknesses below 5 nm. Thinning the body further greatly enhances surface scattering and therefore degrade mobility. Mutiple-gate structures are beneficial because the required body thickness for short-channel control is twice that of single-gated FDSOI devices. For these body-thickness-related mobilities, sufficient short-channel behavior with reduced doping is achieved. Two levers for the phonon-limited mobility are crystal orientation and a deformation of band structure by strain. For electrons, strain in silicon splits the degenerate Δ states and lowers the energy of a sub-band with lower effective transport mass and high-density off-state mass, thus improving drive current. For holes, a complicated deformation of the bands can occur, and scattering rates are also affected. Stress is applied perpendicular to the wafer and in the plane of current flow. In that plane (wafer surface), it is convenient to separate stress in the direction of current flow and the direction perpendicular to that. Furthermore, the stress is uniaxial, biaxial, compressive, or tensile. Figure 9.60 shows wafer surface, in-plane direction line up. Figure 9.61 shows the phonon-limited mobility behaviour. Scattering rates are calculated for the silicon band structure for electrons and holes and phonon dispersion relations at a strain level of 1%, compressive and tensile, respectively. Figure 9.61(a) shows the results for electrons and Figure 9.61(b) the results for holes. The change of mobility is normalized to the standard (100) surface in <110> directions on relaxed silicon substrates. Now, for electrons the optimum surface is the standard (100) surface. For unaxial strain, only tensile provides some moderate gain in electron mobility. There is apparently no benefit for uniaxial compressive strain in the case of electron. These gains are independent of wafer and current flow orientations. Only for biaxial compressive strain, the electron mobility improves in a more significant way if the wafer surface is (110) and the current direction is in the <0-11> direction. The situation looks more promising for holes. There is mobility improvement, in the relaxed case even if a different wafer surface other than (100) is selected. A (110) wafer surface orientation with current flow in the <110> direction gives mobility that is more than two times higher than that in a standard wafer. Further improvement is obtained if the good surface orientation is subjected to compressive strain in <110> current flow. A combination of these two brings the hole mobility close to the electron mobility on standard wafers. Biaxial compressive strain for holes is best for the (110) surface. With sufficiently high levels of tensile strain, hole mobility also increases significantly on a (100) surface if current flow is restricted to the <110> direciton.

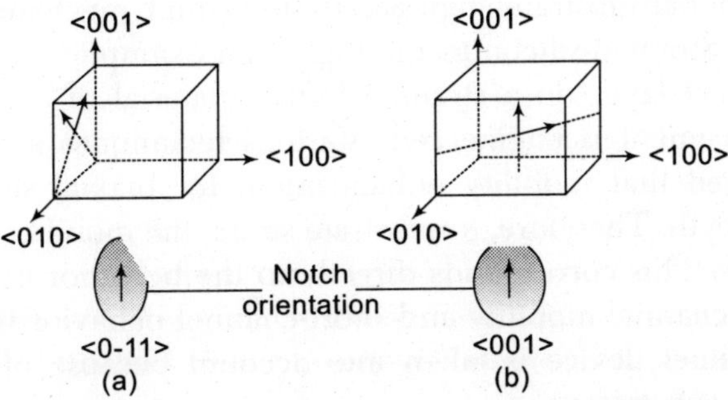

Fig.9.60 *Crystal orientation (wafer surface: gray) and possible current flow strain directions: (a) (100) surface; (b) (110) surface.*

ELectron phonon-limited mobility

Uniaxial						
	Surface	Stress direction	Current direction	Rel.	Com.	Ten.
	100	100	100			
	100	100	010			
	100	010	110			
	011	100	100			
	011	100	0-11			
	011	110	100			
	011	110	110			
	011	110	1-10			
	011	110	11-2			
Biaxial						
	001		100			
	011		100			
	111		100			
	001		110			
	011		0-11			
	111		011			

(a)

Hole phonon-limited mobility

Uniaxial						
	Surface	Stress direction	Current direction	Rel.	Com.	Ten.
	100	100	100			
	100	100	010			
	001	110	110			
	001	110	−110			
	011	100	100			
	011	100	0-11			
	011	110	100			
	011	110	110			
	111	110	11-2			
	011	110	1-10			
Biaxial						
	001		100			
	001		110			
	011		100			
	011		0-11			
	111		11-2			
	111		1-10			

(b)

Severe degradation
Degradation
Little impact
Enhancement
Strong enhancement
Extreme enhancement

Fig. 9.61 *Phonon-limited mobility (a) electron and (b) holes for various crystallogrpahic surfaces, current, and strain directions. Impact on mobility is normalized to standard surface (100) and <110> current direction and relaxed (rel.) substrate. Calculations are done for 1% tensile (ten.) and compressive (com.) strain. Mobility change quoted at inversion density 3×10^{12} cm^{-2} and 4×10^{12} cm^{-2} for uniaxial and biaxial strain respectively.*

Mobility is a long-channel property, and its impact is only in directly measurable for short devices. Figure 9.59(b) and (c), shows mobility improvement by ring-oscillator performance. It shows saturation of performance gain with higher mobility, which is due to velocity saturation. The mobility gain is achieved with combination of various stress techniques. In principle, there is distinction between global-substrate-engineered stress and mechanisms that create a local stress field in the device. Strained-silicon technology is an example of the former. By growing a thin silicon device channel layer on a strained buffer material of a Si_{1-x} Ge_x composition, a channel mobility enhancement is engineered. Various techniques are employed to generate local strain. It is observed that mobility enhancement for biaxial substrate strain is weakly dependent on channel length. Therefore, in substrate strain, the mobility properties are measured on a long-channel device. This corresponds directly to the behavior of a short-channel device. The relationship of long-channel mobility and short-channel behavior is established only if self-heating in the short-channel device is taken into account because of the decreased thermal conductivity in Si_{1-x} Ge_x substrates.

In the case in which the stress is created locally, there is no relationship between the long-channel mobility and short-channel behavior of the device, since the strain field is usually self-aligned with the gate edges. The mobility for this second choice of local strain has to be inferred indireclty from the elctrical data of the shortchannel device compared with proper controls. The situation becomes complicated if there is a superposition of different strain sources. These sources are linear stress, which is mostly longitudinal with respect to the channel current flow; trench-isolation-induced stress, which is both longitudinal and transverse to the direction of current flow (depending on the device width); or embedded SiGe, which is longitudinal to current flow. All of these components depend on details in the device layout and processing conditions, and have either a cumulative or a compensating effect on the total relevant strain.

Scaling Limits for Silicon Devices

High-permittivity insulators are a means of decreasing the electrical insulator thickness without decreasing the physical thickness for improved scalability. However, the net gain is limited, since other two-dimensional effects come into play for thick insulators caused by drain field penetration into the insulator, which modulates the channel charge. There is a possible way around this conundrum (Figure 9.62). When the high-permittivity region is confined to the channel area alone, this path is cut off. This does not change the scale length, rather, it introduces an extra attenuation into the equation describing the drain-induced potential along the channel. This attenuation is rapidly increasing function of the permittivity of the dielectric. If it is large, it will give FET the needed electrostatic integrity. Here, the gate insulator is regarded as polarizable conduit, transferring the potential from the gate to the channel region. The rest of the FET is scaled so that alternate paths for drain potential feed through are minimized. Assume, a geometry with insulator height H and gate length L where the insulator has s permittivity ε_G, and the surrounding ε_{sw}. Consider a square of the dielectric closest to the channel (as shown in Figure 9.62). This couples the dielectric square to the source and drain electrodes with a capacitance $\sim\varepsilon_{sw}$ and to the gate with $\sim\varepsilon_G\ L/H$, per unit width. So, if a drain vs. gate coupling factor of 0.1 is desired, along with $H/L > 10$, then $\varepsilon_G/\varepsilon_{sw} > 100$ is needed. Therefore, materials with very high, anisotropic permittivity are explored.

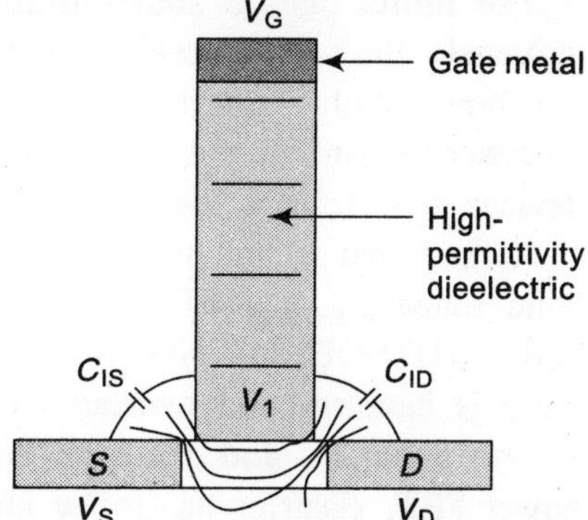

Fig. 9.62 *Ultrahigh-permittivity, high-aspect-ratio gate insulator showing sketched equipotentials (thin lines), and derivation of the potential, V_v of the electric adjacent to the channel region.*

The introduction of this new paradigm removes the gate insulator from the channel-length scaling considerations. So for a double-gated FET a minimum gate length comparable to (~1.5x greater than) the body thickness is required. Since FETs with a body thickness <1 nm are achieved a channel length <1.5 nm is needed. Gate length may approach zero, if the potential is allowed to drop in the external source/drain regions. For instance, Likharev's simulated 4-nm-gate-length FET actually had a channel length closer to 7 nm when depletion into the contacts was considered. This depletion effect characterizes all experimental devices with gate lengths of less than 10 nm. Therefore limit on gate length is removed. So the short gate length, for constant channel length, give degraded performance because of their higher resistance. The limit on channel length is determined by direct drain-to-source tunneling. A simple estimate of this limit is obtained by approximating the potential along the channel in the "off" state by a parabola (Figure 9.63). Such a smooth curve is expected in ultrashort devices, where abrupt transitions are not easy to achieve. The height of the barrier, V_b, determines the "off" current, and the parabola is terminated at "metallic" source and drain contacts where the conduction band is assumed to be pinned at the source and drain potential. The leakage current is dominated by intraband tunneling when

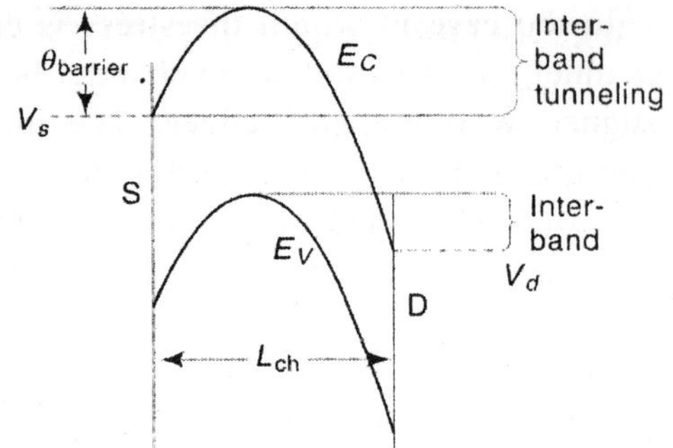

Fig. 9.63 *Schematic band diagram along channel of FET, assuming a parabolic profile, showing schematic S-D (intra-band) and band-to-band (inter-band) tunnel-current energy range.*

$$\frac{kT}{e} \le \frac{\eta}{\pi}\sqrt{\frac{a}{2me}}$$

where a is the curvature of the parabola, related to the geometry through the source/drain distance L_{ch}, barrier height$\theta_{barrier}$, and drain voltage V_d; m is the tunneling mass, e the electric charge, and kT the thermal voltage. Assuming a value of $0.19m_0$ for m, 0.2 eV of $r\theta_{barriers}$, and 0.5 V for V_d requires L_h > 7 nm to suppress tunneling.

The limits due to source/drain tunneling, which are not even a factor in current designs, ultimately limit FET scaling even before electrostatic limits are reached. Power dissipation does not limit scaling on the device level. This is partially because local heat removal from a microscopic part of the device into the 3D surrounding is relatively efficient, and partly because devices operate at a low duty factor, so that local temperature rise is not excessive.

A limit that is not often considered, but is intimately related to power dissipation, is band-to-band tunneling. For FETs on bulk silicon substrates, but even for SOI devices with ultrathin bodies (UTSOI) and silicon wire devices, this tunneling is important. Figure 9.63 shows that there is band-to-band overlap when $V_{ds} > E_g$ barrier, where V_{ds} is the drain-to-source voltage, E_g the bandgap, and θbarrier the barrier height of the turned-off device. For a low-standby-power FET, θbarrier has to be large in order to limit th leakage current so that the condition of above Equation is mostly met. As shown in band-to-band tunneling becomes large when the tunneling distance is ~4 nm. This is typically about 1/3 of the channel length, i.e. ≈12 nm. It is found that channel lengths greater than 20 nm are needed for the ultralow-power options.

The use of strain to enhance mobilities enhances band-to-band tunneling. The bandgap of silicon is very sensitive to strain because of the X-valley symmetry of the conduction band; therefore, strain of either sign always decrease the bandgap. The interaction between strain and band-to-band tunneling leakage must therefore be carefully monitored.

VI. CONTINUOUS MOSFET DEVICE SCALING

Impact of Velocity on MOSFET Switching Time

An increase in channel carrier velocity results form backscattering reduction, and while some of it results from scaling of the characteristic length for backscattering, the channel mobility increase is, from this point in time onward, the main lever for continuing decrease of transistor delay τ in proportion to L_g.

Divided below is the relationship among carrier velocity, MOSFET drain current, and switching time. The drain saturation current in a MOSFET normalized to the channel width, I_D/W, is as follows:

$$(I_D/W) = Q_s'v = C'_{\text{oxinv}}(V_G - V_t)v, \tag{1}$$

where Q_s' is the channel charge areal density at the virtual source, C'_{oxinv} is the gate-to-channel capacitance per unit area at inversion, and v is the effective carrier velocity at the virtual source, which is defined as the point in the channel at which the charge density is given by $C'_{\text{oxinv}}(V_G - V_t)$. V_G is the applied gate-source voltage, and V_t is the effective threshold voltage in saturation, i.e., obtained from linear exatrapolation of the I_D–V_G curve to $I_D = 0$; V_t is given by

$$V_t = V_{t0} - \delta V_D, \tag{2}$$

where V_{t0} is the effective threshold voltage at drain-to-source voltage V_D, equal to zero, and δ is the drain-induced barrier-lowering (DIBL) coefficient.

The intrinsic MOSFET switching delay, τ, is given by

$$\tau = \frac{\Delta Q_G}{I_{\text{eff}}} \tag{3}$$

where ΔQ_G is the charge difference at the gate electrode between the two logic states, including both channel charge and intrinsic gate electrode fringing capacitance charge, and is given by

$$\Delta Q_G = C'_{\text{oxinv}} W(V_{dd} - V_t) + C_f^* \tag{4}$$

and I_{eff} is the effective MOSFET switching current given by

$$\begin{aligned} I_{\text{eff}} &= [I_D(V_G = V_{dd}/2, V_D = V_{dd}) + I_D(V_G = V_{dd}; V_D = V_{dd}/2)]2 \\ &= [Q_s'(V_G = V_{dd}/2, V_d = V_{dd}) + Q_s(V_G = V_{dd}, V_D = V_{dd}/2)]v/2 \\ &= C'_{\text{oxinv}} W[(3 - \delta)V_{dd}/4 - V_t]v. \end{aligned} \tag{5}$$

Using equations (4) and (5) in (3) results in

$$\tau = \frac{V_{dd} - V_t + (C_t^* V_{dd}/C'_{\text{oxinv}} L_g)\, L_g}{[(3-\delta)/4](V_{dd} - V_t)\quad v} \tag{6}$$

where V_{dd} is the supply voltage and C_f^* is the total effective gate fringing capacitance, including all internal and external fringing capacitance. It is interesting to note that C_f^* is independent of technology generation for properly scaled device, with a value of about 0.5 fF/μm.

Because of the existance of finite resistance between the source contact (and the drain contact) and the channel R_s, the actual carrier velocity at the virtual source v_{xo}, is given by

$$v_{xo} = \frac{v}{[1 - C'_{oxinv} R_s W (1 + 2\delta) v]} \tag{7}$$

The virtual source point is located near the top of the potential barrier between source and channel, and v_{xo} is related to the so-called source injection velocity, $v\theta$, or ballistic-limit velocity. Figure 9.64 illustrates these concepts. The ballistic velocity $v\theta$ for electrons vs. effective electric field in the channel is plotted in Figure 9.65, for both relaxed and strained, Si. For the strained Si it is assumed that only the Δ_2 valleys are populated, with no-longer changes relative relaxed Si. The assumption that all electrons occupy the Δ_2 valley corresponds roughly to an energy splitting between the Δ_2 and Δ_4 valleys greater than 140 meV, or a biaxial tensile strain level higher than approximately 1%.

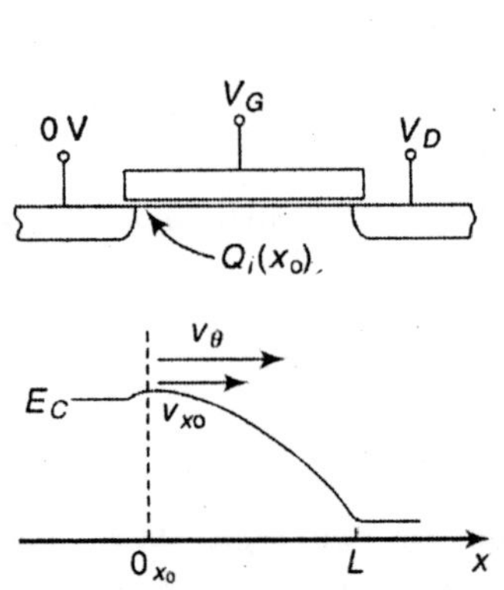

Fig. 9.64 *Illustration of the virtual source position x_o in the current.*

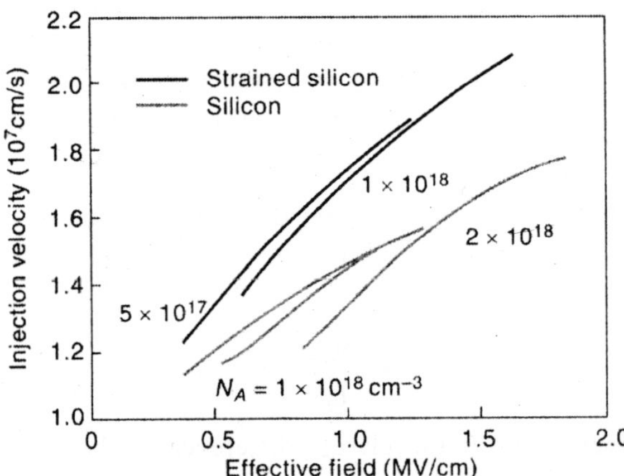

Fig. 9.65 *Injection velocity $v\theta$ v. channel effective field for relaxed and ~1% biaxially strained bulk MOSFETs with different channel dopings.*

While Equation (1) is used to model transistors in the ballistic limit, it actually fits state-of-the-art transistor v_{xo} for both electrons and holes are tracked for the same period of time, for gate length L_g ranging from 480 nm to 35 nm. C_{oxinv} is given, δ is extracted from the data, and R (or at least its upper bound) is reasonably estimated from the output I–V charateristics near $V_D = 0$. The values of the denominator in Equation (7) are higher than 0.89, indicating no more than ~20% correction velocity, v. The values of v_{xo} extracted from the literature data and required hole velocities at $L_g = 10$ nm are shown in Figure 9.66. It is observed that for both electrons and holes, the carrier velocity v_{xo} increases with decreasing L_g. For all gate lengths down to 60 nm, the channel is unstrained (100) silicon, and therefore the carrier mobility vs. effective field relationship is unchanged. The main reason for the velocity increase with scaling is understood from the scattering theory. It is due to reduction of the characteristics length of the potential barrier near the source, as L_g is scaled, and therefore reduction of backscattering.

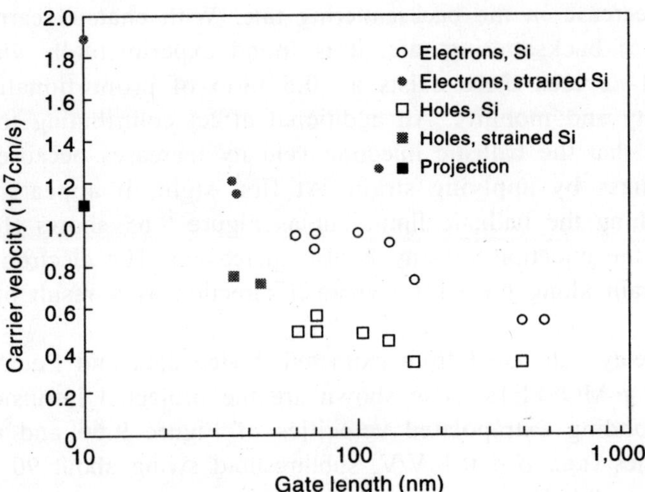

Fig. 9.66 *Source velocity v_{xo} of eletrons and holes vs. gate length. Also shown are velocities at $L_f = 10$ nm that are required to continue the relation of intrinsic FET switching delay time to gate length.*

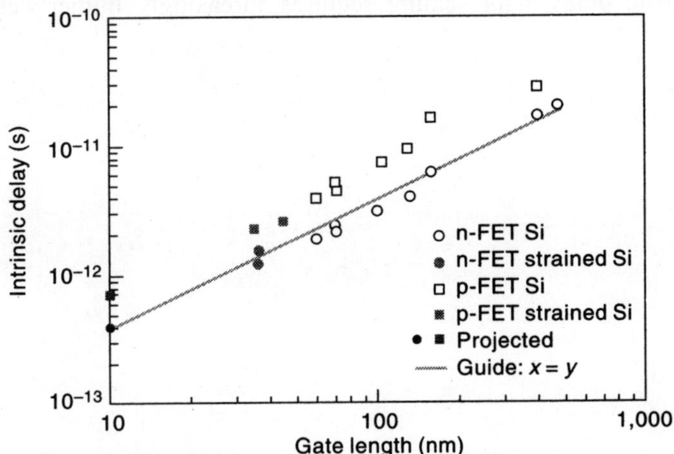

Fig. 9.67 *Calculated historical intrinsic FET switching delay time and gate length.*

The reduction in the characteristic length of backscattering is due to scaling of the electrostatic design of MOSFETs, and is achieved via innovations in source/drain and channel doping. Now, v_{xo} should increase (provided v_{xo} is smaller than vθ) when there is a reduction in backscattering, either by reduction of the length over which backscattering occur, or by reduction of the scattering rate. However, data shows that from L_g ~ 130 nm to 60 nm there is a saturation of velocity increase that is due to the increase of coulombic scattering near the source; with increase in doping that counterbalances the decrease in the backscattering effective length.

Below 60 nm, the increase in velocity observed in Figure 9.66 is due to the introduction of well-known strain-induced mobility increases in the channel via innovative process step, which has brought about a decrease in the backscattering rate. With channel carrier mobility being a proxy for the inverse of backsattering rate, it is found experimentally via the application of strain in short-channel devices there exists a ~0.5 ratio of proportionality between channel electron or hole velocity and mobility. An additional effect contributing to this proportionality ratio is due to the fact that the ballistic injection velocity increases because of the reduction in the carrier effective mass by applying strain. At first sight, it appears that both n-and p-MOSFETs are approaching the ballistic limit, but as Figure 9.65 shows (for electrons), this is not necessary, because the injection velocity is also increasing. The electron injection velocity is higher with uniaxial strain along the <110> channel direction as a result of decreased effective mass.

MOSFET intrinsic delay calculated from extracted device data and Equation (5) is shown in Figure 9.67 for n- and p-MOSFETs. Also shown are the projected intrinsic delays at $L_g = 10$ nm, using the corresponding extrapolated velocities of Figure 9.66 and making assumptions about device electrostatics (i.e., $\delta = 0.1$ V/V, subthreshold swing about 90 mV/decade resulting in $I_{off} \sim 300$ nA, at $V_{dd} = 0.8$ V.

The increase in velocity and therefore decrease of the L_g/v term, counterbalances a parallel increase in Equation (6), resulting in near-perfect proportionality between τ and L_g. Continuous improvement in intrinsic delay with scaling requires inreasingly higher velocities, (at least for electrons in silicon).

Atomic-Level Manipulation of CMOS

For continued progress in the semiconductor industry, the introduction of new process steps such as ion implantation, spacer formation, self-aligned devices or salicide has enabled progress in the past. The new process technologies and materials have emerged to facilitate continued device scaling and performance improvement. The process issues that affect replacement of conventional device structures and to determine where new process elements will be indispensable are discussed below.

The semiconductor revolution began, long bipolar devices fabricate discussed on slabs of polycrystalline Ge. Single-crystalline materials were later introduced making possible the fabrication of grown junction transistors. Migration to Si-based devices was initially hindered by the stability of the Si/ SiO_2 materials system. The stability and low interface-state density of Si/ SiO_2 materials system provided passivation of junctions which in turn effects the migration from bipolar devices to field-effect devices. Both complementary metal-oxide-semiconductor devices (CMOS) and poly silicongate technology allowed self-alignment of the gate to the source/drain of the device to develop. These innovations permitted a significant reduction in power dissipation and a reduction of the device overlap capacitance, improving frequency performance, and changes involved replacing the whole device gate electrodes. Since the advent of the self-aligned, poly-silicon-gated transistor, CMOS scaling is carried out with this fundamental device at the core. Subsequent changes have taken place at the periphery of the device. Figure 10.1 shows the electron microscopy (TEM) image of the device for the 90-mm technology node. The device is built on thin silicon-on-insulator (SOI) substrates and has self-aligned silicide and dual spacer.

The first process module that reached the atomic limit is the gate dielectric module. Figure 10.2 is a high-resolution TEM lattice image of a silicon oxynitride gate dielectric film developed for the 90-nm-technology node. The dielectric is only a few monolayers in thickness. Continued scaling requires a reduction in this thickness by 70% to maintain short-channel control as the

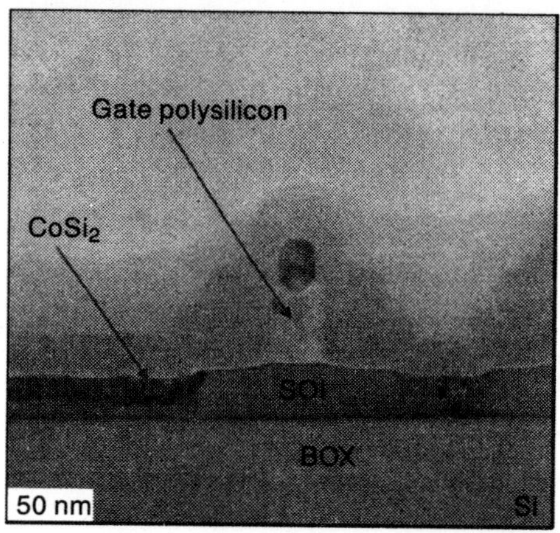

Fig. 10.1 *Cross-sectional TEM (XTEM) image of a 90-nm-technology-node device, illustrating sub-50-nm polysilicon gate.*

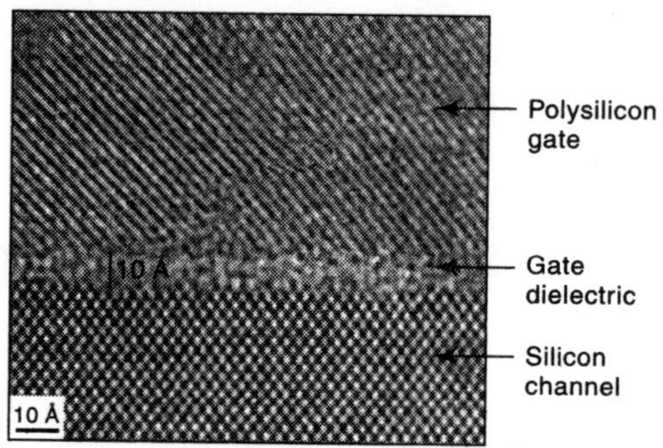

Fig. 10.2 *XTEM lattice image of the Si substrate, gate dielectric, and poly-silicon for a 90-mm technology-node device.*

physical gate length is scaled. Though, changes (new device structure and new materials set) have occurred, but the change in materials set for a fixed device structure (Ge bipolars migrating to Sibipolars) occurred only before the Si/SiO_2 system. In the Si/SiO_2 interface, example, planar SiO_2-passivated bipolars are supplanted by AI-gated SiO_2 FETs and Si CMOS devices are improved by fabrication on SOI substrates. Thus, migration to high-k gate dielectrics or other materials for the channel of the FET is difficult. The new device based on quantum effects and interactions required, an advantage of the Si technology infrastructure and understanding SiO_2/Si interface.

THIN-FILM DEPOSITION

As devices are scaled, there is a need to control the thickness of thin-film depositions to the atomic-layer scale. Figures 10.3 and 10.4, respectively, show the plots of gate dielectric and sidewall spacer requirements. The requirements for gate dielectric i.e. equivalent oxide thickness (EOT); tolerance approach will be 0.02 nm. Since scaling of the conventional device necessitates that the gate dielectric move away from SiO_2 based films, the physical tolerance requirements are relaxed by the ratio of the dielectric constant of the new material to that of SiO_2. But for the sidewall spacer, thickness-control requirements will approach the 2-nm level. Given that gate dielectric tools can achieve tolerances of less than 0.01 nm for 1.5-nm films, these requirements are too stringent in the basis of current capability. However, tolerance is achieved by thermally grown films whose depositions are highly sub linear with growth time and are much more reproducible than chemical-vapor-deposited (CVD) films. Many of the films in FEOL processing are deposited by CVD or plasma-enhanced CVD (PECVD), and achieving the required film thickness control is a challenge.

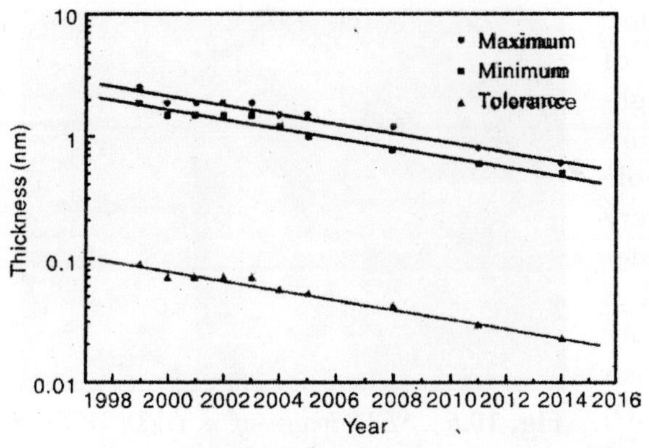

Fig. 10.3 *Minimum and maximum projected requirements for the gate dielectric equivalent-oxide thickness and the projected tolerance.*

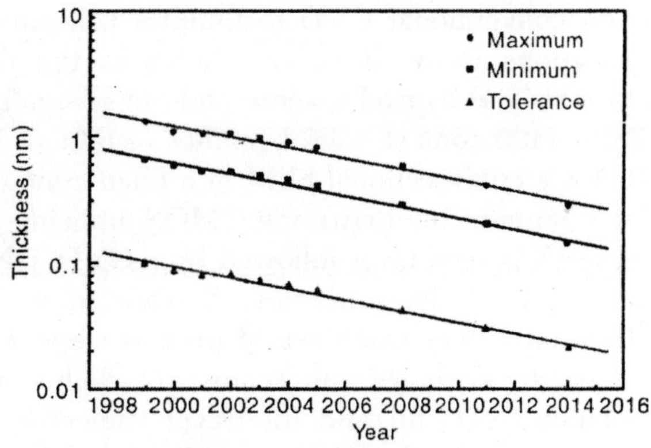

Fig. 10.4 *Minimum and maximum projected requirements for the sidewall spacer thickness and the projected tolerance.*

CVD processes are governed by any one of two mechanisms: 1) kinetically controlled 2) mass-transport-limited. At low growth temperatures, the rate of growth is low and there is an overabundance of reactant species. The growth rate is determined by the rate of thermally activated surface reactions with activation energies ranging from 1 to 4 eV per molecule. It is dependent on the growth surface and the crystal orientation, (in the case of epitaxy). The high activation energies make temperature control extremely critical to growth uniformity. As the temperature increases the rate of surface reaction increases to the point at which the supply of reactants or the transport of reaction byproducts limits the growth rate.

The mass-transport is characterized by a lower activation energy, (a few tenths of an eV per molecule) due to the temperature dependence of the gas–phase diffusion constant. Since the supply of reactants at the wafer surface is limited the local growth rates across a wafer vary because of macroscopic depletion of reactant in the reaction chamber as well as micro loading over three different length scales: wafer level, device level, and an intermediate scale.

Both spacer sidewall film depositions and selective raised source/drain contacts are good examples of structures which CVD technology has reached. Sidewall spacer deposition thickness is a function of the gate pattern density because the growth rate is limited by the reactant diffusion rate. This results in a slower deposition rate in the local vicinity of high-pattern-density fate features due to the increased surface area for deposition with a fixed supply of reactant.

Contact liners are another examples of physical vapor deposition (PVD) and CVD. Contact liners have progressed from PVD to collimated PVD to ionized PVD to ionized PVD (i-PVD) and finally to CVD. Coverage of the sidewalls at the bottom of the hole is limited to a few percent of that deposited on the upper rim of the contact hole because of the angular distribution of sputtered species from the target.

Collimated and ionized PVD techniques provide a more directional flux at the substrate surface. It also provides more material to the bottom of high-aspect-ratio contact holes. In spite of these advances in PVD technology, some applications have moved to CVD liner processes.

Even conventional CVD techniques will encounter limitations in small contact holes as the flux of reactants and byproducts into and out of small, high-aspect-ratio contact holes becomes restricted. Figure 10.5 is a cross-sectional SEM of a filled contact hole for 130-nm-technology-node CMOS, utilizing a very thin i-PVD seed layer followed by a CVD TiN liner and CVD W fill processes. A seam in the fill is observed. Other examples of process steps that are subject to keyholes and seams are shallow-trench isolation (STI) fill and interlayer dielectric (ILD) dielectric deposition that precedes the first contact level. The pressure only be reduced by a certain amount before the flux of reactant species becomes too small to be practical, and in PECVD reactions, charging problems can result from plasma damage at low pressures. Sequences of deposition and etching (dep-etch processes) provide some improvement in filling high-aspect –ratio features, but uniformity over all pattern factors in not easily achieved in a process that involves balancing competing reactions.

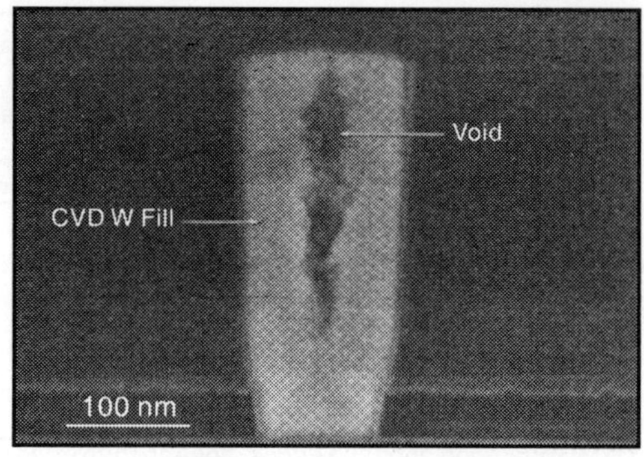

Fig. 10.5 *SEM image of a CVD W-filled contact hole for 130-nm-technology-node CMOS. A CVD TiN liner was deposited on a thin i-PVD seed layer preceding the W CVD fill.*

EXTENSION JUNCTION AND CONTACT RESISTANCE

Another concern for continued scaling of CMOS devices is external resistance (R_{ext}). As devices are scaled, the on resistance, $R_{on,}$ of the intrinsic devices is reduced. It is important to keep the parasitic source/drain resistance a small fraction of R_{on}; in order to maintain good trans conductance and overall performance at the device terminals. Figure 10.6 is a graph of R_{on} of high-performance n-FET devices by year of volume manufacturing buildup for high-performance logic technologies. A similar graph is be made for p-FETs. Usually the total

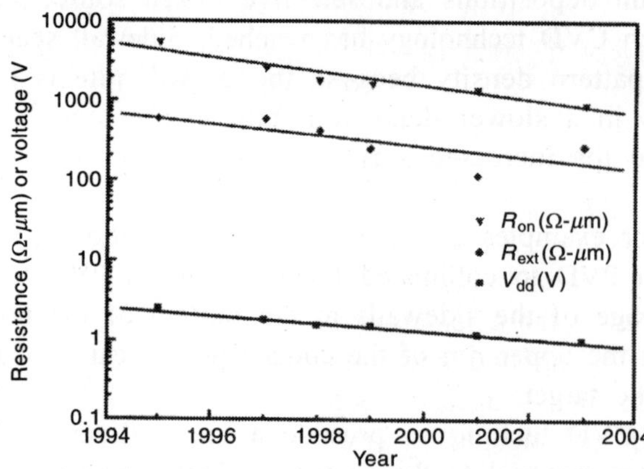

Fig. 10.6 *R_{on}, R_{ext}, and supply voltage (V_{dd}) trends by year of technologies in or nearing production.*

external resistance, R_{ext} is kept to about 10% of R_{ON} to maintain good performance. For 90-nm-technology node, *n*-FET R_{ext} less than 100 W-mm is required. Further scaling for the 60-nm node and beyond, requires R_{ext} significantly less than 100 W-mm.

The total external resistance equals

$$R_{ext} = 2(R_{ac} + R_{sp} + R_{sh} + R_{co}), \tag{1}$$

where R_{ac} is the accumulation layer resistance, R_{sp} is the spreading resistance of the junction, R_{sh} is the sheet resistance of the junction, and R_{co} is the contact resistance between the contact metal and the silicon. The contact resistance R_{co} is a function of the specific contact resistivety of the silicide/silicon interface ρ_c, as well as the sheet resistance $\rho_\square$ of the doped silicon, which determines the geometric flow of current. In this case where $\rho_\square$ is low with respect to ρ_c, current flows over the entire silicide length L_s into the silicon. R_{co} is given by

$$R_{co} = \rho_c / L_s W \tag{2}$$

for a device of width W. However, for the case in which silicide consumes much of the deep junction or for SOI on thin Si, such that the $\rho_{\square\square}$ under the silicide is high, the effective length for current flow is reduced, and the contact resistance saturates at

$$R_{co} \approx \sqrt{(\rho_c \rho_\square)} / W \tag{3}$$

The length L_s drops out of this expression, and current flows out of an effective length l'. The specific contact resistivety dependence is:

$$\rho_c \approx \exp[(4\pi\sqrt{(\varepsilon_s m^*)}/h)(\phi_b/\sqrt{N_d})], \tag{4}$$

where the two important parameters are the substrate doping concentration N_d and the barrier height at the silicide/silicon interface ϕ_b.

The sheet-resistance term of the external resistance is dominated by the extension sheet-resistance component in present devices' Approximately,

$$R_{sh} = \rho_e L_e / (X_j W), \tag{5}$$

where ρ_e is the specific resistively of the extension junction, L_e is the length of the extension junction, and X_j is its depth L_e is typically less than the spacer width L_{sp} because the thermal cycle and lateral straggle of the extension junction are less than those of the deep junction the value of R_{sh} is currently from 40 to 10 Ω-μm and scales lower with continued scaling. However, in order to keep R_{sh} a small fraction of R_{ext}, continued scaling of ρ_c is required even as the junction depth is reduced.

By reducing R_{ext}, reduced sheet resistance, R_{sh}, requires heavy doping, but shallow junction depth degrades the sheet resistance [see equation (5)]. Reduced contact resistance R_{co}, also requires heavy doping, but reduced junction depth compromises the $\rho_\square$ term in Equation (3) while reducing the amount of Si that can be consumed for silicide formation without having degraded contact resistance due to the silicide intersecting the junction at low carrier concentration [equation (4)]. Reduced spreading and accumulation resistance, R_{sp} and R_{ac}, require heavy doping with very abrupt profiles. For junctions formed by implantation and RTA, profile abruptness can be degraded with deeper junctions, thus putting minimization R_{sh} and R_{co} at odds with minimizing R_{sh} and R_{co}.

Laser annealing is utilized to achieve ultrashallow junctions with high activation due to the high solubility of dopants at or near the melting point of Si. In this technique the laser pulse is absorbed in the gate conductor and results in damage. Utilizing laser annealing with a replacement gate flow removes this problem. Care is required to prevent deactivation of the dopants with subsequent processing. There are several variants to the sequence of extension junction formation, deep junction formation, and selective epitaxy that is utilized, but requires a shallow, abrupt, and highly doped junction to be formed by implantation and annealing. So care must be taken to ensure that the epitaxial growth does not deactivate the dopants or lead to transient-enhanced diffusion. Therefore, growth at low temperatures is desired, but growth rates on n-FET and p-FET regions of the CMOS circuit are different because the growth rate is kinetically controlled by surface reaction.

ATOMIC LAYER DEPOSITION/EPITAXY

Due to the limitation of conventional PVD, CVD, and RIE techniques, processes with atomic-layer control of thin-film deposition and sequence of self-limiting surface reaction, are required for continued technology scaling. The mechanism for the deposition of compounds, is extended to elemental films. If the film is grown epitaxially on a crystalline substrate, the technique is known as atomic layer epitaxy (ALE); otherwise it is known as atomic layer deposition (ALD). ALD and ALE are used to deposit a wide range of materials. Commercial tools are available for the ALD of high-k gate dielectrics and W liner materials, but ALD is used for many other process steps. It enables controlled fabrication of many other parts of the device, to tight tolerances, if tooling and processes for the ALD of materials, (such as Si for raised sourced/drain contacts and SiO_2 and Si_3N_4 for spacer materials) is brought to a production-worthy level. The status of ALD and AlE deposition of a range of materials gives a sense of extending ALD beyond W and high-k gate dielectrics.

For the deposition of ZnTe films, it is shown that layer-by-layer growth of II-VI compounds are achieved, and that the adsorption coefficients of the impinging atoms depend strongly on the atomic species of the uppermost layer of the film. If, only one elemental flux is provided, one monolayer is adsorbed but additional species are adsorbed because of the lack of formation of a II-VI chemical bond. Also, if the vapor pressures of the elements exceed that of the compound, excess reactant atoms are made to re-evaporate before a beam of the next species is allowed to react with the surface.

ALE provides a process in which films are insensitive to temperature and the flux of reactant species. The reactants are provided in sequential pulses: -A-B-A-B etc. As long as at least one monolayer of reactant can adsorb in each pulse, and there is sufficient time between pulse to desorb excess reactant, layer-by-layer growth is achieved. The dependence of growth rate on temperature is also reduced. For temperatures below the temperature for congruent evaporation, but still high enough for sufficient surface mobility of adsorbed species and desorption of excess species, each A-B sequence provides on monolayer of compound growth.

Temperature-insensitive ALE growth of ZnTe is obtained between 593 and 673K, with a growth rate of approximately one ZnTe monolayer per –A-B cycle. Above 673K, the growth rate drops to near zero as evaporation of the compound occurs. Sequential exchange reactions are used to obtain layer-by-layer growth of Ta_2O_5 and ZnS film, from a combination of pulsed molecular beams of the metal chloride (from effusion ovens) interleaved with gas pulses of either H_2O or H_2S to complete the reaction cycle. To know about the exchange reactions taking place, Auger spectroscopy of the growth surfaces is performed after each pulse. In both cases, the weekly chemisorbed Cl present on the surface after the metal chloride adsorption is completely removed by the adsorption of either H_2O or H_2S. This reaction produces HCl vapor and either Ta_2O_5 or ZnS films, respectively, with no traces of Cl present. After multiple pulses, films >300 nm thick are produced

This technique is used in the III-V materials system for the growth of GaAs and is described as a gas-source MBE apparatus. Arsine and trimethylgallium used as the reactant, While self-limited monolayer-by-monolyaer growth is achieved poor-quality material is produced, with mobilities of ~100 $cm^2 V^{-1}s^{-1}$ and carrier densities greater than 1×10^{18} cm^{-3}. The technique is extended to CVD growth at more conventional operating pressures for both GaAs and AIAs. Using the same reactants, but in an atmosphere of hydrogen gas. Layer-by-layer growth is achieved by placing the substrate on a rotating susceptor. The material quality obtained by this method is much better, with 77 K photoluminescence peak widths of 11 meV are obtained. GaAs is grown by similar means; room-temperature electron mobilities exceeding 5500 cm^2 $V^{-1}s^{-1}$ are obtained, with residual carrier densities less than 1×10^{15} cm^{-3}. Thus, in order to ensure good materials quality by ALE, it is important to maintain a passivated surface during growth. High-quality films are more difficult to obtain by MBE in the ALE mode as the growth has to be interrupted to allow desorption of reaction byproducts or switching of species. So there is significant incorporation of impurities. In contrast, AlE by CVD means is manufacturable, at least when there is a suitable –A-B- reaction sequence, as in the growth of GaAs and AIAs.

The ALE growth technique when extended to an elemental semiconductor such as Si, like adsorption of Si_2H_6 at cryogenic substrate temperatures followed by laser irradiation by an ArF excimer source at 193 nm and evacuation to desorb reaction byproducts is utilized. Deposition occurs only on the Si surface and not on SiO_2, unless the adsorption time is longer than that used for deposition on Si. SiH_2Cl_2 and H_2 are used as precursors. At 1123 K SiH_2Cl_2 decomposes to $SiCl_2$ on the surface with the desorption of H_2. Subsequent evacuation of the reactor and pulsing with H_2 gas at 1123 K leads to the desorption of the remaining surface Cl as HCl. Layer-by-layer growth proceeds as long as the substrate temperature and the partial pressure of the SiH_2Cl_2 are kept low. Growth by this technique is also selective to SiO_2 masked regions. A disadvantage with these approaches is that the Si surface, provides more impurities.

The adsorption of $SiClH_3$ and SiH_2Cl_2 on Si (100) show that while $SiClH_3$ is not a suitable precursor for ALE growth due to substantial surface coverage of hydrogen, SiH_2Cl_2 has much lower coverage of hydrogen and leads to single-monolayer coverage of Cl at temperatures near 773 K. However, desorption of HCl at 773 K is significant. So, SiH_2Cl_2 adsorption is not strictly self-limiting. As a result, other approaches are attempted to achieve controlled layer-by-layer growth. By using exposures of Si_2H_6 and Si_2Cl_6 chlorine and hydrogen surface termination is maintained. At 738 K the film growth rate is two monolayers per cycle (one cycle equals one Si_2H_6 and one Si_2Cl_6 exposure), and the desorption of surface hydrogen by Si_2Cl_6 dosing is a self-limiting process. Desorption of surface Cl by dosing with Si_2H_6, is not strictly self-limiting and is kinetically controlled in this temperature regime. Atomic hydrogen is used to complete the Cl-H exchange reaction in a self-limiting way with 673 K Si_2Cl_6 exposure. This results in Si adsorption until Cl fully terminate the surface, making the Si adsorption step self-limiting. The terminating Cl layer is removed by exposure to atomic hydrogen. At 673 K, H_2 desorbs rapidly from the surface, regenerating the surface dangling bonds for the next Si_2Cl_6 adsorption. Si_2Cl_6 and SiH_2Cl_2 are not ideal for ALE because they adsorb dissociatively, and one of the Cl atoms of these fills a surface site, resulting in less than one monolayer coverage of adsorbed Si-Cl. By using, SiH_2Cl_2 and increasing its residence time in the reactor, monochlorosilance-SiClH- is produced. SiClH is desirable for ALE growth of Si because when it adsorbs on a free Si surface, it produces H_2 and leaves a monolayer of Si-Cl, i.e., a perfectly Cl-terminated surface. Atomic hydrogen is again used to complete the cycle by desorption of HCl. Monolayer growth per cycle is achieved over a range of temperatures from 823-883 K. This growth technique may not lend itself to the production of electronic-quality material, since it relies on keeping the temperature high enough to desorb hydrogen from the surface, leaving free Si sites, which lead to the incorporation of contaminants.

At present, the exchange reaction useful for ALE growth technique for Si is not as clear as for III-V compounds for example, a hydrogen-terminated surface and a Si precursor can adsorb (without throwing off excess) Cl that will react with some of the hydrogen-terminated sites, blocking full monolayer coverage. Additionally, most of the current options require either a source of atomic hydrogen or a photodissociative step. The ALE of Si, and SiGe, can provide significant leverage for device fabrication. Raised source/drain fabricated without pattern-loading effects and controlled channel region of strained Si FETs are elements that are enable by ALE of Si and SiGe.

Atomic-layer deposition of silicon nitride is achieved by nitridization of Si alternated with Si monolayer deposition from $SiCl_2H_2$. Remote plasma nitridization is carried out with NH_3 as the nitrogen source, and layer growth occurs over a range of substrate temperatures from 523 K-673 K, for plasma powers greater than 40 W. The growth rate saturates at one-half monolayer per cycle. Films are used to create a stacked dielectric consisting of a 2-3-nm-thick thermal oxide followed by two monolayers of ALD SiN. The introduction of nitrogen in the gate dielectric lead to incorporation at the lower interface, where degradation of device performance results. The advantage is that nitrogen is incorporated away from the Si interface. Capacitors demonstrate good resistance to boron diffusion through the dielectric and low flat-band shift due to fixed charge, reduced tunneling currents, and good dielectric breakdown.

The atomic-layer deposition of SiO_2 at room temperature, through the use of catalyzed sequential surface reaction breaks up the $SiCl_4 + 2H_2O \rightarrow SiO_2 + 4\ HCl$ reaction into two half-reactions.

$$\text{(A) Si-OH* + } \rightarrow \text{SiO-Si-Cl}_3^* + \text{HCl} \quad (6)$$

and

$$\text{(B) SiCl* + H}_2\text{O} \rightarrow \text{Si-OH* + HCl} \quad (7)$$

where * represents the surface functional group. At room temperature and low pressures, the reaction rate between $SiCl_4$ and H_2O is negligible, but when the reaction is carried out in the presence of strong Lewis base such as NH_3, catalysis of the reaction at low temperatures occurs because of the interaction of the Lewis base with the surface functional groups. The NH_3 forms a salt that complexes with the HCl reaction products in both half-reactions. Multiple pulses are used for half-reaction (A) in order to remove the reaction byproducts. A number of individual SiCl4 pulses are provided in the presence of a of a 75-mm H_g pressure of NH_3. Half-reaction (B) is carried out with a direct mixture of H_2O and NH_3, which is allowed to react with the surface for several minutes and then removed from the chamber by means of a vacuum pump. At high enough H_2O exposure time, NH_3 pressure, and number of $SiCl_4$ pulses, the reaction saturates at one monolayer per –A-B- cycle at room temperature. The catalysis of surface reactions by gas-phase reagents represents a significant breakthrough that applied to other ALD and ALE systems.

The gate dielectric is the film with the most stringent thickness and uniform control. Thermal oxidation of Si delivers extremely good uniformity and reproducibility because of the nonlinear growth rate with time. The demise of thermal SiO_2 and oxynitride as gate dielectric materials is not due to tolerance; rather, the total film-thickness requirement approach a few monolayers, as the requirement for decreasing equivalent-oxide thickness is maintained.

Figure 10.7 is a plot of the number of monolayers of deposited film for different choices of gate dielectric and gate conductor in order to fulfill the gate-dielectric equivalent-oxide thickness.

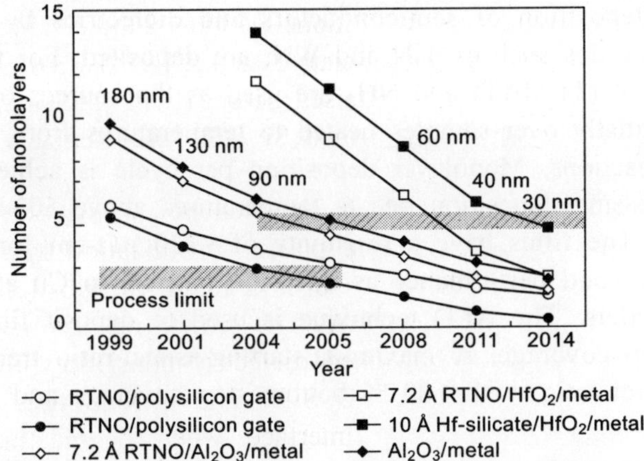

Fig. 10.7 *Equivalent number of monolayers of gate dielectric for several different gate dielectric and gate conductor combinations. The left-hand and right-hand crosshatched regions respectively indicate the practical limits for processing of single- and multiple-layer dielectric films, for dual-layer films, the thickness of the underlayer is considered fixed at a minimum value for the rapid thermal nitrided oxide (RTNO) or Hf-silicate films, respectively.*

The open circles are for an oxynitride film with a dielectric constant 1.3 times that of pure oxide and a polysilicon gate with 10% depletion of dopants. The filled circles are for the same materials set, but the gate activation degrades due to increased band bending at the thinner EOT. For either set the graph shows that polysilicon/oxynitride gates are suitable up to the 90-nm node, but not at the 60-nm node because films of less than two to three monolayers are required.

The open diamonds are for Al_2O_3 as the dielectric material, with a silicon-oxynitride film (as an underlayer) and a metal gate electrode. Since this is a bilayer dielectric, the minimum thickness is about five monolayers, (two to three monolayers of each film). For Al_2O_3, negative results are obtained when no interfacial film is used, but if Al_2O_3 is to be used, it must be done without the benefit of an underlayer. In that case, the filled diamonds in the graph can be compared against the 2-3-monolayer limit. On this basis, Al_2O_3 /metal gate stacks are useful until the 40- or even 30-nm nodes are reached.

The open squares represent the case of HfO_2 on silicon-oxynitride with a metal electrode. The filled squares represent the case of Hf silicate that is obtained by direct interaction of HfO_2 with the Si substrate. The curves show that while HfO_2 on oxynitride is suitable for the 60-nm node, continued scaling is requires the use of a silicate or other higher-dielectric-constant interlayer.

More advanced structures, such as replacement gates or horizontal double-gated devices, can be used for the deposition process to severe micro loading. The technique of ALD is ideally suited. The gate-dielectric deposition may not be used for implementing ALD techniques into high-performance CMOS. As an example, consider the deposition of Al_2O_3 by ALD. Monolayer control of the growth process is achieved by sequential self-limited adsorption of trimethylaluminum and H_2O at a substrate temperature of 573 K ALD films are utilized in fabrication of 80-nm FETs.

In addition to the deposition of semiconductors and dielectrics by ALE and ALD, metals such as W and metal nitrides such as TiN and WN, are deposited. For the case of TiN, tetrakis-ethylmethylamino-titanium (TEMAT) and NH_3 are used as the sources of Ti and N, respectively. They are passed sequentially over samples heated to temperatures from, with N_2 gas in between to prevent gas-phase reactions. Monolayer deposition per cycle is achieved for temperatures up to 493 K with an increasing deposition rate at temperatures above 503 K due to the prevalence of gas-phase reactions. The films have a resistivity of <230 μΩ-cm for deposition temperatures below 463 K and show good performance as diffusion barriers to Cu at temperatures up to 873 K using unpatterned wafers. The ALD technique is used to deposit films into high-aspect-ratio trenches to observe step coverage. A maximum-starting-aspect-ratio trench of 6:1 was filled to the point of >60:1 aspect ratio, with >80 % bottom step coverage and no signs of pinch-off of the trench.

As ALE and ALD are surface-sensitive, it is possible to conceive the reactions that proceed on some surfaces example, GaAs gets deposited selectively on patterned GaAs areas opened through a Ga_2O_3 mask. The condition in which there is no GaAs deposition on the masking material is a function of the hydrogen pressure in the reactor. AlGaAs, in contrast, deposits on the mask over the entire range of pressures. In another example, is high crystallographic selective observed during ALE of GaAs to grow quantum wires. At high growth temperatures and with long hydrogen purge times after exposure to AsH_3, no GaAs growth on the GaAs (111) A and (110) planes is observed, while GaAs growth occurs on the GaAs (100) plane.

Other concern about ALD is low growth rate and contamination. As the dimensions of structures are decreased, so are the thicknesses of deposited films required in their fabrication. As a result, the deposition time required for an ALD will depend on the needed film thickness.

Thermal or other assisted desorption is utilized to complete one or more of the half-reactions and leave an unterminated surface. Thus, deposition sequences are devised so that they always leave the growing surface terminated with a species that is not easily removed, expect by the second half-reaction. Suitable half-reaction and surface catalysts must be found to move ALD and ALE into the production line.

ATOMIC LAYER CONTROL ETCHING

Figure 10.8(a) is a cross- sectional SEM image of Si trench etching for shallow-trench isolation in a 180-nm-node. Wide trench profiles and depth are noted for different linewidths. The RIE depth for the center trench is about 90% of the depth of the outer trenches, which have a larger lateral dimension. The sidewall profile of the center trench is also less vertical than the sidewalls of the outer trenches. Figure 10.8(b) is a cross- sectional SEM of a contact hole adjacent to a local interconnect hole for 130-nm-node CMOS. The contact hole is etched to shallower depth because of the smaller area and uni-dimentionality of the feature compared to the local interconnect on the right. The sidewall profile of the contact hole is also less vertical than is the sidewalls of the interconnect. Reactive ion etching (RIE) is microloaded due to distortion of electric fields near the top and mask charging limitations. The solution for this is the migration to lower etching pressures. This may not be acceptable because of increased charging damage observed at lower etch pressures. Therefore microloading, anisotropy, and selectivity compete with one another for RIE processes. Another, approach is neutral-beam etching. A wide range of techniques are used to generate a neutral beam of molecular and radical etchant species with translational energies from 2 to 600 eV.

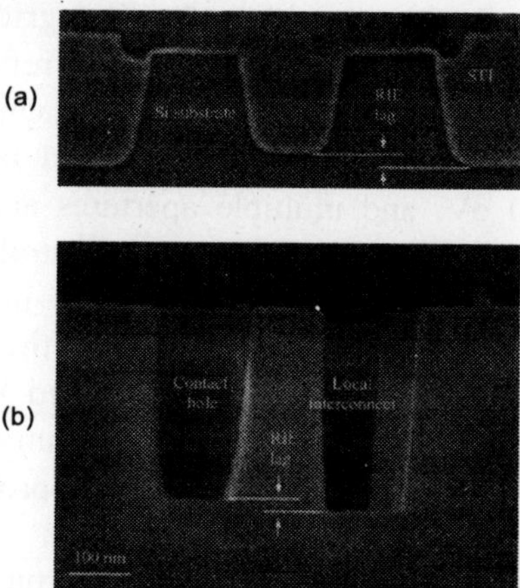

Fig. 10.8 *(a) Cross-sectional SEM image of the STI trench for 180-nm-technology-node CMOS.. (b) Cross-sectional SEM image of a contact hole adjacent to a local interconnect hole for 130-nm-technology-node CMOS.*

Anisotropic etching of Si and GaAs is performed by hyperthermal molecular and radical beams produced by heating etchant such as Cl_2 or SF_6 in resistively heated jets. The adiabatic expansion reduces the rotational temperature of the molecules and increases their translational energy. Jet temperatures from 1073K to nearly 3273K are used. Also used are larger-diameter nozzles, and the beam energy is generated thermally. As the nozzle diameter is reduced to 100 μm and the ratio of Cl_2 in He is reduced, Cl_2 molecular beams with translational energies up to 3 eV are produced with a nozzle temperature 1183 K. Beam divergence is small; therefore, anisotropic etching is possible, but the major challenge is etching of the sidewalls by scattered species. By reducing the substrate temperature, reaction with scattered molecules is reduced, and anisotropic etching by a Cl_2 hot molecular beam is achieved in the case of Si but scattered Cl radicals etch the sidewall. Both Cl_2 molecular and Cl radical beams produce anisotropic etching of GaAs structures because the reaction probability of Cl radicals with the GaAs is nearly unity. Thus, etch rates of RIE are achieved, although over beam areas are only about 1 cm^2. Scaling for full wafer-scale processing requires an array of such beams, combined with wafer rastering.

Higher translational energies are obtained by the use of a pulsed laser. This technique utilizes a 1-mm nozzle with a 125-psig source of SF_6 pulsed into a vacuum chamber with a precisely timed CO_2 laser pulse, producing a neutral beam consisting of F and S radicals. The F translational energy is tuned to 18 eV. Etching of Si substrate at room temperature with 4.8-eV, F radicals results in significant undercutting of masked regions. When the energy is raised to 18eV, undercutting is reduced, but microtrenching becomes more severe.

Still higher translational energies are used for etching by neutralizing energetic ions, generated by plasma sources. Energies for this approach range from tens of eV to KeV levels. Neutralization is accomplished by electrostatically deflecting charged species, through the use of multi- aperture electrodes at the beam source combined with reflection grids over the sample, by utilizing cusped magnetic fields to deflect ions and electrons, or by reflecting the beam off a surface.

SiO_2 is selectively etched to polysilicon, by utilizing a magneto-microwave source to generate ionized Ar and CHF_3 as well as ionized radicals produced by the dissociation of CHF_3. The ions are extracted at 400 to 600 eV, and multiple apertures are used to facilitate the transfer of the ion's charge but not its kinetic energy to thermal neutral species. In this system, etching occurs due to the kinetic energy of neutral beam of Ar impinging on the substrate where neutral radicals are adsorbed. The neutral radicals and the energetic neutral Ar beam are generated in two separated coaxial plasmas. This is termed neutral-beam-assisted etching. By this means, anisotropic SiO_2 etch rates of, more than 50 nm/min are obtained over a 200-mm wafer. Though uniformity is limited to about 8% by the production of a uniform beam, this technique can be scaled to large area with reasonable etch rates and reasonably good angular distribution within the beam. However, because of the apertures and reflector grids in the path of the beam, contamination is there.

A magnetic field is used to remove electrons and positive substrate bias, to repel ions and still obtain neutral-beam fluxes of 10-20eV species which are capable of Si etch rates of 10nm/ min. Lower beam densities of $2 \times 10^{14}/cm^2/s$ are obtained. Each such collision results in as much as 50% loss of the beam energy, but the benefit of reducing UV irradiation is true. It is believed that through modification of the source, beam densities upto $1 \times 10^{18}/cm^2/s$ can be achieved, enabling resist removal at rates exceeding 1 μm/min.

ATOMIC–LAYER ETCHING (ALET)

Neutral-beam etching, holds promise for avoiding some of the current challenges faced in the RIE of structures, but while etchant species are produced with sufficient low translational energies that provide low, controlled etch rates. But, etching by this technique does not proceed in an atomic-layer-by-layer. In contrast, atomic-layer etching (ALET), or digital etching, is demonstrated below for both GaAs and Si.

Etching of the (100) surface of GaAs at room temperature by ALET is achieved using a sequence consisting of a Cl_2 gas pulse, a purge cycle to remove excess gas-phase Cl_2 and bombardment with 100-eV electrons followed by another purge to remove reaction byproducts. The (100) face of GaAs consists of alternating planes of either Ga or As. Two cycles are required for the removal of one GaAs layer. The etch rate is independent of the Cl_2 exposure dose, with a value of about 0.1 nm per etch cycle, is lower than the value 0.14 nm per cycle expected for single-layer removal. Replacing the 100-eV Ar^+ ions produces a slightly different behavior. The etch rate increases with Cl_2 exposure dose up to a self-limiting value of about 0.2 nm per cycle. Above that Cl_2 dose, a longer incubation time for etching is required because sufficient Ar' irradiation is needed to remove the excess adsorbed Cl_2 before ClM_x etching occurs. Though this reaction is self-limiting, the etch rate of 0.2 mm per etch cycle is higher than for single –layer removed, and due to sputtering.

ALET of Si is achieved either by using Cl_2 gas adsorbed on the Si at room temperature, similar to the etching of GaAs or from the cryogenic adsorption of F atoms. For cryogenic etching, separate adsorption, reaction, and desorption steps are provided in a three- station reactor. First, wafers on an electrically floating sample holder are passed under a microwave discharge of $CF_4 + O_2$ which produces F. Temperatures lower than 213K are required in order to prevent spontaneous etching. The sample is then passed under a source of UV irradiation in order to enable the reaction of F with the Si surface. Finally, the sample is passed under an Ar+ irradiation time is allowed. However, etch rates for both lower or higher than the value for monolayer removal per cycle are obtained depending on the F-atom adsorption time, implying that the Ar+ beam is responsible for reaction as well as desorption when excess F is available. Anisotropically etched trenches with depths of the order of 200 nm produced by 1400 cycles of etching a patterned substrate. A substrate temperature of 113K is required because undercutting of the mask is observed at 213K. The sidewalls produced at are quite vertical.

ALET reactions with GaAs and Si do not leave the etched surfaces, with any terminating species (at the end of each etch cycle). The surface is open to reaction with other etchants in the reactor thereby it lead to more than one monolayer of etching per cycle; and, in the case of Si reaction with O_2 or H_2O in the reactor the etch is micro masked to further Si etching. The progression of ALE from ultrahigh-vacuum deposition systems which leaves no surface termination, at high pressures with hydrogen there is surface termination. Thus, reactions meeting these goals are possible to devise. Further, it is possible to combine the concepts of ALET with neutral beam processing to achieve new, more powerful processes. For example,

ALET which utilizes an Ar^+ ion beam or an electron beam should be able to replace it with Ar neutral. Greater gains come through the use of neutral beams to provide the correct sequence of radicals to accomplish a suitable layer-by-layer etch sequence.

In addition to ALET and neutral-beam etching, two processes are developed in Si technology that are self-limiting but do not employ strictly atomic-layer-by-layer removal mechanisms. One is for the etching of Si and the other of SiO_2 removal. In the first example, a process consisting of sequential ozone oxidation followed by ex situ aqueous HF etching is developed for the purpose of ultra shallow depth profiling with X-ray photon and Auger electron spectroscopy. The Si removal per cycle is determined by the oxidation of the Si surface by ozone exposure. Removal of 0.5 nm per cycle is achieved which shows that more than a monolayer is removed from the surface. Therefore for depth profiling, total film removals are limited to <10 nm. Also, since wet etching is used, this process is non-anisotropic. Consider a self-limiting reaction for removing SiO_2. HF and NH_3 are reacted with oxide at room temperature in vacuum to form ammonium hexafluorosilicate, $(NH_4)_2$ SiF_6, which remains on the surface and serves as a diffusion barrier to further reaction of the HF with the SiO_2 surface. After the reaction is stopped, the reaction byproduct is removed by thermal desorption at 373k or by dissolution in H_2O. Thus by this results in the reaction approximately 12-nm SiO_2 is removed. If removal of thicker films is desired, repeated sequencing is to be carried out, or the reaction is carried out at heigher temperatures to produce a thinner byproduct layer. Another very interesting feature of this reaction sequence comes from the fact that the byproduct undergoes a volume expansion of factor of 3 with respect to the SiO_2 film removed. The byproduct fills in cracks, scratches, or grooves in SiO_2 during etching, thereby reducing the reaction rate and thus reducing surface roughness.

SUBSTRATE ENGINEERING-NEW MATERIALS

A move away from SiO_2-based dielectrics as dictated by gate dielectric scaling is a move away from one of the fundamental advantages of Si-based CMOS over other materials systems—the Si/SiO_2 interface. That may warrant a re-evaluation of the materials set for future devices. The continuation of current scaling trends, require the semiconductor industry to develop a new gate dielectric material so that physical gate lengths in the vicinity of 20 nm are produced.

Below (Table 1) is a comparison of some mechanical and electrical properties of a selected group of semiconductor materials.

Table 1 *Mechanical and electrical properties of some semiconductor materials.*

	Si	*Ge*	*GaAs*	*InP*
T_m (*C)	1415	937	1238	1062
Lattice constant (A)	5,4309	5,6461	5,6532	5,8687
Thermal expansion coefficient (10^{-6}/ K)	2.5	6.1	5.4	4.6
Microhardness (Nmm^2)	11270	7644	7500	4100
Mobility at 298 K (cm^2/ V-s)	n – 1500	n – 3800	n – 8800	n – 4600
	p – 450	p – 400	p – 400	p – 150
E_g (eV)	1.1	0.67	1.35	1.27

Because the mobilities of materials other than Si are attractive, consider some of the alternatives to Si. The industry is faced with a choice: direct fabrication of substrates having diameters of 300 mm or greater for new materials, or epitaxial growth of a new material on Si substrates. The requirements for a large wafer size will be set by the level of integration already achievable in Si technology. Direct growth of other semiconductors from the melt are carried out, but wafer sizes are limited, and material quality is poor because of the loss of column V elements below the melting point for the III-V compounds. As indicated in Table 1, none of these materials has a hardness value as high as that of Si, so handling them is more difficult. So, the introduction of new materials will utilize the Si substrate, but heteroepitaxy of these materials is difficult because of the large lattice mismatch and the differences in thermal expansion coefficients.

Thus, the homoepitaxial growth of active layers of these materials on wafers of the right size, (at epitaxial growth temperatures, lower than the melt), the equilibrium concentration of vacancies is greatly reduced. The film is then transferred to a wafer of another material such as Si, using bonding and etchback. Point defects are reduced by this means; but extended defects are transferred into the active device layer. *limits of heteroepitaxial growth* strain relaxation.

For heteroepitaxial growth of a film the epitaxial layer grows with perfect crystalline alignment, but the grown crystal lattice is teragonally distorted. Elastic strain will not result in the generation of dislocations until a critical layer thickness is exceeded. There are two well-known models for the prediction of critical layer thickness. One is the bending force required to bend threading dislocations in the plane of the interface. The other the self-energy of creation of a dislocation. In both models, the force or energy is provided by the mismatch strain in the system, which is proportional to the natural log of the layer thickness. As the energy for creation of dislocations is greater than that required to move pre-existing threading dislocations, the resulting critical layer thickness is slighty greater than predicted by the dislocation movement model. Films with a thickness above the dislocation motion model but below the dislocation creation model are metastable, because any imperfection leads to the relaxation of the layer. The substrate is, in effect, of infinite thickness. Ideal misfit dislocations are pure edge dislocations at the growth interface. However, in both Si and GaAs systems,60° dislocations are common and result in the relief of mismatch and a threading component into the epitaxial film. The aerial density of misfit dislocations, N_m, required to fully relax a heteroepitaxial film on the basis of the relaxed lattice constants of the films being considered, a_1, a_2 is given by:

$$N_m \approx 4/(1/a_1^2 - 1/a_2^2), \tag{8}$$

which gives about 2×10^{12} dislocaitions/cm^2 for a mismatch of 1×10^{-3}. For system with a few percent mismatch, such as SiGe on Si, relaxed buffer layers are grown with dislocation densities below 1×10^5 cm^{-2}. Achieving these dislocation densities, which are still too high for most device applications, require film thicknesses greater than a micron.

Another concept is the growth of a 200-nm-thick. In GaAs layer on 80-nm-thick GaAs. This growth thickness is about double the critical layer thickness. X-ray diffraction show that, the samples grown on GaAs did not have misfit dislocations in the epitaxial film. Cross-sectional TEM reveal that dislocations are pulled down into the substrate film. For substrates thinner than the epitaxial film, the image force pulls dislocations into the substrate.

The relaxation of layers of $Si_{0.85}Ge_{0.15}$ grown on thin SOI that exceeded the classical equilibrium critical layer thickness by a significant amount, but not the critical layer thickness are examined. A SOI film is first thinned to 50 nm, and 180 nm of SiGe is grown at 773K. The layers are subsequently annealed at temperatures from 973-1323K.X-ray diffraction show that the layer are fully attained after growth, with no dislocations detected, but are partially relaxed by annealing at 973K. TEM analysis show that the misfit dislocations that are present at the Si-SiGe interface after annealing thread down into the Si substrate rather than up into the epitaxial layer, with a density exceeding 1×10^7 cm^{-2}, while no dislocations within the detection limit are detected in the SiGe. Further annealing at 1173K has no effect on the strain or dislocations, indicating that the film is relaxed to the equilibrium condition, where the strain can no longer push threading dislocations. The growth and annealing temperatures are too low to allow the tensile strain in the SOI to be accommodated by viscous reflow of the buried oxide, thus showing plastic deformation at the Si/SiO_2 interface.

Also, 1 μm of $Si_{0.6}$ $Ge_{0.4}$ is grown at 773K on thin SOI. It is "relaxed" at growth, without requiring post-growth annealing. The TEM_S showed a dense array of dislocations in the SOI film with none threading into the SiGe layer, whereas in a sample of the same thickness and composition grown on bulk Si > 2×10^{11} dislocations cm^{-2} are there in the SiGe. Thus, for mismatched growth on thin SOI, the dislocations are directed downward toward the substrate by the image force, leading to misfit accommodation even at low growth temperature.

By adding boron to the buried oxide of SOI wafers, viscous reflow is achieved at growth temperatures. SiGe epitaxial films on such SOI substrates are from 38% to 64% relaxed. Relaxation is a function of boron content in the buried oxide (BOX). Further relaxation, up to 95%, is obtained by post-growth annealing at 1273 K. Most of the relaxation is accommodated by threading of misfit dislocations downward into the substrate. Misfit dislocation densities as low as 1×10^3 cm^{-2} are observed in partially relaxed films. This approach demonstrates that strain in the epitaxial film can be relaxed beyond the lower limit for motion of misfit dislocations. So there are at least two classes of "compliant" substrates; the first relieves the mismatch strain in the substrate (utilizing the motion of dislocations). Since finite strain is required for dislocations to relieve strain; it provides a mechanism for complete relaxation. Viscous bonding interfaces are used in the GaAs system, where an In/Ga bond and a glass diffusion barrier are used to bond a thin GaAs substrate to a GaAs wafer. Thin (10-nm) GaAs substrates are used to bond a wafer with thin borosilicate glass to grow strained $Ga_{0.91}$ $In_{0.09}$ As films having thickness five times greater than the critical layer thickness; is carried out by ogranometallic CVD(OMCVD) at 973K. Here, there is a reduction in misfit dislocations threading up into the epitaxial layer as compared to growth on bulk GaAs; very little strain relaxation occurs. However, these films are not subject to post-growth anneals.

SiGe layer are also grown on porous silicon substrates. Substrates consist of a two-layer porous structure with higher porosity at the bottom followed by a lower-porosity film beneath an epitaxially grown thin Si buffer layer are used. $Si_{0.8}$-$Ge_{0.2}$ films grown on these substrates do not have the regular array of misfit dislocations.

$In_{0.35}$ $Ga_{0.65P}$ are grown to a thickness exceeding critical layer thickness on GaAs substrates by a factor of 30. This is achieved by utilizing a GaAs substrate of 10-nm thickness bonded to thick GaAs bulk substrate with a >10° twist angle between the (110) directions while keeping the surface normals parallel. This twist angle introduces a dense square array of screw dislocations confined to the substrate with a spacing (d) given by

$$d = |b|/2 \sin(\theta/2) \approx |b|/\theta \tag{10}$$

for small θ, where (b) is the Burgers vector and θ is the twist angle. This results in less than 2-nm dislocation spacings in GaAs for twist angles of > 10°. OMCVD of 300-nm InGaP films at 913 K on θ = 17° twist-bonded substrates and bulk substrates show threading dislocations, while growth in the twist-bonded substrates show no dislocations, despite exceeding the critical layer thickness by more than 30 times.

The growth of InSb with a 14.7% lattice mismatch on similar substrates is observed. This twist angle is increased to 40°, and the thickness of the bonded substrate is reduced to <2 nm in order to accommodate the increased strain. A 650-nm film of InSb is grown by MBE on bulk GaAs and twist-bonded substrates. XTEM analysis show dislocation densities exceeding $1 \times 10^{11} cm^{-2}$ on bulk substrates, whereas no dislocations are present for growth on twist-bonded substrates.

While a significant reduction in the misfit dislocation density threading into heteroepitaxially grown films is achieved for a wide range of materials systems, the mechanism for long-range accommodation of the mismatch into the substrate layer is not well understood. Though TEM observations confirm limited misfit relaxation, yet characterization of the degree of relaxation in the heteroepitaxially grown layers is lacking. It is claimed that fully relaxed films are grown by "free slipping" at the substrate interface; It is likely that in some of the systems like strain is present in the heteroepitaxial film, and misfits occurred in the substrate.

In spite of the uncertainties in the physical mechanisms these new techniques provide a significant reduction in the generation and propagation of dislocations in heteroepitaxial systems. They also provide a pathway for production of superlatives thereby modifying the electronic transport and resulting in new "designer" materials for high-performance ultralarge-scale integration.

The TEM images in Figure 10.9(a) and (b) show a SOI wafer fabricated by oxygen implantation and annealing (SIMOX), before and after the annealing step. The large number of defects that ate introduced by high-dose oxygen implantation are almost completely

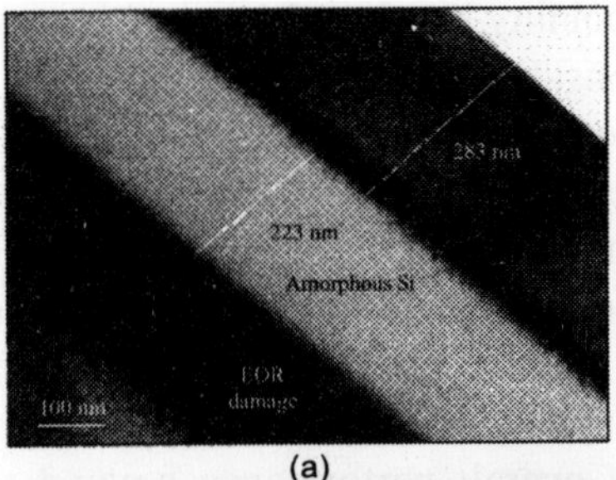

(a)

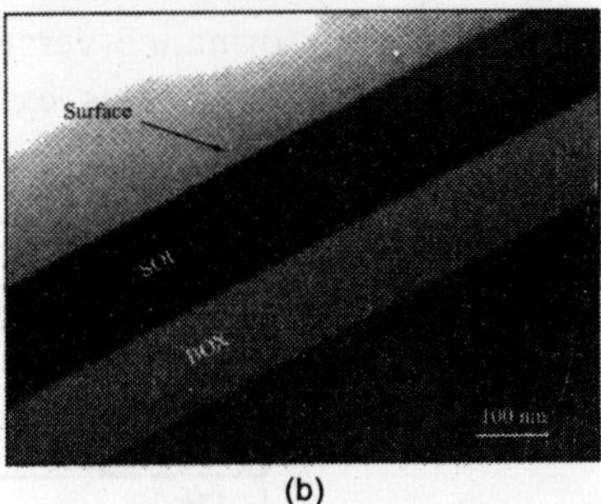

(b)

Fig. 10.9 *(a) XTEM image of an SOI wafer in the process of being fabricated. Bulks Si substrate received a high-dose oxygen implant, resulting in an amorphous region and end-of-range damage. Crystal-line defects above the amorphous region are observed. (b) XTEM image of completed SOI substrate, formed by annealing a wafer processed as in part (a) under optimal conditions to coalesce the oxygen in the substrate into a buried SiO_2 layer (BOX) and repair the crystalline damage.*

removed by the annealing step. With none present in the field of view of the TEM image. The defect density is not zero; however, the few defects that remain are benign with respect to device performance and yield.

OPTIMIZING CMOS

Since power dissipation is becoming a dominant limitation on the continued improvement of CMOS technology, the best way to design transistors in the presence of power constraints should be done, to obtain maximum performance for a fixed amount of power. It is chip performance, not device performance, that matters. By using, simplified models, the basic elements for determining chip performance, including intrinsic transistor characteristics, circuit delay, tolerance issues, basic microprocessor composition, and dissipation and heat removal considerations are achieved.

The optimal energy consumption conditions, including high-k gate insulators, metal gates, high-mobility semiconductors, improved heat removal, and the use of multiple layers of circuitry are considered.

When device dimensions are large and threshold voltages are high (at the left side of curves), dissipation caused by leakage currents are low. The shrinking of dimensions reduces capacitance and increases performance at fixed power. However, device scaling very leads to increasing leakage current, due to quantum-mechanical tunneling and subthreshold current. Figure 10.10. Show that when the total power is constrained, leakage dissipation dominates the power consumption. This leaves very little power left over for active circuit switching. So there is a loss of overall performance beyond some point in the scaling process.). The height and position of the maximum performance-versus-scaling curve depends on the maximum on the power constraint, but a maximum does exist.

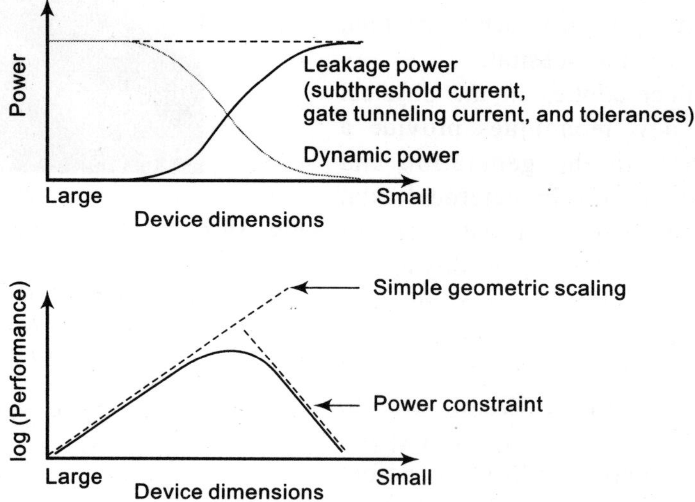

Fig. 10.10 *Schematic illustration of the existence of an optimal device miniaturization with maximal processing performance.*

Optimization Results

Figure 10.11 shows the detailed results of optimizations for the 90-nm to 32-nm-nodes, using the node characteristics shown in Table 2. These optimizations are performed for a dual-core processor chip with aggressive air cooling. Seven variables optimized are: gate length, oxide thickness, halo doping, mean width, mean repeater spacing, mean repeater width, and V_{DD}. The peak in performance versus power in Figure 10.11(a) occurs because the heat-sink technology is fixed and the temperature rise is constrained. The peak corresponds to the power at which the maximum temperature is first reached, as is shown in the constant temperature contours in the figure. In this case, the maximum temperature rise is 333K. The only way to increase power further, without increasing temperature, is to spread the chip out. This lengthens wires and slows down the chip even though the power level is higher. Low-power designs require larger, less scaled devices [Figures 10.11, 10.12] in order to reduce leakage currents, indicating that only the highest-power applications can utilize extremely scaled devices. Figure 10.12 shows the optimal allocation of power dissipation among the various mechanisms for a processor using water cooling (which allows higher power dissipation), from which gate leakage dissipation should not exceed a few percent, but optimal subthreshold leakage can exceed 50% for very high-power designs. It is found that the gate lengths [e.g., Figure 10.11(b)] agree reasonably well at the power levels, but supply and threshold voltages tend to be lower rising only slowly as the lithographic dimension increase. Two main reasons exist for this voltage discrepancy: (1) process variations (which are not included would slightly increase our optimized voltages; and (2) voltages used are not optimal. As the supply voltages are determined by external considerations, there is much resistance to lowering voltages. Furthermore, standby power should be quite low (unlike the optimized results in Figure 10.12), which requires higher V_T, and commensurately higher $V_{DD.}$

Table 2 *Valuesfor various constant model parameters.*

Description	*Symbol*	*Value*
Activity factor over logic depth	α_s/ℓ_D	0.012
I-V curve power law	α	1,462
I-V formula gate-length exponent	β	0.405
I-V formula mobility exponent	s	0.430
Mobility calibration parameter	μ_0	132.3 cm^2/V_{-s}
Critical field	E_C	2.5×10^4V/cm
Halo exponent 1	n_1	–0.574
Halo exponent 2	n_2	2.18
Maximum logic depth	ℓ_{Dmax}	10
Number of logic stages in typical instruction	n_{LI}	60
Latency penalty weighing factor	γ	0.1

Consider future technology options. Metal gates are simulated by removing the poly- Si depletion effects and adjusting the work function. High-K fate stacks are simulated using a double-layer bandgap-dependent tunneling model. Figure 10.13 shows high-*k* combined with metal- gate potentially yield excellent chip-level performance enhancement, as is also seen in

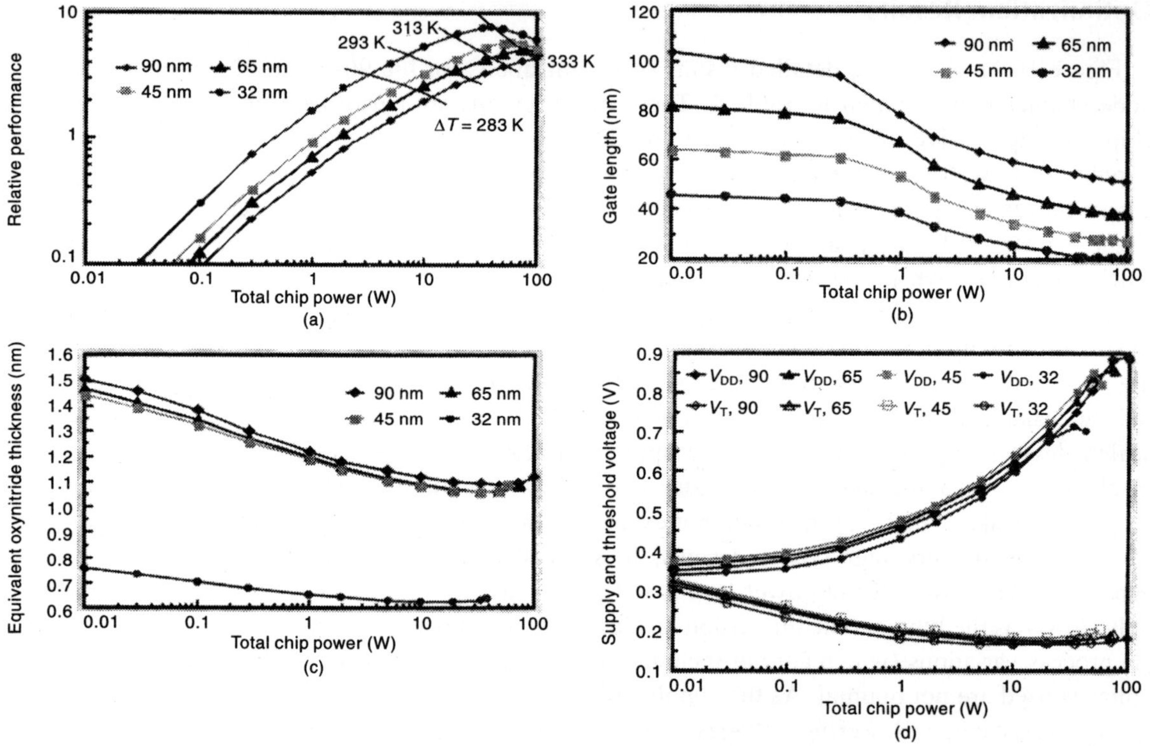

Fig. 10.11 *(a) Optimized equivalent versus power for 90-nm to32-nm nodes. Junction temperature rise is indicated by the added contours. (b) Optimum gate lengths corresponding to part (a). (c) Optimum equivalent oxynitride (gate insulator) thickness corresponding to part (a). The 32-nm- node case a high-k , material. (d) Optimum supply and threshold voltage (V_{TSAT}) corresponding to part (a).*

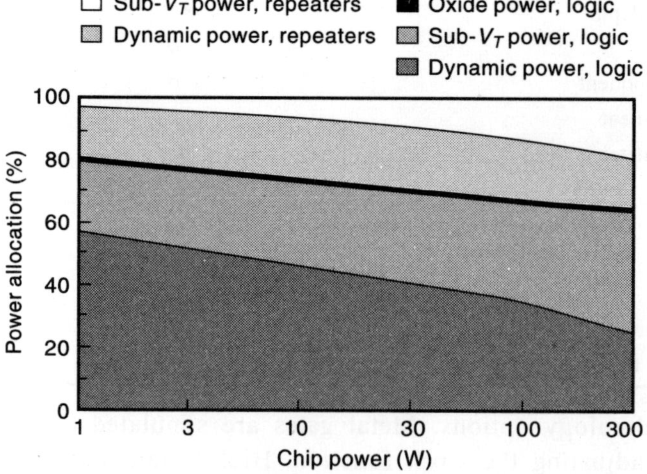

Fig. 10.12 *Cumulative power allocation fraction for the logic in processor cores, based on 45-nm-node optimizations. (At less than 0.5%, the graph segment for oxide power in repeaters is too small to be visible.)*

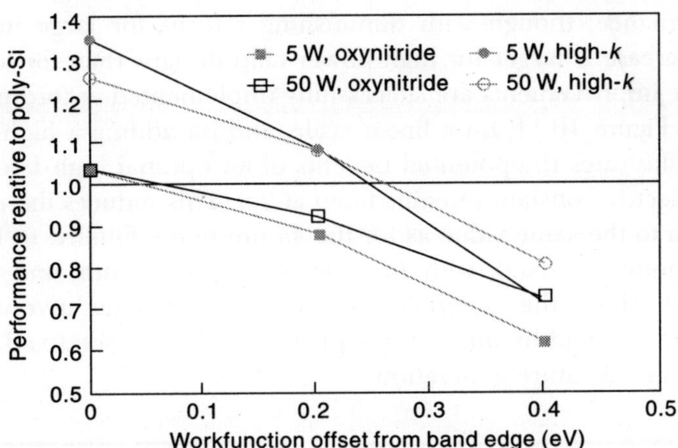

Fig. 10.13 *Relative chip performance versus workfunction offset from the band edge, for oxynitride and high-k gate dielectrics at the 45-nm node, and for two different chip power levels (5W and 50 W). Workfunction offset values are measured from the Si band edge toward midgap, so that 0 is equivalent to the poly-Si offset. Performance is relative to a poly-Si gate, oxynitride case.*

the 32-nmnode in Figure 10.11, but metal gates by themselves do not offer much benefit over poly-Si, even for work functions that are equivalent to poly-Si. As the work functions shifts from band edge toward midgap, a significant loss of performance occurs for both metal-gate. This loss occurs because the optimizations compensate for midgap workfunctions by lower doping (which increases depletion depth) and by raising the supply voltage (which necessitates thicker oxide). According to these optimization, the benefits of the use of high-*k* are lost for PD-SOI by the time the workfunction reaches quarter-gap.

Many future technology options involve increasing mobility, such as the use of strain, hybrid-orientation substrates, and SiGe layers. Figure 10.14 shows that mobility increases can

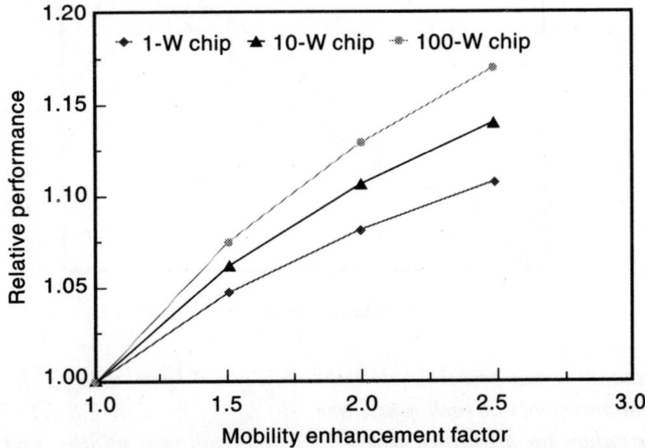

Fig. 10.14 *Relative chip performance increase versus mobility enhancement for the 45-nm-node, for three different chip power targets. Mobility is normalized to its value in unstrained FETs.*

increase chip performance, though with diminishing returns for large increases in mobility, This performance increase is larger for high-power chip designs than for low-power designs.

Suppose that some improvements are successfully implemented according to Table 2. This is the same data as for Figure 10.11, on a linear scale, and on adding a high-k option for the 45-nm node, it clearly illustrates the potential benefits of an optimal high-k solution. Figure 10.15 shows the wiring dielectric constant remains fixed at 2.8. This reduces the peak performance of the 32-nm generation to the same value as for the 45-nm node. Finally, in Figure 10.15 assume that the device technology is fixed, with L_G = 36 nm, t_{ox} = 1.1 nm, and the wiring dielectric constant k_{wiring} = 2.8. Here, the generation-to-generation changes involve only the packing density, as driven by the widths and wiring pitch, which are not fixed. A significant peak performance loss occurs in future generations.

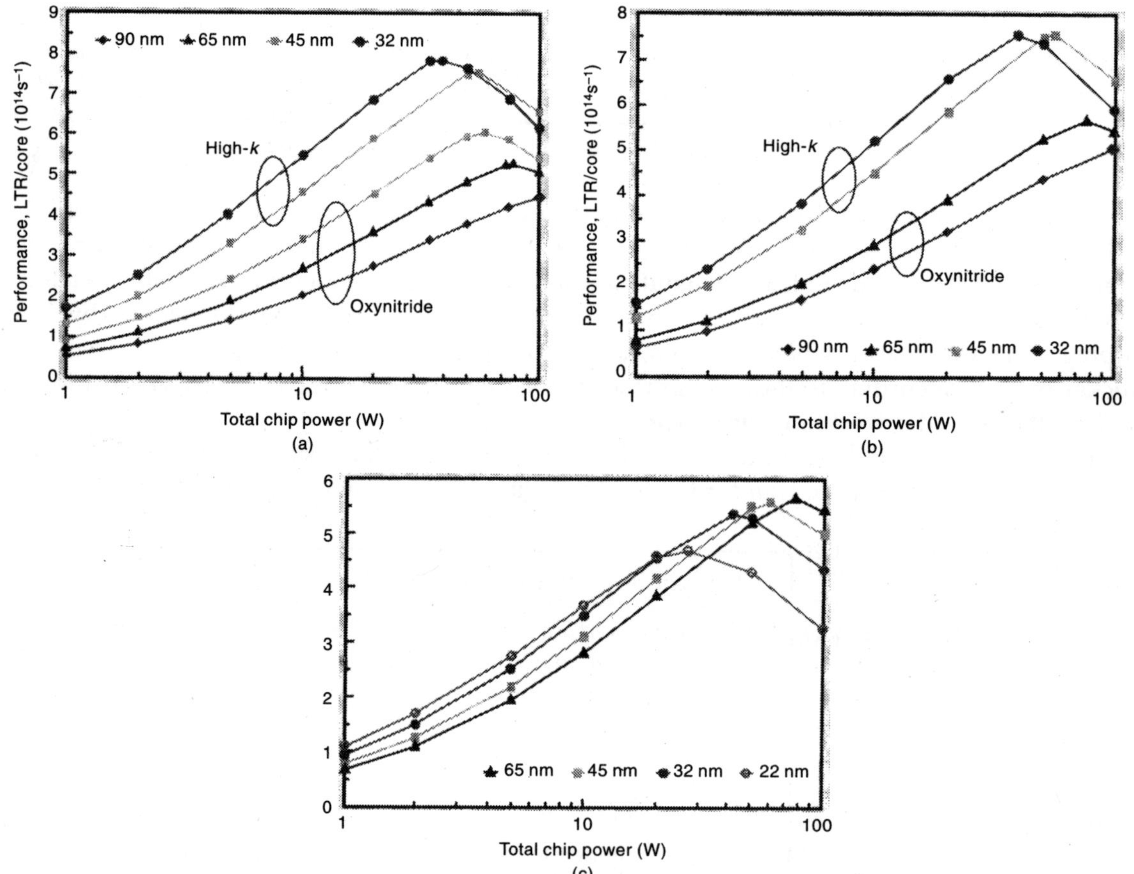

Fig. 10.15 *Chip performance versus power across technology generations, using different assumptions: (a) Full technology enhancements at each generation. (b) k_{wiring} fixed at 2.8 (c) L_G, t_{ox}, and k_{wiring} fixed at 45-nm-node values for all future generations; only voltages, widths, and wire sizes vary. Assumes dual-core processor with aggressive air cooling. (LTR: logic net transition rate.)*

In Figure 10.16, the optimizer is used to assess and compare the potential performance gain. In this figure, the "base" case is the "baseline" 65-nm-node technology to which the other cases are compared. Each point plotted is the peak performance that is possible the given heat sink and the specified temperature rise, corresponding to the highest points on curves points on curves similar to those Figure 10.11. Because the power level at which the peak occurs varies, the data points are somewhat scattered along the *x*-axis. Among the options compared, the following appear to be effective for improving performance: reducing the wiring permittivity by 0.64x, using 3D integration with two layers of active circuitry, and turning off the supply to inactive logic. Reducing variability by 0.7x, while simple making the wiring smaller is not beneficial. Overall, technology changes that truly lower the switching energy appear useful, while changes that only make devices faster or denser at the same switching energy are not valuable when the circuits are power-limited. Because power is the controlling factor, improved heat removal is also quite effective. Note that maximum performance should is achieved by implementing all of the favorable changes simultaneously.

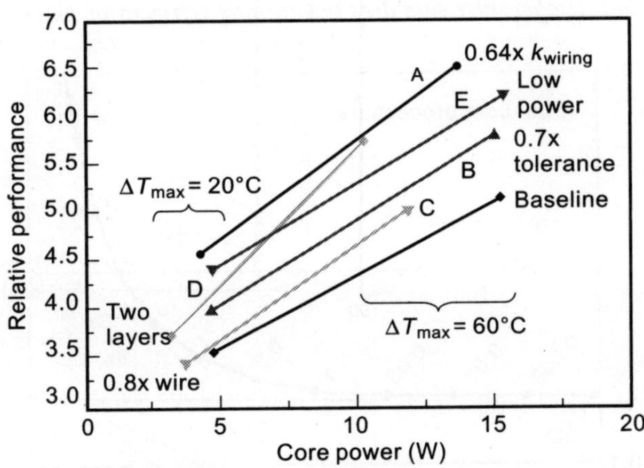

Fig. 10.16 *Peak performance versus power for aggressive air cooling and two different maximum allowed temperature rises, comparing various technology options (different lines) with the 65-nm-node technology baseline. Options compared are improvements in k_{wiring} (A), tolerance (B), and wire size (as indicated on plot) (C), 3D integration using two layers of active circuitry (D), and use of low-power circuit techniques to eliminate two thirds of passive power by turning off inactive circuit blocks (E).*

Thus, improved heat-sink technology offers a direct path to larger performance gains, as shown in Figure 10.17. Also there is a modest performance increase by decreasing the heat-sink temperature, but this performance gain disappears if the refrigerator power is taken into account. Thermal solutions are difficult because such high-power processors are very inefficient, and performance is only increasing as roughly the log of the power. Note that is not actually be possible to deliver high power levels to the chip.

This efficiency challenge Figure 10.18, shows average energy dissipated per logic transition (total logic power divided by LTR) versus overall performance, for optimizations that cores technology generations (including the wire pitch and the halo behavior among the optimized

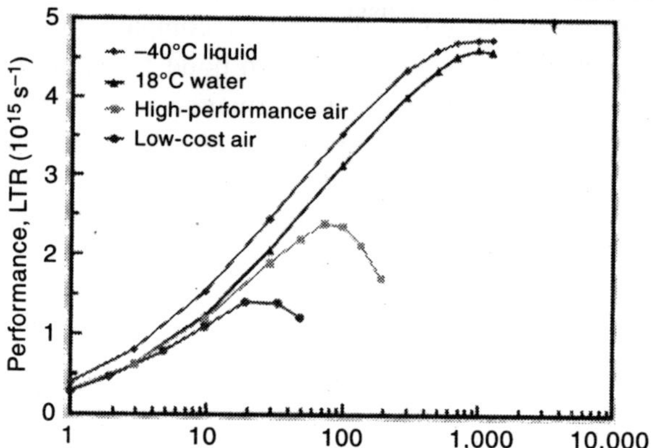

Fig. 10.17 *Dependence of total optimized chip performance on cooling technology, for 45-nm-node four-core processor. The four cooling cases are very low-cost air cooling, high-performance air cooling, chilled water cooling through a microchannel heat sink, and 233 K liquid cooling through a microchannel heat sink. The low-temperature case does not include refrigerator power.*

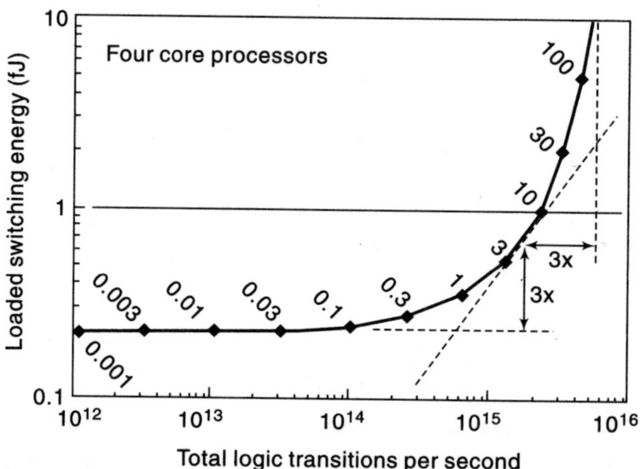

Fig. 10.18 *Average loaded switching energy versus performance for cross-generational optimization (involving nine parameters). Number label on each point gives total chip power for that point.*

variables); yielding a total of nine variables being optimized. These optimizations are considered for four-core processor chips. The very high-power designs are quite energy-inefficient, on a logic transition basis, compared to what is possible at lower power. The knee in this curve is very interesting, because it turns out to be within a factor of ~3 from both the lowest-energy designs and the highest-performance designs.

One way to address the energy inefficiency of the high-power design points involves the use of smaller, lower-power cores in parallel. Figure 10.19 shows a fixed number of transistors is divided into the higher numbers of cores; have higher performance because the smaller cores result in relatively less wiring capacitance due to the shorter wires. This basic performance increase must of course be adjusted for architecture and system effects associated with increased parallelism.

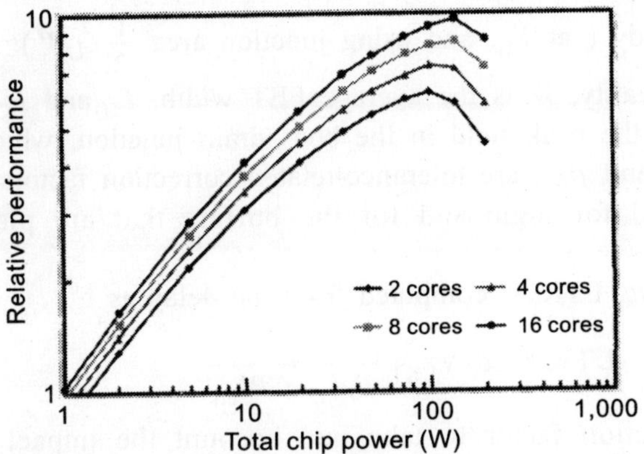

Fig. 10.19 *Dependence of performance on power for different numbers of cores for a 32-nm-technology node.*

Only the logic devices in the processor core are actually optimized; it is assumed that the power and speed of the clock and latch circuits, registers, memories, and I/O can ultimately all be optimized with essentially the same power/performance result as the logic part of the processor. Use of actual chip data, to set the processor core complexity, and use of the fraction of the chip area and power that is used for logic is made. The optimization controls the actual size of the chip, and hence the power density, by adjusting the device and wire sizes.

Because memory actually occupies the majority of the chip in modern processors, it may seem unreasonable not to include it in the optimizations.

The basic optimization methodology starts with definitions for power and delay as functions of the underlying technology parameters. In an inner programming loop, one degree of freedom (usually the supply voltage, V_{DD}) is used to satisfy the power constraint, and then in the outer loop, the remaining variables are optimized to find the maximum possible performance.

The total power (P_{TOT}) calculation includes dynamic switching power (P_{DYN}), power due to subthreshold leakage current (P_{subVT}), power due to gate oxide tunneling current (P_{ox}), and power due to drain-to-body tunneling current (P_{B2B}), as defined in the following equations:

$$P_{\mathrm{TOT}} = (P_{\mathrm{DYN}} + P_{\mathrm{subVT}} + P_{\mathrm{OX}} + P_{\mathrm{B2B}})_{\mathrm{logic}} + (P_{\mathrm{DYN}} + P_{\mathrm{subVT}} + P_{\mathrm{OX}} + P_{\mathrm{B2B}})_{\mathrm{repeaters}}, \quad (1)$$

$$P_{\mathrm{DYN}} = \alpha_S N_{\mathrm{CKT}} \frac{1}{2} \langle C \rangle (V_H - V_L) V_{\mathrm{DD}} \frac{1}{\beta_\tau \ell_D \tau} \quad (2)$$

$$P_{\mathrm{subVT}} = 1.7\ \beta_{I\mathrm{off}} \cdot N_{\mathrm{CKT}} \cdot V_{\mathrm{DD}} \cdot W \cdot J_{\mathrm{off}}(V_T, V_{\mathrm{DD}}, t_{\mathrm{ox}}, \eta, L_{\mathrm{CH}}), \quad (3)$$

$$P_{\mathrm{OX}} = N_{\mathrm{CKT}} \cdot FI \cdot V_{\mathrm{DD}} \cdot L_G \cdot W \cdot J_{\mathrm{ox}}(V_T, V_{\mathrm{DD}}, t_{\mathrm{ox}}, \eta), \quad (4)$$

$$P_{\mathrm{B2B}} = N_{\mathrm{CKT}} \cdot FI \cdot V_{\mathrm{DD}} \cdot \frac{L_G}{2} \cdot W \cdot J_{\mathrm{B2B}}(F_{\max}, V_{\mathrm{DD}}) \quad (5)$$

Where V_{DD}, V_{H}, V_{L}, and V_{T} are the supply voltage, high logic-level, and threshold voltage, respectively, and τ is the average switching delay of a loaded logic stage (a NAND gate, with average fan-in, *FI*, usually set to 2 and average fan-out of 1.65) N_{CKT} is the number of logic gates, $\langle C \rangle$ is the average total load capacitance, α_s is the switching activity factor, ℓ_{D} is the logic depth, J_{off} is off-current density (at V_L) of a typical logic FET, J_{ox} is the oxide tunneling current density from drain to body (at V_H, and using junction area $\frac{1}{2}L_G W$). t_{ox} is the oxide thickness, η is the subthreshold ideality, W is the average FET width, L_G and L_{CH} are the gate length and channel length, F_{MAX} is the peak field in the body-drain junction, which depends on the doping and the voltage, and β_τ and β_{Ioff} are tolerance-related correction factors. The power contributions are separately computed for logic and for the buffers that are placed in long wires (i.e., repeaters).

The *performance metric*, LTR, is computed from the delay as

$$\mathrm{LTR} = \alpha_S N_{\mathrm{CKT}} \frac{1}{\beta_\tau \ell_D \tau CPI_{\mathrm{eff}}} \tag{6}$$

where CPI_{eff} is a correction factor to take into account the impact of long wires and their repeaters. The loaded logic delay computation proceeds in steps. First, the basic device delay is computed using a modified *CV/I* form to by taking into account output conductance effects.

$$\tau_1 = \frac{V_{\mathrm{DD}}(C_{\mathrm{parasitic}} + C_{\mathrm{wire}} + C_{\mathrm{gateload}})}{2I_{\mathrm{eff}}} \tag{7}$$

where $I_{\mathrm{eff}} = \frac{1}{2}\left[I_{\mathrm{DS}}\left(V_{\mathrm{DD}}, \frac{1}{2}V_{\mathrm{DD}}\right) + I_{\mathrm{DS}}\left(\frac{1}{2}V_{\mathrm{DD}}, V_{\mathrm{DD}}\right)\right]$, the C_S are average capacitances, as indicated by their subscripts, and $I_{\mathrm{DS}}(V_{\mathrm{DS}}, V_{\mathrm{GS}})$ is drain current as a function of drain and gate voltages, which is described in the next section. Next, the wire *RC* and time-of-flight delays are computed, and combined using an empirical formula:

$$\tau_2 = R_{\mathrm{wire}}\left(\frac{1}{2}C_{\mathrm{wire}} + C_{\mathrm{gateload}}\right), \tag{8}$$

$$\tau_3 = L_{\mathrm{wire}}/(c/2), \tag{9}$$

$$\tau_4 = (\tau_2^{4/3} + \tau_3^{4/3})^{3/4} \tag{10}$$

where R_{wire} is the temperature-dependent wire resistance, L_{wire} is the average wire length, and c is the speed of light. Finally, these delays are combined and divided by a rise-time correction factor (due to Sakurai and Newton):

$$\tau = \frac{\tau_1 + \tau_4}{0.5 + (1 - V_T / V_{\mathrm{DD}})/(1 + \alpha)} \tag{11}$$

where α is the power-law exponent.

Organic-inorganic Electronics

For the past many years inorganic and gallium arsenide semiconductors, silicon dioxide insulators, and metals such as aluminum and copper have been the backbone of semiconductor industry. However, effort is made in "organic electronics" to improve the semiconducting, conducting, and light-emitting properties of organic (polymers, oligomers through novel synthesis and self-assembly techniques.

Organic–inorganic hybrid materials enable the integration of useful organic inorganic characteristics within a single molecular-scale composite. Unique electronic and optical properties are observed.

That innovative organic materials are essential to the performance increase in semiconductors, storage, and displays. However, the majority these organic materials are either used as photoresists or insulators. They do not conduct current to act as switches or wires and they do not emit light.

For semiconductors, two major classes of organic materials have made possible logic chips; phtoresists and insulators. Photoresists are the key materials that define chip circuitry and enable the constant shrinking of device dimensions. Available, photoresist materials limited the obtainable resolution of the optical tools to ~5.0 μm (~500 transistors cm^{-2}). But, owing to unique lens design and light sources, new resists are developed for lithographic scaling. The increased resolution capability of photoresists combined with optical tool enhancements has enabled the fabrication of 1.2 million transistors cm^{-2} with feature size of 180 nm, significantly smaller than the 248-nm exposure wavelength.

Polymeric insulators are essential to the performance and reliability of semiconductor devices. They are used in the packaging of semiconductor chips, where low-cost epoxy materials found applications as insulation for wiring in the fabrication of printed wiring boards and as encapsulants to provide support/protection and hence reliability for the chips. Although the polymeric dielectrics are used in the packaging of chips, but a polymer that replaces the silicon dioxide dielectric typically used on-chip throughout the industry as an insulator is there. The seven levels of metal wiring required to connect the millions of transistors on a chip effect chip performance because of signal propagation delay and crosstalk between wiring.

Improvement in interconnect performance requires reduction of the resistance (R) and capacitance (C) copper is replaced for aluminum wiring because it is a low-resistivity metal, Also as a low-k polymeric material, SiLK is used for oxide insulators, thereby improving the total interconnect wiring performance by ~37%.

Organic materials have provided performance and reliability for storage products and displays. The density of magnetic storage has increased at a faster rate than even semiconductor devices. Innovations in lubricants, thin-film head materials, and magnetic media have led to these advances. Similarly, the resolution of active-matrix liquid crystal displays (AMLCDs) is approaching photographic quality. The development of liquid crystals, fundamental understanding of alignment layers, and color filters using innovative dyes have contributed to the "Bertha" display with 200-dpi resolution and wide viewing angle.

Nontraditional materials such as conjugated organic molecules, short-chain oligomers, longer-chain polymers, and organic-inorganic composites are being developed that emit light, conduct current, and act as semiconductors. The ability of these materials to transport charge (holes and electrons) due to the π-orbital overlap of neighbouring molecules provides their semiconducting and conducting properties. The self-assembling or ordering of these organic and hybrid materials enhances this π-orbital overlap and is key to improvements in carrier mobility. The recombination of the charge carriers under an applied field leads to the formation of an exciton that decays radiatively to produce light emission. (Schematic of semiconducting and light-emission devices are provided in Figure 11.1 and 11.2.) In addition to their electronic and optical properties, many of these thin-film materials posses good mechanical properties flexibility and (toughness) and are processed at low temperatures using techniques familiar to the semiconducting and printing industries, such as vacuum evaporation, solution casting, ink-jet printing, and stamping. These properties lead to new form factors in which roll-to-roll manufacturing is used to create products such as low-cost information displays on flexible plastic, and logic for smart cards and

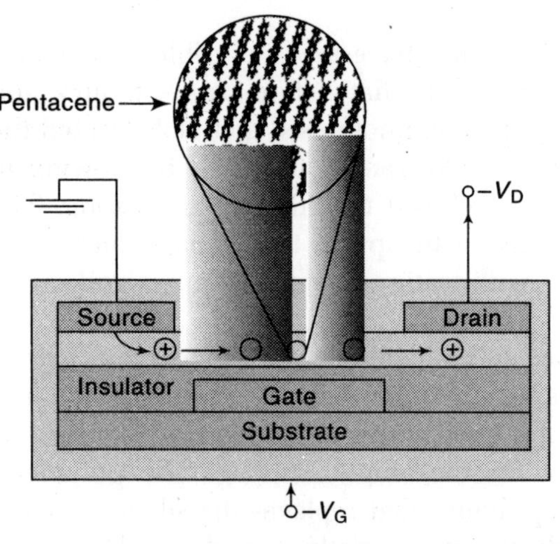

Fig. 11.1 *Organic semiconducting p-type thin-film transistor with top contacts.*

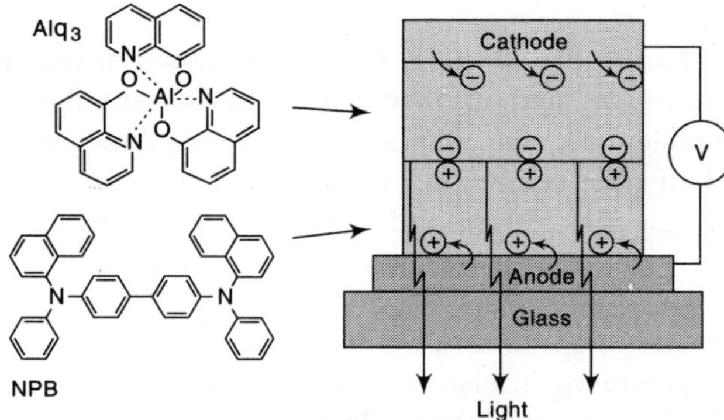

Fig. 11.2 *Organic light-emitting diode which uses Alq_3, tris(8 hydroxyquionolinato)aluminum, as the electron transport and emitting layer, and NPB, N,N′-di(napthalene-1-yl)-N,N′diphenyl-benmzidine, as the hole transport layer.*

Semiconductor	Representative chemical structure	Mobility ($cm^2C^{-1}s^{-1}$)
Silicon	Silicon crystal	300–900
	Polysilicon	50–100
	Amosphous silicon	~1
Pentacene	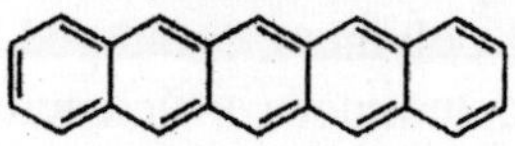	~1
α,ω-dihexyl-sexithiophene	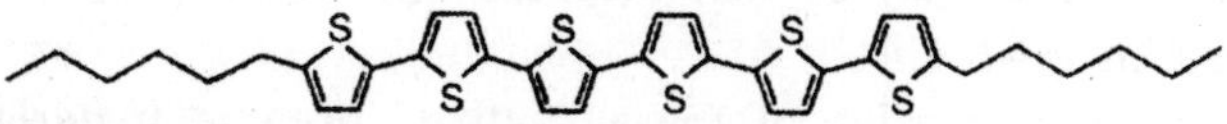	10^{-1}
α,ω-dihexylanthra-dithiophene	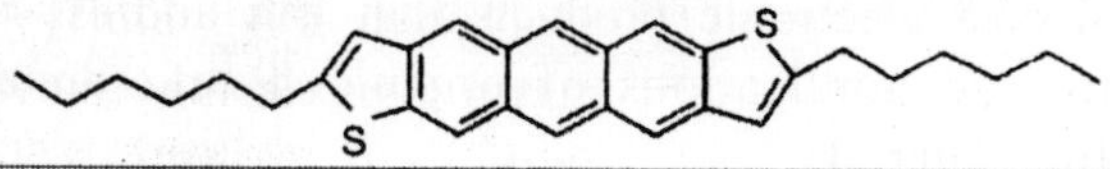	10^{-1}
Regioregular poly (3-hexylthio-phene)	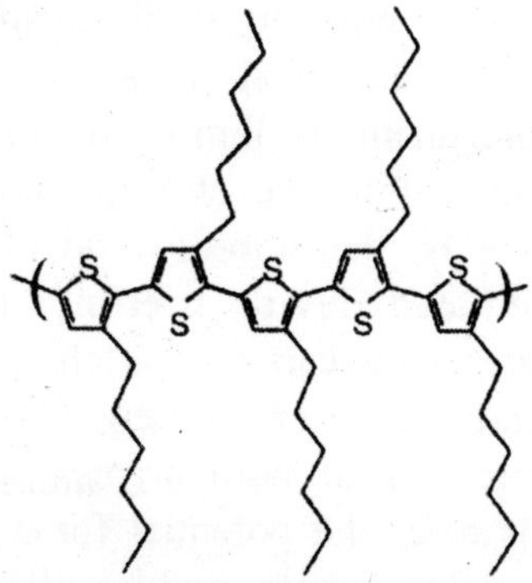	10^{-1}
Organic–inorganic hybrid	Phenethylamine–tin iodide	~1

Fig. 11.3 *Chemical structures and comparison of mobilities of classes of organic and inorganic semiondutors.*

radio-frequency identification (RFID) tags. Effect on semiconducting conjugated organic thiophene oligomers thiophene polymers and the small pentacene molecule have led to improvements in the mobility of these materials. Figure 11.3 shows the chemical structure and mobilities of representative classes of organic materials compared to those of inorganic silicon materials. In Figure 11.3, evaporated films of pentacene have achieved mobilities comparable to that of the amorphous silicon used to fabricate the thin-film transistors (TFTs) which drive the liquid crystal pixels in AMLCD flat-panel; displays. Recently a new class of materials, organic-inorganic perovskites has achieved the mobility of amorphous silicon. These carrier mobilities are now useful for applications that do not require high switching speeds, all of the previously reported materials operated at high voltages. Low-voltage operation of these materials enables applications for portable electronics where battery lifetime is a concern.

Since these organic materials and hybrids are polycrystalline, it is difficult to achieve the mobility of the single-crystal silicon used in high-performance microprocessor. Measurement on single organic crystals of *p-type pentancene* and an *n-type perylene* show mobilities of 2.7 $cm^2 V^{-1}s^{-1}$ and 5.5 $cm^2V^{-1}s^{-1}$—orders of magnitude lower mobility than single-crystal silicon. However, the organic-inorganic perovskites have shown a Hall mobility of 50 $cm^2V^{-1}s^{-1}$, providing a possible path to increased performance.

The majority of these semiconducting organic materials are p-type, transporting holes (h^+) rather than electrons.

In organic light-emitting diodes (OLEDs). There is dramatic increase in luminous efficiency of light-emitting molecular solids and polymers compared to typical inorganic LEDs.

It is found that the highest observed luminous efficiencies of derivatives of solution-processed semiconducting polymers exceed that of incandescent lightbulbs, thus eliminating the need for the backlight that is used in AMLCDs.

The electronic and optical properties of these "active" organic materials are not suitable for some low-performance, low-cost electronic products that can address the needs for lightweight portable devices the status and applications of organic hybrid conducting, semiconducting, light-emitting, and magnetic materials.

Organic compounds generally have a number of disadvantages, including poor thermal and mechanical stability. In addition, while electrical transport in organic materials has improved, the room-temperature mobility is fundamentally limited by the weak van der Waals interactions between organic molecules (as opposed to the stronger covalent and ionic forces found in extended inorganic systems). In OLEDs, the stability and electrical transport characteristics of organic materials contribute to reduced device lifetime. For OFETs, the inherently upper bound on electrical mobility translates to a cap on switching speeds.

Organic-inorganic hybrid materials represent a new class of materials that may combine desirable physical properties characteristic of both organic and inorganic components within a single composite. Inorganic materials offer the potential for a wide range of electronic properties (enabling the design of insulators, semiconductors, and metals), magnetic and dielectric transitions, substantial mechanical hardness, and thermal stability. Organic molecules, provide high fluorescence efficiency, large polarizability, plastic mechanical properties, ease of processing, and structural diversity. The hybrid materials discussed below encompass organic–inorganic composites that alternate chemically and electronically at the molecular level.

Modulating the electronic structure of materials on the nanometer length scale evolves unique and potentially superior electronic and optical properties in hybrid materials that are not characteristic of their organic and inorganic building blocks. For example, thermal evaporation of amorphous multilayers of copper pthalcoyanine (CuPc) and TiO_x, with a periodicity of ~50 Å, forms a composite material with a modulated electronic structure analogous to that of type-II quantum well structures. The lowest energy states for holes is the highest occupied molecular orbital (HOMO) of CuPc, while that for electrons in the conduction band of TiO_x. Photogenerated electron–hole pairs are therefore separated at the organic/inorganic interface. The charge separation induced by modulated structure, coupled with the higher mobility of the TiO_x component, enhances the photoconductive gain of the composite by 40 times compared to that of single-component CuPc layers.

Sol-gel methods, are used to prepare organic-inorganic silicates for optical applications. The organic components helps mechanically to prevent cracking during thin-film and monolithic solid formation and adds optically interesting functionality. Selectively (e.g., through masks) bleaching photosensitive organic molecules embedded in a silicate matrix modify the refractive index of the composite and are used to photopattern waveguiding structures for integrated optics. Nonlinear, optically active (NLO) organic dye molecules are oriented within silicate networks by applying an electric field (known as "poling"). Confining the molecules in the inorganic framework improves the nonlinear optical response of the composite, since the inorganic framework restricts the motion of the organic molecules and prevents randomization of molecular orientation due to thermal relaxation. The restrictive environment of the silicate network is also useful as a matrix for organic LED and solid-state laser materials. The optically clear silicate framework limits rotations/vibrations that decrease the photoluminescence quantum yield of dyes. Grafting of both organic chromophores and charge transporters into the silicate structure is considered for LED applications. It improves the lifetime and stability of the active organic component by preventing segregation and crystallization during device operation. External quantum, efficiencies as high as 0.21% are achieved—but with relatively high turn-on voltages of ~18 V.

Wet-chemical techniques are also used to prepare monodisperse organic–inorganic quantum dot (QD) samples of a variety of semiconductor and metallic materials. Organic ligands, used as the solvent and to control the QD growth rate during preparation, coordinate the surfaces of the QDs and may be exchanged for alternate organic molecules with different lengths and electronic structures. The organic ligands also act to sterically stabilize the inorganic QDs in solutions and enable them to be incorporated into polymeric hosts and to be manipulated into two- and three-dimensional organic-inorganic QD superlattices and colloidal crystals. Interactions between neighbouring QDs in the solids, whose strengths are mediated by the intervening organic molecules, give rise to novel electronic and optical properties. As the distance between quantum dots is reduced to <100 Å, electronic energy transfer processes shift the emission of semiconducting QD solids, and exchange interactions (for distances <5 Å) in metal QD systems give rise to an insulator-metal transition. Semiconductor QDs, such as those containing CdSe and CdS cores with trialkylphosphine chalcogenide capping layer, are good chromophores and are incorporated as emissive layers in LEDs. The increased photoluiminescence quantum yield of CdSe/CdS core/shell QDs enables LEDs with external quantum efficiencies as high as 0.22% and operating voltages of ~4 V.

ORGANIC-INORGANIC PEROVSKITES: STRUCTURES AND PROPERTIES

A class of *crystalline* organic-inorganic hybrids is the organic-inorganic pervokites.

The basic layered perovskites structures, $(R\text{-}NH_3)_2MX_4$ and $(NH_3\text{-}R\text{-}NH_3)MX_4$ ($X = Cl^-$, Br^-, or I^-), are shown in Figure 11.4. Each of the inorganic layers consists of sheets of corner-sharing metal halide octahedra. For the charge-balancing requirements, the M cation is generally a divalent metal that can adopt an octahedral anion coordination, such as Cu^{2+}, Ni^{2+}, Co^{2+}, Fe^{2+}, Mn^{2+}, Cr^{2+}, Pd^{2+}, Cd^{2+}, Ge^{2+}, Sn^{2+}, Pb^{2+}, Eu^{2+} or, Yb^{2+} and trivalent metals Bi^{3+} and Sb^{3+}. The charge difference due to 3+ ions is accommodated by vacancies on the metal site. These inorganic layers are often referred to as "perovskite sheets" because they are conceptually derived from the three-dimensional AMX_3 perovskite structure by taking a one-layer-thick cut

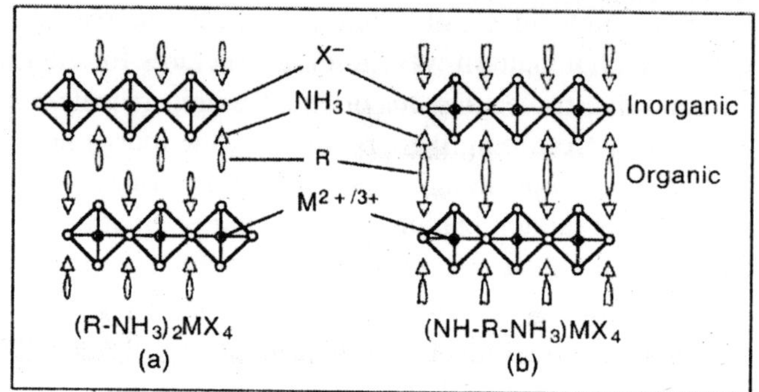

Fig. 11.4 *Representation of single-layer ($n = 1$) <100>-oriented perovskites with (a) monoammonium ($R\text{-}NH_3^+$) or (B) diammonium ($^+NH_3\text{-}R\text{-}NH_3^+$) organic cations. Note that divalent (M^{2+}) metals generally occupy the metal site. For certain systems, however, a mixture of trivalent (M^{3+}) cations and vacancies can occupy the metal site, yielding $M_{2/3}X_3^{2-}$ layers.*

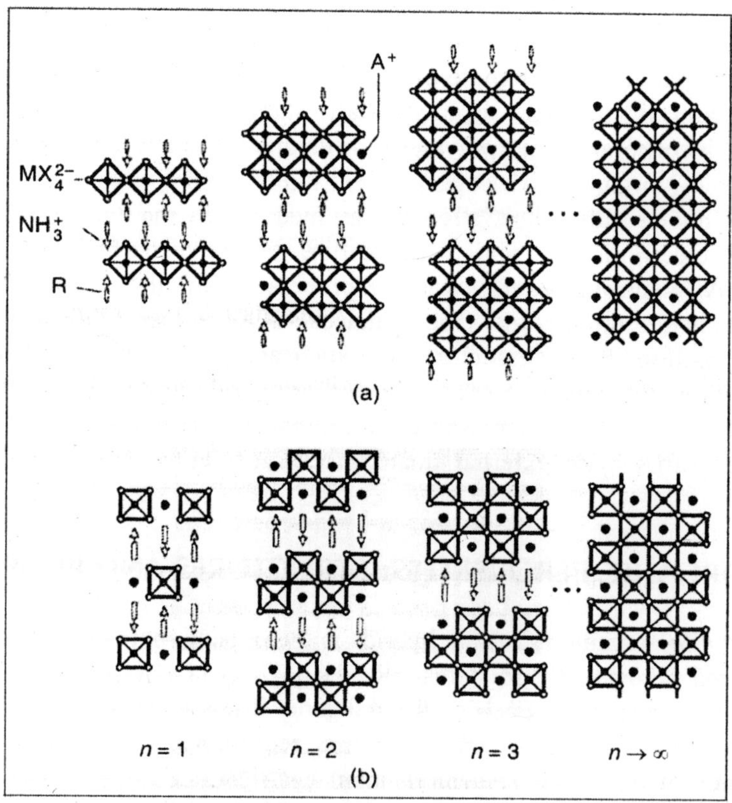

Fig. 11.5 *Representation of (a) <100>-and (b) <110>-oriented families of layered organic-inorganic perovskites. For the <100>-oriented $[(R\text{-}NH_3)_2A_{n-1}M_nX_{3n+1}]$ and <110>-oriented $[(R\text{-}NH_3)_2\ A_nM_nX_{3n+2}]$ structures, R is an organic group, A^+ is a small organic (e.g., $CH_3NH_3^+$) or inorganic (e.g., Cs^+) cation, M is generally a divalent metal cation, X is a halide (Cl^-, Br^-, I^-), and n defined the thickness of the perovskites sheets. The R-group can be used to control which of the two structural families forms. Note that the $n \to \infty$ compound, AMX_3, is the same for each family.*

along the ⟨100⟩ direction of the three-dimensional crystal lattice. In addition to single-layer perovskite sheets, the structures can also be conveniently synthesized with multiple ⟨100⟩-oriented sheets [Figure 11.5(a)], enabling control over the dimensionality for the inorganic framework, or with differently oriented cuts (e.g., ⟨110⟩ vs. ⟨100⟩) of the three-dimensional perovskite structure [Figure (11.5b)]. These structural modifications are simply achieved by altering the combination of organic and inorganic salts in the starting solution from which the hybrids are crystallized. The large repertoire of inorganic frameworks enables tailoring of the electronic, optical, and magnetic properties.

The organic component of the ⟨100⟩-oriented structures consists of a bilayer (for monoammonium) or monolayer (for diammonium) of organic cations (see Figure 11.4). In the case of monoammonium cations, R-NH_3 [Figure (1.4a)], the ammonium head of the cation hydrogen/ionically bonds to the halogens in one inorganic layer, and the organic R-group.extends into the space between the inorganic layers. For diammonium cations, $^+NH_3$-R-NH_3^+ [Figure 1.4(b)], the molecules span the distance between adjacent inorganic layers; therefore no van der Waals gap exists between the layers. The organic R-group consists of an alkyl chain or a single-ring aromatic group. These simple organic layers help to define the degree of interaction between and the properties arising in the inorganic layers.

The unique structural and chemical characteristics of the organic-inorganic perovskites provide useful physical properties. Many of the layered hybrid perovskites, especially those containing germanium (II), tin (II), or lead (II) halide sheets, are analogous or multilayer quantum well structures, with semiconducting, or even metallic, inorganic sheets alternating with organic layers having a relatively large HOMO-LUMO (lowest unoccupied molecular orbital) energy gap [Figure 11.6)]. These self-assembling structures share many of the interesting properties of quantum well structures prepared by molecular beam epitaxy (MBE). In fact, the organic-inorganic perovskites are grown in single-crystalline form.

Note that in Figure (11.6a) the conduction band of the inorganic layers is substantially below that of the organic layers, and the valence band of the inorganic layers is similarly above that of the organic layers. Therefore, the inorganic sheets act as quantum wells for both electrons and holes. In principle, other arrangements of organic and inorganic energy levels are possible. If metal halide sheets having a larger band gap are integrated with more complex, conjugated (i.e., smaller HOMO-LUMO energy gap) organic cations, the well and barrier layers are reversed [Figure 11.6(b)]. Alternatively, if the bandgaps for the organic and inorganic layers are offset [Figure (11.6c)], it leads to a type II heterostructure in which the wells for the electrons and holes are in different layers. The width of the barrier and well layers are adjusted by changing the length of the organic cations and the number of perovskite sheets between each organic layer.

The perovskite quantum well structures, with optically inert (i.e., large HOMO-LUMO energy gap) organic cations and semiconducting inorganic sheets, exhibit sharp resonances (Figure 11.7) in their room-temperature optical absorption spectra due to an exciton state associated with the inorganic semiconducting layers. As excitons are associated with the bandgap of the inorganic framework, the spectral positions of the transitions are substituted by different metal cations or halides with the inorganic framework. The strong binding energy of the

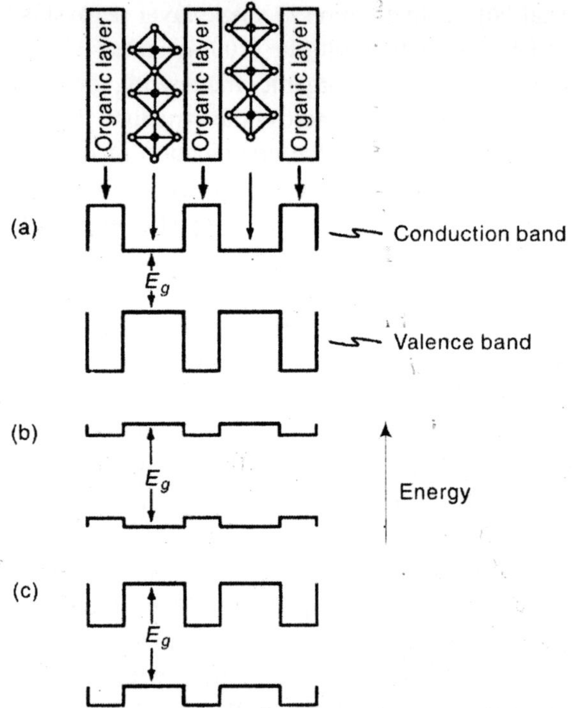

Fig. 11.6 *Organic-inorganic perovskites structure and several energy-level schemes that arise within these structure. (a) semiconducting in-organic sheets alternate with organic layers having much wider bandgaps, result in a type I quantum well structure. In (b), wider-bandgap inorganic layers and organic cations with a smaller HOMO-LUMO gap result in the well/barrier roles of the organic and inorganic layers being switched. In (c), by shifting the electron affinity of the organic layers relative to the inorganic layers, a staggering of the energy levels leads to a type II quantum well structure.*

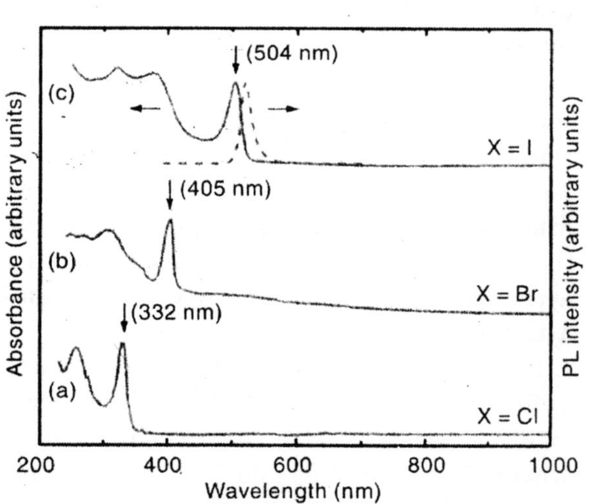

Fig. 11.7 *Room-temperature UV-vis absorption spectra for thin films of $(C_4H_9NH_3)_2PbX_4$ (a) $X = Cl$, (b) $X = Br$, (c) $X = I$. In each spectrum, the narrow indicates the position of the exciton absorption peak (with the wavelength in parentheses). In (c), the corresponding photoluminescence (PL) spectrum (λ_{ex} = 370 nm) is indicated by the dashed curve. Note that small (~15 nm). Stokes shift between the absorption and emission peaks for the excitonic transition.*

excitons, enables the optical features to be observed at room temperature. This is due to two-dimensionality of the inorganic structure, coupled with the dielectric modulation between the organic and inorganic layers. In addition to the sharp transition in the absorption spectra, the large exciton binding energy and oscillator strength leads to strong photoluminescence, nonlinear optical effects and tunable polarition absorption.

The electrical transport properties of the layered perovskites are examined, with an semiconductor-metal transition (Figure 11.8) $(R\text{-}NH_3)_2(CH_3NH_3)_{n-1}Sn_nI_{3n+1}$ and $[NH_2(I){=}NH_2]_2(CH_3NH_3)_nSn_nI_{3n+2}$, as a function of increasing perovskite layer thickness (i.e., increasing *n*— see Figure 11.5). While most metal halides are good insulators, the tin (II) halide-based perovskites exhibit high carrier mobilities and, in some cases, even metallic conduction. The three-dimensional perovskite, CH_3NHSnI_3, for example (i.e., the n $\rightarrow \infty$), is a low-carrier-density p-type metal with a Hall hole density of $1/R_he \approx 2 \times 10^{19}$ cm^{-3} and a Hall Mobility of $\mu \approx$ 50 cm^2 V^{-1}s^{-1} at room temperature.

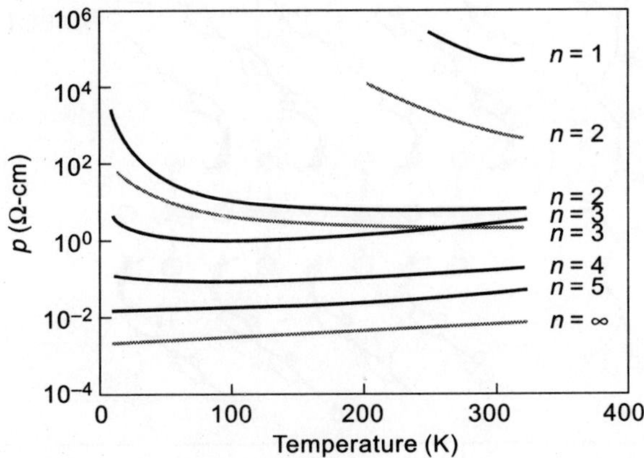

Fig. 11.8 *Resistivity (ρ) as a function of temperature for pressed pellet samples of the layered perovskite familites* $(C_4H_9NH_3)_2(CH_3NH_3)_{n-1}Sn_nI_{3n+1}$ *(black curves) and* $[NH_2C(I) = NH_2]_2(CH_3NH_3)Sn_nI_{3n+2}$ *(red curves), as well as for the n* $\rightarrow \infty$ *end-member of both families* $(CH_3NH_3SnI_3)$.

Most of the perovskites contain relatively simple (large HOMO-LUMO gap) organic cations, such as alkylammonium or phenethylammonium cations. More complex organic molecules can be incorporated, subject to certain chemical and structural constraints. First, the organic molecule must contain one or two terminal cationic groups that ionically interact with the extended inorganic anion. These groups should effectively hydrogen-bond to the inorganic framework. A layered perovskite structure is the tethering group; is a protonated primary amine. Second, the organic molecule must fit within the "footprint" provided by the inorganic framework, with the lateral dimensions of the molecule being limited by the potential for steric interactions between nearest-neighbor molecules. So the length of the molecule (which should extend away from the perovskite sheets) should take a wide range of values, as the distance between perovskite sheets vary. Therefore, long and narrow molecules are favored over molecules with a large cross-sectional area. Finally, interaction between the organic R-groups can either stabilize or destablize the perovskite structure. These interactions include hydrogen bonding, aromatic-aromatic, or van der Walls interactions.

The oligothiophene derivative, 5,5‴-bis(aminoethyl)-2,2′:5′, 2″:5″, 2‴-quaterthiphene (AEQT), shown in Figure (11.9a), is incorporated within a layered pervoskite framework [Figure (11.9b)] consisting of $MX_4{}^{2-}$ (M = Cd, Pb, Sn X = Cl, Br, I) or $M_{2/3}X_4{}^{2-}$ (M = Bi^{3+}, Sb^{3+}) sheets. The

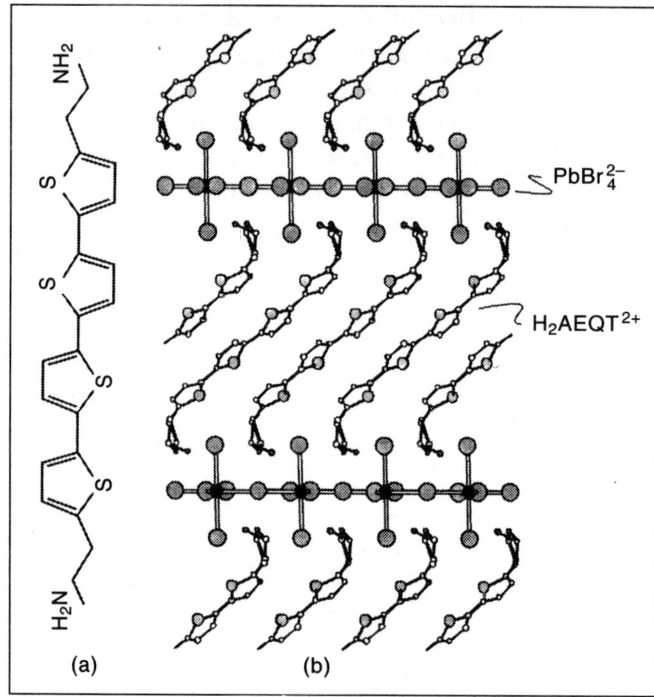

Fig. 11.9 *(a) The 5,5‴-bis(aminoethyl)-2,2′:5′,2″:5″,2‴-quaterthiophene molecule (AEQT). (b) Perovskite structure of* $(H_2AEQT)PbBr_4$ *[i.e.,* $(C_{20}H_{22}S_4N_2)PbBr_4$*], with the doubly protonated AEQT cation occupying the organic cation site of the structure.*

diprotonated AEQT cation has an appropriate tethering group (ethylammonium) and long, narrow profile for incorporating within the layered perovskite framework. Each quaterthiophene oligomer is ordered between the metal halide sheets in a herringbone arrangement with respect to neighbouring quaterhitophenes. This arrangement is partially templated by the relatively rigid inorganic framework.

Annealed thin films of the $(H_2AEQT)PbX_4$ (XC = Cl, Br, I) materials exhibit the characteristic exciton absorption no emission from the inorganic exciton state for nay of the quaterthippnet-containing pervoskites. Rather, for X = Cl, strong photoluminescence is observed only from the quaterthiophene moiety. The QEST luminescence is progressively quenced across the X = Cl → Br → I series, as the bandgap of the inorganic framework decreases in energy compared to the HOMO-LUMO gap of the chromophore molecule (e.g., see Figure 11.6). This quenching is associated with energy transfer and/or charge separation between the organic and inorganic components of the structures.

Other chromophore-containing perovskites have also recently been considered, including the compounds $(R\text{-}NH_3)_2PbCl_4$ (R = 2-phenylethyl, 2-napthylmethyl, and 2-anthrylmethyl). This is a particularly interesting serties because, for R = 2-phenylethyl, the singlets and triplet states of the organic cation are higher in energy than the exciton state of the inorganic sheets; therefore, emission, from the inorganic exciton is observed in the photoluminescence spectrum.

For R = 2-napthlylmethyl, the inorganic exciton state falls between the organic singlet and triplet states, and phosphorescence from the organic molecules dominates the emission spectrum. Finally, for R = 2-naphthylmethyl, the inorganic exciton state is higher in energy than both the organic singlet and triplet states, and chromophore singlet emission is dominant.

An interesting feature of the oligomer- or dye-containing organic-inorganic perovskites is that the orientation of the organie cation is controlled or templated as a result of the inorganic framework. This has a substantial impact on physical properties or potential applications of these materials. For example, if organic chromophores can be held in an orientation that minimizes electronic interaction with neighboring chromophores, concentration quenching of the fulorescence efficiency is reduced. Furthermore, in hybrids containing organic molecules with an extended π system, templating is used to affect the electrical transport properties of the organic layer. In JFET devices, the mobility of the organic channel layer is strongly affected by the degree of molecular ordering of the oligomer layer. In vacuum-deposited organic films, some control over oligomer ordering is attained by varying the substrate temperature during

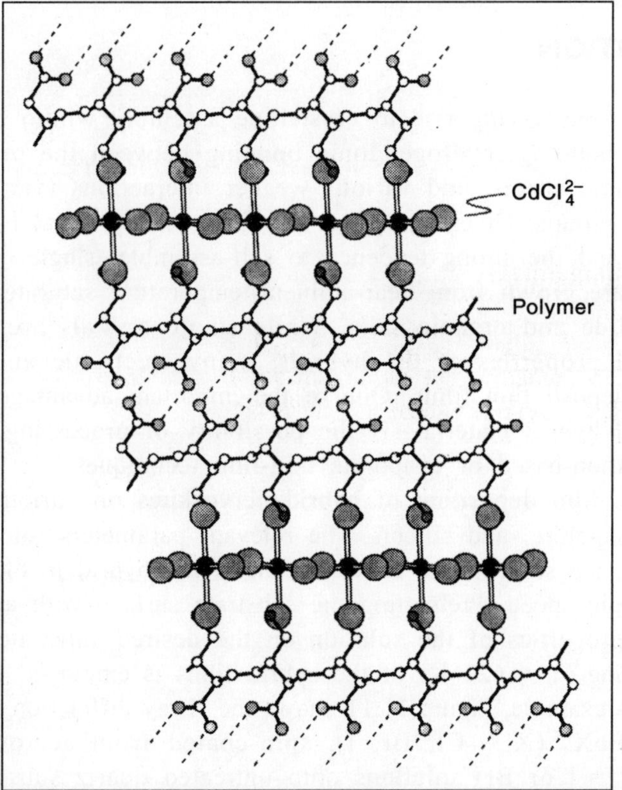

Fig. 11.10 *Crystal structure of ordered polymerized product formed by subjecting (HOOC–CH=CH–CH=CH–CH$_2$NH$_3$)$_2$CdCl$_4$ to UV or γ irradiation. Polymerization yields the 1,4-disubstituted trans-polybutadiene [–CH(COOH)–CH=CH–CH(CH$_2$NH$_2$)–]$_\infty$. Hydrogen bonding between COOH groups on adjacent layers is shown by dashed lines.*

sublimation or through the appropriate attachment of functional groups to the oligomer. The organic-inorganic pervoskite systems show that the inorganic sheets can be used to template the formation of single-crystalline layers of organic molecules, potentially providing a pathway to higher mobilities within oligomer layers.

Another interesting example of the templating influence of thin organic sheets involves solid-state polymerization within the organic layer of perovskite structures. Here, instead of R-groups in the $(R\text{-}NH_3)_2MX_4$ structures, more reactive groups (such as diene or diyne groups) are employed. In one example, the hydrochloride salt of 6-amino-2,5-*trans*, *trans*-hexadienoic acid, within a cadmium(II) chloride pervoskite framework, polymerizes under ultraviolet (UV) or γ irradiation. The polymerization occurs through a 1,4 addition mechanism, leading to a well-ordered polymer layer (Figure 11.10). The same photoreactive organic cation in a copper chloride framework does not undergo polymerization. The structural differences in the inorganic framework, due to distortion of the copper (II) chloride octahedre, lead to an unfavorable configuration of the monomer within the organic layer. Consequently, the inorganic framework is used to control a monomer-containing hybrid system that is susceptible to polymerization.

THIN-FILM DEPOSITION

The relevant interactions giving rise to crystalline assembly within the hybrid perovskites include covalent/ionic bonding, hydrogen/ionic bonding between the organic cations and the halogens in the inorganic sheets, and various weaker interactions (van der Walls, π-π, etc.) between the organic R-groups. Given the good solubility of any metal halides in polar organic and aqueous solvents, and the strong tendency to self-assemble; single crystals of the organic-inorganic perovskites are grown from near-ambient-temperature saturated solutions containing the relevant metal halide and organic salts. While single crystals are useful for examining structure and physical properties of the hybrids, many electronic and optical applications require the ability to deposit thin films. One of the important advantages of organic-inorganic perovskites and related hybrid materials is the possibility of processing the materials using a number of simple solution-based or evaporate thin-film techniques.

Spin-coating enables film depositing of hybrid pervoskites on various substrates, including glass, quartz, plastic, sapphire, and silicon. The relevant parameters for this technique include the choice of substrate, the solvent, the concentration of the hybrid in the solvent, the substrate temperature, and the spin speed. Pretreating the substrate surface with an appropriate adhesion improves the wetting properties of the solution on the desired substrate. Also, postdeposition low-temperature annealing ($T < 623$ K) of the hybrid films is employed to improve crystallinity and phase purity. As an example, Figure 11.11 shows the X-ray diffraction patterns for unannealed films of $(C_4H_9NH_3)_2PbX_4$ (X = Cl, Br, I), spin-coated from acetone (X = I) or N, N-dimethylformamide (X = 1 or Br) solutions onto untreated quartz substrates. The spin-coated films are generally smooth (~1–2 nm RMNS roughness) and suitable for optical studies or device fabrication. In addition to spin-coating, other possible solution-based deposition techniques for the organic-inorganic perovskites include ink-jet printing, stamping, and spray coating. All of these techniques require a suitable solvent for the organic-inorganic hybrid.

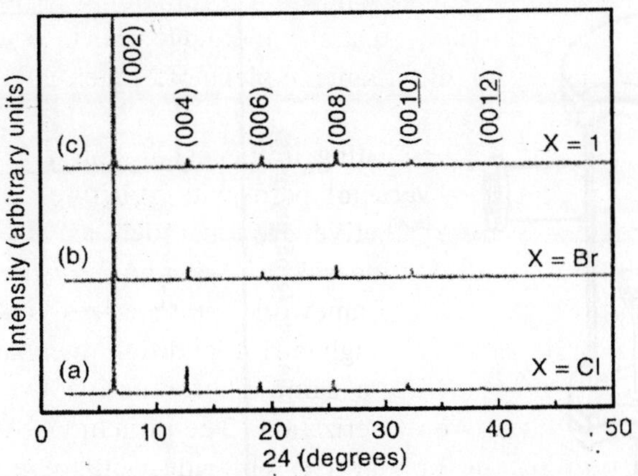

Fig. 11.11 *Provide X-ray patterns for spin-coated thin films of $(C_4H_9NH_3)_2PbX_4$ where (a) X = Cl, (b) X = Br, and (c) X = I. The X-ray reflection indices are given above the data for X = I, and are the same for each sample.*

A solvent cannot be identified for a solutionbased deposition technique. This results either from a lack of solubility of the hybrid or from problems with substrate wetting. Furthermore, for some applications, vacuum-compatible thin-film techniques are appropriate. The problems with evaporative film deposition for the hybrid materials is that the organic component of the structure generally dissociates or decomposes from the material at substantial lower temperatures or in a shorter time period relative to the inorganic component of the structure. Two approaches are employed to overcome this thermal or incompatibility between the two components during evaporation. In two-source thermal evaporation, the organic ammonium salt and the metal halide components of the hybrid are contained in separate evaporation source, and each is heated to an appropriate temperature to enable a nominally controlled evolution. Multiple substrates are placed above the two sources with in the vacuum system. Assuming that the evaporation rate for the two components is controlled and calibrated, a stoichiometric and often crystalline film of the organic–inorganic composite is prepared.

While the metal halide salt evaporates, the organic salt deposition is difficult to control, thus resulting in poor film homogeneity and irreproducible film characteristics. In addition, each distinct organic-inorganic hybrid system requires new heater parameters to balance the evolution rates of the two components. A single-source thermal ablation (SSTA) technique is used that employs, a single evaporation source and very rapid heating. A deposit of the organic-inorganic hybrid is placed on a thin tantalum heater within a vacuum chamber (Figure 11.12). When a large current is passed through the heater, the hybrid ablates form the source over a very short interval (<1 s) and deposits on the substrates positioned above the heater. Since the ablation process is instantaneous, both the organic and inorganic components evaporate from the heater simultaneously, without significant decomposition of the organic component. In many areas, the as-deposited films are single-phase and crystalline (Figure 11.13), indicate that the organic-inorganic hybrids can reassemble on the substrate even at room temperature. For more complex organic cations, a short (<20 min), low-temperature (<473K) annealing is sometimes required to improve the film crystallinity.

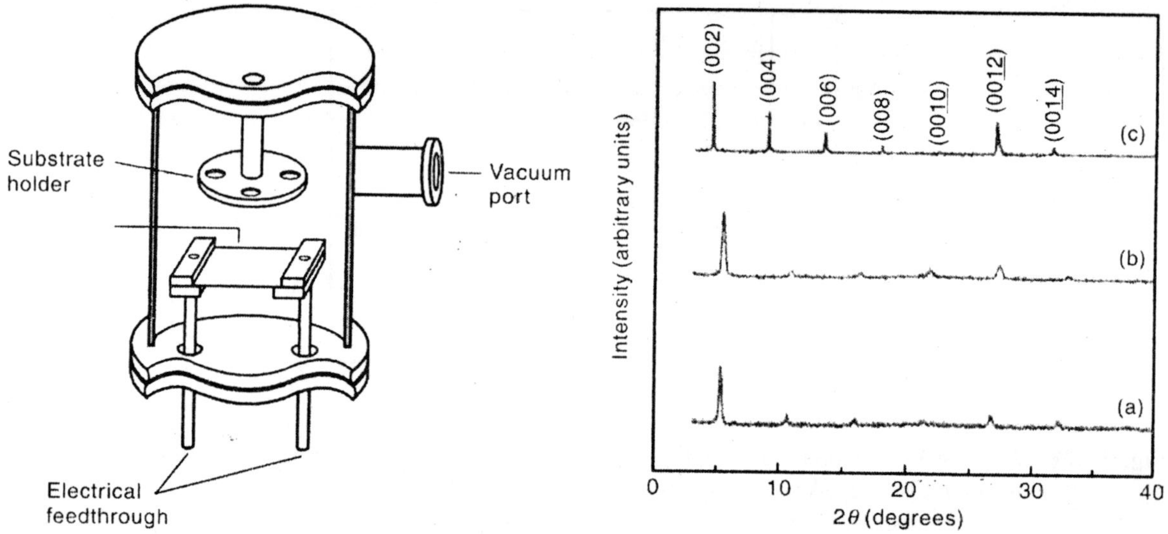

Fig. 11.12 *Cross-sectional view of a typical single-source thermal ablation (SSTA) chamber.*

Fig. 11.13 *X-ray diffraction patterns for hybrid perovskite thin films deposited using the SSTA technique: (a) $(C_6H_5C_2H_4NH_3)_2PbBr_4$; (b) $(C_6H_5C_2H_4NH_3)_2PbBI_4$;* **(c)** *$(C_4H_9NH_3)_2(CH_3NH_3)Sn_2I_7$. The X-ray reflection indices for pattern (b) are the same as those for pattern (a).*

Some organic components do not withstand the heating required for evaporative or sublimation processes. As an alternative, dip-processing technique in which metal halide films, predeposited on a substrate, are dipped into an ambient-temperature solution containing the organ cation is achieved. The solvent for the dipping solution is chosen so that the organic salt is substantially soluble, but the starting metal halide and the final organic-inorganic hybrid are not. The organic cations react with the metal halide on the substrate and form a single-phase crystalline film of desired hybrid. For the family $(R\text{-}NH_3)_2(CH_3NH_3)_{n-1}M_nI_{3n+1}$ (R = butyl, phenethyl; M = Pb, Sn; n = 1, 2) toulene/2-propanol mixtures work well as the solvent, and the dipping timies are relatively short (several seconds to several minutes, depending on the system and film thickness). The advantages of the dip-processing technique include that a simultaneously suitable solvent for both the organic and inorganic components of the hybrid is not required and that the organic cations need not be heated.

ORGANIC-INORGANIC DEVICES (APPLICATIONS)

The electrical and optical properties of the hybrid perovskites, makes it possible to build devices with these materials, example; the use of organic-inorganic hybrids in LEDs. The excitonic transitions, of $(R\text{-}NH_3)_2MX_4$ (M = Ge, Sn, Pb; X = Cl, Br, I) perovskites, give rise to strong photoluminescence that is tuned by incorporating different metal or halogen atoms in the structure. These useful optical haracteristyics make the perovskites attractive as potential emissive materials in electroluminescent (EL) devices. EL is induced in these materials by attaching silver paint contacts to single crystals of $(C_6H_5C_2H_4NH_3)_2$ $(CH_3NH_3)Pb_2I_7$. By applying an electric filed (>10 kV-cm^{-1}) in the plane of the pervoskite sheets and cooling it below 200 K, orange emission is observed, due to avalanche breakdown. Also, EL devices are prepared, with a structure analogous to that of traditional OLEDs, but with hybrid $(R\text{-}NH_3)_2PbI_4$ perovskite light-emitting layers. These perovskite organic bilayer consist of optically inert (in the visible spectrum) phenethylammonium ($C_6H_5C_2H_4NH_3$ $^+$) or cyclohexenylthylammonium ($C_6H_9C_2H_4NH_3^+$) cations. The heterostructure device consists of an indium tin oxide (ITO) anode, a spin-coated $(R\text{-}NH_3)_2PbnI_4$ emissive layer, an evaporated 1,34-oxadiazole, 2,2′-(1,3-phenylene)bis[5-[4-(1,1-dimethylethyl)phenyl]] (OXD7) electron transport layer, and a MgAg cathode. When the OILED (Organic-Inorganic Light-Emitting Diode) is driven at liquid nitrogen temperature, it exhibits intense and efficient EL at 24 V. Unfortunately, the emission intensity drops rapidly with increasing temperature, rendering such devices impractical for display applications. The reduction in EL efficiency near room temperature results, from thermal quenching of the excitons.

In these initial devices, light emission arises from excitons within the inorganic component of the structure. To obtain room-temperature EL from the hybrids, the dye molecule AEQT (Figure 11.9) is synthesized, to replace the optically inert alkyl or aromatic component (used in layered perovskite structures), incorporate additional functionality (fluorescence efficiency) within the organic layer of the perovskite, thus increasing efficient room-temperature emission. Films of $(H_2AEQT)PbCl_4$ are deposited using the single-source thermal ablation (SSTA) method, followed by a short (5 min), low-temperature 338 K, inert-atmosphere anneal, using a digitally controlled hot plate. The UV-vis absorption spectra of the annealed SSTA-deposited films exhibit the characteristic exciton absorption (331 nm) for the lead-chloride-based perovskite (Figure 11.14), demonstrating that the correct structure is formed. However, the films have a featureless X-ray powder diffraction pattern, indicating a small grain size of the perovskite structure. The smooth film morphology and fine-grain structure are also confirmed with atomic force microscope (AFM) and scanning electron microscope (SEM). Higher annealing temperatures result in a gradual increase of the grain size.

The device structure in Figure 11.15, consists of a circular (3/4-in.-diameter), optically polished quartz substrate on which ITO is e-beam-deposited (1500 Å, 14 Ω/□) as an anode. To avoid shorting between the anode and cathode, 1300 Å of SiO_2 is e-beam deposited on top of the ITO through contact masks, thereby defining four rectangular areas (3 mm × 1 mm) of exposed ITO. After cleaning the substrate using solvent-based and oxygen plasma processes, a patterned 3000-Å $(H_2AEQT)PbCl_4$ film is deposited on the exposed ITO areas using SSTA at 10^{-7} mm Hg, followed by the short low-temperature annealing cycle. Subsequently, 200 Å of OXD7 is vacuum-deposited by resistive heating at 10^{-7} mm Hg. An Mg:Ag [20:1] alloy is coevaporated (600 Å, 10^{-7} mm Hg) through a contact mask with four rectangular 2-mm × 7-mm openings. These openings are at a 90° angle to the exposed ITO, thus providing an active device area of

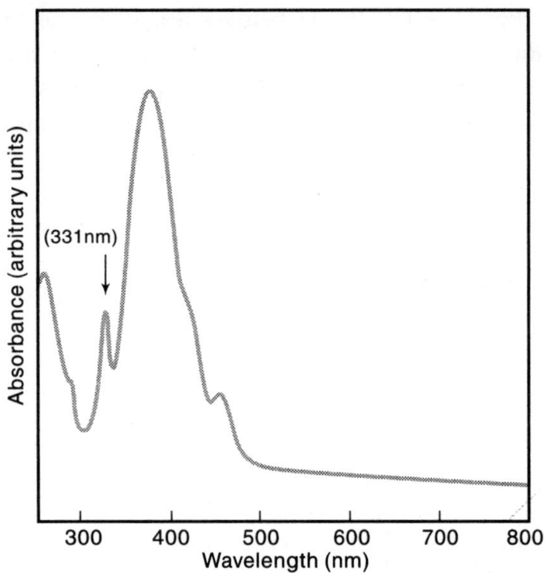

Fig. 11.14 *Room-temperature UV-vis absorption spectrum for an annealed SSTA-deposited film of $(H_2AEQT)PbCl_4$ on a quartz substrate. The peak from the inorganic sheet exciton is marked with an arrow. The broader features at a wavelengths above 350 nm corresponds to the AEQT absorption.*

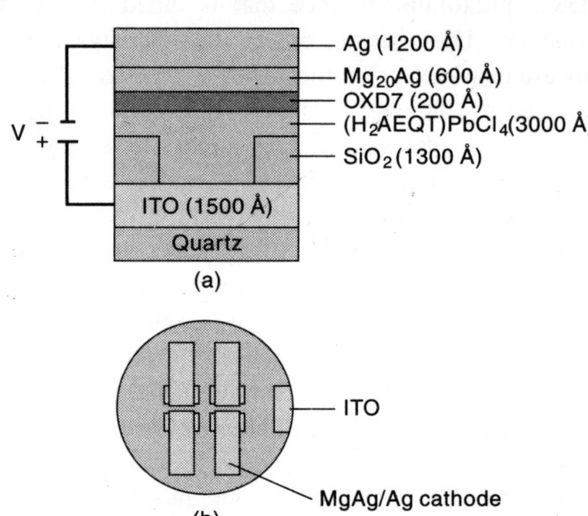

Fig. 11.15 *(a) Cross section of the OILED device structure (not to scale); (B) top view of the circular substrate containing four devices. For clarity, the OXD7 layer (on top of the patterned hybrid perovskite layer) is not shown.*

1 mm × 2 mm [Figure 11.15(b)]. Another 1200 Å of AG is deposited on top to inhibit oxidation. The vacuum system is connected to the antechamber of a nitrogen-filled glovebox, for inert conditions. The devices are encapsulated using a cover glass and epoxy before.

Bright green –yellow light (λmax ≈ 530 nm) is observed when the devices are forward-biased under ambient conditions (Figure 11.16). Figure 11.7 shows the electroluminesecnce–voltage characteristics for the OILED structure. A low turn-on voltage of about 5.5 V is observed. The IV and ELV curves superimpose, indicating a balanced electron and hole injection. The eletroluminescence spectrum corresponds well to the photoluminescence spectrum of $(H_2AEQT)PbI_4$ [Figure 11.17(b)], and to the dye salt AEQT 2HCl. The maximum efficiency is 0.1 lm/W at 8 V and 0.24 mA. Devices with a 600-Å $(H_2AEQT)PbCl_4$ film has a lower turn-on voltage of 4.8 V compared to the devices with a 3000-Å emission layer. Even with the thicker film,

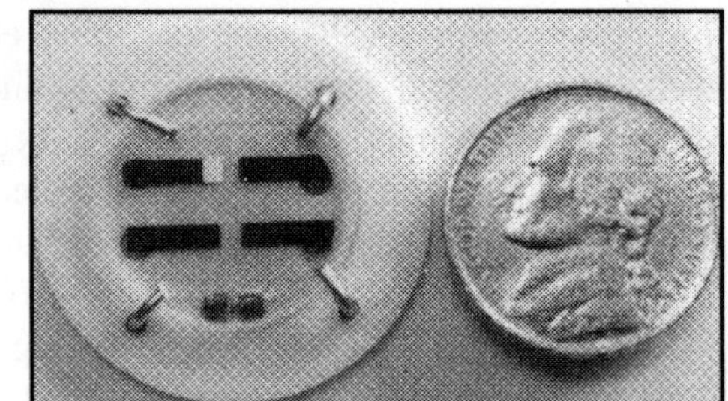

Fig. 11.16 *Photograph of an operational OILED device based on $(H_2AEQT)PbCl_4$ as the emitting material. The nickel on the right provides a scale for size comparison.*

the turn-on voltage remains relatively low. The thicker organic-inorganic emission layers are in conventional OLEDs (typically <600 Å) thereby improving the device reliability as it renders the devices less prone to pinholes and shorts.

The advantages of molecular-level sequencing of organic and inorganic layers is: First, the artificial layering (i.e., by successive evaporation) of organic dye molecules (e.g., Alq_3) and wide-bandgap inorganic compounds (e.g., LiF) has shown to enhance LE efficiency and device lifetime compared to systems with a single layer of the emitting dye. The improved device performance is attributed to better carrier injection, transport, and electron-hole recombination. The self-assembling hybrid perovskites provide a single evaporation step of achieving this type of alternating structure. Hybrid perovskite devices designed with lead bromide inorganic sheets exhibit efficiencies 20 to 30 times lower than the ones with lead chloride sheets, whereas with cadmium chloride (wider bandgap) layers produced have higher efficiencies of ~0.16 1m/W. Therefore, the OILEDs made with laternating Alq_3/LiF layers, the self-assembling organic-inorganic structures can have a large impact on the luminous efficiencies exhibited by the emitting species within the structure. The choice of inorganic framework also has a substantial effect on the electronic properties of the perovskite and hence on the electrical characteristics of the device. As is seen in Figure 11.17(a), the higher bandgap of cadmium chloride gives rise to a higher turn-on voltage of about 9 V (compared to 5.5 V), despite the use of a thinner emissive layer (1200 Å versus 3000 Å).

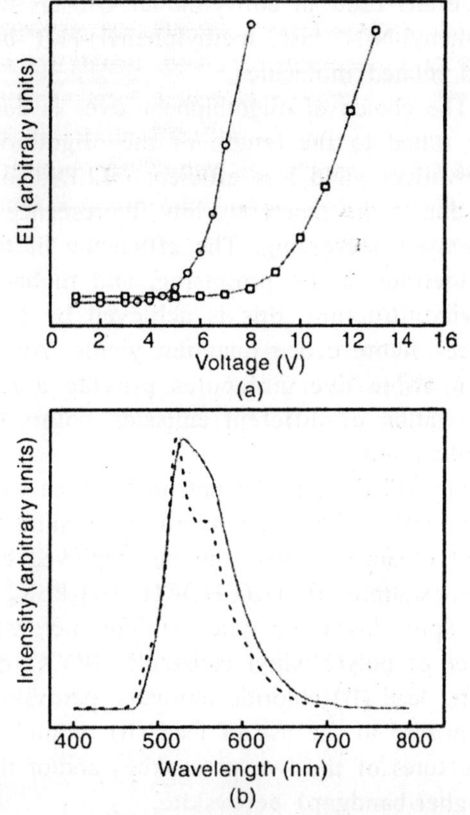

Fig. 11.17 *(a) Room-temperature electroluminescence–voltage (EL–V) characteristics of an OILED with a 3000-Å layer of (H_2AEQT)PbI_4 (filled circles), and one with a 1200-Å layer of (H_2AEQT) $CdCl_4$ (open squares). (b) Room-temperature electroluminesecnce (solid curve and photoluminescence (dashed curve, excited at 306 nm) spectra of an (H_2AEQT) PbI_4 film.*

Another, feature of the hybrid perovskites is that it acts as a template, inducing different molecular packing than that achievable in a purely organic film. This control over packing has the potential for reducing the quenching interactions between the dye molecules thereby increasing their overall luminescence quantum efficiency. Finally, the hybrid perovskite films exhibit a high thermal stability (393–473 K depending on the inorganic framework) against grain growth, enabling more robust films. For comparison, glass transition temperature of materials used in conventional OLEDs range from 333K of common molecules such as N-N′ diphenyl-N-N′ bis(3-methylphenyl)-[1-1′-biphenyl]-4-4′-diamine (TDP) to > 423 K for starburst and related molecules.

The choice of oligothiphene dyes is due to the electronic properties of the organic cation that are tuned to the length of the oligothiophene moiety. Thin, organic films of oliogothiphene derivatives yield less efficient OLEDs, so EL quantum yields range from 10–9 to 10–2%. This us due to the inherently low fluorescence quantum yield of oligothipophenes caused by effective intersystem crossing. The efficiency of the preliminary hybrid perovskite devices (~0.2 lm/W) is therefore quite promising, and higher values are expected with further fine tuning of the device structure, this is achieved by the incorporation of chromophores with fundamentally better fluorescence quantum yields. Appropriate modification of a wide range of structurally compatible dye molecules provide a repertoire of candidate cations and also enables the realization of different emission colors (e.g., blue, red)—a requirement for full-color display applications.

For OILEDs, if the potentially high carrier mobilities and the tunable electronic structure are achievable within the hybrid perovskite framework, then, these materials (even with optically inert organic cations) can be employed as carrier transport layer within LED device structures. For example, $[C_6H_5(CH_3)CHNH_3]_2PbX_4$ (X = Cl, Br) perovskites are demonstrated as hole transport layers (p-type semiconductors) in an OILED with a conventional organic emitter layer of poly(N-vinyl carbazole) (PVK) doped with the laser dye Coumarin 6. It is found that using lead (II) chloride inorganic perovskite sheets results in a tenfold improvement in efficiency compared to the use of lead (II) bromide layers. This improvement is due to the different band structures of the two perovskites and/or the better electron-blocking abilities of the lead chloride (higher-bandgap) perovskite.

Semiconducting hybrid perovskites are also attractive as a new class of channel materials for thin-film field-effect transistors (FETs) because they combine the higher carrier mobilities of ionic and covalently bonded inorganic semiconductors with low-temperature thin-film techniques that make organic semiconductors exciting as alternative channel materials. The first organic-inorganic hybrid material is shown as the semiconducting channel in a TFT, using the hybrid peorvskite $(C_6H_5C_2H_4NH_3)_2SnI_4$. Figure 11.18 depicts a typical organic-inorganic TFT device structure. Thin films of the hybrid perovskite are spin-coated from a methanol solution, in an inert atmosphere, onto thermally oxidized degenerately n-doped silicon wafers. The SiO_2 layer acts as the gate insulator and is 400, 1500, or 5000 Å thick. The silicon wafer is used as the gate electrode, and high-work-function metals, such as Pd, Au, or Pt, are deposited by e-beam evaporation through silicon membrane masks and serve both as the source and drain electrodes and to define the channel dimensions. While the organic-inorganic perovskite are deposited either before or after metallization, deposition the source and drain electrodes before spin-coating eliminates exposing the hybrid perovskite films to potentially harmful metal deposition conditions (e.g., high temperatures).

Spin-coating single-layer (n = 1) tin (II) iodide perovskites with simple monoammonium organic cations results in oriented polycrystalline thin films having ~300-nm in-plastic grain size. Figure 11.19 shows the X-ray diffraction pattern and scanning electron micrograph (SEM)

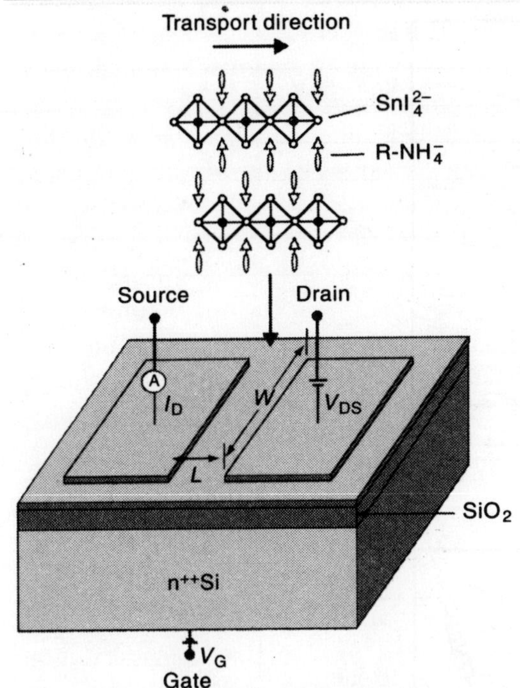

Fig. 11.18 *Diagram of a TFT device structure employing a layered organic-inorganic perovskite as the semiconductor channel.*

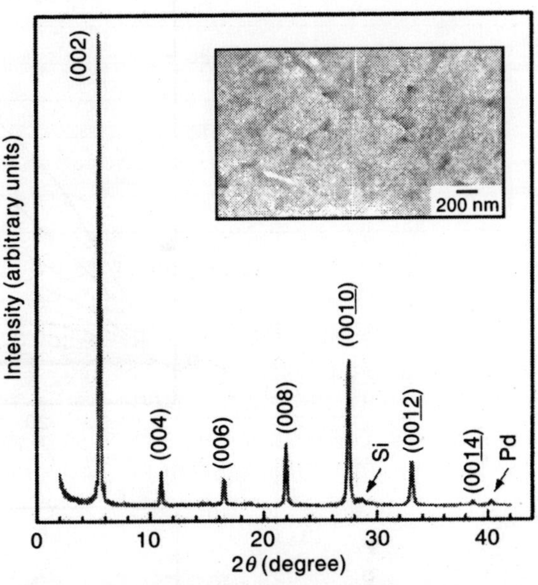

Fig. 11.19 *X-ray diffraction pattern for a completed TFT with $(C_6H_5C_2H_4NH_3)$ $2SnI_4$ as the semiconducting channel and Pd source and drain electrodes. Inset is an SEM image of the hybrid perovskite film.*

of a completed $(C_6H_5C_2H_4NH_3)_2SnI_4$ TFT. The sharp X-ray diffraction pattern of only $(0\ 0\ \ell)$ reflections and the smooth morphology in the SEM image are typical of spin-coated tin (II) iodide perovskites with simple monoammonium organic cations. The formation of extended inorganic sheets is also reflected in sharp exciton peak in the optical absorption spectra at a wavelength ($\lambda_{max} \approx 608$ nm) characteristic of the tin (II) iodide perovskite.

Field-modulated conductance in tin (II) iodide perovskites is observed for both simple aliphatic and aromatic organic cations, $R\text{-}NH_3^+$. Each tin (II) iodide-based perovskite forms a p-channel transistor. Figure 11.20 show representative device characteristics for a TF with the organic-inorganic perovskite $(C_6H_5C_2H_4NH_3)_2SnI_4$ as the channel layer and a 5000-Å-thick SiO_2 gate insulator layer. Application of a negative bias to the gate electrode ($V_G < 0$) increases the number of majority holes in the semiconducting channel contributing to the drain current (I_D). These devices show typical transistor like behavior, as I_D increases linearly at low source-drain voltage (V_{DS}) and then saturated as V_{DS} increases and the holes in the channel are pinched off near the drain electrode. Application of a positive gate bias ($V_G > 0$) depletes the holes in the channel, turning the device off.

The device characteristics of these organic-inorganic field-effect transistor (OIFETs) have the standard equations used for both inorganic and organic semiconducting channel materials. Figure 11.20(b) show the dependence of I_D and $I_D^{1/2}$ versus V_G, at $V_{DS} = -100$ V. As determined

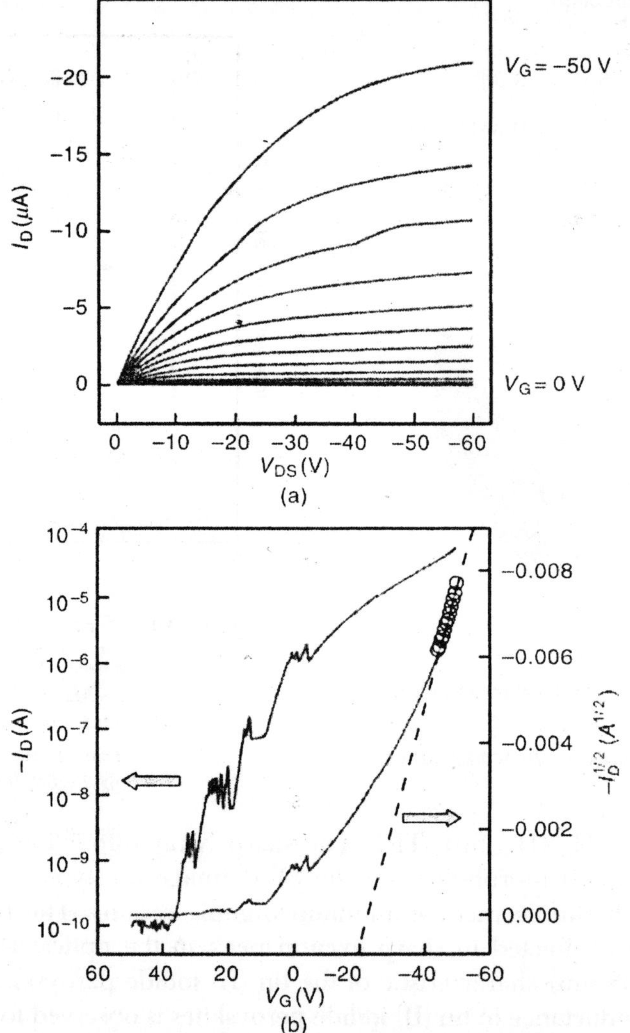

Fig. 11.20 *(a) Drain current I_D versus source-drain voltage V_{DS} as a function of gate voltage V_G for a TFT with a spin-coated channel of $(C_6H_5C_2H_4NH_3)2SnI_4$, a channel length L = 28 μm, and a channel width W = 1000 μm. The gate dielectric is 5000 Å SiO_2. (b) Plots of I_D and $I_D^{1/2}$ versus V_G at constant $V_{DS} = -100$ V, as used to calculate current modulation (I_{on}/I_{off}) and field-effect mobility (μ).*

from these curves, the field-effect mobility for the device is μ = 0.61 $cm^2 V^{-1}s^{-1}$, and the I_{ON}/I_{OFF} ratio is ~10^6. These device characteristics are representive of $(C_6H_5C_2H_4NH_3)_2SnI_4$ and are comparable to the performance of amorphous silicon and the best organic semiconductors deposited in high vacuum. The I_{ON}/I_{OFF} ratio is achieved by patterning the semiconduction perovskite. The leakage current between source/drain electrodes and the back gate is effectively avoided. Unpatterned device structures show the same field-effect mobilities, but leakage current between gate and drain electrodes reduces the magnitude of current modulation. Similar device characteristics (μ and I_{ON}/I_{OFF}) are achieved at substantially lower applied voltages for devices fabricated using a thinner gate insulator.

Although $(C_6H_5C_2H_4NH_3)_2SnI_4$ channels show the best device characteristics of the organic-inorganic TFTs prepare do date, and tin(II) iodide perovskites with other organic molecules also show field-modulate conductance. Devices prepared with the alkylammonium tin iodides, $(C_nH_{2n-1}NH_3)_2SnI_4$ (n = 4 to 12) exhibit transistor-like behavior but have lower field-effect mobilities (~10^{-3} to10^{-2} $cm^2\ V^{-1}s^{-1}$) compared to the phenethylammonium tin(II) iodide devices. They also do not completely turn off under positive gate bias. The infereior device performance stems from differences in trap characteristics and/or in film and grain morphology after spin-coating.

Three-dimensional Integrated Circuits

Three-dimensional (3D) integrated circuits (ICs), which contain multiple layers of active devices, have the potential dramatically enhanceship performance, functionality, and device packaging density. They also provide for microchip architecture and may facilitate the integration of heterogeneous materials, devices, and signals.

CHALLENGE OF CMOS TECHNOLOGY

The development of IC technology is driven by the need to increase both performance and functionality whole reducing power and cost. The goal is achieved by the use of two solutions: (1) scaling devices and associated interconnecting wire through the implementation of new materials and processing innovations, and (2) introducing architecture enhancements to reconfigure routing, hierarchy, and placement for critical circuit building blocks.

- *Front-end-of-lime (FEOL) scaling:* An accelerated gate-length scaling pushes the gate-dielectric and junction to its physical limits, therefore, continued conventional bulk-Si CMOS device scaling of the diode thickness junction depth, and depletion width has become quite difficult, and necessitates the replacement of bulk MOSFETs with CMOS device structures. Silicon-on-insulator (SOI) technology, which offers higher performance because of junction capacitance reduced and lack of body effects, has been developed. Further, scaling of SOI thickness reduces short-channel effect and eliminates most of the leakage paths, but it rapidly degrades mobility, thereby limiting the extent of SOI scaling. Strained Si channels offering mobility enhancement are achieved, but future structures which combine the benefit of SOI and strained silicon technology are to be constructed, using devices geometry and technology developed for double-gate FETs and FinFETs. A key challenge for these novel integration and device options is the increasing difficulty in their fabrication and the incompatibility of various designs with planar structures.

- *Back-end-of-line (BEOL) scaling:* CMOS scaling trends result in a design in which billions of transistors are interconnected by tens of kilometers of wires packed into an area of square centimeters. Wires deliver power to each transistor and provide a low-skew synchronizing clock. However, increasing wiring complexity and challenges in improving wire delay to keep up with intrinsic gate delay are key issues for BEOL technology. Although many new materials and processes have been introduced to meet metal conductivity and dielectric permittivity requirements, it is expected that interconnect metallization of long wires with resulting RC delay, two yield, and high cost of fabrication will limit the performance of ICs beyond the 45-nm-technology node.
- *Architecture:* The conventional planar IC has limited floorplanning choices, and these in turn limit system architecture performance improvements. This leads to issues related to the interconnect loading in the network of long wires and the need for signal repeaters used for clock distribution. However, repeaters are responsible for a significant fraction of the total power consumption on a chip. Also, existing two-dimensional (2D) IC designs may not be suitable for the integration of disparate signals (digital, analog, or rf) or technologies (SOI, SiGe heterojunction bipolar transistors or HBTs, GaAs, etc.) In addition, because of IC scaling trends, traditional computer-aided-design (CAD) practices and tools have required an increased number of design cycles, raising time to market and cost per chop function. Therefore, a solution is required that both alleviates the interconnect bottleneck and provides new avenues for the advanced device and architectural innovation.

BENEFITS OF 3D INTEGRATED CIRCUITS

One of several promising solutions being explored is the 3D integration and packaging technology (also known as vertical integration), in which multiple layers of active devices are stacked with vertical interconnections between the layers (Figure 12.1) to form 3D integrated circuits (ICs). Even in the absence of continued device scaling, 3D ICs provide potential performance advances, since each transistor in a 3D IC accesses a greater number of nearest neighbors, and each circuit functional block has higher bandwidth. Other benefits of 3D ICs include improved packing density, noise immunity, improved total power due to reduced wire length/lower capacitance, superior performance, and the ability to implement added functionality. These features are described below in more detail.

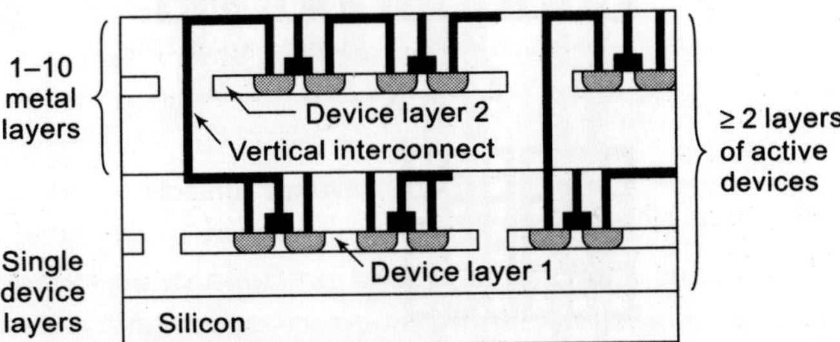

Fig. 12.1 *Schematic diagram of three-dimensional integrated circuit (3D1C) showing two stacked device layers with their corresponding metallization levels and inter-device-layer connections (vertical interconnects).*

Power

Initial analyses of 3D wire-length reduction shows that 3D integration provides a smaller wire-length distribution, with the largest effect associated with the longest paths. These shorter wires decrease the average load capacitance and resistance and decrease the number of repeaters needed for long wires. Since interconnect wires with their supporting repeaters consume a significant portion of total active power, the reduced average interconnect length in 3D IC, compared with that of 2D counterparts, will improve the wire efficiency (~15%) and significantly reduce total active power by more than 10%.

Noise

The shorter interconnects and consequent reduction of load capacitance in 3D ICs reduce the noise due to simultaneous switching events. The shorter wires have lower wire-to-wire capacitance, resulting in less noise coupling between signal lines. The shorter global wires with reduced numbers of repeaters should have less noise and less jitter, providing better signal integrity.

Logical Span

Because MOSFET fan-out is limited to a fixed amount of capacitive gain per cycle, the increasing intrinsic gate load is significantly constrained by extrinsic load capacitance (wires). Since 3D IC provides a lower wiring load, it makes it possible to drive a greater number of logic gates (fan-out).

Density

In three dimensions, active devices are stacked and the size of a chip foot print is reduced. This added dimension improves the transistor packing density, since circuit components stack on top of each other, as in Figure 12.2, where an n-FET is placed over a p-FET. When the total layout area (the sum of the device area and the metal routing area) is compared for 2D and 3D

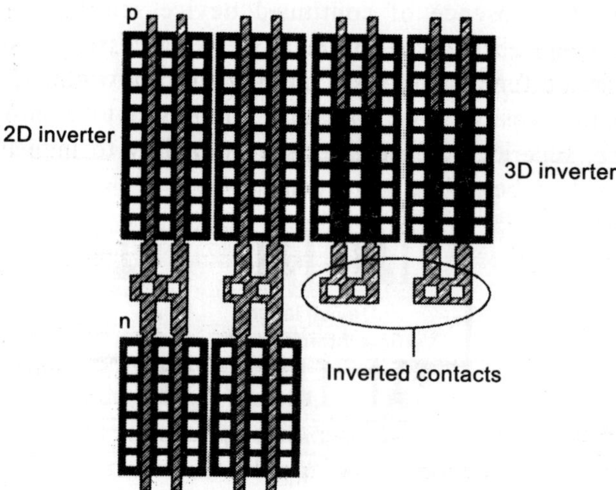

Fig. 12.2 *Layout designs of the 2D and 2D inverters with fan-in equal to 1, showing large (30%) areal gain for the 3D case.*

standard cells with difference inverter designs, a 30% areal benefit for the 3D cells is achieved. The ability to stack circuit elements, thus shrink the footprint and potentially reduce the volume and/or weight of a chip. Hence, it is of great interest for wireless, portable electronics, and military applications.

Higher-density and hence higher-speed SRAM circuits are created. For example, the pull-up p-MOS device is stacked over the n-MOS in a 3D approach to save device area. However, since metal routing occupies a large portion of the total layout area, the total cell area reduction depends strongly on the chip architecture and the metal routing design. Successful stacked CMOS SRAM cell technology is achieved, but it is limited by extremely tight alignment tolerance requirement for interlayer contacts.

Performance

3D technology enables the memory arrays to be placed above or under logic circuitry, resulting in an increased bandwidth and thus a significant performance gain in communication between memory and microprocessor. In particular, as the amount of on-chip memory increases the latency of the path from logic to memory becomes a limiting factor. The ability to stack logic and memory is achieved.

In addition, the maximum system performance depends on the number of device layers. Maximum performance depends on power dissipation constraints. In the presence of power constraints, there are scaling optima that yield maximum computation (for example, if devices are scaled too far, leakage consumes too much of the power). Simple layering models of device and system dependencies are developed, and optimization is performed. These layering models ignore the impact of blockage due to signals passing through a layer. As depicted in Figure 12.3, the results show significant potential advantage for 3D integration, with performance increasing roughly as the square root of the number of circuit layers that are stacked. For these data points; device characteristics (such as V_{dd}, V_T, t_{ox}; gate length, mean FET width, wire half-pitch, and repeater spacing) are optimized for maximum performance, performance is calculated as *Performance = total number of logic switching events per second in a processor core.*

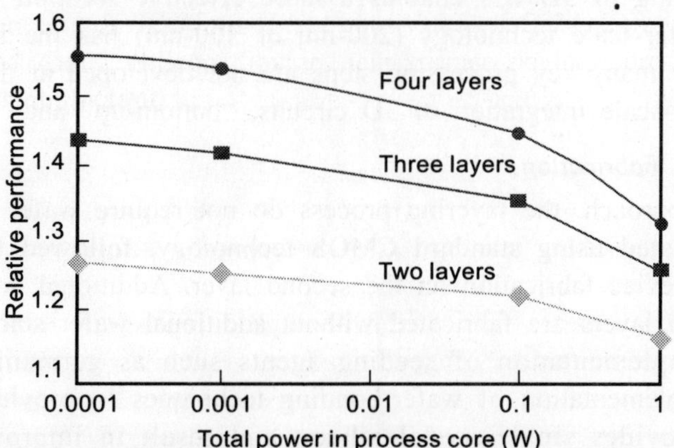

Fig. 12.3 *Relative performance for different numbers of stacked layers vs. the pre-set total power in the process core, showing performance increase as the square root of the number of layers stacked.*

Functionality

3D integration allows the incorporation of new elements. It enables the implementation of related design flexibility, including new system architectures. The application is the combination of dissimilar technologies (memory, logic with extension to rf, analog, optical, and microelectromechanical systems) to create hybrid circuits.

3D IC FABRICATION TECHNOLOGY

3D IC fabrication technology is accomplished by the implementation of diverse processing sequences. The simplest way to distinguish among various methods is by differentiating between chip-level and wafer-level processing during the layering of key circuit components. Then the process is further differentiated by determining whether the layer stacking is done using a face-to-face or face-to-back approach. A detailed description of some of the most promising 3D assembly methods is presented below:

Chip Stacking

3D stacking technology for packaging is focused mainly on chip-stacking methods. Today many 3D packaging systems are manufactured, but high-density memory modules are a key application. Typically a 3D package stacks bare dies or multichip modules (MCMs), securing the full chips by using epoxy or glues and creating electrical connections by wire-bonding techniques. Novel 3D packages utilizes peripheral interconections that are several millimeters long, but higher interconnect density with shorter links (hundreds of micron) between stacked layers is achieved by incorporating conducting vertical through hole across the chip 3D packaging. Hence, a key process technology element for optimizing 3D IC; is a methodology for high-density, smaller-dimension interlayer connections. Chip-to-chip and chip-to-wafer methods, are utilized to accomplish this goal.

Wafer-scale Fabrication

A wafer-level stacking of 3D ICs enables a more effective solution than the chip-stacking techniques. 3D IC wafer-scale technology (200-nm or 300-nm) has the advantage of increased design flexibility, since many key processing steps are not developed at the die level. There are two schemes for wafer-scale integration of 3D circuits: "bottom-up" and "top-down" fabrication.

Bottom-up Wafer-scale Fabrication

In the bottom-up approach, the layering process do not require wafer stacking. The bottom-most layer is first created using standard CMOS technology, followed by the formation of a second Si layer, and device fabrication on the second layer. Additional layers, are added on the top. The subsequent Si layers are fabricated without additional wafer stacking using solid-phase crystallization, the implementation of seeding agents such as germanium or nickel, lateral overgrowth or the implementation of wafer-bonding techniques to provide a new Si substrate. The latter method provides single-crystal silicon and result in improved device quality in comparison with the first method. However, thermal constraints, facilitate to maintain good performance in the underlying IC layer.

Top-down Wafer-scale fabrication

In the top-down method, multiple 2D IC circuits are fabricated in parallel and then "assembled" to form 3D IC. Such an approach enables the performance optimization of each layer and its functional verification prior to stacking, and results in acceptable yield. It is inattractive for applications in which layers of disparate technologies are closely stacked. Key process challenges of the top-down 3D IC technology include high-quality, low-temperature bonding (<673 K), as back-end materials (metals and low-*k*) are a part of the structure, tight alignment tolerance, the integrity of contacts between device layers, and high process reliability.

3D IC Stacking

As shown in Figure 12.4, 3D IC structures are characterized according to the parts of the circuit design that are layered. More specifically the 3D integration is application-specific, and conceptually it is partitioned as stacking layers of devices, circuits, macros, circuit functional units, or chips. As depicted in Figure 12.4, depending on 3D application or partition level, a specific input/output (I/O) or interlayer via density is achievable.

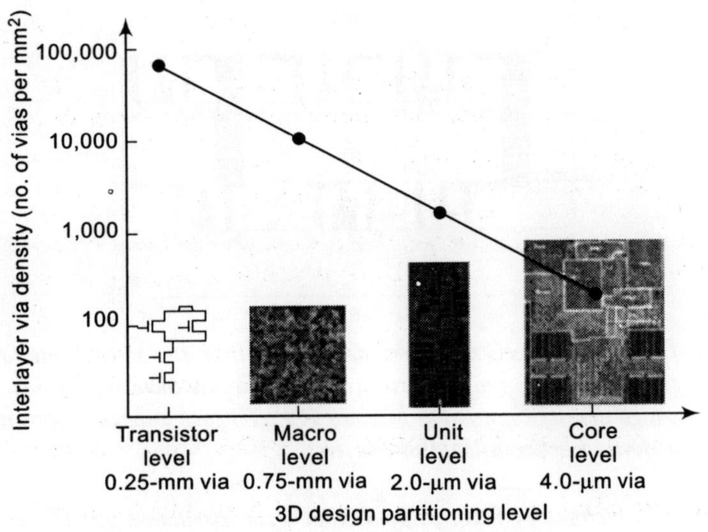

Fig. 12.4 *Diagrams for 3D integration (based on 3D partitioning level).*

Further, depending on the position of the top of the second layer with respect to the top of the first layer after stacking, the process is described as "face-to-face" if the two tops are facing each other, or "face-to-back" if they are not. The most promising methods for creating 3D ICs using face-to-face and face-to-back options are shown in Figure 12.5, and their assembly technology features are listed in Table 1. These options are used to build chip-to-chip, wafer-to-wafer, and chip-to-wafer 3D ICs, but a specific process flow is easier for a particular chip- or wafer-level technology.

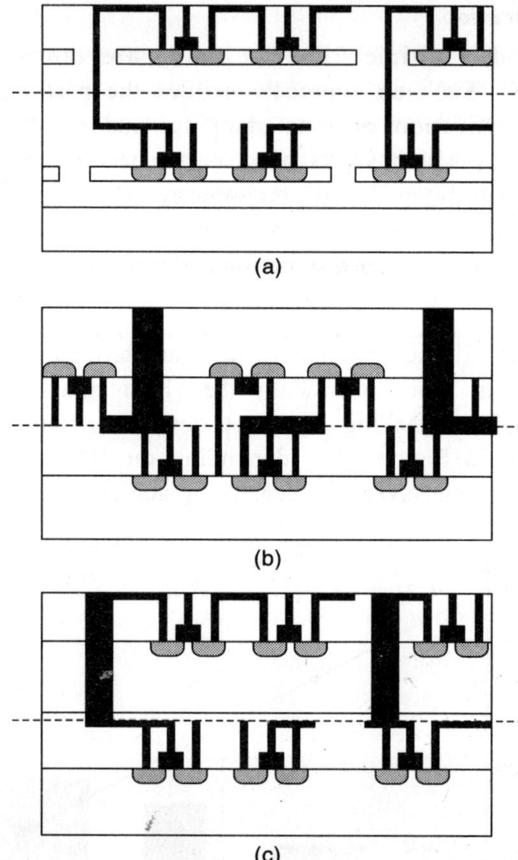

Fig. 12.5 *Diagrams of assembled 3D IC structures; dashed line indicates bonded interface: (a) SOI-based face-to-back process with closely coupled layers; (b) face-to-face bonding (avoids need for glass substrate and achieves high density connections between ICs); (c) face-to-back process with some Si remaining and deep vias formed between the device layers.*

Table 1 *Technology features associated with assembled 3D IC structures*

Process feature	*(a) SDI-based face-to-back process*	*(b) Face-to-face process*	*(c) Face-to-back process*
Bonding medium	Fusion or adhesive	Cu-Cu	Cu-Cu
Distance between device layers	Smallest	MiddleLargest	
Glass substrate needed	Yes	No	Yes
Alignment required	Aggressive (sub-μm)	Few μm	More relaxed
Minimum via pitch	Very tight (~0.4 μm)	~10 μm	20–50 μm
Interlayer via density	Very high (~10^8 cm^{-2})	High (~10^6 cm^{-2})	Lower
Suitability for SOI vs. bulk wafer	SOI	Either	Either
Chip vs. wafer bonding	Wafer/wafer only	Either	Either
Directly extendable to >2 layers	Yes	No	Yes
Connection to package	Standard	Deep via	Standard

Figure 12.5(a) shows a structure in which the distance between device layers is minimized by removing the entire Si substrate between the layers. Bonding between the device layers is achieved through blanket dielectric fusion bonding or the use of an adhesive interlayer, after which interlayer electrical connections are formed. Figure 12.5(b) shows the face-to-face bonding option, which is effective for creating high-density Cu-Cu bonded links between layers by requires deep vias for bringing signals out to the package. The structure in Figure 12.5(c) has the largest interlayer via dimensions and the lowest via density, along with the most relaxed alignment tolerances. The choice of structure and fabrication method of depends on the specific goal and application of the 3D IC technology.

The 3D assembled structure [Table 1, column (a)] is described as having the shortest distance between stacked device layer, the highest interconnection density, and extremely aggressive wafer-to-wafer alignment requirements. A unique n-FET and p-FET layer is stacked to derive full benefit from the 3D IC process. The process flow to fabricate such structures is depicted in Figure 12.6. With stringent design requirements, the key process optimization is focused on development of state-of-the-art interdevice layer connections.

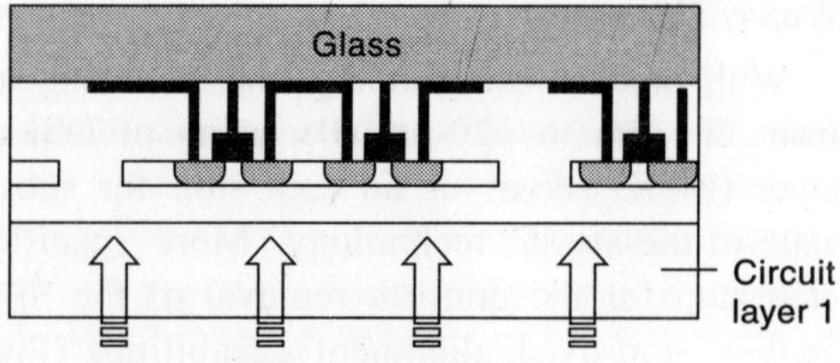

• Attach circuit to glass substrate
• Remove original substrate

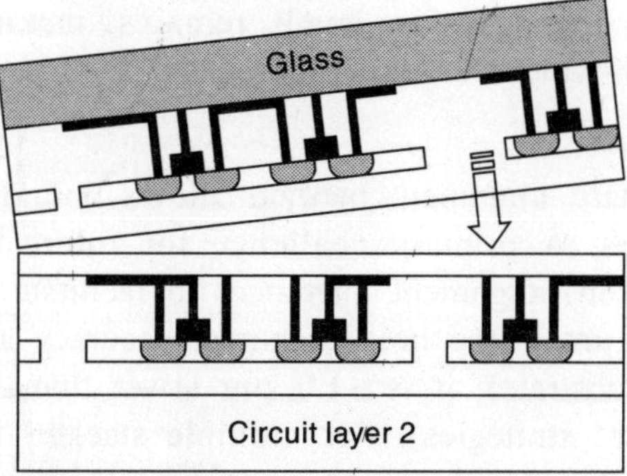

• Align and bond top circuit to bottom circuit
•

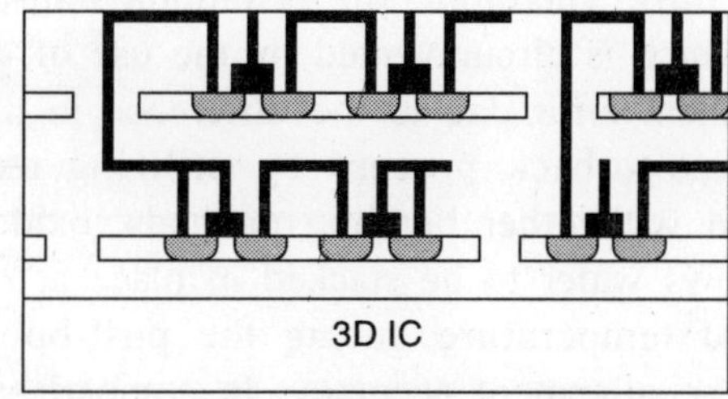

• Remove glass substrate and sdhesives
• Form vertical interconnects

Fig. 12.6 *Diagrams of assembly process (layer transfer methodology to fabricate 3D ICs).*

KEY 3D IC TECHNOLOGY

Independent of the final 3D IC structure, the assembly method always involves the integration of four key technology areas: thinning of the wafers, interdevice-layer alignment, bonding, and interlayer contact pattering. An additional challenge in achieving high-density I/O signal through the stack layers arises from thermal mismatch between the bonded layers, affecting alignment tolerance. Also, thermal dissipation of high-performance CMOS devices is already a concern in 2D ICs; for 3D circuits, heat spreading and self heating become critical issues. All of these 3D IC integration challenges require new materials and process innovations.

Wafer Thinning

With mechanical grinding and polishing and plasma or wet etching silicon wafers are thined from 200 nm to ~20-μm. By using of OSI and glass substrates bulk Si is removed. The buried layer (BOX) serves as an etch stop for substrate thinning, enabling the use of high-performance state-of-the-art IC technology. More specifically, the BOX in SOI wafers provides a selective etch stop for the uniform removal of the Si substrate; combined with use of a glass substrate, it enables improved alignment capabilities (Figure 12.6). Both features greatly simplify the layer-transfer process, providing a means of obtaining the shortest distance between devices. The final "decal" structure on a glass carrier has all of the bulk SI removed; only the device layer with its metallization levels remains, making the stack transparent and hence enabling the "through-wafer" alignment process.

Alignment

Standard alignment method allows both front-side (through-wafer) and back-side alignment strategies. A primary challenge for future high-density 3D ICs is the requirement for high (submicron) alignment tolerances to facilitate higher-level circuit designs. As 3 sigma value (3σ) of ~1.0 μm is the best alignment accuracy achieved using the through-wafer alignment strategy (glass substrate); it is ~1.0 μm lower than from nontransparent alignment methods (back-side alignment strategies). For multiple stacked fully thinned IC device layers, good alignment is easily achieved. If a nontransparent carrier is used, the wavelength-dependent signal attenuation through Si degrades alignment accuracy (especially for layers in which remaining Si is thicker than 40 μm). Therefore, the resolution transparency in Si poses a challenge for nontransparent wafers which is circumvented by the use of a glass flow.

Alignment error due to the difference in the CTE of the two layers is minimized in the SOI-based face-to-back process by utilizing oxide-fusion bonding at room temperature. When compared with other bonding methods, oxide-fusion bonding shows clear superiority (Table 2) as it allows wafer to be stacked in place at room temperature during alignment. It is shown that increased temperature during the post-bonding anneal strengthens the bond but does not change the alignment accuracy. In comparison, since Cu bonding occurs at higher temperatures, extremely good temperature control is maintained. Accuracy using bonding with adhesive layers degrades, as adhesive becomes viscous during the bonding process, thus causing alignment patterns to shift. It is important to notice that the placement error of state-of-the-art lithography tools is <0.02 μm and so does not limit alignment precision.

Table 2 *Technology features various bonding methods*

Critical aspects	*Oxide fusion bonding*	*Thermo-compression bonding*	*Bonding with adhesive layers*
Minimum bonding temperature	Room temperature	Depends on metal; for Cu 573–673K	Mostly 473–573K
State of material during bonding	Solid	May temporarily be viscous if metals are alloyed	Viscous
Special requirements	None	Good temperature control	Good temperature control
Ability to preserve alignment during bonding	High	Low	Low

Large alignment errors are induced by bowing for the wafers. Every processing step changes the bow of a wafer, sometimes by hundreds of microns. To achieve optimal alignment, the bow should be less than 20 μm for 200-nm wafers during alignment. To maintain this bow target, compensation methods, such as the deposition of counter-pre-stressed films, are implemented prior to the bonding step. Similarly, surface smoothness and local planarity are critical for high-accuracy alignments, as they affect the ability of the optics of an alignment tool to focus on alignment mark structures.

Bonding

For all types of bonding methods, the quality of the bonded interface depends strongly on surface roughness and cleanliness. In particular, a fusion bonding requires atomically smooth surfaces. The combination of chemical-mechanical polishing (CMP) and wet chemical surface treatment is often used prior to bonding to ensure clean and reactive bonding surfaces. Cleaning procedures and a port-deposition annealing sequence control bond strength so that they reduce the formation of voids at the bonding interface. Thus, for the oxide-fusion bonding process, reduction of the bulk concentration of —OH groups in oxide (post-deposition) before bonding enhances the ability of the oxide to absorb by products released during the bonding anneal and is critical in obtaining defect-free bonded interfaces Figure 12.7 shows a cross-sectional TEM image of two SOI CMOS device layers bonded by oxide fusion.

Fig. 12.7 *Cross-sectional TEM image of two metallized, stacked, and oxide-fusion-bonded SOI CMOS device layers.*

Since surface roughness requirements for fusion-bonded surfaces are very stringent (<1.0 nm) and not easily achieved, metal-to-metal bonding options because their roughness specifications are higher (<20 nm). However, in metal-to-metal low-temperature bonding process, high pattern density is required to provide high bond strength and interface stability during further processing steps. Bonding using polymeric or dielectric glue layers has the least stringent surface planarity requirements, but the use of viscous glue leads to shift of these layers during bonding, thereby limiting alignment tolerance (Table 2). Temperature for all of these bonding approaches is comparable with the thermal constraints of each functional layer, 723K for post-CMOS FEOL processes.

Inter-device-level 'via' Fabrication

For all three structures depicted in Figure 12.5, the 3D IC technology requires the formation of high-aspect ratio vias. The patterning and metallization process for the creation of such vias (e.g., plasma etch, metal fill, and CMP) is compatible with other BEOL process strategies. All metallization techniques place specific limitation on the maximum aspect ratio of vias and lead to design limitations with respect to the layout of active and passive devices on each layer. The

BOX layer in a SOI substrate is used to control the transferred device layer thickness to very tight tolerances. This minimizes the effective aspect ratio of the interwafer via by enabling vertical stacking of the layers spaced on a few microns apart. To utilize the full potential of 3D IC, vias of submicron diameter dimensions are required to be compatible with state-of-the-art FEOL technology. Hence, the performance and viability of 3D ICs built by stacking high-performance CMOS devices depends on bonding alignment tolerance and on the structural and electrical integrity of the submicron high-aspect-ratio vias connecting device layers.

Figure 10.8 shows the capability to fabricate small (sumbmicron) interconnecting 3D IC copper-filled vias with high aspect action (6:1 < 11:1) using a single-damascene process. The via profile, metal liner, and Cu plating processes are modified slightly from a standard back-end-of-line via formation sequence to achieve proper fill of these high-aspect ratio structures. The smallest vias, with a bottom diameter of ~0.14 μm, height >1.6 μm, and sidewall angle of approximately 86 degrees, are formed on a 0.4-μm pitch, equivalent to an extremely high via density $>10^8$ via per cm^2.

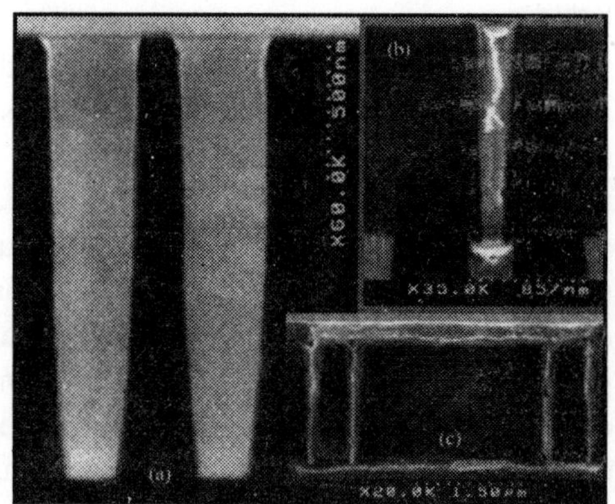

Fig. 12.8 **(a)** *Polished cross-sectional SEM images of Cu-filled vias with a 6:1 aspect ratio and height ~1.,6 μm; (b) cleaved SEM image of isolated via; (c) cleaved SEM image of via structure with diameter ~175 nm and high aspect ratio.*

Vias with bottom critical dimensions of ~01.4 μm × 0.14 μm correspond to a 0.13-μm CMOS BEOL technology, but owing to the higher aspect ratio of the interlevel vias in 3D ICs, their resistance is two or three times higher than that of a typical back-end via. Measurements of resistance per link of 3D via chains connecting the first metal level of top and bottom wafers indicate resistance value of ~2–4 Ω per link and good yield for via chains with 100–10,000 vias. This confirms a successful metallization process through the bonded interface. Further process optimization is required to achieve acceptable yields for the longer chain lengths. 'VIAS' with such high density are rarely used because of the space taken up by active circuitry on the upper device layer and alignment challenges. Nevertheless, by this process building ultrahigh-density, low-parasitic links between layers using materials and processes compatible with pre-fabricated circuitry are observed. The alignment accuracy required to reliably interconnect the various device circuits fabricated ranges from 0.5 to 2.5 μm and is successfully achieved

Thermal Dissipation

Device temperature increase is a major concern in 2D SOI technology. Because of the poor heat conductivity of the BO layer, temperature increases of 353–393 K/mW/μm of width in transistors are there. In addition, a rise in temperature causes device performance variation and is critical for matching in analog circuits. Also, the performance of the clock buffer is affected by device temperature increases. For SOI and bulk devices, every 10K increase in junction temperature degrades buffer performance by 1.2% and 1.32%, respectively. Various tests, including pulsed *I–V*, body-contact diode, polySi resistance, and subthreshold slope methods are used to measure temperature in 2D SOI transistors and are utilized to test 3D ICs. The reduced surface-area-to-volume ratio of 3D structures inevitably lead to increases in power density and affects the intrinsic heating of high performance chips. Therefore, for some applications the use of heat-dissipating structures to minimize thermal gradients and local heating is required, but it affects the interlevel interconnect layout and the design of the 3D chip.

Useful Acronyms

μCP	Microcontact printing
aa	amino acid
Ab	antibody
ADMET	Adsorption, distribution, metabolism, excretion and toxicology
AES	Auger electron spectroscopy
AFM	Atomic Force Microscope (Microscopy)
Ag	antigen
ALD	Atomic Layer Deposition
ALE	Atomic Layer Epitaxy
ASIC	Application-specific integrated circuit
ATMP	Advanced Therapy Medicinal Product
ATP	adenosine triphosphate
BBB	blood brain barrier
BCC	Body-centred Cubic (Crystal structure)
BEEM	Ballistic Electron Emission Microscopy
BET	Brunauer, Emmett and Teller method of measuring surface-area
BioNEMS	biofunctionalised nanoelectromechanical systems
BMR	Ballistic Magnetoresistance
BSF	Back surface field
BZ	Brillouin Zone
CARL	Chemically amplified resist lithography
CCD	Charge-coupled device
Cermet	Ceramic-Metal Composite
CFC	Chloroflourocabrons
CMOS	Complementary Metal-Oxide Semiconductor
CMP	Chemical Mechanical Polishing
CNS	central nervous system
CNT	Carbon Nano Tube
CPC	Condensation Particle Counter
CRD	chronic respiratory disease
CT	computer tomography
CVD	cardiovascular disease
CVD	Chemical Vapour Deposition
CW	Continuous wave
CWA	CEN Workshop Agreement
Cz	Czochralski

DBQW	Double-barrier quantum-well
DFB	Distributed feedback (QCL)
DFT	Density Function Theory
DLC	diamond-like carbon
DLTS	Deep level transient spectrocopy
DMA	Differential Mobility Analyzer
DNA	dexoyribonucleic acid
DOF	Depth of focus
DPN	dip-pen nanolithography
DRAM	Dynamic random access memory
DUV	Deep ultraviolet
EBIC	Electron beam induced current
ECL	Emitter-coupled logic
ECM	extracellular matrix
ECR	Electron cyclotron resonance (CVD, plasma etching)
EDP	Ethylene diamine/pyrocatechol
EDX	Energy Dispersive X-ray analysis
EELS	Electron Energy Loss Spectroscopy
EEPROM	Electrically erasable programmable read-only memory
EFA	Field emitter cathode array
EL	Electroluminescence
ESCA	(also called XPS) Electron Spectroscopy for Chemical Analysis
ESEM	Environmental Scanning Electron Microscope
ESR	Electron spin resonance
ESTOR	Electrostatic data storage
Et	Ethyl
EUV	Extreme ultraviolet
EUV	Extreme Ultra-Violet
EUVL	Extreme ultraviolet lithography
EXAFS	Extended X-ray Absorption Fine Structure
Fab	fragment antigen binding (of an antibody molecule)
FCC	Face Centred Cubic (Crystal structure)
FEG-SEM	Field Emission Gun Scanning Electron Microscope
FET	Field-Effect Transistor
FIB	Focused ion beam
FP	Fabry-Perot
FRET	fluorescence resonance energy transfer
FTIR	Fourier transform infrared
FWHM	Full Width, Half Maximum
GDOES	Glow Discharge Optical Emission Spectroscopy
GMR	Giant Magneto Resistance
HBT	Hetero bipolar transistor
HEL	Hot-embossing lightography
HEMT	High electron mobility transistor
HIT	Heterojunction with intrinsic thin layer
HOMO	Highest occupied molecular orbital
HREELS	High-Resolution Electron Energy Loss Spectroscopy
HREM	High-Resolution Electron Microscopy
hTEP	human tissue-engineered product
HVOF	High Velocity Oxygen Fuel
IC	Integrated circuit
ICP	Inductively coupled plasma
IMPATT	Impact ionization avalanche transit time
IPG	In plane gate
IR	Infrared

ITO	Indium-tin-oxide
ITRS	International technology roadmap for semiconductors
Laser	Light amplification by stimulated emission of radiation
LBIC	Light beam induced current
LCD	Liquid crystal display
LDD	Light-emitting diode
LED	Light-emitting diode
LED	Light Emitting Diodes
LEED	Low energy electron diffraction
LISA	Lithographically induced Self-Assembly
LMIS	Liquid metal ion source
LOC	Lab-on-a-chip
LPE	Liquid phase epitaxy
LSS	Lindhardt, Scharff, SchiØtt (Researchers)
LUMO	Lowest unoccupied molecular orbital
M	Metal
MAL	Mould-assisted lithography
MBE	Molecular Beam Epitaxy
MCT	Mercury cadmium telluride
Me	Methyl
MEMS	Micro Electro Mechanical Systems
MEMS	Microfabricated Electro Mechanical Systems
MFH	Magnetic Fluid Hypothermia
MFM	Magnetic Force Microscopy
MIS	Metal-insulator-semiconductor
MMIC	Monolithic microwave integrated circuit
MNT	(Forum)-Micro and nano technologies
MNT	Molecular Nanotechnology
MOCVD	Metallo-organic chemical vapor deposition
MODFET	Modulation-doped field-effect transistor
MOLCAO	Molecular orbitals as linear combinations of atomic orbitals
MOS	Metal-oxide-semiconductor
MOSFET	Metal-oxide-semiconductor field effect transistor
MPU	Microprocessor unit
MQW	Multi wall nanotubes
MRI	Magnetic Force Microscopy
MST	Micro Systems Technology
MWCNT	Multi-Walled Carbon NanoTube
MW	Molecular weight
MWNT	Multi wall nanotubes
NA	Numerical aperture
NAND	Not and
Nd:YAG	Neodymium yttrium aluminum garnet (laser)
NDR	Negative differential resistance
NEMS	Nano Electro Mechanical Systems
NEMS	Nanofabricated ElectroMechanical Systems
NIL	Nanoimprint lithography
NIL	Nanolmprint Lithography
NM	Nuclear Medicine
NMOS	n-Channel metal oxide-semiconductor (transistor)
NMR	Nuclear magnetic resonance
NMR	Nuclear Magnetic Resonance
NOR	Not or
NOx	Nitrogen Oxides
NQR	Nuclear Quadrupole Resonance

NSOM	Near-field Scanning Optical Microscopy
OCL	Quantum cascade laser
OLEDs	Organic Light Emitting Diodes
PADOX	Pattern-dependent oxidation
PCR	Polymerase Chain Reaction
PDMS	Polydimethylsiloxane
PE	Plasma etching
PECVD	Plasma-enhanced chemical vapor deposition
PEEM	PhotoEmission Electron Microscopy
PEG	Polyethylene glycol
PET	Polyethyleneterphthalate
PET	Position Emission Tomography
PL	Photoluminescence
PLAD	Plasma doped
PMMA	Polymethylmethacrylate
PREVAIL	Projection reduction exposure with variable axis immersion lenses
PTFE	Polytetrafluorethylene (Teflon®)
PVD	Physical vapor deposition
QDNs	Quantum Dot Nanotcrystals
QSE	Quantum size effect
QWIP	Quantum well infrared photodetector
RAM	Random access memory
RBS	Rutherford backscattering spectrometry
RCA	Radio Corporation of America (Company)
RF	Radio frequency
RGD	arginine, glycine and aspartic acid motif
RHEED	Reflection high-energy electron diffraction
RIE	Reactive ion etching
RITD	Resonant interband tunneling diode
RLS	Resonance Light Scattering
RNA	Ribonucleic acid
RTA	Rapid thermal annealing
RTBT	Resonant tunneling bipolar transistor
RTD	Resonant tunneling diode
SAMMS	Self Assembled Monolayer on Mesoporous Supports
SAM	Self-Assembled Monolayer
SCALPEL	Scattering with angular limitation projection electron beam lithography
SCM	Scanning Capacitance Microscopy
SCZ	Space charge zone
SECM	Scanning Electro Chemical Microscopy
SEM	Scanning Electron Microscope (Microscopy)
SET	Single Electron Transistor
SFIL	Step and flash imprint lithography
SFM	Scanning Force Microscope
SHT	Single hole transistor
SIA	Semiconductor Industry Association
SICM	Scanning Ion Conductance Microscopy
SIMOX	Separation by implantation of oxygen
SIMS	Secondary Ion Mass Spectrometry
SL	Superlattice
SMD	Surface-mounted device
SMPS	Scanning Mobility Particle Sizer
SNP	Single Nucleotide Polymorphisms
SOI	Silicon on insulator
SOS	Silicon on sapphire

SPM	Scanning Probe Microscope (Microscopy)
SPR	Surface Plasmon Resonance
STM (STEM)	Scanning Transmission Electron Microscope
STM	Scanning Tunnelling Microscope (Microscopy)
SWCNT	Single-Walled Carbon NanoTube
SWNT	Single-Wall NanoTube
TA	Thermal analysis
TED	Transferred electron device
TEM	Transmission Electron Microscope (Microscopy)
TEOS	Tetraethylorthosilicate
TFT	Thin film transistor
TMAH	Tetramethylammonium hydroxide
TOFSIMS	Time of Flight SIMS
TSI	Top surface imaging
TTL	Transistor-transistor logic
TUBEFET	Single carbon nanotube field-effect transistor
UHV	Ultrahigh vacuum
ULSI	Ultra large scale integration
UPS	UltraViolet Photoelectron Spectroscopy
UV	Ultraviolet
VHF	Very high frequency (30–300 MHz; 10–1 m)
VLSI	Very large scale integration
VMT	Velocity-modulated transistor
V-PADOX	Vertical pattern-dependent oxidation
VPE	Vapor phase epitaxy
XOR	Exclusive or
XPS	X-Ray Photoelectron Spectroscopy
XRD	X-ray diffraction
ZME	Zeolite modified electrode
ZVI	zerovalaent ion

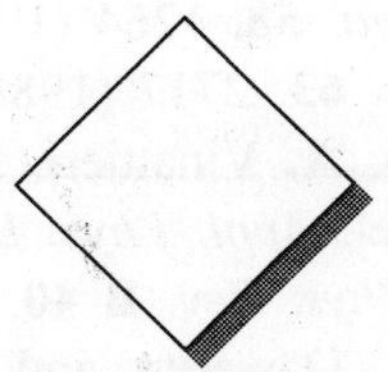

Bibliography

Park, S. H.; Yan, H; Reif, J. H; LaBean, J. H.–Nano Technology, 2004, 15, 525.

Fundamentals of Semiconductors–P. Y. Yuand M. Cardona (Springer, Berlin, 2001).

The Physics of Semiconductors with application to Optoelectronic Devices by K. F. Brennan (CUP, Cambridge, 1999).

Nanoelectronics and Information Technology: Advanced Electronic Materials and devices edited by R. Waser (Wiley VCH, Weinheim, 2003).

P. M. Brosenberger and S. Weiss; Organic Photoreceptors for Imaging System, Marcel Dekker, Newyork.

IEEE Trans Magn 38, 285 (2002),

Phys. Rev. **B 70**, 035310 (2004),

Phys. Rev. **B 70**, 195208 (2004),

Appl. Phys Lett. **80**, 1150 (2002),

Phys. Rev. Lett. **90**, 47204 (2003).

Molecular Electronis–Science and technology, American Institute of Physics, N. Y. (1991), Vol. 852.

Molecular Electronis—Aviram, A. and M. A. Ratner, Annals of the Newyork Academy of Sciences, New York, Vol. 852, (1998).

Bradley D.D.C., Molecular Electronics—Aspects of the Physics, Chem. Britain 719 (1991).

A. Richter, G. Behme, M. Sptitz, Ch. Lienau, T. Elsässer, M. Ramsteiner, R. Nötzel, and K. H. Ploog, *Phys. Rev. Lett.* 2145 (1997).

R. Nötzel, U. Jahn, Z. C. Niu, A. Trampert, J. Fricke, H. P. Schönherr, T. Kurth, D. Heitmann, L. Däweritz and K. H. Ploog, *Appl. Phys. Lett.* 72, 2002 (1998).

U. Jahn, R. Nötzel, J. Ringling, H. P. Schönherr, H. T. Grahn, K. H. Ploog, and E. Runge, *Phys. Rev.* **B 60**, 11038 (1999).

F. Aslina, P. V. Santos, H. P. Schönherr, W. Seidel, R. Nötzel, and K. H. Ploog, *Phys. Rev.* **B 67**, 161305 (2003).

R. Nötzel, M. Ramsteiner, Z. C. Niu, H. P. Schönherr, L. Däweritz, and K. H. Ploog, *Appl. Phys. Lett.* **70**, 1578 (1997).

D. Smith, and C. Mailhiot, *Phys. Rev. Lett.* **58**, 1264 (1987).

D. Smith, and C. Mailhiot, J. *Appl. Phys.* **63**, 2717 (1988).

D. Alderighi, M. Zamfirescu, M. Gurioli, A. Vinattieri, S. Sanguinetti, M. Povolotskyi, J. Gleize, A. Di Carlo, P. Lugli, and R. Nötzel, *Appl. Phys. Lett.* **84**, 786 (2004).

M. Ilg, K. H. Ploog, and A. Trampert, *Phys. Rev.* **B 40**, 17111 (1994).

R. Nötzerl, M. Ramsteiner, Z. C. Niu, L. Däweritz, and K. H. Ploog, *Physica*, **E 2**, 979 (1998).

L. De Caro, and L. Tapfer, *Phys. Rev.* **B 51**, 4374 (1995).

Balestra, D., T. Matsumoto, M. Tsuno, H. Nakabayashi, Y. Inoue, and M. Koyangi, Moderate kink effect in fully depleted thin-film SOI MOSFETs, *Electron. Lett.* **31**, 326 (1995).

Bruel, M., Silicon on insulator material technology, *Electron. Lett.* **31**, 1201 (1995).

Chan, M., S. K. H. Fung, K. Y. Hui, C. Hu, and P. K. Ko, SOI MOSFET design for all-dimensional scaling with short channel, narrow width and ultra-thin films, *IEDM Tech. Dig.*, 631 (1995).

Chen, J., S. Parke, J. King, F. Assaderaghi, P. Ko, and C. Hu, A high-speed SOI technology with 12 ps/18 ps gate delay operating at 5V/1.5V, *IEDM Tech. Dig.*, **35** (1992).

Choi, J. Y., and J. G. Fossum, Analysis and control of floating body effects in fuly depleted SOI MOSFET's, *IEEE Trans. Electron Devices* **38**, 1384 (1991).

Colinge, J-P., SOI Technology: Materials to VLSI, 2nd ed, Kluwer, Boston (1997).

Cristoloveanu, S., Silicon films on sapphire, *Rep. Prog. Phys.* **3**, 327 (1987).

Cristoloveanu, S., A review of the elecrical properties of SIMOX substrates and their impact on device performance, *J. Electrochem. Soc.* **138**, 3131 (1991).

Cristoloveanu, S., Hot-carrier degradation mechanisms in silicon-on-insulator MOSFETs, *Microelectron. Reliab.* **37**, 1003 (1997).

Cristoloveanu, S., SOI: A metamorphosis of silicon, IEEE Mag.: Circuits & Devices 99–18, **15** (1) 26 (1999).

Cristoloveanu, S., and S. S. Li, Electrical Characterization of SOI Materials and Devices, Kluwer, Norwell (1995).

Cristoloveanu, S., and G. Reichert, Recent advances in SOI materials and device technologies for high temperature, *1998, High Temperature Electronic Materials, Devices and Sensors Conf. Proc.* (1998).

Cristoloveanu, S. D. Munteanu, and M. Liu, A review of the pseudo-MOS transistor, *IEEE Trans. Electron Devices* **47** (5), 1018 (2000a).

Cristoloveanu, S., T. Ernst, D. Munteanu, and T. Ouisse, Ultimate MOSFETs on SOI: ultra thin, single gate, double gate, or ground plane, *Int. J. High Speed Electron. Systems* **10** (1), 217 (2000b).

Ernst, T., and S. Cristoloveanu, The ground-plane concept for the reduction of short-channel effects in fully-depleted SOI devices, *SOI Technology and Devices IX*, Electrochem. Soc., Pennington, p. 329 (1999).

Ernst, T., D. Munteanu, S. Cristoloveanu, J-L. Pelloie, O. Faynot, and C. Raynaud, Detailed analysis of short-channel, SOI Dt-MOSFET, *Proc. ESSDERC'99*, Ed. Frontiers, Neuilly, p. 380 (1999a).

Ernst, T. *et al.*, Investigation of SOI MOSFETs with ultimate thickness, *Microelectron, Eng.* **48** (1-4), 339 (1999b).

Franck, D., S. Laux, and M. Fischetti, Monte Carlo simulations of a 30 nm dual gate MOSFET: how short can Si go? *IEDM Tech. Dig.*, 553 1992).

Hafez, I. M., G. Ghibaudo, and F. Balestra, Analysis of the kink effect in MOS transistors, *IEEE Trans. Electron Devices* **37**, 818 (1990).

Hisamoto, D., T. Kaga, and E. Takeda, Impact of the vertical SOI 'DELTA' structure on planar device technology, *IEEE Trans. Electron Devices* **38**, 1419 (1991).

Hisamoto, D. *et al.*, A folded-channel MOSFET for deep-sub-tenth micron eta, *IEDM Tech. Dig.*, 1032 (1998).

Jomaah, J., F. Balestra, and G. Ghibaudo, Experimental investigation and numerical simulation of low frequency noise in thin film SOI MOSFETs. *Phys. Stat. Sol.* (a) **142**, 533 (1994).

Koh, Y-H. *et al.*, 1 Gigabit SOI DRAM with fully bulk compatible process and body-contacted SOI MOSFET structure, *IEDM Tech. Dig.*, 579 (1997).

Lagnado, I., and P.R. de la Houssaye, Silicon on sapphire technology: Quo vadis II, *Silicon-On-Insulator Technology and Devices X*, Electrochemical Soc., Pennington, Vol. 2001–3, p. 265 (2001).

Kesan, V. P., S. Subbanna, P. J. Restle, M. J. Tejwani, J. M. Aitken, S. S. Iyer, and J. A. Ott, High performance 0.25 μm *p*-MOSFETs with silicon-germanium channels for 300 K and 77 K operation, 1991 *IEDM Technical Digest*, pp. 25–28 (1991).

Lo, S.-H. D. A. Buchanan, Y. Taur, and W. Wang, Quantum-mechanical modeling of electron tunneling current from the inversion layer of ultra-thin-oxide *n*-MOSFETs, *IEE Electron Device Lett.*, **18**, 209 (1997).

R. Nötzel, N. Ledentsov, L. Däweritz, M. Hohenstein, and K. Ploog, *Phys. Rev. Lett.* **67**, 3812 (1991).

R. Nötzel, J. Temmyo, and T. Tamamura, *Nature* (London) 369, 131 (1994).

R. Nötzel, J. Menniger, M. Ramsteiner, A. Ruiz, H. P. Schönherr, and K. H. Ploog, *Appl. Phys. Lett.* **68**, 1132 (1996).

W. T. Tsang, and A. Y. Cho, *Appl. Phys. Lett.* **30**, 293 (1977).

E. Kapon, D. M. Hwang, and R. Bhat, *Phys. Rev. Lett.* **63**, 430 (1989).

X. L. Wang, M. Ogura, and H. Matsuhata, *J. Cryst. Growth* **171**, 341 (1997).

A. Hartmann, L. Loubies, F. Reinhardt, and E. Kapon, *Appl. Phys. Lett.* **71**, 1314 (1997).

M. Walther, T. Röhr, G. Böhm, G. Tränkle, and G. Weimann, *J. Cryst. Growth* **127**, 1045 (1993).

S. Tsukamoto, Y. Nagamune, M. Nishioka, and Y. Arakawa, *Appl. Phys. Lett* **63**, 355 (1993).

H. Fujikura, and H. Hasegawa, *Jpn. J. Appl. Phys.* **35**, 1333 (1996).

K. C. Rajkumar, A. Madhukar, K. Rammohan, D. H. Rich, P. Chen, and L. Chen, *Appl. Phys. Lett.* **63**, 2905 (1993).

Q. Gong, R. Nötzel, H. P. Schönherr, and K. H. Ploog, *Appl. Phys. Lett.* **77**, 3538 (2000).

R. Nötzel, M. Ramsteiner, J. Menniger, A. Trampert, H. P. Schönherr, L. Däweritz, and K. H. Ploog, *J. Appl. Phys.* **80**, 4108 (1996).

R. Nötzel, J. Menniger, M. Ramsteiner, A. Trampert, H. P. Schönherr, L. Däweritz, and K. H. Ploog, *J. Cryst. Growth.* **175/176**, 1114 (1997).

R. Nötzel, M. Ramsteiner, J. Menniger, A. Trampert, H. P. Schönherr, L. Däweritz, and K. H. Ploog, *Jpn. J. Appl. Phys.* **35**, L297 (1996).

Appenzeller, J., R. Martel, P. Solomon, *et al.* Scheme for the fabrication of ultrashort channel MOSFETs, *Appl. Lett.* **77**, 298 (2000).

Assaderaghi, F. *et al.*, A 7.9/5.5 psec room/low temperature SOI CMOS, *IEDM Tech. Dig.*, 415 (1997).

Baccarani, G., M. R. Wordeman, and R. H. Dennard, Generalized scaling theory and its application to a 1/4 micrometer MOSFET design, *IEEE Trans. Electron Devices* **31**, 452 (1985).

Balestra, F., Performance and physical mechanisms in deep submicron SOI MOSFETs, *Electron Technol.* **32**, 50 (1999).

Balestra, F., M. Benachir, J. Brini, and G. Ghibaudo, Analytical models of subthreshold swing and threshold voltage for thin and ultra-thin film SOI MOSFETs, *IEEE Trans. Electron Devices (USA)* **37**, 2303 (1990).

Balestra F., S. Cristolveanu, M. Bènachir, J. Brini, and T. Elewa, Double-gate silicon on insulator transistor with volume inversion: a new device with greatly enhanced performance, *IEEE Electron, Device Lett.* **8**, 410 (1987).

Balestra F., J. Jomaah, H. Ghibaudo, O. Faynot, A-J. Auberton-Hervé, and B. Giffart, Analysis of the latch and breakdown phenomena in N and P channel thin film SOI MOSFET's as a function of temperature, *IEEE Trans. Electron Devices*, **41**, 109 (1994).

Damle, P. S., A. W. Ghosh, and S. Datta, Theory of nanoscale device modeling, in *Molecular Nanoelectronics*, edited by, M. Reed (2003).

Datta, S., *Electronic Transport in Mesoscopic Systems*, Cambridge University Press, Cambridge, UK (1997).

Datta S., W. Tian, S. Hong, R. Reifenberger, J. I. Henderson, and C. P. Kubiak, Current–voltage characteristics of self-assembled monolayers by scanning tunneling microscopy, *Phys. Rev. Lett.* **79**, 2530 (1997).

Desjoqueres, M. C., and D. Spanjaard, *Concepts in Surface Physics*, 2nd edn, Springer-Verlag, Berlin (1996).

Di Ventra, M., S. T. Pantelides, and N. D. Lang. First-principles calculation of transport properties of a molecular device, *Phys. Rev. Lett.* **84**, 979 (2000).

Emberly, E. G., and G. Kirczenow, Multiterminal molecular wire system: a self-consistent theory and computer simulations of charging and transport, *Phys. Rev.* B **62** (15), 10451 (2000).

Ferry, D. K., and S. M. Goodnick, *Transport in Nanostructures*, Cambridge University Press, Cambridge, UK (1997).

Frisch, M. J., G. W. Trucks, H. B. Schlegel, G. E. Scuseria, M. A. Robb, J. R. Cheeseman, V. G. Zakrzewski, J. A. Jr. Montgomery, R. E. Startmann, J. C. Burant, S. Dapprich, T. T. J. M. Berendschot, H. A. J. M. Reinen, H. J. A. Bluyssen, C. Harder, and H. P. Meier, *Appl. Phys. Lett.* **54**, 1827 (1989).

Y. Arakawa, K. Vahala, A. Yariv, and K. Lau, *Appl. Phys. Lett.* **47**, 1142 (1985).

Y. Arakawa, K. Vahala, A. Yariv, and K. Lau, *Appl. Phys. Lett.* **48**, 384 (1986).

K. Vahala, Y. Arakawa, and A. Yariv, *Appl. Phys. Lett.* **50**, 365 (1987).

G. P. Agarwal, and N. K. Dutta, *Long-Wavelength Semiconductor Lasers* (Van Nostrand Reinhold, New York, 1986).

W. Wegscheider, L. N. Pfeiffer, K. W. West, P Littlewood, O. Narayan, M. Hagn, M. M. Dignam, and R. E. Leibenguth, *Solid-State Electron*, **40**, 1 (1996).

M. Z. Maialle, E. A. de Andrada Silva, and L. J. Sham, *Phys. Rev.* **B 47**, 15776 (1993).

A. Vinattieri, J. Shah, T. C. Damen, D. S. Kim, L. N. Pfeiffer, M. Z. Maialle, and L. J. Sham, *Phys. Rev.* **B 50**, 10868 (1994).

W. Wegscheider, G. Schedelbeck, G. Abstreiter, M. Rother, and M. Bichler, *Phys. Rev. Lett.* **79**, 1917 (1997).

G. Schedelbeck, W. Wegscheider, M. Bichler, and G. Abstreiter, *Science* **278**, 1792 (1997).

W. Wegscheider, G. Schedelbeck, M. Bichler, and G. Abstreiter, "Atomically Precise, Coupled Quantum Dots Fabricated by Cleaved Edge Overgrowth." In *Festkörperprobleme/Advances* in *Solid State Physics.* Vol. **38**, edited B. Kramer (Vieweg. Wiesbaden, 1998), p. 153.

D. Gammon, E. S. Snow, B. V. Shanabrook, D. S. Kratzer, and D. Park, *Phys. Rev. Lett.* 76, 3005 (996); *Science* 273, 87 (1996).

P. Platzman unpublished.

Dennard, R. H., F. H. Gaensslen, H. N. Yu, V. L. Rideout, E. Bassous, and A. R. LeBlanc, Design of ion-implanted MOSFETs with very small physical dimensions, *IEEE J. Solid-State Circuits*, **SC-9**, 256 (1974).

Fair, R. B., and H. W. Wivell, Zener and avalanche breakdown in as-implanted low-voltage silicon N–P junctions, *IEEE Trans. Electron Devices*, **ED-23**, 512 (1976).

Fischetti, M.V., and S. E. Laux, Band structure, deformation potentials, and carrier mobility in strained Si, Ge, and SiGe alloys. *J. Appl. Phys.*, **80** (4), 2234 (1996).

Frank, D., S. Laux, and M. Fischeti, Monte Carlo simulation of a 30 nm dual-gate MOSFET. How far can Si go? 1992 *IEDM Technical Digest*, 553 (1992).

Frank, D. J., Y. Taur, and H.-S. Wong, Generalized scale length for two-dimensional efects in MOSFETs, *IEEE Electron Device Lett.*, **19**, 385 (1998).

Frank, D. J., *et al.*, Monte Carlo modeling of threshold variation due to dopant fluctuations. *VLSI Technology Symposium*, Kyoto, Japan, June (1999).

Gaensslen, F. H., V. L. Rideout, E. J. Walker, and J. J. Walker, Very small MOSFETs for low temperature operation, *IEEE Trans. Electron Devices*, **ED-24**, 218 (1997).

Kahng, D., and M. M Atalla, Silicon dioxide field surface devices, presented at *Device research Conference*, Pittsburgh (1960).

Index